Transmission, Distribution, and Renewable Energy Generation Power Equipment

Aging and Life Extension Techniques

Transmission, Distribution, and Renewable Energy Generation Power Equipment

Aging and Life Extension Techniques

Second Edition

Bella H. Chudnovsky

CRC Press
Taylor & Francis Group
Boca Raton London New York

CRC Press is an imprint of the
Taylor & Francis Group, an **informa** business

CRC Press
Taylor & Francis Group
6000 Broken Sound Parkway NW, Suite 300
Boca Raton, FL 33487-2742

First issued in paperback 2020

© 2017 by Taylor & Francis Group, LLC
CRC Press is an imprint of Taylor & Francis Group, an Informa business

No claim to original U.S. Government works

ISBN-13: 978-1-4987-5475-0 (hbk)
ISBN-13: 978-0-367-73639-2 (pbk)

Visit the Taylor & Francis Web site at
http://www.taylorandfrancis.com

and the CRC Press Web site at
http://www.crcpress.com

Contents

SECTION I   Transmission and Distribution Electrical Equipment

SECTION II Renewable Energy Equipment Challenges

SECTION III Testing, Monitoring, and Diagnostics of Electrical Equipment

Preface

Every electrical generation, transmission, and distribution apparatus is a complex engineering system of electrical and mechanical components made of various conductive and insulating materials. When in service, these systems are exposed to multiple environmental stresses (atmospheric corrosive gases, contaminants, high and low temperatures); mechanical stresses (vibrations, shocks, handling); electrical stresses and electrostatic discharges; and many other internal and external impacts. The effect of stresses is cumulative, leading to progressive damage and significant deterioration (aging) of the electrical systems. Continuous aging sooner or later results in disruption or even complete depletion of the ability of the electrical apparatus to function properly and safely. The second edition of this book presents, extends, and updates a thorough analysis of the factors that cause and accelerate the aging of conductive and insulating materials of which electrical apparatus is made.

The updated version of this book also includes additional parts and chapters that summarize the issues of the reliability and safety of electrical apparatus and supporting equipment in the expanding field of renewable energy generating technologies: solar, wind, tide, and wave power. The review of aging factors and mitigating means allowing to extend an equipment useful life also covers the structural elements and mechanical parts of the wind, wave and tidal turbines, as well as the specific issues of solar panels deterioration. The goal is to provide the knowledge and understanding of the importance of preventing equipment failure that frequently results from the aging, negligence, and unique outdoor environments such as seas and deserts, and severe climatic conditions. In the modern world of "green energy," the equipment providing clean electrical energy needs to be properly maintained to prevent a premature failure.

A thorough analysis of the factors that accelerate aging and cause the failure of various materials in electrical apparatus allows to suggest multiple techniques for diminishing the impact of deteriorating factors, thus preventing a premature failure. Various aging-mitigating procedures extending the life of the electrical and structural equipment have emerged and became available since the publication of the first edition of this book. The author's purpose is to help finding proper ways to improve a performance and extend the life of generating, transmission, and distribution electrical equipment by recognizing and slowing down the aging processes.

This book is designed to serve as a reference manual for engineering, maintenance, and training personnel to aid in understanding the causes of equipment deterioration. Under one cover, it makes available extensive information, which is very hard to obtain since it is scattered among many different sources such as manufacturers documentation, journal papers, conference proceedings, and general books on plating, lubrication, insulation, and so on.

The book is an important source of practical knowledge for different audiences, including electrical and maintenance engineers and technical personnel responsible for utilization, operation, and maintenance of transmission and distribution electrical equipment at virtually every power plant and industrial facility. College instructors and professors may use this source as supplemental material for teaching classes on electrical equipment maintenance concepts and procedures. Industrial training personnel may use this book to develop manuals on proper maintenance procedures and choice of materials. It teaches electric maintenance personnel to identify the signs of equipment aging and recommends various techniques for the protection of electrical apparatus from deterioration and damage. This book combines research and engineering material with practical maintenance recommendations given in layman's terms, which makes it useful for audiences of various levels of education and experience.

Author

Bella Helmer Chudnovsky earned her PhD in applied physics at Rostov State University (RSU) in Russia. For the first 25 years, she was working as a successful scientist for the Institute of Physics at RSU and since 1992 at the University of Cincinnati. During the last 12 years of her career, she worked as an R&D engineer for Schneider Electric-Square D Company, where her principal areas of activities were aimed at resolving multiple aging problems and developing means of mitigating deteriorating processes in power distribution equipment. In this field, she has published 40 papers in national and international technical journals and conference proceedings on topics that are summed up in the book *Electrical Power Transmission and Distribution: Aging and Life Extension Techniques*, published in 2012. In the second edition of the book, she included the review of the issues of aging and means of life extension techniques for renewable energy power equipment.

Acronyms

I. TRADITIONAL TRANSMISSION AND DISTRIBUTION EQUIPMENT

A

AAAC	All-aluminum alloy conductor
AAC	All-aluminum conductor
AB	Alkylbenzenes
AC	Alternate current
ACAR	Aluminum conductor alloy reinforced
ACGIH	American Conference of Governmental Industrial Hygienists
ACSR	Aluminum conductor steel reinforced
AGMA	American Gear Manufacturers Association
AIS	Air-insulated substations
AMG	Aging management guidelines
AMS	Aerospace material specification
ANSI	American National Standards Institute
ASTM	American Society for Testing of Materials
ATH	Alumina trihydrate

C

CB	Circuit breaker
CBM	Condition-based maintenance
CCF	Common-cause failures
CCT	Continuous current test
CD	Current density
CIC	Cable in conduit
CIGRÉ	International Council on Large Electric Systems (Conseil International des Grands Réseaux Électriques)
CM	Corrective maintenance
COTS	Commercial off-the-shelf
CR	Contact resistance
CSPE	Chlorosulfonated polyethylene (synthetic rubber)
CT	Current transformer

D

DC	Direct current
DDF	Discharge dissipation factor
DES	Disconnectors and earthing switches
DGA	Dissolved gas analysis
DP	Degree of polymerization
DPC	Diphenylcarbazide (test)
DOD	Department of Defense
DOE	Department of Energy
DS	Disconnect switch
DTS	Distributed temperature sensing
DWV	Dielectric withstanding voltage

E

EC	Electrical conductor (grade of aluminum)
EDS	Energy-dispersive x-ray spectroscopy
EIM	Electrical insulating material
EIS	Electrical insulating system
EMAT	Electromagnetic acoustic transducers
EMI	Electromagnetic interference
EN	Electroless nickel
ENIG	Electroless nickel immersion gold
EP	Electrode potential
EP	Extreme pressure
EPDM	Ethylene propylene diene monomer
EPM	Electrical preventive maintenance
EPR	Ethylene propylene rubber (type of cable insulation)
EPRI	Electric Power Research Institute
ES	Earthing switch
ETFE	Modified ethylene tetrafluoroethylene (type of cable insulation)

F

FA	Fatty acid
FAA	Federal Aviation Administration
FEP	Fluorinated ethylene propylene (type of cable insulation)
FFA	Furfural analysis
FRA	Frequency response analysis
FOTC	Fiber-optic transmission conductor
FOV	Field of view
FTR	Fiber-optic transceiver
FxHy	Fluorinated thiol
FxOHy	Fluorinated ether thiol

G–H

GCA	General condition assessment
GIS	Gas-insulated substation or switchgear
GRP	Glass-reinforced plastic
GTPP	Geothermal power plant
HASL	Hot air solder leveled
HDG	Hot-dip galvanizing
HF	High frequency
HK	Knoop hardness
HMWPE	High-molecular weight polyethylene (type of cable insulation)
HNBR	Hydrogenated nitrile butadiene rubber
HP	High potential
HPEN	Electroless nickel with high content of phosphorus
HPLC	High-performance liquid chromatography
HSLA	High-strength, low-alloy (steel)
HV	High voltage
HVAC	Heating, ventilation, and air conditioning
HVIC	High-voltage insulator coating

I

IACS	International Annealed Copper Standard
IC	Integrated circuit

ICPC	Inductively coupled plasma spectroscopy
IDLH	Immediately dangerous to life or health
IDT	Interdigital transducers
IEC	International Electrotechnical Commission
IEEE	Institute of Electrical and Electronics Engineers
IMC	Intermetallic compound
iNEMI	International Electronics Manufacturing Initiative
IR	Insulation resistance
IR	Infrared
ISM	Industrial, scientific, and medical (radio bands)
ISO	International Standards Organization
IT	Intellectual technology
IT	Instrument transformer
ITAA	Information Technology Association of America

L–M

LDM	Laser distance meter
LPEN	Electroless nickel with low content of phosphorus
LTC	Load tap changer
LV	Low voltage
MCC	Motor control center
MCCB	Molded case circuit breaker
MCW	Microcrystalline wax
MEMS	Micro electro-mechanical system
MF	Major failure
mf	Minor failure
MFG	Manufacturer
MFG	Mixed flowing gas (test)
MSDS	Material safety data sheet
MTTF	Mean time to failure
MV	Medium voltage

N

NASA	National Aeronautics and Space Administration
NCI	Nonceramic insulator
NEC	National Electric Code
NEI	Nuclear Energy Institute
NEMA	National Electrical Manufacturers Association
NEPP	NASA Electronic Parts and Packaging
NFPA	National Fire Protection Association
NIOSH	National Institute for Occupational Safety and Health
NLGI	National Lubricating Grease Institute
NOX	Nitrogen oxides
NRS	Nuclear Regulatory Commission
NUMARC	Nuclear Management and Resources Council

O

OCB	Oil circuit breaker
OEM	Original equipment manufacturer
OHTL	Overhead transmission line
OLTC	On-load tap changer

OQA	Oil quality analysis
OSHA	Occupational Safety and Health Administration
OSP	Organic solderability preservative

P

PAG	Polyalkylene glycol
PAO	Polyalphaolefins
PBB	Polybrominated biphenyl
PBDE	Polybrominated diphenyl ether
PCB	Printed circuit boards
PCB	Polychlorinated biphenyls
PD	Partial discharge
PdM	Predictive maintenance
PE	Polyethylene
PF	Power frequency
PFAE	Perfluoropolyalkylether
PFPE	Perfluorinated polyether
PILC	Paper insulated with lead sheath (type of cable insulation)
PM	Periodic maintenance
POE	Polyolesters
PPE	Polyphenyl ether (type of cable insulation)
PPLP	Laminate of paper with polypropylene (type of cable insulation)
PPP	Paper with polypropylene (type of cable insulation)
PRD	Pressure relief device
PSTM	Point source to tower measurement
PTFE	Polytetrafluoroethylene
PVC	Polyvinyl chloride

R

RCM	Reliability-centered maintenance
RF	Radio frequency
RFI	Radio-frequency interference
RH	Relative humidity
RIV	Radio-influence voltage
RLC	Electrical circuit consisting of resistance (R), inductance (L), and capacitance (C)
ROHS	Reduction of hazardous substances
RR	Red rust
RTB	Reactor trip breakers
RTD	Resistive temperature detector
RTI	Relative temperature index
RTV	Room temperature vulcanization (silicone)

S

SAE	Society of Automotive Engineers
SAW	Surface acoustic wave
SCC	Stress-corrosion cracking
SEM	Scanning electron microscopy
SHC	Synthetic hydrocarbons
SIR	Silicon rubber
SS	Salt spray (test)

T

TAN	Total acid number
TBM	Time-based maintenance
TDCG	Total dissolved combustible gases
TDS	Technical data sheet
TDS	Total dissolved solids
TEV	Transient earth voltage
THI	Transformer health index
TLV	Threshold limit value
TPPO	Thermoplastic polyolefin (type of cable insulation)
TPR	Thermoplastic rubber
TR	Temperature rise
TRS	Total reduced sulfur
TR-XLPE	Tree-retardant cross-linked polyethylene (type of cable insulation)

U–V

UHF	Ultra-high frequency
UL	Underwriters Laboratories Inc.
UPS	Uninterruptible power supply
URD	Underground residential distribution
UV	Ultraviolet
VCB	Vacuum circuit breakers
VCI	Vaporized corrosion inhibitors
VI	Viscosity index
VLF	Very low frequency
VOC	Volatile organic compound
VT	Voltage transformer

W–X

WEEE	Waste Electrical and Electronic Equipment
WHS	Winding hot spot
WI	Whisker index
WR	White rust
WTMS	Wireless temperature monitoring system
WWTP	Wastewater treatment plant
XLPE	Cross-linked polyethylene (type of cable insulation)
XLPO	Cross-linked polyolefin (type of cable insulation)
XRD	X-ray diffraction

II. RENEWABLE ENERGY EQUIPMENT

A–B–C

ACORE	American Council on Renewable Energy
AE	Acoustic emission
ALWC	Accelerated low water corrosion
ASCE	American Society of Civil Engineers
ASES	American Solar Energy Society
AWEA	American Wind Energy Association
AZ	Atmospheric zone
BCSE	Business Council for Sustainable Energy
CanWEA	Canadian Wind Energy Association

CMS	Condition monitoring system
CSA	Canadian Standard Association
CSP	Concentrated Solar Power

D–E–F

DIBt	Deutsches Institut für Bautechnik (DIBt) (German Länder Governments)
DIN	Deutsches Institut für Normung (German Institute for Standardization)
DTF	Dry film thickness
EMEC	European Marine Energy Centre
EP	Epoxy paint
EPT	Energy payback time
EVA	Ethylene vinyl acetate (encapsulation)
EWEA	European Wind Energy Association
EWTS	European Wind Turbine Standard
FIT	Feed-in tariff
FOD	Foreign object debris
FRP	Fiberglass reinforced plastic

G–H–I

GL	Germanische Lloyd
GWEC	Global Wind Energy Council
HAWT	Horizontal axis wind turbines
ICCP	Impressed current cathodic protection
IWES	Fraunhofer Institute for Wind Energy and Energy System Technology
IZ	Immersion zone

L–M–N–O

LCA	Life cycle assessment
LCIA	Life cycle impact assessment
LEC/LCOE	Levelized energy cost/Levelized cost of energy
MIC	Microbiologically influenced corrosion
NACE	National Association of Corrosion Engineers
NERC	North American Electric Reliability Corporation
NORSOK	Norsk Sokkels Konkuranseposisjon (Norwegian Technology Centre)
NREL	National Renewable Energy Laboratory
O&M	Operations & Maintenance
OTEC	Ocean thermal energy conversion
OWC	Oscillating water column

P–R

PET	Polyethylene terephtalate film (Mylar)
PID	Potential-induced degradation
PRO	Pressure-retarded osmosis
PSA	Pressure-sensitive adhesives
PSP	Pneumatically stabilized platform
PUR	Polyurethane resin
PV	Photovoltaic
REC	Renewable energy credits
RED	Reverse electrodialysis
RES	Renewable Electricity Standard
RPS	Renewable Portfolio Standard
RUM	Reactive Unscheduled Maintenance

S–T–U

SCADA	Supervisory Control and Data Acquisition
SEPA	Solar Electric Power Association
SHM	Structural health monitoring
SPI	Solar Power International
SREC	Solar Renewable Energy Certificates
SWCC	Small Wind Certification Council
SWT	Small wind turbine
SZ	Splash zone
TCM	Turbine condition monitoring
UAV	Unmanned aerial vehicle
USP	Utility Scale Power
UT	Ultrasonic testing
UWIG	Utility Wind Integration Group
UZ	Underwater zone
VAWT	Vertical axis wind turbines

W

WEC	Wave energy converter
WEC	White etching crack
WFMS	Wind Farm Management System
WPA	Wind Powering America
WRA	Wind Resources Area
WREZ	Western Renewable Energy Zone
WSF	White structure flaking
WT	Wind turbine
WTF	Wind turbine farm
WTG	Wind turbine generator

Section I

Transmission and Distribution Electrical Equipment

1 Electrical Contacts
Overheating, Wear, and Erosion

1.1 ELECTRICAL CONTACTS

Electrical contacts provide electrical connections, and the primary purpose of an electrical contact is to provide the passage of electrical current across the contact interface. There are four categories in contact design: make–break contacts, sliding contacts, fixed contacts, and demountable contacts.

1.1.1 CLASSIFICATION OF ELECTRICAL CONTACTS

1.1.1.1 Make–Break Contacts

Make–break contacts are designed either to maintain the connection between the conducting materials or break apart during overload or short circuit.

Make–break contacts come in several types, which depend on the level of power load. Higher current types of make–break contacts are almost always made out of copper, due to its high conductivity. Smaller make–break contacts are often made of copper on the inside and a silver alloy on the outside. Different interrupting medium (compressed air, oil, SF_6, vacuum, etc.) is used for insulation and arc quenching. Separable and breaking contacts control the electric circuit in plug connectors and circuit breakers (CBs), and are used for periodical closing and opening of an electrical circuit in different contactors, switches, relays, and so on.

Make–break contacts include *stationary* contact and *moving* contact.

Stationary contacts provide the permanent joint and also are divided into two categories: *clamped* (bolted, screwed, and wrapped) and *nonseparable* (welded, soldered, and glued). In moving contacts, at least one contact member is rigidly or elastically connected to the moving part of the device. The contacts may be tipped with an arc-resistant material to resist erosion from the high-power arc, and the surfaces may be plated with silver to improve conductivity. *Moving* contacts are also of several kinds: *separable, breaking,* and *sliding.*

1.1.1.2 Sliding or Rolling Contacts

Sliding contacts are made of the conductors that slide over each other without separation. The sliding contacts are used in electrical machines, current pickoffs of transport and lifting machines, where they commutate currents of high and moderate intensity. Sliding contacts are used also in radio-electronic devices, in control and automatic systems operating on low-current level [1].

Sliding or rolling contact design is based on a different concept than make–break contact. These contacts never actually break contact with one another during relative movement which is breaking a flow of electricity. These contacts will move fast during this movement, and therefore must be extremely durable. Sliding contacts provide very little resistance, so they work quickly and accurately.

As a rule, low-current sliding contacts should combine high wear resistance and low friction with high stability of the contact voltage drop. To meet these requirements, contacts with good oxidation and wear resistances made of gold, platinum, rhodium, and palladium alloys are widely used. In addition to high cost of these components, they are not always suitable to meet the service requirements such as vibration stability, impact resistance, and reliability under severe weather and extreme conditions (e.g., low-current sliding contacts of ships and airplanes).

1.1.1.3 Fixed Contacts

Fixed contacts or connections must be clamped or bolted together at the beginning of their use and remain that way for years without being moved. They are used to maintain the overall integrity of the component of which they are a part. These contacts are not stopping or starting electrical flow. Fixed contacts include a wide range of *bolted* and *crimped (clamped)* contacts. Terms *crimp* and *clamp* are often used to describe the method used to attach a connector to coaxial cable. With a crimp connector, a ring is crimped around the outer conductor (shield) to a slotted or knurled stem. A clamp connector uses a V-shaped wedge ring to secure the outer conductor (shield) to the connector body.

Crimped joints employ the ultimate extreme force of contact making, causing the metal to flow, and make a permanent connection. These types of joints are very attractive for permanent connections due to the trouble-free nature of these joints, and the simplicity and rapidity of the crimping operation. Bolted or crimped contacts are used in interrupting chambers to secure and to maintain the integrity of the electrical component. Bolting is used because it is cheap and convenient.

A *clamped joint* avoids the reduction in cross section caused by drilling to insert bolts, and gives a more uniform distribution of the contact force, making the contact more efficient and hence running cooler. Clamped contacts are made by mechanically joining conductors directly using intermediate parts, specifically, clamps. These contacts may be assembled or disassembled without damaging the joint integrity. The simplest case of a clamped contact is the joint of two massive conductors with flat contact surfaces, such as busbars.

1.1.1.4 Demountable (Detachable) Contacts

Demountable (detachable) contacts are used to detach a breaker unit from an electrical network, which allows the breaker to be repaired or maintained. This type of electrical contacts is usually found in medium-voltage (MV) metalclad CBs. It helps in taking the breaker off the network by easily sliding it off the busbars for maintenance purposes, which should be done off load.

These contacts, like the make–break contacts, may be carrying high currents at high voltages (HVs; e.g., HV isolators or high or MV fuse contacts). They have to carry current reliably for long periods, without overheating or loss of contact, but do not control whether the current continues to flow through the breaker, like a make–break contact.

Demountable (detachable) contacts are not subjected to the stress of arcing; hence, they do not get the inherent cleaning action associated with it. They are frequently designed to have some frictional action on closing to remove superficial oxide or corrosion films which might impede contact. These contacts have a high contact force but not so high as the contact force in a bolted contact because of the excessive mechanical wear which would be caused when separating the contacts.

Demountable contacts are almost always attached and detached while there is no electricity running through the breaker. *Copper and its alloys* are the most frequently used materials for the bulk of demountable contacts.

1.1.2 Parameters of Electrical Contacts Affecting the Performance

In all types of electrical contacts, the conducting materials join together and the performance of electrical contact often depends on quality of the joint and conditions of the materials in mating parts. Additionally, the mechanical ability to break the electrical current efficiently during an overload or a short circuit is a very important factor in make–break and sliding contacts.

In automotive application, the electrical contact reliability is an essential issue as the cars experience severe operating conditions such as the engine vibration, operating temperature, and outdoor environmental conditions. The performance of electrical systems and devices depends on the reliability of electrical contacts; therefore, a thorough understanding of what may cause the failure in electrical contacts is extremely important.

To improve a performance and extend, the life of electrical contact and of electrical equipment in whole, multiple parameters and conditions should be considered and analyzed. The reliability and longevity of electrical contact depend on many factors determined by design (electrical and mechanical), and may be enhanced by improving environmental and service conditions [2].

Electrical and mechanical designs play an important role for the contacts, which functions is to interrupt an electrical current, to provide a required function during the life of equipment.

In *electrical* design the following parameters should be considered:

- Voltage (circuit voltage, voltage at the contacts during circuit interruption)
- Current (amperes, direct or alternating current, frequency)
- Type of load (inductive, capacitive, resistive, motor load, and overload requirements)
- Contact protection and arc suppression

In *mechanical* design to provide proper contact when the electricity is flowing or should be interrupted, the following factors should be considered:

- Contact force
- Frequency of operation
- Speed of opening and closing
- Wipe or slide between contacts on closure, or butt type closure
- Chatter during opening, or bounce during closing
- Contact gap when fully open

It is important to take into account the method of operating contacts: mechanical, electromagnetic, thermostatic, or manual.

Environmental conditions should always be considered for all types of electrical contacts to control the quality of metallic surfaces connected together in the contact whether they are in open air or sealed. It is important to take into account a *composition of atmosphere* in which contacts will operate: air, other gases such as nitrogen, and various corrosive gases. It is important to evaluate the *presence of salt in air, dust, and particles* from wear of adjacent parts as well as to know what would be the range of *ambient temperatures and humidity*. Several service requirements and conditions should be under control to provide safe performance of the electrical contacts.

Contact resistance (CR) can affect both reliability and operating life. Specific resistance must permit carrying of required current without producing a destructive overheating. Interface resistance must be low enough to provide safety of the circuit.

Contact erosion may result from burning under arcing, or mechanical wear. Erosion rate must be evaluated in choosing both contact material and design.

Material transfer (a deposit of material on one contact from another, with corresponding peaks and craters) may interfere with proper operation or calibration of many devices.

Contact sticking (or welding) occurs when contacts fuse together due to heat from electrical arcing or excessive electrical overloads. Material selection, provision for heat dissipation, and opening forces must be sufficient for the device.

In the following sections some of the important factors, which may strongly affect the quality and performance of electrical contacts are analyzed.

1.2 ELECTRICAL RESISTANCE AND TEMPERATURE RISE ON ELECTRICAL CONTACTS

The principal function of an electrical connection is to satisfactorily carry the electrical load over its entire service life. The electrical load can be expected to have daily fluctuations from no load to full load and frequently to very heavy overloads, thus causing wide fluctuations of operating temperature. In addition, the ambient temperature can be expected to fluctuate between daily extremes and between seasonal extremes. The effect of this heat cycling on a poorly designed or improperly installed connection is frequently progressive deterioration and ultimate failure of the connection or associated equipment.

Multiple aging processes result in gradual degradation of electrical connectors. This degradation may be generated by combination of contact wear, corrosion, decrease of the pressure, and other processes observed in the service. Degradation of contact surfaces induces conduction problems, which in turn result in local temperature rise (TR).

Rise in contact temperature is one of the major factors that pose a big threat to the stability of electrical contacts. Continuous exposure of electrical contact to elevated temperatures contributes to the downgrading of contact quality. Therefore, measuring contacts TR provides an important and quite convenient method of monitoring the condition of electrical connections [3].

1.2.1 DEFINITION OF CONNECTOR TEMPERATURE RISE

TR on electrical connector depends on multiple factors, such as thermal and electrical properties of the connector material; mechanical configuration of the connector; electrical properties of connection interface; ability of surrounding structures to transfer heat to and from a connector, and so on.

An attempt to develop a simple approximation for TR in an electronic connector requires many simplifications and assumptions [4,5], such as (1) the connection is made of two contacts of equal conducting length and equal and constant cross-sectional area; (2) the fixed ends of the contacts are assumed to function as heat sinks at ambient temperature; (3) resistance at the contact interface is assumed negligible; (4) thermal energy is generated at a constant rate in the material, based entirely on resistive heating; (5) the maximum temperature occurs at the contact interface; (6) all excess thermal energy is carried away by conduction through the base metal; (7) there is no heat transfer by convection or radiation.

The model of the connection TR based on these assumptions [4] was suggested in 1976 by R.E. Colin. The simplified equation shows that TR grows proportionally to the square of current and length of conductor, and is inversely proportional to the square of the cross-sectional area of the conductor and its electrical and thermal conductivity.

The usefulness of this equation as a predictive tool is limited because of multiple assumptions and simplifications used to create it. In other words, any resemblance to the actual TR of a real contact is often coincidental. In fact, to calculate TR on busbars or plug-in connections in particular installation, one should consider many factors effecting TR. These factors are heat transfer via convection, radiation and conduction, heat dissipation and absorption in the structure surrounding connection, and so on. Also, the variables used in the equation will affect the TR regardless of contact shape [4].

However, this simple approximation maybe helpful in some cases. For example, it allows to make comparisons between different materials used in the same design serving as a measure of a relative performance rather than an actual performance.

According to the simplified assumptions, connector TR is proportional to the square of the current passing through it. This is negligible in signal contacts but more important in power contacts and switches.

The change in temperature is also proportional to the square of the conducting length. As length increases, bulk resistance increases, and the heat sinks are farther away from the hottest point on the connector. With the current miniaturization trend in connectors, this becomes an advantage. Correspondingly, the TR is inversely proportional to the square of the cross-sectional area. With a larger cross section, the current can pass through more efficiently and can better carry away heat. The smaller cross-sectional areas associated with miniaturization will thus offset the gains made by decreasing the length.

TR is inversely proportional to both thermal and electrical conductivity of the material. Higher conductivity materials will perform better than lower conductivity materials in any given contact. Additionally, higher conductivity materials like copper beryllium will allow for thinner and narrower cross sections and smaller contacts, resulting in cost savings for the connector manufacturer.

1.2.2 Thermal Condition Causing Connector Failure

Excessive temperature causes connectors to fail by breaking down the insulation or the conductivity of the connector material. Failures usually occur in an avalanche—type style, which often happen as a sequence of the following events. If operating temperature increases, insulation tends to become more conductive. Simultaneously, conductor resistance increases, which results in further TR. This effect leads to the temperature raising above the maximum designed operating temperature, resulting in damage to the insulation material and the conductor. The resulting failure can be either partial or complete.

Complete failures occur if the operating temperature reaches the point where the conductor begins to melt, breaks down electrical conductivity, or where the insulation fails. A graph of service life versus operating temperature is shown in Figure 1.1. As shown, service life is directly proportional to how close operating temperature is to the maximum rated insert temperature of the device [5].

Prolonged operation, at elevated temperatures and humidity, within rated values, can result in overall degradation of connector performance, for example, increases in CR, corrosion of the shell, deterioration of the insert material, lessening of locking spring effectiveness (resilience aging is accelerated), jamming of corroded threaded coupling mechanisms, and so on.

Low temperatures can also cause conductors to fail, but such failure mechanisms are relatively rare. Low temperatures actually tend to slow the detrimental chemical effects, increase conductivity,

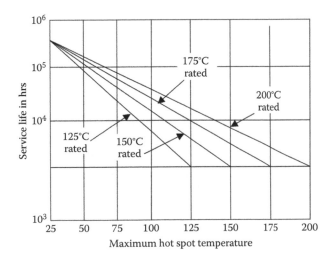

FIGURE 1.1 Service life versus hot spot temperature. (Adapted from *Connectors: Failure Mechanisms and Anomalies*, Online. Available: http://www.navsea.navy.mil/Home/WarfareCenters/NSWCCrane/Resources/SD18/Products/Connectors/FailureMechanismsAnomalies.aspx)

and result in a longer life. However, extremely low temperatures can cause damage to the nonmetallic portions of the connector [5].

Contaminant ingress generally leads to corrosion, which also impedes current flow and increase energy loss in the electrical interfaces of the connector. Eventually, the electrical CR increases to the point where the connector temperature becomes very high, ultimately leading to catastrophic failure [6].

1.2.3 Overheating of Connectors and Detection Techniques in Transmission and Distribution Systems

Two techniques are traditionally available for live-line inspection of connections in transmission and distribution lines: infrared thermography (IRT) and electrical resistance measurements. To determine thermal conditions of electrical connections, utility companies often rely solely on temperature measurements based on IRT surveys to manage maintenance activities.

A number of utility companies have developed empirical guidelines defining which maintenance action is required at specific electrical connector thermal condition. An example of such guidelines [6], adopted by a North American company, is given in Table 1.1.

One major limitation of IRT in surveying outdoor power lines is the effect of convective cooling by wind. Even for connections in the late stages of degradation, convective cooling may mask the deteriorating connection. Difficulties also arise in the absence of significant convective cooling. In this case, deteriorating connections maybe appreciably warmer than IRT measurements indicate due to a number of inaccuracies in calibration and environmental factors. Utility companies are aware of the limitations of IRT-based guidelines and are continually reviewing and updating them.

Measurement of connection resistance across the splice and across a conductor length identical to the splice length allows to determine connector resistance rise, which results in thermal runaway and consequent failure. Several devices are available commercially that allow live-line resistance measurements which provide earlier detection of connection overheating than measured by IRT technique.

One such device is the Ohmstik™ manufactured by SensorLink Corporation of Acme, WA, available worldwide. This device may be operated from an electrically insulated hot stick and allows rapid resistance measurements at line voltages as large as 500 kV [7]. It measures the micro-ohm resistance of conductors, connectors, splices, and switching devices positioned directly on an energized, HV lines. This measurement is much more direct than IRT, and is not

---

TABLE 1.1

Thermal Condition of Electrical Connector and Required Maintenance Action

Class	Temperature Rise	Stage of Overheating	Maintenance Action
I	>0.5°C	Minor	Connection should be closely monitored and repaired at next scheduled maintenance
II	5–30°C	Significant	Repair at first opportunity with due consideration to loading environment
III	>30°C	Severe	Repair immediately with due consideration to minimizing power disruption

Source: Modified from Timsit R. S. Origin, detection, and cost of connector degradation in electrical transmission and distribution systems, *Proceedings of the 1st International Conference on Reliability of Electrical Products and Electrical Contacts*, Suzhou, China, p. 1, 2004.

[a] Temperature rise based on IRT measurements (50% maximum load) and assessment of connector condition.

FIGURE 1.2 Overhead robot for inspection of HV power lines. (From *Expliner-Robot for Power Line Inspection*, HiBot, Online. Available: http://www.hibot.co.jp/en/products/robots_1/expliner-robot-for-power-line-inspection_12. With permission.)

subject to emissivity, weather, current loading, background, and other influences that cause infrared (IR) errors.

The most advanced development for deterioration detection of power line connections are inspection robots, such as The Expliner robot by HiBot (Hitachi High-Technologies Corporation, Japan), Québec's LineScout, and the Electric Power Research Institute (EPRI)'s robot are the latest techniques in providing the services of inspecting HV power lines.

Japanese Expliner (Figure 1.2) is a self-propelled robot [8] that moves along overhead high-voltage transmission lines to perform inspection of the lines by checking their external conditions, measuring the diameter, and even detecting internal corrosion. Expliner is also used to obtain detailed images of cable spacers, jumpers, insulation discs, and other components.

1.3 WEAR AND WIPE OF ELECTRICAL CONTACT

Proper performance of electrical contact depends strongly on mechanical damage of the contact's surface which is called wear. Connector life is affected by cyclic wear out failure mechanisms. As the connector contacts are repetitively engaged and disengaged, plating surfaces are eroded and exposed to corrosive elements of the surrounding atmosphere.

Repeated mating and un-mating also results in physical wear of the contact material, affecting the integrity of the connecting interfaces, the connector shell engagement interfaces, and the mounting/cable attachment hardware.

Surface contact points become worn, making unsymmetrical contacts and corrosion results in nonconducting films on the contact surface. The result can be a significantly increased interface resistance. In power connection applications, an increase in temperature at the interface can accelerate further contact interface deterioration [5].

The metal surface of any new electrical contact is usually covered with asperities, which are slight protrusions from a surface such as the bumps or points. As two mechanical surfaces slide past each other these asperities collide, resisting motion. As the force increases, asperities are sheared off or deform eventually becoming loose wear particles.

Electrical contacts are subjected to wear because of shock, vibration, and thermo-mechanical stresses resulting in fretting, increase in CR, and eventual failure over the lifetime of the product. For example, in high speed sliding contacts the excessive heat due to the intense electrical and frictional heating can result in high temperatures near the electrical contact interface [6]. This may degrade the performance of materials and result in severe wear. The wear process in high-speed sliding electrical contact is a big concern.

Wear is a very complex phenomenon. Based on the failure mechanism, wear can be defined in many ways and categorized into several types such as abrasive wear, adhesive wear, surface fatigue and corrosion, and so on, as presented in Refs. [9–11].

Abrasive wear is described by a mechanism when a surface is moving against a harder material. The harder material could exist in the form of particles, which may enter the contact system externally or they can be internally generated by oxidation or other corrosive processes. The surface material is removed or displaced by abrasion due to rubbing against hard particles. The mechanical damage generated by abrasive wear generally has a deleterious effect on electrical contact reliability.

Adhesive wear occurs when asperities interact, leading to transfer of metal from one surface to another, which usually occurs at high speeds and temperatures. A severe form of adhesive wear is scuffing. Adhesive and abrasive wear are among the dominant mechanisms under electrical sliding conditions.

Scuffing wear is defined as surface damage characterized by the formation of local welds between sliding surfaces. In scuffing, there is a tendency for material to be removed from a slower surface and deposited on the faster surface, or material is removed from the hotter surface and deposited on the cooler surface.

Surface fatigue is a form of wear which is predominant in rolling contact bearings. These bearings are subjected to repeated intense loadings producing the shearing stress within the surface. This finally may result in pitting failure. Pitting wear usually occurs during the relative motions where stresses exceed the endurance limit of the material.

Fretting wear occurs as a result of very small oscillatory displacement between surfaces. In electrical contacts, fretting occurs when two loaded surfaces in contact undergo relative oscillatory tangential motion, known as slip, as a result of vibrations or cyclic stressing [1,12,13]. The amplitude of relative motion is very small. When relative motion exists between two surfaces, the surfaces can be attacked by a variety of wear modes; they can be damaged in different ways depending on various factors like the thermal and chemical environment at the point of contact and materials of the mating surfaces and surface properties.

As fretting proceeds, the area over which slip is occurring usually increases due to the incursion of debris. The amount of debris produced depends on the mechanical properties of the material and its chemical reactivity. This debris produced by fretting is mainly the oxide of the metal involved, and the oxide occupies a greater volume than the volume of metal destroyed. If space is confined, this will lead to seizure of the contact.

Since fretting wear actually consists is a combination of several forms of wear, initiated by adhesion, amplified by corrosion and having its major effect by abrasion or fatigue, it is impossible to predict which mode is dominant. It is possible, however, to select the dominant wear mode based on the type of system, the nature of relative motion between the contacting surfaces, and the application. More about fretting processes can be found in Chapter 4 of this book.

Contact wipe is an action designed into a contact so that the contact motion exceeds the initial electrical contact touch point. Using a flexible contact arm causes the contact surface to wipe against its counterpart as the arm bends slightly. Contact wipe is measured as the distance of travel (electrical engagement) made by one contact with another during its engagement or separation or during mating or un-mating of the connector halves. The wipe removes oxide and contaminants from the electrical contact surface improving electrical resistance. The so-called "wiping motion" has a strong surface cleaning effect, which is desirable; it is even a necessary condition for good metal to metal contact. However, it causes also deformation and wear [2].

A correlation between contact *wiping motion* and *vibration frequency* effecting CR was studied in Ref. [14] based on a comprehensive model. It was shown that debris builds as two surfaces in electrical connectors wipe against each other in a cyclical way. The distance at which two surfaces rub affects CR. The longer was the distance (longer wipe), the more rapidly electrical resistance increased because more material was exposed to air than at shorter wipe. It also was demonstrated that low-frequency vibrations increased debris buildup more than high-frequency vibrations.

1.3.1 MEANS TO INCREASE WEAR RESISTANCE

1.3.1.1 Choice of Contact Material and Plating

Electrical contacts are usually plated in order to prevent corrosion and increase resistance to wear caused by corrosion. Plating of detachable electrical contacts experiences wear because of the motion between contacts or it may corrode itself. Once the protecting plating has been worn out, electrical contact will fail rapidly due to corrosion or fretting corrosion. Therefore, the wear resistance of the plating is a very important parameter for the long lifetime of electrical contacts [15].

Many measures which improve the wear resistance can diminish the conductivity of the plating. Due to the fact that plating of electrical contacts must have both a high wear resistance and a high-electrical conductivity, the manufacturing of high-performance plating of electrical contacts creates a great challenge.

1.3.1.1.1 Noble Metals

Using metals which have little or no oxide film-forming tendency, generally known as noble or precious metals, as plating materials is considered as an effective measure to avoid failures caused by wear by fretting. *Gold* is one of the most commonly used noble plating materials for high-performance electrical contacts. Since pure gold is very soft, using gold as coating material may diminish the contact lifetime caused by low wear resistance of the plating material [16].

In order to improve the wear resistance of gold plating, hard gold is usually used. The high degree of hardness is achieved by alloying gold with such elements as cobalt, iron, or nickel. Another technique of producing hard gold plating is modifying gold with nanoscale particles, such as aluminum and titanium oxides, or polytetrafluoroethylene (PTFE). However, the effect of alloying elements is limited by the galvanic process and other surface properties, which are also required for electric contacts [15].

The design requirements for contacts made of noble metals or with noble metal plating are fundamentally different from those for nonnoble metals. Important aspects with *noble* metals are a clean and smooth surface and a well-designed geometry. When corrosion is a concern, a pore-free layer should be provided over noble metal. A normal force of 0.3–0.5 N is sufficient to make good contact on clean noble surface. However, for a good reliability the commonly required minimum of 1 N and redundancy of two parallel contacts is still a good recommendation, particularly when short interruptions are of concern. At longer term, the major concern with noble metal contacts is to avoid pore corrosion and keep the surface free from contamination. It may be achieved by using pore-free underplating for gold plating [17].

With the gold price rising lately, the search of possible alternate plating for the applications where gold was usually applied was conducted in Ref. [18]. *Silver* has a unique combination of material properties such as the highest thermal and electrical conductivity of any metal and a relatively low hardness, which results in superior joule heating thermal performance and making it attractive for power applications.

Silver also does not have the "noble" character of gold. It will readily form surface tarnish films when exposed to atmosphere containing even minor concentration of hydrogen sulfide [19] in the presence of chloride and elevated humidity, which may accelerate formation of silver sulfide film. However, still silver has a long history of performing well even in the tarnished state in higher normal force/lower durability power applications and similar signal applications.

The fact that silver will tarnish in most connector environments and is not a durable finish may present a problem for silver used in signal connectors which require significantly lower normal force and higher durability, where gold plating is usually used. There are potential risks associated with applying and testing silver finished contacts in applications appropriate for hard gold.

In Ref. [18], it was determined that applying silver plating with nickel underplating in a hard gold finished connector design/application will most likely degrade these performance factors, though it may be a viable cost-effective finish for some situations by adjusting performance requirements of a connector to levels more appropriate for a silver finish (e.g., lower mating cycles), for example, by reducing the number of required cycles.

An effective way to reduce or prevent silver tarnish films from forming could be environmental shielding. Examples of environmental shielding methods are the use of connector housings (open, closed, or actively sealed), equipment enclosures, environmental controls, greases or gels, reduced sulfur, and/or sulfur absorbing packing material. They can all serve to limit the flow of corrosive elements to the silver contact surfaces [19].

It was suggested in Ref. [18] to use an appropriate environmental shielding as well as silver surface treatment, proprietary products capable of providing both environmental exposure resistance and lubricity. Surface treatments can be used to mitigate corrosion processes as well improve wear, durability, and insertion force performance. However, there are the potential risks associated with using a surface treatment outside of its proper operating range. If the surface treatment is exposed to a temperature at which it is not stable, it may cease to function depending on the duration of the excessive thermal exposure. Any selected surface treatment would have to be appropriate for the "lifetime" exposure of the intended application.

1.3.1.1.2 Nonnoble Metals

Contacts behavior is fundamentally different if they are made from or are plated with nonnoble metals. Copper and copper alloys (brass and bronze) are the most widely used base materials for electrical contacts. Lately, shape memory alloy (SMA) was also developed for electrical connector to prevent fretting corrosion.

Both copper and copper alloys are prone to corrosion and oxidation. Therefore, electrical contacts are usually coated with protecting plating. The most extensively used coating materials are tin, silver, gold, palladium, nickel, and various alloys of these metals. Since the surface of these metals may oxidize, much more force (5–10 N) is needed to break the oxide film and make the good connection.

The exception is tin, which is a very soft metal with a very thin and hard oxide layer. A relatively low force is enough to break a tin oxide layer and create a large contact area. However, if tin is used as plating over copper, it transforms into CuSn intermetallics, which has significantly higher electrical resistance than both tin and copper. When intermetallics is formed, still higher forces (5–10 N) are needed to make better connection for tin-plated contact like for other harder nonnoble metals. More about various types of plating and their properties are presented in Chapter 3 of this book.

Advantages of nonnoble metals are that they are not prone to pore corrosion and that the surface roughness and contamination are less important because of the higher force. With tin plating, a higher roughness can be even beneficial; the valleys can act as sources of fresh tin during the wear process. The higher force also makes relative motion less likely to occur, however, when relative motion does occur then fretting corrosion will take place. Fretting corrosion is the major failure (MF) mechanism for nonnoble contacts [13].

The effect of fretting on aluminum conductors coated with various contact materials was studied in Refs. [16,17]. It was observed that as fretting corrosion progresses, the degradation characteristics such as melting/arcing, abrasion, adhesion, and delamination wear occur in aluminum power applications using a variety of contact materials, both fretting amplitude and contact force affect the rate of fretting degradation, and the larger amplitudes produce greater degradation rates while the larger forces produce the lower rates.

There are typically two phases of contact wear in plated contacts. The first phase is the wear of the plating, which normally takes place faster than the second phase. The second phase is the wear of the interlayer or the base metal. There is usually a very good correlation between phase I of wear and lifetime of electrical contacts, since the time of phase I of wear corresponds to the time of partially worn-out plating which leads to first instabilities of electrical contacts. The rate of wear in phase I actually represents the wear performance of plating [15].

In high-speed sliding electrical contacts, self-lubricating contact materials are used, such as novel materials based on graphite. It was found that wear of these contact materials in the absence of an electrical current is a combination of fatigue and abrasive forms of wear. Without current, the wear rate is proportional to load and sliding speed, and increases when hardness of contact material decreases (Archards' wear equation). When electrical current is applied, wear processes are more complicated with different morphology of wear debris [9].

1.3.1.2 Current, Voltage, and Frequency

The electrical conditions determine how critical the different parameters are in a certain application. For high currents, the contact and constriction resistances need to be low to avoid overheating of the contact point. For low currents there is no special contact problem. With voltages, it is the other way around: with HVs the connection is not critical because fretting will improve the contact if the dry circuit resistance is too high, while with low voltages (LVs) high resistances may cause problems.

With the increasing use of high frequencies, several new problems have arisen. Short interruptions may switch equipment off or distort a data flow. Also, contacts for shielding and grounding, for example, in coax connections, become very critical and do no longer allow resistive and unstable connections from grounds and shields as it leads to common path distortion problems.

1.3.1.3 Lubrication

Lubrication as a technique to provide better performance and protection of electrical contacts is a very complex subject [19–26]. Solid and liquid lubricants are often used to enhance the performance of electronic and electrical connectors. In connector industry, one of the solid lubricants often used in power contacts is lamellar graphite, particularly with electrical brushes in machinery and electric trains, to take advantage of the unique lubrication and electrical conductivity properties of that material. Liquid lubricants reduce friction and minimize mechanical wear, they also must mitigate fretting corrosion and protect against atmospheric corrosion in contact areas of connectors [26].

In most applications, lubrication is advantageous to electrical contacts because it provides lower insertion forces and less wear. Lubricants with special additives also protect the contacts from corrosion. An important requirement for effective lubricant action in connectors is chemical inertness toward the metal surfaces to be lubricated. For example, lubricants used on copper or silver-plated surfaces should not contain sulfur-containing compounds because these may react with the metals to generate resistive metal sulfide films and increase CR.

However, lubrication has also disadvantages. For example, a lubricated surface may retain more dust, the wiping motion may get less effective, and lubricant may enable sliding micromotion where this would not take place without lubrication. Lubricant can polymerize and form insulating films; it can form a varnish and can creep to places where it is not wanted.

Proper and controlled application of lubricants is a difficult task. When lubricant is applied before assembly, it may be partly removed during the assembly process, but after assembly the contacts are mostly less accessible. Also, housings and packaging materials may become contaminated with lubricant.

The thickness of lubricating film is known to be important for the electrical function; however, it is hard to measure and hard to control initially and particularly after a few mating cycles has taken place. Lubricants must still be considered to be a part of the product surface finish, to be specified on the product drawing and subjected to the procedures of testing and quality control, like any other finish. More information about lubrication techniques and lubrication materials used for electrical contacts is given in Chapter 6.

1.3.2 Means to Evaluate a Degree of Electrical Contact Wear

The stability of electrical properties is one of the most important of the materials used for electrical contact. Wear of contact surfaces may adversely affect these electrical properties, especially in designs where the contact surfaces are relatively thin coatings.

American Society for Testing and Materials (ASTM) Standard B794 "Standard Test Method for Durability Wear Testing of Separable Electrical Connector Systems Using Electrical Resistance Measurements" [27] provides means to compare various material systems on a basis relevant to their application in electrical connector contacts. Repeated insertion and withdrawal of a connector may cause wear or other mechanical damage to the electrical contact surfaces, rendering those surfaces more susceptible to environmental degradation.

This test method is intended to detect degradation of the electrical properties of the connector by such processes. This Standard describes procedures for conducting wear and durability testing of electrical connectors; the procedures produce quantitative results.

The Standard specifies the test in which sample connectors are wired for precision resistance measurements of each test contact. The samples are divided into two groups; then resistance measurements are made of each test contact. The connectors in one group undergo a number of insertion/withdrawal cycles appropriate for the particular connector under test, and the resistances of these connectors are measured again. The connectors in the other group are not disturbed.

All samples are subjected to an accelerated aging test; then the resistances are measured again. All samples are separated (withdrawn), exposed to an accelerated aging test in the un-inserted condition, removed from the test, reinserted, and resistances measured again. The various resistance measurements are compared to detect effects of the wear and aging on electrical performance.

A special environmental test chamber is used to provide an accelerated aging test in the environment with ascending and descending temperature in the range 25–65°C and relative humidity (RH) in the range 87%–92% during 24 h. The samples should be positioned in the environment test chamber for 10 days between each measurement of resistance.

The results of the tests may be used to compare the performance of different connector designs so that meaningful design choices can be made. Such results may also be used to compare the performance of a connector to a previously established standard to evaluate the quality of the samples under test.

This Standard also specifies the techniques of measuring the effects of repeated insertion and withdrawal of separable electrical connectors which are harmful to the electrical performance of the connector. This method is limited to low-current electrical connectors designed for use in applications where the current through any one connection in the connector does not exceed 5 A, and where the connector may be separated a number of times during the life of the connector. This method is limited to electrical connectors intended for use in air ambient temperatures where the operating temperature is less than 65°C.

1.3.3 Effect of Wear Resistance on Life of Electrical Contacts

In applications where electrical contacts are subjected to wear because of shock, vibration, corrosion and fretting, and other deteriorating factors, the life of the contacts is significantly shorten

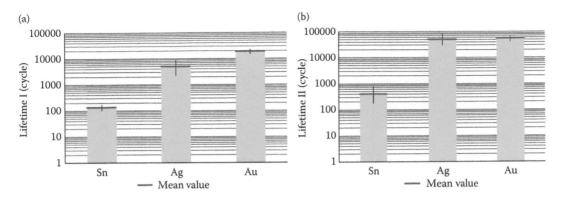

FIGURE 1.3 Lifetime I (a) and II (b) of contacts with different coating materials: tin, silver, and pure (soft) gold. (Adapted from Song J., Koch C., and Wang L. Correlation between wear resistance and lifetime of electrical contacts, *Advances in Tribology*, 2012, Article ID 893145, 9pp., Online. Available: http://www. hindawi.com/journals/at/2012/893145/.)

compared with those contacts which are exposed to lesser degree of the conditions that lead to increase in CR.

An interesting study in Ref. [15] shows the correlation between the lifetime of electrical detachable contact and wear resistance of plating protecting the base metal of the contact from corrosion. Different criteria may be used to determine the lifetime of electrical contacts. Useful life of the contact comes to end when electrical resistance increases so much that excessive heat produced when the current is flowing through the contact area may result in thermal failure of the contact. The number of cycles resulting in the rapid increase of CR was used as lifetime, since it has the strongest impact on the behavior of electrical contacts.

Two limits are set for the definition of the lifetime. Lifetime I is defined as the number of cycles which leads to a 300% increase of CR. Lifetime II is the number of cycles when the CR of 300 mΩ is measured. At the end of lifetime I, first damages on plating occur and the contacts can normally still work properly. At the end of lifetime II, a normal function of electrical contacts cannot be expected. The wear of the surfaces was evaluated by confocal laser scanning microscope (CLSM) which measured the surface topography with a precision of up to 10 nm. A scanning electron microscope (SEM) with a Focused Ion Beam System (FIB) and an energy-dispersive X-ray spectroscopy (EDX), an X-ray fluorescence spectrometer, and an optical microscope were also used for the material and surface analysis of the plating.

The study in Ref. [15] was performed on electrical contacts with tin, silver, and pure gold (soft gold) and hard gold plating (Figure 1.3). Lifetime I of tin plating was very low, only about 100 cycles. The reason for this was the softness of tin plating, which was partially worn out after a very small number of motion cycles. Contacts with both silver with the thickness 3 μm and soft gold plating (0.6 and 1 μm) displayed much longer lifetimes I and II. With increasing rate of wear (decreasing wear resistance), the lifetime of electrical contacts decreases.

The lifetime of gold plating can be further improved by means of modification of gold plating with hard alloying elements or nanoscale particles (hard gold), the rate of wear can be decreased by two to three decades and consequently the lifetime of electrical contacts can also be increased by more than two decades (Figure 1.4).

1.3.4 Effect of the Thickness of Plating on Wear Resistance

The study in Ref. [11] showed that the wear of gold plating is proportional to the cycles of motion; therefore, the number of cycles to the wear out of the plating is expected to be proportional to the

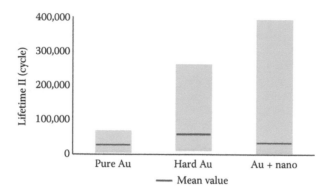

FIGURE 1.4 Effect of using hard gold and gold plating with nonoparticles on contact lifetime compared with pure (soft) gold. (Adapted from Song J., Koch C., and Wang L. Correlation between wear resistance and lifetime of electrical contacts, *Advances in Tribology*, 2012, Article ID 893145, 9pp. Online. Available: http://www.hindawi.com/journals/at/2012/893145/.)

thickness of the plating. At the end of the plating life, the plating is completely worn out and the fretting corrosion of the interlayer can begin.

The results of study in Ref. [12] showed that a minimum thickness of gold plating approximately 0.7 μm is required to achieve a long lifetime. However, thickness is a necessary but not sufficient condition for a long lifetime. Some samples with a thick plate also displayed a short lifetime.

Nickel underplating has a strong effect on the wear resistance of gold plating extending the lifetime of electrical contacts. Samples with nickel underplating of 2.3 μm or less survived less than 10,000 cycles. All samples which survived an exceptionally large number of cycles of more than 200,000 have a nickel interlayer of more than 3 μm. The detailed analysis of the contacts with gold plating modified with nanoscale particles and Ni underplating showed the longest average lifetime of about 120,000 cycles.

A very good correlation between the rate of wear and lifetime of electrical contacts specifically for hard gold plating modified with hard alloying elements and nanoscale particles previously studied for using in bearings in Ref. [15]. These types of gold plating with increased wear resistance applied to electrical contacts may result in large increase in the lifetime of electric contacts. However, it is very important to use optimal amount, size, and combination of nanoparticles in gold plating, the thickness of both the gold plating and the nickel interlayer in order to reach the desired lifetime improvement.

Effect of plating thickness on wear caused by fretting corrosion of tin-plated automotive connectors was studied in fretting motion experiments among other control parameters which included temperature, humidity, load, frequency, and other factors in Ref. [21]. It was found that load and tin thickness have a higher impact on the increase in CR caused by fretting corrosion than other factors. The thicker plating (5 μm) at lower load (1 N) caused much higher CR then thinner plating (1 μm) at higher load (3 N) due to thicker layer of tin oxide film on the contact surface.

1.4   WEAR AND EROSION OF ARCING CONTACTS

In separable and breaking contacts (or make–break contacts) during the opening of electrical circuit in different breakers, contactors, switches, and relays, the arcing or sparking occur. During these processes, ejection of contact material in the discharge takes place, which results in significant wear of contact surfaces. Damage to material resulting in electrical discharge is classified as a variety of erosion, specifically, *electroerosion wear* [1].

Electrical arc erosion plays a crucial role in the reliability and life of power switching devices. Depending on the contact material's behavior in response to an electrical arc, surface damage can induce severe changes in contact material properties that will impact the power switching device's functioning. Because serious accidents have been caused by contact failures, contact reliability is one of the most important factors for the design of such apparatuses.

Arc erosion induced by the electrical arc at each contact opening and contact closing plays a major role in reliability and safety defining life of power switching devices. Surface damages resulting from arcing can lead after a certain number of operations to contact welding, destruction, and failure, after which the power switching device cannot fulfill anymore one of the functions it has been designed for.

When the switch contacts start to separate, CR increases because of the fewer contact points until contact voltage reaches melting potential (in the range 0.4–0.7 V for most metals). Molten metal bridges the separating surfaces and finally breaks. Soon after the rupture of the molten bridge, the gap between the electrodes is filled by metal vapor coming from the explosion of the molten bridge. During this process, some material may transfer between contacts, especially if direct current (DC) is interrupted.

Rupture of the molten bridge opens the circuit at LV. When source voltage exceeds 14 V, an arc may form as the molten bridge ruptures, delaying circuit opening, and pitting the switch contacts. In summary, during one breaking operation of electrical contacts, arc erosion results from the material removed and ejected contact material particles from the electrode and its redeposition on the contact surfaces.

Arc erosion behavior of the contact and their performance depend significantly on the materials used for the contacts. The materials chosen for arcing contacts should be arc resistant. Among them are various alloys and oxides of copper (Cu), silver (Ag), tungsten (W) as well as metal composites with carbon (C). For example, arcing contacts in CBs are usually made from alloys AgW, CuW, AgWC. At higher switching current, the arc is hotter and the contact erosion increases. Also, contacts that switch inductive currents erode quicker than those with resistive loads.

1.4.1 ARCING EROSION IN SF$_6$ CIRCUIT BREAKERS

In LV, MV, and HV CBs, the process of arcing and the damage (erosion) of the arcing contacts depends on the type of the breaker and type of insulating media used in a specific design (air, SF$_6$, vacuum, or oil). Example of fixed contact damage caused by the arcing in SF$_6$ HV CB is shown in Ref. [27] (Figure 1.5).

To evaluate the condition of the arcing contacts, an internal inspection can be done; however, time-consuming and costly maintenance procedures must be followed in order to securely handle the SF$_6$ gas and arc byproducts. It should be remembered that excessive arcing contact wear and/or misalignment may result in a decrease of the CB's breaking capacity. A new dynamic-CR measurement method was developed in Refs. [28,29] and validated by field tests which were performed on air-blast and SF$_6$ gas CBs.

The new method is based on the breaker CR measurement during an opening operation at low speed. After reviewing the characteristics of the dynamic resistance curve and the measuring system and parameters, the relevant values that can be extracted from the resistance curve for detecting contact anomalies wear and/or misalignment.

1.4.2 WEAR OF THE CONTACTS IN HV AND MV AIR-BLAST CBs

Wear of moving contacts may cause a significant change in CR depending on the level of wear. In Figures 1.6 and 1.7, a new moving contact is compared with three moving contacts, one of them is slightly worn, the second one has higher degree of wear, and the third one is seriously worn [28].

FIGURE 1.5 Arcing spots on the fixed contacts of HV 230-kV reactor SF$_6$ circuit breaker. (From Landry M. et al. New measurement method of the dynamic contact resistance of HV circuit breakers, *CIGRE Paris Session*, 2004. Photo courtesy of Hydro-Québec.)

FIGURE 1.6 Moving contacts of HV air-blast circuit breaker: new contact (left) and slightly worn contact (right). (Modified from Landry et al. New measurement method of the dynamic contact resistance of HV circuit breakers, *Proceedings of the IEEE PES Transmission and Distribution Conference and Exposition*, pp. 1002–1009, Dallas, TX, May 2006.)

It was shown that by measuring dynamic resistance of the contact using technique developed in Refs. [27–29] allows to detect contact anomalies and evaluate the level of the contact wear. Dynamic CR of seriously damaged contact (right picture in Figure 1.7) is twice larger than that of the new contact (left picture in Figure 1.6).

Arcing damage of MV air-magnetic CB is shown in Figure 1.8.

FIGURE 1.7 Moving contacts of HV air-blast circuit breaker: worn contact (left) and seriously worn contact (right). (Modified from Landry et al. New measurement method of the dynamic contact resistance of HV circuit breakers, *Proceedings of the IEEE PES Transmission and Distribution Conference and Exposition*, pp. 1002–1009, Dallas, TX, May 2006.)

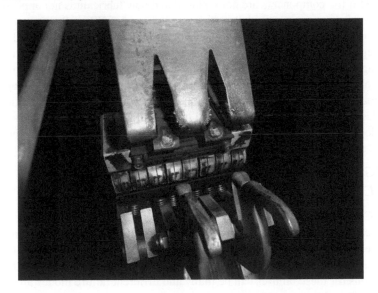

FIGURE 1.8 Damaged stationary and moving arcing contact with copper splatter in MV air-magnetic CB. (From Online. Available: http://www.circuitbreakersblog.com/2012/07/image-of-circuit-breaker-main-and-arcing-contacts/. With permission.)

1.4.3 Wear of CB Contact Due to the Lack of Lubrication

The electrical contacts in CB may experience wear due to wrong lubrication. In LV CB, the primary points of wear are (a) at the arcing contact/tensioner contact area and (b) at the contact roller/tensioner contact area.

An example of the contact wear in LV CB is given in Ref. [30]. The wear of the arcing contact and roller from an ITE breaker due to the lack of lubrication is shown in Figure 1.9. Lack of lubricant at

FIGURE 1.9   Contact wear in ABB/ITE K-600 LV circuit breaker. (From White J. Critical maintenance for circuit breakers, *NETA World Magazine*, pp. 1–8, Summer 2007, Online. Available: http://www.netaworld. org/sites/default/files/public/neta-journals/NWsu07-White.pdf. With permission.)

point (a) causes wear across both contact faces, while lack of lubricant at point (b) causes the breaker to slow down and grooving to be cut into both the roller and the tensioner contact. Eventually, the breaker will seize if the components are not replaced and new lubricant is not applied.

One of the most important rules of lubrication maintenance of CBs is that penetrating oils should never be used for lubrication [19], because it results in old lubricant being flushed out of the pivot points and not replaced with the real lubricant. When the breaker heats up due to current flow through it, the penetrating oil quickly dissipates and a contacts' wear begins. Original equipment manufacturers (OEMs) are always very specific about the lubricants used in their breakers giving that the breakers have to perform under a variety of conditions, many times in very adverse conditions. Using other lubricants not designed or proven to function well in these conditions will eventually lead to a failure. More on the rules of lubrication, maintenance of CBs is given in Chapter 6 of this book.

1.5   ARCING DAMAGE, COKING, AND FILMING OF ELECTRICAL CONTACTS IN TRANSFORMER LTC

The Load Tap Changer (LTC) or On-Load Tap Changers (OLTC) contacts of the transformer are exposed to switching operations in oil. Concerns were raised about the arc erosion behavior of the contact immersed in oil. Since, the OLTC is the only moving component of the transformer, it is particularly vulnerable. Approximately, half of all transformer failures are caused by failures in the OLTC. The reasons for these failures can be many and different failure modes are important under different operating conditions.

Common LTC failures result from increased CR, material failure, breakdown of insulating oil (coking), contact wear, and so on. A metallurgical investigation was conducted in Ref. [31] to assess the influence of the metallurgical characteristics of the contact such as microstructure and hardness on arc erosion of its surface in oil.

The variations in arc erosion damage may be caused by the differences in characteristics of the contact material that contains tungsten and copper, such as the hardness and microstructure. The hardness of the material influences arc erosion of the contact, but its influence was found much stronger on the contact microstructure.

Heating and *coking* of LTC contacts occur when the LTC is inactive for long periods or subjected to loading that exceeds maximum current rating of the transformer. When the utilities are

FIGURE 1.10 Heavy coking of LTC in ABB transformer UZC. (Courtesy of Dr. M. Duval, Hydro Quebec IREQ. With permission.)

experiencing LTC failures, failure analysis often reveals that the carbon was precipitating out of the oil onto the heated contact surfaces and forming a granulated, carbon-based deposit or "coke."

Coking is caused by carbonization of oil at high temperature. Also, particles of coke are very erosive to the contact surfaces and may cause premature wear. The carbon deposit acts as an insulator causing further heating. Temperature eventually exceeds the thermal operating limits of the contact assembly, resulting in thermal failure. Examples of failed LTC due to coking in different transformers are presented in Refs. [32,33]. A picture of heavy coking of LTC in ABB Transformer is shown in Figure 1.10. Moving contacts in LTC show different stages of wear during inspection (Figure 1.11).

FIGURE 1.11 Damaged LTC moving contact. (Courtesy of Rick Youngblood, Doble Engineering.)

Another phenomenon that may result in LTC failure is "filming," a formation of the films on the surfaces of the contacts, which are made of polymerized insulating fluid. High temperature and stress of the oil or other insulating fluids are the factors that lead to polymerizations of the oil, and bonding of long chains polymer structures to the contact surfaces.

With time polymerized oil forms a resinous, varnish-like film over the contacts and mechanism. Quenching of the arc by the oil becomes less effective if the contacts have a heavy layer of film on them, so the arc is sustained longer on the contact surface. This may cause hot spot overheating of contacts and result in premature wear. In extreme cases, overheating of the contacts may lead to contact failure. Filming of the mechanism will cause the device to work harder during normal changing of the taps under load. These conditions can cause an increase in operating temperature of the device. In extreme cases, filming can become so advanced that it may lead to binding and subsequent mechanical failure of the LTC.

1.6   CONTROL OF MAIN AND ARCING CONTACTS WEAR

In LV air CBs and magnetic air CBs, the arcing contacts are expected to evaporate at a rate that depends on the number of interruptions and the magnitude of current that is interrupted. As arcing contacts erodes, the dimensions between the contacts are changing. Evaluation of contact erosion based on measuring these distances in closed and open positions is an important part of the maintenance inspection to reveal wear in components well before a CB loses its ability to interrupt an overload current. An example how to measure and, therefore, control the wear of arcing contacts is given in Ref. [34].

There are several simple instru ments called *Contact Wear Indicator*, which is a device that allows the user to establish if the main contacts need replacement. In LV air CBs, if maintenance or inspection is necessary, the arc chambers are easily removed so one may visually inspect the contacts and wear indicator groove [35].

To check the wear value maintenance, personnel has to remove the arc chutes and visually check contact wear at least once a year and following each short circuit. The contact wear indicators constitute an absolute minimum value that must not be overrun. To plan and reduce the number of shutdowns, an electronic wear counter is available. A visual check is required when the counter reaches 100. When the counter reaches 300, the contacts must be replaced [36].

Interrupters in modern vacuum CBs are also equipped with a contact wear indicator or other simple contact measurement means that require no special tools. Many MV outdoor CBs are equipped with contact wear indicators accessible via a bolted panel for ease of maintenance.

The contact erosion analyzer is used to determine the extent of wear to the CBs contacts [37]. In this device, the breaking currents of the CB are used to determine the contact wear.

The contact burn off is calculated using the integral of the CB breaking currents. The result is compared with the configured reference values. If the result of this comparison exceeds the warning or alarm value, the corresponding signal contact is activated, and the signal light on the contact wear analysis device shows the status. With this warning or alarm, the device shows whether the CB contact system has already been worn down to a specific level, which allows servicing work to be scheduled in good time.

A different approach is used in HV Gas-Insulated Switchgear (GIS), which is equipped with CB monitoring system. Beside multiple values obtained from the sensor directly, among them contact and/or drive travel, switching times (opening, closing, close open), and so on, there are values which are calculated by mathematical methods. CB calculations give the main and control circuits operating times, contact travel, and contact velocity during closing and opening operations, arc integral (I^2t) during opening operations. These quantities are a good indicator of CB wear which has to be resolved to prevent the catastrophic consequences [38].

REFERENCES

1. Braunovic M., Konchits V. V., and Myshkin N. K. *Electrical Contacts: Fundamentals, Applications and Technology*, CRC Press, Boca Raton, FL, USA, 645pp., 2007.
2. Dijk Piet van. Critical aspects of electrical connector contacts, *Proceedings of the 21st International Conference on Electrical Contacts*, ICEC, Zurich, 2002, Online. Available: pvdijk.com/pdf/21thiceccriticalaspects.pdf.
3. *Electrical Connections for Power Circuits, Facilities Instructions Standards, and Techniques*, Volume 3-3, United States Department of the Interior Bureau of Reclamation (USBR), pp. 1–20, November 1991, Online. Available: http://www.usbr.gov/power/data/fist/fist3_3/vol3-3.pdf.
4. Gedeon M. Connector temperature rise, *Brush Wellman Technical Tidbits*, 23, November 2010, Online. Available: http://materion.com/~/media/Files/PDFs/Alloy/Newsletters/Technical%20Tidbits/Issue%20No%2023-%20Connector%20Temperature%20Rise.pdf.
5. *Connectors: Failure Mechanisms and Anomalies*, Online. Available: http://www.navsea.navy.mil/Home/WarfareCenters/NSWCCrane/Resources/SD18/Products/Connectors/FailureMechanismsAnomalies.aspx.
6. Timsit R. S. Origin, detection, and cost of connector degradation in electrical transmission and distribution systems, *Proceedings of the 1st International Conference on Reliability of Electrical Products and Electrical Contacts*, Suzhou, China, p. 1, 2004.
7. *Ohmstik Plus – Live-Line Ohmmeter*, SensorLink Corporation, Acme, USA, Online. Available: http://www.sicame.com.au/product_pdf/Testing%20Equipment%20-%2003%20-%20Ohmstik%20&%20Qualstik%20Plus.pdf.
8. *Expliner-Robot for Power Line Inspection*, HiBot, Online. Available: http://www.hibot.co.jp/en/products/robots_1/expliner-robot-for-power-line-inspection_12.
9. Zhao H., Barber G. C., and Lui J. Friction and wear in high speed sliding with and without electrical current, *Wear*, 249(5), 409–414, 2001.
10. Moran J., Sweetland M., and Suh N. P. Low friction and wear on non-lubricated connector contact surfaces, *Proceedings of the 50th IEEE Holm Conference on Electrical Contacts and the 22nd International Conference on Electrical Contacts*, Seattle, WA, pp. 262–266, September 2004.
11. Varenberg M., Halperin G., and Etsion I. Different aspects of the role of wear debris in fretting wear, *Wear*, 252, 902–910, 2002.
12. Dijk Piet van and Meijl Frank van. Contact problems due to fretting and their solutions, *AMP Journal of Technology*, Vol. 5 June, 1996, Online. Available: http://www.te.com/documentation/whitepapers/pdf/5jot_2.pdf.
13. Dijk Piet van, Rudolphi A. K., and Klaffke D. Investigations on electrical contacts subjected to fretting motion, *Proc. 21st International Conference on Electrical Contacts ISEC*, Zurich, 2002, Online. Available: http://www.pvdijk.com/pdf/21thicecinvestigationsonfretting.pdf.
14. Bryant M. D. Resistance buildup in electrical connectors due to fretting corrosion of rough surfaces, *Proceedings of the 39th IEEE Holm Conference on Electrical Contacts*, Pittsburgh, Pennsylvania, pp. 178–190, September 1993.
15. Song J., Koch C., and Wang L. Correlation between wear resistance and lifetime of electrical contacts, *Advances in Tribology*, 2012, Article ID 893145, 9pp., Online. Available: http://www.hindawi.com/journals/at/2012/893145/.
16. Braunovic, M. Effect of fretting in aluminum-to-tin connectors. *IEEE Trans. Components, Hybrids and Manufacturing Technology*, 13(3), 579–586, 1990.
17. Braunovic, M. Fretting in nickel-coated aluminum conductors, *Proceedings of the 36th IEEE Holm Conference on Electrical Contacts and 15th International Conference on Electrical Contacts ICEC*, Seattle, WA, pp. 461–471, 1990.
18. Myers M. The performance implications of silver as a contact finish in traditionally gold finished contact applications, *Proceedings of the 55th IEEE Holm Conference on Electrical Contacts*, Vancouver, Canada, pp. 307–315, 2009.
19. Chudnovsky B. H. Lubrication of distribution electrical equipment, in *Electrical Power Transmission and Distribution: Aging and Life Extension Techniques*, Chapter 4, CRC Press, Boca Raton, FL, USA, 411pp., 2012.
20. Ito T., Matsushima M., Takata K. et al. Factors influencing fretting corrosion of tin plated contacts, *SEI Technical Review*, April 2007, Online. Available: http://www.autonetworks.co.jp/en/r_and_d/report/pdf/64-01.pdf.

21. Zhang J. The application and mechanism of lubricants on electrical contacts, *Proceedings of the 40th IEEE Holm Conference on Electrical Contacts*, Chicago, IL, pp. 145–154, 1994.
22. Antler M. Electronic connector contact lubricants: The polyether fluids, *Proceedings of the 32th IEEE Holm Conference on Electrical Contacts*, Boston, MA, pp. 35–44, 1986.
23. Timsit R. Connector lubricants enhance performance, *Connector Specifier*, 28–31, June 2001.
24. Dijk, Piet van. Some effects of lubricants and corrosion inhibitors on electrical contacts, *Proceedings of the 16th International Conference on Electrical Contacts (ICEC)*, Loughborough, UK, pp. 67–72, 1992.
25. Achanta S. and Drees D. Effect of lubrication on fretting wear and durability of gold coated electrical contacts under high frequency vibrations, *Tribology*, 2(1), 57–63, 2008.
26. ASTM Standard B794-97. *Standard Test Method for Durability Wear Testing of Separable Electrical Connector Systems Using Electrical Resistance Measurements*, American Society for Testing and Materials ASTM, Philadelphia, 6pp., 1997.
27. Landry M. and Brikci F. Dynamic contact resistance measurements on HV circuit breakers, *IEEE/PES Switchgear Committee Meeting*, Montreal, October 2005, Online. Available: http://www.ewh.ieee.org/soc/pes/switchgear/presentations/2005-2_Lunch_Landry.pdf.
28. Landry M., Mercier A., Ouellet G. et al. New measurement method of the dynamic contact resistance of HV circuit breakers, *CIGRE Paris Session 2004*, Online. Available: http://www.zensol.com/Articles/A_Complete_Strategy_DRM_IEEE.pdf.
29. Landry M., Turcotte O., and Brikci F. A complete strategy for conducting dynamic contact resistance measurements on HV circuit breakers, *IEEE Transaction on Power Delivery*, 23(2), 710–716, 2008, Online. Available: http://www.zensol.com/Articles/A_Complete_Strategy_DRM_IEEE.pdf.
30. White J. Critical maintenance for circuit breakers, *NETA World Magazine*, pp. 1–8, Summer 2007, Online. Available: http://www.netaworld.org/sites/default/files/public/neta-journals/NWsu07-White.pdf.
31. Schellhase H.-U., Pollock R. G., Rao A. S. et al. Load tap changers: Investigations of contacts, contact wear and contact coking, *Proceedings of the 48th IEEE Holm Conference on Electrical Contacts*, Orlando, FL, pp. 259–272, 2002.
32. Duval M. Application of duval triangles 2 to DGA in LTCs, *IEEE Meeting WG C57.139 Committee*, Nashville, March 2012, Online. Available: http://grouper.ieee.org/groups/transformers/subcommittees/fluids/C57_139/S12-DuvalPresentation.pdf.
33. Griffin Paul J., Lewand Lance R., Peck Richard C. et al. Load tap changer diagnostics using oil tests— A key to condition-based maintenance, *Proceedings of the 2005 Annual International Conference of Doble Clients*, Boston, 21pp., 2005.
34. Sprague Michael J. Service-life evaluations of power circuit breakers and molded-case circuit breakers, *IEEE IAS Pulp and Paper Industry Conference*, Atlanta, GA, 2000, Online. Available: http://www.eaton.eu/ecm/groups/public/@pub/@electrical/documents/content/ct_129977.pdf.
35. Masterpact™ NT and NW, *Universal Power Circuit Breakers, Instruction Manual*, Schneider Electric, Bulletin Catalog #0613CT0001 R 11/12, p. 7, 2012, Online. Available: http://static.schneider-electric.us/docs/Circuit%20Protection/0613CT0001.pdf.
36. Masterpact™ NT and NW, *Maintenance Guide*, 11/2009, Bulletin LVPED508016EN, Schneider Electric Industries SAS, Online. Available: http://www2.schneider-electric.com/resources/sites/SCHNEIDER_ELECTRIC/content/live/FAQS/12000/FA12084/en_US/Masterpact%20Maintainence%20guide.pdf.
37. *High Voltage Circuit Breakers from 72.5 kV up to 800 kV*, Siemens AG Energy Sector, Erlangen, Germany, 2012, Online. Available: http://www.energy.siemens.com/hq/pool/hq/power-transmission/high-voltage-products/circuit-breaker/Portfolio_en.pdf.
38. *HV Circuit Breaker Monitoring System, KONČAR—Electrical Engineering Institute*, Zagreb, Croatia, Online. Available: http://www.koncar-institut.com/solutions/switching_apparatus_switchgear_monitoring_systems/high_voltage_circuit_breaker.

2 Bolted, Plug-On, and Busbar Connections
Materials and Degradation

2.1 ELECTRIC CONNECTIONS MATERIALS

Electrical connection in power system is defined as one which exhibits minimum resistance, both at initial assembly and during the long service life. Defective electrical connections in MV and HV power systems are involved in many circuit or equipment failures. Correct assembly and preventive maintenance of efficient electrical connections for power circuits allow minimizing the risk of such failures.

Properties of conductor metals and recommended methods of making connections are discussed in Ref. [1] to illustrate the need for the proper selection of connectors; the proper preparation of conductors; and the proper application of fusion, compression, and bolted connectors. Materials used in conductors and connections are discussed in many sources and will be shortly summarized in this section.

2.1.1 COPPER AND COPPER ALLOYS CONNECTIONS

Copper (Cu) is abundant in nature and its cost is relatively low. Copper was the most widely used electrical conductor (EC); however, when the cost of copper increased considerably, it was replaced in many applications with aluminum (Al). High ductility of copper allows casting, forging, rolling, drawing, and machining it. Copper hardens when worked, but annealing restores soft state.

As soon as Cu is exposed to the atmosphere, copper oxide is formed, and once stabilized it prevents further oxidation of the copper. Copper oxide is generally broken down by reasonably low values of contact pressure. Unless the copper is very badly oxidized, good contact can be obtained with minimum cleaning. More oxides are formed when the copper is heated, for example, by bad connection. High-resistant Cu oxides will generate more heat until the conductor breaks. At elevated temperatures copper oxidizes in dry air, it is oxidized in an ammonium environment, and is also affected by sulfur dioxide.

As copper forms heavy oxide films of relatively high resistance, especially when arcing occurs, it is not suitable for low current and voltage contacts where the contact forces are low. Copper or copper alloys are used for the current-carrying parts on to which the contacting tips are welded or riveted.

The copper alloy used for the bulk of the moving contact member needs to be able to retain its mechanical properties when running hot, due to the light construction required, and thus special copper alloys are often selected which combine mechanical stability with good electrical conductivity. Copper may be alloyed with manganese, nickel, zinc, tin, lead, gold, aluminum, chromium, beryllium, and so on [2].

2.1.2 ALUMINUM AND ALUMINUM ALLOYS CONNECTIONS

Aluminum (Al) and various alloys of Al are ductile metals which allow rolling, drawing, spinning, extruding, or forging. Al has high corrosion resistance which allows using Al in saline atmosphere.

Al forms a tough, very hard, invisible, and high-resistance oxide, which is both impermeable and protective in nature.

Al oxide forms quickly on exposed aluminum, and once stabilized, the oxide prevents further oxidation. This tough film gives aluminum its good corrosion resistance. After a few hours, the oxide film formed is too thick to permit a low-resistance contact with cleaning. The film is so transparent that the bright and clean appearance of an aluminum conductor is no assurance of a good contact. After cleaning the oxide film from aluminum, a compound must immediately be applied to prevent the oxide from reforming.

For highly corrosive environments, a proper selection of alloy or anodization may be used. Aluminum resists pitting from arcs better than any other metal.

Aluminum is one of the most anodic metals. When aluminum is connected to copper or steel in an electrolyte (e.g., water) galvanic corrosion takes place. The potential difference between the conductors varies proportionally with the distance between them in the galvanic series [3]. This difference causes corrosion of the aluminum but leaves the cathodic material unharmed. More information on galvanic corrosion in connection made of dissimilar metals is presented in Chapter 4, "Detrimental Processes and Aging of Plating."

There are multiple advantages of using Al in conductors. For equal ampere ratings, an aluminum conductor must have 1.66 times the cross-sectional area of a copper conductor. At this gauge, aluminum has 75% of the tensile strength and 55% of the weight of copper. In spite of its higher resistance, the larger size of aluminum conductors keeps it cooler and also helps keep the corona loss down. However, required larger size of Al conductor restricts its use in applications, where there are space limitations such as in enclosed switchgear.

For joining aluminum-to-aluminum conductors, it is usually recommended that an aluminum-bodied connector is the proper choice, since this obviously eliminates the galvanic corrosion of dissimilar metals. However, even for this case, care must be taken to prevent crevice corrosion and to select an alloy of aluminum for the connector body that is free from cracking due to stress corrosion [1,4].

For joining aluminum-to-copper conductors, an aluminum-bodied connector is the best choice since it prevents galvanic corrosion of the aluminum conductor, the most vulnerable element to attack in the connection.

By making the aluminum connector massive in comparison to the copper conductor, where the copper conductor emerges from the connector, the electrolytic current density (CD) over the exposed face of the aluminum connector is greatly reduced. In addition, because the aluminum connector body is massive in the region where the corrosion occurs, the small loss of metal caused by corrosion is insignificant even after long periods of service.

2.1.3 Steel in Connections

Steel is used for current conductors. Because of low thermal and electrical conductivity, steel is used primarily as a strengthening agent for conductors. Some of its applications are aluminum cable steel reinforced (ACSR), copper cable steel reinforced (CCSR), copper weld, and alumaweld. The main advantage of using steel is its strength.

When any of these three metals (aluminum, copper, and steel) are connected together, significantly different coefficients of expansion and allowances for expansion of copper, aluminum, and steel must be considered when the busways and connectors are designed.

2.2 PLATING AND ANODIZING OF THE CONDUCTORS

Plated busbars outperform unplated (bare) busbars by providing stable contact resistance (CR) and a low maximum operating temperature that increases the service life of the bus joint. More importantly, stable CR joints will reduce the need for frequent maintenance, decrease overall downtime of equipment and maintenance costs, and greatly reduce the risk of catastrophic failures. Industry practices recommend that all bus contacts be plated.

Most government, IEEE, and insurance provider specifications require that all bolted bus connections be plated in accordance with applicable standards.

Silver plating is recommended for copper contact surfaces which must be operated at elevated temperatures. Tin and nickel plating are also used on copper and aluminum connectors to prevent the formation of oxides. However, when wet, these plated metals can cause galvanic corrosion. More information on plating of electrical equipment and deteriorating processes and aging of plating can be found in Chapters 3 and 4 of this book.

Anodizing is an electrolytic process which increases the corrosion and abrasion resistance of aluminum (primarily bolts). The aluminum is made anodic with respect to a dilute acid which causes an additional porous oxide layer to form on the aluminum. This layer can be dyed and/or coated.

There are downsides of anodizing, which by definition requires building an oxide film on the surface. If the surface has an oxide film or is bare, a more strenuous approach is needed to clean it. Wire brushes, steel wool, fine sandpaper, or a scuff pad may be used but it should be done with caution. If the surface was abraded during the cleaning, it should still remain even. The surrounding surfaces should be protected from loose particles and cleaned thoroughly after abrading. Conductive and insulating particles can find their way into insulating systems and cause serious problems later.

2.3   TECHNIQUES TO JOIN ELECTRICAL CONDUCTORS

To distribute electrical power from a supply point to a number of output circuits copper or aluminum busbars are used. It is necessary that a conductor joint shall be mechanically strong and has a relatively low resistance which must remain substantially constant throughout the life of the joint. Efficient joints in busbar conductors can be made very simply by bolting, clamping, riveting, soldering, or welding. Bolting and clamping are used extensively, and copper welding is now more generally available through improvements in welding technology.

2.3.1   ELECTRICAL CONNECTIONS MADE BY FUSION

Fusion involves joining by melting of the two parent metals or a third metal as in soldering and brazing.

Soldering is one of the oldest means of joining electrical connections. A well-made soldered joint is a good joint but the quality is directly dependent on skill, and inspection is nearly impossible. Soldering increases the probability of corrosion by the introduction of different metals and flux. Also, soldering may weaken the conductor by annealing and the heat may damage adjacent cable insulation. The low melting point of solder may permit a joint failure under overload conditions.

Brazing makes an excellent connection when it is properly done. It is essentially a soldering process with a high-temperature solder such as silver solder. Its high temperature requirement of 650–700°C (1200–1300°F) may damage insulation and anneal the conductor. Soldered or brazed joints are rarely used for busbars unless they are reinforced with bolts or clamps, since heating under short-circuit conditions can make them both mechanically and electrically unsound.

Welding provides one of the best electrical connections; it fuses the parent metals of the two conductors being joined and theoretically produces a perfect joint. Because of the special skill required, welding has not been widely used as a field method of making connections. However, the widespread use of aluminum pipe bus has brought about a limited application for welding since soldering or brazing aluminum is difficult to control.

2.3.2   ELECTRICAL CONNECTIONS MADE BY PRESSURE

Pressure connections are made by physical contact between coincident peaks in the parent metals. With suitable pressure, as from a connector, the peaks are deformed creating a greater contact and conducting area. Because of unequal expansions of different conductors and/or connectors, voids may be reformed between conductor and connector. If the connection is not airtight, the voids provide a place for oxides to form.

Pressure connectors have become widely accepted as the most suitable method of making electrical connections in the field because of their reliability, convenience, and economy. Pressure connectors apply and maintain pressure between the contact surfaces by means of a clamp-type or compression-type fitting.

Clamp-type connectors apply and maintain pressure by means of clamping bolts, wedges, springs, or a combination of these. Clamped joints are formed by overlapping the bars and applying an external clamp around the overlap. As there are no bolt holes, the current flow is not disturbed resulting in lower joint resistance. The extra mass at the joint helps to reduce temperature excursions under cyclic loads. Well-designed clamps give an even contact pressure and are easy to assemble, but take up more space than a bolted joint and are more expensive to manufacture.

Compression-type connectors apply and maintain pressure by compressing the connector about the conductor by means of suitable tools. Wedge-type connectors (for taps, jumpers, etc.) installed by a special tool which utilizes a powder cartridge to supply the wedging force. Internally fired connectors (terminal lugs, dead ends, and line splices) utilize internal wedges actuated by powder charges fired by impact or an external electrical circuit.

2.4　BOLTED CONNECTIONS

In electrical industry, sections of the bus structure are typically spliced together with bolted connections. Efficient joints in copper busbar conductors are formed by overlapping the bars and bolting through the overlap area. Bolted connections are compact, reliable, and versatile but have the disadvantage that holes must be drilled or punched through the conductors, causing some distortion of the current flow in the bar.

Bolted splice connections allow for modularization of the rigid bus system components. The bolted splice joints can be assembled with inexpensive steel bolts, thus eliminating the need for expensive welding processes, which is another economical drive to use bolted connections.

Proper design of busbar joints guarantees a long equipment life. A good bolted busbar connection provides good conductivity, so that the bus system will meet the temperature rise (TR) requirements in the American National Standards Institute (ANSI) standards. The joint must withstand thermal cycling, so that the low-resistance joint will be maintained for the life of the equipment.

Busbar systems with bolted connections are used in power stations and high-power buildings. Various factors such as dust, humidity, the oxidation film resistance, and operating temperature affect the lifetime of electric contacts. The coated connections provide an excellent stability and low initial CR.

Dust contamination formed on contact surface is one of the main reasons of the electric contact failure. Films on the surface of contacts create an excess of the electric resistance which also can cause failures in contact applications. Humidity in the atmospheric environment also influences the surface of contacts [5].

To provide good conductivity, the joint pressure should be high, but not too high to cause a cold flow of the bus material, which would result in deterioration of the joint with time. In normal installation environments, the joint should have good resistance to corrosion. It must be able to withstand the mechanical forces and thermal stresses associated with short-circuit conditions.

2.4.1　BOLTING ARRANGEMENTS

Both electrical and mechanical considerations have to be taken into account in deciding the number, size, and distribution of bolts required to produce the necessary contact pressure to provide high joint efficiency [6]. A schematic bolted busbar joint arrangement is presented in Figure 2.1.

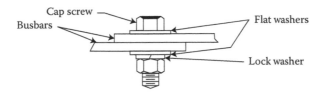

FIGURE 2.1 Bolted busbar joint. (From Olsen T. W. *Bus Joint Fundamentals*, Siemens Technical Topics, #16, April 2001, Online. Available: http://w3.usa.siemens.com/us/internet-dms/btlv/PowerDistributionComm/PowerDistribution/docs_MV/TechTopics/ANSI_MV_TechTopics16_EN.pdf. With permission.)

A joint normally decreases in resistance with an increase in the size and number of bolts used. Bolt sizes usually vary from M6 to M20 (metric bolt size). Four and six bolts are used in each joint with a preference for four bolts in narrow conductors and six in large conductors.

The torque chosen for each bolt size depends on the bolt material and the maximum operating temperature expected. Because of the strength of copper, deformation of the conductor under the pressure of the joint is not normally a consideration. The bolts in bolted connectors should be tightened with a torque wrench to the torque value recommended by the connector manufacturer [7]. Table 2.1 shows typical bolting arrangements including size and number of bolts and washers for various busbar sizes.

The recommended torque settings are applicable to high-tensile steel or aluminum bronze fasteners with unlubricated threads of normal surface finish. In the case of stainless steel bolts, these torque settings may be used, but the threads must be lubricated prior to use. In addition to the proof or yield stress of the bolt material and the thread characteristics, the correct tightening torque depends on the differential expansion between the bolt and conductor materials.

Galvanized steel bolts are normally used, but brass or bronze bolts also have been used because their coefficients of expansion closely match the copper conductor and hence the contact pressure does not vary widely with operating temperature.

TABLE 2.1
Typical Busbar Bolting Arrangements (Single Face Overlap)

Bar Width (mm)	Joint Overlap, (mm)	Metric Bolt Size/ Number of Bolts	Hole Size (mm)	Washer Diameter/ Thickness (mm)
16	32	M6/2	7	14/1.8
20	40	M6/2	7	14/1.8
25	60	M8/2	10	21/2.0
30	60	M8/2	10	21/2.0
40	70	M10/2	11.5	24/2.2
50	70	M12/2	14	28/2.7
60	60	M10/4	11.5	24/2.2
80	80	M12/4	14	28/2.7
100	100	M12/5	15	28/2.7
120	120	M12/5	15	28/2.7
160	160	M16/6	20	28/2.7
200	200	M16/8	20	28/2.7

Source: Modified from David C. and Toby N. Copper for busbars. *Guidance for Design and Installation*, Copper Development Association Publication No 22, European Copper Institute Publication No Cu0201, 108 pp., Revision May 2014, Online. Available: http://www.copperalliance.org.uk/docs/librariesprovider5/pub-22-copper-for-busbars/copper-for-busbars-all-sections.pdf?sfvrsn=2.

Copper alloy bolts also have the advantage that the possibility of dissimilar metal corrosion is avoided. As these alloys do not have an easily discernible yield stress, care has to be taken not to exceed the correct tightening torque. Because of their nonmagnetic properties, copper alloys may also be preferred to mild or high-tensile steel where high magnetic fields are expected. Alternatively, a nonmagnetic stainless steel may be used. In most cases, however, high-tensile steel is used for its very high yield stress.

It was shown in Refs. [8,9] that the CR of overlapping bolted/pad joints may be decreased by using various designs, such as slanting the edges of the busbars/pads under 45° and making slots in the overlapping areas. Such design may significantly reduce the CR of a joint and improve its mechanical integrity. These improvements are results of enlarged contact area and creation of a uniform current distribution at the contact interface.

2.4.2 RECOMMENDED MATERIALS AND HARDWARE FOR BOLTED CONNECTIONS

When connections are made between different metals, the type of the plating of either connection should be considered to make a correct choice of the material of connector, bolts, and type of washer. Table 2.2 shows the choices for materials and plating for connections, bolts, and washers, which will allow reducing the effects of corrosion [1].

It is important to use nuts, flat washers, and lock washers when recommended, which are made of the same material as the bolt. Bronze connector bodies are usually made of silicone bronze. Steel bolts used for connection may be zinc or cadmium plated or stainless steel. When steel bolts are used for bolting flat aluminum busbar or aluminum to copper, the use of Belleville washers in place of flat washers to compensate for the differences in expansion of steel, aluminum, and copper is recommended. General recommendations of choosing hardware for joints made from different materials are as follows:

- Copper-to-copper joints: silicon bronze or stainless steel hardware
- Copper-to-aluminum joints: tin-plated silicon bronze hardware or stainless steel hardware
- Aluminum-to-aluminum joints: aluminum or stainless steel hardware
- Copper-to-steel joints: silicon bronze, stainless steel, or galvanized steel hardware
- Aluminum-to-steel joints: tin-plated silicon bronze hardware or stainless and galvanized steel hardware

TABLE 2.2
Recommendations for Materials, Plating, and Hardware for Electrical Connectors

Connection (1)	Plating on (1)	Connection (2)	Plating on (2)	Connector Body	Bolt	Washer
Al	None	Al	None	Al	Al	Lock washer
Al	None	Al	Sn	Al	Al	Same
Al	None	Cu	Sn	Al	Al	Same
Cu	None	Al	Sn	Bronze	Bronze or steel	Same
Cu	None	Cu	Sn	Cu or bronze	Bronze	Bronze lock washer
Cu	None	Cu	None	Same	Same	Same
Bronze	None	Bronze	None	Same	Same	Same
Cu	Sn	Al	Sn	Al or bronze	Al, bronze or steel	Belleville washer

Source: Modified from *Electrical Connections for Power Circuits, Facilities Instructions Standards, and Techniques,* Volume 3-3, United States Department of the Interior Bureau of Reclamation (USBR), pp. 1–20, November 1991, Online. Available: http://www.usbr.gov/power/data/fist/fist3_3/vol3-3.pdf.

Some manufactures' require the busbars silver plated (standard) or tin plated (optional) to improve the resistance to corrosion. The recommended bolt is a high-strength grade 5 cap screw, while the nut is a grade 2 (heavy wall) nut. These requirements are based on the fact that the grade 2 nut is more ductile than the grade 5 bolt, so that when the nut is torqued in place, the threads in the nut will tend to be swaged down and burnished to a degree, which results in a more equal distribution of load on all threads. This spreads the force more evenly and avoids unacceptable stress levels in the bolt and the nut. The joint includes a large-diameter, thick flat washer on both sides of the joint, adjacent to the busbars. A split lock washer is installed under the nut to assure that the joint stays tight over the life of the equipment [6].

It is important to notice that nuts, flat washers, and lock washers (if used) must be of the same material as the bolt. Bronze, when recommended, should be silicone bronze. Steel bolts may be zinc or cadmium plated, or stainless steel. When steel bolts are used for bolting flat aluminum busbar or aluminum to copper, the use of Belleville washers in place of flat washers to compensate for the differences in expansion of steel, aluminum, and copper is recommended in Ref. [1].

2.4.3 Aging Mechanisms of Bolted Joint

Whenever currents are transmitted in the order of a few hundred amperes to a few thousand amperes—or even tens of thousands of amperes—problems arise at the busbar joints as a result of excessively high joint resistance. Several variables affect this resistance, which increases with time because of aging. The heat losses rise at the same time. Ultimately, excessive heating can lead to total failure of the joint.

Service life can vary widely depending on the ambient conditions. Progressive *film buildup* increases the joint resistance, which results in excessive heat losses and therefore the temperature of the joint. High temperature of the joint accelerates the chemical reaction producing the film buildup. All these factors together can ultimately cause the busbar joint to fail totally as a result of *overheating*. A review of factors that affect the connector performance and detailed analysis of the degradation mechanisms of power connections are given in Refs. [10,11].

One of the factors that may lead to bolted connection failure is the *difference in coefficient of thermal expansion*. Differential expansion between the bolt and bar materials results in an increase or decrease in bolt tension (and therefore contact pressure) as temperature changes. Both aluminum and copper alloys expand at a greater rate than steel. The use of steel bolts, without the benefit of properly sized and applied Belleville washers will result in rapid creep of softer material of the connectors, and the joint will loosen over time. As it loosens, the resistance will rise, and with that given rise in resistance, the thermal rise for a given current will increase. Galvanized steel bolts are often used with copper busbars but copper alloy bolts, for example, aluminum bronze (CW307G), are preferred because their coefficients of expansion closely match that of copper, resulting in a more stable contact pressure [7].

The difference in coefficient of thermal expansion could be one of the factors that may lead to bolted Al-to-Cu connection failure, which for aluminum is 1.36 time that of copper. When a bolted aluminum–copper joint is heated by the passage of current, aluminum will tend to expand relative to copper, causing displacement of the contact interface. The shearing forces generated by differential thermal expansion will rupture the metallic bonds at the contact interface and cause significant degradation of the joint.

Published experimental evidence and reports of trouble in service show that bolted joints may not be resistant to degradation and failure (see also Section 2.4.8). More information of possible mechanisms of bolted joins deterioration (oxidation, corrosion, etc.) is given in Chapters 3 and 4 of this book.

2.4.4 Condition and Preparation of Connection Surfaces

The condition of the contact surfaces of a joint has an important effect on its efficiency. The surfaces of the copper should be flat and clean but need not be polished. Machining is not usually required.

Perfectly flat joint faces are not necessary as very good results can be obtained in most cases merely by ensuring that the joint is tight and clean. This is particularly the case where extruded copper bars are used. Where cast copper bars are used, however, machining may be necessary if the joints are to obtain a sufficiently flat contact surface.

Oxides, sulfides, and other surface contaminants have, of course, a higher resistance than the base metal. Copper, like all other common metals, readily develops a very thin surface oxide film even at ordinary temperatures when freely exposed to air, although aluminum oxidizes much more rapidly, and its oxide has a much higher resistivity.

The negative temperature coefficient of resistance of copper oxide means that the joint conductivity tends to increase with temperature. This does not, of course, mean that a joint can be made without cleaning just prior to jointing to ensure that the oxide layer is thin enough to be easily broken as the contact surface peaks deform when the contact pressure is applied.

2.4.4.1 Surface Preparation

Contact surfaces should be flattened by machining if necessary and thoroughly cleaned. A ground or sand-roughened surface is preferable to a smooth one. Surface contamination, specifically surface oxide, must be expected on all conductors. These oxide films are insulators and must be broken down to achieve the metal-to-metal contact required for efficient electrical connections. Aluminum oxide, in particular, forms very rapidly; therefore, aluminum must be thoroughly cleaned immediately prior to making the connection. In addition to cleaning, the surface should be covered with a good joint compound to exclude moisture, thus preventing the reoxidation.

Joint compounds are used when making electrical connections to (1) prevent formation of oxides on the cleaned metal surfaces and (2) to prevent moisture from entering the connection and reduce chances of corrosion. Among widely used joint compounds are petrolatum (trade name Vaseline), Burndy Pentrox "A," NO-OX-ID, and many others. Joint compound should be applied immediately after cleaning the contact surfaces. The joint surfaces should then be bolted together, the excess compound being pressed out as the contact pressure is applied. The remaining compound will help to protect the joint from deterioration.

Another way of protecting contacts from exposure to atmosphere is using so-called contact *sealants* [12]. Various sealant formulations have been developed to provide improved electrical and mechanical performance as well as environmental protection to the contact area. The use of sealants is recommended for aluminum-to-aluminum or aluminum-to-copper connections.

Sealants are also recommended for copper-to-copper joints which are subject to severe corrosive environments. Nongritted sealants are recommended for flat connections and as a groove sealant in bolted connectors such as parallel groove clamps. A gritted sealant is primarily used in compression connectors. The sharp metallic grit particles provide multicontact current-carrying bridges through remaining oxide films to ensure superior electrical conductivity.

It should be noted that in cases where joints have to perform reliably in higher than normal ambient temperature conditions, it may be advisable to use a high-melting point joint compound to prevent it from flowing out of the joint, leaving it liable to attack by oxidation and the environment.

The surface preparation plays a very important role in the performance of the overlapping bolted joints. The significance and effect of surface preparation and lubrication on the CR was studied and reported in Refs. [13,14].

2.4.4.2 Plating of the Joints

In the electric power industry, optimizing power flow is a primary concern. In the past (beginning of twentieth century) copper or aluminum busbars were not plated, which resulted in significantly inefficient joints. The effects of busbar surface irregularities, the formation of nonconductive surface films caused by the local environment and temperature, and the formation of copper oxides led to a high joint resistance and voltage drop, resulting in hot spots, and consequently to catastrophic failure.

The use of powders and greases can help, but only for a short period of time. Industrial practices recommend that all bus contacts be plated. Most government, IEEE, and insurance provider specifications require that all bolted bus connections be plated in accordance with applicable specifications. One of the key elements to effective busbar contact plating is applying a uniform deposit of sufficient thickness to provide corrosion protection.

Plating with a soft metal such as *silver* or *tin* effectively forms a compressive layer on the surfaces of the bus. The force applied when bolting the surfaces together squeezes the conductive material into the low areas, effectively increasing the contact area and decreasing the overall joint resistance. Some conductive "wipe-on" coatings such as Cool Amp® may be quick and inexpensive, but the actual silver deposits are very thin and yield only temporary results due to degradation by heat and environmental factors over much shorter periods of time [15].

Other processes may not provide the thickness and durability required to ensure adequate "compression" of the joint and long term corrosion protection required to mitigate the effects of oxidation. Brush plating using pure silver provides a simple, cost-effective solution for in-place plating of bus systems during routine maintenance and can also be of value in upgrading bus, rather than replacing existing bus, when generator or system capacity increases are desired.

The use of electroplated silver later on copper bus joints results in improvement in the joint efficiency over its life. The application of a 5- (0.0002″) to 1.3-μm (0.0005″)-thick deposit such as silver, nickel, or tin can improve the lifetime performance of the joint and reduce required maintenance substantially.

These days plating usually used on the joints is silver or *nickel*, particularly where equipment is manufactured to American standards which require plated joints for high temperature operation. Nickel plating provides a harder surface than silver and may therefore be preferable. Nickel plating is expensive to apply and must be protected prior to the final jointing process as they are always very thin coatings and can therefore be easily damaged. There is also some doubt as to the stability of these joints under prolonged high-temperature cycling. Very high CRs can be developed some time after jointing.

The plating of the contact surfaces of a bolted or clamped joint with pure tin or a *lead–tin alloy* is normally unnecessary, although advantages can be gained in certain circumstances. For example, if the joint faces are very rough, *tin plating* may result in some improvement in efficiency. In most cases, however, tin may prevent oxidation and hence subsequent joint deterioration. It may therefore be recommended in cases where the joints operate at unusually high temperatures or current densities or when subjected to corrosive atmospheres. More information on plating is given in Chapter 3, "Plating of Electrical Equipment."

2.4.4.3 Procedures for Making Connections

The procedures of making connection may depend of material of the connectors [1]. For example, when connecting Al-to-Al bus it is recommended after coating surface with joint compound to abrade the contact surfaces through the joint compound with a wire brush. Aluminum oxide forms immediately upon exposure; therefore, joint compound during brushing should not be removed.

When connecting Al-to-Cu or Al-to-bronze bus with or without tin plating it is recommended to place the aluminum member above the tinned copper member, particularly in outdoor installations where mixed metals are used. Aluminum conductors must be installed above copper conductors whenever possible.

Moisture on copper conductor surfaces will accumulate copper ions. If positioned above aluminum, these copper salts will wash onto the aluminum and cause galvanic corrosion. In the event the aluminum conductor is located below the copper, a "drip loop" should be provided on the copper conductor. The drip loop redirects copper conductor around and below the aluminum conductor for attachment. The loop formed allows corrosive moisture to drip from the copper conductor safely below the aluminum.

An aluminum surface should never be connected to a silver plated surface. Aluminum in contact with silver results in a highly corrosive joint, which will further result in a high resistance connection.

It is also recommended to clean the tinned copper or bronze surface by a few light rubs with fine steel wool. When connecting Cu-to-Cu or Bronze-to-Bronze connections, it is recommended to clean bare (unplated) contact surfaces to bright metal with emery cloth. If there are nicks and ridges on the surfaces they should be removed by filling, after which all copper particles should be wiped.

2.4.5 Effect of Pressure on the Resistance of Bolted Contact

It has been shown that the CR is dependent more on the total applied pressure than on the area of contact. If the total applied pressure remains constant and the contact area is varied, the total CR of bolted joint remains practically constant. The greater the applied total pressure the lower will be the joint resistance and therefore for high efficiency joints high pressure is usually necessary. This has the advantage that the high pressure helps to prevent deterioration of the joint.

Electrical connections are often made up of materials that have various coefficients of thermal expansion. In addition, the joint materials also carry more current than the bolts. This causes the joint to heat up more than the bolts. The resultant differential thermal expansion (DTE) results in an increase in bolt load, possibly causing the joint material to yield. This yielding of the material causes a decrease in load holding the joint together which in turn increases the electrical resistance during the next thermal cycle.

Eventually enough heat may be generated to result in what is referred to as a "hot spot." "Hot spots" are produced at bolted electrical joints when there is more heat generated by current passing through the joint than can be effectively dissipated. This can result in catastrophic failure of the joint.

As a bar heats up under load, the contact pressure in a joint made with steel bolts tends to increase because of the difference in expansion coefficients between copper and the steel. It is therefore very important that the initial contact pressure is kept at the level that the contact pressure is not excessive when it is at operating temperature. If the elastic limit of the bar is exceeded, the joint will have a reduced contact pressure when it returns to its cold state due to the joint materials having deformed or stretched. To avoid this, it is helpful to use disc spring washers whose spring rating is chosen to maintain a substantially constant contact pressure under cold and hot working conditions.

This type of joint deterioration is very much more likely to happen with soft materials, such as aluminum, where the material elastic limit is low compared with that of high conductivity copper. Several aging mechanisms have been studied in Refs. [16,17] for bolted electrical busbar connections, such as chemical changes in the constriction area (chemical aging) and the decrease of the joint force caused by creep in the metallic conductor material (creep aging). Both aging mechanisms take place simultaneously and depend on temperature.

2.4.6 Use of Belleville Washers

Another method to keep bolted connection tight after differential thermal expansion is the use of Belleville springs [18], particularly when steel bolts are used for aluminum-to-copper connection, with aluminum having thermal expansion coefficient 1.36 times higher than that of copper. Belleville springs counteract the effects of differential thermal expansion by maintaining sufficient load on bolted electrical connections to prevent "hot spots" during and after temperature cycles. A typical electrical assembly between aluminum or copper bus and aluminum lug includes flat washer with Belleville spring.

A bolted joint without a Belleville spring relies on the bolt stretch to produce the load at the contact joint. Bolt stretch produces small amounts of movement at very high spring rates. Creep of the material making up the joint can cause significant loss of load at the joint with little movement of the material. A Belleville spring's deflection to flat is seven to ten times the stretch of a bolt for the same load. The combined deflection of the Belleville spring and the stretch of the bolt produce a much lower spring rate for the same load on the contact joint.

Throughout the industry, opinions vary as to what the "correct" method is to install Belleville washers. According to Framatome Connectors International (FCI), the following is a successful, time-tested procedure [19]:

1. Place a flat washer between the concave side of the Belleville washer and the surface of the member you are joining. By doing so, you capture the Belleville between the bolt head and large flat washer. Make sure you use a flat washer having an outside diameter greater than that of the flattened Belleville so there is no overhang. Also, choose a flat washer that is twice the thickness of the Belleville washer.
2. Fit the assembly (bolt, Belleville washer, and flat washer) into its hole. There should be no interference with washers of adjacent bolts and no overhang over surface edges.
3. Tighten the nut (with a washer of its own) onto the bolt until you feel a sudden and noticeable increase in torque. The Belleville washer is now flat; it is not necessary to "back off" the nut after you have tightened to this point.

2.4.7 IMPORTANCE OF RETIGHTENING OF BOLTED JOINTS

One of the simplest and most widely used methods of maintaining the contact force in a bolted joint is retightening. Retightening has a very beneficial effect on the overall performance of bolted joints as manifested by a considerable stabilization of the contact force and resistance. The effect is more pronounced in the joints assembled with a combination of a spring lock (Grower) and thin flat washers. The initial pronounced loss of the contact force in this joint configuration is substantially reduced by subsequent retightening.

In a joint with a combination of disc spring (Belleville) washer and thick flat washers, the initial contact loss of contact force is considerably lower due to the beneficial action of the Belleville washer. Nevertheless, retightening improves even further the integrity of this joint. As the creep and stress relaxation are responsible for the loss of contact force, from these results it can be inferred that retightening lowers the rates of creep and stress relaxation in bolted joints. It should be pointed out, however, that despite the beneficial effect of retightening on the mechanical and electrical stabilities of bolted joints, this method of mitigating the loss of contact force in bolted joints is rather impractical.

2.4.8 DISADVANTAGES AND DEGRADATION OF BOLTED CONNECTIONS

There are multiple issues with bolted connections, which should be taken into account and addressed during maintenance. Prior to making bolted connections, each connecting part requires the removal of surface films and oxides if the busbar is tin or silver plated. When the surfaces of connecting parts are not properly prepared, high resistance connections are often the result. Each bolted connection is made by assembling multiple loose parts including nuts, bolts, and Belleville washers. Properly assembling each connection can be difficult and time consuming.

Many bolted connections are made by driving a thread-forming screw or bolt into an aluminum or copper busbar. Proper care must be taken during assembly to assure the screw or bolt is not cross-threaded or stripped. If either occurs, it may result in a loose connection. Even when properly installed, annual inspection and retightening of bolted connections is necessary due to the loosening effects of vibration and thermal cycling.

In a properly assembled bolted joint, the stresses in connection increase very rapidly with increase in bolt diameter and also with the yield strength of the bolt material. Therefore, the use of bolts of too small a diameter is insufficient to provide adequate contact area that will lead to unsatisfactory joint performance. As the areas of low resistance in connection are developed only under the bolt head or flat washer (when used), and increase rapidly in area with increasing bolt/washer size, this should be taken into account when deciding the number and the size of the bolts to be used in a connection.

Conventional bolted connections require special on-site surface preparation and tightening tools to achieve proper connection pressure. Making good bolted connections with sufficient low-resistance contact spots requires the use of calibrated tightening tools, careful preparation of the contact surfaces, correct assembly of the connection pieces, followed by applying the appropriate amount of torque according to tightening recommendations. Inadequate preparation and installation reduces the number of contact spots, increases current constriction, and can result in overheating at the connection.

Degradation mechanisms in bolted connection include the effect of different coefficient of thermal expansion for different materials discussed in the Section 2.4.3. The shearing forces generated by differential thermal expansion will rupture the metallic bonds at the contact interface and cause significant degradation of the joint.

The study of fretting damage in real-life tin-plated aluminum and copper connectors, commonly used for distribution transformers, revealed the presence of four distinct forms of fretting damage: electrical erosion by melting and arcing, accumulation fretting debris, delamination, and abrasion [20].

Corrosion is also recognized as one of the most significant reliability concerns for bolted connections. The corrosion problem is particularly severe in some areas, since connections must often be made between dissimilar metals, particularly, aluminum to copper, in which case deterioration is caused by the action of galvanic corrosion and build-up of corrosion products leading to the loss of the mechanical integrity at the conductor–connector interface and ultimately, failure of a joint.

Copper-to-copper joints in most circumstances are highly resistant to corrosion and do not normally need additional protection. However, in environments containing ammonia, sulfur, and chlorine compounds, especially where the humidity may be high, protection is required [7]. More information of deteriorating processes in connectors and means of protection will be presented in Chapters 4 and 5 of this book.

2.4.9 RECOMMENDED ARRANGEMENTS FOR BOLTED JOINT FOR AL AND CU BUSBAR CONDUCTOR

The preferred configurations for jointing aluminum or copper busbar conductors are shown in Figure 2.2 [20]. To have a more uniform stress distribution in the busbars under the washer and to avoid its buckling, it is recommended to use flat washers at least 3–4 mm (~1/8″ to 5/32″) thick.

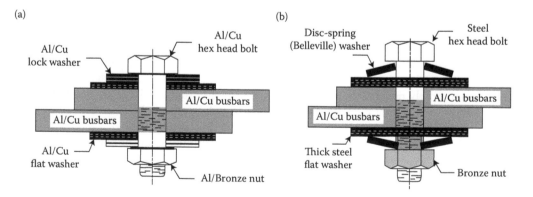

FIGURE 2.2 Recommended bolted joint configuration for Al and Cu busbar joints: (a) without Belleville washers and (b) with Belleville washers. (Adapted from Braunovic M., Myshkin N. K., and Konchits V. V. Power connections, in *Electrical Contacts: Fundamentals, Applications and Technology*, Chapter 7, CRC Press, Taylor & Francis Group, Boca Raton, FL, 2006.)

When steel bolts are used, it is recommended to use disc spring washers with a high spring constant and flat washers with thickness at least twice that of the disc spring washer.

Furthermore, the disc spring has to have a very high spring constant to provide the required elasticity of a joint during temperature excursions caused by joint overheating, short circuit conditions, and other causes impacting on the mechanical integrity of a bolted joint.

One method of avoiding failures due to differential thermal expansion consists of using the bolting hardware of the same material as that of conductor. For instance, when the conductors and hardware are made of aluminum, the coefficient of their thermal expansion is essentially the same and thus a high contact pressure is maintained during both heating and cooling cycles. The same principle applies to all copper-bolted joints.

As the compressive stress in a bolted joint is concentrated under the head and nut of the bolt, bolts made of high-strength aluminum alloys, such as 7075 T-72 alloy, are recommended. Aluminum bolts are usually anodized and covered with a thin layer of a lanoline lubricant.

For jointing the copper busbars, bolts made of a high-strength bronze (Everdur) are preferable over those of steel. Bronze bolts should be assembled dry. Aluminum alloy and bronze bolts are nonmagnetic and therefore not subject to heating from hysteresis losses in AC fields, as is the case with steel bolts.

2.4.10 FAILURES CAUSED BY INADEQUATE ELECTRICAL CONNECTIONS

Several failures reported in the United States Nuclear Regulatory Commission (NRC) Information Notice [21] caused by inadequate electrical connections. For example, a link assembly failed that allowed a short circuit between phase conductors. The failure of the link was attributed to thermal cycling that caused the bolted connections on the central flexible link to loosen. It was determined that a preventive maintenance procedure omitted torque verification of the bolts.

In another document [22], it was found that loose bolted connection could cause the fault of the breaker that provides charging current for the battery. Inadequate electrical connections can lead to unanticipated plant transients and the failure or unavailability of safety related equipment. They can also affect equipment important for safety or can potentially challenge safety-related equipment. NRC review of operating experience has determined that the following items are important to ensure the integrity of electrical connections:

* Torque verification
* Visual inspections
* Periodic measurements of the bolted connection temperature using thermography
* Resistance measurements
* Adherence to vendor recommendations
* Use of proper lubricants for switch contact surfaces
* Identification of single point connection vulnerabilities

2.5 PLUG-ON ELECTRICAL CONNECTIONS

Plug-on connections are used extensively in residential, commercial, and industrial applications. They can be found in many everyday electrical devices such as in switchgear. The design of the plug-on connection is relatively simple. A conductor is inserted between hardened metal jaws (Figure 2.3) and the spring force from the jaws maintains the proper contact pressure for good connections.

The plug-on design allows to avoid many common connectivity problems, including those due to vibration and thermal cycling. During installation, the connector force from the plug-on jaws and beveled fingers wipe the nonconducting surface films from the busbar and jaw surface. This "in-line" wiping action develops a line of electrical contact spots to maximize the connection's

FIGURE 2.3   Beveled multiple fingers in a jaw. (From *Square D® I-Line® Plug-On Connections*, Data bulletin 600DB1004, Schneider Electric USA, December 2010, 8 pp., Cedar Rapids, IA, Online. Available: http://static.schneider-electric.us/docs/Circuit%20Protection/0600DB1004.pdf. With permission.)

electrical performance. Line contact spots actually produce lower resistance connections than circular contact spots of the same area [23].

A significant advantage of the plug-on connection is that it creates multiple current paths on each side of the busbar. Each of the fingers on a plug-on jaw provides an individual current path with an equal contact area. Devices with higher current-carrying capacity are designed with more contact fingers to minimize electrical resistance. In contrast, the electrical resistance values in bolted connections often measure much higher than expected when considering the apparent connection surface area. The actual electrical contact spots are only within the thin ring around the bolt hole.

Oxidation and corrosion can erode electrical contact areas if the connections are not properly protected. To protect plug-on connection from corrosion and oxide development, the components are usually plated with tin or silver. Additionally, a protective compound is applied to the connections to maintain low-resistance contact areas.

However, the use of tin-plated Al busbars in a plug-on jaw to busbar connection may eventually lead to arcing and thermal failure. The history of elastic connections across the industry indicates a substantially higher occurrence of connection degradation in elastic connections to aluminum busbars than to copper busbars. The connection failure happens as a result of unstable CR growth. This instability in a CR can be a result of chemical or physical changes in the plating, which are caused and accelerated by the high temperatures, which the connection may experience in carrying different loads. The degradation of the contact can lead to high CR, which can become large enough to cause contact arcing and melting.

The history of the failures of aluminum-to-copper connections is extensive, although tin is still one of the most common commercial plating for aluminum to improve the stability and decrease the galvanic corrosion of aluminum-to-copper connection [3]. Though the mechanisms of failure are not fully understood and explained, numerous processes such as oxidation, stress relaxation, differential thermal expansion, and galvanic and fretting corrosion are attributed to degradation of the contacting interface and a corresponding change in CR [24].

2.6   BUS-STAB SEPARABLE ELECTRICAL CONNECTIONS

Field experience with separable power contacts involving aluminum busbars has indicated failures in certain industrial locations. These failures were mechanical erosion and electrical burning of metal at the contact areas, resulting ultimately in an open-circuit condition. Configuration of bus-stab separable electrical connection is shown in Figure 2.4.

Bus-stab contacts are subjected to three modes of mechanical motions [20]. One is slow slide motions with respect to the stab contact (along busbar) that causes elongation of the stab contact area

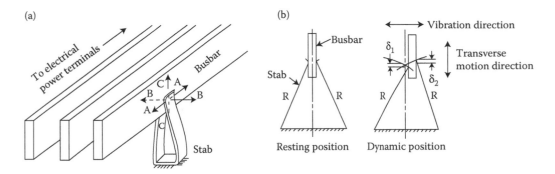

FIGURE 2.4 (a) Direction of mechanical motion acting on bus-stab contacts and (b) transverse motion of stab contact along busbar width. (Adapted from Braunovic M., Myshkin N. K., and Konchits V. V. Power connections, in *Electrical Contacts: Fundamentals, Applications and Technology*, Chapter 7, CRC Press, Taylor & Francis Group, Boca Raton, FL, 2006.)

along the busbar (direction A–A, Figure 2.4). The motion occurs when the busbars are subjected to variations in electrical load, changing their lengths due to thermal expansion.

The second mode of motion (direction B–B, Figure 2.4) is attributed to electromagnetically induced vibrations. The driving force for these motions is created by currents flowing in adjacent bus members. In common busway configurations, the forces between busbars can be as large as 100 Nm of busbar length under rated current conditions. Such forces can cause busbar displacement of the order 20 mm perpendicular to those induced by thermal expansion.

The third mode of bus-stab contact motion is transverse displacement perpendicular to both thermal and electromagnetically induced motions (direction C–C, Figure 2.4). It is due to rigid stab mounting and the vibratory motions. Although such motion displacements are small, significant fretting wear of the bus or the plating may occur because of increasing number of slides.

The resulting degradation of protective plating would lead to the formation of high-resistance films, surface damage, and arcing, which, in turn, favors more chemical attack, burning, and failure of the contact metals.

The best performance is obtained when the contact members are plated with silver, rather than tin, and mechanically held together with forces of at least 70 N. Nevertheless, when the contact members are subjected to 120-Hz relative slide motions with amplitudes greater than 5 mm, the time to failure is greatly shortened. Hence, unless transverse motions can be minimized below the critical amplitude level, bus-stab protective plating will wear at appreciable rates and eventually be worn out, resulting in higher CR and temperature and final "burn" failure.

2.7 GAS-INSULATED BUSBAR CONNECTIONS

SF_6-gas has been used for insulating purposes in the MV and HV field for several decades. The electrical strength of SF_6 is considerably higher than that of air. SF_6-gas is chemically neutral, non-toxic, and non-inflammable. Because of these characteristics and the increasing demand for totally enclosed MV equipment, SF_6-insulated busbar connections have been developed by Swiss company MGC Moser-Glaser [25].

These systems may be used for connections between GIS, transformers, and cables, as well as between GIS and non-SF_6-insulated equipment (such as short-circuit current reactors). The system is designed for rated voltage up to 40.5 kV and rated current: 2500–3150 A (Figure 2.5).

In such system, modular construction elements allow individual solutions for any compact MV equipment. The SF_6-gas-insulated system consists of aluminum or copper conductors interconnected by type-tested high-current plugs with special HV insulators used to center these conductors inside aluminum-protecting tubes. Minimum phase distance is suitable for adaptation to any

FIGURE 2.5   Gas-insulated busbar system. (Modified from Gaslink® SF6 busbar system. *MGC Moser-Glaser*, Online. Available: http://www.mgc.ch/en/products/gaslinkbusbarsystem.)

compact SF_6-switchgear. Busbar system allows separating two different gas chambers with different pressures using SF_6-gas partition.

Universal angle-type cast-aluminum housings allow changes of direction and branch-offs in all three dimensions. Gas sealing is provided by special SF_6-proof O-rings inserted in specially designed grooves with a maximum leakage of no more than 0.5% per year.

2.8   RELIABILITY OF ELECTRICAL CONNECTORS

A reliability of a particular part or product is a probability that a part will perform as required for a desired period of time during which it is subjected to a given set of environmental and/or mechanical stresses. This definition suggests that the failure criteria are defined, and there is an expected lifetime of the part. Stresses that are considered could be environmental, such as temperature, humidity, corrosive agents, and so on, and mechanical, such as vibration, shock, abrasion, and so on. All these conditions are defined by application of the part or product, and in different applications the reliability of product will be different.

Various methods developed to predict reliability are not useful for electrical connectors which are susceptible to many environmental and mechanical stresses, which have failure criteria that vary depending on the system in which they are used, and for which little reliable field failure analysis data exist [26]. Other methods (physics-of-failure) require that potential failure mechanisms of a design be identified and then a test program which will force these mechanisms be developed. The goal is to cause, at an accelerated rate, the types of failure inducing degradations which would be expected to occur in actual operation.

By using these methods one may produce a quantitative estimate of reliability for a specific use (e.g., a particular system). These reliability parameters are *mean time between failures* (MTBF), *mean time to failure* (MTTF) or average service life of connectors, *failure in time* (FIT), failure rates, and service life [27].

MTTF and MTBF are difficult to obtain for electrical connectors. Determination of MTTF or MTBF requires that:

1. Large percentage of contacts and connectors actually fail during testing
2. Time of each failure is known

However, most often few, if any, failures (however defined) occur during typical connector tests. Determining MTTF or MTBF for electrical connector is found difficult and impractical, which makes the use of these reliability parameters for connectors not justified [28].

In the case of complex, multiple application components such as connectors, which are susceptible to multiple and sometimes interacting environmental and mechanical stresses, the method which can provide good estimates of qualitative or quantitative reliability at a reasonable cost and in a reasonable time is the physics-of-failure method.

2.9 NATIONAL AND INTERNATIONAL STANDARDS AND SPECIFICATIONS ON MATERIALS, TESTING, AND DESIGN OF BUSBARS AND BUSWAYS

American Specifications:

C 29.1 Test methods for electric power insulators.
C 37.20 Switchgear assemblies including metal-enclosed bus.
C 37.30 Definition and requirements for HV air switches insulators and bus supports.
C 37.31 Indoor apparatus insulators, electrical and mechanical characteristics.

British and European Standards:

BS 23 Copper and copper–cadmium trolley and contact wire for electric traction.
BS 7884 Copper and copper–cadmium stranded conductors for overhead electric traction and power transmission systems.
BS 159 Busbar and busbar connections.
BS 6931 Glossary of terms for copper and copper alloys.
BS 1432 Copper for electrical purposes, strip with drawn or rolled edges.
BS 1433 Copper for electrical purposes, rod and bar.
BS 1434 Copper for electrical purposes—commutator bar.
BS 1977 High conductivity copper tubes for electrical purposes.
BS EN 1652 Copper and copper alloys. Plate, sheet, and circles for general purposes.
BS EN 12165 Copper and copper alloys. Wrought and unwrought forging stock.
BS EN 12166 Copper and copper alloys. Wire for general purposes.
BS EN 12163 Copper and copper alloys. Rod for general purposes.
BS EN 1652 Copper and copper alloys. Plate, sheet, strip, and circles for general purposes.
BS 4109 Copper for electrical purposes—wire for general electrical purposes and for insulated cables and cards.
BS 4608 Copper for electrical purposes—rolled sheet, strip, and foil.
BS 5311 High-voltage alternating-current circuit-breakers.
BS EN 60439-2 Specification for low-voltage switchgear and controlgear assemblies. Particular requirements for busbar trunking systems (busways).
BS EN 1976 Copper and copper alloys. Cast unwrought copper products.
BS EN 1978 Copper and copper alloys. Copper cathodes.
BS EN 60439-1 Specification for low-voltage switchgear and controlgear assemblies. Specification for type-tested and partially type-tested assemblies.

IEC Specifications:

IEC 28 International standard of resistance for copper.
IEC 137 Bushings for alternating voltages above 1000 V.
IEC 273 Dimensions of indoor and outdoor post insulators and post-insulator units for systems with nominal voltage greater than 1000 V.
IEC 344 Guide to the calculation of resistance of plain and coated copper conductors of low-frequency wires and cables.
IEC 349 Factory-built assembler of low-voltage switchgear and controlgear.

IEC 60439-1:1999 MOD Low-voltage switchgear and controlgear assemblies. Part 1: Type-tested and partially type-tested assemblies (Edition 4:1999 consolidated with amendment 1:2004).

Australian/New Zealand Standards [29]:

AS/NZS 3000:2007 Electrical installations.
AS/NZS 3012: Electrical installations—construction and demolition sites.
AS/NZS 3017 Electrical installations verification guidelines.
AS/NZS 3439.1:2002 Low-voltage switchgear and controlgear assemblies. Part 1: Type-tested and partially type-tested assemblies.
AS 60890:2009 A method of temperature-rise assessment by extrapolation for Partially Type Test Assemblies (PTTA) of low-voltage switchgear and controlgear.

2.10   ELECTRICAL CONNECTIONS GLOSSARY

Online Sources:

1. www.glenair.com, Connector and Contact Glossary of Useful Terms
2. www.connectpositronic.com, Positronic Industries, White paper "Connector Industry Lexicon"
3. http://www.osha.gov/SLTC/etools/electric_power/glossary.html "Glossary"

B–C

Bayonet coupling: A quick coupling device for plug and receptacle connectors, accomplished by rotation of a cam operating device designed to bring the connector halves together.

Closed entry contact: A female contact designed to prevent the entry of a pin or probing device having a cross-sectional dimension (diameter) greater than the mating pin.

Compatible connectors: Connectors that are intermountable, intermateable, or interchangeable.

Connection: Intentional electric contact between conductors or intentional junction between waveguides including optical fibers.

Connector contact: The primary electrically conductive element of connectors. The contact system is comprised of a male contact and a female contact. In general, contacts are available in a wide variety of sizes. Contacts can be provided with multiple termination types, including wire crimp and solder, printed circuit board (PCB) solder, straight and right angle mount, and straight mount compliant press-in.

Connector front: Side of a connector which is the mating face.

Connector housing (shell): Part of a connector into which the connector insert and contacts are assembled.

Connector insert (insulator, molding, dielectric): The connector insulating element which also supports and positions the contacts in the connector system. Connectors can be supplied as a free connector or a fixed connector.

Connector interface: Two surfaces of mating connectors which face each other when mated.

Connector rear: Side of a connector in which termination of the connector occurs.

Contact: The conductive element in a connector that makes actual contact for the purpose of transferring electrical energy.

Contact area: The area in contact between two conductors, two contacts, or a conductor and a contact permitting the flow of electricity.

Contact arrangement: Number, spacing, and arrangement of contacts in a connector.

Contact arrangement (contact variant, contact layout): Number, spacing, and configuration of contacts in a component.

Contact, bifurcated female (split tine): Contact containing forked/branched contact members with one contact point on each branch.

Contact, closed entry female: Contact which has an opening that is an unbroken ring. Closed entry contacts provide a higher degree of reliability.

Contact, coaxial: A contact with two conductive surfaces—a center contact and a surrounding coaxial sleeve.

Contact extraction force: Axial force required to extract a removable contact from a connector insert.

Contact, female (socket): Contact gender in which mechanical and electrical engagement is made on the inner surface of the contact.

Contact float: Permitted free movement of a contact in a connector insert.

Contact force: Normal force (90°) which exists between engaged contact surfaces providing contact pressure.

Contact, male (pin): Contact gender in which mechanical and electrical engagement is made on the outer surface of the contact.

Contact resistance: Electrical resistance of a pair of engaged contacts. Resistance may be measured in ohms or millivolt drop at a specified current over the engaged contacts.

Contact retainer: A device either on the contact or in the insert to retain the contact in an insert or body.

Contact retention: The axial load in either direction which a contact can withstand without being dislodged from its normal position within an insert or body.

Contact sequence: Sequence in which contacts with a different mating length mate, with the purpose of providing stepped insertion forces and/or stepped electrical continuity.

Contact, size: A designation to differentiate one contact from another. Numbers are commonly used for this purpose. The designator numbers are associated with a specific male contact diameter; the smaller the designator, the larger the contact size.

Contact, shielded: A contact which carries alternating current shielded from unwanted signals (EMI/RFI) by one or more outer (protective) conductors. These contacts are not generally matched to the impedance of the cable they terminate.

Contact shoulder: The flanged portion of a contact which limits its travel into the insert.

Contact, thermocouple: A contact made of special materials, used in connectors as a means of measuring temperature electrically. Materials often used for these contacts are alumel, chromel, constantan, iron, copper, and others.

Coupler: Intermediate device that can be used to attach accessories or mounting mechanisms, making two nonmatching connectors intermateable.

Creepage distance: Shortest distance along the surface of a solid insulating material between two conductive parts.

Creepage path: Path that electricity must follow across a dielectric to bridge cross two conductors; longer creepage paths reduce the likelihood of arc damage or tracking.

Crimp: The physical compression (deformation) of a contact barrel around a conductor in order to make an electrical connection.

Crimp barrel: Conductor barrel designed to accommodate one or more conductors and to be crimped by means of a crimping tool.

Crimp contact: A contact, pin or socket, whose back portion (wire barrel) is a hollow cylinder into which a stripped wire (conductor) is inserted. The sidewalls of the wire barrel are then mechanically compressed (uniformly deformed) using a crimping tool to captivate the conductor.

Crimp termination: Type of connection that secures a metal sleeve to a conductor by mechanically crimping the sleeve with pliers or a similar tool.

Crimped connection: Permanent connection made by the application of pressure inducing the deformation or reshaping of the barrel around the conductor of a cable.

Crimping: Method of permanently attaching a termination to a conductor by pressure deformation or by reshaping the barrel around the conductor to establish good electrical and mechanical connection.

Crimping dies: Portion of the crimping tool that shapes the crimp.

Crimping tool: Mechanism used for crimping.

Crimping zone: Part of a crimp barrel where the crimped connection is achieved by pressure deformation or reshaping the barrel around the conductor.

D–E

Depth of crimp: The distance the indenter penetrates into the barrel.

Die closure: The gap between indenter dies at full handle closure. Usually defined by Go/No-Go dimensions.

Effective press-in length: Length of contact between the press-in zone of a press-in termination and the metal plating of the plated-through hole in a printed board in which the press-in termination is inserted.

Electrical engagement length (wipe length, contact wipe, electrical engagement): Distance a contact glides on the surface of its mating contact during engagement or separation.

Electromagnetic interference (EMI): Electrical or electromagnetic energy that causes unwanted responses in electronic equipment.

Engaging force: Force required to engage fully a pair of mating components including the effect of a coupling, locking, or similar device.

Extraction tool (removal tool): Device for extracting removable contacts from a component.

F

Feed-thru: Connector that has double-ended terminals to facilitate simple distribution and bussing of electrical circuits.

Female connector: A connector which utilizes female contacts.

First mate/last break contacts: Contacts which provide different contact levels and sequencing. NOTE: During connector module engaging (insertion), at least one contact of the higher level shall connect before any contact of a lower contact level; whereas during connector module disengaging (withdrawal), at least one contact of the higher level will not disconnect before any contact of a lower contact level.

Fixed board connector: Connector mounted on a mother board or back plane, for engagement with a free cable connector and/or free board connector.

Fixed cable connector (bulkhead mount, panel mount): Connector mounted to a rigid surface for attachment to wire or cable.

Fixed connector: The portion of connector system designed for attachment to a rigid surface.

Fixed jackscrew (jack post): The portion of a jackscrew system that does not rotate.

Float mounting: Mounting method permitting movement to facilitate alignment of two mating components.

Float-mounting connector: Fixed connector with mounting means permitting movement to facilitate alignment with the mating connector.

Free board connector: Connector mounted on a printed board which can be separated from a mother board or back plane, for engagement with a free cable connector and/or free board connector.

Free cable connector (free connector): Connector for attachment to the free end of a wire or cable.

Front mounted: Pertaining to a component with its mounting flange in front of the mounting surface when looking at the mating face or front side of the component.

Front release: A term indicating the direction the contact removal tool must enter the connector to allow for the removal of contacts. On a front release connector, the contact removal tool must be inserted in the contact cavity from the front or face of the connector to release the contact retention clip. Whether front release or rear release, the contacts are inserted from the rear of the connector.

Front release contact: Removable contact in which release is effected from the mating face.

G–H

Gasket: Also called a grommet; a component that forms an environmental seal by surrounding a connector interface with an elastic polymer.

Grommet: A resilient elastomeric seal bonded to the rear of a connector. It is designed with internal sealing barriers that grasp and seal on the wire's insulation to prevent contaminants from entering into the rear of the connector.

Ground: Any zero-voltage point. Earth is considered a zero voltage grounding point.

Ground wire (drain wire): An extra conductor (usually a bare wire) added to a cable for connection of the grounding path.

Guide pin: A pin, rod, or projection extending beyond the mating face of a component designed to guide the mating of the component to ensure proper alignment of the contacts.

Guide socket: A socket which receives a guide pin designed to guide the mating of the component to ensure proper mating alignment of the contacts.

High density connector: A connector having its pins arranged close together without compromising system performance.

Holding strength: Connector's ability to remain assembled to a cable under tension.

I–J

Indenter: That part of a crimping die, usually the moving part, which indents or compresses the contact barrel.

Inner conductor: Central conductive structure in a coaxial structure, such as the center contact in a coaxial connector.

Insert: Connector part that holds the contacts in the proper arrangement and insulates them from each other and from the outer shell.

Insert retention: Axial load in either direction that an insert must withstand without being dislocated from its normal position in the connector shell.

Insertion force: Force required to fully insert a set of mating components without the effect of a coupling, locking, or similar devices.

Insertion loss: Loss of load power due to cable or component insertion; expressed in decibels as the ratio of power received at the load before insertion to the power received at the load after insertion.

Insertion tool: Device used to insert contacts into a component.

Inspection hole: A hole placed at the bottom end of a contact wire barrel to permit visual inspection to see that the conductor has been inserted to the proper depth in the barrel prior to crimping.

Installing tool: A device used to install contacts into a connector. A device used to install taper pins into taper pin receptacles.

Interchangeable: Pertaining to a component when all elements guarantee compliance of electrical, mechanical, and climatic performance of mated connectors when individual connector halves are from different sources. This is the highest level of compatibility. See also "Intermateable" and "Intermountable."

Interconnect: A conductor within a module or other means of connection which provides an electrical interconnection between the solar cells (UL 1703).

Interface: The two surfaces on the contact side of a mating plug and receptacle. The surfaces will face each other and interface when mated.

Interfacial seal: The sealing of mating connectors over the entire area of the interface and around each contact. This is accomplished when resilient material, with raised barriers around each cavity on the pin interface, displaces into the hard recessed (chamfered) cavities on the socket interface. This creates what is commonly called "cork and bottle seal."

Intermateable: Pertaining to each of two components when they feature identical dimensions for electrical and dimensional interfaces.

Intermountable: Pertaining to each of two components when their overall dimensions on printed board or panel cut-out, and cable termination are identical.

Jackscrew system: A locking device which uses the mechanical advantage of threaded coupling to couple and uncouple connector pairs. The system consists of a fixed jackscrew and a rotating jackscrew. Jackscrew systems can also be used as a coding device for connectors.

L–M

Locator: Device for positioning terminals, splices, or contacts into crimping dies, positioner, or turret heads.

Locking device: An accessory that provides mechanical retention of mated connectors.

Lockout: A feature of fixed, female, threaded hardware that minimizes the possibility of rotation of male rotating hardware once the system is coupled.

Lower limiting temperature: Value within a connector specification, smallest admissible value of the temperature at which the connector, as defined by the climatic category, is intended to operate.

Male connector: A connector which utilizes male contacts.

Mate: The joining of two connectors.

Mated pair: A plug and receptacle joined or to be joined together.

Millivolt drop test: A test designed to determine the voltage loss due to resistance of a crimped joint.

Mounting bracket: Connector accessory used to mechanically fix a connector to a mounting surface.

Mounting flange: Projection from a component for the purpose of attaching the component to a rigid surface.

Mounting plate: Plate for mounting electronic and electrical devices.

O–P

O-ring: Also referred to as peripheral seal; it is used around the periphery of a connector shell and is compressed internally between the plug and receptacle shells when mated to prevent contaminants from entering the connector.

Over-voltage: A condition in which voltage exceeds a specified limiting value of a component.

Panel cut-out: Hole or group of holes cut in a panel or chassis for the purpose of mounting a component.

Pin contact: A contact having an engagement end that enters the socket contact.

Plating: The overlaying of a thin coating of metal on metallic components to improve conductivity, provide for easy soldering or prevent rusting or corrosion.

Plug: Part of the two mating halves of a connector that is free to move when not fastened to the other mating half.

Polarization: Integral feature within a connector system to ensure corresponding male and female contacts are engaged when the connectors are mated.

Polarizing plug: An accessory to a connector which allows a connector pair to be polarized by blocking specific contact positions.

Positioner: A device when attached to a crimping tool locates the contact in the correct position.

Power connector: A connector in which part or all of its contacts are used to supply power to a component or device. Current rating is relevant to the application.

Power contact: Type of contact used in multicontact connectors to support the flow of rated current.

Press-fit contact: Electrical contact that can be pressed into a hole in an insulator, printed board, or metal plate.

Press-in connection: Solderless connection made by inserting a press-in termination into a plated through hole of a printed board.

Press-in connection, compliant: A press-in connection in which the termination "complies" to the shape of a plated-through hole.

Press-in zone: Specially shaped section of a press-in termination which is suitable for the press-in connection.

Printed-board connector: Connector specifically designed to facilitate connections to printed boards.

Pull-out force: Force necessary to separate a conductor from a contact or terminal, or a contact from a connector, by exerting a tensile pull.

Pull test: Tensile test: a controlled pull test on the contact crimp joint to determine its mechanical strength.

Push fastener (board lock): An accessory to a PCB mount connector allowing rapid insertion and securing of the connector prior to the soldering operation.

Q–R

Quick disconnect connector: Connector fitted with a coupling device which permits relatively rapid unmating.

Quick disconnect coupling: Type of coupling device which permits relatively rapid uncoupling.

Quick disconnect locking device: Device which allows for rapid connects and disconnect of connector pairs. Many systems consist of fixed lock tabs and actuation levers.

Radio frequency contact (RF contact): An impedance-matched shielded contact.

Range, wire: The sizes of conductors accommodated by a particular barrel. Also the diameters of wires accommodated by a sealing grommet.

Rated current: The amount of electrical current flowing through a connector or connector contacts expressed in terms of a specified set of conditions or test method. A component's current rating is used as an aid in specifying the components use as part of an electrical system.

Rear mounted: Connector installed from the inside of a box onto a panel that can only be removed from inside the equipment.

Rear release contact: Removable contact in which release is effected from the rear.

Receptacle: Fixed or stationary half of a two-piece multiple contact connector, usually the mounting half that contains the socket contacts.

Rectangular connector: A connector that has a housing and/or insert that is rectangular in shape.

Removal tool: A device used to remove a contact from a connector. A device used to remove a taper pin from a taper pin receptacle.

Right angle PCB mount: The relative position of PCB termination when the terminations are at a right angle to the mating position of the contact.

Rotating jackscrew (screw lock): The portion of a jackscrew system that rotates.

S

Sealing plug: A plug which is inserted to fill an unoccupied contact aperture in a connector insert. Its function is to seal all unoccupied apertures in the insert, especially in environmental connectors.

Separating force: Force required to separate fully a pair of mating components including the effect of a coupling.

Socket contact: A contact having an engagement end that will accept entry of a pin contact.

Solder contact: Contact designed for the attachment of the conductor by solder. NOTE: Solder contacts may assume but are not limited to the following types: (a) the wire is surrounded by the terminal, barrel, bucket, cup, well, and so on, (b) the wire surrounds the terminal: post, standoff, tag, and so on, (c) the wire passes through the terminal: eyelet, hook, and so on, (d) no wire is involved, as in printed circuit applications: solder pin post, SMT terminations (e.g., gull wings) spill, tail, and so on.

Soldered connection: Connection made by soldering.

Solderless connection: The joining of two metals by pressure means without the use of solder, braze, or any method requiring heat.

Spacer: An accessory that is used to control the distance from and secure a connector to its mounting surface.

Straight PCB mount: The relative position of PCB terminations when the terminations are a linear extension of the mating portion of the contacts.

Strip: To remove insulation from a conductor.

T–U

Terminal: Conductive part of a device, electric circuit, or electric network, provided for connecting that device, electric circuit, or electric network to one or more external conductors.

Termination point (termination, termination type): Part of a contact, terminal of a contact, terminal, or terminal end to which a conductor is normally attached.

Threaded coupling: Means of coupling by engaging screw threads present on the mating components.

Turret head: A device that contains more than one locator which can be indexed by rotating a circular barrel, and when attached to a crimping tool, positions the contact.

Under load: The condition of a component when current is flowing through the component during normal use.

Upper limiting temperature: Value within a connector specification, greatest admissible value of the temperature at which the connector, as defined by the climatic category, is intended to operate.

V–W

Wiping action: Action used to mate contacts with a sliding motion to remove small amounts of contamination from the contact surfaces and establish better conductivity.

Withdrawal force: Force required to fully withdraw a set of mating components without the effect of a coupling, locking, or similar device.

Withstand voltage (breakdown voltage, dielectric strength): Value of the test voltage to be applied under specified conditions in a withstand test, during which a specified number of disruptive discharges is tolerated. The withstand voltage is designated as: (a) conventional assumed withstand voltage, when the number of disruptive discharges tolerated is zero. It is deemed to correspond to a withstand probability PW = 100% (this is in particular the case in the low-voltage technology); (b) statistical withstand voltage, when the number of disruptive discharges tolerated is related to a specified withstand probability, for instance PW = 90%.

Working voltage: Highest permissible AC value (specified in volts, r.m.s.) or highest DC value (specified in volts) across any particular insulation which can occur when the equipment is supplied at rated voltage.

REFERENCES

1. *Electrical Connections for Power Circuits, Facilities Instructions Standards, and Techniques*, Volume 3-3, United States Department of the Interior Bureau of Reclamation (USBR), pp. 1–20, November 1991, Online. Available: http://www.usbr.gov/power/data/fist/fist3_3/vol3-3.pdf.

2. David C. *Copper in Electrical Contacts, Copper Development Association Publication No 223*, European Copper Institute Publication No Cu0169, 40 pp., July 2015, Online. Available: http://www.copperalliance.org.uk/docs/librariesprovider5/resources/tn-23-copper-in-electrical-contacts-pdf.pdf.

3. Chudnovsky B. H. Detrimental processes and aging of plating, in *Electrical Power Transmission and Distribution: Aging and Life Extension Techniques*, Chapter 2, CRC Press, Boca Raton, FL, USA, 411 pp., 2012.

4. Suvrat B., Dulikravicha G. S., Murtyb G. S. et al. Stress corrosion cracking resistant aluminum alloys: Optimizing concentrations of alloying elements and tempering, *Materials and Manufacturing Processes*, 26(3), 363–374, 2005, Online. Available: http://web.eng.fiu.edu/agarwala/PDF/2011/7.PDF.

5. Farahat M. A. Factors Affecting the Life Time of the Electric Joints, *Proc. 14th International Middle East Power Systems Conference (MEPCON'10)*, Paper ID 145, Cairo, Egypt, December 19–21, 2010, Online. Available: www.sdaengineering.com/MEPCON10/Papers/145.pdf.

6. Olsen T. W. *Bus Joint Fundamentals*, Siemens Technical Topics, #16, April 2001, Online. Available: http://w3.usa.siemens.com/us/internet-dms/btlv/PowerDistributionComm/PowerDistribution/docs_MV/TechTopics/ANSI_MV_TechTopics16_EN.pdf.

7. David C. and Toby N. Copper for busbars. *Guidance for Design and Installation*, Copper Development Association Publication No 22, European Copper Institute Publication No Cu0201, 108 pp., Revision May 2014, Online. Available: http://www.copperalliance.org.uk/docs/librariesprovider5/pub-22-copper-for-busbars/copper-for-busbars-all-sections.pdf?sfvrsn=2.

8. Braunovic M. Effect of connection design on the contact resistance of high power overlapping bolted joints, *IEEE Transactions on Components, Packaging and Manufacturing Technology*, 25(4), 642–650, 2002.

9. Yanko S., Nikos M., and Valery M. Thermal field distribution in bolted busbar connections with longitudinal slots, recent researches in circuits, *Systems and Signal Processing*, CIRCS-26, 2011, Online. Available: www.wseas.us/e-library/conferences/2011/Corfu/CIRCS/CIRCS-26.pdf.

10. Slade, P. G. (Ed.). *Electrical Contacts: Principles and Applications*, Marcel Dekker, New York, p. 155, 1999.

11. Timsit R. S. The technology of high power connections: A review, *20th International Conference on Electrical Contacts*, Zurich, Switzerland, p. 526, 2002.

12. *Electrical Jointing*, Tyco Electronics, Online. Available: http://www.gvk.com.au/pdf/electrical_jointing.pdf.

13. Braunovic M. Effect of different types of mechanical contact devices on the performance of aluminum-to-aluminum joints under current-cycling and stress relaxation conditions, *Proc. 32nd Holm Conference on Electrical Contacts*, Boston, MA, pp. 133–141, 1986.

14. Braunovic M. Evaluation of different types of contact aid compounds for aluminum-to-aluminum connectors and conductors, *IEEE Transactions on Components, Packaging and Manufacturing Technology*, 8, 313, 1985.

15. *A Look at Bus Connections: Why a Localized Silver Deposit Makes Sense*, SIFCO Selective Plating, White Paper, February, 2010, Online. Available: http://www.dokor.com/reference/plated-business/1-2_bus_bar_white_paper.pdf.

16. Braunovic M. Evaluation of different contact aid compounds for aluminum-to copper connections, *IEEE Transactions on Components, Packaging and Manufacturing Technology*, 15, 216, 1992.

17. Stephan S., Helmut L., Kindersberger J. et al. Creep Ageing of bolted electrical busbar joints, *Proc. 19th International Conference on Electric Contacts, Nuremberg*, Germany, pp. 269–273, 1998.

18. George D. *Maintain Bolt Preload on Electrical Connections Using Belleville Springs*, Solon Manufacturing Company, ElecConn White Paper, Online. Available: www.solonmfg.com/springs/pdfs/ElecConnWhitePaper.pdf.

19. Di Troia Gary, Kenneth W., and Gaylord Z. *Connector Theory and Application. A Guide to Connection Design and Specification*, FCI USA Inc., 43 pp., Revised 4th Edition, 2007, Online. Available: https://www.academia.edu/9463837/BURNDY_PRODUCTS_Connector_Theory_and_Application.

20. Braunovic M., Myshkin N. K., and Konchits V. V. Power connections, in *Electrical Contacts: Fundamentals, Applications and Technology*, Chapter 7, CRC Press, Taylor & Francis Group, Boca Raton, FL, 2006.

21. NRC Information Notice 2010-25: Inadequate Electrical Connections, *US NRC office of Nuclear Reactor Regulation*, Washington, DC, Nov 17, 2010.

22. LER 361/2008-006-00, Loose connection bolting results in inoperable battery and TS violation, *US NRC Office of Nuclear Reactor Regulation*, Washington, DC, issued 2008, Online. Available: pbadupws.nrc.gov/docs/ML0826/ML082660036.pdf.

23. *Square D® I-Line® Plug-On Connections*, Data bulletin 600DB1004, Schneider Electric USA, December 2010, 8 pp., Cedar Rapids, IA, Online. Available: http://static.schneider-electric.us/docs/Circuit%20Protection/0600DB1004.pdf.

24. Braunovic M. Evaluation of different platings for aluminum-to-copper connections, *IEEE Transactions on Components, Hybrids, and Manufacturing Technology*, 15(2), 204–215, 1992.

25. GASLINK® SF6 busbar system. *MGC Moser-Glaser*, Online. Available: http://www.mgc.ch/en/products/gaslinkbusbarsystem.

26. Mroczkowski R. S. *Connector Design/Materials and Connector Reliability*, Technical paper, P351-93, AMP Incorporated, 14 pp. 1993, Online. Available: http://www.te.com/documentation/whitepapers/pdf/p351-93.pdf.

27. Georg S. Reliability assurance and service life of electrical contacts, *Harting Tech News*, 14, 68–73, 2006, Online. Available: http://www.newark.com/pdfs/techarticles/harting/reliabilityAssuranceAndServiceLife.pdf.

28. Pascucci V. C. A brief overview of reliability in general and for electrical connectors in particular, *TYCO Electronics*, 7 pp.

29. Roland B. Operating temperature of current carrying copper busbar conductors, dissertation, University of Southern Queensland Faculty of Health, *Engineering & Sciences*, Australia, 117 pp., October 2013, Online. Available: eprints.usq.edu.au/24629/1/Barrett_2013.pdf.

3 Plating of Electrical Equipment

3.1 ELECTROPLATING FOR CONTACT APPLICATIONS

3.1.1 SILVER PLATING

Silver (Ag) plating has many different uses in an industrial setting. It can be used as an engineering coating owing to its superior conductivity and corrosion resistance. When used in plating, silver's conductivity allows for extensive use in electronics and semiconductor industries. It is also used extensively in the aerospace, telecommunications, military, and automotive industries.

3.1.1.1 Physical Properties of Silver Plating

Silver plating is considered to be one of the most highly conductive plated surfaces. It is widely applied to copper conductors of any kind, including wires.

In electrical power distribution, a busbar—a thick strip of copper or aluminum—conducts electricity within a switchboard, distribution board, substation, or other electrical apparatus. Busbars may be connected to each other and to electrical apparatus by bolted or clamp connections. Often, joints between high-current bus sections have matching surfaces that are silver plated to reduce contact resistance (CR).

Offering conductivity and corrosion resistance, silver plating creates a surface that can be soldered and exhibits low electrical resistance. It can be used as engineering coating as well as for bearing surfaces and antigalling applications. Silver plating should conform to Mil QQ-S-365D and ASTM B 700 Standards, as well as to ISO 4521, "Metallic Coatings—Electrodeposited Silver and Silver Alloy Coatings for Engineering Purposes."

Silver resists oxidation by air but is attacked by compounds containing sulfur. Industrial and urban atmospheric environments as well as certain materials contain sulfides. Under these conditions, the tarnishing of silver becomes inevitable. Tarnishing can have various degrees of severity. For more about silver corrosion, see Chapter 5.

Silver and silver-plated components can yellow slightly and sometimes do not discolor any further. The electrical conductivity of silver is not affected by a light yellowing.

In other cases, tarnishing can lead to a dark brown or black color. This discoloration can be partial or total, depending on the conditions of storage or use (finger marks, opened packing, etc.). In addition to aesthetics, the effects of excessive tarnishing at the electric level may be significant (Section 5.3). Silver sulfides are unstable with a rise in temperature.

Technical characteristics of silver layers:

- Specific electrical resistance: $16-18.8 \times 10^{-9}\ \Omega$
- Electrical conductivity: up to $62.5 \times 10^6\ \Omega^{-1}\ m^{-1}$ (at 20°C)
- Hardness: 70–160 HV
- Melting point: 960°C
- Coefficient of linear expansion: $19.3\ \mu m°C^{-1}\ m^{-1}$

3.1.1.2 Silver Plating Thickness for Electrical Applications

The thickness of plating strongly depends on the application and environment to which the silver will be exposed. It was found that at thicknesses <2 μm silver plating is porous and provides no proper protection of base metal from corrosion. For industrial applications, when electrical equipment is serving in a corrosive environment, the thickness of the silver plating should be in the range of 2–40 μm. The lowest thickness, ~2 μm, may be applied only for bolted contacts assembled in a factory and inaccessible to customers. A thin silver plating such as this is usually used in plating copper. For contacts assembled on site, the thickness of the silver plating should be no less than 5 μm. This is also the minimum thickness of the silver plating on aluminum and aluminum alloys, as well as on ferrous alloys.

3.1.1.3   Use of a Nickel Underplate for Silver Plating

It is recommended that a nickel underplate (a minimum of 1.25 μm) be used whenever possible in plating silver. The nature of the tarnish film will change significantly if copper alloy elements from the substrate reach the surface of the silver. This can occur through mechanisms such as diffusion or corrosion creep at breaks in the silver electrodeposit. At higher temperatures, oxygen will diffuse through silver to the copper alloy interface at a relatively high rate and can lead to blistering if no nickel underplate is used. A nickel underplate will also prevent a relatively weak layer of silver–copper intermetallics from forming at temperatures greater than 150°C, which could lead to adhesion problems [1].

In most separable contact interface applications that use a nickel underplate, silver plating thickness is typically in the range of 2 μm or greater. What silver thickness is appropriate depends on application factors such as environmental severity, the time at a specific temperature, durability requirements, nickel underplate, and surface treatment. If no nickel underplate is used, a greater thickness of silver may be required to prevent substrate corrosion products from getting to the surface [2]. These higher plating thicknesses also provide more silver material between the atmosphere and the substrate material(s), possibly leading to more wear cycles before any substrate material is exposed.

Silver may be used in plating various base metals. In many cases, the application of underplating prior to plating silver is necessary to provide proper adhesion and corrosion resistivity [3]. Depending on the base metal, different underplating metals should be applied and the thickness of underplating varies (Table 3.1). The data presented in Table 3.1 are a summary of information given in multiple recourses.

On the other hand, constant evolution in a silver plating's appearance is a sign of the presence of sulfur in the immediate vicinity of the electrical apparatus. In such conditions, antitarnishing

TABLE 3.1
Underplating Types and Thicknesses of Silver Platings for Various Base Metals

Base Metal	Underplating Type	Underplating Thickness (μm)
Copper	None	–
Brass (CuZn alloy)	Cu	4
	Ni	5
Bronze (CuSn alloy)	Cu	4
Ferrous alloys	Cu	8
Aluminum and aluminum alloys	Cu[a]	8
	CuSn (bronze)[b]	3
	CuSn	2
	Ni (electroplating)	5
	Ni (electroless)	5

[a] Zincade process (be described in more detail in Section 3.6) and plating process consisting of direct deposit of electrolytic bronze and copper developed by PEM in Europe [4].
[b] Alstan plating processes (be described in more detail in Section 3.6).

treatment of the silver plating is recommended. For electrical applications, only those antitarnishing compounds or techniques which do not contain lacquer and do not experience discoloration should be considered. These compounds should be easily stripped without any damage, should provide antitarnishing protection, and most importantly, must not affect electrical conductivity.

It is important to note that antitarnish and passivation treatments on silver coatings should be compliant with Reduction of Hazardous Substances (RoHS) requirements, which prohibit the use of hexavalent chromium (CrVI). Chromium-free solutions exist and are appropriate for the antitarnish protection/treatment of conductive parts.

3.1.1.4 Types of Silver Platings

ASTM B700 Standard [5] establishes the requirements for electrodeposited silver coatings that may be matte, bright, or semibright finishes. Silver plating is usually employed as a solderable surface and for its electrical contact characteristics, as well as for its high electrical and thermal conductivity, thermocompression bonding, wear resistance on load-bearing surfaces, and spectral reflectivity.

Coatings shall be classified into types according to minimum purity, grade according to surface appearance (bright, semibright, or matte), and class according to whether any surface treatment has been applied. Silver coatings shall undergo preplating operations such as stress relief treatment, strike, and underplating, as well as postplating embrittlement relief.

Silver plating, therefore, may be from white matte to very bright in appearance. Corrosion resistance may depend on the base metal. Hardness varies from about 90 Brinell to about 135 Brinell, depending on the process and plating conditions. Solderability is excellent, but decreases with age. Silver plating has excellent lubricity and smear characteristics for antigalling uses on static seals, bushing, and so on. According to Mil QQ-S-365D Standard, the thickness of silver plating should be no less than ~13 μm (0.0005″) unless specified otherwise.

Silver is an excellent conductor and is therefore widely used in contacts and joints in switchgear assemblies [6]. Popular methods for silver plating of contact surfaces are electroplating and the dip emersion process. A plating thickness of 5–15 μm is considered adequate for plating a copper or aluminum busbar. However, a thicker deposit may be needed because of the porosity of silver plating, particularly where environmental conditions may be harsh. For a disconnect switch (DS), plating must be thick enough to prevent exposure of the base metal to contamination, because the joint will be connected and disconnected many times during the service life of the equipment.

There is a serious disadvantage in using silver plating when a connection is made between silver-plated Al or Cu, because silver, like Cu, is cathodic to aluminum, and may cause galvanic corrosion of Al (for more about galvanic corrosion, see Section 4.4).

3.1.2 Tin Plating

Tin or tin alloy coating is a cost effective and reliable alternative to noble metal finishes (gold, silver) thanks to tin's low cost, low CR, and good solderability.

3.1.2.1 Physical Properties of Tin Plating

The following standards define the tin plating intended for engineering purposes: Mil-T-10727C, "Tin," and ASTM B-545, "Standard Specification for Electro-Deposited Coatings of Tin," and Standard ISO 2093-1987, "Specifications and Test Methods for Tin Electroplated Coatings." Tin plating is used for corrosion protection, to facilitate soldering, and to improve antigalling characteristics. Copper alloys containing more than 5% Zn should have a copper or nickel underplating.

According to Tin Military Standard Mil-T-10727C, there are two types of tin plating: Type I—electrodeposited plating and Type II—hot-dipped plating. Plating color is gray-white in plated conditions; it is soft but very ductile. Corrosion resistance is good (coated items should meet 24-h 20% salt spray (SS) requirement). Tin is not suitable for low-temperature applications, it changes structure and loses adhesion when exposed to temperatures below −40°C (−40°F).

However, the use of tin plating is limited by tin's low durability characteristics and susceptibility to fretting corrosion, which will be discussed in Section 4.5. These limitations may be avoided by using tin plating only in applications where a relatively low number of mating cycles are required, and by using appropriate contact design and lubrication (as needed) to reduce susceptibility to fretting corrosion.

Technical characteristics of tin layers:

- Hardness: 20–27 HV
- Melting point: 205–232°C
- Specific electrical resistance: $115 \times 10^{-9}\ \Omega$
- Coefficient of linear expansion: 17–27 $\mu m°C^{-1}m^{-1}$
- Electrical conductivity: up to $8.7 \times 10^{6}\ \Omega^{-1}m^{-1}$ (at 20°C)

From July 1, 2006, the implementation of the RoHS directive, which bans lead in products, requires using RoHS-compliant components. Some component manufacturers already propose that pure tin (or high-tin content) surface finishes replace Sn-containing lead (Pb). These finishes present a major reliability risk which can reduce their use because pure tin plating is prone to tin whiskers (tiny tin filaments that may cause short circuits). The effect of RoHS compliance on the tin whiskers phenomenon will be discussed in Section 3.5.

Table 3.2 shows acceptable techniques and types of underplating for the better adherence of tin plating for electrical applications.

For electronic components, such as connectors, various types of tin plating are used: matte Sn and SnCu, bright Sn, and hot-dipped Sn and SnCu, all usually with Ni underplating. Electronic components with a lead frame are plated with matte Sn alloys either with 2%–4% of Bi or with 1.5%–4% of Ag. Tin plating could be applied with or without underplating. However, when tin plating is applied to electronic components without a lead frame, Ni underplating is used, and "postbaking" or postplating annealing at 150°C for 1 h, is recommended. Other types of platings used for electronic components are hot-dipped Sn, SnAg, SnCu, and reflowed tin.

TABLE 3.2
Tin Plating Techniques for Electrical Applications

Base Metal	Plating Technique	Underplating Type	Underplating Thickness (μm)
Copper	Matte, bright, or bright hot-dipped	None	–
Copper–zinc alloys (brass)	Same	Ni or Cu	2.5–4
Copper–tin (bronze)	Same	Cu	4
Copper–aluminum alloys	Same	Cu	4
Aluminum and aluminum alloys	Zincate[a]	Cu	8
	Alstan[b]	CuSn bronze	3
	PEM process[c]	CuSn bronze and Cu	2
			8
	Ni underplating	Ni	5[c]
Ferrous alloys		Cu	8

[a] These techniques will be presented in more detail in Section 3.6.3.

[b] Plating process consisting of direct deposit of electrolytic bronze and copper developed by PEM in Europe [2].

[c] While plating tin on aluminum base, Ni underplating should not exceed 10 μm in order to avoid increases in CR, so usually it is recommended to be not more than 5 μm for electrical engineering applications.

TABLE 3.3
Minimal Thicknesses of Tin Platings for Electrical Applications

Environment	Min. Thickness (μm) (Base Metal Contains Cu)	Min. Tin Thickness (μm) (Other Base Metals)
Indoor, dry atmosphere	5	5
Indoor, with condensation	10	12
Outdoor, extreme T	15	20
Outdoor, corrosive	30	30

Source: Multiple industrial and OEMs Plating Standards.

3.1.2.2 Tin Plating Thickness for Electrical Applications

Tin plating thickness depends on which base metal tin is going to be deposited. Tin electroplating on copper is recommended to be not less than 5 μm, and tin electroplating on aluminum is recommended to be at least 8 μm (Table 3.3). The specifics of tin plating on Al will be discussed later in Section 3.6.

To comply with the requirements of the European directive 2002/95/CE (RoHS directive) of the European Parliament and the European Union Council of January 27, 2003 regarding the restriction of hazardous substances in electric and electronic equipment and, in particular, regarding lead substitution in tin–lead coatings, some substitute lead-free tin alloy platings may be considered for electrical applications. Among those are tin–bismuth alloy (SnBi), tin–copper alloy (SnCu), and tin–silver alloy (SnAg). Pure tin (matte or bright), as well as silver or nickel plating, may also be used as substitutes; however, the use of pure tin plating does not provide protection from tin whisker growth and fretting corrosion. Some properties of tin alloy platings, the substitute for tin–lead, are shown in Table 3.4 (for more about fretting corrosion and tin pest, see Chapter 4).

3.1.3 NICKEL PLATING

Electroplated Ni is not traditionally used in plating conductive parts of electrical equipment; more often, it is applied as an undercoating for other metals such as gold, tin, or palladium. It acts as a barrier layer to prevent the diffusion of the base metal to the surface. In the case of tin-coated contacts, it prevents the formation of copper–tin intermetallics (discussed in Sections 4.2 and 4.3). Nickel plating passivates pores and bare edges, reducing the potential for pore corrosion and creep corrosion.

3.1.3.1 Applications of Nickel Plating in the Electrical Industry

There are some special applications of nickel plating, such as for the aluminum used for electrical products such as wires, cables, or terminal connectors. In some industrial applications, the surface

TABLE 3.4
Properties of Some Sn–Pb Substitute Platings

Property	SnCu	SnAg	SnBi
Electrical resistance	Fair	Fair	Fair
Anticorrosion	Good	Good	Good
Antifretting	Fair	Fair	Fair
Antitin pest	Fair	Good	Good
Antiwhiskers	Fair	Good	Good

of the metal has to be modified in order to promote a good and durable electrical contact. This is particularly the case for the aerospace industry, where nickel electroplating on aluminum has proved to be a good means of significantly improving the surface conductivity of the substrate.

However, traditional processes of nickel plating on aluminum are complex to achieve as they require chemical pretreatments and several sublayers. That is why an original direct nickel electroplating method has been developed, which requires only two steps to cover the substrate. The resulting nickel layer not only exhibits a good adhesion but also allows meeting specifications in terms of electrical contact [7,8]. It was also shown that a nickel plating layer is stable up to high working temperatures (about 400°C) [7]. In studies of the effect of fretting in the contact properties of nickel-plated aluminum, it was shown that contact zones are considerably degraded by fretting, which could be mitigated with the application of lubricants and by using heavier loads [9,10].

Nickel plating was tested for copper busbars bolted to aluminum [11] and demonstrated an excellent stability and low initial CR. As proved in Ref. [12], nickel plating on copper-to-aluminum connections provides a better performance than silver- and tin-plated connections under different operating and environmental conditions. Nickel can be directly plated over several metals, typically steel, brass, and copper. In some cycles, appropriate immersion treatments or preplate deposits precede nickel. Aluminum, because of its unique electropositive nature, must first be conditioned by immersion zincating before plating either electrolytic or electroless nickel (EN).

3.1.3.2 Physical Properties and Thickness of Nickel Plating

Properties of nickel electrodeposited plating are defined in Federal Specification QQ-N-290 (1997) [13] and SAE AMS-QQ-N-290 (2000) 14, which cover the requirements for electrodeposited nickel plating on steel, copper and copper alloys, and zinc and zinc alloys. Electrodeposited nickel plating covered by these specifications is determined to be one of the following two classes: Class 1—corrosion protective plating and Class 2—engineering plating.

Class 1 plating is divided into seven different grades with different thicknesses, from Class A with the thickest plating (40 μm) to Class G with the thinnest coating (3 μm). According to these specifications, Class 1 plating shall be applied on a plating of copper on steels, and on copper and copper-based alloys (Table 3.5). Class 1 plating shall be applied to an underplating of copper or yellow brass on zinc and zinc-based alloys. Copper alloys containing zinc equal to or greater than 40% shall have a copper underplate of ~7.5 μm (0.0003 in.) minimum thickness.

Class 1 has three processing grades: SB—single-layer coating in a fully bright finish; SD—single-layer coating in a dull or semibright finish, containing <0.005% sulfur and having an elongation greater than 8%; and M—multilayer coating, either double layer or triple layer. When applied on copper and copper alloys, electrodeposited nickel plating with a thickness of 5–20 μm protects the base metal from

TABLE 3.5

Minimum Thicknesses of Class 1 Nickel Plating on Copper and Copper Alloys

Coating Grade	Ni Thickness (μm)	Coating Grade	Ni Thickness (μm)
A	40	E	10
B	25	F	5
C	20	G	3
D	15		

Sources: From Ni Plating Federal Specification QQ-N-290 (1997), SAE AMS-QQ-N-290 (2000), and Aerospace Material Specification/ Nickel Plating (Electrodeposited) AMS-QQ-N-290.

corrosion in indoor and outdoor environments; the same protection of aluminum and aluminum alloys is provided by electrodeposited nickel plating with a thickness of 10–30 μm. On aluminum and aluminum alloys, an immersion plating of zinc or tin and electrodeposited copper and other undercoatings are recommended in order to guarantee adhesion before the application of nickel coating.

Electroplated nickel is often used in plating copper conductors. There are multiple standards which define the thickness of nickel plating on copper conductors, depending on the diameter of the conductors, such as ASTM B 3, ASTM B 5, ASTM B 170, ASTM B 286, ASTM B 355, MIL-W-16878, MIL-W-22759, MIL-W-27038, and so on.

Nickel-plated copper has wide applications in aircraft wiring for special wind and moisture problem areas, as well as for aircraft engine and fire zone safety wiring, automotive engine and exhaust monitoring, high-temperature industrial and home appliances, electronic components, electric furnaces, heat tracing wiring, and downhole oil logging.

Nickel-plated copper combines the desired features of both metals: the 100% IACS conductivity of copper necessary for efficient electrical and thermal energy transfer and the high-temperature stability and corrosion resistance provided by nickel. Bare (unplated) copper conductors are generally rated for military applications operating at up to 105°C. Above this temperature, bare copper will oxidize excessively, become brittle and less ductile, and lose conductivity. Various plating thicknesses of nickel increase the useful continuous operating temperature of the plated copper conductor to as high as 450°C (842°F) [15].

3.2 ELECTROLESS PLATING

Electroless plating is a process for chemically applying metallic deposits onto substrates using an autocatalytic immersion process without the use of electrical current. It differs from electroplating, which depends on an external source of direct electrical current to produce a deposit on the substrate material. Since electrical current cannot be distributed evenly throughout the component, it is very difficult to obtain uniform coatings with electrolytically applied deposits. Electroless deposits, therefore, are not subject to the uniformity problems associated with electroplated coatings.

EN is one of the most used plating materials in the electronic and electrical industries. Major industry specifications for EN are AMS 2404, "Electroless Nickel Plating"; AMS 2405, "EN, Low-Phosphorus"; ASTM B656, "Guide for Autocatalytic Nickel–Phosphorus Deposition on Metals for Engineering Use"; ASTM B733, "Standard Specification for the Autocatalytic Nickel–Phosphorus Coatings on Metal Coatings"; and so on.

3.2.1 ELECTROLESS NICKEL (EN): PHYSICAL PROPERTIES

EN can be applied with excellent adhesion to many different substrates: steels, stainless steels, aluminum, copper, bronze, and brass.

Deposits may contain different amounts of phosphorus or boron depending on the chemicals in the autocatalytic process. The microstructure of an EN–phosphorus deposit strongly depends upon the alloy content of the deposit.

3.2.1.1 Chemical Composition and Structure of EN Plating

Various EN systems with phosphorus are formulated to codeposit from 1% to 13% phosphorus. The phosphorus content in the alloy is the most significant parameter to control the properties of plating. Generally, as plated, the higher phosphorus (above 9%) alloy deposits are often softer and tend to be nonmagnetic. Phosphorus levels can vary, ranging from 3% to 12% by weight. The industry identifies EN coatings according to their phosphorus content, for example:

- Low-phosphorus EN: 1%–3%
- Low–medium phosphorus: 4%–6%

- Medium-phosphorus EN: 7%–9%
- High-phosphorus EN: 10%–13%

At low and medium phosphorus levels (<7% by weight), the EN deposit is microcrystalline, consisting of many small grains, ~2–6 nm in size. As the amount of alloyed phosphorus increases, the microstructure changes to a mixture of amorphous and microcrystalline phases and finally to a totally amorphous phase (>10% by weight).

3.2.1.2 Physical Properties of EN Plating

When using EN in electrical applications, it is important to remember that its physical properties, such as hardness, electrical resistance, and corrosion resistance, are related to the plating composition.

3.2.1.2.1 Hardness

The *hardness* of an EN deposit is inversely related to its phosphorus content. As the phosphorus content increases, the as-plated hardness decreases. In all cases, EN deposits can be hardened through heat treatment, which causes the formation and precipitation of nickel phosphide (Ni_2P). Typical heat treatment conditions for full hardness are 400°C (752°F) for 1 h in an inert atmosphere. If there is no access to an inert atmosphere oven and discoloration is objectionable, or there is a need to harden without affecting the hardness of the substrate, full hardness can be obtained at lower temperatures with longer times.

3.2.1.2.2 Wear Resistance

EN coatings have good *wear resistance* because of their high hardness and natural lubricity. This, coupled with the uniformity of the EN deposit, makes it an ideal wear surface in many sliding wear applications. Relatively soft substrates with poor abrasion resistance, such as aluminum, can be given a hard, wear-resistant surface with EN. EN has also found widespread use in antigalling applications, where the use of certain desirable materials could not otherwise be used due to their mutual solubility and propensity to gall and seize. A deposit's hardness can be increased through heat treatment to further enhance wear properties, rivaling the wear properties of hard chromium.

All the physical properties of EN depend on the amount of phosphorus or boron in the deposit (density, hardness, corrosion resistance, etc.) [16–19]. Some physical properties of EN with phosphorus are shown in Table 3.6.

3.2.1.3 EN Film Thickness

Another important parameter which significantly influences the properties of the deposited film is the thickness of the film. One particularly beneficial property of EN is its uniform coating thickness, which can affect the ultimate performance of the coating and can also eliminate additional finishing after plating.

With electroplated coatings, the thickness can vary significantly depending upon the part's configuration and its proximity to the anodes. With EN, the coating thickness is the same on any section of the part exposed to a fresh plating solution and can be controlled to suit the application. Grooves, slots, blind holes, and even the inside of the tubing will have the same amount of coating as the outside part.

3.2.2 EN: Corrosion Resistance

A very important property of EN plating is its strong corrosion resistance, which makes it a very attractive plating for use in the electrical industry for equipment exposed to corrosion. Corrosion resistance of EN is stronger for coatings with a higher content of phosphorus, based on the amount of time before the first signs of corrosion appear after the exposure of plating to SS (Table 3.7).

TABLE 3.6

Physical Properties of EN Plating with Phosphorus

Property	Low Phosphorus, 1%–3% P	Low–Medium Phosphorus, 4%–6% P	Medium Phosphorus, 7%–9% P	High Phosphorus, 10%–13% P
Deposit density range (g/cm³)	8.6–8.8	8.3–8.5	8.0–8.2	7.6–7.9
Hardness as plated (HK_{100})	725–800	625–750	500–600	450–425
Hardness after heat treatment (HK_{100})	900–1100	850–1100	850–1000	850–950
Coefficient of thermal expansion (mμ/m °C)	12–15	11–14	10–15	8–10
Electrical resistivity (μΩ-cm)	10–30	15–45	40–70	75–110
Thermal conductivity (W m • K⁻¹)	6.28	6.70	5.02	4.19
Tensile strength (MPa)	200–400	350–600	800–1000	650–900
Elongation (%)	0.5–1.5	0.5–1	0.5–1	1–2.5
Melting range (°C)	1250–1360	1100–1300	880–980	880–900

Source: Based on data from "Electroless Nickel Plating" by Mike Barnstead and Boules Morcos. http://www.pfonline.com/articles/electroless-nickel-plating.

TABLE 3.7

Hours to First Appearance of Rust Pit for EN Plating with Different Phosphorus Contents

Thickness of Coating (μm) (Mils)	Low–Medium Phosphorus, 4%–5% P	Medium Phosphorus, 6.5%–8% P	High Phosphorus, 10%–12% P
12.7 (0.50)	24	24	250
22.4 (0.88)	96	96	1000
38.6 (1.52)	96	96	1000
50.8 (2.00)	96	96	1000

Source: Based on data from ASTM Standard B117 SS Test.

Optimizing film thickness will control corrosion resistance in engineering applications. ASTM B733 specifies the thickness of the Ni alloy film depending on the service condition, which can vary from 60 μm (2.4 mil or 0.0024″) for very severe conditions to 5 μm (0.2 mil or 0.0002″) for mild ones.

In general, choosing a proper composition and thickness of the coating will control its corrosion resistance. Corrosion properties of electroless coatings depend not only on the thickness of the film and its nonmetal content [20,21] but also on different parameters characterizing the process of plating, such as the coating time [22], solution age [23], and heat treatment [21].

3.2.3 EN: Electrical Resistivity

The use of EN in the electrical industry for application to electrical conductors strongly depends on the electrical resistivity of EN. Pure metallurgical nickel electrical resistivity is about 6.05 μΩ-cm. The electrical properties of electroless coatings vary with the plating composition. It is very important for EN alloy films deposited on the contact surface to have a *low electrical resistivity*.

For high-phosphorus deposits, coatings are significantly less conductive than conventional conductors such as copper. EN deposits containing 6%–7% of P have resistivity in the range of 52–68 $\mu\Omega$-cm. When P content is about 13%, the electrical resistivity is 110 $\mu\Omega$-cm. Heat treatment of EN reduces its electrical resistivity [22]. Alloying elements such as phosphorus as well as the presence of amorphous phases increase the electrical resistivity of the deposit.

For low-phosphorus deposits containing 2.2% P, the electrical resistivity is about 30 $\mu\Omega$-cm. It was found [24–26] that nickel alloys containing <2 at% of P are the most promising contact materials for replacing precious metals. Nickel alloy deposits with a minimum thickness of 2.5 μm exhibit a favorably low CR (2–7 $\mu\Omega$) after exposure to several accelerated environmental tests [20].

Because of the relatively thin layers used, however, for most applications, the resistance of EN is not significant. Heat treatments precipitate phosphorus from the alloy and can increase the conductivity of EN by two to four times. The formulation of the plating solution can also affect conductivity. EN with boron yields the lowest resistivity of any commercial EN type. Phosphorus content also has a strong effect on the thermal expansion of EN.

Because of their electrical and magnetic properties, EN alloys are widely used by the electronic industry for a variety of functions [26,27]. For electronic components, EN is used for its combination of superior corrosion resistance and solderability.

Other important considerations with these components are the coating's uniform thickness and its consistent electrical, thermal, and physical properties. Coaxial connectors, headers, housings and cases, heat sinks, diode cans, shutters, interlocks, and memory disks and drums are among the many electronic applications currently plated with EN.

The major high-volume component plated is the aluminum memory disk found in many computer storage devices. It consumes a huge amount of EN. The high-phosphorus EN used in this application has to be exceptionally smooth and nonmagnetic to serve as a base for the subsequent magnetic layers.

Another large electronic application is the coating of both aluminum and zinc connectors. Here, the coating serves to provide corrosion protection and wear resistance and is especially important because of the complex shape of these components. These are for both commercial and military uses.

3.3 EN AS A PLATING ALTERNATIVE FOR ELECTRICAL APPARATUSES IN CORROSIVE ATMOSPHERE

The coating deposited on the surface of copper parts of electrical contacts must display two important properties. First, it must withstand a harsh environment—mostly high humidity and a high concentration of hydrogen sulfide, chlorine, and/or other corrosive gases and particles in the air. This specific environment is generally typical of wastewater or sewage treatment plants. Hydrogen sulfide is also widely used in metallurgy, paper mills, at phosphorus mills and plants, and in the process of producing oil additives and other chemicals.

For electrical connectors working in areas with harmful atmospheric components (chlorine, hydrogen sulfide, chloric and sulfuric acid vapors, water vapor, salts, etc.), the use of silver and tin is often unacceptable. These coating materials corrode very quickly, leaving base metals unprotected. In a harsh environment, a different type of plating which can better withstand a high concentration of corrosive gases and vapors may be used.

For electrical applications, CB primary parts/disconnects/stabs and LV buses need to be plated with the plating that will protect current-carrying parts from resistance deterioration in a heavy corrosive industrial environment. Such a plating with high corrosion resistance must have a relatively low electrical resistance in order to successfully compete with the silver and tin plating traditionally used in finishing the conductive parts of electrical apparatus.

EN may be considered as an alternative plating for current-carrying parts in some specific applications because some formulations of EN have a relatively low electrical resistance and at the same time provide excellent corrosion protection. However, the application of EN plating to electrical conductive parts requires careful and thorough testing.

3.3.1 Testing of EN for Use in Electrical Applications

Testing the possibility of applying EN to current-carrying parts of a specific electrical equipment is needed to verify that the anticorrosion properties of EN are superior to that of various traditional electroplating types (silver and tin) in identical conditions. It must also be determined whether the electrical resistance of EN plating is comparable to traditional silver and tin plating.

If testing confirms that EN plating provides not only its already well-known perfect anticorrosion protection but also an acceptable electrical resistance in specific electrical applications, then it becomes necessary to run field testing of EN-plated electrical equipment in energized conditions exposed to industrial corrosive environments for extended periods of time. This testing will allow, if successful, the possibility of using EN as an alternative plating for the corrosion protection of current-carrying parts of specific electrical apparatus.

3.3.1.1 Testing the Anticorrosion Properties of EN Plating

The best way to test the anticorrosion properties of EN plating and compare these properties with those of traditional electroplating is to run a test using identical copper coupons in identical laboratory conditions. To receive statistically reliable results, three coupons were plated with one of the five platings of interest: two EN platings with different phosphorus contents and three traditional electroplatings: silver (Ag), tin (Sn), and nickel (Ni). Three coupons were left unplated (bare copper). In as-plated coupons, the amount of deposited metal (Sn, Ag, Ni, or Ni–P alloys) in each plating before the corrosion test was 100 at%, as determined using an x-ray elemental analysis.

The corrosive effect of environmental gases is strongly enhanced by other environmental parameters such as high humidity and high ambient temperature, which altogether rapidly destroy the integrity of metal finishes and produce extremely heavy tarnish films on some base metal surfaces. Conditions of corrosion testing were designed to represent an atmospheric environment similar to the one which causes failures in CB service conditions. These conditions include a gaseous mixture of air with H_2S and water vapor. Another predefined condition was elevated temperature. According to many reviews, the most concentrated corrosive environment among all industries has been observed at wastewater treatment plants (WWTPs) and older paper mills. The choice of corrosion test conditions has been based on standards that define the limit of corrosive gases concentration in the environment of such industrial sites.

The Occupational Safety and Health Administration (OSHA), the American Conference of Governmental Industrial Hygienists (ACGIH), and the National Institute for Occupational Safety and Health (NIOSH) Standards describe environments that could be observed at paper or sewage treatment plants both indoors and outdoors [28–30]. The OSHA ceiling for H_2S is 20 ppm, and the peak is 50 ppm for 10 min. The NIOSH ceiling for H_2S is 10 ppm for 10 min, and evacuation is over 65 ppm. The ACGIH Standard defines threshold limit value (TLV) and the concentration which could be "immediately dangerous to life or health" (IDLH), and represents the maximum level from which one could safely escape within 30 min for H_2S equal to 300 ppm. For the pulp and paper industry, total reduced sulfur (TRS) levels, including H_2S and elemental sulfur vapor, vary considerably for each point source and are related to the age of the mill [31].

Newer paper mills with efficient recovery boilers typically produce TRS levels of <20 ppm. At some older Kraft mills, TRS levels approaching 2000 ppm have been observed. Hydrogen sulfide levels in the vapor above the condensate were found to be 450 ppm at some paper mills. H_2S could also be liberated from the sludge at the presses, with H_2S levels from 40 to 100 ppm. However, levels spike to 750 ppm, depending on throughput volumes and feed composition. Besides, according to Ref. [30], the actual changes in concentration from outdoors to indoors are really not that significant, amounting to an average decrease of <50% for reduced sulfur.

Based on definitions from these tests, the following conditions for corrosion tests have been selected: H_2S concentration—2000 ppm; RH—100%, temperature—40°C (104°F; highest indoor possible); test duration—30 days. These conditions are much more severe than the conditions of a Class IV environment, according to the classification system developed at the Battelle Institute in Columbus, OH, for mixed flowing gas (MFG) testing conditions. Class IV represents the most severe level of corrosive environment [32,33] with T at 40°C (104°F), RH 75%, and H_2S level at 200 ppb (compared with 2000 ppm in our test, which is 10,000 times higher). Exposure time (30 days) is also longer than specified for MGF testing (20 days). The corrosion test was run in a 4-L heated kettle which served as the test chamber; the coupons were hung in the kettle's upper portion. The kettle was partially filled with deionized water through which a mixture of 2000 ppm of H_2S with air was continuously bubbled at a slow rate while the water temperature was maintained at about 43°C (110°F). Thus, the coupons were exposed to 2000 ppm of H_2S environment that was nearly saturated with water vapor.

After the 30-day exposure, the samples were removed, washed with deionized water, air dried, and examined visually for scaling, discoloration, and other signs of corrosion attack. The coupons of bare copper and tin-plated coupons were covered with heavy black scale. The silver plating was heavily tarnished. Pure Ni-electroplated coupons had dull luster with slight tarnish from some filming. The best in appearance were the coupons plated with EN plating—they still had a bright luster with a slight tint and no apparent scaling.

Further analysis included elemental composition of the surface/deposits, morphology of the plated surfaces (microphotographs of the cross section of the coupons) and microphotographs of the surface at different magnifications using the SEM/energy-dispersive x-ray spectroscopy (EDS) technique. The chemical composition of the deposits after the corrosion test helped to determine which corrosion products were formed on the surface. The result of the elemental analysis of the surfaces of the coupons with two different pure metal electroplatings (silver and tin) is compared with the corroded bare copper coupon as shown in Table 3.8.

From the visual appearance of the surfaces, tin plating is almost entirely destroyed, and chemical data show that the black scale on the surfaces consists of copper oxides and sulfides. The silver plating after the corrosion test is also heavily damaged, with less than 15% of silver plating remaining; in some spots, there is up to 30% silver in the form of oxides and sulfides. Both silver and tin plating had significant corrosion products on the surface with discontinuous plating observed in the microphotographs of the sample cross section. Very unusual formations covering the surface were found on the surface of silver-plated coupons (Figure 3.1). Based on chemical analysis of the "Corrosion Sunflower," no silver is present in the "flower"—only copper, oxygen, and sulfur. Microphotographs of cross sections of both silver and tin platings show discontinuity, or even complete loss, of the plating.

TABLE 3.8

Surface Compositions of Copper Coupons with Electrolytic Tin, Silver, and Nickel Plating before and after Corrosion Tests

Corrosion Test	Cu, at%		Ag, at%		Sn, at%		Ni, at%		O, S, at%	
	Before	After	Before	After	Before	After	Before	After	Before	After
Ag plating	0	70–83	100	29–15	–	–			0	1–2
Sn plating	0	0	–	–	100	78–81			0	21–17
Ni plating	0	0	–	–	–	–	100	100	0	0
Bare copper	100	90	–	–	–	–			0	10

Source: Chudnovsky B. H. Degradation of power contacts in industrial atmosphere: Plating alternatives for silver and tin, *Proceedings of the 49th IEEE Holm Conference on Electrical Contacts*, pp. 98–106, 2003. © 2003, IEEE.

FIGURE 3.1 Sulfide-oxide "flowers" made of Cu, O, and S (×8000) cover the surfaces of silver-plated copper coupons after corrosion test.

Compositions of the corroded coupon surfaces plated with two different EN plating compositions (1) and (2) and Ni electroplating are shown in Table 3.9.

The coupons with pure electrolytic Ni and one EN plating type did not show any damage or inclusions after the corrosion test, and one EN plating type was found partially oxidized.

According to the results of the corrosion test, the electroplated nickel and EN-plated coupons showed much better corrosion resistance, with less corrosion product formed on the surface than for traditional silver and tin electroplating. This corrosion test confirmed that EN plating withstands a very concentrated corrosive atmosphere much better than traditional silver and tin plating, which are virtually destroyed during corrosion tests.

It is important to note that the results obtained in the corrosion test characterize the corrosion resistance of the plating in predetermined conditions in a specific accelerated test. Also, the compositions of the coupon surfaces present the range of composition of the surface layer rather than its exact composition. The area of the spot under the electronic beam is very small (from 1 to 100 μm^2) and the relative composition of the film at this spot strongly depends on the set of elements which

TABLE 3.9

Surface Compositions of Copper Coupons with Electroless Ni (EN) Plating before and after Corrosion Test

Corrosion Test	Ni–P Alloy, at%		O, at%	
	Before	After	Before	After
EN(1) plating	100	93–94	0	7–6
EN(2) plating	100	100		

Source: Chudnovsky B. H. Degradation of power contacts in industrial atmosphere: Plating alternatives for silver and tin, *Proceedings of the 49th IEEE Holm Conference on Electrical Contacts*, pp. 98–106, 2003. © 2003, IEEE.

are covering this tiny part of the surface. This composition varies in a very wide range at different spots. Analytical data may vary in different measurements and also depend on experimental conditions.

These data provide an overall picture of a sulfuric environment's effect on silver, tin, and nickel platings and compare their corrosion resistance with that of EN alloy plating. However, the results of the corrosion test do not necessarily correlate directly with the actual corrosion resistance of each plating material tested in a sulfide-rich environment. It is very important to consider plating quality, such as porosity and the continuity of the plating, as well as the specific surface treatment of the base material, which, in addition to the chemical nature of the plating, may also contribute to corrosion resistance.

3.3.1.2 Testing of the Electrical Properties of EN Plating

In general, EN is not universally superior to silver or tin. EN plating has a significantly higher electrical resistance than traditional silver and tin plating. Therefore, when in electrical applications, EN plating is considered as an alternative plating for electrical current-carrying parts; it is essential to confirm that EN-plated equipment could be applied at the same rate as originally designed by the OEM.

There are several electrical tests designed to measure electrical resistance of electrical conductive parts according to the ANSI Standards. One of the most important tests is the continuous current test (CCT) setup for CBs in accordance with ANSI Standard C37.09-1979, Section 4.4. Therefore, prior to using EN plating in refurbished electrical equipment, it is strongly recommended that a series of life tests be performed, including a CCT.

CCT was run on equipment specifically designed for such a testing: six manufactured model sets made of copper busbars with bolted connections to test the electrical resistance of current-carrying parts coated with the same plating types, which have been tested for corrosion resistance. Each model set was plated with one of the five different finishes: two compositions of EN plating, one with a low content of phosphorus (LPEN) and another with high content of phosphorus (HPEN), and three traditional electroplatings (silver, tin, and nickel), with one set left with no plating (bare Cu). A model set of parts was designed to fit the CCT setup for CBs in accordance with ANSI Standard C37.09-1979, Section 1.4.4. CCT was performed on each of the six model sets of copper parts, plated or bare, assembled, enclosed in a cardboard box and connected to a power supply.

According to the test conditions, resistance (R) was measured on two pairs of contacts (upper and lower) twice, at the beginning and at the end of each test at 1600 A. The ambient temperature (T) and contact temperature were measured every half an hour as well as R cold (CR before the test) and R hot (CR measured at the end of the test). Resistance rise (ΔR) was calculated as the difference between these two values. Resistance rise for various platings during the CCT is shown in Table 3.10.

TABLE 3.10

Resistance Rise on Energized Electrical Contacts with Five Various Platings and Bare Copper during CCT

ΔR ($\mu\Omega$)	Cu	Ag	Sn	Ni	EN1	EN2
Cold contacts	40	38	40	42	44	43
Hot contacts	40	40	41	53	55	43

Source: Chudnovsky B. H. Degradation of power contacts in industrial atmosphere: Plating alternatives for silver and tin, *Proceedings of the 49th IEEE Holm Conference on Electrical Contacts*, pp. 98–106, 2003. © 2003, IEEE.

TABLE 3.11

Temperature Rise on Energized Electrical Contacts with Five Various Platings and Bare Copper during CCT

Time in Energized Condition (h)	Bare Cu ΔT (°C)	Ag ΔT (°C)	Sn ΔT (°C)	Ni ΔT (°C)	EN1 ΔT (°C)	EN2 ΔT (°C)
0.5	0	0	0	0	0	0
2.5	103	85	90	84	90	96
5.0	110	92	96	92	92	96

Source: Chudnovsky B. H. Degradation of power contacts in industrial atmosphere: Plating alternatives for silver and tin, *Proceedings of the 49th IEEE Holm Conference on Electrical Contacts*, pp. 98–106, 2003. © 2003, IEEE.

The TR (the difference between the ambient temperature and the temperature of the contact) for various platings and bare copper after 0.5, 2.5, and 5 h in energized conditions is shown in Table 3.11.

The result of the tests shows that neither the electrical resistance rise nor the TR for EN-plated contacts is significantly higher than these values for traditional plating.

3.3.2 FIELD TESTING OF EN-PLATED ELECTRICAL EQUIPMENT IN ENERGIZED CONDITIONS

Two previous tests proved that an EN plating of a carefully chosen formulation and thickness may offer not only the best anticorrosive properties but also a relatively low electrical resistance, not significantly exceeding the resistance of the best conductive platings such as silver and tin. EN's inherent hardness means that it does not require additional hardness treatments. The combination of these properties could make EN plating a good alternative to traditional plating for the electrical connectors of CBs exposed to corrosive atmospheres.

The last stage of the testing should provide proof that the real electrical apparatus' current-carrying parts coated with EN plating will properly function while being exposed to extreme corrosive conditions without the extensive deterioration of electrical resistance and failure.

The industrial environment of a WWTP has been chosen as a testing site for the third stage of EN testing. As the electrical unit for the field test, a MV electrical apparatus (MV electrical contactor) was chosen for reconditioning with EN plating after serving for a relatively short period of time at the same WWTP in heavily corroded conditions. The original ("as-is") condition of the unit after serving for a relatively short time at WWTP is easily seen in Figure 3.2. The contacts of the contactor are covered with heavy black scale, which previously was silver plating.

3.3.2.1 Live Electrical Tests

According to the industry standards, there are several tests for determining the condition of electrical apparatuses. One of the tests is the high potential (Hipot or HP) test to measure the quality of insulation between electrical parts; this test is also called the dielectric withstanding voltage (DWV) test. The theory behind the test is that if a deliberate overapplication of test voltage does not cause the insulation to break down, the product will be safe to use under normal operating conditions—hence the name DWV test. This test is helpful in finding nicked or crushed insulation, stray wire strands or braided shielding, conductive or corrosive contaminants around the conductors, terminal spacing problems, and so on.

The setup and test guidelines of the HP test are defined in IEC Standard 61010, "Safety Requirements for Electrical Equipment for Measurement, Control, and Laboratory Use" and in the

FIGURE 3.2   Heavily corroded electrical contacts of MV contactor after being exposed to the corrosive atmosphere of a WWTP.

Underwriters Laboratories Inc. (UL) Standard 61010A-1, "Electrical Equipment for Measurement, Control, and Laboratory Use; Part 1: General Requirements."

This test is usually applied to verify that the insulation of a product or component is sufficient to protect the operator from electrical shock. In a typical HP test, HV is applied between a product's current-carrying conductors and its metallic shielding. The resulting current that flows through the insulation, known as leakage current, is monitored by the Hipot tester.

The *insulation resistance (IR) test* is used to provide a quantifiable resistance value for all of the product's insulation. The test voltage is applied in the same manner as in a standard Hipot test, but is specified to be DC. The voltage and measured current value are used to calculate the resistance of the insulation. According to the requirements, the contactor in "as-is" conditions was thoroughly tested.

The *HP test* was performed with the contactor in the closed position, a phase-to-ground and phase-to-phase for each pole. The contactor should sustain test voltage for 1 min. Before and after the Hipot test, *IR* was measured and recorded at DC in closed position between phase-to-ground and phase-to-phase.

The contactor also passed through the CCT in which temperature was measured with thermo-couples installed at all significant points along the current-carrying path. CR was measured with the unit in closed position across each pole before and after CCT. The results of the test of the contactor in "as-is" condition are shown in Figure 3.3, which show that CR (top chart) and TR (bottom chart) on corroded contacts have extremely high values, which resulted in thermal failure of the unit.

3.3.2.2   Electrical Properties of the Contactor Reconditioned with EN Plating

After completion of the corroded unit testing, the same apparatus was refurbished with EN plating. Out of two previously tested EN platings, the one that showed the best anticorrosion properties has been chosen for plating. The contacts of the MV contactor refurbished with EN-type plating are shown in Figure 3.4. All required electrical tests have been performed to determine and compare the electrical properties of the apparatus before and after reconditioning with selected EN plating.

CR is compared in Figure 3.3a for the unit in "as-is" conditions and after refurbishment with EN plating. CR in CR1 was measured between primary disconnect fingers and CR2 was measured

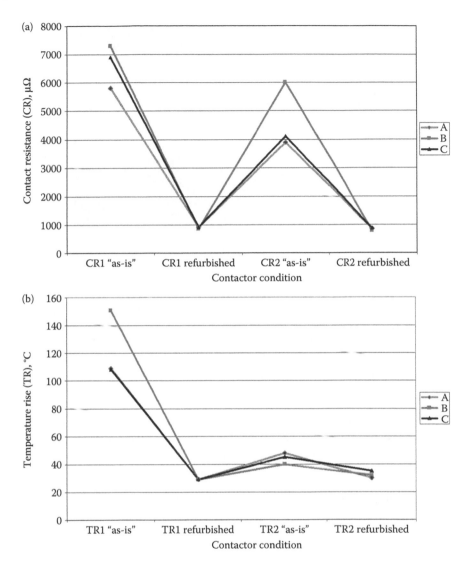

FIGURE 3.3 Electrical CR on three phases of corroded ("as-is") and refurbished contactor (CR1 measured between primary disconnect fingers and CR2 measured by passing them) (a); TR on disconnect fingers on three phases of corroded ("as-is") and refurbished contactor (TR1 measured on load primary disconnect fingers TR2 measured on line primary disconnect fingers) (b). (Chudnovsky B. H. Degradation of power contacts in industrial atmosphere: Plating alternatives for silver and tin, *Proceedings of the 49th IEEE Holm Conference on Electrical Contacts*, pp. 98–106, 2003. © 2003, IEEE.)

bypassing them. TR values are compared also for corroded unit and for the contactor refurbished with EN plating in Figure 3.3b. TR (difference between measured T and ambient T) was measured on load primary disconnect fingers (CCT1) and on line primary disconnect fingers (CCT2).

CCT was conducted according to standard requirements with applied rated current 400 A, with a maximum TR (ΔT) of 65°C above the ambient temperature. The test of the "as-is" unit was prematurely interrupted because of extreme overheating of heavily corroded connections. The CR test showed that the resistance of the current-carrying path refurbished with EN plating is higher than the resistance of the same path plated with silver (~100 $\mu\Omega$). However, based on the results of TR measurements, according to CCT, it was concluded that this specific apparatus with EN plating may be used at the same rating as originally designed by OEM.

FIGURE 3.4   MV contactor reconditioned with EN plating.

After the series of life tests were successfully completed, the refurbished unit was installed in the control room of the WWTP facility to determine the rate of plating deterioration at full load in a heavily corrosive environment. The result of the field testing of EN-plated contactors is remarkable; after 20 months of service, no visible signs of corrosion could be detected (Figure 3.5).

This is an example of the EN plating use for electrical application, demonstrating that EN plating of carefully chosen composition may withstand a corrosive, hot, and humid atmosphere much better than traditional silver and tin plating, which have been practically destroyed in the H_2S atmosphere of a WWTP.

The EN finish used in the presented example provided not only the best anticorrosion properties but also relatively low and stable electrical resistance and high hardness as plated. The combination of these properties makes this plating a good alternative for traditional tin and silver plating for electrical connectors of CBs working in a corrosive atmosphere.

3.3.2.3   Precaution in Electrical Applications of EN Plating

In general, EN is not universally superior to silver or tin. EN plating has a higher electrical resistance than traditional plating. The test results presented in this chapter show that EN plating could be used

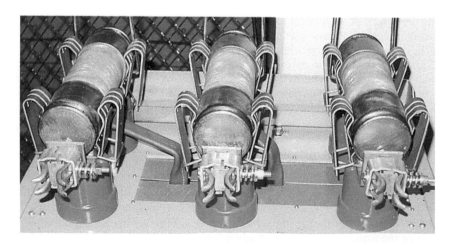

FIGURE 3.5   No signs of corrosion are seen on electrical apparatus with corrosion protective plating after 20 months in service at WWTP.

in specific electrical installations without derating the equipment. However, these results cannot be applied blindly to any other type of electrical equipment. Prior to refurbishing a unit or installation, a series of life tests, like the ones described in the example, should be run to demonstrate that the EN-plated equipment could be applied at the same rating as originally designed by OEM.

Considering the higher electrical resistance of EN plating, the heat dissipation and evacuation capabilities of every unit and installation must also be evaluated to determine whether the plating still can be applied at the OEM's original rating. However, in most applications, electrical equipment is rarely used at full-rated current. If a unit suffers extremely extensive corrosion, which causes multiple equipment failures, it may be derated according to the results of a CCT to protect the entire installation from corrosion-related thermal accidents.

3.4 ZINC ELECTROPLATING AND GALVANIZATION

Zinc is more corrosion resistant than steel in most natural atmospheres, the exceptions being ventilated indoor atmospheres, where the corrosion of both steel and zinc is extremely low, and certain highly corrosive industrial atmospheres. Coatings of metallic zinc are generally regarded as the most economical means of protecting against corrosion. Seven methods of applying a zinc coating to iron and steel are in general use: hot-dip galvanizing, continuous-line galvanizing, electro-galvanizing, zinc plating, mechanical plating, zinc spraying, and painting with zinc-bearing paints.

3.4.1 ZINC ELECTROPLATING

Zinc plating is a common sacrificial coating used in finishing steel parts to provide protection of iron (steel) from corrosion (red rust [RR]). Zinc is usually applied electrolytically to a typical thickness of 5–7.5 μm (0.0002″–0.0003″). The zinc plating protects the underlying steel by the formation of a "galvanic cell," which results in the zinc corroding preferentially to the steel. RR will not start forming until all zinc has been converted into white rust (WR; zinc oxide). Normally, zinc plating is used for indoor applications, but can also be used as a base for painting [34].

By itself, zinc plating with a thickness of 5–7.5 μm will probably get not more than 12 h of SS protection per ASTM-B117. With a clear chromate topcoat, this is increased to 24–36 h; while a yellow chromate top coating can achieve protection up to ~96 h. Even though it is mostly used as a functional coating, zinc plating does have some decorative appeal. Other properties of zinc plating include moderate appearance, excellent abrasion resistance, and excellent paint adhesion.

Yellow zinc plating is applied as above, with added yellow dichromate passivate, which greatly improves corrosion resistance. Yellow zinc plating provides SS protection (ASTM-B117) for ~96 h.

Black zinc plating is applied as above with black silver nitrate passivate, giving a matte black appearance and a similar corrosion resistance as yellow zinc plating.

3.4.2 ZINC GALVANIZATION PROCESSES

There is usually at least one process that is applicable to any specific purpose. Because these processes are complementary, there are rarely more than two processes to be seriously considered as the best choice for a particular application [35].

3.4.2.1 Hot-Dip Galvanizing

In hot-dip galvanizing, the steel or iron to be zinc coated is usually completely immersed in a bath of molten zinc. This is by far the most widely used of the zinc-coating processes and has been practiced commercially for almost two centuries. The modern hot-dip galvanizing process is conducted in carefully controlled plants by applying the results of scientific research, and is far removed from that conducted years ago, although it is still dependent on the same basic principles.

3.4.2.2   Continuous Galvanizing

This process, known as the *Sendzimir process*, uses a small amount of aluminum in the zinc bath and produces a coating with essentially no iron–zinc alloy and with sufficient ductility to permit deep drawing and folding without damage to the coating.

3.4.2.3   Electrogalvanizing

The additional development of continual electrogalvanizing lines added another dimension to zinc-coated steel. In this process, very thin, formable coatings of zinc, ideally suited for deep drawing or painting are electrodeposited on a variety of milled products by the steel industry: sheet, wire, and in some cases, pipe. Electrogalvanizing at the mill produces a thin, uniform coat of pure zinc with excellent adherence. Electrogalvanized steel is produced by electrodepositing an adhering zinc film on the surface of sheet steel or wire. These coatings are not as thick as those produced by hot-dip galvanizing and are mainly used as a base for paint.

3.4.2.4   Process of Galvanizing

Before the iron or steel pans are dipped in the molten zinc, it is necessary to remove all scale and rust. This is usually done by pickling in an inhibited acid. To remove molding sand and surface graphite from iron castings, shot or grit blasting is generally used, usually followed by a brief pickling operation. In the dry galvanizing process, the work is prefluxed by dipping in a flux solution of zinc ammonium chloride and then passed to a low-temperature drying oven. It is then ready for dipping into the molten zinc bath, the surface of which is kept relatively clear of flux. In contrast, in wet galvanizing, the pickled articles are dipped into the molten zinc bath through a substantial flux blanket.

3.4.3   Conversion Zn Plating: Passivation with CrIII or CrVI

Conversion Zn plating which includes dichromate is a sealer that is deposited on zinc plating. It is very thin and in itself does not provide much corrosion resistance. However, when it is deposited onto a metal, such as zinc plating, it enhances the corrosion resistance and is an excellent base for paint that can be applied over the plating. CrVI (hex chrome) passivation is currently prohibited in compliance with the RoHS directives and replaced with CrIII (tri chrome) passivation. Some of the major differences between the two chromium conversions are in corrosion resistance, color variability, self-healing, and identification.

3.4.3.1   Corrosion Resistance

Under SS tests, corrosion resistance achieved by hexavalent chromiums is 120+ h. Under similar levels of SS tests, trivalent chromate provides ~36 h of protection. A topcoat can be added to increase corrosion resistance. The amount of topcoat coverage directly affects the corrosion resistance of the product, and currently no industry standards exist that mandate topcoat thickness. This means the topcoat could vary and so will the corrosion resistance. The topcoat also creates other barriers, such as increased cost and the inability to paint.

The Standard ISO 4042-E25-009, "Fasteners—Electroplated Coatings," provides requirements in the climatic test, which is a corrosion test in neutral salt spray (NSS) described in ISO 9227. Conditions of the test correspond to the required classes of the environment. So for indoor service conditions, WR withstand should last from 48 to 72 h, and RR withstand from 72 to 120 h. WR indicates a corrosion of the zinc coating (object of the CrIII passivation) and RR indicates the corrosion of the substrate (object of the Zn layer). (More about zinc corrosion is given in Chapter 5.) Any other requirement regarding resistances to RR and WR is the subject of another technological solution aiming at increasing the resistance to corrosion (Zn-alloyed nickel, Zn lamellar coating, the addition of a film forming—an additional layer of finishing: mineral [topcoat], organic, or organometallic). These solutions will have to be determined by the specifications. To use zinc plating in outdoor applications, the RR withstand for zinc-plated stainless steel should be 600 h.

3.4.3.2 Color Variability

In many cases, customers tend to purchase the yellow or gold version of zinc dichromate. This color change is created during the hexavalent chromate conversion process, which also greatly improves corrosion resistance. With the move to trivalent chrome, the conversion process will no longer generate a color change and color must be added by using a dye. The use of a dye makes it difficult to maintain a uniform color and UV exposure can eventually cause the color to fade. The corrosion properties are also sacrificed as the die has no corrosion resistance properties. The product becomes nothing more than a clear zinc finish that has been dyed yellow.

3.4.3.3 Self-Healing Properties

Trivalent chromate alone has no self-healing properties. This can be improved with the addition of a topcoat, but still only provides about 33% of the self-healing protection of the hexavalent chromium. The topcoat also creates other barriers, such as increased cost and the inability to paint.

3.4.3.4 Identification

There is no easy way to verify whether the products supplied are trivalent or hexavalent. Both are considered a yellow zinc finish; they are similar in color and it is difficult to distinguish the two. The only true way to determine the finish is through chemical testing. The absence of Cr VI must be checked by the diphenylcarbazide (DPC) test called the "drop test."

3.4.3.5 Cost Issue

Tri chrome by itself is not much more expensive than hex chrome (chemicals are more expensive, disposal is cheaper), although the difference emerges when corrosion resistance is considered, since tri chrome is markedly inferior in corrosion resistance. Thus, to achieve improvement in the corrosion resistance of tri chrome with some kind of more exotic conversion coating would inherently cost more, would usually involve extra steps and more labor, and would probably have a higher reject rate (also driving up costs).

All these differences should be taken into consideration when performing and using the parts/fasteners with conversion zinc plating.

3.5 METAL WHISKERS ON PLATING (NONCORROSIVE PHENOMENON)

3.5.1 WHISKER PHENOMENON AND CHARACTERISTICS

Metal whiskers are the filaments that grow on various metals and cause electrical short circuits and electronic component damage. There are at least two different phenomena identified for different growth conditions. One type of whisker that grows in any environment, including a vacuum, is found on galvanized or electroplated surfaces of tin, zinc, silver, gold, cadmium, indium, aluminum, lead, indium, antimony, and so on. First described in 1946, highly conductive whiskers can reach several millimeters in length with a typical diameter between 1 and 5 μm and growth rates around 0.1 Å/s under ambient conditions.

The precise mechanism for such growth remains unknown, but theories suggest that whisker formation is encouraged by compressive mechanical stresses developed during the electroplating process. The corrosion-related phenomenon of gigantic whiskers growing on silver is described in Section 5.5. Such whiskers are found only on silver plating, alloys, and pure metal, which have been exposed to an industrial atmosphere rich in hydrogen sulfide. There is no explanation of this phenomenon.

In this section, a historical review of risks and failure mechanisms and the development of various mitigation techniques associated with metal whiskers is presented.

The tendency of pure tin plating to form tin whiskers has been known for many years. Tin whiskers have been found to form on a wide variety of tin-plated component types under a range of environmental conditions. These whiskers are composed of nearly pure tin and are therefore electrically conductive and can cause shorting of electronics. The growth of whiskers has caused reliability problems for electronic systems that employ components plated with tin [36–39].

Other metal galvanized or electroplated deposits are also prone to growing whiskers. Metal whiskers are completely different from the more commonly known dendrites, not only by their shape and orientation but also by the mechanism that forms them. Dendrites form in damp environments when metal ions dissolve and are then redistributed in the presence of an electromagnetic field due to a phenomenon called electromigration.

As to metal whiskers, there is not a single theory that can explain their growth. What causes tin whiskers to grow is still under debate, although it is generally accepted that stresses involved in plating play a major role [40].

3.5.1.1   Conditions and Characteristics of Growth

Most metal whiskers grow in any environment including a vacuum and do not require dissolution of the metal, the presence of electromagnetic field, or a specific range of ambient temperature. Whiskers are found on galvanized or electroplated surfaces of tin, zinc, silver, gold, lead, cadmium, and other metals. These electrically conductive growths can reach several millimeters in length; infrequently, whiskers as long as 10 mm have been reported. Most studies are focused on tin whiskers since most electronic components use tin plating (Figures 3.6 and 3.7).

Extensive accounts of dozens of studies may be found in Refs. [39,41], which study various mechanisms for whisker growth and the effects of multiple process parameters. The effects of plating process parameters such as CD, temperature, substrate preparation, substrate materials, and

FIGURE 3.6   Tin whiskers. (a) On tin-plated housing on variable air capacitor built in 1959, found in 2006; (b) on a 1960s era variable air capacitor; (c) on connector pins; and (d) on printed circuit card guides from Space Shuttle. (Courtesy of the NEPP Program, nepp.nasa.gov/whisker/)

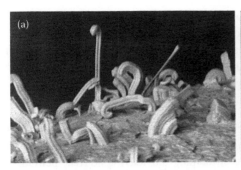

FIGURE 3.7 SEM pictures of tin whiskers: (a) straight and hooked tin whiskers; (b) spiral shaped tin whiskers. (Courtesy of the NEPP Program, nepp.nasa.gov/whisker/)

bath components have been studied. The influence of plating thickness, underplating, postplating annealing, plating structure, and alloying agents on whisker growth has been evaluated.

None of the theories proposed for the explanation of the phenomenon can be considered as proven, although some of them have allowed the development of methods for growth prevention. One of these theories is that whiskers form in response to the compressive stresses that develop during the electroplating process; another theory suggests that mechanical stresses produce an environment favorable for whisker formation. Nonetheless, the precise mechanism for metal whisker formation remains unknown.

There is much debate in the industry regarding which tin plating processes are prone to whisker formation. Most of the literature agrees that "pure tin" electroplated surfaces (especially those that employ brighteners in the plating process) are the most susceptible to whisker formation. There are also reports that tin–lead plating can also grow whiskers; however, such whiskers are generally reported to be <50-μm long.

Some of the observed characteristics of tin whiskers [42] are

1. *Shape*: Whiskers may be straight, kinked, hooked, or forked with outer surfaces often grooved. Some growths may form as nodules or pyramidal structures.
2. *Incubation period*: The incubation time before whiskers form is very unpredictable and can range from days to years. Under high stress conditions, whiskers have been found to grown very rapidly.
3. *Growth rate*: Growth rates from 0.03 to 9 mm/year have been reported. Growth is highly variable and is likely to be determined by a complex range of factors including plating chemistry, plating thickness, substrate materials, grain structure, and environmental storage conditions.
4. *Whisker size*: Whiskers as long as a few millimeters are most common, with some whiskers as long as 10 mm in length (Figure 3.7). Typical diameters are a few microns, while some are found to be as large as 10 μm.

3.5.1.2 Environmental Factors

There is contradictory information regarding the influence of various environmental factors on whisker formation and the ability of these factors to accelerate or retard the growth.

Environmental factors that need to be considered are

- *Temperature*: Some reports state that ~50°C ambient temperature is optimal for whisker formation; while others observe that at room temperature, whiskers grow faster. Whisker growth usually ceases at temperatures above 150°C.
- *Pressure*: Whiskers will grow in a vacuum as well as at atmospheric pressure.

- *Moisture*: Some observations show that whiskers grow more readily in high humidity, while others do not find moisture to be a contributing factor.
- *Electric field*: Whiskers grow spontaneously without requiring an applied electric field. However, it was found that whiskers can bend due to electrostatic forces, thus increasing the possibility of tin whisker shorts.

3.5.1.3 Historical Account of Metal Whisker Hazards

Metal whiskers, highly conductive filaments reaching several millimeters in length, can cause electrical short circuits or electronic component damage. The specific causes of failure are particularly difficult to define because whiskers may vaporize or disintegrate, leaving no trace behind.

The most serious hazard posed by metal whiskers is a metal vapor arc in which the whisker vaporizes into a cluster of extremely conductive metal ions capable of carrying hundreds of amperes of current. Metal whiskers can also cause voltage to jump across electrical circuitry and cause short circuits. The history of these failures began during World War II, when cadmium whiskers were found in electrical equipment.

When tin and zinc began to be used instead of cadmium later in the 1940s, Bell Labs reported shorting caused by the whiskers. When investigating failures on channel filters in telephone transmission lines, Western Electric and Bell Laboratories first documented the existence of metal whiskers in the 1950s and 1960s [36,37]. Bell Labs investigated whether the addition of some element to tin coating would prevent whiskering. After it was found that adding lead prevented the growth, they discontinued research on the metal whisker phenomenon.

Since then, hundreds of scientific and technical papers have been published and reviewed. The best and the largest collection of papers and images on whiskers are posted on the National Aeronautics and Space Administration (NASA) website, the page nepp.nasa.gov/WHISKER/. A complete account and review of these studies is given in Refs. [36–40]. The most comprehensive account of documented failures caused by metal whiskers is given in Refs. [36,37]; see Table 3.12.

3.5.2 Tin Whisker Mitigation Techniques

Development of mitigation techniques [43] that could greatly enhance resistance to whisker formation for tin-based films is very important, since tin whiskers are the most dangerous growth for electric and electronic devices. However, none of these techniques necessarily prevents the whiskers' ultimate formation.

Since compressive stresses are thought to be fundamental to whisker formation, mitigation approaches should somehow alter the stress state within the tin. Underplating, alloying, plating chemistry, and heat treatment, all have an effect on the stress state of the tin and encompass a variety of commercially available mitigation methods. The choice of the method also depends on the application.

3.5.2.1 Underplating

Underplating may be effective by forming a barrier at the tin–substrate interface. Therefore, adding a nickel (Ni) layer between tin plating and a copper (Cu) base metal may mitigate whisker formation by forming a barrier to the diffusion of Cu into the Sn. The thickness, porosity, and ductility of the nickel plating are very important to ensure an effective barrier layer for the copper that will not crack during lead formation. A minimum porosity-free layer thickness of 0.5 μm has been shown to mitigate the growth of Sn whiskers for devices subjected to Pb-free reflow assembly conditions [44].

It has also been shown that Ni of this thickness is not fully converted and consumed to intermetallic compounds (IMCs) after $3 \times 260°C$ reflows, and that the nickel layer is sufficiently ductile and will not crack during a subsequent lead-forming operation [45]. On the other hand, Ni underplating may not be beneficial in conditions where the compressive stress-inducing mechanisms originate at

TABLE 3.12
Metal Whiskers Failures: 2000s

Year	Application	Industry	Failure Cause	Whiskers On?
2000	Galaxy VII (Side 2)	Space (complete loss)	Tin whiskers	Relays
2000	Missile program "D"	Military	Tin whiskers	Terminals
2000	Power Mngm Modules	Industrial	Tin whiskers	Connectors
2000	Solidaridad I (Side 2)	Space (complete loss)	Tin whiskers	Relays
2001	Galaxy MR (Side 1)	Space	Tin whiskers	Relays
2001	Hi-Rel	Hi-Rel	Tin whiskers	Ceramic chip caps
2001	Nuclear power plant	Power	Tin whiskers	Relays
2001	Space ground test equipment	Ground support	Zinc whiskers	Bus rails
2002	DirecTV 3 (Side 1)	Space	Tin whiskers	Relays
2002	Electric power plant	Power	Tin whiskers	Microcircuit leads
2002	GPS receiver	Aeronautical	Tin whiskers	RF enclosure
2002	MIL aerospace	MIL aerospace	Tin whiskers	Mounting hardware (nuts)
2002	Military aircraft	Military	Tin whiskers	Relays
2002	Nuclear power plant	Power	Tin whiskers	Potentiometer
2003	Commercial electronics	Telecom	Tin whiskers	RF enclosure
2003	Missile aerogram "E"	Military	Tin whiskers	Connectors
2003	Missile aerogram "F"	Military	Tin whiskers	Relays
2003	Telecom equipment	Telecom	Tin whiskers	Circuit breaker
2004	Military	Military	Tin whiskers	Waveguide
2005	Communications	Radio (1960s vintage)	Tin whiskers	Transistor TO package
2005	Millstone nuclear power plant	Power	Tin whiskers	Diode (axial leads)
2005	Optus B1	Space	Tin whiskers	Relays
2005	Telecom equipment	Telecom	Tin whiskers	RF enclosure
2006	Galaxy MR (Side 2)	Space	Tin whiskers	Relays

Source: Courtesy of the NASA Electronic Parts and Packaging (NEPP) Program, nepp.nasa.gov/whisker/.

the tin surface or within the tin matrix itself. Ni does not prevent the corrosion and oxidation of Sn in a high-humidity environment.

Adding silver (Ag) underplating between tin plating and copper (Cu) base metal has been proposed as a method of mitigating whisker formation, similar to Ni as noted above [46]. Components using silver underplating should have a minimum silver thickness of 2 μm. However, there is limited whisker test data supporting the effectiveness of the Ag underplating in whisker mitigation.

3.5.2.2 Addition of Lead

Tin plating with >3% lead (Pb) has been the industry-accepted finish for over 50 years and has been shown to mitigate Sn whisker formation [47]. It has become the standard used when comparing and evaluating Sn whisker mitigation practices. Since the 1990s, most U.S. MIL specs require adding Pb to any tin coatings used around electronics, with the concentration usually as 2%–3% Pb by weight. However, adding lead to tin plating became no longer an option for most products after July 1, 2006.

In response to campaigning from environmental groups to stop the use of lead, two regulations—the Waste Electrical and Electronic Equipment (WEEE) and RoHS of the European Parliament and the European Union Council—have compelled European and the U.S. electronics manufacturers to produce lead-free components and protective coatings, which have a HP for whiskers formation [48].

This has, in turn, resulted in the increased incidence of electrical failures, although the use of Pb is not 100% guarantee against the formation of whiskers. A further discussion on the influence of RoHS and WEEE initiatives on dealing with tin whiskers is presented in Section 3.5.3.

3.5.2.3 Heat Treatments

Heat treatments may be subcategorized as annealing, fusing, and reflow. Annealing is a heating and cooling process typically intended to soften metals and to make them less brittle. Fusing and reflow are similar in that both melt and resolidify tin plating under relatively slow cooling conditions.

3.5.2.4 Hot-Dip Tin Plating

Hot-dip tin is a molten tin bath process that is intended for electronic components. It has primarily been used for structural steel parts, connectors and devices, such as relays, but not for lead frame components. Hot dipping with $SnAg_4$ or SnAgCu is generally an effective mitigation practice and considered to be whisker free. However, there is evidence that when pure tin is used, this process may not be effective [49].

3.5.2.5 Thicker Tin Finish

Industry data indicate that thicker tin finishes show a lower propensity for tin whiskers and/or a greater incubation time before tin whiskers occur. It is recommended that the tin thickness for components without a nickel or silver underplating be a minimum of 7 μm with 10 μm nominal [50,51]. When nickel or silver underplating is used, the minimum tin thickness should be of appropriate thickness to ensure component solderability over the intended shelf life and to prevent complete consumption of Ni in the intermetallic Ni/Sn phase.

3.5.2.6 Conformal Coating

The use of conformal coatings (materials applied in thin layers, often by dipping, spraying, or flow coating) on a circuit board assembly has been shown to reduce the electrical shorting risks from whiskers [6,52]. Thicker coatings (≥100 μm or 3.9 mil) have been shown to prevent or delay whisker penetration, while thinner coatings act as a dielectric layer for whiskers that may break off in an assembly.

Conformal coating could be used in conjunction with other mitigation methods previously discussed. Conformal coating also reduces the corrosion of Sn. There is an opportunity for more study in this area and the mitigation results will vary with the specifics of the assembly and its sensitivity to whisker problems.

When applied on top of a whisker-prone surface, conformal coat can sometimes keep whiskers from pushing through. When applied to a distant conductor, conformal coat can block whiskers from electrically shunting distant conductors. It also provides an insulating barrier against loose conductive debris.

3.5.2.7 NonTin Plating and Coating

Nontin platings, such as nickel/palladium/gold, nickel/gold, or nickel/palladium, do not have Sn whisker problems and should be considered for lead-frame applications. This plating has a more than 10-year history of field use, 1992 to the present [53]. Nontin plating, however, does have other potential issues, including adhesion to mold compounds that need to be evaluated prior to product conversion. PCBs that were previously hot air solder-leveled (HASL) SnPb will be finished with either electroless nickel immersion gold (ENIG) or HASL (tin/silver/copper Alloy) as the application demands. Organic solderability preservative (OSP) may be used if the application permits [54].

3.5.3 Tin Whiskers and the RoHS Initiative

3.5.3.1 Lead-Free Solders

The RoHS and WEEE legislations set a deadline of June 2006 to achieve the goal of Pb-free solder. It was found that Pb-free solder is, to some extent, more brittle than the traditional solder. Substitute solders that have been developed may be applied too thinly or with too little heat, which results in stressing of the circuit board laminate (a contributing factor in whisker growth). An example would be a recalled product from the Swiss watch-making company Swatch, which reportedly cost $1 billion. The company was later granted a permanent exemption from the RoHS directive for its exports within the European Union (EU) and was allowed to put Pb back in the solder.

Many experts doubt that a reliable lead-free process will be implemented soon, and debate among professionals continues. However, companies such as IBM and National Instruments say that they are now achieving RoHS-compliant techniques even for exempt products. So far, the last source to count on for information is the manufacturers [55]. Currently, the United States has not made Pb-free solder mandatory, but does offer tax benefits for reducing or eliminating its use [56].

3.5.3.2 "Pure" Tin Finishes

Some types of electronics were exempted from the law since its inception. Among these exceptions were military and other national security equipment and certain medical devices. There is no pending U.S. legislation mandating lead-free electronic products, and should such legislation arise, military, aerospace, and medical equipment manufacturers would likely be exempt.

Nevertheless, the Department of Defense (DoD) and NASA believe that the use, and therefore the risk, of pure tin finish (tin which contains <3% Pb by weight shall be considered "pure" for this purpose) on electronic components will increase because: (1) the commercial industry has stated initiatives to eliminate lead (Pb) from electronics; (2) defense and aerospace industry trends show increasing usage of commercial components; and (3) continuing reductions in circuit geometry and power may result in even small whiskers causing catastrophic failures.

However, integrators and designers of high-reliability systems exert little or no control over component-level plating processes that affect the propensity for tin whiskering. Challenges in how to ensure long-term reliability, while continuing to use commercial off-the-shelf (COTS) parts plated with pure tin, continue to arise [57–59]. The National Electronics Manufacturing Centre for Excellence, sponsored by the U.S. Navy, did find that modifying the temperatures at which soldered items are bathed and stored diminished whiskering, but nevertheless recommends the "use of lead in conflict with future industrial practice" [55].

3.5.4 Whisker Mitigation Levels Classification

Currently, there is still no commonly used industry standard for pure tin finishes on applications. Most individual companies are developing their own standard of acceptability of tin and zinc finishes used on parts. Some companies have responded to current industry pressure by completely changing to pure tin and zinc finishes, although they have not yet incorporated new part numbers to reflect this change in material. This practice only leads to confusion in the procurement process. Parts should be inspected on arrival, but this costs a lot of time and money.

In response to this situation, a system of whisker mitigation-level classification standards was proposed in Ref. [57] to match potential risk with appropriate mitigation for tin and zinc whisker issues. It was suggested to define the whisker mitigation level for each hardware system or subsystem. Five proposed whisker mitigation levels (I, II, III, IV, and V) determined the type of control and the recommendation of system types for each level.

In 2006, the ANSI/ITAA published GEIA-STD-0005-2, "Standard for Mitigating the Effects of Tin Whiskers in Aerospace and High Performance Electronic Systems," which established processes for documenting the mitigating steps taken to reduce the harmful effects of tin finishes in electronic products. In this Standard, the nomenclature of the five levels has been changed from I, II, III, IV, and V to 1, 2A, 2B, 2C, and 3.

The Standard describes processes for mitigating the detrimental effects of tin whiskers in electronic systems used in military, aerospace, and other high-reliability applications. The Standard is intended for use to document processes and to ensure performance, reliability, airworthiness, safety, and certifiability of electronic systems in which harmful forms of tin may be introduced into the production or maintenance process.

The highest level of control is provided by Level 3 (or V in Ref. [57]), where no pure tin or zinc finishes are permitted to be used. Typical systems where Level 3 would be appropriate include space-based systems, strategic missile vehicles, and implanted medical devices. However, application-specific risk assessments for the use of tin would, of course, typically not be a part of a standard process for Whisker Control Level 3 because the use of tin is totally banned.

A high level of control is provided by Level 2C (IV), according to which pure tin and zinc finishes are allowed only very limited uses. Special exemptions may be granted at this level of control to permit pure tin or zinc use on a case-by-case basis. Typical end uses where Level 2C would be appropriate include tactical missiles, safety-critical avionics, and so on.

Level 2B (III) provides a moderate level of control. Pure tin and zinc finishes are only permitted when the risk of failure can be reasonably determined to be minimal. Level 2B would apply to a wide range of high-reliability systems. Such systems would typically permit repair and replacement of failed assemblies as a normal part of their logistic plan. Typical systems include surface-based military radar and communications systems, and critical commercial applications.

Application-specific risk assessment will generally be applicable to Whisker Control Levels 2B and 2C. Under Whisker Control Level 2C, each and every incidence of tin usage must be analyzed, justified, and documented. A standard risk assessment algorithm offers one method of performing these analyses in a well-defined and easily documented manner.

Under Whisker Control Level 2B, tin may be used under a set of predefined conditions, or blanket exemptions. These conditions must be defined and justified. A standard risk assessment algorithm again can provide a well-defined and documentable method for justifying exemptions. Here, the use of tin may also be justified on a case-by-case base exactly as with Level 2C.

Level 2A (II) provides for a minimal level of control where pure tin and zinc finishes are restricted from use only in specific applications. Unless otherwise indicated, pure tin and zinc are considered acceptable for use. Typical systems where this level could properly apply include military test equipment, industrial electronics, and transportation. Level I classification provides no special controls for pure tin or zinc finishes. All components may be used without regard to issues of tin or zinc whiskering. This is essentially the "best commercial practices," or COTS approach.

Typical systems where Level 1 (I) would be appropriate for use include consumer electronics and nondeliverable prototypes. The use of pure tin is currently prohibited by the military and NASA. The summary of procurement specifications can be found in Ref. [60].

Whisker Control Levels 1 and 2A do not generally restrict the use of tin in any particular application. Therefore, application-specific risk assessments would typically not be performed as part of the standard process.

The Standard establishes a framework within which manufacturers and their customers can establish guidelines for the control of tin usage. One popular method for determining the suitability of pure tin in a given application is the use of a standard tin whisker risk assessment algorithm; the last revision (D) was accepted in January 2011 [61]. The use of an algorithm facilitates standardization of tin risk decision-making across and between organizations.

Based on mitigation techniques suggested by industrial recommendation and standards, International Electronics Manufacturing Initiative (iNEMI) proposed the recommendations for busbar plating [62] which may be free of metal whiskers (Table 3.13).

When applying tin (Sn) on busbars as a finish, tin whiskers are a concern, particularly when busbar connections result in mechanical stresses on the finish. This finish has been used on busbars in many products for years, so the application may still be acceptable even with tin whiskers. Success highly depends on application and it is up to the user to make the final decision as to acceptance of this finish. iNEMI Tin Whisker User Group concludes [62] that if Sn finishes are used, a tin whisker mitigation practice is recommended.

When chromium finishes are applied, the plating should not contain CrVI (hexavalent chromium).

TABLE 3.13

Recommendations for Busbars Plating

Plating	On Copper Alloys	On Aluminum
None (unfinished)	OK	OK
Nickel	OK	OK over copper strike plating
Silver (immersion or electroplate)	OK	OK over copper strike plating
Chromium	OK	OK over copper strike plating
Matte Sn	Not recommended	Not recommended

Source: From Smetana J. iNEMI recommendations on lead-free finishes for components used in high-reliability products, *iNEMI Tin Whisker User Group IPC/APEX*, February 2006. Online. Available: http://thor.inemi.org/webdownload/projects/ese/tin_whiskers/User_Group_mitigation_May05-REMOVED_5–12–05.pdf. With permission.

3.5.5 WHISKERS ON OTHER METAL PLATINGS

Whiskers have been found on many other plating materials other than tin: zinc, lead, aluminum, cadmium, gold, and so on (Figure 3.8).

Zn whiskers have been well studied [27] and have, in general, the same physical characteristics as tin whiskers:

- *Length:* a few millimeters or less
- *Diameter:* a few microns (thousandths of a millimeter)
- *Shapes:* straight or kinked filaments, nodules

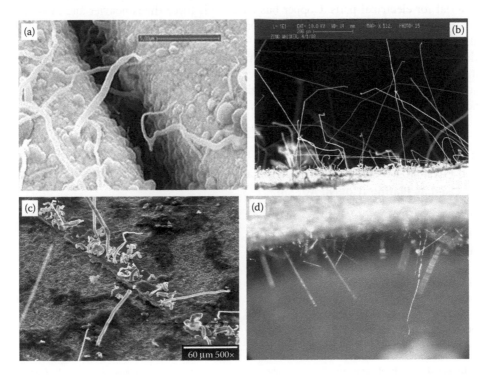

FIGURE 3.8 SEM pictures of whiskers on gold (a), zinc (b,c), and cadmium (d). (Courtesy of the NEPP Program nepp.nasa.gov/whisker/)

- *Texture:* fluted and/or striated along axis of growth
- *Conductive:* can carry tens of milliamperes before melting.

The mechanism of Zn whiskers growth is unconfirmed, as is the case for tin whiskers. However, the most plausible mechanism for tin whiskers is that they grow to relieve compressive stress within the film, and this may also be the case for zinc whiskers. The incubation period ranges from days to years, with growth rate typically <1 mm/year.

The sources of Zn whiskers are the parts with Zn plating, such as mechanical hardware (nuts, screws, and washers) and equipment racks (cabinets, chassis, bus rails, etc.).

Zinc whiskers may grow on electroplated or hot-dip galvanized zinc plating. However, electroplated zinc is far more susceptible to whisker growth [63–65]. Zinc whiskers, which may sometimes grow to be 1–2-mm long, threaten instrument transformers equipment when they become dislodged and airborne, which could happen when the tiles are disturbed during their removal or when pulling or removing underfloor cables. If zinc whiskers are ingested by information technology (IT) hardware equipment (computers, data centers, and so on), a circuit with voltages higher than about 25 V may suffer electrical short circuiting, arcing, signal perturbations, or catastrophic failures [66]. If zinc whiskers are present, remediation involves replacing the contaminated raised-floor tiles and hiring professionals to clean the data center.

3.6 PLATING ON ALUMINUM

3.6.1 Use of Aluminum in Electrical Industry

Most electrical applications of Al use this metal for the bus instead of copper. Physical, mechanical, and chemical properties are compared for copper and aluminum, as well as for three plating materials (nickel, silver, and tin) in Table 3.14 [67]. There are many reasons why aluminum is often chosen as a material for electrical parts. Copper bus material is three times heavier and more expensive than aluminum. The dollars per amp is less for aluminum than it is for copper (current-carrying capacity).

Although there is a definite benefit in cost from using aluminum bus, it is not as conductive as copper. Aluminum has about 1/2 the conductivity of copper. With more conductivity, comes less resistance; therefore, you have less heat generation. Hence, for applications where more heat could be generated and longer life of the equipment is important, copper bus is more often used.

TABLE 3.14

Selected Physical, Mechanical, and Chemical Properties of Base Metals (Al or Cu) and Metals, Used as Plating (Cu, Ni, Ag, Sn)

Physical Property	Al	Cu	Ni	Ag	Sn
Coating thickness (μm)	–	10–15	3–5	3–5	10–15
Electrical resistivity ($\mu\Omega$-cm)	2.96	1.72	6.84	1.59	11.0
Thermal conductivity (W/m · K)	210	380	70	418	64
Microhardness (MPa)	170	500	8000	700	60
Tensile strength (σ_y) (MPa)	172	195	59	55	12
Coefficient of thermal expansion (10^{-6}/°C)	23.6	16.8	13.3	19.6	23
Temperature coefficient of resistivity (10^{-3}/°C)	4.03	3.97	6.90	4.01	4.70
Melting temperature (°C)	660	1084	1455	1482	232

Source: Braunovic M. Evaluation of different platings for aluminum-to-copper connections. *IEEE Transactions on Components, Hybrids, and Manufacturing Technology*, 15(2), 216–224, April 1992. © 1992, IEEE.

Usually, the decision as to which material to choose for the bus is taken based on the specific application. The aluminum bus should be plated to protect it from corrosion and ensure long life because when corrosion or oxidation occurs on aluminum, it decreases its conductivity considerably. The choice is usually made between silver and tin.

3.6.1.1 Choice of Plating

In some cases, when aluminum and copper are joined together in bolted electrical connections, it is important to choose the plating both for Cu and Al that will provide better electrical properties and stability [67]. Tin plating of copper busbars has long been used for joining copper to copper. With the increased use of aluminum in the electrical industry and the rising number of aluminum-to-copper interfaces, the use of tin plating on copper became an established practice.

Silver is an excellent choice for plating. As the best conductor, it is less sensitive to deteriorating processes such as fretting, but silver plating is expensive. Therefore, the use of silver is justified only in specific cases and in environments free of sulfurized contaminants.

Nickel is also a very good plating material; it has superior resistance to corrosion and oxidation and does not form IMC with aluminum, but it also has some disadvantages. It does not protect aluminum from galvanic corrosion, and forms subsurface corrosion when plated over aluminum. It is also sensitive to fretting. A study, however, shows [67] that nickel plating enhances the stability of Al-to-Cu connections.

Tin plating provides better joint connections because it is not as hard a material as silver. The harder the plating material, the more contact surface area is lost. Silver provides for better conductivity. Each material has its benefits and drawbacks, and is chosen depending on the application. Although both materials are acceptable in most installations, there are some atmospheres where silver plating is not desirable. These conditions exist in wastewater treatment and pulp and paper facilities (Section 5.3).

3.6.1.2 Difficulties with the Plating of Aluminum

The main difficulty connected with the plating of aluminum is the presence of the oxide film on the surface of Al. An extreme reactivity of bare aluminum results in rapid reformation of the film while exposed to air or to aqueous solutions.

Additionally, the difference in the linear expansion coefficient of the aluminum alloys and most of the metals commonly deposited on Al may cause adhesion problems. In applications in which considerable temperature changes occur, the differential expansion of aluminum and the metal coating may cause sufficient strain to rupture the bond between the deposit and the basic metal [68].

3.6.2 Metals Used to Plate Aluminum

For nonelectrical applications, nickel and chromium are the metals most commonly deposited on aluminum; however, electrical applications require the application of other metal deposits. Aluminum is anodic to most of the metals commonly deposited on it, and hence, it is essential that the deposited metal be largely pore free if good protection is to be achieved.

Various metals have been applied to aluminum to provide a solderable surface. These include tin, copper, zinc, silver, and nickel. Tin and sometimes silver deposits are used on aluminum busbar ends to improve electrical conductivity at the joints. In the past, lead–tin alloy electrodeposits have been applied to bearing surfaces prior to the ban on lead. In electrical devices, there have been many cadmium-plated aluminum components such as switch boxes and instrument panels where electrical earthing needed to be protected from corrosion, although nowadays cadmium plating is also banned due to toxicity.

Aluminum wires with an extruded copper coating are being used for electrical installations in place of solid copper wires. Other metallic coatings for aluminum conductor wires are in use. When

TABLE 3.15

Plating for Aluminum with Acceptable Physical Properties

Metal	Thermal	Electrical	Wear	Friction	Anticorrosion	Solderability
Cr			Good		Good	
Ni			Good		Good	
Cu	Good	Good				Good
Ag	Good	Good		Good		
Sn		Good		Good	Good	Good
Zn					Good	
Au		Good			Good	Good

Source: Modified from Moller P. *Plating on Aluminum.* TALAT lecture 5205, Online. Available: core.materials.ac.uk/repository/eaa/talat/5205.pdf.

plating has to be chosen for aluminum for a specific application [68], it is important to consider the various physical properties of different metals available for Al plating (Table 3.15).

3.6.3 METHODS FOR PLATING ON ALUMINUM

Pretreatment processes for plating aluminum are presented in Ref. [68]. Pretreatment processes begin with degreasing, followed by etching, deoxidation, and ion-exchange Zn plating (or tin plating). Whenever tin plating is used in the pretreatment process, it is necessary to provide underplating (usually electrolytical bronze strike).

The plating, then, may be either electroless (e.g., Ni or other metals) or electrolytic plating (Cu or Ni). If needed, additional plating may be applied, such as gold (Au) or silver (Ag). These procedures are described in ASTM Standard B253-87(2010), "Standard Guide for Preparation of Aluminum Alloys for Electroplating" [69].

3.6.3.1 Plating Classifications

Generally, methods for producing deposits on aluminum may be classified as follows:

1. Processes based on the zincate immersion technique
2. Processes based on the stannate immersion technique
3. Direct plating methods
4. Processes based on mechanical preparation
5. Chemical etching procedures
6. Processes based on preparation by anodic or chemical oxidation

To obtain a good surface finish of the plated aluminum, a proper pretreatment and jigging of the profiles has to be carried out. The usual process steps are as follows (there is at least one rinsing step after each main process step): polishing, jigging, degreasing/cleaning, etching, zincating (zinc immersion), or double zincating (double zinc immersion). After these steps, electrolytic plating (Cu or Ni) or nickel (electroless) is applied.

Of the preplating techniques, the zincate process is by far the most widely used. The stannate process is recommended for specific applications. The zincate and stannate processes have mainly superseded the other methods for plating on aluminum. Some of these methods are still in use and some continue to be investigated, although with little to no commercial exploitation.

3.6.3.2 Pretreatment by Zincating

To obtain good adhesion when plating on aluminum, it is necessary to start with a precoat of zinc or tin. This precoating is the pretreatment needed prior to the main metal plating. Most plating on aluminum is currently carried out on top of chemically deposited zinc produced by immersion in modified zincate solutions. The zincate process has been developed into a relatively simple, cheap, and reliable technique for plating on aluminum and a wide range of aluminum alloys. It is less sensitive to alloy composition than processes based on anodizing pretreatments, does not need the applied power requirements of anodizing, and is cheaper than the stannate immersion process owing to the high price of tin.

The basic reactions in the alkaline zincate solution are the dissolution of aluminum and the deposition of zinc. The zincate solution is normally made up of zinc oxide and caustic soda, and it has been shown that the ratio of these constituents has a profound effect on the adhesion of the subsequent deposit. Physical conditions, as well as time of immersion and temperature, can also affect the degree of adhesion. It was indicated that the presence of nickel in the zincate solution improves the adhesion of nickel plated directly onto the zinc deposit. Copper also assists the adhesion to some alloys, and the substitution of zinc oxide by zinc sulfate was found to give further improvement.

3.6.3.3 Tin Plating Techniques on Al

Various techniques of plating are used in the electrical industry. Several of them are based on different versions of the same process, called Alstan (Table 3.16), which utilizes proprietary solutions including cyanide-containing chemicals. Alstan plating techniques are supposed to provide tin plating on aluminum with low surface-to-surface CR and to minimize surface corrosion.

For example, the Alstan 80 process consists of an immersion or electrolytic activation step in the proprietary Alstan 80 solution followed immediately by a bronze alloy strike without an intermediate water rinse. Following this processing sequence and a subsequent appropriate acid dip and water rinse, the work is ready to be plated, as desired. The primary advantage of the Alstan 80 process is an improved corrosion resistance of the plated parts. It is particularly important for using tin on aluminum busbars, connectors, and so on in electrical equipment exposed to a corrosive environment.

Other activation techniques use noncyanide chemicals, which are considered more environment friendly, such as the plating technique called Bondal, which uses a surface activation procedure with a cyanide-free zincate solution and copper strike prior to tin plating. Either with or without

TABLE 3.16
Alstan Processes for Plating Tin on Aluminum Alloys

Process	Type	Description
Alstan 60	Immersion	Matte tin plating process for aluminum alloy pistons
Alstan 65	Immersion	Matte tin plating process for aluminum alloy pistons. This process runs with lower tin content
Alstan 70	Immersion	A process for the preparation of aluminum and its alloys for plating with good adhesion to the base metal
Alstan 80	Electrolytic	A process for the preparation of aluminum and its alloys for plating with good adhesion to the base metal
Alstan 88	Electrolytic	A process for the preparation of aluminum and its alloys for plating with good adhesion to the base metal

Source: Adapted from industrial standards for tin plating on aluminum.

cyanide-containing solutions, these processes prepare the surface of the aluminum busbar for tin plating through multiple-step procedures including underplating of a thin layer of bronze or copper.

3.6.4 Quality of Tin Plating on Al for Different Plating Techniques

The plating process itself, particularly the plating technique, is an important factor that may affect the quality of plating. The age of the bath and the position of the part in the bath may also significantly affect the quality of the plating. To determine whether the influence of these factors is significant [70], copper busbars were plated using three different techniques of plating tin on aluminum: Alstan 80, Alstan 88, and Bondal.

Each of two Alstan techniques is based on two different versions of the same process with proprietary solutions including cyanide-containing chemicals. Both techniques are supposed to provide tin plating on aluminum with a low surface-to-surface CR and to minimize the surface corrosion; they are often used for plating Al busbars in the electrical industry. The noncyanide plating technique, Bondal, uses a surface activation procedure with a cyanide-free zincate solution and copper strike prior to tin plating.

The samples also have been plated in three different periods during the life cycle of each bath solution, with a periodicity depending on the frequency of the suppliers' changing or readjustment of his baths. For example, if the supplier used a new bath every Monday, two bars have been taken every Monday, two bars every Wednesday, and two bars every Friday. Samples have also been plated in three different positions in the bath.

SEM testing of the samples in these batches was used to define how well the quality of plating is reproduced with regard to bath age and positions of the parts in the bath. There are two tests to study the quality of the plating: the adhesion test and the thermal shock test.

3.6.4.1 Adhesion Test

The adhesion test should be conducted according to ISO 2819:1980, "Metallic Coatings on Metallic Substrates. Electrodeposited and Chemically Deposited Coatings. Review of Methods Available for Testing Adhesion." The adhesion test involves crosscut and tape tests. It is first required to scribe down to the metal. A cutting blade with a straight edge is used to draw a square of 1 cm² divided into 100 squares (10 horizontal lines separated by 1 mm and crossed by 10 vertical lines). Then, an adhesive tape with peeling force >180 g/cm is applied on the surface and pulled off quickly. If there is any coating peeled away with the tape, the adhesion is considered as nonsatisfactory. Indeed, if the plating fails, the adhesion test at a nonaged status, further tests are no longer necessary and the plating technique is considered as not meeting the specifications.

3.6.4.2 Thermal Shock Test

The tests should be conducted according to the International Standard ISO 2819, which determines the technique of the test. According to the Standard, the samples must be heated at the recommended temperature. They are then quickly thrown in room-temperature water. A temperature of 150°C is the recommended for tin plating over Al, Cu, Zn alloys, and steel substrates. Samples are then examined. If the coating is blistering, peeling, or exfoliating, then the adhesion is considered as not satisfactory. According to the thermal shock test, the adhesion of tin plating on aluminum should not be strongly affected by short exposure to temperatures up to 200°C. For samples exposed to elevated temperature for a longer period of time, the adhesion of the plating begins to deteriorate. According to the Standard, scribing is applied to every sample before and after thermal shock.

3.6.4.3 Plating Techniques and Adhesion of Tin Plating on Al

All three plating techniques provide a good plating quality, according to the tests conducted in Ref. [70]. All three plating types succeeded adhesion and thermal shock tests. Original plating adhesion was found to be satisfactory for both Alstan techniques and the cyanide-free zincate technique Bondal. According to the results of scribing tests after thermal shock at 150°C, all samples plated using all three techniques passed the test and the coatings can be considered properly adhesive. In order to study the quality of the plating beyond the limit of 150°C, the thermal shock test was performed also at 200°C. After thermal shock at 200°C, the surface had an iridescent purple tint. No change in the adhesiveness of the coating was observed.

3.6.4.4 Plating Techniques and the Quality of Tin Plating on Al

The quality of tin plating on aluminum is best determined by using SEM/EDS metallographic analysis. This analysis allows the measurement of the thickness of the tin plating and preplating layer (copper or bronze) and determines if pores, voids, and gaps are present along each interface (Al–Cu/bronze, Cu/bronze–Sn) of tin plating across the layer. With EDS chemical analysis, the chemical composition of copper/bronze preplating layer can be determined. For SEM study, the samples are usually encapsulated with an epoxy resin polymerizing at room temperature. It is to be noted that the polymerization process involves heating samples at 70°C for a few minutes, which is not supposed to affect the quality of the plating.

3.6.4.4.1 *Tin on Al Plating Thickness*

An example of using the SEM technique for measuring the plating and underplating thickness is given in Ref. [70]. Tin and bronze thickness measurements have been made on each SEM capsule, containing three parts of each bar: one part from the middle of the bar and two parts from the edges of the bar. Considering the differences between coating on the curved and linear sides, measures were taken on each side in three different places.

The average value of tin plating and underplating was determined for all three types of plating based on multiple measurements. It was found that in Alstan 80 plating, the average thickness of tin plating was 13 µm and the bronze underplating thickness was 1.1 µm. In the case of Alstan 88 plating, these values were 11 and 1.5 µm, correspondingly. In the case of Bondal plating, the average thickness of tin plating is 5 µm and the thickness of the copper underplate is 1.0 µm.

A slight difference in thickness was observed between curved and linear sides. This well-known phenomenon is the result of differences in the electrical field. For all samples, including different plating bath life cycles, different positions in the bath, and different sides of the samples (linear or curved), tin plating thickness varied in the range between 7.1 and 13.9 µm. Bronze underplating thickness ranged between 0.7 and 1.5 µm. The thicknesses of both plating and underplating were in agreement with OEM specifications.

3.6.4.4.2 *Morphology and Composition of Tin Plating on Aluminum*

The composition and quality of samples for all three types of tin plating on Al were determined using SEM/EDS analysis in Ref. [69]. It was found that for Alstan 80 plating, the upper layer of the plating consists of pure tin, and the underlayer is composed of Sn and Cu with the proportion corresponding to 70Sn30Cu-type bronze (70% Sn and 30% Cu; Figure 3.9). An SEM microphotograph of the cross section of freshly plated Alstan 80 samples also shows that the quality of nonaged plating is good: no separation, voids, or cracks have been found. SEM analysis showed that Alstan 88 samples also had no pores, voids, or gaps along each interface (Al–bronze, bronze–Sn; Figure 3.10). The quality of Bondal samples was also good: no pores, voids, or gaps were detected along each interface (Al–Cu, Cu–Sn).

The results of the tests in Ref. [70] show that all the three techniques of tin plating often used in the electrical industry provide good plating adhesion and a perfect quality of plating with no pores, voids, or gaps detected along each interface and with the thickness of both plating and underplating in agreement with plating standards. It was also shown that the position of the samples in the bath and the age of the bath solution during the bath life cycle have little or no influence on the quality of tin plating on Al.

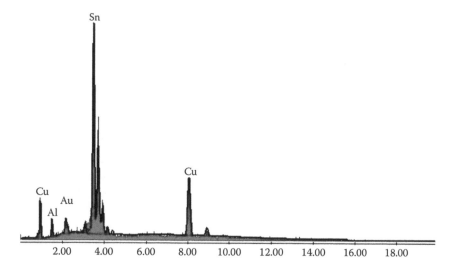

FIGURE 3.9 Composition of copper–bronze underplating in a freshly plated Alstan sample (70% Sn and 30% Cu).

FIGURE 3.10 Microphotograph of freshly plated tin plating on Al with bronze underplating.

3.7 PLATING STANDARDS AND GLOSSARY

3.7.1 National and International Standards and Regulations on Plating

The following national and international standards specify the requirements for various types of plating and different substrates and describe the conditions of the plating procedure that provides the best quality of the plating. Other Standards describe the procedures of testing the plating quality. Included are international regulations focused on environmental safety.

An example of what issues are covered on plating standards is given for ASTM B-545-97 (2009). "Standard Specification for Electro-deposited Coatings of Tin" (The International Classification for Standards [ICS] Number Code 25.220.40 [Metallic coatings], ASTM International, 8pp., 2009):

This specification covers the requirements for electrodeposited tin coatings applied to metallic articles to provide a low CR surface, to protect against corrosion, to facilitate soldering, to provide antigalling properties, and to be a stop-off coating in the nitriding of high-strength steels.

This specification does not cover hot-dipped tin or other nonelectrodeposited coatings, and mill products. Coatings shall be grouped into six service classes, which are based on the minimum thickness and severity of service required for the coating, and three surface appearance types, which are based on the type of electroplating process employed.

The surface appearance types are matte tin electrodeposits, bright tin electrodeposits, and flow-brightened electrodeposits. Coatings shall be sampled, tested, and conform accordingly to specified requirements as to appearance, purity, local and mean thickness, integrity (including gross defects, mechanical damage, and porosity), adhesion, solderability, and hydrogen embrittlement relief.

Standards for Silver Plating:

- ASTM B 700-08, "Standard Specification for Electrodeposited Coatings of Silver for Engineering Use"
- MIL QQ-S-365D, "Federal Specification: Silver Plating, Electrodeposited, General Requirements for (03 June 1985) [S/S by ASTM-B700]"
- ISO 4521, "Metallic coatings—Electrodeposited silver and silver alloy coatings for engineering purposes"
- SAE AMS 2411G-2008 (SAE AMS 2411G-2008), "Plating, Silver for High-Temperature Applications"

Standards for Tin Plating:

- Mil-T-10727C Tin, "Plating: Electrodeposited or Hot-dipped (This specification has been cancelled, but is still in wide use)"
- ASTM B-545-97 (2009), "Standard Specification for Electro-deposited Coatings of Tin"
- ISO 2093-1986, "Electroplated Coatings of Tin: Specification and Test Methods"

Standards for Nickel Plating:

- SAE AMS-QQ-N-290B-2009 (SAE AMS-QQ-N-290B-2009), "Nickel Plating (Electrodeposited)"
- SAE AMS 2403, "Plating, Nickel, General Purpose (Reaffirmed: June 2009)"
- SAE AMS 2423D-2002 (SAE AMS 2423D-2002), "Plating, Nickel Hard Deposit (Reaffirmed: April 2007 and June 2009)"
- MIL-P-18317, "Plating, Black Nickel (Electrodeposited) On Brass, Bronze, or Steel"
- SAE AMS 2404F-2008 (SAE AMS 2404F-2008), "Plating, Electroless Nickel"

Standards for EN Plating:

- AMS 2404, "Electroless Nickel Plating"
- AMS 2405, "Electroless Nickel, Low-Phosphorus"
- ASTM B656, "Guide for Autocatalytic Nickel–Phosphorus Deposition on Metals for Engineering Use"
- ASTM B733, "Standard Specification for the Autocatalytic Nickel–Phosphorus Coatings on Metal Coatings"

Standards for Zinc Plating:

- ASTM F2329-05 (2011), "Standard Specification for Zinc Coating, Hot-Dip, Requirements for Application to Carbon and Alloy Steel Bolts, Screws, Washers, Nuts, and Special Threaded Fasteners"

- Phosphorus ASTM B852-08, "Standard Specification for Continuous Galvanizing Grade (CGG) Zinc Alloys for Hot-Dip Galvanizing of Sheet Steel"
- SAE AMS 2402J-2007 (SAE AMS2402J-2007), "Zinc Plating"

Standards for Galvanizing:

- ASTM A123/A123M-09, "Standard Specification for Zinc (Hot-Dip Galvanized) Coatings on Iron and Steel Products"
- ASTM A143/A143M-07, "Standard Practice for Safeguarding Against Embrittlement of Hot-Dip Galvanized Structural Steel Products and Procedure for Detecting Embrittlement"
- ASTM A384/A384M-07, "Standard Practice for Safeguarding Against Warpage and Distortion During Hot-Dip Galvanizing of Steel Assemblies"
- ASTM A780/A780M-09, "Standard Practice for Repair of Damaged and Uncoated Areas of Hot-Dip Galvanized Coatings"

Standards for Plating on Aluminum:

- ASTM Standard B253-87(2010), "Standard Guide for Preparation of Aluminum Alloys for Electroplating"
- SAE AMS 2420D-2002 (SAE AMS 2420D-2002), "Plating of Aluminum for Solderability Zinc Immersion Pre-treatment Process (Reaffirmed: November 2007)"

Standards for Plating Testing:

- SEMI G62-95 (Reapproved 0302), "Test Method for Silver Plating Quality"
- ISO 2819:1980, "Metallic coatings on metallic substrates. Electrodeposited and chemically deposited coatings. Review of methods available for testing adhesion"
- SAE AIR 4789B-2003 (SAE AIR 4789B-2003), "Aerospace Information Report on Evaluating Corrosion Testing of Electrical Connectors and Accessories for the Purpose of Qualification"

Regulations:

- Waste Electrical and Electronic Equipment (WEEE)
- RoHS, European directive 2002/95/CE (RoHS directive) of the European Parliament and the European Union Council
- ANSI/ITAA GEIA-STD-0005-2, "Standard for Mitigating the Effects of Tin Whiskers in Aerospace and High Performance Electronic Systems"

3.8 PLATING GLOSSARY

Plating terminology is very extensive, and having an explanation of plating terms at hand when reading a paper or book on plating is very useful. The following glossary may not cover all the terms used in the plating industry and technology, but will be very helpful while reading this book. The source of this glossary is "Glossary of Finishing Terms" [71], reprinted with permission.

A

Abrasive blasting: A process for cleaning or finishing by a means of an abrasive directed at the work piece at differing pressures and velocities.

Activation: Elimination of a passive condition on a surface.

Activity (ion): The ion concentration corrected for deviations from ideal behavior. Concentration multiplied by activity coefficient.

Addition agent: A material added in small quantities to a solution to modify its characteristics. It is usually added to a plating solution for the purpose of modifying the character of a deposit.

Adhesion: The attractive force that exists between an electrodeposit and its substrate that can be measured as the force required to separate an electrodeposit and its substrate.

Amorphous: Noncrystalline or devoid of regular structure.

Ampere: The current that will deposit silver at the rate of 0.0011180 g/s. Current flowing at the rate of 1 Coulomb/s.

Angstrom unit (Å): 10^{-8} cm.

Anion: A negatively charged ion.

Anode: The electrode in electrolysis, at which negative ions are discharged, positive ions are formed, or other oxidizing reactions occur.

Anode efficiency: Current efficiency of a specified anodic process.

Anode film: (1) The layer of solution in contact with the anode that differs in composition from that of the bulk of the solution. (2) The outer layer of the anode itself consisting of oxidation or reaction products of the anode metal.

Anode polarization: See polarization.

Anodic coating: A protective, decorative, or functional coating, formed by the conversion of the surface of a metal in an electrolytic oxidation process.

Anodizing: An electrolytic oxidation process in which the surface of a metal, when anodic, is converted into a coating having desirable protective, decorative, or functional properties.

Anolyte: The portion of electrolyte in the vicinity of the anode; in a divided cell, the portion of electrolyte on the anode side of the diaphragm.

Antipitting agent: An additional agent for the specific purpose of preventing gas pits in a deposit.

Automatic machine (or conveyor): A machine for mechanically processing parts through treatment cycles, such as cleaning, anodizing, or plating.

Automatic plating: (1) Full: Plating in which the cathodes are automatically conveyed through successive cleaning and plating tanks. (2) Semi: Plating in which the cathodes are conveyed automatically through only one plating tank.

B

Back electromotive force (EMF): The potential setup in an electrolytic cell that opposes the flow of current, caused by such factors as concentration polarization and electrode films. See EMF.

Barrel burnishing: The smoothing of surfaces by means of tumbling the work in rotating barrels in the presence of metallic or ceramic shot, and in the absence of abrasive. In a ball burnishing, the shot consists of hardened steel balls.

Barrel finishing (or tumbling): Bulk processing in barrels, in either the presence or absence of abrasives or burnishing shot, for the purpose of improving the surface finish.

Barrel plating (or cleaning): Plating or cleaning in which the work is processed in bulk in a rotating container.

Base metal: A metal that readily oxidizes or dissolves to form ions. The opposite of noble metal.

Basis metal (or material): Material upon which coatings are deposited.

Bipolar electrode: An electrode that is not directly connected to the power supply but is so placed in the solution between the anode and the cathode that the part closest to the anode becomes cathodic and the part closest to the cathode becomes anodic.

Blue dip: A solution, once widely used, containing a mercury compound used to deposit mercury upon a metal by immersion, usually prior to silver plating.

Bright dip (nonelectrolytic): A solution used to produce a bright surface on a metal.

Bright plating: A process that produces an electrodeposit having a high degree of specular reflectance in as-plated condition.

Bright plating range: The range of current densities within which a given plating solution produces a bright plate.

Brightener: An additional agent that leads to the formation of a bright plate or that improves the brightness of the deposit.

Brush plating: A method of plating in which the plating solution is applied with a pad or brush, within which is an anode and which is moved over the cathode to be plated.

Buffer: A compound or mixture that, when contained in the solution, causes the solution to resist changes in the pH. Each buffer has a characteristic limited range of pH values over which it is effective.

Buffing: The smoothing of a surface by means of a rotating flexible wheel to the surface of which fine, abrasive particles are applied in liquid suspension, paste, or grease stick form.

Building up: Electroplating for the purpose of increasing the dimensions of an article.

Burnishing: The smoothing of surfaces by rubbing, accomplished chiefly by the movement rather than removal of the surface layer.

Burnt deposit: A rough, noncoherent, or otherwise unsatisfactory deposit produced by the application of an excessive current density (CD) and usually containing oxides or other inclusions.

Bus (busbar): A rigid conducting section for carrying current to the anode and cathode bars.

Butler finish: A finish composed of fine, uniformly distributed parallel lines, having a characteristic luster usually produced by rotating wire brushes or cloth wheels with applied abrasives.

C

Cathode: The electrode in an electrolysis at which positive ions are discharged, negative ions are formed, or other reducing actions occur.

Cathode efficiency: The current efficiency of a specified cathodic process.

Cathode film: The layer of solution in contact with the cathode that differs in composition from that of the bulk of the solution.

Catholyte: The portion of the electrolyte in the vicinity of the cathode; in a divided cell, the portion on the cathode side of the diaphragm.

Cation: A positively charged ion.

Chemical polishing: The improvement in smoothness of a metal by simple immersion in a suitable solution. See Bright dip (nonelectrolytic).

Cleaning: The removal of grease or other foreign material from a surface.

Alkaline cleaning: Cleaning by means of alkaline solutions.

Anodic or reverse cleaning: Electrolytic cleaning in which the work is the anode.

Cathodic or direct cleaning: Electrolytic cleaning in which the work is the cathode.

Diphase cleaning: Cleaning by means of a solution that contains a solvent layer and an aqueous layer. Cleaning is effected both by solvent and by emulsifying action.

DC cleaning: See Cathodic or direct cleaning.

Electrolytic cleansing: Alkaline cleaning in which a current is passed through the solution, the work being one of the electrodes.

Immersion: see Soak cleaning.

Reverse current cleaning: See Anodic or reverse cleaning.

Soak cleaning: Cleaning by immersion without the use of current, usually in an alkaline solution.

Solvent cleaning: Cleaning by means of organic solvents.

Spray cleaning: Cleaning by means of spraying.

Ultrasonic cleaning: Cleaning by any chemical means aided by ultrasonic energy.

Colloidal particle: An electrically charged particle, generally smaller in size than 200 mμ, dispersed in a second continuous phase.

Coloring: (1) The production of desired colors on metal surfaces by appropriate chemical or electrolytical action. (2) Light buffing of metal surfaces for the purpose of producing a high luster.

Composite plate: An electrodeposit consisting of two or more layers of metal deposited successively.

Conductance: The capacity of a medium, usually expressed in mhos, for transmitting electric current. The reciprocal of resistance.

Conducting salt: A salt added to the solution in order to increase its conductivity.

Conductivity: Specific conductance—the current transferred across unit area per unit potential gradient. In the metric system, $K =$ amperes per squared centimeters divided by volts per centimeter. The reciprocal of resistivity.

Contact plating: Deposition of a metal by the use of an internal source of current by immersion of the work in solution in contact with another metal.

Contact potential: The potential difference at the junction of two dissimilar substances.

Conversion coating: A coating produced by chemical or electrochemical treatment of a metal solution that gives a superficial layer containing a compound of the metal, for example, chromate coatings on zinc and cadmium, oxide coating on steel.

Coulomb: The quantity of electricity that is transmitted through an electric circuit in 1 s when the current in the circuit is 1 A. The quantity of electricity that will deposit 0.0011180 g of silver.

Critical current density: A CD above which a new and sometimes undesirable reaction occurs.

Current density (CD): Current per unit area.

Current efficiency: The proportion, usually expressed as a percentage, of the current that is effective in carrying out a specified process in accordance with Faraday's law.

Cutting down: Polishing or buffing for the purpose of roughness or irregularities.

D

Deburring: The removal of burrs, sharp edges, or fins by mechanical, chemical, or electrochemical means.

Decomposition potential: The minimum potential, exclusive of IR drop, at which an electrochemical process can take place at an appreciable rate.

Degreasing: The removal of grease and oils from a surface.

Solvent degreasing: Degreasing by immersion in liquid organic solvent.

Vapor degreasing: Degreasing by solvent vapors condensing on the parts being cleaned.

Deionization: The removal of ions from a solution by ion exchange.

Depolarization: A decrease in polarization of an electrode at a specified CD.

Depolarizer: A substance or a means that produces depolarization.

Detergent: A surface active agent that possesses the ability to clean soiled surfaces.

Anionic detergent: A detergent that produces aggregates of negatively charged ions with colloidal properties.

Cationic detergent: A detergent that produces aggregates of positively charged ions with colloidal properties.

Nonionic detergent: A detergent that produces aggregates of electrically neutral molecules with colloidal properties.

Diaphragm: A porous or permeable membrane separating the anode and cathode compartments of an electrolytic cell from each other or from an intermediate compartment.

Diffusion coating: An alloy coating produced by applying heat to one or more coatings deposited on a basis metal.

Dispersing agent: A substance that increases the stability of a suspension by retarding the flocculation.

E

Electrochemical equivalent: The weight of an element, compound, radical, or ion involved in a specified electrochemical reaction during the passage of unit quantity of electricity, such as a Faraday, ampere-hour, or coulomb.

Electrochemistry: The branch of science and technology that deals with transformations between chemical and electrical energy.

Electrode: A conductor through which current enters or leaves an electrolytic cell, at which there is a charge from conduction by electrons to conduction by charged particles of matter, or vice versa.

Electrodeposition: The process of depositing a substance upon an electrode by electrolysis.

Electrode potential (EP): The difference in potential between an electrode and the immediately adjacent electrolyte referred to some standard EP as zero.

> *Dynamic EP*: The EP measured when a current is passing between the electrode and the electrolyte.
>
> *Equilibrium EP*: A static EP when the electrode and the electrolyte are in equilibrium with respect to a specified electrochemical reaction.
>
> *Static EP*: The EP measured when no net current is flowing between the electrode and the electrolyte.
>
> *Standard EP*: An equilibrium EP for an electrode in contact with an electrolyte in which all of the components of a specified chemical reaction are in their standard states. The standard state for an ionic constituent is unit ion activity.

Electroforming: The production or reproduction of articles by electrodeposition upon a mandrel or mold that is subsequently separated from the deposit.

Electrogalvanizing: Electrodeposition of zinc coatings.

Electroless plating: Deposition of a metallic coating by a controlled chemical reduction that is catalyzed by the metal or alloy being deposited.

Electrolyte: (1) A conducting medium in which the flow of current is accompanied by the movement of matter. Most often an aqueous solution of acids, bases, or salts, but includes many other media, such as fused salts, ionized gases, some solids, and so on. (2) A substance that is capable of forming a conducting liquid medium when dissolved or melted.

Electrolysis: Production of chemical changes by the passage of a current through an electrolyte.

Electrolytic cell: A unit apparatus in which electrochemical reactions are produced by applying electrical energy, or which supplies electrical energy as a result of chemical reactions and which includes two or more electrodes and one or more electrolytes contained in a suitable vessel.

Electromotive series: A table that lists in order to the standard electrode potentials of specified chemical reactions.

Electrophoresis: The movement of colloidal particles produced by the application of an electric potential.

Electroplating: The electrodeposition of an adherent metallic coating upon an electrode for the purpose of securing a surface with properties or dimensions different from those of the basis metal.

Electropolishing: The improvement in surface finish of a metal by making it anodic in an appropriate solution.

Electrorefining: The process of anodically dissolving a metal from an impure anode and depositing it cathodically in a purer form.

Electrotyping: The production of printing plates by electroforming.

Electrowinning: The production of metals by electrolysis with insoluble anodes in solutions derived from ores or other materials.

Embrittlement, hydrogen: See Hydrogen embrittlement.

Electromotive force (EMF): An electrical potential.

Emulsion: A suspension of small droplets of one liquid in another in which it is insoluble. For the formation of a stable emulsion, an emulsifying agent must usually be present.

Emulsifying agent: A substance that increases the stability of an emulsion.

Energy efficiency: The product of the current efficiency and the voltage efficiency for a specified electrochemical process.

Equivalent conductivity: In an electrolyte, the conductivity of the solution divided by the number of equivalents of conducting solute per unit volume, that is, the conductivity divided by the normality of the solution.

Etch (n): A roughened surface produced by a chemical or electrochemical means.

Etch (v): To dissolve unevenly (or uniformly) a part of the surface of a metal.

F

Faraday: The number of coulombs (96,490) required for an electrochemical reaction involving one chemical equivalent.

Ferritic stainless steel: Ferritic stainless steels are chromium alloys with a low carbon content. They are magnetic and have a good ductility, plus resistance to corrosion and oxidation (they are especially resistant to stress corrosion cracking).

Filter aid: An inert, insoluble material, more or less finely divided, used as a filter medium or to assist in filtration by preventing excessive packing of the filter cake.

Flash (or flash plate): A thin electrodeposit, <1 μm. See Strike.

Flocculate: To aggregate into larger particles, to increase in size to the point where precipitation occurs.

Formula weight: The weight, in grams, pounds, or other units, obtained by adding the atomic weights of all elemental constituents in a chemical formula.

Free cyanide: (1) True: The actual concentration of cyanide radical or equivalent alkali cyanide not combined in complex ions with metals in the solution. (2) Calculated: The concentration of cyanide, or alkali cyanide, present in the solution in excess of that calculated to be necessary for forming a specified complex ion with a metal or metals present in the solution. (3) Analytical free cyanide content of a solution, as determined by a specified analytical method.

G

Galvanic cell: An electrolytic cell capable of producing electrical energy by electrochemical action.

Galvanic series: A list of metals and alloys arranged according to their relative potentials in a given environment. See Electromotive series.

Galvanizing: Application of a coating of zinc.

Gassing: The evolution of gases from one or more of the electrodes during electrolysis.

Glass electrode: A half cell in which the potential measurements are made through a glass membrane.

Grinding: The removal of metal by means of a rotating rigid wheel containing abrasive.

Grit blasting: Abrasive blasting with small irregular pieces of steel or malleable cast iron.

H

Half cell: An electrode immersed in a suitable electrolyte. It may be designed to yield a known constant potential, in which case unknown potentials may be measured against it, for example, the calomel half cell.

Hard chromium: Chromium plated for engineering rather than decorative applications. Not necessarily harder than the latter.

Haring cell: A rectangular box of nonconducting material, with principal and auxiliary electrodes so arranged as to permit estimation of the throwing power of electrode polarizations and potentials between them.

Highlights: Those portions of a metal article most exposed to buffing or operations and, hence, having the highest luster.

Hull cell: A trapezoidal box of nonconducting material with electrodes arranged to permit observation of cathodic or anodic effects over a wide range of current densities.

Hydrogen embrittlement: Embrittlement of a metal or alloy caused by absorption of hydrogen during a pickling, cleaning, or plating process.

Hydrogen overvoltage: Overvoltage associated with the liberation of hydrogen.

Hydrophilic: (1) Tending to absorb water. (2) Tending to concentrate in an aqueous phase.

Hydrophobic: (1) Tending to repel water. (2) Lacking affinity for water.

I

Immersion plate: A metallic deposit produced by a displacement reaction in which one metal displaces another from the solution, for example, $Fe + Cu^{2+} \rightarrow Cu + Fe^{2+}$ (copper replacing iron).

Indicator (pH): A substance that changes color when the pH of the medium is changed. In the case of most useful indicators, the pH range within which the color changes is narrow.

Inert anode: An anode that is insoluble in an electrolyte under the conditions prevailing in the electrolysis.

Inhibitor: A substance used to reduce the rate of a chemical or electrochemical reaction, commonly corrosion or pickling.

Interfacial tension: The contractile force of an interface between two phases. See Surface tension.

Ion: A charged portion of matter of atomic or molecular dimensions.

Ion exchange: An exchange of ions between a solution and a solid.

IR drop: The voltage across a resistance in accordance with Ohm's law: $E = IR$, where: E = potential (voltage), I = current, and R = resistance.

K–L

Karat: A 24th part by weight; thus 18-karat gold is 18/24 pure.

Lapping: Rubbing two surfaces together, with or without abrasives, for the purpose of obtaining an extreme dimensional accuracy or superior surface finish.

Leveling action: The ability of a plating solution to produce a surface smoother than that of the substrate.

Limiting current density: (1) Cathodic: The maximum CD at which satisfactory deposits can be obtained. (2) Anodic: The maximum CD at which the anode behaves normally, without excessive polarization.

M

Mat finish (matte finish): A dull finish.

Matrix: A form used as a cathode in an electroforming, a mold or mandrel.

Metal distribution ratio: The ratio of the thicknesses of metal upon two specified areas of a cathode. See Throwing power.

Metalizing: (1) The application of an electrically conductive metallic layer to the surface of nonconductors. (2) The application of metallic coatings by nonelectrolytic procedures such as spraying of molten metal and deposition from the vapor phase.

Microinch: One-millionth of an inch, 0.000001 in. = 0.001 mil.

Micron (μ): One-millionth of a meter, 0.001 mm

Microthrowing power: The ability of a plating solution or a specified set of plating conditions to deposit metal in pores or scratches.

Mil: One-thousandth of an inch, 0.001 in. = 25.4 μ.

Mill scale: The heavy oxide layer formed during hot fabrication or heat treatment of metals.

Motor–generator (MG set): A machine that consists of one or more motors mechanically coupled to one or more generators. In plating such a machine, the generator delivers DC of appropriate amperage and voltage.

N–O

Noble metal: A metal that does not readily tend to furnish ions, and therefore does neither dissolve readily nor easily enters into reactions such as oxidation and so on. The opposite of base metal.

Note—**Since there is no agreement on the sign of electrode potentials, the words "noble" and "base" are often preferred because they are unambiguous.**

Nodule: A rounded projection formed on a cathode during electrodeposition.

Orange peel: A finish resembling the dimpled appearance of an orange peel.

Overvoltage: The irreversible excess of potential required for an electrochemical reaction to proceed actively at a specified electrode, over and above the reversible potential characteristics of that reaction.

Oxidation: A reaction in which electrons are removed from a reactant.

Oxidizing agent: A compound that causes oxidation, thereby itself becoming reduced.

P

pH: A unit of measure depicting the hydrogen concentration of a solution: Scale 1–14, where 7—neutral; <7—acidic; >7—basic.

Passivity: The condition of a metal that retards its normal reaction in a specified environment and is associated with the assumption of a potential more noble than its normal potential.

Peeling: The detachment or partial detachment of an electrodeposited coating from a basis metal or undercoat.

Periodic reverse plating: A method of plating in which the current is reversed periodically. The cycles are usually no longer than a few minutes and may be much less.

Pickle: An acid solution used to remove oxides or other compounds from the surface of a metal by chemical or electrochemical action.

Pickling: The removal of oxides or other compounds from a metal surface by means of a pickle.

Pit: A small depression or cavity produced in metal surface during electrodeposition or by corrosion.

Plastisol: A mixture of resins, plasticizers, and other minor additives, such as pigments and so on, that can be converted into a continuous film by the application of heat. Distinct from baking enamels and so on, in which all of the original mixture becomes a part of the film; there is no significant evaporation of the solvent. The films are usually much thicker than obtainable from coatings which depend on the evaporation of a volatile solvent.

Plating range: The CD range over which a satisfactory electroplate can be deposited.

Polarization: The change in the potential of an electrode during electrolysis, such that the potential of an anode always becomes more noble and that of a cathode becomes less noble than their respective static potentials. Equal to the difference between the static potential and the dynamic potential.

Polarizer: A substance or a means that produces or increases polarization.

Polishing: The smoothing of a metal surface by means of the action of abrasive particles attached by adhesive to the surface of wheels or endless belts usually driven at a high speed.

Primary current distribution: The distribution of the current over the surface of an electrode in the absence of polarization.

R

Rack, plating: A frame for suspending and carrying current to articles during plating and related operations.

Rectification: The conversion of alternating current (AC) into direct current (DC).

Rectifier: A device that converts AC into DC by virtue of a characteristic permitting appreciable flow of current in only one direction.

Reducing agent: A compound that causes reduction, thereby itself becoming oxidized.

Reduction: A reaction in which electrons are added to a reactant. More specifically, the addition of hydrogen or the abstraction of oxygen. Such a reaction takes place, for example, at the cathode in electrolysis.

Relieving: The removal of material from selected portions of a colored metal surface by mechanical means to achieve a multicolored effect.

Resist, n: (1) A material applied to a part of a cathode or plating rack to render the surface nonconductive. (2) A material applied to a part of the surface of an article to prevent reaction of metal from that area during chemical or electrochemical processes.

Ripple (DC): Regular modulations in DC output wave of a rectifier unit, or a motor–generator set, originating from the harmonics of the AC input system in the case of a rectifier, or from the harmonics of the induced voltage of a motor–generator set.

S

Sacrificial protection: The form of corrosion protection wherein one metal corrodes in preference to another, thereby protecting the latter from corrosion.

Sand blasting: Abrasive blasting with sand.

Saponification: The alkaline hydrolysis of fats, whereby soap is formed; more generally, the hydrolysis of an ester by an alkali with the formation of an alcohol and a salt of the acid portion.

Satin finish: A surface finish that behaves as a diffuse reflector and which is lustrous but not mirror like.

Scale: An adherent oxide coating that is thicker than the superficial film referred to as tarnish.

Sealing of anodic coating: A process that, by absorption, chemical reaction, or other mechanisms, increases the resistance of an anodic costing to staining and corrosion, improves the durability of colors produced in the coating, or imparts other desirable properties.

Sequestering agent: A sequestering agent forms soluble complex ions with, or sequesters, a simple ion, thereby suppressing the activity of that ion. Thus, in a water treatment, the effects of hardness can be suppressed by adding agents to sequester calcium and magnesium. See *Chelating Agent*.

Shield (n): A nonconducting medium for altering the current distribution on an anode or cathode.

Shield (v): To alter the normal current distribution on an anode or cathode by the interposition of a nonconductor.

Slurry: A suspension of solids in water.

Spotting out: The delayed appearance of spots and blemishes on plated or finished surfaces.

Stalagmometer: An apparatus for determining the surface tension. The mass of a drop of liquid is measured by weighing a known number of drops or by counting the number of drops obtained from a given volume of the liquid.

Stray current: Current through paths other than the intended circuit, such as through heating coils or the tank.

Strike (n): (1) A thin film of metal to be followed by other coatings. (2) A solution used to deposit a strike.

Strike (v): (1) To remove a coating from the basis metal or undercoat. (2) To plate for a short time, usually at a high initial CD.

Strip: A process or solution used for the removal of a coating from a basis metal or an undercoat.

Substrate: See Basis metal (or material).

Superimposed AC: A form of current in which an alternating current component is superimposed on the direct plating current.

Surface active agent (surfactant): A soluble or colloidal substance having the property of affecting markedly the surface energy of solutions even when present in a very low concentration.

Surface tension: That property, due to molecular forces, that exists in the surface film of all liquids and tends to prevent the liquid from spreading.

T

Tank voltage: The total voltage between the anode and cathode of a plating bath or electrolytic cell during electrolysis. It is equal to the sum of: (1) the equilibrium reaction potential, (2) the IR drop, and (3) the electrode potentials.

Tarnish: The dulling, staining, or discoloration of metals due to superficial corrosion. The film so formed.

Thief (or robber): An auxiliary cathode so placed as to divert to itself some current from portions of the work which would otherwise receive too high CD.

Throwing power: The improvement of the coating (usually metal) distribution over the primary current distribution on an electrode (usually cathode) in a given solution, under specified conditions. The term may also be used for anodic processes for which the definition is analogous.

Total cyanide: The total content of cyanide expressed as the radical CN⁻, or alkali cyanide, whether present as simple or as complex ions.

Transference (or transport or migration): The movement of ions through the electrolyte associated with the passage of the electric current.

Transference number (transport number): The proportion of the total current carried by the ions of a given kind.

Trees: Branched or irregular projections formed on a cathode during electrodeposition, especially at edges and other high-CD areas.

Tripoli: Friable and dust-like silica used as an abrasive.

Tumbling: See Barrel finishing.

V–W

Voltage efficiency: The ratio, usually expressed as a percentage, of the equilibrium reaction potential in a given electrochemical process to the bath voltage.

Water break: The appearance of a discontinuous film of water on a surface signifying nonuniform wetting and usually associated with a surface contamination.

Wet blasting: A process for cleaning or finishing by means of a slurry or abrasive in water directed at high velocity against the work pieces.

Wetting agent: A substance that reduces the surface tension of a liquid, thereby causing it to spread more readily on a solid surface.

Whiskers: Metallic filamentary growths, often microscopic, sometimes formed during electrodeposition and sometimes spontaneously during storage or service, after finishing.

Work (plating): The material being plated or otherwise finished.

REFERENCES

1. Myers M. Overview of the use of silver in connector applications, *Interconnection and Process Technology*, Tyco Electronics, Harrisburg, PA, February 5, 2009.
2. Imrell T. The importance of the thickness of silver coating in the corrosion behavior of copper contacts, *Proceedings of the 37th IEEE Holm Conference*, Chicago, IL, pp. 237–243, 1991.
3. Dunn B. D., de Rooij A., and Collins D. S. Corrosion of silver-plated copper conductors, *ESA Journal*, 8, 307–335, 1984.
4. *PEM Plating Process*, Online. Available: http://www.pem.fr/eng/Bright_reflowed.htm.
5. ASTM B700-08 Standard Specification for Electrodeposited Coatings of Silver for Engineering Use.
6. Braunovic M. Coating (Plating), Section 4.5.4, in *Electrical Contacts: Principles and Applications*, edited by Slade, P. G. Marcel Dekker, Inc., New York, pp. 237–239, 1999.
7. Lefebvre J., Galand J., and Marsolais R. M. Electrical contacts on nicked-plated aluminum: The state of the art, *IEEE Transactions on Components Hybrids, and Manufacturing Technology,* 14(1), 176–180, March 1991.
8. Safrany J. S. Surface treatment on aluminium for electrical applications, *ATB Métallurgie*, 42(1–2), 47–50, 2002.
9. Braunovic M. Fretting of nickel-coated aluminum conductors, *Proceedings of the 36th IEEE Holm Conference on Electrical Contacts*, Montreal, Canada, pp. 461–471, 1990.
10. Bruel J. F. and Carballeira A. Durabilite des contacts electriques en aluminum, [Durability of aluminum electrical contacts], *Proceedings of the SEE (Société de l'Electricité, de l'Electronique et des TIC)*, pp. 139–145, December 1986.
11. Jackson R. L. Electrical performance of aluminum/copper bolted joints, *Proceedings of the IEEE:* Part C, IEEE Publishing, New York, 129(4), 77–85, 1982.
12. Braunovic M. Evaluation of different contact aid compounds for aluminum-to-copper connections, *IEEE Transactions on Components Hybrids, and Manufacturing Technology,* 15(2), 216–224, 1992.

13. Federal Specification QQ-N-290, "Nickel Plating (Electrodeposited)," 1997.
14. *Nickel Plating (Electrodeposited)*, SAE Standard AMS-QQ-N-290, July 2000.
15. Selcuk, H. Fields of application of nickel plated copper conductor, *Wire Journal International*, 39(12), 65, December 2006, Online. Available: http://www.sarkuysan.com/Upload/Document/docu ment_3456096f90eb4f61a619f1376fdee351.pdf.
16. Duncan, R. N. Electroless nickel: Past, present and future. *Proceedings of the Electroless Nickel (EN) Conference*, Orlando, FL, November 1993.
17. Parkinson R. Properties and applications of electroless nickel, *NiDI Technical Series*, 10081, 37 pp., 1997, Online. Available: http://www.nickelinstitute.org/en/TechnicalLiterature/Technical%20Series/ PropertiesandApplicationsofElectrolessNickel_10081_aspx.
18. *Electroless Nickel Plating. A Guide*. Artistic Plating Company, Milwaukee, WI. Online. Available: http://www.artisticplating.net/pdf-files/metal-finishing/Nickel%20Plating/Electroless%20Nickel%20 Plating%20-%20A%20Guide.pdf.
19. Barnstead M. and Morcos B. *Electroless Nickel Plating*, Online. Available: http://www.pfonline.com/ articles/electroless-nickel-plating.
20. Zeller R. L. and Salvati Jr. L. Effect of phosphorus on corrosion resistance of electroless nickel in 50% sodium hydroxide, *Corrosion*, 50(6), 457–467, 1994.
21. Rajam K. S., Rajagopal I., and Rajagopalan S. R. Phosphorus content and heat treatment effects on the corrosion resistance of electroless nickel, *Plating and Surface Finishing*, 63–66, September 1990.
22. Singh D., Balasubramaniam R., and Dube R. K. Effect of coating time on corrosion behavior of electroless nickel–phosphorus coated powder metallurgy iron specimen, *Corrosion*, 51(8), 581–585, 1995.
23. Duncan R. N. The effect of solution age on corrosion resistance of electroless nickel deposit, *Plating and Surface Finishing*, 4–68, October 1996.
24. Holden C. A., Law H. H., Mattoe C. A., and Sapjeta J. Nickel-phosphorus alloy as a low-cost electrical contact material, *Plating and Surface Finishing*, 58–61, April 1987.
25. Holden C. A., Law H. H., and Opila R. L. Factors for producing low contact resistance of nickel-phosphorus alloys, *Plating and Surface Finishing*, 46–49, August 1989.
26. Duffec E. F., Baudrant D. W., and Donaldson J. G. Electroless nickel applications in electronics, in *Electroless Plating: Fundamentals and Applications*, edited by Mallory G. O., Haidu J. B., Inc., Chapter 9, *American Electroplaters and Surface Finishers Society*, Noyes Publications/Williqam Andrew Publishing, LLC, New York, 1990, 539pp.
27. Baudrand D. Electroplating/electroless plating for electronic applications, *Product Finishing*, December, 1999, Online. Available: http://www.pfonline.com/articles/electroplating-electroless-plating-for-electronic-applications.
28. *Occupational Safety and Health Administration*, Online. Available: http://www.osha.gov/pls/oshaweb/ owadisp.show_document?p_table=STANDARDS&p_id=9993.
29. Trauffer E. A. Aminal scrubbing compounds cut TRS levels with no CO_2 reaction, *Pulp and Paper*, 69(5), 121–125, May 1995.
30. Rice D. W., Cappell R. J., Kinsolving W., and Laskowski J. J. Indoor corrosion of metals, *J. Electrochem. Soc.*, 127, 891–896, 1980.
31. Trauffer E. A. A new high efficiency, low cost TRS scavenging system, *Proceedings of the 1993 TAPPI Pulping Conference*, pp. 1165–1172, Publication of Technical Association of the Pulp and Paper Industry (TAPPI), Norcross, GA, 1993, On-line. Available: http://www.tappi.org/Downloads/unsorted/ UNTITLED—pulp931165pdf.aspx.
32. Abbott W. H. The development and performance characteristics of mixed flowing gas test environments, *Proceedings of the IEEE Holm Conference on Electrical Contacts*, Chicago, IL, pp. 63–68, 1987.
33. Krumbein S. J. *Environmental Testing of Connectors for Performance in Service*, Publication of AMP Incorporated, AMP Publication No. 148-91.
34. *Material and Finish Guide*, Online. Available: https://www.scribd.com/document/67031657/ Material-Finish-Guide.
35. *Key to Metals. Nonferrous Knowledge Base*, Online. Available: http://nonferrous.keytometals.com/ default.aspx?ID=Articles.
36. Brusse J. A., Leidecker H., and Panashchenko L. Metal whiskers: Failure modes and mitigation strategies, *2nd International Symposium on Tin Whiskers*, Tokyo, Japan, April 24, 2008.
37. Leidecker H. and Brusse J. A. Metal whiskering, *CALCE International Symposium on Tin Whiskers*, College Park, MD, April 2007.
38. Leidecker H. and Brusse J. A. Tin whiskers: A history of documented electrical system failures, *Technical Presentation to Space Shuttle Program Office*, April, 2006.

39. Galyon G. A history of tin whisker theory: 1946–2004, *SMTAI International Conference*, Chicago, IL, September 2004, pp. 26–30.
40. Brusse J. A., Ewell G. J., and Siplon J. P. Tin whiskers: Attributes and mitigation, *22nd Capacitor and Resistor Technology Symposium Proceedings*, New Orleans, LA, March 25–29, pp. 67–80, 2002.
41. Woodrow T. Evaluation of conformal coatings as a tin whisker mitigation strategy, *IPC/JEDEC 8th International Conference on Pb-Free Electronic Components and Assemblies*, San Jose, CA, pp.1–25, April 18–20, 2005.
42. *Basic Information Regarding Tin Whisker*, Online. Available: http://nepp.nasa.gov/whisker/background/index.htm#q6.
43. Current Tin Whiskers Theory and Mitigation Practices Guideline, *JEDEC/IPC Joint Publication*.
44. Osenbach J. W., Shook R. L., Vaccaro B. T., Amin A., Potteiger B. D., Hooghan K. N., Suratkar P., and Ruengsinsub P. Tin whisker mitigation: Application of post mold nickel underplating on copper based lead frames and effects of board assembly reflow, *Proceedings of the Surface Mount Technology Association*, Chicago, IL, pp. 724–734, 2004.
45. Choi W. K., Kang S. K., Sohn Y. C., and Shih D. Y. Study of IMC morphologies and phase characteristics affected by the reactions of Ni and Cu metallurgies with Pb-free solder joints, *Electronic Components and Technology Conference*, pp. 1190–1196, 2003.
46. Oberndorff P. J., Dittes T. L. M., and Petit L. Intermetallic formation in relation to tin whiskers, *Proceedings of the IPC/Soldertec International Conference: Towards Implementation of the RHS Directive*, Brussels, Belgium, pp. 170–178, June 11–12, 2003.
47. Arnold S. M. The growth of metal whiskers on electrical components, *Proceedings of the IEEE Electrical Components Conference (ECC)*, Philadelphia, pp. 75–82, 1959.
48. Rickett B., Flowers G., Gale S., and Suhling J. Potential for whisker formation in lead-free electroplated connector finishes, *Proc. SMTA International Conference*, Chicago, IL, pp. 707–716, September 2004.
49. *INEMI Tin Whisker Test Committee Results*, ETCT NEMI Workshop, June 1, 2005. Available: http://thor.inemi.org/webdownload/newsroom/Presentations/ECE/Tin%20Whisker%20Workshop%20May%202005/Results.pdf.
50. Osenbach J. W., Shook R. L., Vaccaro B. T., Potteiger B. D., Amin A. N., Hooghan K. N., Suratkar P., and Ruengsinsub P. Sn whiskers: Material, design, processing, and post-plate reflow effects and development of an overall phenomenological theory, *IEEE Transactions on Electronic Packaging Manufacturing*, 28(1), pp. 36–62, January 2005.
51. Schetty R., Brown N., Egli A., Heber J., and Vinckler A. Lead-free finishes—Whisker studies and practical methods for minimizing the risk of whisker growth, *Proceedings of the AESF SUR/FIN Conference*, Nashville, TN, pp. 1–5, June 2001.
52. Kadesch J. and Brusse J. The continuing dangers of tin whiskers and attempts to control them with conformal coating, *NASA's EEE Links Newsletter*, Online. Available: http://nepp.nasa.gov/whisker/index.html.
53. Park S. C. and Abbott D. C. *Nickel-Palladium based Component Terminal Finishes*, HDP User Group International, Inc.
54. Nardone K. Combating the military's tin whisker threat: No-lead strategies for power products. *Military and Airspace Electronics*, Online. Available: http://mae.pennnet.com/display_article/346546/32/ARTCL/none/EXCON/1/Combating-the-military%27s-tin-whisker-threat:-no-lead-strategies-for-power-products/.
55. Jacobsen K. Within a whisker of failure, *Guardian*, Online. Available: http://www.guardian.co.uk/technology/2008/apr/03/research.engineering.
56. Derek B. When making things better makes things worse, *Uptime Magazine*, Online. Available: http://www.uptimemagazine.com/uptime/2009/02/tin-whiskers.html.
57. Pinsky D. and Lambert E. Tin whisker risk mitigation for high-reliability systems integrators and designers, *Proceedings of the IPC/JEDEC 5th International Conference on Lead Free Electronic Components and Assemblies*, Raytheon Technical Library, March 2004.
58. Pinsky D. Tin whisker risk algorithm spreadsheet, Paper presented at *15th Annual International Military & Aerospace/Avionics COTS Conference*, Newton, MA, 2003.
59. Lloyd C. Aerospace lead-free reliability, Boeing, *IPC APEX Reliability Summit*, Los Angeles, CA, February 23, 2007.
60. *Summary of the Current Pure Tin Prohibition for EEE Part Procurement Specifications (Military NASA)*, Online. Available: http://nepp.nasa.gov/WHISKER/reference/eee_specs/spec_summ.html.
61. *Tin Whisker Risk Assessment Algorithm-Revision D*, Online. Available: https://www.reliabilityanalysislab.com/ral_TechLibrary_TinWhisker.asp

62. Smetana J. iNEMI recommendations on lead-free finishes for components used in high-reliability products, *iNEMI Tin Whisker User GroupIPC/APEX*, Online. Available: http://thor.inemi.org/webdownload/projects/ese/tin_whiskers/User_Group_mitigation_May05-REMOVED_5-12-05.pdf.

63. Brusse J. and Sampson M. NASA, Zinc whisker: Hidden cause of equipment failure, *IT Professional*, 6(6), 43–46, November/December 2004.

64. Lahtinen R. and Gustafsson T. The driving force behind whisker growth, *Metal Finishing*, 103(11), 25–29, and 103(12), 33–36, 2005.

65. Reynolds H. L. and Hilty R. H. Investigation of zinc (Zn) whiskers using FIB technology. *Presented at IPC/JEDEC Lead Free North America Conference,* Online. Available: http://www.tycoelectronics.com/customersupport/rohssupportcenter/pdf/Zn_whisker_IPC_2004_paper.pdf.

66. Miller S. K. Whiskers in the data center, *Processor*, 29(30), 1–24, July 27, 2007, Online. Available: http://www.processor.com/editorial/article.asp?article=articles/P2930/31p30/31p30.asp.

67. Braunovic M., Myshkin N. K., and Konchits V. V. Coating materials, Chapter 4.2.3 in the book *Electrical Contacts: Fundamentals, Applications and Technology*, CRC Press, Boca Raton, FL, December 2007, 645pp.

68. Moller P. *Plating on Aluminum*, TALAT lecture 5205, Online. Available: core.materials.ac.uk/repository/eaa/talat/5205.pdf.

69. *ASTM B253-87(2010) Standard Guide for Preparation of Aluminum Alloys for Electroplating*, Online. Available: http://www.astm.org/Standards/B253.htm.

70. Chudnovsky B., Pavageau V., Rapeaux M., Zolfaghari A., and Bardollet P. Thermal aging study of tin plating on aluminum, *Proceedings of the 53rd IEEE Holm Conference on Electrical Contacts*, Pittsburgh, PA, pp. 124–131, September 2007.

71. Techplate. *Glossary of Finishing Terms,* Online. Available: http://www.techplate.com/platingglossary1.htm.

72. Chudnovsky B. H. Degradation of power contacts in industrial atmosphere: Plating alternatives for silver and tin, *Proceedings of the 49th IEEE Holm Conference on Electrical Contacts*, pp. 98–106, 2003.

4 Detrimental Processes and Aging of Plating

4.1 ISSUES OF TIN PLATING PERFORMANCE

4.1.1 GENERAL PRECAUTIONS IN USING TIN PLATING

Tin is a very good electrical conductor, but when applying tin to electrical contacts and other conductive parts it is recommended to take several precautions due to the specific physical properties of tin (Section 3.1). These precautions are summarized in the so-called "Tin Commandments" [1,2]:

1. Tin-coated contacts should be mechanically stable in the mated condition
2. Tin-coated contacts need at least 100 g contact normal force
3. Tin-coated contacts need lubrication
4. Tin coating is not recommended for continuous service at high temperatures
5. The choice of plated, reflowed, hot air leveled, or hot tin dipped coating does not strongly affect the electrical performance of tin or tin alloy-coated contacts
6. Electroplated tin coatings should be at least 2.5 μm (~100 μin.) thick
7. Mating tin-coated contacts to gold-coated contacts is not recommended
8. Sliding or wiping action during contact engagement is recommended with tin-coated contacts
9. Tin-coated contacts should not be used to make or break current
10. Tin-coated contacts can be used under dry circuit or low level conditions

4.1.1.1 Tin and Fretting Corrosion

Tin plating is susceptible to a so-called fretting corrosion, which will be described later in Section 4.5. However, protective lubrication is necessary when it is not possible to achieve a mechanical stability. Application of a thin liquid lubricant film protects the contact surfaces from detrimental effects of disturbance or fretting motions. When a lubricant is applied, friction and generation of wear particles due to the motion are reduced and the surface is protected from an atmospheric oxidation, in and around the contact interface.

A lubricant can act to reduce wear, seal off the interface from the environment, and/or inhibit corrosion of the interface material when it is exposed to the fretting action. It becomes more important for lubrication to be used to prevent fretting corrosion when contact forces approach the minimum recommended level of 100 g (3.53 ounces). On the other hand, with higher contact forces lubrication may be necessary to reduce friction and wear during insertion cycling of the connector [3]. In this respect, lubrication would permit the use of higher contact forces.

Lubrication may be applied to only one side of a connector pair. However, it is recommended that both sides be lubricated whenever possible. Sufficient lubricant is normally transferred during engagement of the connector to provide fretting corrosion resistance to a contact.

Lubricant formulations vary in complexity and effectiveness. They include pure mineral oil (marginal in life and effectiveness) and sophisticated mixtures of natural oils, synthetic oils, and additives for optimized performance. When possible, it is recommended to choose a lubricant with proven antifretting characteristics. A more detailed account of thoroughly tested lubricants that may reduce a negative effect of fretting corrosion is given in Section 6.4.

There are instances when the presence of a lubricant cannot be maintained throughout the life-time of the connector. In these cases an alternative to lubrication such as a mechanically stable design or a noble metal coating should be considered.

4.1.1.2 Tin and Intermetallic Compounds

Tin plating is not recommended for continuous service at high temperature [4]. At elevated temperature, the performance of tin-coated electrical contacts is degraded by the aging effects due to a rapid increase in the diffusion rate of copper and tin. This results in a change in the composition and effective thickness of the tin coating as a hard, brittle, nonuniform resistive layer of copper–tin intermetallic compound (IMC) that grows between the tin layer and the copper base metal. More information about IMC is given in Section 4.5.

The use of nickel underplating is recommended for elevated temperature applications. It does not prevent the formation of IMC because nickel–tin IMC may grow as well. However, the growth rate of the nickel–tin IMC is lower than that of the copper–tin IMC, which results in lesser degree of degradation. The same underplating was found useful in mitigating deterioration of tin plating on aluminum at elevated temperatures (see Section 4.3).

In addition to the detrimental effects of IMC, tin loses a significant portion of its mechanical strength, resulting in substantial creep effects at temperatures >100°C (212°F). Special design parameters such as a high contact force, thick coatings, or limited time–temperature exposure are needed to successfully avoid contact problems.

High-temperature operation would impose special requirements on contact lubricants, and under this condition, the lubricant must withstand the time and temperature requirements representative of the applications. While tin-coated contacts have been used at higher temperatures on an intermittent basis, for example, high force contacts for automotive applications, using tin at temperatures >100°C (212°F) should be based on evaluation to verify the acceptable performance for the intended application.

4.1.2 Thermal Deterioration of Tin Plating on Aluminum

Tin-plated aluminum busbars are often used in electrical applications as a cost-effective replacement of more expensive tin-plated copper. Elastic connections made to tin-plated aluminum conductors are very commonly used in electrical distribution equipment. These connection jaws or clips are usually made of plated copper, which plug on to tin-plated aluminum or copper conductors.

However, the use of tin-plated Al busbars in a plug-on jaw to busbar connection may eventually lead to arcing and thermal failure. The history of elastic connections across the industry indicates a substantially higher occurrence of connection degradation in elastic connections to aluminum busbars than to copper busbars. The connection failure happens as a result of unstable contact resistance (CR) growth. This instability in a CR can be a result of chemical or physical changes in the plating, which are caused and accelerated by the high temperatures, which the connection may experience in carrying different loads. The degradation of the contact can lead to high CR, which can become large enough to cause contact arcing and melting.

The history of the failures of aluminum-to-copper connections is extensive, although tin is still one of the most common commercial plating for aluminum to improve the stability and decrease the galvanic corrosion of aluminum-to-copper connection [5]. Though the mechanisms of failure are not fully understood and explained, numerous processes such as oxidation, stress relaxation, DTE, galvanic and fretting corrosion are attributed to degradation of the contacting interface and a corresponding change in CR [6,7].

4.1.2.1 Accelerated Aging Study of Tin Plating

Degradation of tin and tin alloy plating on copper was studied in many aspects due to the formation of copper–tin (CuSn) IMC, particularly at elevated temperatures [8,9]. On the other hand, the behavior of tin plating on aluminum was not studied well enough to identify the longevity of the plating exposed to elevated temperatures in electrical connections. The possible causes of occasional resistance growth in elastic connections to tin-plated aluminum busbar conductors may be determined by studying the quality of tin plating on aluminum in the as-plated condition and after exposure to an elevated temperature in an accelerated aging study.

The first step in the study, however, should be learning about all aspects of the plating quality before elevated temperatures would be applied. The details of all three tin plating techniques on Al (Alstan 80, Alstan 88, and Bondal) used in this study are given in Section 3.6. According to the results of Ref. [10], all three tested plating techniques produce plating with good adhesion, and demonstrate the absence of the voids, gaps, or cracks in the plating. Strong adhesion quality was found after a short exposure to 200°C (below the Sn melting temperature) with a thermal shock test that reveals any adhesion issues. These results are consistent with a strong adhesion of both the plating and the underplating over the aluminum busbar. Therefore, freshly plated tin on aluminum usually does not show any defects that may lead to the failure.

However, various types of tin plating on aluminum traditionally used in electrical industry may deteriorate and eventually fail if exposed to elevated temperatures for prolonged periods of time. Multiple testing techniques, such as the adhesion test, thermal shock test, SEM metallography, EDS x-ray analysis, and thickness measurements [10], have been used to investigate three different types of tin plating over Al. This study allowed finding out whether exposure to high temperature affects the adhesion and the quality of tin plating on aluminum. Aluminum busbars plated with tin using three different traditional plating techniques have been exposed to elevated temperature for 1–8 weeks. The plated Al busbars have been positioned in a classical oven at the temperature of ambient air stabilized at 150°C. After a specified period of time, one randomly selected bar was removed from the oven and the samples were cut and prepared for further testing. Aging test for Alstan 80 plating lasted for 8 weeks and the testing of plating quality was then performed after 1, 4, and 8 weeks of thermal exposure. For two other types of plating (Alstan 88 and Bondal), a more detailed aging process was studied, with the bars removed from the chamber after each week of exposure during the first 4 weeks.

4.1.2.2 Quality of Thermally Aged Tin Plating

Two tests may help us to determine the quality of the plating: the adhesion test and the thermal shock test. A description of these tests is given in Section 3.6. According to thermal shock test results, the adhesion of tin plating on aluminum was not significantly affected by short exposure to temperatures up to 200°C.

However, for samples exposed to elevated temperature for a longer period of time, the adhesion of the plating begins to deteriorate. Adhesion tests were performed on the samples of the Al busbar with three different tin plating techniques after the aging tests. Alstan 80 samples failed adhesion tests after just 1 week of thermal exposure (Figure 4.1) and Alstan 88 samples failed adhesion tests after 4 weeks of thermal exposure. However, Bondal samples passed adhesion tests after 4 weeks of thermal exposure. According to the results of the scribing test the Bondal plating survived a longer exposure to high temperature than both Alstan 80 and Alstan 88 plating.

The quality of thermally aged tin plating on aluminum was compared to the original quality of the plating based on SEM metallographic analysis in Ref. [10]. This technique allows one to measure the thickness of tin plating and preplating layers (copper or bronze) and to find out whether pores, voids, and gaps are present along each interface (Al–Cu/bronze, Cu/bronze–Sn) of tin plating across the layer. Chemical analysis using the EDS technique defines the chemical composition of every layer on the sample surface, including copper or bronze preplating layers.

In freshly plated Alstan 80 tin, the upper layer of the plating consists of pure tin, and the underlayer is composed of Sn and Cu with the proportion corresponding to 70Sn30Cu-type bronze (70% Sn and 30% Cu). The quality of nonaged plating is good without separation, voids, or cracks. Nonaged Alstan 88 samples also had no pores, voids, or gaps along each interface (Al–bronze,

FIGURE 4.1   Appearance of tin plating on Al that failed the adhesion test.

bronze–Sn). The same was found for Bondal samples; the plating quality was good as well: no pores, voids, or gaps were detected along Al–Cu and Cu–Sn interfaces (see Section 3.6).

To determine how aging time affects the quality of the plating applied using three different plating techniques, an SEM study was conducted on the samples after different periods of the exposure to elevated temperatures [10]. It was found that after 1 week of thermal exposure none of the three types of tin plating had signs of deterioration; there were no pores, voids, or gaps.

However, all three types of tin plating on aluminum demonstrated serious deterioration after a longer exposure to elevated temperature. The first signs of aging appeared in Alstan plating after 2 weeks of exposure: some voids and gaps have been detected along both interfaces Al–bronze and bronze–Sn, and the coating is peeling in a few locations. After 3 weeks at 150°C, multiple pores, voids, and gaps were found along interfaces, and the coating was peeling in more locations. After 4 weeks of thermal exposure, both types of Alstan plating had multiple pores, voids, and gaps along each of the interfaces Al–bronze and bronze–Sn, and multiple locations with coating peeling (Figure 4.2).

Finally, after 5 weeks of exposure, Alstan plating was found to be in a very bad condition with cracks along each interface, separation and fractures of both plating and underplating.

As to Bondal plating, it survived longer; after 2 weeks of thermal exposure the plating was still in good condition without pores, voids, or gaps. However, after 3 weeks it also began to show signs of aging: some voids and gaps appeared along the interfaces Al–Cu and Cu–Sn. The coating was peeling in a few locations.

The test revealed that after 4 weeks of thermal exposure, Bondal plating was no better than both Alstan types; it had multiple pores, voids, and gaps along the interfaces, and multiple locations with coating peeling. The longest time of thermal exposure was 8 weeks, after which the cracks have progressed along each interface leading to separation of long parts of the plating. At the bronze–tin interface, cracks progression led to the fractures of long parts of tin plating (Figure 4.3).

From the results of the aging test in Ref. [10], it appears that the details and rate of deterioration may slightly differ for different plating procedures, but none of the plating was able to withstand the aging test. It was found that even though the Bondal plating survived the first 2 weeks of thermal

FIGURE 4.2   Microphotograph of tin plating on Al after 4 weeks of thermal aging: multiple pores and gaps along interfaces.

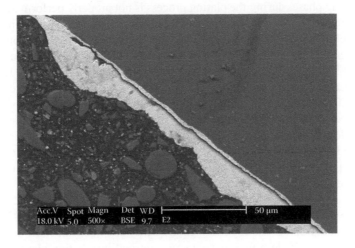

FIGURE 4.3   Microphotograph of tin plating on Al after 8 weeks of thermal aging: plating separated from base metal along interface.

exposure without significant deterioration, the aging process proceeded very fast for all three types of tin plating.

4.1.2.3   Mechanisms of Thermal Deterioration of Tin Plating on Al

It was suggested that theoretically at least three different mechanisms could cause deterioration of tin plating on aluminum during exposure to high temperature for an extended period of time [10]. These three mechanisms are the different ability of the three materials (Sn, Cu, or bronze, and Al) to expand at elevated temperatures, possible preexisting mechanical stress in the plating, and poor quality of tin plating on aluminum.

4.1.2.3.1   Difference in Thermal Expansion of Three Layers

Plating on aluminum (base metal) consists of two types of coating: the lower layer (underplating) is made of copper or bronze and the upper layer is made of tin. If the difference in thermal expansion between the coating material and the substrate is large enough, then thermal shock or temperature elevation can produce mechanical stress for coating different from that for the substrate. Mechanical stress may cause plating delamination and/or produce cracks through the plating. However, the values of thermal expansion coefficients for Sn ($27 \times 10^{-6}/°C$) and Al ($24 \times 10^{-6}/°C$) are quite close, and it is not likely that the thermal expansion coefficient of bronze with Sn as a major component will be significantly different.

4.1.2.3.2 Mechanical Stress

It is known that mechanical stress in the plating is higher on the curved and angular sides of a busbar than on the linear sides. In processing, mechanical stress is absorbed and contained in existing defaults. Heating helps to release this stress, leading to defaults' propagation, creation, and multiplication. Therefore, heating and mechanical stresses are supporting each other in the process of material deterioration. An important observation was made in Ref. [10]: more cracks, voids, and coating peeling are seen on the curved and angular sides than on the linear sides of the aged samples, which may be seen as a proof of the important role of mechanical stress in plating deterioration.

4.1.2.3.3 Quality of the Plating Procedure

During multiple steps of the plating process, including surface activation and cleaning, plating interfaces are exposed to several chemicals, some of them with proprietary composition. If any of the activation and rinsing phases during the plating process is not properly performed, then the residues of the chemicals may pollute the plating interfaces. A detailed SEM/EDS study in Ref. [10] revealed small amounts of elements such as potassium (K), sodium (N), and carbon (C) on interfaces. There could also be other light elements included in the composition of nondisclosed chemicals and in the materials used in the bath rubber lining, such as sulfur (S) and fluorine (F). Even chlorine may be found on the surfaces rinsed with chlorinated water. These impurities, if present, do not cause visible damage to the fresh plating. Over time (longer than 1 week), the chemicals' residues on interfaces could produce bubbles of pressurized gases that tear the plating from inside. This process can be accelerated at elevated temperatures. Therefore, it is possible that, when the tin-plated aluminum busbars have been heated for a period of just 1 week, the plating has been weakened enough to fail the adhesion test. The poor plating quality may be one of the major causes of tin plating failure on Al busbars exposed to elevated temperatures.

4.1.2.4 Tin Plating on Aluminum as a Possible Cause of Connection Overheating

Freshly plated tin on aluminum usually has a good quality and does not show any defects that may lead to failure. Therefore, if aluminum parts in an electrical installation plated with tin could not be overheated, there should be no problems. It is important to evaluate if there is a possibility of tin-plated aluminum parts be exposed to elevated temperature during extended periods of time in service.

If there is a possibility that a conductive part made of tin-plated aluminum may get exposed to elevated temperatures for prolonged periods of time, there is also a possibility that various types of tin plating on aluminum, traditionally used in electrical industry, may deteriorate and eventually fail. The study [10] has shown that after a tin plating is exposed to a temperature of 150°C for just 1 week, adhesion of the plating begins to deteriorate. If thermal exposure is continued, all three traditional types of tin plating displayed poor plating adhesion after the aging test. Aging-related cracks, voids, and separation developed at the plating interfaces although in different concentrations and progression rates depending on the plating technique.

When in service, the electrical current flowing through the connection generates heat, which eventually raises the temperature of the connection. If one part of the connection is made of a tin-plated aluminum busbar, continuous exposure to elevated temperature causes the tin plating to lose its adhesion and develop pores and cracks along conductive surfaces.

The longer the thermal exposure, the more the defects appearing in the plating. This causes the electrical resistance on the connection to rise, which in turn causes more thermal energy to be generated. The process continues until heat runaway leads to the connections' thermal failure. To continue using tin plating on Al in applications, where electrical aluminum parts may be exposed to elevated temperatures, special attention should be paid to the improvement of the tin plating's thermal stability. The use of nickel underplating may provide such an improvement (see Section 4.3).

4.1.3 TIN PEST

4.1.3.1 Definition of Tin Pest

Tin pest is an autocatalytic, allotropic transformation of the element tin, which causes deterioration of tin objects at low temperatures. At 13.2°C (about 56°F) and below, pure tin transforms from a silvery and ductile allotrope of β-form white tin to brittle, α-form gray tin.

The transformation is slow to initiate due to its high activation energy, but the presence of germanium (or crystal structures of similar form and size) or very low temperatures aids the initiation. There is also a volume increase associated with the phase change. Eventually, it "decomposes" into a powder named "tin pest" [11]. The photograph in Figure 4.4 is likely the most famous modern photograph of tin pest. It shows the phenomenon of "tin pest" quite clearly [12].

The decomposition of tin will catalyze itself. As soon as the tin began decomposing, the process speeds up; the mere presence of tin pest leads to more tin pest. The tin objects at low temperatures will simply disintegrate decomposing tin into powder. Tin pest has also been called *tin disease*, *tin blight*, or *tin leprosy (Lèpre d'étain)*. It was observed to occur in medieval Europe, where the pipes of pipeorgans in churches were affected in cool climates.

4.1.3.2 Effects of Alloying Elements and the Environment on Tin Pest

Tin pest was thought to be a problem of pure tin, because tin–lead alloys do not decompose into tin pest. It was found that several alloying metals retard or eliminate the formation of tin pest [13].

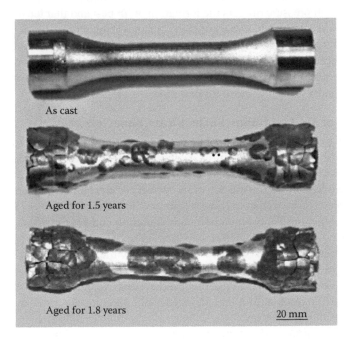

FIGURE 4.4 Formation of β tin into α tin in Sn–0.5Cu at $T < 10°C$. (From Karlya Y., Gagg C., and Plumbridge W. J. *Surf. Mount Technol.* 13(1), 39–40, 2000. With permission.)

TABLE 4.1
Tin Pest Retarding Effects of Lead-Free Alloying Metals

Alloying Metal	Tin Pest Retardant	% Concentration for Effective Inhibition
Bi	Strong	0.3
Sb	Strong	0.5
Pb	Strong	5.0
Cu	None–Weak	>5.0
Ag	Weak–Moderate	>5.0

Source: Lasky R. C. Tin pest: A forgotten issue in lead-free soldering? *Proceedings of the SMTA International Conference*, Chicago, IL, pp. 838–840, September 2004. Online. Available: http://www.indium.com/images/blogs/drlasky/files/TinPestPaper0723Final.pdf. With permission.

The most effective alloying elements retarding tin pest are bismuth, antimony, and lead. However, while bismuth and antimony will suppress tin pest at concentrations of <1 weight%, five times higher concentration of lead is needed [12]. A "rule of thumb" is that alloying metals that are highly soluble suppress tin pest.

Those elements that are not highly soluble in tin and tend to form IMC with tin are less likely to suppress tin pest. Examples of such metals are copper and silver. There are limited data that suggest that silver has a retarding effect on tin pest formation; however, the effect appears to be moderate at best. Silver suppresses tin pest but at a much reduced level. Hence, many people assume that it has not been proven that silver significantly retards the formation of tin pest, especially in the field or use conditions. Copper is considered by most researchers to have a very little effect. Table 4.1 is a summary of the effects of some lead-free alloying elements on tin pest [13].

It is known that high pressure hinders the formation of tin pest and that tensile stress enhances it. These phenomena make sense when one considers that anything that makes it easier to expand the lattice (e.g., tensile stress) should enhance the formation of tin pest, and anything that compresses the lattice (e.g., compressive stress) will retard tin pest formation. It is also reported that tin oxide formation at the surface will retard tin pest formation. Some unknown additions in soldering such as organic contaminants probably retard tin pest also.

4.1.3.3   Example of Tin Pest Failure in Electrical Connectors

A very interesting case study [14] described a tin pest failure that occurred on plated electrical connectors exposed to low temperatures during storage. Several large, tin-plated copper electrical connectors, used in an electric transit system in northeastern US, were subjected to analysis. The connectors were located on the outside of the transit car and helped conduct the electric supply from overhead cables to the transit car engine circuit. The connectors were ordered with a tin–bismuth coating, containing 2% bismuth. Bismuth is used to combat tin pest (Table 4.1) and it was the reason why the coating was selected.

Some of the coated connectors were put into service while the remainder were boxed up and placed in a storage yard over the winter. The following spring, additional connectors were required to enter the service. On opening the storage box, workers discovered that the once light-gray-colored connectors were now covered with a fine "cigarette ash"-like powder, with regions of the plating clearly separating from the substrate.

Using the SEM technique and a more sensitive spectroscopic method, namely inductively coupled plasma spectroscopy (ICPC), it was found that the gray powder and unaffected areas of the

plating contained tin and <0.1% bismuth. With such low levels of bismuth, plating would be susceptible to tin pest if the right conditions are present.

The storage environment for the connectors was assessed and the data revealed that the average monthly temperature from October to May was below 13°C (55°F), and during the winter
months December, January, and February it was nearly 0°C (below 30°F). As noted earlier,
at temperatures below 13°C (55°F), tin pest propagates. Therefore, the conditions were suitable for tin pest to initiate, and the tin plating was in a susceptible condition (i.e., essentially
bismuth-free).

X-ray diffraction (XRD) was used in Ref. [15] to determine if tin pest had developed. Since tin
pest is the transformation of tin from a body-centered tetragonal structure (β tin) to a cubic structure
(α tin), the use of XRD on the affected areas provided conclusive evidence of the phase constituents
present. In the unaffected areas the tin plating appeared to consist of metallic β tin. However, in the
affected, powdery areas the tin was present in the α-form—tin pest.

4.1.3.4 Impact of RoHS on Possible Tin Pest Failures

On July 1, 2006, a new directive, the RoHS Directive, took effect [16]. The RoHS Directive restricts
the use of six hazardous materials in the manufacture of various types of electronic and electrical
equipment. It is closely linked with the WEEE Directive which sets collection, recycling and recovery targets for electrical goods and is part of an initiative to solve the problem of huge amounts of
toxic electronic waste.

RoHS is often referred to as the lead-free directive, but also restricts the use of six other substances: lead, mercury, cadmium, hexavalent chromium (CrVI), polybrominated biphenyls (PBB),
and polybrominated diphenyl ether (PBDE).

Although RoHS is a step in the right direction toward dealing with toxic electronic waste, it
could lead to a greater number of tin pest failures. Tin pest had largely become a problem in the
past, which brought to life the wide use of tin alloys to combat it. One of the popular and effective
tin alloys is the one with lead, the use of which is restricted by RoHS. Since RoHS bans most uses
of lead the problem has returned, as lead-free alloys contain 95%–99% tin. With the adoption of the
RoHS directive regulations in Europe, and similar regulations elsewhere, traditional lead/tin solders
have been replaced by solders containing primarily tin, rendering prevention of tin pest (and related
problems such as tin whiskers) a modern technological challenge [14,15]. Tin pest could danger-
ously affect the safety and functionality of electronic products used across many manufacturing
sectors, such as the avionics and electronic industry.

For example, the leads of some electrical and electronic components are plated with pure tin. In
cold environments, this can change to α-modification gray tin, which is not electrically conductive,
and fall off the leads. After heating backup, it changes back to β-modification white tin, which is
electrically conductive, and can cause electrical short circuits and failure of equipment. Such problems can be intermittent as the powdered particles of tin move around.

Tin–lead solders have been used in the electronic industry for decades because of their good
conductivity, fluidity, and solderability. The lead is added to the tin solders to improve fluidity, with
the unexpected side benefit that it is the most common alloying addition used for combatting tin
pest. With the introduction of RoHS, and its restrictions on lead, new lead-free solders had to be
developed. Tin pest can be avoided by alloying with small amounts of electropositive metals soluble
in tin's solid phase, for example, antimony or bismuth, which prevent the decomposition. Silver,
indium, and lead have also been used, but lead is not soluble in tin's solid phase.

The most common replacements are tin–copper, tin–silver, and tin–silver–copper. All three
alloys have a similar solderability property as the lead–tin solders they are replacing. However, one
property they do not possess is an inherent resistance to tin pest, since they no longer contain the
pest-mitigating lead. Testing of these lead-free solders has shown that they can be susceptible to tin
pest [17,18].

So these new solder materials may be susceptible to tin pest when exposed to low temperatures. For most consumer electronics there should not be any issue, since they are rarely exposed to temperatures below 13°C (~55°F). The impact that RoHS may have in relation to tin pest failures would be electronic products operating in cold climates, such as telecommunication base stations (cell phone towers), automotives (vehicles for snow conditions such as Sno-Cats), military (weapons systems), space (satellites), and outdoor recreation electronics.

4.2   USE OF UNDERPLATING FOR PLATING LONGEVITY

4.2.1   MITIGATING ROLE OF UNDERPLATING

In conjunction with various plating types, underplating is often used to enhance the performance of the plating. Without underplating, the prime finish would be plated directly over base metals, which results in a number of failure mechanisms, such as

1. Formation of IMC
2. Diffusion of base metal (copper) through plating (gold)
3. Increased wear of the plating
4. Thermal deterioration of the plating (tin on aluminum)
5. Diffusion of zinc through base metal (certain brass alloys)
6. Through pore base metal corrosion

There are three basic underplating types, namely, silver, copper, and nickel, which could be used in contact systems.

Silver was used as underplating in the early 1960s. However, it was found that silver migrated through gold and formed disruptive films on the contact surfaces or at the very best cosmetic discoloration. Silver as an underplate is no longer in use.

Copper underplating is only used in a few applications due to diffusion tendency and tin/copper intermetallic formation. Copper as underplating covers substrate surface defects, which reduces porosity by creating a smoother surface. It does not, however, enhance wear resistance or increase corrosion resistance. The downside of copper underplating is its ability to diffuse through gold and its tendency to form IMC with tin (alloys).

Nickel is the most common and important type of underplating currently used.

4.2.2   ADVANTAGES OF NICKEL AS UNDERPLATING

Tin plating on copper tends to form SnCu IMC, whose electrical properties affect the conductivity of electrical contact and play a detrimental role in longevity and performance of the contacts. For more about the formation of IMC, see Section 4.5. One of the techniques used to minimize or slow down the process of CuSn IMC formation is application of Ni underplating.

Ni underplating provides the following functions for the plating:

• Covers defects in the substrate surface, which reduces porosity
• Acts as an internal diffusion barrier preventing the diffusion of zinc and copper from penetrating through gold
• Serves as an external diffusion barrier preventing pollutants from reacting with base metals; therefore it is increasing corrosion resistance
• Increases wear resistance
• Eliminates Cu/Sn intermetallics
• Improves resistance to thermal deterioration of the plating

4.2.2.1   Ni Underplating Provides a Diffusion Barrier

Ni underplating is used often to plate gold over other metals, such as copper. Copper atoms can diffuse through precious metal plating, and will oxidize once they reach the surface. This diffusion is time and temperature dependent. Higher temperatures provide a greater number of atoms with the necessary energy to move through the lattice. Therefore, diffusion will occur more readily at higher temperatures.

Ni underplating provides a very effective diffusion barrier, preventing copper from migrating to the surface. It can slow down creep corrosion from bare edges, and can passivate pores in the overlying layers, minimizing pore corrosion. Even if there are no pores in the plating and there are no bare edges, it is still possible for the base metal to corrode.

4.2.2.2   Ni Underplating Prevents the Formation of Intermetallics

When applied under tin plating on copper, nickel underplating helps to prevent copper–tin intermetallic formation in tin and tin–lead-coated contacts [19–21]. This increases the service life if the nickel underplate has low residual tensile stress. Otherwise, the nickel underplate may reduce the fatigue life of the contact.

4.2.2.3   Ni Underplating Improves Wear Resistance

If the electrical contact is going through many mating and unmating cycles, increasing contact wear resistance becomes very important. The wear resistance of a surface depends on the hardness of every layer, not just the outermost layer. Nickel is typically much harder than the base metals and the noble metal contact finishes commonly in use. This means that the use of nickel underplate increases the overall hardness of the contact surfaces and makes the contact interface more durable.

4.2.2.4   Ni Underplating Increases Corrosion Resistance

Application of Ni layer underplating provides an effective means for corrosion protection of base metal from pore and creep corrosion.

4.2.2.4.1   Pore Corrosion

Nickel can significantly reduce the possibility of pore corrosion. Nickel underplating will have fewer pores than the top layer because it is usually applied in a thicker layer than the overlying precious metal. In order for the base metal to be exposed, the pores in every coating layer would have to coincide. It means that the probability of a nickel underplating to be exposed to the atmosphere is very low, and therefore the base metal would not be exposed as well. A protective nickel oxide layer will form in any such pores. Since the oxide does not creep, the pores are effectively passivated, and the noble metal contact surface will remain unblemished.

4.2.2.4.2   Creep Corrosion

The passive oxide formed by nickel will also help to limit the amount of creep corrosion from bare edges. The exposed copper at the bare edges will oxidize, and the corrosion product will tend to spread outward. However, the nickel underplating will oxidize as well. This passive layer of nickel oxide will inhibit the spread of the active copper corrosion product across the underplate. Unfortunately, it will not entirely prevent the eventual creep of corrosion product onto the precious metal, but it will significantly reduce the rate of spreading. This will increase the useful life of the contact. However, for contacts that operate in aggressively corrosive environments, it is always best to eliminate bare edges.

4.2.2.5   Other Advantages of Ni Underplating

Nickel layer between base metal and plating serves to enhance the adhesive and abrasive sliding wear resistance of base metal. Ni underplating improves base metal durability during fretting.

It increases the nobility of the contact if the base metal is porous and if sulfur, hydrogen sulfide, and some other pollutants are present.

4.2.3 RECOMMENDED THICKNESS OF NICKEL UNDERPLATING

The thickness of Ni underplating depends on application. Both electroplated Ni and EN are used for plating gold over other metals, such as copper or aluminum. Electroplated Ni is used as underplating for tin plating on copper and aluminum.

The use of ductile nickel underplate for hard gold <5 µm thick is nearly universal. A typical range of nickel thickness in gold-plated connectors is 1.25–4.0 µm. The lower limit is a sufficient underplating thickness to provide the basic benefits to the nickel underlayer. The upper limit is determined by both cost–benefit and mechanical considerations. The cost–benefit issue is obvious; more nickel means more plating time and material cost. The mechanical considerations are more complicated. On the one hand, as the nickel plating thickness increases, the ductility of the nickel tends to decrease, and reduced ductility can lead to cracking of the plating. On the other hand, the roughness of the thicker plating tends to increase, and the increased roughness may compromise porosity and wear performance.

Nickel underplating is used for plating gold on aluminum for aerospace applications [22]. The process includes an electrolytic copper underplating (5 µm) and then electrolytic nickel deposit (10 µm), and after that electrolytic gold deposit (5 µm). Some OEM specifications for plating tin over aluminum recommend applying an Ni underplating of 5 µm. Some suppliers of distribution and transmission voltage power system products and solutions specify the technique of plating of wrought or cast aluminum for hot tinning applications as the Alstan 70 [23] process with 5 µm of nickel under 5 µm of tin.

However, there are some precautions that must be taken when using nickel underplating.

- Ductile nickel should be used. Some nickel baths result in brittle conditions to exist. This has implications for crimp applications and when preplated materials are used.
- Some nickel baths are stress-related, increasing potential residual stresses in contacts.
- The thickness of the underplating must be controlled. It should not be less than 1.25 µm (50 µin.) for porosity reasons. It should not be more than 4 µm (150 µin.) as it can start having an impact on the "composite" modulus of elasticity. This could affect spring rates and other mechanical features such as mating forces.

Additionally, with all the advantages that nickel underplating provides, it is important to take into account that nickel, when used for gold plating, will also eventually diffuse through gold. It may also form nickel–tin intermetallics when used as underplating for tin (Section 4.3). The diffusion and intermetallic growth rates for nickel are very slow and, under normal application environments, usually pose little or no potential functional problems.

4.3 APPLICATIONS OF Ni UNDERPLATING

4.3.1 USE OF Ni UNDERPLATING FOR TIN PLATING ON COPPER

Application of Ni underplating plays a twofold role when used for tin plating on copper [19–21]. First, it prevents copper–tin intermetallic formation in tin and tin–lead-coated contacts. Another important role of nickel underplating is to slow down the formation of tin whiskers. However, there is a chance that nickel itself may form IMC with tin. More detailed explanation of the formation of IMC will be given in Section 4.5.

4.3.1.1 The Formation of Ni–Sn Intermetallics

The use of Ni underplate for plating Sn over Al is in practice, but studies of the behavior of two interfacing metals (nickel and tin) are conducted generally on the solder joints. It turns out that Ni also forms an intermetallic phase with tin, although the process of Ni–Sn intermetallics formation is different from that of Cu–Sn intermetallics.

It was determined that the formation of IMC between tin-based solder and electroplated Ni substrate is inevitable. However, the thickness of the Ni–Sn IMC layer is approximately only one-quarter as thick as the Cu–Sn IMC formed in the same conditions (Figure 4.5) [24].

The growth of Ni–Sn IMC is linear with the temperature. The Ni–tin IMC layer usually consists of Ni_3Sn_4 brittle phase, since other phases (Ni_3Sn and Ni_3Sn_2) can be formed only on a very rough surface or under temperatures higher than 200°C. The Ni–Sn intermetallic layer is peaky and spiky, and Ni_3Sn_4 whiskers come into solder along the interface, which is another difference to Cu–Sn intermetallics that is usually smooth.

It was shown in Ref. [25] that brittle Ni_3Sn_4 IMC contributes mainly to the fatigue failure of the solder joint and also is the root cause of fracturing near the IMC layer. Another difference between two types of IMCs is that the volumeric shrinkage of Ni_3Sn_4 layer is more, about 10.7% during the transformation from solid phases of Sn and Ni to the Ni_3Sn_4 compound, compared with volumeric shrinkages to form Cu_6Sn_5 and Cu_3Sn, which are only 5% and 8.5%, respectively.

As the thickness of IMC layer increases, internal strain and intercrystalline defects are formed and increase gradually in severity at the grain boundary of the Ni_3Sn_4 IMC and IMC/solder interface. They play a significant role in mechanical fatigue, when train accumulation around IMC results in cracks initiating and propagating. Therefore, the thicker the layer the shorter the fatigue lifetime of the solder joint under mechanical stresses.

In Ref. [25] the formation of Cu–Sn and Ni–Sn intermetallics has been studied in Sn–Pb plating. The data on resistivity of IMCs compared with that of tin and copper given in Ref. [25] are used in Figure 4.6 to show how significant is the growth of resistance in intermetalics, particularly in Ni–Sn IMC. Hardness is also high in IMCs (~420 BHN). It was also shown [25] that the total growth of Cu–Sn IMC is almost twice faster than the growth of Ni–Sn IMCs (Table 4.2).

Intermetallics tend to grow faster in bright tin plating compared with matte tin–lead plating. Slightly different data on the thickness of the IMC are given in Ref. [24], but they still show that Ni–Sn IMC layer is much thinner than that of Cu–Sn growing in the same conditions. It is interesting to compare the thickness of Ni–Sn IMC growing at different temperatures. Since Cu–Sn

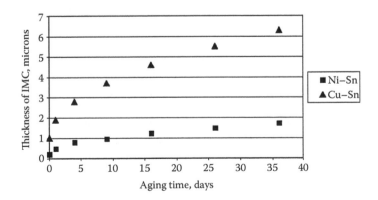

FIGURE 4.5　Thickness of IMC after aging at 120°C for 0–36 days. (Chan Y. C. et al. Reliability studies of BGA solder joints—effect of Ni–Sn intermetallic compound, *IEEE Transactions on Advanced Packaging*, 24(1), 25–32, February. © 2001, IEEE.)

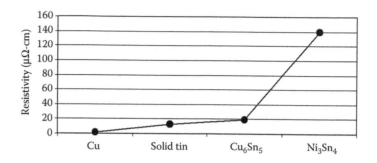

FIGURE 4.6   Electrical resistivity of Cu–Sn and Ni–Sn IMC compared with copper and tin. (Chan Y. C. et al. Reliability studies of BGA solder joints—effect of Ni–Sn intermetallic compound, *IEEE Transactions on Advanced Packaging*, 24(1), 25–32, February. © 2001, IEEE.)

TABLE 4.2

Growth of Cu–Sn and Ni–Sn Intermetallics in SnPb Plating at 150°C

Tin Plating Type	Aging Time (weeks)	Cu_6Sn_5 (µm)	Cu_3Sn (µm)	Cu–Sn Total (µm)	Ni_3Sn_4 (µm)
Matte	1	1.95	1.3	3.25	1.46
Bright	1	3.9	1.79	5.69	2.11
Matte	4	2.2	2.0	4.2	2.28
Bright	4	3.74	2.93	6.67	3.41

Source: Chan Y. C. et al. Reliability studies of BGA solder joints—Effect of Ni–Sn intermetallic compound, *IEEE Transactions on Advanced Packaging*, 24(1), 25–32, February 2001, IEEE.

intermetallics tends to grow faster than Ni–Sn IMC, it allows to suggest that usage of nickel barrier plating may retard the growth of the intermetallics [26].

4.3.2   Nickel Underplating as a Tin Whisker Mitigation Technique

Using a nickel or silver underplate has been proven to be excellent in preventing the formation of Cu_6Sn_5 intermetallics. It has been suggested that the formation of intermetallics in the base metal below the pure tin plating creates stresses that promote the growth of tin whiskers. Based on this assumption, in the development of several tin whisker mitigation strategies [27], it has been suggested that using barrier layer such as nickel between the base metal and the pure tin finish can reduce the likelihood of tin whisker growth. The results of studies on the effect of nickel underplating on tin whiskers growth are controversial.

To provide a quantitative relationship between the size of the whiskers and various factors, the concept called whisker index (WI) was introduced in Ref. [28]. WI is defined as a function of the number, length, and diameter of the whiskers. The effect of applying Ni underplating was studied in Ref. [28] by comparing the WI for satin bright tin, satin bright tin after reflow and satin bright tin over a nickel underlayer. It was determined that both reflow and nickel underlayer further reduce the propensity of whisker formation. A nickel barrier layer of 1.5 µm over a copper base material significantly reduced the growth of tin whiskers on a low stress tin finish.

By utilizing a nickel underlayer with the satin bright tin finish effectively eliminates whisker formation. WI as a function of time measured in months decreased from 10 for satin bright tin to

3 for reflowed satin bright tin, and no whiskers have been found in satin bright tin with a nickel barrier layer after 6 months of aging. After 10 months of aging, the difference between WI in two satin bright tin platings without Ni underplating became smaller (9.5 and 8), but there are still no whiskers found in the plating with a Ni barrier layer.

In Ref. [29], lead frame samples have been tested for over 3400 h in different temperature and RH conditions: (a) 105°C and 100% RH and (b) 50°C and 25% RH. Both lead frames were on bronze substrates and had 1–2 μm underplating (silver or nickel) with 5–7 μm tin on the top. Results have shown hillock and whisker growth differences in the samples depending on the environmental conditions, which is attributed to the different stress-causing mechanisms. Growth differences have also been found in the different underplating materials.

On the other hand, the data presented in Ref. [30] indicate that tin whiskers can still grow on the parts with nickel underplating. However, in this study a nickel undercoat was applied over silver, which was in turn applied over a ceramic material. In another extensive study [31] the conclusion was that a nickel underlayer is not effective in preventing tin whisker growth.

In the latest survey available on different mitigation techniques [32], application of Ni underplate fell under the category of "weak mitigators," while the silver underplate is in the category of "very weak or nonmitigators," which means that the use of nickel underplating on copper substrates is considered to reduce tin whiskers' growth. For the manufacturer, it is important to have part suppliers providing Ni underplating of the desired quality.

However, the base material has been shown to contribute to whisker growth despite the application of a barrier layer. For example, analysis of the data in Ref. [31] indicated that both whisker density and lengths are related to tin thickness. The effectiveness of the underplating appears to depend on the base material as well as the process parameters used to deposit the pure tin finish. At present, it is not clear whether the underplating could permanently stop whisker formation. However, nickel is reported to be the underplating of choice when using a pure tin finish.

4.3.3 Ni UNDERPLATING FOR TIN PLATING ON ALUMINUM

In Section 4.1 it was shown that in applications, where electrical aluminum parts may be exposed to elevated temperatures, using tin plating on Al may lead to severe deterioration of the plating, resulting in a significant increase of electrical resistance. The energized parts with increased resistance generate extra heat, which eventually causes thermal failure.

Because tin-plated aluminum conductive parts and connections remain in wide use in the electrical industry, it is necessary to further improve the thermal performance of tin plating. The role of Ni underplating in improving the ability of tin plating on aluminum to perform well at elevated temperatures has been studied in Ref. [33].

4.3.3.1 Plating, Sample Preparation, and Testing Techniques

The properties of tin plating on Al with Ni underplating have been studied for the specific technique of the plating process currently used in the electrical industry, based on OEM specifications for indoor electrical application (bolted connection) tolerating water condensation.

4.3.3.1.1 Plating Process

Aluminum connection bars have been tin plated using nickel underplating. The thickness of bright tin plating was a minimum of 12 μm. Since tin exhibits good corrosion resistance just as Ni does, 5–6 μm of Ni is usually enough to protect the Al busbar. The thickness of the applied Ni underplating was a minimum of 5 μm, with a maximum thickness of 10 μm to limit CR. In addition, the tin plating process consisted of the following sequence with each step of the process completed with rinsing: degreasing (alkaline etching); deoxidation (acid etching); double activation (zincate); electrolytic nickel plating; and electrolytic tin plating.

4.3.3.1.2 Samples

Ten plated aluminum parts (alloy EN AW-6082) have been randomly taken from one batch of the automatic production line after having passed production tests including visual inspection and fluorescence × thickness measurement. Each of the selected six parts was cut into three parts, two parts for the adhesion test and one for thickness measurement (before and after the accelerated aging test).

4.3.3.1.3 Testing Techniques

To determine thermal properties of tin plating with Ni underplating, the accelerated aging test and the adhesion test have been used in the same sequence as was done in Ref. [10] and described in Section 4.1.

4.3.3.2 Quality of the Plating and Interfaces

SEM/EDS metallographic analysis was performed in Ref. [32] to determine

- Thickness of tin plating and preplating Ni layer
- Presence of pores, voids, and gaps along each interface (Al–Ni, Ni–Sn)
- Morphology of tin plating across the layer
- Chemical composition of plating and preplating layers
- Formation of intermetallics at the Ni–Sn interface

Average thicknesses measured on freshly plated samples was 15 μm for tin plating and 6 μm for nickel underplating [33] according to OEM tin coating specifications.

Nonaged samples demonstrated very good adhesion: tin plating and nickel underlayer are applied onto each other with a perfect fit. There are no voids, pores, or cracks along all interfaces (Al–Ni and Ni–Sn) (Figure 4.7). In addition, all nonaged samples passed the adhesion test: the original coating can be considered to be properly adhesive, and its quality is good.

Samples aged up to 8 weeks at 150°C still exhibited a strong adhesion. The quality of the adhesion of the aged samples was confirmed by the absence of any peeling during the tape adhesion test. According to SEM analysis, there is no evidence of any cracks or separation at the two plating interfaces (Al–Ni and Ni–Sn) (Figure 4.8).

Acc.V	Spot	Magn	Det	WD	Exp	100 μm
15.0 kV	4.6	250×	BSE	9.9	1	

FIGURE 4.7 Microphotograph of aluminum (at the left side of the picture) with Ni underplating layer and tin plating.

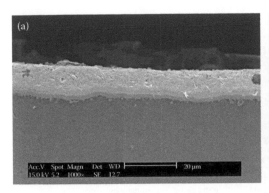
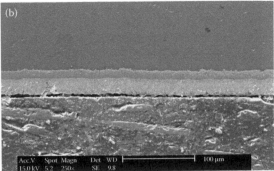

FIGURE 4.8 Microphotograph of Sn plating on Al with Ni underplating: (a) after 3 weeks of thermal aging; (b) after 4 weeks of thermal aging.

4.3.3.3 Formation of Ni–Sn Intermetallics

SEM metallography revealed the formation of an additional layer at the Ni–Sn interface, and x-ray chemical analysis confirmed that this layer is a Ni–Sn IMC. Ni–Sn IMC layer is also present in non-aged (as-received) samples as well, but to a lesser degree. Actually, the IMC layer of freshly plated samples is discontinuous, and its maximum thickness is about 1 μm.

After 1 week of aging, a continuous layer appears at the interface of the tin plating and the Ni underlayer. The thickness of the IMC layer increases with increasing high-temperature exposure. It is found that after 3 weeks of aging the IMC layer stops growing: the thickness of Ni–Sn IMC did not change during the following weeks of thermal exposure [33]. IMC thickness stabilization has already been observed in some studies, and is considered as the result of a diffusion-controlled growth for thick layer [34].

The maximum thickness of the intermetallic layer ranges between 3 and 4 μm, which is comparable with IMC thicknesses measured after about 4-week aging at 120–150°C [22,24]. In addition, no evidence was found that the IMC growth induces a reduction in the thickness of the Ni underplating.

Formation of Al–Ni IMC is also possible on another interface—between Ni layer and the base metal Al. Aluminum–nickel IMC was found in Al to Ni-plated Cu connectors [35], as well as in electronic materials annealed at high temperatures (400°C)—both Ni_2Al_3 and $NiAl_3$ IMCs [36]. However, after 8 weeks of aging, no traces of Al–Ni IMC on the Al–Ni interface could be found [10].

4.3.3.4 Comparison of Aging Behavior of Sn Plating with Ni, Bronze, or Cu Underlayer

A study of the quality of tin plating on aluminum using either bronze or copper as the underplating revealed poor adhesion characteristics after 2 or 3 weeks of exposure at 150°C [10]. The purpose of the study in Ref. [33] was to determine whether nickel underplating might improve the thermal performance of tin plating on aluminum. It was found that Ni underplating outperforms bronze or copper used as underplating in traditional tin plating techniques (Table 4.3). All layers (including Ni–Sn IMC layer) show a very good adhesion after 8-week aging.

The influence of the Ni–Sn formation on the CR (especially for bolted connection) is expected to be quite weak. Although the resistivity of Ni–Sn IMC is about 3–4 times higher than that of nickel [37], a significant increase in CR is more likely to come from plating defects, including cracks and delamination, than from a change in material resistivity, as was suggested in the investigation on tin plating with either bronze or copper underplating [10]. Besides, the results of Refs. [10,34] demonstrate that the accelerated aging test is a very good way of discriminating different plating techniques.

Formation of Ni–Sn IMC is observed in nonaged samples as a discontinuous layer, although the thickness of the IMC is growing with time. However, after 3 weeks at 150°C, the IMC thickness is roughly 3–4 μm and it no longer increases during the next 5 weeks. It appears that the formation of

TABLE 4.3

Results of the SEM Study and the Adhesion Test of Aged Samples Using Bronze, Cu, and Ni as Underplating

Plating Facts			Period of Thermal Aging (weeks)			
Type of Tin Plating and Underplating	Average tin Plating Thickness (μm)	Underplating Thickness (μm)	2	3	4	8
Alstan 80/88, Bronze	11	1.5	Some voids and gaps, adhesion test passed	Multiple pores and gaps along interfaces, adhesion test failed	Multiple voids and gaps along interfaces, coating peeling, adhesion test failed	Failed
Bondal, Cu	5	1	No voids or gaps, adhesion test passed	Some voids and gaps along interfaces, adhesion test passed	Multiple voids and gaps along interfaces, coating peeling, adhesion test failed	Failed
OEM specs, Ni	15	6	No voids or gaps, adhesion test passed	No voids or gaps adhesion test passed	No voids or gaps, adhesion test passed	No voids or gaps, adhesion test passed
Ni–Sn IMC thickness (μm)			2–2.5	3–4	3–4	3–4

Source: Modified from Chudnovsky B., Pavageau V., and Rapeaux M. The quality of tin plating on aluminum exposed to elevated temperatures: The role of underplating, *Proceedings of 24th International Conference on Electrical Contacts (ICEC)*, Saint-Malo, France, pp. 495–500, 2008.

Ni–Sn IMC does not reduce the thickness of the Ni underplating. Stabilization of the IMC thickness was observed in other studies, but is not yet fully understood.

As the data in Table 4.3 show, tin plating on aluminum with Ni as underplating provides an excellent quality of the plating in terms of adhesion and stability even after 8-week thermal aging at 150°C. For this temperature, the thicknesses of tin plating and Ni underplating are high enough to prevent depletion that might induce significant degradation of the plating structure. This is a great improvement in plating quality compared to traditional tin plating techniques on aluminum using either bronze or copper as underplating.

4.3.4 Ni Underplating for Gold Plating

Gold is a noble metal; hence it is very stable; it does not chemically interact with any chemical compound normally found in a connector environment to form an electrically insulating film. Gold is also a good thermal conductor as it is an excellent electrical conductor. For these reasons gold is recommended for use as contact plating in high reliability applications, particularly when the contacts are operating in corrosive environments. Another good quality of gold is that it is not susceptible to fretting degradation and gold contact performance can be enhanced with lubrication.

The protection offered by gold can be compromised if there are pores or cracks in the coating, through which atmospheric pollutants can attack the substrate metal to form insulating corrosion products. A limited amount of porosity can be tolerated depending upon the severity of the contact's service environment.

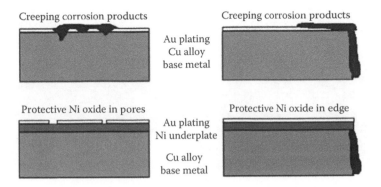

FIGURE 4.9 Passivation of pores and bare edges by the use of a nickel underplating for gold plating. (From Gedeon M. Nickel as an undercoating, *Brushwellman Technical Tidbits*, 4(6), June 2002, Online. Available: https://materion.com/~/media/Files/PDFs/Alloy/Newsletters/Technical%20Tidbits/June02.pdf. With permission.)

Because gold is a noble metal and because thin gold plating tends to be porous, gold coatings are susceptible to the creep of base metal corrosion products across the surface of the gold after formation at pore sites and edge boundaries. Corrosion creep can be inhibited by applying a nickel underplating prior to application of the gold. The nickel underlayer serves as a diffusion barrier to most base metal constituents, and nickel corrosion products are self-limiting, passive, and not susceptible to corrosion creep. A nickel underlayer can act as an inhibitor of the pore and creep corrosion, and a diffusion barrier for gold [38,39].

An illustration of how Ni underplating works against pore and creep corrosion for gold plating is shown in Figure 4.9 [38].

Pore-Corrosion Inhibitor—Nickel underplating forms passive oxides at the base of pores. The underplating serves as a pore-corrosion inhibitor provided that the environment does not contain significant amounts of acidic pollutants, such as chlorine.

Corrosion Creep Inhibitor—Nickel acts as a barrier to corrosion product migration.

Diffusion Barrier—Nickel underplating slows down the diffusion of base metal constituents of the contact spring (such as copper and zinc) to the gold surface, where they could oxidize and form an insulating film of corrosion product.

Mechanically Supporting Underlayer for Contacting Surfaces—A relatively hard nickel underplating serves as a mechanical support for the gold coating, and increase its durability.

The nickel underplating must be continuous and have sufficient thickness to perform the particular function for which it is intended. As a general rule, a minimum thickness of 1.25 μm (50 μin.) of nickel should be used for plating under gold.

4.4   GALVANIC CORROSION: CONNECTIONS MADE OF DISSIMILAR METALS

4.4.1   HAZARD: GALVANIC CORROSION

The corrosion characteristics of a metallic material are very important in electrical design. A number of different mechanisms cause corrosion, such as when the metal enters a chemical reaction with other elements such as oxygen, chlorine, or sulfur. One of the most common mechanisms of corrosion, namely *electrochemical* or *galvanic corrosion*, can develop when two *dissimilar metals* are electrically connected and exposed to an *electrolyte*.

An electrolyte is an electrical conducting solution in which positively charged and negatively charged ions can move freely. Acids, alkalis, and salt solutions, such as sea water, make good

FIGURE 4.10 Galvanic corrosion of dissimilar metals. ((a) From http://www.interconnectionworld.com/index/display/article-display/194291/articles/connector-specifier/volume-19/issue-12/features/electrical-connections-what-you-can-do-to-prevent-corrosion.html. With permission. (b) From http://corrosion.ksc.nasa.gov/images/gal3.jpg, courtesy of NASA.)

electrolytes. The electrolyte may be present in bulk as when immersed in sea water or as a thin condensed film of water on the metal surface. In moist industrial areas harmful electrolytes are formed by the absorption of gases by rain and fog to form acids and salts, while in seaside areas they are formed by the combination of salt with the moisture to form saline solutions.

Electrical interconnect components are subject to galvanic corrosion due to the incompatible coupling of plating materials and base metals. Examples of galvanic corrosion when two dissimilar metals connected together are shown in Figure 4.10.

4.4.2 Definition of Dissimilar Metals

Each metal is characterized by its electrical potential. The varying potential indexes of some metals and alloys in the electrolyte are shown in the *galvanic series* in Table 4.4. Those metals which are on the top of the table with small values of electrical potential are considered as noble or *cathodic*, and the metals and alloys in the lower part of the table are less noble or *anodic* than the metals above them. The *anodic* metal will corrode and the *cathodic* metal (more noble) is protected from corrosion.

Whenever dissimilar metals are in the presence of an electrolyte, a difference in electrical potential is developed. One metal becomes the cathode and receives a positive charge. The other becomes the anode and receives a negative charge.

In the presence of an electrolyte a galvanic cell is thus set up where one of the metals will act as an anode and the other will act as a cathode according to the electrical potential of each metal. When these metals are in contact, an electrical current will flow, as in the case of any short circuited electric cell. This electrolytic action causes an attack on the anodic metal, leaving the cathodic metal unharmed.

Generally, the closer one metal is to another in the series, the more compatible they will be; that is, the galvanic effects will be minimal. On the other hand, the farther one metal is from another, the greater the corrosion will be.

The extent of the attack is proportional to the strength of the electrolytic current, which in turn is proportional to the electrical potential difference developed. The magnitude of the potential difference generated between two dissimilar metals can be seen from the position of these metals in the galvanic series. When two metals are in contact in an electrolyte, the one higher up in this series is the anode, the corroded metal, whereas the one lower is the cathode, the protected metal.

The further apart the metals are in this series, the greater the electrolytic potential difference and the greater the attack on the anodic metal. General rankings of the galvanic corrosion susceptibility

TABLE 4.4

Electrical Potential Index (V) for Metals and Alloys

Metals, Alloys	Index (V)
Cathodic or Noble End	
Gold, solid and plated, gold–platinum alloy	0.00
Silver, solid or plated	0.15
High nickel–copper alloys	0.15
Nickel, solid or plated, Monel	0.30
Copper, solid or plated; low brasses or bronzes; copper–nickel alloys; nickel–chromium alloys	0.35
Brass and bronzes	0.40
High brasses and bronzes	0.45
18% chromium-type corrosion-resistant steels	0.50
Chromium plated; tin plated; 12% chromium-type corrosion-resistant steels	0.60
Tin-plate; tin–lead solder	0.65
Lead, solid or plated; high lead alloys	0.70
Aluminum, wrought alloys of the 2000 Series	0.75
Iron, wrought, gray or malleable, plain carbon, and low alloy steels	0.85
Aluminum, wrought alloys other than 2000 Series aluminum, cast alloys of the silicon type	0.90
Aluminum, cast alloys other than silicon type, cadmium, plated and chromate	0.95
Hot-dip-zinc plate; galvanized steel	1.20
Zinc, wrought; zinc-base die-casting alloys; zinc plated	1.25
Anodic or Active End	

Source: Adapted from Army Missile Command Report RS-TR-67–11, *Practical Galvanic Series.*

of different contacting material combinations show that silver-to-gold and silver-to-tin are satisfactory combinations in many environments; with silver-to-tin presenting the most risk in harsher environments.

Galvanic series relationships are useful as a guide for selecting metals to be joined. It will help to select metals as close as possible to each other in galvanic series. These metals which will have a minimal tendency to corrode if connected. The positions of the coupled metals in galvanic series will also indicate the need or degree of protection to be applied to lessen the expected galvanic interaction (Military Standard 889B [Notice 3], "Dissimilar Metals," May 1993).

4.4.3   GALVANIC CORROSION OF COPPER-TO-ALUMINUM CONNECTIONS

The two types of corrosion that exhibit the greatest influence on the electrical properties of a metal are oxidation/atmospheric corrosion and galvanic corrosion. Both copper and aluminum have a high resistance to oxidation because of the manner in which their surfaces oxidize. In the case of these two metals a thin oxide film is formed on their surfaces. This film becomes a protective coating, which offers resistance to any further oxidation. In corrosive atmospheres copper is usually protected by corrosion-resistive plating such as silver or tin.

Since most of the commonly used electrical conductors are resistant or protected from oxidation, galvanic corrosion becomes very important when bare copper conductor is connected to bare aluminum conductor. Copper and aluminum are quite far apart in the series (Table 4.4), copper being *cathodic* and aluminum being *anodic*. Hence, when aluminum and copper are in contact in an electrolyte, the aluminum can be expected to be severely attacked.

Aluminum will, in contact with copper and in the presence of an electrolyte, corrode progressively and the process will continue as long as the electrolyte is present or until all the aluminum has been consumed, even though the build-up of corrosion products may limit the rate of corrosion at the surface. The performance of aluminum-to-copper connection is affected by corrosion in two ways. Because of corrosion, the contact area is drastically reduced, causing an electrical failure. Severely corroded connector may cause a mechanical failure as well. In most cases, failure is due to a combination of both effects.

Several conditions must be observed in selecting the most appropriate connection, where it consists of a junction between aluminum and copper conductors. Since aluminum has a high electrical potential compared with that of copper, it will corrode progressively in contact with copper and in the presence of an electrolyte until contact between the two metals is destroyed. It is an undesirable practice to employ a copper connector for use with aluminum because of the possibility of galvanic corrosion.

It is worth noting that there is another issue with copper-to-aluminum connections in addition to possible galvanic corrosion. The use of a copper-to-aluminum connector could result in a resistance increase and premature joint deterioration, and the reason for the insufficient performance of such a connector is that there is a difference in thermal expansion between these two metals.

Aluminum has a higher coefficient of thermal expansion than copper, and for a given TR, aluminum will expand more than an equal size copper member. Therefore, as the temperature rises, both copper and aluminum expand, but to a different degree.

Copper, not being able to expand as much as aluminum, restricts aluminum and causes it to flow. Upon cooling, a dimensional difference begins to create a greater electrical resistance and higher TR. As the repeated cycles continue, resistance and temperature become progressively greater until ultimate failure occurs.

In general, the use of a copper-to-aluminum connector could result in a resistance increase and early joint deterioration: (1) due to the presence of electrolytic fluids in the environment and (2) due to variations in the connection temperature. Therefore, a bare copper-to-aluminum connection may be used only if there is absolutely no other option available.

4.4.4 Protection of Copper-to-Aluminum Connections from Galvanic Corrosion

There could be situations when the use of copper-to-aluminum connections in unavoidable. In those cases the contact area should be coated with either protective plating or a protective compound/lubricant. Either option will slow down galvanic corrosion, which otherwise will cause a connection failure.

The most important factor for galvanic corrosion to exist is humidity. In order to limit the detrimental effect of galvanic action in corrosive environments and maintain a low CR, various palliative measures may be recommended, such as plating of Cu and Al buses to suppress the galvanic corrosion of the aluminum-to-copper joint and utilization of proper metallic fasteners. Application of contact aid compounds prevents ingress of oxygen and humidity inside the contact interface [40].

4.4.4.1 Plated Aluminum Connections

To minimize galvanic corrosion of aluminum-to-copper connectors, a protective coating at the contacting interface must be introduced. Plating has been used as a means of making an aluminum connector suitable for a copper conductor. The plating of aluminum is much more critical than the plating of a more noble metal such as copper.

To be effective in reducing galvanic corrosion between the copper conductor and the aluminum connector, the plated metal must be closer in the electrolytic series to copper than aluminum is. Therefore, it must be cathodic to aluminum. Since porosity and minor scratches are always present, galvanic action can be expected in the presence of moisture, resulting in an attack on the aluminum under the plating. Corrosion tests reveal an attack in the form of a mottled appearance and flaking of the plating.

In addition, the presence of plated metal can cause a galvanic attack of the aluminum conductor, thus reducing the protection offered to this conductor in an aluminum connector. The question of what is the most effective coating for protection of aluminum from galvanic corrosion is still being discussed and studied, but many recommend performing tin plating of a uniform thickness over the entire surface of the contact. Tin serves as a suitable intermediate metal between copper and aluminum. Copper and zinc platings have also been in use.

4.4.4.2　Fasteners

A plated copper bar or a plated copper terminal fitting may be connected to a plated aluminum bar. Metallic fasteners shall be chosen to minimize dissimilar metal corrosion. When it is not possible to avoid joining dissimilar metals, the metal of the fastener shall be from a higher potential group than the structure [41,42]. The connection should be made with a plated steel bolt in conjunction with plated Belleville spring washers and wide series plated steel washers [43,44].

4.4.4.3　Corrosion Protective Compound for Copper-to-Aluminum Connections

Some products, such as petrolatum-type compound containing zinc dust, effectively protect contacts made of dissimilar metals from galvanic corrosion. However, some lubricants may induce galvanic corrosion; therefore, the products for this application should be chosen very carefully. For example, the use of graphite, which has a high noble potential, is not recommended for copper-to-aluminum connections, because it may cause severe galvanic corrosion of copper alloys in a saline environment [45].

A detailed study was performed in Ref. [46] to evaluate the protective properties of various compounds available commercially and recommended for electrical connections. Some compounds provide a good corrosion protection for copper-to-aluminum connections in a wide working temperature range from –25 to +130°C.

When aluminum connectors or conductors are involved, proper cleaning of the aluminum as well as the use of a good connector lubricant or compound are essential for trouble-free service. Hard and nonconductive aluminum oxide, developing on the surface of bare aluminum conductors exposed to air, must be removed from the Al surface prior to making a connection. In all cases, the aluminum conductor should be cleaned by means such as scratch brushing and immediately coated with the connector compound [40].

A generous amount of joint compound should be applied to all joint surfaces before assembly to seal out air and improve corrosion resistance. A bead of compound should appear all around the edges of the joint when the connection is tightened. Excess sealant squeezed out of the joint may be left as is or removed [43].

4.4.5　Galvanic Corrosion in Steel Connections with Aluminum and Other Metals

The connection between metallic parts made of aluminum and steel alloys is possible in electrical installations. A chart that shows electrical potentials for different aluminum and steel alloys can be found in Ref. [47]. Stainless steel alloys can significantly change their potential and become much more active if exposed to stagnant or poorly aerated water. The result of these stagnant conditions is oxygen depletion and the low noble potential, which can make the stainless steel susceptible to corrosion in conditions which might otherwise be considered noncorrosive.

To initiate galvanic corrosion, the two alloys of steel or aluminum and steel must be in electrical contact with each other. The two metals can be bolted, welded, or clamped together, or they can even be just resting against each other. The metal junction must be bridged by an electrolyte, and almost any fluid falls under this category, except for distilled water. Even rain water is likely to become sufficiently conducting after contact with common environmental contaminants.

If the conductivity of the liquid is high (a common example is sea water) the galvanic corrosion of the less noble metal will be spread over a larger area; in low-conductivity liquids the corrosion will be localized to the part of the less noble metal near the junction. An example of such a connection exposed to galvanic corrosion is shown in Figure 4.11.

FIGURE 4.11   Galvanic corrosion in connection between steel and brass plate (a) and stainless-steel bolt and aluminum (b). (From http://corrosion.ksc.nasa.gov/galcorr.htm, courtesy of NASA.)

In this example, a stainless steel bolt is connected to an aluminum plate (Figure 4.11). If it collects water in the corner at the edge of the bolt but the remainder of the plate remains dry, the effective area of the less noble aluminum is only the wetted region; it is quite possible for the aluminum plate to be attacked by galvanic corrosion in the region immediately surrounding the bolt.

To prevent galvanic corrosion in aluminum-to-steel connections [48] one should be particularly aware of zinc-plated or galvanized fasteners in stainless steel sheets—a common substitution because of apparent cost saving or better availability. These less noble fasteners are likely to be rapidly attacked by galvanic corrosion. It is often practical to prevent electrical contact between the dissimilar metals. This may be achieved by the use of nonconducting (e.g., rubber or plastic) spacers, spool pieces or gaskets, perhaps in conjunction with sleeves around bolts.

4.4.6   General Precautions to Minimize Galvanic Corrosion in Connections

The following precautions and joining methods are recommended for galvanic corrosion mitigation, whenever it becomes necessary to assemble relatively incompatible metals.

Recommendations for electrical connections:

- The metal couples design should provide that the area of the cathodic member of the connection is considerably smaller than the area of the anodic member. For example, bolts or screws of stainless steel should be used for fastening aluminum sheet, but not vice versa.
- A compatible metallic gasket or washer should be inserted between the dissimilar metals prior to fastening.
- The cathodic member should be plated with a metal compatible with the metal of the anodic member.
- A proper electrically conductive sealant should be selected for the connection.

Recommendations for nonelectrical connections: A nonabsorbing, inert gasket material or washer should be inserted between the dissimilar materials prior to connecting them.

Other precautions:

- Seal all edges of a pair of surfaces made of metals far from each other in galvanic series that are in contact, or in close proximity, and are to be connected, to prevent the entrance of liquids.
- Apply corrosion-inhibiting pastes or compounds under heads of screws or bolts inserted into dissimilar metal surfaces whether or not the fasteners had previously been plated or otherwise treated.

- In some instances, it may be sufficient to apply an organic coating to the faying surfaces prior to assembly. This would be applicable to joints which are not required to be electrically conductive.
- In the cases where it is practical or will not interfere with the proposed use of the assembly, the external joint should be coated externally with an effective paint.

4.5 OTHER DETRIMENTAL PROCESSES AFFECTING PLATING PERFORMANCE

4.5.1 INTERMETALLIC COMPOUNDS

IMC, by definition, is a combination of solid phases containing two or more metallic elements, with optionally one or more nonmetallic elements, whose crystal structure and properties differ from that of the other constituents.

The growth of IMC is often found in those parts having a metallic layer plated over a base metal. The formation of IMC may significantly affect the electrical properties and reliability of electrical connections, mostly in electronic devices [49], particularly when it is found in the solder joints. However, in some cases the resistance of plated electrical parts in electrical equipment may also be increased due to the formation of IMC. Many dissimilar metals in close contact will diffuse and form IMC. An IMC consists of two or more elements, always with the same precise ratio of atoms.

4.5.1.1 Copper–Tin Intermetallic Compounds

Every tin-plated copper alloy experiences the formation of two types of copper–tin IMC—Cu_6Sn_5 and Cu_3Sn—at the interface of the tin and the base metal. For these two copper–tin IMCs, the ratios between Sn and Cu atoms are 6:5 and 3:1, correspondingly.

The compound initially forms at the interface between the plating and the base metal and grows until eventually all the tin is consumed. The rate at which intermetallic Cu–Sn compounds are growing depends on time and temperature. Copper–tin IMC are hard and brittle; at the surface they are easily oxidized. The oxidized IMC can adversely affect CR and solderability.

Because the different types of tin plating are produced in different ways, with different thickness, their IMC reach the metal surface at different times [45]. Several studies have shown that, during the manufacturing process, commercial *hot-dipped tin* quickly forms an intermetallic layer with the thickness of 0.5–1 μm (20–40 μin.). Pockets or layers of intermetallic have been found on rather "fresh" commercial *hot-dipped tin*. A stringent connector performance requires thicker coatings.

With *air leveled tin* in which the tin is applied from a molten bath, there is an immediate intermetallic layer at the tin interface of about the same thickness as commercial *hot-dipped tin*. But the air jets allow a thicker tin layer to be produced; the *air leveled tin* surface has an abundance of residual ("pure") tin just after it is produced.

Because electroplated tin is not produced with hot molten tin, little or no intermetallic is formed at the copper–tin interface during the plating process. The structure of *electroplated tin* exhibits a finer grain size and tends to be more porous than the essentially "cast" structure of the "molten method" products. This affects the growth rate of the intermetallic. *Reflow tin* starts as electroplated tin so there is essentially no intermetallic at this step. But the furnace treatment to melt the tin initiates a thin layer of intermetallic about the same thickness as that initially generated by the molten tin processes.

4.5.1.2 Effects of Temperature and Time on the Formation of Cu–Sn IMC

Copper–tin IMCs form at the interface of the tin plate and the base metal, for all the different processes used for the tin-coated copper alloy strip. With increasing time and temperature, the intermetallic grows to the surface, becomes oxidized, and can affect contact integrity [50]. A schematic illustration of the process [51] is shown in Figure 4.12.

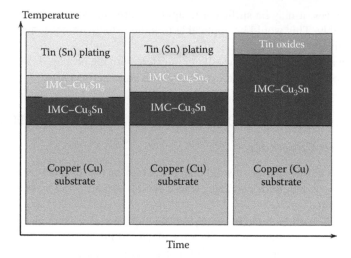

FIGURE 4.12 Intermetallic growth on tin-coated copper alloy strip. (Modified from *Copper–Tin Intermetallic Compounds*, Online. Available: https://www.copper.org/applications/industrial/DesignGuide/performance/coppertin03.html.)

Because commercial *hot-dipped tin* already has some IMC at the surface when produced, its service life is likely to be the shortest. After 3 months in storage at summer temperatures, practically all of the surface could be composed of intermetallics.

The contacts with electroplated tin, relatively free of intermetallic when plated, offer a longer service life. But after 4 weeks (without elevated temperature) IMC of about 25 μin. thickness can be detected at the copper–tin interface. This too will eventually grow to the surface. The growth rate of the intermetallic on electroplated tin is more rapid than that on *air leveled tin* and *reflow tin* (and commercial *hot dipped tin*). Their cast micro structures with big grains inhibit intermetallic growth at moderate temperatures. This provides longer service time or a thinner tin plating can be used for an equivalent performance.

Intermetallic growth can be retarded by using a "barrier metal" (a metal that diffuses much more slowly with the base alloy and the tin). Nickel has long been plated under gold to act as this diffusion barrier and is similarly effective in tin plating (see Section 4.3). Nickel underplating is often considered as means of slowing down or reducing the formation of CuSn IMC. However, as shown in Section 4.3.3, it is not a complete solution to the problem since other IMC may grow on the interfaces, such as Ni–Sn IMC.

A barrier layer may significantly slow down the diffusion of elements other than copper, for example, zinc (zinc oxides at the contact surface are resistive). It was found that a copper underplating will prevent or impede zinc from reaching the surface; 0.5 μm (20 μin.) is often used on *reflow tin* products. For brasses, 2.5 μm (100 μin.) of copper is effective. Diffusion barriers are often used on *electrotinned* and *reflow tin*. Incorporation of a barrier for commercial *hot-dipped tin* and *air leveled tin* would require a separate operation [50].

4.5.1.3 Resistance of the Contacts with Tin Coating

Tin is characterized by high electrical conductivity, excellent solderability, and relatively low cost. Tin plating is accepted as a method of preventing contamination from the base metal. Tin softness makes it easy to break up films with minute sliding action and relatively light loads. The different kinds of tin coatings have different rates of degradation; some have limited tin thickness, some are less porous, some diffuse faster, and some embody diffusion barriers. Some of these characteristics and all of these differences affect CR.

CR of tin coatings generally increases with exposure to elevated temperatures. For example, 50 h of exposure of tin plating to a temperature of 125°C does not significantly change the resistance of the coating, but exposure to the same temperature for ~120 h increases the resistance up to 12 µΩ, and after 250 h of exposure the resistance of tin plating reaches 105 µΩ. After 250 h, the tin on all the samples had formed IMC [52]. Yet there were distinct differences; oxides, films, and surface contaminants from the tin and/or base metal are likely causes. An effective underplating (such as nickel) with a "thick" tin plate would provide better performance.

4.5.2 FRETTING CORROSION AND A MEANS OF PROTECTION

Fretting is a term to describe a small-amplitude rubbing motion between two surfaces. By definition, the amount of relative movement between the contacting surfaces is about 10–100 µm for fretting. Since 100 µm is a thickness of a sheet of paper, fretting motion is pretty small and hard to see. Vibration and thermal expansion/contraction are the main sources that create fretting motion.

Fretting corrosion is a form of accelerated atmospheric oxidation which occurs at the interface of contacting materials undergoing slight, cyclic relative motion. Fretting and fretting wear can occur on nonelectrical things such as bearings, but fretting corrosion is often associated with electrical contacts [49,53].

4.5.2.1 Fretting Corrosion of Electrical Contacts

A major reason for contact failure with tin surfaces is because of fretting corrosion, a deteriorating process caused by small but repeated micro-motion between contacting surfaces. A higher contact force is effective in minimizing this motion (and thus the fretting corrosion), but cannot be used unless a subsequent increase in the insertion/withdrawal force is acceptable.

In electrical contacts involving nonnoble metals, fretting action can cause rapid increases in CR, which in the worst cases proceeds to virtual open circuits in a matter of minutes.

Fretting corrosion is a build-up of insulating, oxidized wear debris that can form when there is small-amplitude fretting motion between electrical contacts. The oxidized wear debris can pile up sufficiently at the electrical contact spots that the electrical resistance across the connection increases significantly. Figure 4.13 shows a magnified, cross-sectional view of the fretting corrosion sequence on a tin-plated terminal contact interface [54].

At the beginning, the thin tin oxide layer cracks under pressure when contact is first made (Step 1). The tin oxide layer is represented as a thin black line (A). Nice, clean tin (B) penetrates through

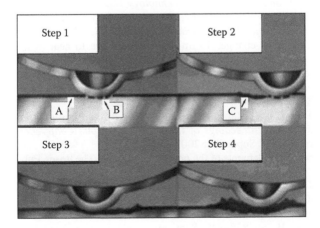

FIGURE 4.13 Fretting corrosion sequence on a tin-plated electrical contact. (From *What is Fretting Corrosion*, Delphi, Online. Available: http://www.hlinstruments.com/RoHS_articles/Reliability%20articles/ What%20is%20Fretting%20Corrosion%20-%20Delphi.pdf. With permission.)

the cracks in the tin oxide layer, making a very stable electrical connection. If the contact spot moves, a microscopic amount from fretting motion (Step 2), the clean tin is exposed to the air and quickly forms an insulating tin oxide film (C).

Every time there is motion at the contact spot, the cycle repeats and more tin oxide wear debris builds up, as shown in Step 3. With continued microscopic fretting movement between contacting surfaces, sufficient insulating tin oxide wear debris can build up so that high resistance and/or intermittent connections can develop (Step 4).

Fretting action seems to produce a surface condition that depends on the hardness of the material. The softer materials suffer more surface damage, pitting, material loss, and generation of wear debris. The harder materials suffer very little damage and not much wear debris is generated. It was shown in Ref. [54] that CR increases at very rapid rates under conditions of fretting corrosion.

All of the more commonly used nonnoble contact materials, such as tin (dull and bright tin, tin alloys) and nickel, experience large increases in resistance during fretting action (small-amplitude cyclic motion). The use of such materials should always be analyzed in the applications where fretting corrosion could occur on electrical contacts as a result of vibration, mechanical motion, or DTE.

Lubrication has been identified as a very good way of preventing fretting corrosion [43]. The use of lubricants helps in wear reduction; it aids in the self-cleaning and flushing of wear debris from contacts. Finally and perhaps most importantly, the advantage of lubrication is that liquid lubricants have the ability to exclude oxygen from the contact, thus preventing oxidation of wear debris and freshly exposed contact material. A more detailed description of the use of lubrication for protecting electrical contacts from fretting corrosion is given in Chapter 6.

REFERENCES

1. Whitley J. H. The tin commandments, *Plating and Surface Finishing*, 68(10), 38–39, October 1981.
2. *The Tin Commandments: Guidelines for the Use of Tin on Connector Contacts*, AMC Technical Report, July 1996.
3. Abbott W. H. and Whitley J. H. The lubrication and environmental protection of alternatives to gold for electronic connectors, *Proceedings of the 21st Annual Holm Seminar on Electrical Contacts*, Chicago, IL, pp. 9–16, 1975.
4. Whitley J. H. *Connector Surface Plating: A Discussion of Gold and the Alternatives,* AMP Publication No. EN114, AMP Incorporated, 1980, 14pp.
5. Braunovic M. Effect of fretting on the contact resistance of aluminum with different contact materials, *IEEE Transactions on Components, Hybrids, and Manufacturing Technology* 2(1), 25–31, March 1979.
6. Braunovic M. Fretting damage in tin-plated aluminum and copper connectors, *IEEE Transactions on Components, Hybrids, and Manufacturing Technology* 12(2), 215–223, June 1989.
7. Braunovic M. Evaluation of different platings for aluminum-to-copper connections, *IEEE Transactions on Components, Hybrids, and Manufacturing Technology* 15(2), 204–215, April 1992.
8. Geckle R. J. Metallurgical changes in tin-lead platings due to heat aging, *IEEE Transactions on Components, Hybrids, and Manufacturing Technology* 14(4), 691–697, December 1991.
9. Braunovic M. Effect of intermetallic phases on the performance of tin-plated copper connections and conductors, *Proceeedings of the Forty-Ninth IEEE Holm Conference on Electrical Contacts*, Chicago, IL, pp. 124–131, September 2000.
10. Chudnovsky B., Pavageau V., Rapeaux M., Zolfaghari A., and Bardollet P. Thermal aging study of tin plating on aluminum, *Proceedings of the 53rd IEEE Holm Conference on Electrical Contacts*, Pittsburgh, PA, pp. 124–131, September 2007.
11. Levy J. Understanding the elements of the periodic table, *Tin*, The Rosen Publishing Group, New York, 48 pp., 2009.
12. Karlya Y., Gagg C., and Plumbridge W. J. Tin pest in lead-free solders, *Solder. Surf. Mount Technol.* 13(1), 39–40, 2000.
13. Lasky R. C. Tin pest: A forgotten issue in lead-free soldering? *Proceedings of the SMTA International Conference,* Chicago, IL, pp. 838–840, September 2004. Online. Available: http://www.indium.com/images/blogs/drlasky/files/TinPestPaper0723Final.pdf.
14. Burns N. D. A tin pest failure, *J. Failure Anal. Prev.* 9(5), 461–465, 2009.

15. *Tin Pest Control*, National Physics Laboratory, Online. Available: http://www.npl.co.uk/science-technology/electronics-interconnection/tin-pest-control.

16. Restriction of the Use of Certain Hazardous Substances in Electrical and Electronic Equipment (RoHS), Directive 2002/95/EC of the European Parliament and of the council of 27 January 2003, *Official Journal of European Union*, 02/13/2003, Online. Available: https://ec.europa.eu/growth/single-market/european-standards/harmonised-standards/restriction-of-hazardous-substances_en.

17. Plumbridge W. Recent observations on tin pest formation in solder alloys, *J. Electron. Mater.* 37(2), 218–223, 2008.

18. Plumbridge W., Gagg C., Williams N., and Karlya Y. Tin pest in Sn-05wt.% Cu lead-free solder, *J. Met.* 53(6), 39–41, 2001.

19. Antler M. Gold plating contacts, effect of thermal aging on contact resistance, *Proceedings of the IEEE Holm Conference on Electrical Contacts*, Philadelphia, PA, pp. 121–131, 1997.

20. Pinnel M. R. and Bennett J. E. Qualitative observation on the diffusion of copper and gold through a nickel barrier, *Met. Trans. A.* 7619–7635, 1976.

21. Lau J. H. and Liu K. Global trends in lead-free soldering, solid state technology, Part 1, *Solid State Technology*, 13(1), January 2004, Online. Available: http://electroiq.com/blog/2004/01/global-trends-in-lead-free-soldering/.

22. Moller P. D. Plating on aluminum, *TALAT Lecture 5205*, Dansk Technisca Hoogschol, Lyngby, 19pp., 1994, Online. Available: http://core.materials.ac.uk/repository/eaa/talat/5205.pdf.

23. Gowman L. P. Plating on aluminum by the Alstan 70 process, *Electroplating and Metal Finishing*, 20(3), 81–85, 1967.

24. Chan Y. C., Tu P. L., Tang C. W., Hung K. C., and Lai J. K. L. Reliability studies of BGA solder joints—Effect of Ni-Sn intermetallic compound, *IEEE Transactions on Advanced Packaging*, 24(1), 25–32, February 2001.

25. Geckle R. J. Metallurgical changes in tin-lead platings due to heat aging, *IEEE Proceedings of the forty first Conference on Electronic Components and Technology*, Atlanta, GA, pp. 218–228, 1991.

26. Danielsson H. Lead-free soldering causes reliability risks for systems with harsh environments, *Adv. Microelectron.* 29(3), 6–11, May/June, 2002.

27. *Mitigation Strategies for Tin Whiskers, Prepared by* M. Osterman, CALCE-EPSC, update 8-28-02, Online. Available: http://www.calce.umd.edu/tin-whiskers/TINWHISKERMITIGATION.pdf.

28. Zhang Y., Xu C., Fan C., Abys J. A., and Vysotskaya A. Understanding whisker phenomenon whisker index and tin/copper, tin/nickel interface, *Proceedings of the APEX Expo*, San Diego, CA, pp. 506-1-1–506-1-10, January 2002.

29. Horvath B., Illes B., and Harsanyi G. Investigation of tin whisker growth: The effects of Ni and Ag underplates, *Proceedings of the thirty second International Spring Seminar on Electronics Technology*, Brno, Czech Republic, pp. 1–5, 3–17, May 2009.

30. Brusse J., Ewell G., and Siplon J. Tin whiskers: Attributes and mitigation, *Capacitor and Resistor Technology Symposium (CARTS)*, New Orleans, LA, pp. 68–80, March 25–29, 2002.

31. Panashchenko L. and Osterman M. Examination of nickel underlayer as a tin whisker mitigator, *IEEE ECTC Conference*, San Diego, CA, pp. 1037–1043, May 2009.

32. Touw Anduin (Ed). *Results of Mitigation Effectiveness Survey & Plans for GEIA-STD-0005-2 Revision*, Boeing Company, 6-24-2010, Online. Available: http://www.calce.umd.edu/tin-whiskers/MitigationSurvey.pdf.

33. Chudnovsky B., Pavageau V., and Rapeaux M. The quality of tin plating on aluminum exposed to elevated temperatures: The role of underplating, *Proceedings of 24th International Conference on Electrical Contacts (ICEC)*, Saint-Malo, France, pp. 495–500, 2008.

34. Jeon Y. D., Paik K.-W., Bok K.-S., Choi W.-S., and Cho C.-L. Studies of EN under bump metallurgy solder interfacial reaction and their effects on flip chip solder joint reliability, *J. Electron. Mater.* 31(5), 520–528, 2002.

35. Oberg A., Gustafsson R., Saksvik O., Stomberg H., and Olsson K.-E. The ageing physics of electrical contacts subjected to DC current, *Proceedings of the Forty Second IEEE Holm Conference Joint with the 18 International Conference on Electrical Contacts*, Chicago, IL, pp. 189–194, 1996.

36. Liu W. C., Chen S. W., and Chen C.-M. The Al/Ni interfacial reactions under the influence of electric current, *J. Electron. Mater.* 27(1), L-6–L-9, 1998.

37. Frear D. R., Burchett S. N., Morgan H. S., and Lau J. H. (Eds). *The Mechanics of Solder Alloy Interconnects*, Van Nostrand Reinhold Publishing, New York, Chapter 3, p. 60, 1994.

38. Gedeon M. Nickel as an undercoating, *Brushwellman Technical Tidbits*, 4(6), June 2002, Online. Available: https://materion.com/~/media/Files/PDFs/Alloy/Newsletters/Technical%20Tidbits/June02.pdf.

39. *Golden Rules: Guidelines for the Use of Gold on Connector Contacts*, AMP Incorporated, Tech. Report, Contact Physics Research, Tyco Electronic Corporation, 11pp., 1996, Online. Available: www.te.com/documentation/whitepapers/pdf/aurulrep.pdf.

40. Basic connection principles, In *Burndy Reference*, pp. O2–O5, Online. Available: http://ecat.burndy.com/Comergent/burndy/documentation/section%20o-reference.pdf.

41. NASA Standard KSC-E-165D Specification for electrical ground support equipment fabrication, *NASA Engineering Development Directorate*, January 2, 2009, Online. Available: http://everyspec.com/NASA/NASA-KSC/KSC-SPEC/KSC-E-165D_7116/.

42. *NASA Standard STD PO23 Electrical Bonding for NASA Launch Vehicles, Spacecraft, Payloads, and Flight Equipment*, Online. Available: http://everyspec.com/NASA/NASA-NASA-STD/NASA-STD-P023_DRAFT_16586/.

43. Aluminum Bus Bar, *US Department of Transportation, US Coast Guard Circular 02-79,* March 1979, Online. Available: http://www.uscg.mil/hq/cg5/NVIC/pdf/1970s/n2-79.pdf.

44. *Low-Voltage Electrical Power Conductors and Cables (600 Volts and below)*, US Department of Veterans Affair, Technical Information Library, NCA Master Construction Specifications, Electrical, Specs 26-05-21, Online. Available: http://fwcontracting.com/home/wp-content/uploads/2010/11/260521-LOW-VOLTAGE-ELECTRICAL-POWER-CONDUCTORS-AND-CABLES-6.pdf.

45. *Engineering and Design—Lubricants and Hydraulic Fluids*, US Army Corp of Engineers (USACE), Publication No. EM 1110-2-1424, 197pp., Online. Available: http://www.publications.usace.army.mil/Portals/76/Publications/EngineerManuals/EM_1110-2-1424.pdf?ver=2016-02-25-125404-800.

46. Braunovic M. Further studies of different contact aid compounds for aluminum-to-copper connections, *Proceedings of the Forty-Fifth IEEE Holm Conference on Electrical Contacts*, Pittsburgh, PA, pp. 53–62, 1999.

47. LaQue F. L. *Marine Corrosion: Causes and Prevention*, John Wiley and Sons, New York, 179pp., 1975.

48. Courval G., Allin J., and Doyle D. Galvanic corrosion prevention of steel-aluminum couples, SAE Technical Paper 932357, *SAE Automotive Corrosion and Prevention Conference and Exposition*, Warrendale, PA, August 1993.

49. Braunovic M., Konchits V. V., and Myshkin N. K. Intermetallics, in *Electrical Contacts: Fundamentals, Applications and Technology*, CRC Press, Boca Raton, FL, 645pp. 2007.

50. *Effect of Time and Temperature on Copper-Tin*, Online. Available: http://www.copper.org/applications/industrial/DesignGuide/timecompounds03.html.

51. *Copper-Tin Intermetallic Compounds*, Online. Available: https://www.copper.org/applications/industrial/DesignGuide/performance/coppertin03.html.

52. *CR When Using Tin Coatings*, Online. Available: http://www.copper.org/applications/industrial/DesignGuide/contact03.html.

53. Bock E. M. and Whitley J. H. Fretting corrosion in electrical contacts, *Proceedings of the Twentieth Annual Holm Seminar on Electrical Contacts*, Chicago, IL, pp. 154–174, October 29–31, 1974.

54. *What is Fretting Corrosion*, Delphi, Online. Available: http://www.hlinstruments.com/RoHS_articles/Reliability%20articles/What%20is%20Fretting%20Corrosion%20-%20Delphi.pdf.

5 Electrical Equipment in a Corrosive Environment

Electrical equipment installed at industrial or commercial facilities could be exposed to both indoor and outdoor environment. Composed of different kinds of materials, both metals and insulators, electrical equipment are subjected to deteriorating corrosion processes, the rate of which strongly depends on the presence of various gases, vapors, and particles and combinations thereof in the surrounding air [1]. The following sections focus on different corrosive factors and the possible response of each of the materials that constitute the equipment to the atmospheric exposure.

5.1 CORROSION FACTORS IN THE ATMOSPHERE

5.1.1 Types of Corrosive Atmospheres

Atmospheric corrosion has been defined to include corrosion by air at temperatures between −18°C and 70°C (~0–160°F) in the open and in enclosed spaces of all kinds. Deterioration in the atmosphere is sometimes called weathering. This definition encompasses a great variety of environments of differing corrosivity. The factors that determine the corrosivity of an atmosphere include industrial pollution, marine pollution, humidity, temperature (especially the spread between daily highs and lows that influence condensation and evaporation of moisture) and rainfall.

Corrosive atmospheres are usually classified into four basic types, although most environments are mixed without any clear borderlines [2]. The four types are indoor, rural, marine, and industrial, where "indoor" may be considered as the mildest environment. The type of atmosphere may vary with the wind pattern, particularly if corrosive pollutants are present.

5.1.1.1 Indoor Atmosphere

Normal indoor atmospheres are generally considered to be quite mild when the ambient humidity and other corrosive components are under control. However, some combinations of conditions may actually lead to relatively severe corrosion problems. Any enclosed space that is not evacuated or filled with a liquid can be considered as an indoor atmosphere. If not ventilated, such an environment may contain fumes, which could be highly corrosive in the presence of condensation or high humidity.

5.1.1.2 Rural Atmosphere

Rural atmospheres are typically the most benign and do not contain strong chemical contaminants. However, the rural atmosphere can be extremely corrosive to most construction materials near a farm operation, where by-products are made up of various waste materials.

Dry or tropical atmospheres are special variants of the rural atmosphere. In dry climates, there is little or no rainfall, but there may be a high RH and occasional condensation. This situation is encountered along the desert coasts. In the tropics, in addition to the high average temperature, the daily cycle includes a high RH, intense sunlight, and long periods of condensation during the night. In sheltered areas, the wetness from condensation may persist long after sunrise. Such conditions may produce a highly corrosive environment.

5.1.1.3 Marine Atmosphere

A marine atmosphere is full of fine particles of sea mist carried by the wind to settle on exposed surfaces in the form of salt crystals. The quantity of salt deposited can vary considerably with wind velocity and may, in extreme weather conditions, even form a very corrosive salt crust. The quantity of salt contamination decreases with distance from the ocean, and is strongly affected by wind currents.

The marine atmosphere also includes the space above the sea surfaces where splashing and heavy sea spray are encountered. Equipment exposed to these splash zones are indeed subjected to the worst conditions of intermittent immersions with wet and dry cycling of the corrosive agent.

5.1.1.4   Industrial Atmosphere

An industrial atmosphere is characterized by pollution composed mainly of sulfur compounds such as sulfur dioxide and nitrogen oxides. Sulfur dioxide (SO_2), from burning coal or other fossil fuels, is picked up by moisture and oxidized by some catalytic process on the dust particles into sulfuric acid. The acid settles in microscopic droplets and falls on exposed surfaces ("acid rain"). The primary sources of nitrogen oxides (NO_x) are motor vehicles, electric utilities, and other sources that burn fuels. Nitrogen oxides, hydrocarbons and ozone cause smog (smoky fog) in modern cities.

The contaminants in an industrial atmosphere, together with dew or fog, produce a highly corrosive, wet, acid film on exposed surfaces. In addition to the normal industrial atmosphere in or near chemical plants, other corrosive pollutants may be present. These are usually various forms of chloride which may be much more corrosive than the acid sulfates. The reactivity of acid chlorides with most metals is more pronounced than that of other pollutants such as phosphates and nitrates.

Even in the absence of any other corrosive agent, constant condensation on a cold metallic surface may cause an environment similar to constant immersion for which a component may not have been chosen or prepared for. Electrical systems are often installed in confined areas close to ground level or, worse, below ground where high humidity may prevail. For example, the frame and contacts in an electric junction box were found corroded just 4 years after installation [2]. While the junction box in this example was only at the ground level, the wires coming to the box were buried without additional insulation and were in constant contact with the much cooler ground than ambient air in a room; all these factors together caused the corrosion. Additional information on corrosive atmosphere containing smoke is given in Section 5.10.

5.1.2   Factors Affecting Atmospheric Corrosion

Atmospheric corrosion is accelerated by several environmental factors, such as elevated RH, high temperatures, presence of particulate matter and aerosols, corrosive gases, and so on.

5.1.2.1   Relative Humidity

The most important factor in atmospheric corrosion is moisture, in the form of rain, dew, condensation, or a high RH. RH is defined as the ratio of the quantity of water vapor present in the atmosphere to the saturation quantity at a given temperature, and it is expressed in percent.

A fundamental requirement for atmospheric corrosion processes is the presence of a thin film of electrolyte that can form on metallic surfaces when exposed to a critical level of humidity. The critical humidity level is a variable that depends on the nature of the corroding material, the tendency of corrosion products and surface deposits to absorb moisture, and the presence of atmospheric pollutants. It has been shown, for example, that this critical humidity level is 60% for iron if the environment is free of pollutants.

A thin film of electrolyte is almost invisible, but contains corrosive contaminants, which are known to reach relatively high concentrations, especially under conditions of alternate wetting and drying. In the presence of thin film electrolytes, atmospheric corrosion proceeds by balanced anodic and cathodic reactions. The anodic oxidation reaction involves the corrosion attack on the metal, while the cathodic reaction is naturally the oxygen reduction reaction.

In the absence of moisture, most contaminants would have little or no corrosive effect. Rain may have a beneficial effect as it washes away atmospheric pollutants that have settled on exposed surfaces. This effect is particularly noticeable in marine atmospheres. On the other hand, if the rain collects in pockets or crevices, it may accelerate corrosion by supplying continued wetness to such areas.

Marine environments typically have a high percent RH (%RH), as well as salt-rich aerosols. Studies have found that the thickness of an adsorbed layer of water on a zinc surface increases with %RH and that corrosion rates increase with the thickness of the adsorbed layer. There also seems to be a particular thickness of the water layer that, when exceeded, can limit the corrosion reaction owing to limited oxygen diffusion. However, when metallic surfaces become contaminated with hygroscopic salts their surface can be wetted at lower %RH. The presence of magnesium chloride ($MgCl_2$) on a metallic surface can make a surface apparently wet at 34% RH, while sodium chloride (NaCl) on the same surface requires 77% RH to create the same effect.

Dew and condensation are undesirable from a corrosion standpoint if not accompanied by frequent rain washing, which dilutes or eliminates contamination. A film of dew, saturated with sea salt or acid sulfates, and acid chlorides of an industrial atmosphere provide an aggressive electrolyte for the promotion of corrosion. Also, in the humid Tropics where nightly condensation appears on many surfaces, the stagnant moisture film either becomes alkaline from reaction with metal surfaces or picks up carbon dioxide and becomes aggressive as a dilute acid.

5.1.2.2 Temperature

Temperature plays an important role in atmospheric corrosion in two ways. First, there is a normal increase in corrosion activity, which can theoretically double for each 10°C increase in temperature. As the ambient temperature drops during the evening, metallic surfaces tend to remain warmer than the humid air surrounding them owing to their heat capacity. Therefore, metallic surfaces do not collect condensation until sometime after the dew point has been reached.

As the temperature begins to rise in the surrounding air, the temperature of the metal structures makes them act as condensers. A film of moisture is forming on metal surfaces, and the period of wetness may be much longer than the time the ambient air is at or below the dew point. How long the moisture is deposited on metal surfaces varies with the thickness of the metal structure, air currents, RH, and direct radiation from the sun.

Cycling temperature has produced severe corrosion on metal objects in the tropical climate, in unheated warehouses, and on metal tools or other objects stored in plastic bags. It is advisable to maintain the temperature of metal parts 10–15°C above the dew point since the dew point of an atmosphere indicates the equilibrium condition of condensation and evaporation from a surface. This will guarantee that no corrosion will occur by condensation on a surface that could be colder than the ambient environment.

5.1.2.3 Deposition of Aerosol Particles

The behavior of aerosol particles in outdoor atmospheres depends on their formation, movement, and capture. The concentration of these particles depends on a multitude of factors, including location, time of day or year, atmospheric conditions, presence of local sources, altitude, and wind velocity.

The highest concentrations are usually found in urban areas, reaching up to 109 particles per cm³, with the particle size ranging from around 100 μm to a few nm. Size is normally used to classify aerosols, and the other properties can be inferred from size information. The highest mass fraction of particles in an aerosol is characterized by particles having a diameter in the range of 8–80 μm.

Some studies have also indicated that there is a strong correlation between wind speed and the deposition and capture of aerosols; a correlation was also found between chloride deposition rates and wind speeds. Aerosols can be produced either by ejection into the atmosphere (primary aerosol production) or by physical and chemical processes within the atmosphere (secondary aerosol production). Examples of primary aerosols are sea spray and windblown dust. Secondary aerosols are often produced by atmospheric gases reacting and condensing or by cooling vapor condensation (gas to particle conversion).

Once an aerosol is suspended in the atmosphere, it can be altered, removed, or destroyed. An aerosol cannot stay in the atmosphere indefinitely, and average lifetimes are of the order of a few days to a week. The lifetime of any particular particle depends on its size and location. Studies of the migration of aerosols inland of a sea coast have shown that typically the majority of the aerosol particles are deposited close to the shoreline (typically 400–600 m) and consist of large particles (>10 μm diameter), which have a short residence time and are controlled primarily by gravitational forces. The aerosols also have mass and are subject to the influence of gravity, wind resistance, droplet dry out, and the possibilities of impingement on a solid surface, as they progress inland.

5.1.2.4 Pollutants, Corrosive Gases

The ANSI/ISA Standard S71.04-1985, "Environmental Conditions for Process Measurement and Control Systems: Airborne Contaminants," describes how various pollutants contribute to degradation of equipment performance. There are three types of gases that are the prime cause of corrosion of electronics: acidic gases, such as hydrogen sulfide, sulfur and nitrogen oxides, chlorine, and hydrogen fluoride; caustic gases, such as ammonia; and oxidizing gases, such as ozone.

Of the gases that can cause corrosion, acidic gases are typically the most harmful. For instance, it takes only 10 parts per billion (ppb) of chlorine to inflict the same amount of damage as 25,000 ppb of ammonia. Each site may have different combinations and concentration levels of corrosive gaseous contaminants.

Performance degradation can occur rapidly or over many years, depending on the particular concentration levels and combinations present at a site. The following paragraphs describe how various pollutants contribute to equipment performance degradation. See more about effect of pollutants on corrosion in electronic devices in Section 5.9.

5.1.2.4.1 Active Sulfur Compounds

This group includes hydrogen sulfide (H_2S), elemental sulfur (S), and organic sulfur compounds such as the mercaptans (RSH). Even at low concentration levels, they rapidly attack copper, silver, aluminum, and iron alloys. The presence of moisture and small amounts of inorganic chlorine compounds and/or nitrogen oxides greatly accelerates sulfide corrosion. Active sulfurs and inorganic chlorides are considered to be the predominant cause of atmospheric corrosion in the process industries.

5.1.2.4.2 Sulfur Oxides

Oxidized forms of sulfur (SO_2, SO_3) are generated as combustion products of sulfur-bearing fossil fuels such as coal, diesel fuel, gasoline, and natural gas. SO_2 has been identified as one of the most important air pollutants that contribute to the corrosion of metals. At low concentration levels, sulfur oxides can passivate reactive metals and thus retard corrosion. However, at higher levels, they will attack certain types of metals. The reaction with metals normally occurs when these gases dissolve in water to form sulfurous and sulfuric acids.

Precise methods are available to monitor continuously the amount of sulfur dioxide in a given volume of air. However, since only the actual amount of hydrated sulfur dioxide or sulfur trioxide deposited on metal surfaces is important, this is only indirectly related to the effect of sulfur dioxide on corrosion. The pollution levels can also be measured in terms of the concentration of the dissolved sulfate in rain water.

5.1.2.4.3 Nitrogen Oxides (NO_x)

Nitrogen oxides (NO, NO_2, N_2O_4) are formed as combustion products of fossil fuels and have a critical role in the formation of ozone in the atmosphere. They are also believed to have a catalytic effect on corrosion of base metals by chlorides and sulfides. In the presence of moisture, some of these gases form nitric acid that, in turn, attacks most common metals.

5.1.2.4.4 *Inorganic Chlorine Compounds*

This group includes chlorine (Cl_2), chlorine dioxide (ClO_2), hydrogen chloride (HCl), and so on. Chemical reactivity of these compounds depends on the specific gas composition. In the presence of moisture, these gases generate chloride ions which react readily with copper, tin, silver, and iron alloys. These reactions are significant even when the gases are present at low concentration levels. At higher concentrations, many materials are oxidized by exposure to chlorinated gases.

Sources of chloride ions, such as bleaching operations, sea water, cooling tower vapors, and cleaning compounds, and so on, should be considered when classifying industrial environments. They are seldom absent in major installations. Particular care must be given to equipment that are exposed to atmospheres containing chlorinated contaminants.

5.1.2.4.5 *Hydrogen Fluoride*

This compound is a member of the halogen family and reacts like inorganic chloride compounds.

5.1.2.4.6 *Ammonia and Derivatives*

Reduced forms of nitrogen (ammonia (NH_3), amines, ammonium ions (NH_4^+)) occur mainly in fertilizer plants, agricultural applications, and chemical plants. Copper and copper alloys are particularly susceptible to corrosion in ammonia environments.

5.1.3 AIRBORNE CONTAMINATION IN DATA CENTERS

The effect of airborne contaminations becomes a very serious issue in the performance of data centers [3]. Airborne contaminations may affect data center equipment in three different ways: chemical, mechanical, and electrical.

Two common chemical failure modes are copper creep corrosion on circuit boards and the corrosion of silver metallization in miniature surface-mounted components.

Mechanical effects include heat sink fouling, optical signal interference, increased friction, and so on. Electrical effects include changes in circuit impedance, arcing, and so on.

Most IT equipment are not installed in corrosive environments where it can be exposed to higher risk of failure. Most data centers are well designed and are in areas with relatively clean environments with mostly benign contamination. Most data centers should not experience hardware failures related to particulate or gaseous contamination. Still there are some data centers which may have harmful environments arising from the ingress of outdoor particulate and/or gaseous contamination.

In some rare instances, contamination has been known to be generated within the data center.

Reduction of circuit board feature sizes and miniaturization of components are necessary to improve hardware performance. It also makes the hardware more prone to attack from contamination in the data center environment.

The recent increase in the rate of hardware failures in data centers high in sulfur-bearing gases led to recommendations to monitor and control dust and gaseous contamination in addition to temperature–humidity control. These additional environmental measures are necessary to reduce the two most common recent failure modes of copper creep corrosion on circuit boards and the corrosion of silver metallization in miniature surface-mounted components. The failures of data centers due to airborne contamination, the causes of the failures, and the means to protect data centers from corrosion are discussed below in Section 5.9.5.

5.1.4 ZINC WHISKERS

Another form of particulate contamination very harmful to hardware reliability is the zinc whiskers, which are the most common electrically conductive particles found in data centers (Figure 3.8b and c in Section 3.5 of Chapter 3). The undersides of some steel-raised floor tiles are coated with zinc to prevent corrosion. The stringers and pedestals supporting the tiles may also be coated with electroplated or hot-dip galvanized (HDG) zinc. Zinc whiskers may grow on both types of coatings, although electroplated zinc is far more susceptible to whisker growth [4,5].

Zinc whiskers may sometimes grow to be 1–2 mm long and become dislodged and airborne, which could happen when the tiles are disturbed during their removal or when pulling or removing underfloor cables. If zinc whiskers are ingested by IT equipment, circuits with voltages higher than about 25 V may suffer electrical short circuiting, arcing, signal perturbations, or catastrophic failures [6]. If zinc whiskers are present, remediation involves replacing the contaminated raised-floor tiles and hiring professionals to clean the data center.

5.2 EFFECT OF ENVIRONMENT ON BARE METALS

It is important to find out how different materials used in the construction of electrical apparatus behave in a corrosive atmosphere. Steel, aluminum, and copper and their alloys are the three main metals used in electrical applications. Steel is typically used in housing, enclosure, and support systems (struts and cable trays). Copper is often used in conductors. Aluminum can be found most often in conductors, conduits, armor, and supports (cable trays). Each of these metals is susceptible to corrosion in different environments. In general, atmospheric, soil, and galvanic corrosion represent three major types of corrosion that metals are exposed to in the electrical industry. The following sections contain some specific information on the corrosion performance of the most popular materials in normal atmospheric conditions [7–11].

5.2.1 Iron and Steel in Enclosures, Frames, Rails, and So Forth

The iron, as the major component of steel, is exposed to all kinds of environments. It tends to be highly reactive with most of these because of its natural tendency to form iron oxide. When it does resist corrosion, it is due to the formation of a thin film of protective iron oxide on its surface by reaction with oxygen of the air. This film can prevent rusting in air at 99% RH, but a contaminant such as acid rain may destroy the effectiveness of the film and permit continued corrosion. Thicker films of iron oxide may act as protective coatings, and after the first year or so, could reduce the corrosion rate significantly.

The corrosion rates for steel have been determined by the ISO 9223 "Corrosion of metals and alloys—Corrosivity of atmospheres—Classification." While the corrosion rate of bare steel tends to decrease with time in most cases, the difference in corrosivity of different atmospheres for a particular product is tremendous. In a few cases, the corrosion rates of ferrous metals have been reported to increase with time. Analysis of the exposure conditions generally reveals that if there is an accumulation of contaminating corrosive agents, this increases the severity of the exposure. Steels containing low amounts of copper are particularly susceptible to severe atmospheric corrosion. In one test conducted over a 3.5-year period in both a marine and an industrial atmosphere, a steel containing 0.01% copper corroded at a rate of 80 μm/year, whereas increasing the copper content by a factor of five reduced the corrosion rate to 35 μm/year. Other tests comparing gray cast iron, malleable iron, and low-alloy steels indicated that their corrosion resistances were approximately the same.

The corrosion behavior of carbon steels is influenced significantly by small variations in copper and phosphorus content. Plain cast iron appears to have a corrosion rate about half of the 0.2% copper steel in a marine atmosphere. In an industrial atmosphere, structural carbon steel showed a penetration of about 20 μm, copper structural steel about 10 μm, and low-alloy steel about 4 μm after 5 years of exposure.

As indicated in ISO 9223, it is impossible to give a corrosion rate for steel in the atmosphere without specifying the location, composition, and certain other factors. If one can relate exposure conditions to those described in the literature, a fairly good estimate can be made of the probable corrosion behavior of selected material. However, all aspects of the exposure of the metal surface must be considered. A high-strength, low-alloy (HSLA) steel may show a superior corrosion resistance 12 times higher than carbon steel when freely exposed to a mild environment. When the severity or the physical conditions of exposure are changed, the HSLA steel will show less superiority, until in crevices or on the back side of structural forms in a corrosive atmosphere, it will be no better than carbon steel.

Stainless steels (Types 200 and 300), which contain high percentages of nickel and chromium, can keep their shiny aspect without tarnishing for many decades. Steels containing only chromium (Type 400) as the principal alloying constituent tend to rust superficially, but the others are relatively free from surface atmospheric corrosion. Many of them are susceptible to stress-corrosion cracking (SCC) in many common environments.

In electrical applications where steel is usually used as the housing, enclosure, or support system material, corrosion needs to be controlled by measures such as galvanizing, alloying, or painting. Galvanizing steel with zinc provides steel with a sacrificial anode that will continually deplete to protect the more cathodic steel. Alloying steel with chromium allows for the formation of a tightly adhering surface oxide layer, thus turning steel into stainless steel. Painting prevents corrosion by sealing out the agents of corrosion. Various means of protecting steel parts in electrical installations against corrosion are presented in Section 7.7 of Chapter 7.

5.2.2 Copper and Copper Alloys: Parts of the Conductive Path

Bare copper and its alloys are not exposed to the atmosphere in great quantities when compared with steel. However, these materials bring aesthetic value to building construction, in addition to excellent corrosion resistance. Extensive tests have been conducted on the corrosion resistance of copper and its alloys to various atmospheres. Various alloys were exposed to rural, industrial, and marine atmospheres for periods of up to 20 years. A fairly clear picture can be obtained of the corrosion behavior of copper from data accumulated in these tests and the calibrations of relative corrosivity of the test sites. The addition of tin to copper produces bronze, an alloy that is harder and more wear- and corrosion-resistant than either of the pure metals. Brasses, copper alloys with zinc, are the most numerous and the most widely used of the copper alloys because of their low cost, easy or inexpensive fabrication and machining, and their relative resistance to aggressive environments. They are, however, generally inferior in strength to bronzes and must not be used in environments that cause dezincification.

Copper alloys with nickel are called copper-nickels or cupronickels. The alloys' names usually reveal their basic composition, for example, aluminum bronze, nickel–aluminum bronze, silicon bronze, and so on. Like the copper–tin bronzes, the alloys have a light golden color, high corrosion resistance, and excellent mechanical properties. These alloys may contain several additional alloying elements to provide specific alloy properties. Tin bronzes containing more than 10% tin are generally harder and more corrosion resistant than brass and up to 20% tin is often added.

The so-called phosphor bronzes additionally contain small amounts of phosphorus to further increase hardness and wear resistance; hence, they are often specified for sliding contacts and connectors. Some alloys are susceptible to specific types of corrosion such as SCC of brass. These types of corrosion contribute to the failure of the material in mechanical respects without significant weight changes or losses in thickness.

Bare copper easily corrodes in atmospheres containing water and acidic vapors, producing white, blue, and green deposits of copper chlorides, sulfates, carbonates, and other compounds. For example, copper (II) sulfate ($CuSO_4$) is a common salt of copper. Copper sulfates exist as a series of compounds that differ in their degree of hydration. The anhydrous form is a pale green or gray-white powder, while the hydrated form is bright blue. Both forms may be converted into each other: blue powder will turn white when heated and, conversely, white powder will turn blue

FIGURE 5.1   Green and blue corrosion of bushing stud made of bare copper.

when contacted with water. Copper salts have higher electrical resistance than metallic copper, thus degrading conductive properties of copper parts and resulting in failure. An example of extensive corrosion of copper stud inside the bushing is shown in Figure 5.1.

5.2.3   NICKEL AND NICKEL ALLOYS: ELECTRICAL CONTACTS AND PLATING

Electrodeposited nickel and EN are widely used as a protective coating for atmospheric exposure, and some nickel alloys, while selected for other reasons, are also exposed to atmospheric corrosion. Studies of corrosion behavior in different atmospheres showed that nickel tends to be passive in a marine atmosphere. The ratio between the corrosion rate for nickel exposed to the industrial atmosphere and that exposed to rural or marine atmospheres was 28:1. For more about corrosion properties of nickel and nickel alloys in the form of plating and underplating, see Sections 5.1 and 5.2 of Chapter 5, and Section 6.3 of Chapter 6.

5.2.4   ALUMINUM AND ALUMINUM ALLOYS IN ELECTRICAL APPLICATIONS

Bare aluminum is used in the electrical industry to manufacture busbars, transmission and distribution conductors. It is produced in the form of wrought products, extrusions, and castings with a variety of alloying elements to produce the desired mechanical and electrical properties.

Because of technological advances and the favorable price as compared to copper, there has been a continuous growth in the volume, sizes, and varieties of aluminum conductors. Although the increasing use of aluminum bus conductors is credited largely to the economic factors, advance in joining techniques and general experience have prompted its use in many manufacturing, chemical, and utility installations.

Aluminum is used in making overhead transmission lines (OHTL), almost to the exclusion of copper. There are four major types of overhead conductors used in electrical transmission and distribution: all-aluminum conductor (AAC), all-aluminum alloy conductor (AAAC), aluminum conductor steel reinforced (ACSR), and aluminum conductor aluminum-alloy reinforced (ACAR).

The ACSR cable is a specific type of high-capacity, high-strength stranded cable typically used as a bare overhead transmission cable and as primary and secondary distribution cables. All-aluminum alloy cable was developed to retain the mechanical and electrical properties of ACSR while improving the weight and corrosion resistance characteristics. The excellent corrosion resistance of aluminum in AAC has made it a conductor of choice in coastal areas. ACAR is a composite aluminum–aluminum alloy conductor which is specially designed for each application to optimize the properties.

The favorable qualities of aluminum for its use in electrical applications are relatively high electrical and thermal conductivities, low density, nonmagnetic properties, ease of drawing down to smaller wire sizes, and high resistance to weathering. Noncurrent-carrying applications of aluminum are numerous in transformers, capacitors, motors, and other types of electrical equipment.

While pure aluminum has excellent atmospheric corrosion resistance and is used extensively as a cladding material for this very reason, alloys containing copper and silicon as the principal alloying components are often susceptible to atmospheric corrosion and should be used with care. In a rural atmosphere, the corrosion rate for most aluminum alloys is about 0.06 μm/year. For alloys containing large amounts of copper, the rate is about twice as much. Changes in tensile strength because of corrosion vary from 0 to <1% for sheet material.

The aluminum alloys employed for conductor accessories, including drawn, extruded, and cast products, vary with the specific application and the preferences of the individual manufacturer. However, the alloys are generally selected to provide suitable conductivity, high resistance to atmospheric corrosion, galvanic compatibility with EC grade aluminum, and satisfactory mechanical properties. Typical suitable alloys are 6061-TG for wrought forms and 356-T6 for castings.

In a marine environment, the differences between corrosion resistances of aluminum alloys may be very large. Some aluminum alloys develop severe pitting and a voluminous white corrosion product under some exposure conditions in a marine atmosphere. The usual corrosion behavior in the atmosphere involves pitting and roughening of the surface with a fairly large decrease in the corrosion rate after the first 1–3 years of exposure. In a harsh industrial atmosphere, aluminum corrosion can be much greater than in the marine atmosphere. In terms of average corrosion rate, the initial rate would be about 0.08 mm/year, dropping to <0.3 mm/year by 7 years.

Severe pitting has been encountered where aluminum surfaces were contaminated by either alkaline dust or coral dust containing chlorides, followed by condensation. On some of the South Pacific islands, dust collected from the surfaces of sheltered structures, such as those inside aircraft wings, contained 67% chloride by weight.

While designing aluminum equipment, care must be taken to avoid dissimilar metal couples and the attendant galvanic corrosion. Copper and rusty steel are particularly bad when in contact with aluminum. Due to the passive film, stainless steel can be used in contact with aluminum with little expectation of accelerated corrosion, despite the differences in electrolytic potential. In addition, designers should be aware of the possibility that some aluminum alloys may be sensitized to intergranular corrosion by heat treatment. Galvanic corrosion of dissimilar metals connections and a means of protection are described in the Section 6.4 of Chapter 6.

As would be expected, constant exposure to moisture with a limited supply of oxygen to the aluminum surface leads to rapid corrosion of any aluminum apparatus or equipment component. This is due to the highly reactive nature of aluminum that leads to the formation of oxides or hydroxides. In the presence of oxygen, a protective aluminum oxide film develops on any aluminum surface. This oxide film is substantially unreactive with the normal constituents of the atmosphere. If the film is removed by mechanical or chemical means and the aluminum is exposed to water, then a rapid reaction sets in and large quantities of aluminum are converted into hydroxide and subsequently to oxide.

Most of the 1000, 3000, and 6000 series of wrought alloys, the magnesium, and the silicon magnesium casting alloys are relatively immune to SCC in the normal atmosphere. However, other alloys may be susceptible under certain conditions. This aspect should be investigated carefully by the designer or user if stresses are high and the atmosphere is corrosive.

In the design of aluminum structures, the usual precautions of avoiding crevices or pockets and coupling with dissimilar metals must be observed. In some marine or industrial atmospheres, the aluminum was found perforated at the laps within a few months of exposure. In addition, stress concentrations should be avoided, such as those found in some riveted structures, in the vicinity of welds, and at notches or inside corners. Where it is impractical to avoid connection between dissimilar metals, the aluminum should be electrically insulated from the more noble metal by means of washers, sleeves, and so on. In some instances, covering the noble metal with an organic finish is sufficient to greatly reduce galvanic couple corrosion.

5.3 ATMOSPHERIC CORROSION OF SILVER PLATING

Silver is probably the most widely used contact material available. It has the highest electrical and thermal conductivity of any known metals. As a result of silver's high electrical and thermal conductivity, contacts composed of fine silver work well at current in the low to medium range (1–20 A) where light to moderate contact pressure (CR) is available and low CR is a requirement. Silver has the lowest cost of all precious metal contact materials and is readily formed into various contact shapes (rivets, buttons, etc.) due to its ductility. Silver plating is a very popular coating for various contact materials, providing a high quality of corrosion protection against many types of environments.

5.3.1 Silver Plating Corrosion and Tarnish

Silver shows good resistance to oxidation and tarnishing except in the presence of sulfur. Sulfur-containing atmospheres will produce silver sulfide that increases CR. There are various forms and degrees of silver corrosion depending on the environment, application, and silver plating quality and thickness [12].

5.3.1.1 Sulfuric Corrosion

Silver corrodes in an environment containing various sulfuric gases such as H_2S, OCS, CS_2, and SO_2. The corrosive effect on silver of H_2S and OCS gases is about an order of magnitude stronger than that of CS_2 and SO_2. However, a product of silver corrosion, silver sulfide can be formed by contact with SO_2 in moist air, but only at SO_2 concentrations two to three orders of magnitude higher than that typical of ambient environments. Hydrogen sulfide (H_2S) is usually present in chemical plants, oil refineries, production of artificial fibers, steel mills, and the paper and pulp processing industry due to process technologies. It was found that even at a minor concentration of H_2S, corrosion produces silver sulfide on the parts in contact with the environment.

Hydrogen sulfide (H_2S) gas practically always is present in an ambient air. In the United States, it was found that the concentration in urban and industrial areas of it is in the range of 0.02–5.0 ppb and in remote areas it is in the range of 0.005–0.5 ppb. There is no precise definition of the minimum concentration of H_2S required for tarnishing to be observed on silver, in whatever form the silver is exposed, either bulk or plating. Additional risk of sulfuric corrosion for specific electrical apparatus may come from outgassing of hydrogen sulfide or other sulfuric compounds from seals, gaskets, and other polymers within electrical units and closed housing.

5.3.1.2 Silver Tarnish

By definition, "tarnish" means loss of metallic luster, which is caused by a thin layer of corroded metal. The amount of silver tarnishing is a function of the RH, ambient temperature (T), gas concentration, and time of exposure. Silver will react with extremely low levels of H_2S, and the minimum level will be in the part per billion (ppb) range and even in the part per trillion (ppt) range. Silver and H_2S react to form dark brown or black silver sulfide Ag_2S; one of the most insoluble salts of silver, this is what causes the tarnish, which has been progressing with time even in facilities with a mild environment, such as manufacturing or assembling plants.

An example of a tarnished silver-plated copper bus is shown in Figure 5.2. After exposure to a corrosive industrial facility environment, the tarnish film on a silver-plated copper bus grows very thick, and dark brown and black silver sulfides form flakes on exposed surfaces (Figure 5.3).

FIGURE 5.2 Tarnished silver plating on copper bus after various periods of storage at manufacturing facility.

FIGURE 5.3 Heavy silver plating tarnish on copper bus after several months in service at industrial facility.

Some factors could accelerate the development of silver tarnishing during transportation, storage, and exploitation of the equipment:

- Elevated temperature
- High humidity
- Light (UV)
- Ozone rate in the atmosphere
- Presence of salts on the surface, possibly due to handling with naked hands

An additional source of sulfuric gases may be an inappropriate packing material because certain cardboards contain sulfur. The packing material should pass a special silver tarnish test [13] to not cause an additional silver tarnishing.

5.3.1.3 Silver Whiskers

When a thick layer of Ag_2S is formed on silver plating and the parts are exposed to a high temperature and hydrogen sulfide, the new process of growing thin filaments (whiskers) begins (Figure 5.4). Silver whiskers grow practically everywhere on corroded silver-plated parts of switchgear, but more intensely in areas with higher temperature, such as bus joints and sliding contacts, outside edges and corners of the contacts. Temperature gradients in these areas may encourage this phenomenon.

FIGURE 5.4   Finger cluster of corroded breaker after 2 months of service in paper mill facility (pulp processing mill).

Silver corrosion results in a high resistance, which produces more heat, which in turn stimulates further tarnishing and the growth of whiskers [14]. A detailed description of the silver whisker phenomenon is given in Section 7.5 of Chapter 7.

5.3.2   RED-PLAGUE CORROSION

A special type of copper corrosion called "red plague" has been known for decades. Copper is susceptible to the formation of red cuprous oxide when stored or used in a moist or high humidity environment. "Red plague" corrosion was first documented in reports on corrosion dated 1965 or so, and has been found on a regular basis since then both in the United States and Europe.

When silver is plated over copper, there can be an accelerated corrosion of the copper caused by defects in the silver plating through galvanic action (Section 6.4 of Chapter 6). Imperfections such as pinholes, pores, and breaks in silver plating allow moisture and oxygen to penetrate the silver plating. The sacrificial corrosion of copper occurs in a galvanic cell of silver (cathode) and copper (anode), resulting in the formation of red cuprous oxide (Cu_2O). Secondary corrosion products are black cupric oxide (CuO) and copper hydroxides (bluish green). This reaction is promoted by the presence of moisture (H_2O) and oxygen (O_2) at an exposed copper–silver interface, such as an exposed conductor end (crimp terminations). The process is caused by poor plating quality control (pin-hole, porosity, thin coating) or mechanical damage during stranding or handling (scratches, nicks, abrasion). In the cables, copper corrosion is accelerated by penetration of moisture, oxygen, flux residue, and solvents under the insulation jackets.

The color of the deposit may vary depending on the amount of oxygen available, commonly noted as a red and reddish-brown surface discoloration—hence the term "red plague." Once initiated, the sacrificial corrosion of the copper base conductor can continue indefinitely in the presence of oxygen (O_2) [15–17]. The corrosion known as "red plague" is identifiable by the presence of a

FIGURE 5.5   Silver-flash plated copper bus after being shipped overseas. (Courtesy of Finishing.com at http://www.finishing.com/280/03.shtml; the picture is taken by Steve Whatley.)

brown-red deposit on the surface. It was found that copper wires with silver plating <1 μm thick are susceptible to red-plague corrosion. An example of corrosion that may be identified as "red plague" is shown in Figure 5.5. Silver flash plating disappeared from the copper bus after being shipped overseas and left copper unprotected and exposed to oxidation and corrosion.

Occurrences of red-plague corrosion have periodically affected aerospace and defense programs for >40 years. The risk imposed by red plague is related to the extent of corrosion. Sufficiently advanced corrosion can lead to electrical and/or mechanical failure as copper is consumed within affected conductors. A major risk to system performance therefore exists when primary signal or power conductor paths are affected. Wire and cable industry process improvements have not eliminated the problem.

Concerns about galvanic corrosion of silver-coated copper wire are specified in many documents used by NASA, the European Space Agency (ESA), military, and aerospace, for example, ASTM B 961-08 "Standard Specification for Silver Coated Copper and Copper Alloy Stranded Conductors for Electronic Space Application," ECSS-Q-ST-70-20C "Determination of the Susceptibility of Silver-Plated Copper Wire and Cable to 'Red-Plague' Corrosion," SAE AIR4487 "Investigation of Silver Plated Conductor Corrosion (Red Plague)," and many others.

Red Plague Control Plan (RPCP) [17] determines silver coating requirements to mitigate the development of red-plague corrosion. Primary and shield conductors should be plated with not less than 2 μm (80 μin.) silver plating on average. After stranding, the silver coating thickness on each of the individual conductor strands shall not be <1 μm (40 μin.). The ESA responded to the issue of red plague by requiring lot sample verification that the silver coating has a minimum coating thickness of 2 μm (80 μin.) for all critical wiring installations utilizing silver-coated copper wires.

5.3.3   Underplating Corrosion

When silver is plated over copper with some areas of bare copper exposed, underplating corrosion may occur (Figure 5.6). In the example of underplating corrosion [18], the finger clusters of MV CB were silver plated only along the edges, which left large amounts of exposed copper. The CBs were installed in a vented air-conditioned trailer. Normally, these contacts cycled once daily, on in the morning and off at night. When energized, the operating temperature inside the box was 75–90°C.

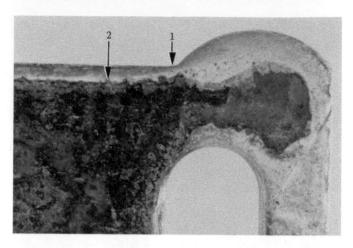

FIGURE 5.6   Underplating corrosion: silver plating on the contact surface (1) and corroding copper under silver plating (major component—copper carbonate hydroxide) (2) (4× magnification). (From Under Coating Corrosion of Silver Plated Copper in a Circuit Breaker. http://www.corrosionlab.com/Failure-Analysis-Studies/20034.under-coating.silver-plated-copper.htm, courtesy of Corrosion Testing Laboratories, Inc., Newark, DE. With permission.)

When cycled off, the box was cooled to ambient temperature overnight. The contacts carried either 4160 or 13,800 V AC.

There are two basic areas on the finger cluster shown in Figure 5.6: area 1 is silvery metallic in color (typical of silver plate), while area 2 is green in color (typical of corroding copper). Corrosion products penetrated under silver plating, causing it to flake off. To prevent underplating corrosion, silver should be plated on the whole finger cluster area, leaving no exposed copper.

5.3.4   Effect of Silver Plating Thickness and Quality on Sulfuric Corrosion

Silver sulfide (Ag_2S) is a major product of silver corrosion in atmospheres containing sulfuric gases. Silver "tarnish" is the very first step in silver sulfuric corrosion, when the corrosion layer forms a dark thin film, which can relatively easily be removed with a tarnish cleaner or scotch-bright material. As soon as a significant layer of silver is consumed, silver sulfide is formed on the surface as a heavy dark gray or black flaking deposit. The formation of silver sulfide is linear with time.

It is well known that corrosion resistance of electroplated silver strongly depends on plating quality. Specific types of plating the bath may produce plating with higher corrosion resistance or lower corrosion resistance. For example, plating without surfactants at lower CD would produce more loose and porous plating [19]. Surface conditions, such as roughness and cleanliness, play an important role in plating corrosion resistance. A combination of two conditions, poor quality of the plating characterized as loose and porous and exposure to humid atmosphere, may result in "red plague" corrosion.

It was found that CR is strongly influenced by the corrosion film thickness [20,21]. It was also concluded that silver coating thickness is the most important factor in corrosion resistance. If the silver coating is too thin or of low quality, the corrosion film will include copper corrosion products and its presence will always be detrimental to the contact. Contacts with a thin silver coating behave almost like bare (unplated) copper contacts.

The thicker the silver plating is, the better the base material is protected from corrosion, since a thicker plating is less porous and does not allow corrosive gases to penetrate through the plating. Silver flash, which usually has a thickness of several tenths or even hundredths of microns, would be consumed in a corrosive atmosphere very fast.

5.3.5 CORROSION OF A COPPER BUS WITH FLASH SILVER PLATING

Copper is often used in various applications together with silver flash, which is a very thin plating that does not provide long-lasting protection from corrosion, particularly in environments with relatively high concentration of corrosive gases or particulate matter. The preferred environment for a silver flash bus is considered to be the office or environmentally controlled light industrial environment. These environments are typically controlled and are relatively free from corrosive contaminants in the area around the equipment. Silver flash plating of about 0.025–1.25 µm may not show any visible signs of tarnish after being stored for many months at the facility with a clean environment (Figure 5.7).

The use of silver-plated/coated bus is not recommended in the industrial facilities with the potential for excessive corrosion rates, such as WWTPs, pulp and paper processing and manufacturing plants, chemical industrial facilities, and so on. The use of silver flash in corrosive environment may result in "red plague" copper corrosion as shown in Figure 5.5. Examples of "blue" and "green" copper corrosion of the CB bus stabs coated with silver flash are shown in Figure 5.8.

FIGURE 5.7 Copper bus with silver flash stored for 12 months at manufacturing facility with clean environment.

FIGURE 5.8 "Blue" and "green" corrosion of copper bus with silver flash.

5.4   EFFECT OF SILVER CORROSION ON CONTACT RESISTANCE

5.4.1   Silver Tarnish and Contact Electrical Resistance

Silver sulfide is not a conductive material and having a contact surface covered with a corrosion layer may significantly decrease the electrical conductivity. The effect of corrosion on the electrical properties of the contact mainly depends on the thickness of the corrosion layer. The focus of this section is on a relatively benign and less harmful stage of silver corrosion—silver tarnish—with a very thin layer of silver sulfide on the contact surface.

5.4.1.1   Thickness of Silver Tarnish

Air quality, ambient temperature, and humidity are two most important factors influencing the thickness of silver tarnish. Usually, the thickness of a silver plating is in the range of 6.25–12.5 μm. The longer the silver plating is exposed to corrosion, the thicker the tarnish film that is formed. The color of the silver plating is changing gradually after various periods, from 1 month to 5 years of exposure to the atmosphere of a common manufacturing facility storage. After only several months in storage, silver tarnish is very well seen as a dark yellow tint. In <3 years, the discoloration of the plating is very strong and in 5 years the plating has a heavy brown deposit (Figure 5.2).

When exposed to the corrosive industrial environment of a paper mill, the tarnish film is very thick, forming a dark brown/black scale in a couple of months of exposure (Figure 5.3). The only silver that will not tarnish would be the one kept in vacuum or has a coating on top to isolate its surface from the atmosphere. Silver sulfide is not an insulating substance; it is a semiconductor and may therefore conduct electricity. However, since the electrical resistivity of silver sulfide (Ag_2S) is ~100,000 times higher than that of silver (15–20 μΩ-cm for Ag_2S vs. 1.6×10^{-4} μΩ-cm for Ag), it reduces the surface conductivity of the silver even if a very thin film is formed.

To what extent silver tarnish will effect electrical conductivity depends on the thickness of the tarnish. When silver contacts are used, one of the factors that make them good is that the silver is soft and when rubbed against the other contact, it cleans itself and smears, so good metal–metal contact can be sustained. However, if the contacts are not designed to do this, there could be a longer term problem with an accumulation of low conductivity sulfides. The influence of sulfide corrosion on electrical contact was studied in detail in Ref. [22] because of a growing concern for the safety of satellites stored on the ground for a long time before launching.

Initially, sulfur rapidly attacks the silver surface. Once the sulfide layer becomes continuous (100% of the surface is covered), growth takes place on the sulfide surface. Thereafter, sulfidation is limited by the rate at which the sulfur-containing gas is supplied to the surface or the rate at which silver, a highly mobile ion, diffuses through the sulfide layer. It was concluded that most electrical connectors can tolerate some sulfide on their surfaces. Silver sulfide grown on a pure silver surface is soft and malleable, being easily pushed aside even under low load. This would allow enough micro contacts to produce a low CR.

Although the thickness at which tarnish begins to noticeably degrade the electrical properties of the circuit is not well established, it was found that the silver sulfide layer becomes continuous at a thickness of about 150 Å. Tests examining the electrical conductivity of silver seem to indicate that 200 Å would be a reasonably conservative estimate of the thickness at which electrical properties begin to deteriorate.

The data presented in Ref. [21] can be used to evaluate the approximate "safe" storage time in an environment with very low concentration of H_2S where the electrical contact performance remains acceptable. For example, in urban and industrial areas the concentration of H_2S is found to be in the range 2–5.0 ppb. At lower concentrations of H_2S contamination, electrical properties of the contact will not be deteriorated by the tarnish film formed. However, at higher end of the H_2S concentration (5.0 ppb), the tarnish film able to deteriorate the electrical properties of the contact will be formed only for 2000 h (83 days or <3 months) at room temperature and >75% RH. At higher temperatures, the tarnish film will grow faster.

The data on the thickness of corrosion film forming on silver in three types of field environments, according to MFG test specifications, are given in Ref. [23]. It was found that in a mild Battelle Class II environment (with 10 ppb of H_2S), a corrosion film with thickness greater than 1000 Å on silver will be formed for <1 year. It was shown that the formation of a 1000 Å thick corrosion film on silver may lead to an increase of the contact electrical resistance to >10 mΩ after exposure to a mild office environment only for 1+ year. For a more harsh environment with 2 ppm H_2S, a corrosion film of 500 Å is formed only for 1000 h (~40 days) [24]. The relationship between the CR and the exposure time to H_2S gas was studied in Ref. [25] as a function of the film thickness and the time of exposure to 3 ppm of H_2S at 40°C and 80%–85% RH. In such conditions and on exposure for less than 100 min, the corrosion film thickness will grow to 500 Å.

5.4.1.2 Effect of the Current Load and Mechanical Load on the Electrical Resistance of Corroded Contact

The CR of the contact covered with corrosion film depends on the contact load (force), which varied from 1 to 100 g. This means that Ag_2S is mechanically weak.

In Ref. [25], the current was found to strongly affect the CR and the size of the contact areas; that is, the CR was lowered with increasing the current load. Also, it was determined that thin corrosion films on silver do not always cause high CR in high current applications despite their high resistivity.

Depending on the thickness of silver tarnishing, it can have various degrees of severity. If the parts are just slightly yellow and the discoloration does not evolve more, then it should not be considered as seriously damaging for contact electrical properties. Conductivity is not affected by a light yellowing. In other cases, tarnishing can lead to dark brown and black colors. This coloring can be partial or total depending on the conditions of the storage or use (finger marks, opened packing, etc.). If the discoloration is in constant evolution, it should be seen as a sign of the presence of a source of sulfuric gas in the immediate environment of the products. In this case, conductivity has a very good chance of being degraded.

The reactions of atmospheric pollutants with silver-based contact materials also produce corrosion products which adversely affect the contact's electrical behavior. In Refs. [25,26], three silver-based contact materials, Ag/Cu/Ni, Ag/Ni, and Ag/MgO/NiO, were exposed to an accelerated corrosion environment containing SO_2, H_2S, NO_2, and Cl_2, with a maintained elevated temperature and high humidity. Both electrical degradation and corrosion film growth characteristics were studied as functions of exposure time. CR increased rapidly with the formation of a growing corrosion film made mostly of silver chlorides and silver sulfides.

5.4.2 TECHNIQUES OF TARNISH CLEANING

Due to the detrimental effect of silver tarnish on electrical contact performance, it is important to protect silver plating from corrosion. If a corrosion film is already formed, it should be timely and properly removed from the contact surfaces. Different techniques have been developed to delay contact corrosion, but whatever be the kind of antitarnish treatment applied, it should not interfere with the electrical properties of the coating.

The techniques of cleaning silver are as old as jewelry made of silver. The first published recommendations came from Theophilus of twelfth century: "If … silver … has become blackened with age, take some black charcoal, grind it very fine, and sift it through a cloth. Take a linen or woolen cloth that has been wet in water and put it on the charcoal. Lift it up and rub the … silver all over until you remove all the black stain. Then wash it with water and dry it in the sun or by a fire or with a cloth. Then take white chalk and scrape it very fine into a pot and, dry as it is, rub it with a linen cloth over the … silver, until it recovers its pristine brilliance" [27].

It would be helpful to correlate the color of a tarnish film with its thickness, and be able to determine when the corrosion film may begin to deteriorate the electrical properties of the contacts. In Ref. [28], it was shown that evaluation of tarnish film color may be used to assess the tarnish formation and to relate the color change to the degree of tarnishing. The results of this study correlate very well with the behavior found for the weight increase (thickness of the tarnish layer) versus either the hydrogen sulfide concentration or exposure time.

Tarnish-cleaning products may have different consistency. They can be solids (such as the common washing soda) that dissolve in water, in which case tarnish is removed using electrolytic action; they could be fluid that requires one to rinse off the chemicals used for tarnish film removal. Some silver cleaning materials (abrasive fluids) have been tested for removing a heavy tarnish (0.5–1 μm) [29].

It was found that calcium carbonate, γ alumina, and chromium oxide suspended in deionized water and containing nonionic surfactant helped remove tarnish successfully and caused the least amount of damage to the silver. The use of cleaning powder is not recommended. Overzealous pressure when cleaning with powders may help remove significant amounts of silver, which ultimately will wear through the plating completely.

The antitarnish products could be semisolid (cream, paste, and emulsion). Some cleaners that are available in paste or emulsion not only remove tarnish but also contain tarnish inhibitors.

The use of chemical dip solutions (often referred to as silver dip) is effective, but requires much attention:

- Silver-plated parts in the dip solution should not stay in solution for >10 s.
- If heavy tarnish is treated, the surface of the restored silver may have a matt finish.
- Dip solutions can stain or even etch the materials other than the silver; any drops of dip solutions that fall on these materials should be rinsed off immediately.
- Silver-plated parts should be washed and dried thoroughly after cleaning with dip solutions.

Several criteria should be applied to select the best tarnish cleaner to use in the service maintenance and refurbishment of silver-plated copper electrical contacts.

The best products among many available in the market are those that satisfy the following conditions:

1. Clean the tarnish instantly
2. Contain no abrasives and therefore do not harm silver plating
3. Do not require rinsing in soapy water or any other fluid
4. Applied by a simple procedure and without spill
5. Can be applied without any special employee protection

Some products provide cyanide-free instant tarnish removal for pure and plated silver, return substrate to as-plated appearance in most cases, and rejuvenate surfaces and eliminate the need for costly stripping and replating.

5.5 SILVER WHISKERS: A MYSTERIOUS AND DANGEROUS PHENOMENON

5.5.1 History of Silver Whiskers

Thick hair-like formations on the conductive parts of electrical equipment have been seen from the 1920s onwards and have caused a number of violent failures, but are not yet practically studied and clearly understood. In the papers published in Germany in the early 60s, similar formations were found on the contactor and were called silver whiskers in Refs. [30,31].

In Ref. [32], a very interesting historical account was presented on the past discoveries and studies of silver whiskers. In early 1970s, a serious problem occurred on one of the electro-mechanical contactors in LV switchgear assembly in Durban, South Africa. During investigation, strange fibers were found on the silver-plated terminal flags. The fibers were very conductive, and further experiments proved that the fibers were pure silver. The circumstances described in Refs. [30,31] were similar to those in Durban, South Africa. The switchgear was installed in a switch house adjacent to a sulfur plant, and sulfur was present in the ambient atmosphere together with a high RH of the ambient air. It was noticed that the high ambient temperature was not the reason of growing these fibers but the TR just originating from a contact subjected to aging.

In mid 70s, another product failure appeared, this time on a refinery on the Caribbean's Curacao Island. During an investigation of the failed switchgear, the same types of fibers as reported in Durban case were detected. Ambient circumstances on the Caribbean island were similar to the Durban refinery case that indicated again that silver whiskers were the cause of the failure. Shortly after this incident, a similar short circuit incident in a Singapore refinery took place which was also most probably initiated by silver whiskers. Actually in all the three plants (Durban, Caribbean's, and Singapore), silver whiskers were found on relatively new switchgears (early 1970 supplies).

Later, a few papers describe the extensive growth of silver whiskers at a synthetic fiber factory in France [33], Kraft linerboard mill in Southeastern United States [34], and Coking plant in the United Kingdom [35]. In all these locations, hydrogen sulfide (H_2S) was present in the atmosphere.

The first and the only extensive study of silver whiskers was prompted by the failures at a synthetic fiber factory [33] in 1982. The atmosphere in the factory had H_2S levels of several dozens of ppm, which is very high by any standard. No other contaminants, such as pure sulfur and SO_2, were found in the environment. As later confirmed by analysis, the contaminated parts were heavily sulfidized. After the failure of electrical equipment, improved ventilation of the premises was incorporated, but did not bring the expected results, and although the level of H_2S was reduced to 0.5 ppm, incidents began to occur again.

In the course of a thorough periodic examination of the suspect equipment, "metallic"-looking filaments, some of them >1 cm (0.4″) long, were observed on the supports of certain fuse holders. In some cases, large numbers of very fine filaments appeared in tufts; in others, they looked like needles growing at right angles to the metallic surface.

This phenomenon was found not only on silver-plated copper parts but also on contacts made of AgNi alloy. The filaments grew most readily on the outside edges and corners near contact zones, but not necessarily on the points thought to be the hottest.

Silver whisker growth was successfully reproduced in a laboratory setup, in which different levels of H_2S (50 and 1 ppm) were maintained. The temperature of the chamber was stabilized at 40°C and the RH was 40%. The test was carried out on a triple-pole fuse holder, whose poles, linked in series, had a current corresponding to the nominal intensity of the apparatus (50 amp) passing through it. The supports of the fuse holders were made of copper, electroplated with silver of 10 μm thickness. The cylindrical ends of the fuses were also silver plated.

In the course of the experiment it was found that the whiskers started growing as soon as a fairly thick layer of silver sulfide is formed on the silver-plated parts. The only difference in H_2S level of 20 or 1 ppm was that the formation of corrosion film took longer for lower level of gas.

When the temperature of the sample remained below about 140°C, typical Ag_2S crystals with length not exceeding a few μm have been growing on the fuse holder, but no filament growth was observed. As soon as the temperature of the support rose above 140°C in a matter of hours, minuscule filaments began to appear. In 24 h, these filaments had grown to 10 mm. They looked exactly like the filaments on the apparatus at the factory.

Each filament was made up of a bunch of fibers of about 1 μm diameter. The diameter of the filaments measured at the base varied between 10 and 70–80 μm. The sections where the filaments seem to grow best were on the outside edges and corners. The speed at which filaments grew in these conditions reached about 1 mm/h or 0.3 μm/s.

The whiskers were made up of Ag and S but with much less sulfur than there is in Ag_2S. It was assumed [33] that silver whiskers have been partially transformed into silver sulfide, which covered filament surfaces exposed to an ambient H_2S.

Another observation was given in Ref. [34] that under proper conditions there is no restriction on the length to which the whisker could grow. There are other observations showing that silver whiskers may grow at room temperature as well. In Ref. [35], the sample was laid on a bench in the laboratory and subjected to heavy concentrations of several corrosive gases, including hydrogen sulfide, at ambient temperature. After some time, it was found that whiskers up to 4 mm in length had developed.

5.5.2   Factors That Affect the Growth of Silver Whiskers

Based on the observations and experiments described in Refs. [21,30–39], several factors initiating and accelerating the growth of silver whiskers may be determined.

5.5.2.1   Environmental Factors

It is obvious that H_2S gas even at very low levels is a primary factor to initiate whiskers growth. A layer of silver sulfide (Ag_2S) of a certain thickness needs to form for whiskers to start growing. Elevated temperature (above 140°C) of the parts seems to accelerate the growth. The influence of elevated humidity is not defined yet. The influence of gases other than H_2S present in the atmosphere on silver sulfide formation has not been thoroughly investigated. It was shown in Ref. [39] that the presence of NO_2 greatly enhances the silver corrosion rate, which was 5 times faster than that expected when the combination of H_2S and NO_2 is present in the atmosphere. However, too fast a corrosion process would consume silver plating completely before the whiskers may develop.

5.5.2.2   Plating Factors

The role of silver plating thickness in silver corrosion was studied in Ref. [19]. The test was performed on three sets of model contacts made from copper electroplated with 2, 5, and 20 μm of silver precorroded for 1, 5, and 10 days, respectively, in an atmosphere containing a mixture of Cl_2, NO_2, and H_2S (Battelle class III). Then precorroded contacts were powered and tested under the normal force. It was found that the whiskers start growing from the corrosion layer formed in an atmosphere containing H_2S concentration as low as 0.1 ppm. With the thinnest silver coating (2 μm) covered with the thickest corrosion film (10 days in the Battelle chamber), silver whiskers formed just outside the conducting area of the anode in the 15 min test. At the time the whiskers were forming, the temperature of the contacts was close to 160°C.

5.5.2.2.1   Thickness of Ag_2S Film

The effect of silver sulfide film thickness on the start of whiskers growth was studied in Ref. [38]. Until the thickness of the silver sulfide film reached 0.075 μm (750 Å), no whiskers were found. Such a thickness of the tarnish film is formed in 1 week on silver exposed to 20 ppm H_2S at room temperature in humid air (75% RH).

5.5.2.2.2   Base Material

The role of base material in whisker growth is not clear yet. Silver whiskers are found growing not only on silver-plated copper but also on Ag90/Ni10 alloy [33]. In Ref. [21], whisker formation was observed on coupons made of silver with purity 99.9%–99.99%. In this experiment, silver coupons were exposed to an atmosphere containing 3 ppm H_2S at 40°C and 80%–85% RH.

5.5.3   Failures in Electrical Equipment Caused by Silver Whiskers

Exposure of silver-plated parts of the CB to hydrogen sulfide results in the formation of a thick dark brown or black layer of silver sulfide and multiple clumps of silver whiskers (Figure 5.4). In most cases, the presence of corroded silver and whiskers led to the thermal failure of the CB.

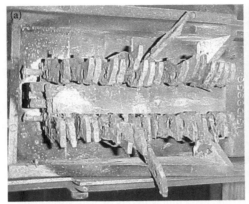

FIGURE 5.9 Images of the parts of circuit breaker failed due to the heavy corrosion and growth of silver whiskers: finger cluster (a) and arc chute (b).

In recent years, the growth of silver whiskers was seen on the corroded silver-plated parts of LV CBs in service in the control room of a pulp recycling plant [12]. This happened 1 year after the factory switched from one pulp processing technology to a different process, which used a chemical reaction that produced H_2S gas. The level of H_2S in the control room was detected to be in the range 0.5–2 ppm.

Since then, every 2–3 months the maintenance personnel at the plant have performed cleaning of the units having extensive corrosion and whiskers growth. Within 2 months after thorough cleaning, two LV CBs failed after violent thermal runaway (Figure 5.9).

After the failure, the failed breakers were disassembled, and clumps of the whiskers were found everywhere inside the current-carrying path of the breakers. Almost 12 g (4 oz) of whiskers were collected and studied using SEM/EDS techniques [14].

5.5.4 Study of the Silver Whisker Phenomenon

5.5.4.1 Visual Appearance of the Whiskers

Most of the whiskers have the shape of a single thick thread, which may eventually split into several thinner filaments. The filaments have many different colors, shapes, and sizes ranging from tiny thin "dawn"-looking formations to thick and strong threads, which are bound together into large tufts and clumps (Figure 5.10).

FIGURE 5.10 Silver whiskers inside failed breakers.

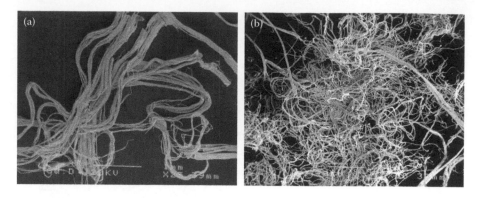

FIGURE 5.11   Extra-long and thick whiskers (a) and extra thin whiskers (b) with silver luster.

The color of the whiskers ranges from a shiny silver and gold luster to dull dark gray or black. The color of some whiskers changes along the length of the filament from bright silver to bright yellow or from metallic luster to dark dull gray or black. The thickness of the whiskers is up to 1 mm (0.04 in.). Some of the whiskers shown in Figure 5.11 grew as long as 5–10 cm (2–4 in.).

5.5.4.2 Morphology

The morphology of the whiskers was studied by using SEM. The samples for SEM analysis were selected from whisker clumps under stereo microscope magnification.

SEM photographs of the whiskers at high magnification in Figures 5.12 and 5.13 show "baby" whiskers at the beginning of the growth. They start extruding from the flakes as a group of many thin straight or curly filaments of different thickness, which may then fuse together into a thick filament.

The color of the whisker surface (with metallic luster or dull and dark) seems to correlate well with the surface morphology. The surface appearance of the whiskers with a metallic luster and dull dark color is significantly different [14]. Under high magnification, various groups made of long threads have been found. Each thread eventually splits into several thinner filaments. Some filaments are bound together, forming agglomerates. The thickness of the whiskers varies from 2–4 μm to almost 1000 μm (1 mm).

5.5.4.3 Chemical Composition

Chemical composition of the whiskers was determined by EDS. X-ray analysis of chemical composition of the whiskers [14] is based on elemental composition of a relatively thin upper layer of the

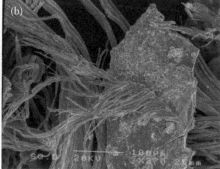

FIGURE 5.12   Silver whiskers, growing from the flakes: "nursery" (a), and pictured together with thicker (older) whiskers (b).

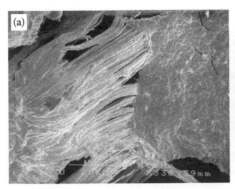 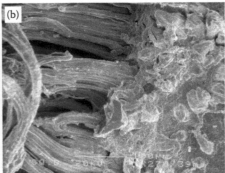

FIGURE 5.13   Silver whiskers at the roots of growth.

whiskers, since EDS techniques provide quantitative analysis with a sampling depth of 1–2 μm. It was found that whiskers having silver luster are made of 94%–99.4% of silver. Some of the whiskers with silver luster also contain up to 4.5% copper. The whiskers with yellow luster have higher content of copper (from 7% to 33%).

The chemical composition of the whiskers also correlates well with the color of the whiskers. Sulfur (S) was found on the surface of the whiskers with the content of S varied from traces (on the whiskers with silver luster) and up to 9% sulfur. That much sulfur was found on the whiskers with dull dark color, which is enough to constitute a pure silver sulfide on the surface. When the whisker has one yellow end and the other end is dark, the content of sulfur along the whisker is growing gradually from 1.8% on the yellow end up to ~4.5% on the dark end.

The difference in color and composition of the long and thick whiskers may be explained by the different temperature conditions of the whisker "birth place." The silvery whiskers could grow much faster in hotter areas and be exposed to H_2S for a shorter period of time than same-size whiskers, which grew slower in cooler locations. The whiskers that have been exposed to corrosive gas for a longer time have a thicker silver sulfide layer on the surface and therefore are dark.

The composition of the whiskers is compared with the composition of the flakes collected from the surfaces of the corroded conductive parts of the breaker. The flakes contain silver, copper, and sulfur in proportions corresponding to the mixture of silver and copper sulfides.

5.5.4.4   Chemical Composition of the Whisker Cross Section

One of the samples for x-ray analysis was prepared to determine the internal composition of the whisker. The sample was cut through the base metal, thick corrosion layer, and small whisker growing from the corrosion layer. All parts of the sample, where chemical compositions have been determined, contained three elements: copper (Cu), silver (Ag), and sulfur (S). The composition was measured along the line starting from the base metal made of pure copper and then across the corrosion layer, which was found to be made of 86.8 at% Cu, 2.9 at% Ag, and 10.7 at% S. This composition corresponds to a mixture of copper and silver sulfides. Chemical analysis along the cross section of the whisker revealed that it was made of pure silver. Silver content changed drastically across the border between the corrosion layer and the whiskers from ~3% in the corrosion layer to 100% in the whisker's body [14].

An important conclusion is that silver whiskers are mostly metallic formations; therefore, they are highly conductive. Taking into consideration the rate of growth, the size, and the amount of whiskers developing practically everywhere but mostly on the corners and edges of conductive parts, of the energized electrical equipment, it is obvious that the consequences of the phenomenon are extremely dangerous.

5.5.5　Silver Whiskers Puzzle

The silver whisker phenomenon cannot be explained today no matter how much we know about the conditions in which these amazing "creatures" grow. The origin and the forces that initiate and support the growth of the whiskers are not yet defined. Not a single reliable reason has been suggested to explain why these filaments are growing on silver under suitable conditions. Perhaps one of this book's readers may find out what makes silver whiskers grow so intensely.

5.5.5.1　What Do We Know?

As was found, silver whiskers spontaneously grow on copper parts plated with silver, but may also grow on silver alloy or pure silver. Whiskers having silver metallic luster are made mostly of silver. When grown on plating, they may contain up to 4% base metal copper. Whiskers with yellow metallic luster contain silver and 8%–30% copper. It seems that in general the make-up of the whiskers and the conditions of growth are known. The presence of H_2S and elevated temperature seem to be two major accelerating factors of the growth.

5.5.5.2　What Do We Not Know or Understand?

It is not clear how to explain the following facts. Silver whiskers may grow from corroded silver coupon or from silver alloy. However, more often they are found on silver-plated copper. The speed at which the whiskers grow reaches 1 mm/h. When growing on silver plating, metal filaments made of almost pure silver are growing from a corrosion layer made of silver and copper sulfides containing less than 3% silver.

The original silver plating thickness is no more than 10 μm on copper substrate. The filaments are up to 10 cm long and 1 mm thick.

A simple calculation shows that to make just one silver whisker with such dimensions from silver contained in the layer of original plating with a thickness of 10 μm, the metal should be collected from the area of ~100 cm². Moreover, in each location the whiskers grow in large tufts (see Figure 5.5) that are made of dozens of such filaments.

5.5.5.3　Questions Not Answered Yet

- What is the mechanism of growth of a 100% silver filament (single crystal?) from a corrosion layer made of CuS and Ag_2S?
- What makes silver reduction from silver sulfide proceed so fast?
- What is the driving force that causes silver from a corroded 10-μm thick plating layer to be drawn into gigantic filaments of 1000 μm thickness and 100,000 μm length made of pure Ag?
- Where are all these amounts of silver coming from?

5.5.5.4　Native Silver Wires

In geology, the phenomenon of whiskers growing on rocks is well known for many years. Many native whiskers have been found all over the world by geologist and miners. These unique growths are appreciated for their authentic beauty and unusual shapes and colors. They were found on rocks containing tiny amounts of precious metals (gold, silver) and also other metals. Many of the samples of these whiskers (called native wires) are offered for sale on the Internet. Surprisingly, the appearance of native silver wires is practically the same as that of silver whiskers shown in the pictures in this chapter (Figure 5.14).

The major difference is in the age of the whiskers in these two groups. Reviewing Figure 5.14, try to find out which of them matured for a couple of months in corroded electrical switchgear and which ones take centuries or millenniums to grow on the rocks? Is the silver whisker phenomenon the same as that observed in nature?

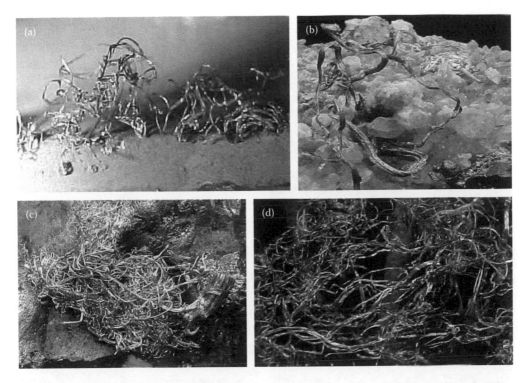

FIGURE 5.14 Corrosion silver whiskers and natural silver wires (a, b, c, and d): is there a difference?

5.5.6 OTHER DISCOVERIES OF SILVER WHISKER GROWTH

The fact that pure silver whiskers grow from silver sulfide is confirmed in the study of *in situ* high-temperature XRD study of silver sulfide Ag_2S in Ref. [40] to better understand the polymorphs of Ag_2S on heating. Powder of Ag_2S was synthesized in a standard chemical reaction, analyzed by XRD and identified as β-Ag_2S, acanthite.

SEM photographs in Figure 5.15 show that during high-temperature XRD thermal processing silver whiskers developed in Ag_2S powder on heating strip. The powder was exposed to 650°C. The presence of silver whiskers in heated Ag_2S powder was confirmed by XRD. The results of this study confirm that pure silver whiskers grow from silver sulfide under elevated temperature but do not provide an explanation of this phenomenon.

Silver whiskers have been found on silver-plated copper contacts during Eaton investigation of the failure in Netherland [41]. The whiskers could cause a short circuit in an assembly, which was found in the petrochemical industry. This finding also confirms that most likely silver whiskers grow on sharp edges of silver coated parts in a sulfur rich, warm and moisture environment.

Laboratory tests were carried out by Eaton and Shell Laboratories in Netherland. Close cooperation between both the laboratories lead to the conclusion that the phenomena of silver whiskers is resulting from silver sulfide on the silver-plated contact surfaces of components used in the LV switchgear assemblies [32]. During these studies, the bundles of silver whiskers have been grown in laboratory setup in the atmosphere with high concentration of H_2S and elevated temperature and RH.

A switchgear failure in Switzerland in 2007 revealed the presence of silver whiskers on electrical contacts shown in Figure 5.16. These silver whiskers are seen at high magnification in SEM photograph in Figure 5.17.

FIGURE 5.15 Silver whiskers formed during HTXRD thermal processing of Ag_2S at 650°C: (a) Ag_2S powder on a HTXRD heating strip, post heating; (b) Silver whiskers attached to Ag_2S removed from the heating strip; (c) SEM micrograph (500×) of a silver whisker growing out of Ag_2S powder;(d) SEM micrograph (500×) of a bundle of silver whiskers. (From Blanton et al. In situ high-temperature X-ray diffraction characterization of silver sulfide Ag_2S, JCPDS-International Centre for Diffraction Data 2011 ISSN 1097-0002, pp. 110–117, *Powder Diffraction*, 26(2), SI (JUN), 114–118, 2011, Online. Available: www.icdd.com/resources/axa/vol54/V54_14.pdf. With permission.)

FIGURE 5.16 Silver whiskers found on circuit breakers installed at paper mill in Europe. (From Personal communication.)

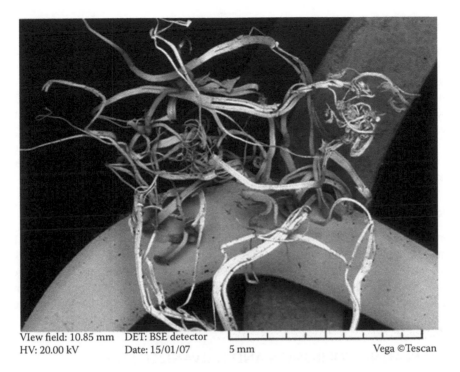

VIew field: 10.85 mm   DET: BSE detector
HV: 20.00 kV   Date: 15/01/07   5 mm   Vega ©Tescan

FIGURE 5.17 SEM photograph of silver whiskers found on switchgear installed at paper mill. (From Personal communication.)

5.6 TIN PLATING CORROSION

Tin does not have great structural strength and so it finds most application as a coating for both ferrous and nonferrous substrates. Use is made of the corrosion resistance, soldering characteristics or ability to prevent galvanic corrosion effects between other metals. During atmospheric exposure, the presence of a substrate, exposed through pores in the coating, has little effect on the corrosion of tin.

Tin is often used as a plating for two metals widely used in the electrical industry: copper and aluminum. Although tin is considered resistant to atmospheric corrosion, there are several negative characteristics of tin plating, such as susceptibility to fretting corrosion, formation of intermetallics when plated on copper, low softening and melting temperature, and softness. These degrading tin properties are discussed in more detail in Chapters 3 and 4, including the formation of tin whiskers, one of the major failure (MF) causes of electric and electronic devices using tin as a plating or solder.

Tin plating withstands relatively harsh environmental conditions with corrosive gases pretty well. However, tin reacts with different gases in the atmosphere forming various nonconductive compounds that may interfere with the electrical conductivity of tin plating [42,43]. CR is relatively stable for tin contacts in the atmosphere containing a mixture of SO_2, NO_2, and Cl_2 gases [43]. This is because relatively few corrosion products are formed on tin contacts. Also, the surface films formed on tin contacts shatter more easily upon plastic deformation.

Corrosion experiments in atmospheres containing different combinations of SO_2, NO_2, and Cl_2 demonstrate that two types of corrosion processes take place in plated contacts. One is surface corrosion, which occurs when the surface finish is severely attacked by a corrosive atmosphere. The other is pore corrosion, which occurs when the underplating materials are attacked to a greater extent than surface corrosion. Both processes occur in tin contacts and the former is dominant in atmospheres containing Cl_2.

5.6.1 TIN OXIDATION

Under most atmospheric exposure conditions, corrosion of tin depends on the formation and stability of an oxide film [42]. The thickness of a tin oxide film is shown to depend on the RH of the air. It was found that for samples with a starting thickness of tin oxide of 2 nm stored in a cupboard, the thickness of tin oxide film more than doubled to 4.5 nm in <45 days of observation, while for the samples stored in a desiccator, the oxide growth was much slower, reaching ~2.5 nm in 25 days of storage.

It was also shown that at room temperature an amorphous film is formed, while at higher temperatures it is apparently crystalline α-SnO, possibly with some SnO_2. The increase of the CR of tin coatings exposed to the atmosphere can be related to the growth of the oxide film on tin.

5.6.2 REACTION OF TIN WITH OTHER GASES

Under normal atmospheric conditions, H_2S and SO_2 have little effect on tin, but above 100°C, tin sulfide (SnS) is formed. High concentrations do, however, produce darkening or tarnish, while at low temperatures liquid or gaseous SO_2 may produce a crystalline solid with tin. The halogens attack tin readily at room temperature except for fluorine, which only reacts at a high rate above 100°C. Similarly, fumes from concentrated acids attack tin. Tin resists corrosion by the vapors of organic acids and other organic compounds such as might be emitted from wood or other packing materials.

5.7 ZINC PLATING CORROSION AND GALVANIZED STEEL

5.7.1 ATMOSPHERIC CORROSION OF ZN

Zinc is found in electrical equipment mostly on nonconductive parts, such as enclosures and hardware in the form of plating or as deposit on steel structures. Corrosion of zinc is possible when these parts are exposed to the atmosphere, either indoor or outdoor [44–46].

The behavior of zinc and zinc coatings during atmospheric exposure has been closely examined in multiple tests conducted throughout the world. The performance of zinc in a specific atmospheric environment can be predicted within reasonable limits. Precise comparison of the corrosion behavior in atmospheres is complex because of the many factors involved, such as the wind direction, the type and intensity of corrosive fumes, the amount of sea spray, and the relative periods of moisture or condensation and dryness.

However, it is generally accepted that the corrosion rate of zinc is low. It ranges from 0.13 μm/year in dry rural atmospheres to 0.013 mm/year in more moist industrial atmospheres. Zinc is more corrosion resistant than steel in most natural atmospheres. For example, in sea coast atmospheres the corrosion rate of zinc is about 1/25 that of steel. However, in ventilated indoor atmospheres the corrosion of both steel and zinc is extremely low.

Zinc is pretty resistant to atmospheric corrosion due to the formation of insoluble basic carbonate films. Environmental conditions that interfere with the formation of such films may attack zinc quite rapidly. Important factors that control the rate at which zinc corrodes in atmospheric exposures are the duration and frequency of moisture, the rate at which the surface dries, and the extent of industrial pollution of the atmosphere. In dry air, zinc is slowly attacked by atmospheric oxygen. A thin, dense layer of oxides is formed on the surface of the zinc, and outer layer then forms on top of it. The outer layer breaks away occasionally, but the underlayer remains and protects the metal, limiting its interaction with the oxygen. Under these conditions, which occur in some tropical climates, the zinc oxidizes very slowly. The rate of drying is also an important factor because a thin moisture film with higher oxygen concentration promotes corrosion.

The influence of multiple atmospheric factors, such as pollution, humidity, and temperature, on the corrosion of zinc is related to their effect on the initiation and growth of protective films. Zinc may be safely used in contact with most common gases at normal temperatures if water is absent, because moisture stimulates corrosion attack on zinc. Dry chlorine does not affect zinc. Hydrogen sulfide (H_2S) is also harmless because insoluble zinc sulfide (ZnS) is formed. On the other hand, SO_2 and chlorides have a corrosive action because water-soluble and hygroscopic salts are formed in a moist environment.

When used indoor, zinc corrodes very little in ordinary indoor atmospheres of moderate RH. In general, a tarnish film begins to form at spots where dust particles are present on the surface: the film then develops slowly. This attack may be a function of the percentage of RH at which the particles absorb moisture from the air.

Rapid corrosion can occur where the temperature decreases and where visible moisture that condenses on the metal dries slowly. This is related to the ease with which such thin moisture films maintain high oxygen content because of the small volume of water and large water/air interface area.

Studies of the corrosion of zinc when it was exposed to the atmosphere in the form of coatings on steel, either hot dipped or electroplated, indicated that there was less than a 10% difference between the corrosion rate of galvanized iron, zinc die castings, and three grades of rolled zinc. It was also found that the corrosion rate tended to be a linear function of time. In some instances, where the rate changed after a period of time, it was concluded that the amount of contamination in the atmosphere had changed.

Corrosion rates for zinc, like many other metals, can vary by as much as two orders of magnitude in atmospheric environments depending on the specific environmental conditions. Therefore, it is important to know the specific corrosion rate in a given application environment in order to effectively use zinc-coated steels in outdoor structures. For example, when compared with steel in the marine environment, corrosion attack on zinc where chloride deposition is important has low rates. Such an excellent resistance is attained by the hard, dense, protective products of corrosion in a chloride atmosphere. However, similar results cannot be achieved in a sulfurous atmosphere where the products are soft, voluminous, and nonprotective.

5.7.2 WHITE RUST ON ZINC

Zinc is a sacrificial barrier to protect the base steel. Zinc will sacrifice itself and eventually disappear before the base metal starts to corrode. Zinc plating protects the underlying steel by the formation of a "galvanic cell," which results in the zinc corroding preferentially to steel. It is usually used as a functional indoor coating. By itself, 5–8 μm of zinc plating will probably get no more than 12 h of SS protection per ASTM-B117. With a clear chromate top coat this is increased to 24–36 h, while a yellow chromate top coating (yellow zinc) can achieve protection up to approximately 96 h.

Where freshly galvanized steel is exposed to pure water (rain, dew, or condensation), the water will react with zinc and progressively consume the coating. Zinc is oxidized and changed to WR that is a basic compound of zinc carbonate and zinc hydroxide; WR has white color and is called in the technical literature as "white rust" or "wet storage staining."

The most common condition in which WR occurs is with galvanized products that are stacked together, tightly packed, or when water can penetrate between the items and remain for extended periods. It is manifested as a bulky, white, powdery deposit that forms rapidly on the surface of the galvanized coating under certain conditions. The appearance of the corroded galvanized steel parts of a CB with "white rust" is shown in Figure 5.18.

"White rust" causes considerable damage to zinc plating and is always detrimental to its anticorrosion function [47–49]. Different types of Zinc plating have different levels of ability to resist the formation of "white rust" (Table 5.1).

5.7.3 GALVANIZED STEEL

Galvanized steel is the most important application of zinc. One of the most popular methods of making steel resist corrosion is by alloying it with zinc. When steel is submerged in melted zinc, the

FIGURE 5.18 "White rust" on corroded galvanized steel parts of a CB exposed to moisture.

TABLE 5.1

Approximate Hours to Formation of White Rust (WR) and Red Rust (RR) on Zinc Plating after Salt Spray (SS) Test

Type of Zinc Plating	Hours to WR	Hours to RR
Zinc	10	70
Zinc with clear chromate	75	200
Zinc with yellow chromate	100	250
Zinc–cobalt with yellow chromate	150	450
Zinc–cobalt with yellow chromate and post sealer	300	600

Source: Based on the data from ASTM B-117.

chemical reaction permanently bonds the zinc to the steel through galvanizing. Coatings of metallic zinc are generally regarded as the most economical means of protecting steel against corrosion.

Seven methods of applying a zinc coating to iron and steel are in general use: hot-dip galvanizing, continuous-line galvanizing, electro-galvanizing, zinc plating, mechanical plating, zinc spraying, and painting with zinc-bearing paints. Most of these are hot-dipped galvanized coatings containing a small amount of aluminum.

The thickness of electroplated coatings is considerably lower than those applied by the hot dip process. Galvanizing produces a zinc coating on the steel surface and is one of the most effective methods for corrosion protection of steel. This is attributed to the excellent corrosion resistance of zinc coatings, particularly in atmospheric environments. Zinc coating/plating processes are described in more detail in Chapter 5.

The most common method for estimating the corrosion life of galvanized steels uses a generalized value to represent the corrosion rates for five predetermined atmospheric environments as a function of zinc coating thickness. Service life is defined as the time to reach 5% rusting of the steel surface. It can be used to estimate the service life of a given coating thickness or to specify a coating for a given environment.

Corrosivity maps specify the thickness of galvanized coating required to achieve a useful life of 50 years in various environments. The method is applicable to zinc coated steel produced by batch or continuous galvanizing, including hot-dip, electrogalvanized, and thermal sprayed coatings. However, it does not apply to coatings containing >1% of alloying elements. The method assumes that the galvanized product is free of significant defects that could accelerate corrosion. For example, the service life of an HDG Zn coating with a thickness of 40 μm is estimated to be ~30 years in an industrial environment, while it could last >50 years in a rural environment. However, the service life prediction does not consider issues of water entrapment that can create conditions for severe corrosion.

The galvanizing process prevents steel from corroding by providing cathodic barrier and patina protection [50].

- Cathodic protection: Zinc is more anodic than steel. Thus, when a corrosion cell forms—when both zinc and steel have an electrolyte and a return current path present—zinc readily gives up electrons that protect the steel from corrosion. Zinc will protect the base steel until all of the galvanized coating is consumed.
- Barrier: Zinc metal is very dense and thus does not allow moisture, or electrolytes, to penetrate the galvanized coating. Therefore, the base steel does not corrode.
- Patina: When exposed to the atmosphere immediately after the galvanizing process is complete, zinc metal reacts with oxygen in the air to form a very thin zinc oxide powder on the galvanized coating surface. After a few days, the zinc oxide reacts with hydrogen in the air to form zinc hydroxide. As the zinc hydroxide is exposed to moisture in the air over a period of months, a thin film of zinc carbonate forms. Zinc carbonate is a passive patina film that is tightly bound to the remaining zinc of the galvanized coating and gives an HDG coating its durability.

5.7.4 Signs of Galvanized Steel Corrosion

Galvanized steel corrosion is usually easily recognized in the presence of rust and pits. The same signatures reveal the corrosion of other types of steel/iron parts.

5.7.4.1 Rusting

Corrosion of steel is easily recognized because the corrosion product is RR. As soon as a protective layer of zinc plating on and around steel parts is destroyed and converted into porous "white rust," there is no more barrier between steel and oxidizing agents of the atmosphere, which reacts with iron to form rust on the surface of steel parts. When iron base alloys corrode, dark corrosion products usually form first on the surface of the metal.

If moisture is present, this ferrous oxide coating is converted into hydrated ferric oxide, which is known as "red rust." This material will promote further attack by absorbing moisture from the air. The shade of iron oxides ranges from a dull yellow through various oranges and reds to a deep black. "Red rust" (Fe_2O_3) and "black rust" (Fe_3O_4) are usually mixed together and the color of corrosion deposit may be different in shade depending on which oxide in the mixture is present in a larger amount.

When an electrical unit such a CB is exposed to moisture, the protective zinc plating corrodes and degrades, leaving the base metal (steel) exposed to corrosion as well. Extensive rusting of the parts made of steel (bolts, fasteners, mounting plates, and laminated arms) then develops.

5.7.4.2 Pitting Corrosion

Pitting is a form of extremely localized corrosion that leads to the creation of small holes in the metal. The driving power for pitting corrosion is the lack of oxygen around a small area. This area becomes anodic while the area with excess of oxygen becomes cathodic, leading to very localized galvanic corrosion. The corrosion penetrates the mass of the metal, with limited diffusion of ions, further pronouncing the localized lack of oxygen. Pitting can be initiated by a small surface defect, being a scratch or a local change in composition, or damage to protective coating. Polished surfaces display higher resistance to pitting. An example of pitting corrosion on galvanized steel parts is shown in Figure 5.19.

FIGURE 5.19   Pitting corrosion on galvanized steel.

The alloys most susceptible to pitting corrosion are usually the ones where corrosion resistance is caused by a passivation layer. Regular carbon steel will corrode uniformly in sea water, while stainless steel will pit. All stainless steels can be considered susceptible to pitting, but their resistances vary widely. Their resistance to attack is largely a measure of their content of chromium, molybdenum, and nitrogen.

The addition of about 2% molybdenum increases pitting resistance of stainless steel.

Another important factor is the presence of certain metallurgical phases (in particular, the grades 303, 416, and 430F of steel containing inclusions of manganese sulfide) which have very low pitting resistances, and ferrite may be harmful at austenitic grades in harsh environments. A clean and smooth surface finish improves the resistance to attack.

5.7.5   Factors Affecting Galvanized Steel Corrosion

Several factors may accelerate the rate of galvanized steel corrosion.

5.7.5.1   Environment

The areas near the ocean coast might be influenced by multiple environmental parameters (airborne salinity, meteorological systems, hydrochemical parameters).

These factors might have a strong effect on the corrosion of mild steel, which is the material that the nonconductive parts of electrical equipment, such as CBs, are often made of. It is a well-known fact that corrosion of steel on a sea coast is 400–500 times greater than it is in a desert. The closer to the coast line the equipment is located, the higher the corrosion rate is. Steel specimens located 80 ft from a coast corroded 12 times faster than specimens located 800 ft away from a coast.

Salt is the chief contaminant. "Black Rust," found in the pits on the surface of corroded steel plates, is more often formed in saline environments with limited access to oxygen.

5.7.5.2   Thickness of Zinc Plating

A factor that could increase the rate of zinc plating deterioration exposed to wet conditions is the thickness of the plating. The rate of corrosion for zinc plating depends on the atmosphere to which the plating is exposed. ASTM standards specify different minimum zinc plating thicknesses according to the severity of different atmospheric conditions determined by the corrosion rate (see Table 5.2).

TABLE 5.2

Mean Corrosion Rates for Zinc Plating and Thicknesses of Zinc Plating for Different Atmospheric Conditions

Atmosphere	Mean Corrosion Rate (μ per year)	Condition	Minimum Plating (μ)
Industrial	5.6	Mild dry indoor	5
Urban nonindustrial	1.5	Moderate corrosive	8
Suburban	1.3	Severe corrosive and outdoors	12
Rural	0.8	Exceptionally severe—marine	25
Indoors	≪0.5		

Source: Based on the data from ASTM B695-85 and ASTM-B633.

According to the ASTM standards, if zinc plating had not been initially exposed to wet conditions and then installed indoors, it will take about 10 years to completely deteriorate even the thinnest plating specified (5 μm).

However, if plating was wetted and the formation of "white rust" had started to develop, it will be only a matter of time how soon it will be all consumed. The deterioration begins when a localized corrosion cell is formed. The activity of such corrosion cell (pit) results in rapid penetration through the zinc coating to the steel.

It will take somewhat longer for converting zinc plating into "white rust" for thicker plating, but as soon as the WR is formed there are no more barriers for oxidizing agents to penetrate through "white rust" to the surface of steel.

5.7.6 CORROSION OF GALVANIZED STEEL IN CIRCUIT BREAKERS

Many parts of power CBs are made of galvanized steel. Corrosion of galvanized steel may lead to the failure of electrical equipment exposed to excessive moisture. In this example, the parts of an MV CB were found to be severely corroded after 2 years of previous refurbishment. Signs of corrosion have been found on plated steel parts, such as bolts, washers, mounting plates, laminated arms, and so on. The parts with corrosion spots are shown in Figures 5.20 and 5.21. To determine the root cause of corrosion, it is very important to carefully examine the parts and locate all signs of corrosion.

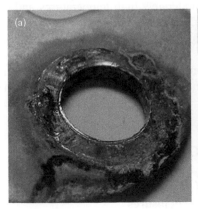

FIGURE 5.20 "White rust" and "red rust" on the mounting plates (a) and around the opening in steel plates (b).

FIGURE 5.21 Pitting corrosion found on the surfaces of steel parts: black and brown pits (a) and green pits (b).

According to the reconditioning program, all steel parts have been zinc-plated, although the type of zinc plating is rarely predetermined in reconditioning procedures. The color of the parts of the corroded breaker allows defining the type of plating. For example, bronze–gold or light gold color is specific for yellow zinc (Zn chromate) plating.

Three major types of defects have been found: white deposits on the surfaces, dark pits on the plates of laminated arms, and brown corrosion product on plates and fasteners. The most extensive brown corrosion or "red rust" specific to the corrosion of iron is observed on mounting plates, bolts, and washers made of steel (Figure 5.20).

In the close vicinity of brown spots and pits on all the parts where steel is rusted, a white deposit is present, which means that Zn plating has been corroded, forming "white rust," leaving the base metal of the parts(steel) unprotected from corrosion by Zn plating and exposed to air and water. Without this protection, not only "red rust" is produced but also the pitting corrosion develops at random in the pits on the surfaces of the parts.

On CB parts, the pitting corrosion is mostly found on those plates of the laminated arms that are tightly stacked together (Figure 5.21). It can be explained by the moisture condensed between the plates, where it accumulates and does not dry out for long periods of time (the so-called "sheltered" corrosion). The color of pits is different in different spots: black, dark brown, and light brown (Figure 5.21a).

Most probably, the corrosion products in the pits consist of a mixture of different iron oxides. Pitting usually occurs under certain conditions, particularly involving high concentrations of chlorides (such as sodium chloride (NaCl) in sea water), moderately high temperatures and also the presence of neutral or acidic solutions containing other chlorides or halides. This explains why the color of some pits has a greenish hue to it (Figure 5.22b), which might be due to the admixture of $FeCl_2$ in some spots.

Since the facility where the breaker has been installed was located near the sea coast, environmental factors such as elevated airborne salinity and humidity could play a destructive role in severe corrosion of steel and lead to failure of the CB.

5.8   MEANS OF CORROSION PROTECTION OF ELECTRICAL EQUIPMENT

5.8.1   Protective Coatings for Conductive Parts, Enclosures, and Frames

5.8.1.1   Metallic Coatings for Conductive Parts and Enclosures

Metallic coating is usually used to protect the base metal from corrosion. As protective coatings, a distinction must be made between those such as zinc, cadmium, and aluminum on steel that protect

FIGURE 5.22 Creep corrosion field failure of PCB in high sulfur environment by corrosion products bridging vias. (From Schueller R. Creep corrosion on lead-free printed circuit boards in high sulfur environments, *SMTA 2007*, Online. Available: http://www.dfrsolutions.com/wp-content/uploads/2012/06/2007_08_creep_corrosion_on_lead-free_pcb_in_high_sulfur_environments.pdf. With permission.)

by sacrificial behavior and those that must provide a substantially continuous protective envelope around the protected metal. In the latter category are metals such as nickel, tin, silver, brass, and chromium. Sacrificial metals corrode as coatings in much the same manner as solid metals do until the base or protected metal is exposed at pores or bare areas. Then the galvanic corrosion action begins to accelerate the corrosion of the protective coating. This galvanic couple effect tends to protect the base metal at the pores or bare spots.

This reaction is different from what happens in the case of the more noble protective coatings that act as a cover and resist the atmosphere due to their inertness, passivity, or protective films. As soon as a pore or bare spot appears, the corrosion of the base metal is accelerated. Die castings of a metal such as zinc covered with a copper–nickel–chromium coating will suffer severe corrosion at large pores or discontinuities in the coating.

Some metal coatings are applied by electroplating, while the others are produced by flame or plasma spraying, hot dipping, electrostatic sputtering, or vapor deposition. The corrosion rate of a metal coating is largely independent of the method of application, except where impurities play a role. In noble types of coating, porosity of metal coatings is very important, because it affects mechanical, electrical, and anticorrosion properties of the plating. There are many factors that influence a coating porosity, but among them, the most important factor reducing porosity seems to be the thickness of electrodeposited coating. Porosity tends to decrease rapidly (by exponential law) as the coating thickness increases.

Tin and tin alloys tend to be inert in the atmosphere unless any specific type of contamination exists. As far as exposure to the atmosphere is concerned, tin appears as a coating on steel containers but it can accelerate the corrosion of the more anodic steel by pitting. Due to their inertness, silver and gold are used as alloys to improve corrosion resistance of the more reactive metals. Corrosion of tin and silver is discussed in more detail in the following sections.

5.8.1.2 Polymeric Coating and Paints for Enclosures and Frames

The exterior surfaces of electrical equipment, such as enclosures and frames, if not metal plated, are protected from the environment by applying paints or polymeric coatings. To protect steel enclosures from corrosive solvents, fumes, and gases, various powder coatings (Polyester, Epoxy) and polyurethane paint are used. Detailed information about the polymeric coatings used in the electrical industry is given in Section 17.7 of Chapter 17.

Paints and coatings are designed to provide a long-lasting barrier between the metallic material they cover and the atmosphere. Various criteria and their order of importance have been established when selecting the best coating for enclosure, which basically perform two functions: protection and decoration.

However, essentially all polymers exposed to environmental factors will change in some manner. The active rays of the sun become potent agents of change in organic materials. Further polymerization of the resin can occur to produce embrittlement. Other types of bonding can be triggered to make polymers more crystalline. Any volatile component of the material, such as a plasticizer, can be evaporated. The polymer chains may be simply oxidized and broken up to destroy the product. Oxygen, ozone, and moisture act with the sunlight to degrade the plastics.

The external evidence for attack may be blushing (loss of gloss), chalking, or change in the color of the product. This is often observed on the epoxy and polyester polymers when they have been boldly exposed to the environment. However, only mechanical tests will reveal the extent of degradation of either thermoplastic or thermosetting resins.

The effect of high atmospheric temperatures or heating from a direct exposure to the sun can be particularly strong on thermoplastic polymers. Creep or distension of the polyvinyl chloride (PVC) and polyethylene (PE) plastics will occur readily unless provision is made to prevent overheating or stressing of the materials. Other thermoplastics can be dimensionally stable under normal atmospheric temperatures.

The strength of these thermosetting resins is not noticeably changed. Polymeric materials should be thoroughly tested if they are to be exposed to the atmosphere. ASTM Recommended Practice D1435 describes the appropriate conditions for such a test exposure and suggests tests that might be used to evaluate changes in the materials. Changes in the mechanical and physical properties of the polymers are determined for definite results. Weight gains or losses can be of interest, but do not provide substantive results.

Accelerated cabinet testing of the materials can be performed with essentially the same validity as when testing metallic materials. Poor materials can be eliminated, but extrapolation of the data to forecast the life of plastic parts in the atmosphere should not be attempted. ASTM D1499, D2565, and G26 may be used to ensure the controlled cabinet testing of the plastics. ASTM D750 should be used for the evaluation of the elastomers.

5.8.2 Means of Protection from Silver Corrosion

There are several ways to protect metals from corrosion which would be applicable to various materials exposed to different corrosive atmospheres.

5.8.2.1 Silver Protection from Corrosion

Any developed antitarnishing methods will only delay silver corrosion but cannot eliminate it. None of these methods satisfy all applications. The most important requirement for any type of silver plating treatment used in the electrical industry is to not affect the electrical resistance. Various methods useful in decorative applications such as the application of various protective films such as lacquers, oils, and inorganic deposits such as metal oxides on silver plating [51] cannot be utilized for electrical contacts.

The application of electrolytic silver alloy deposits and composite coatings is another type of antitarnishing technique that is often used for luxury jewelry items but that cannot be used here due to high electrical resistance of these deposits. The most frequently used antitarnishing technique in the electric and the electronic industry is metal coating applied by immersion or electrochemically.

5.8.2.2 Silver Plating Thickness

The role of silver plating thickness in the corrosion process might be very important. It was shown in Ref. [20] that when the silver coating thickness increases, the total corrosion film thickness decreases and the corrosion film contains less copper. First, these changes lead to a dramatic decrease in CR. Second, the build-up of silver sulfide thick enough to start the growth of the whiskers might be significantly slowed down.

Therefore, the thickness of silver coating is a very important factor in the reliability of contacts in environments that are corrosive for both silver and copper. The copper creep corrosion film is thought to be detrimental to silver-plated contacts, mainly due to a higher growth rate in comparison with silver corrosion films. Thick pore-free silver coating with a thickness of about 20 μm is recommended to prevent the influence of the copper substrate on the corrosion mechanism.

5.8.2.3 Alternate Plating

Corrosion-induced growth of the filaments on metals other than silver has not yet been seen. Therefore, the use of other than silver plating may eliminate the silver whisker problem. Substituting silver with expensive gold plating might not be a good solution. There are difficulties with this strategy, since many control switches and protective relays are not available with gold-plated contacts. Tin plating has other problems, such as a much higher resistance than silver, galling, and softness; these make it less than ideal for sliding contact applications. However, in Ref. [34], it was recommended that tin plating instead of silver be used in electrical equipment for the paper industry, in current-carrying parts of contactors and breakers, and in the stabs and stab mating fingers, everywhere except arcing contacts.

When choosing tin for plating in a very aggressive atmosphere it is useful to follow the "Tin Commandments" developed by Whitley [52]. It is considered to be poor practice to use pure tin plating on contacts because of whiskers, which develop on the surface no matter whether there is atmosphere or vacuum. The use of tin alloys with 5% or less of lead or other tin alloys was suggested as a practical solution to the whisker problems [53]. However, the introduction of the RoHS initiative limited the use of lead for most applications of tin plating. The problems that arise due to tin whiskers growth and other tin plating issues are described in Section 5.5 of Chapter 5 and Section 6.1 of Chapter 6.

5.8.3 Conversion Treatment

The purpose of a conversion treatment (passivation or antitarnishing treatment) is to delay the appearance of tarnishing. Conversion coating is a generic term for any process that converts the surface of a part into a coating. For example, chromating is an immersion process in which metal parts are immersed in a solution containing chromates and other chemicals. The purpose of this process is to convert the surface into a thin film that enhances the corrosion resistance. In this process, Chromium VI is present in solution and if this treatment is applied, CrIV will be present in tarnishing and passivation layers on silver coatings.

Recently, there has come a rule that a conversion treatment should comply with the requirements of the European directive 2002/95/CE (RoHS directive [54]) of the European Parliament and the European Union Council of January 27, 2003. This directive restricts the presence of hazardous substances in electric and electronic equipment, and hexavalent chromium (Chromium VI) is one among them. RoHS principles and reach are global.

The directive indicates that anything covered by RoHS entering the EU must be in compliance; this includes cables made in China, parts molded in the United States and PCBs made in Japan. If it is destined for the EU, it is impacted by RoHS. And it is not just the EU that is taking steps to reduce the toxins in electronic devices, California's Electronics Waste Recycling act of 2003 (SB 20, Chapter 526) echoes the RoHS directive and was supposed to take effect on January 1, 2007.

5.8.4 Chromium-Free Varnish-Preventative Processes

Antitarnish processes without heavy metal ions are designed to add value to silver and silver-plated substrates by preserving the as-plated appearance and minimizing tarnishing of the surface. The parts are immersed into a liquid concentrate, which after treatment leaves a thin, invisible organic film on the plated surfaces that in general has no influence on the surface resistance of electrical contacts. This protection works only for 6–12 months; therefore it would be a suitable treatment to protect electrical contacts from tarnishing in storage. In service protection of the surfaces will be only temporary; it will not prevent corrosion in the long run.

According to technical specifications, some solutions seem appropriate for use in electrical applications. However, since some chemicals provide a top coat (or sealer) on the parts, for specific applications they should be tested to prove that film application does not deteriorate the electric continuity.

5.8.5 Means of Preventing the Corrosion of Zinc-Plated Steel Parts in Electrical Equipment

Corrosion of galvanized steel in electrical equipment is often the result of exposure of the components to moisture, which causes zinc plating oxidation and formation of porous white deposit on the parts ("white rust"). A premature, rapid loss of galvanized zinc coating on steel surfaces results in the corrosion of the underlying steel. The most probable root cause of the rusting of the galvanized steel parts is an exposure of electrical equipment to wet conditions: rain, dew, or condensation.

For those equipment that are usually installed indoors, such as CBs, water could get inside the breakers during the period between shipment and installment. The possibility of moisture getting inside such equipment depends on how the units have been packed for shipment and how long and where they have been stored at the customer facility before being installed.

WR is a postgalvanizing phenomenon. One way to prevent rust is by checking the manner in which the unit is packed, handled, and stored before it is installed and used. The presence of WR is not a sign of the galvanized coating's performance, but rather points to the responsibility of all those involved in the supply chain to ensure that the causes of WR are recognized and the risks of its occurrence minimized on zinc-plated steel.

Every precaution should be taken to prevent electrical apparatus from becoming and remaining wet. During shipping, the unit should be protected to ensure that no water intrusion takes place even when it rains while shipping. Plastic wrapping with tarp could be a solution to prevent water from getting inside the unit, but the wrapping should not be too tight in order to allow the unit to "breathe" by air circulation. If it does get wet, it should be stored in such a manner that it is allowed to dry. The customer should not allow zinc-plated apparatus to be exposed to water or dampness for extended periods of time after receiving the unit. The zinc-plated apparatus should be stored in a warm, dry environment. The unit should not be removed from the cover until it has reached room temperature.

5.8.6 Vaporized Corrosion Inhibitors

Vaporized corrosion inhibitors (VCI) presented in Refs. [55–58] may be used for the temporary protection of silver plating. VCI are widely available in the market; these compounds do not have a significant effect on the electrical CR. VCI emitters can be placed within electrical enclosures to retard internal corrosion, as long as the air exchange with the outside environment is limited. They release a chemical vapor into the enclosed still air space, to form a protective invisible molecular film settling on the metal surfaces as a barrier layer that limits excess of corrosive gases to the plating surface.

A full range of corrosion inhibitors has been developed to protect ferrous and nonferrous metals, including formulations that protect combinations of metals. Providing protection for 1–2 years, each such container is effective within an area of up to 1.8 m (6 ft) in radius. Similar vapor-releasing compounds are also available as tapes or spray-can formulations for use on surfaces in the open air.

For storing apparatus or components, according to the environment of exposure, a special packing containing VCI could be used that would protect the material during transport and storage. Several effective VCI products are widely available in the market. Vapor corrosion inhibitors may slow down the corrosion of silver-plated copper parts in switchgear in atmospheres containing over 200 ppm of mixed gases such as SO_2, H_2S, HCl, and so on.

There is a temperature limit for using this technique, since the higher the temperature is, the faster the VCI evaporates. At an operating temperature of 120–140°F the useful lifetime will start decreasing, and if they are used continuously at elevated temperature, containers emitting VCI should be replaced more often than once every 2 years.

5.8.7 LUBRICATION

Sealing out corrosive gases with lubricants can help to protect metal surfaces against atmospheric corrosion. The use of lubricants is widely accepted as an effective means of protecting silver-plated contacts from corrosion [59]. However, for many years, the benefits of using lubricants for corrosion inhibition were overlooked because prospective users were skeptical that a nonconductive substance could be applied to contacts without interfering with the conduction.

The choice of the lubricant for corrosion protection should be based on the thorough evaluation of the product for survival in long-term use and the ability of such products to provide friction and wear reduction. Corrosion inhibitors/lubricants applied at any sliding contact points and surfaces will do this job very well. Special attention should be given to the correct choice of these products to not induce interference with the electrical properties of conductive parts. An effective lubricant for electrical parts should be chemically inert towards the metal surfaces to be lubricated.

Lubricants used on silver-plated surfaces should not carry sulfur-containing compounds. Lubricants for electrical parts should not collect dust and particulate matter and should be resistant to oxidation and chemically inert to corrosive atmospheric components. These lubricants should not build nonconductive deposits on contact surfaces after multiple operations and retain long-term thermal stability at least up to 200°C (~390°F). Any corrosion inhibitor must be thoroughly tested for electrical application. In fact, very few lubricants have been identified that provide exceptional long-term corrosion protection and yet do not produce any adverse effect on connector surfaces. An example of a lubricant that proved useful for corrosion protection for connectors made or plated with precious metals is six-ring polyphenyl ether (PPE) or its mixture with microcrystalline wax, although the wax content should be controlled.

Multiple products such as greases with a synthetic soap thickener and other commercial products have been tested and proven useful for bolted contact application. These issues will be presented in more details in Chapter 8.

5.9 ENVIRONMENTAL EFFECTS AND CORROSION IN ELECTRONICS

In the electric power industry, multiple *electronic devices* are used such as microprocessor-based controllers of power system equipment in CBs, transformers, and capacitor banks. This electronic devices receive data from multiple sensors and power equipment, and can issue control commands, such as tripping CBs if they sense voltage, current, or frequency anomalies, or raise/lower voltage levels in order to maintain the desired level.

Common types of electronic devices include protective relaying devices, OLTC controllers, CB controllers, capacitor bank switches, recloser controllers, voltage regulators, and so on. Digital protective relays are electronic devices, using a microprocessor to perform several protective, control and similar functions. A typical electronic device can contain multiple protection and control functions controlling separate devices, an auto-reclose function, self-monitoring function, communication functions, and so on. Because of a very important role of electronic devices in performance and safety of electrical equipment, it is essential to provide an environment in which electronic devices will not prematurely deteriorate and fail. However, electrical equipment is often installed in corrosive environment, therefore the role of electronics corrosion resistance and protection cannot be underestimated.

Electronic manufacturing industry uses a number of metallic materials in various forms. Also new materials and technology are introduced all the time for increased performance. In recent years, corrosion of electronic systems has been a significant issue. Electronic components are used in virtually all environmental conditions, both indoors and outdoors, in vehicles, and so on containing multiple metallic components, electronic devices corrode.

Corrosion of component leads or solder joints are often clearly visible. PCB conductors may corrode from exposure to a corrosive environment where the solder resist does not provide sufficient protection [60]. Conductors may also corrode under the influence of current, by electrolytic corrosion or electrochemical migration. In this case, the conductor at the most positive potential corrodes and the metal is redeposited at the more negative conductor. Electrochemical migration may occur also inside the board, producing copper filaments, cathodic anodic filaments (CAF), along the glass fibers of FR4 laminates. Corrosion rates are higher at higher humidity and in the presence of ionic contaminants such as corrosive flux residues.

5.9.1 Factors of Reduced Corrosion Reliability in Electronics

There are several factors leading to reduced corrosion reliability of electronic components [61].

5.9.1.1 Miniaturization

Conductors of integrated circuits (ICs) are getting thinner and thinner, distances between conductors are getting shorter and metal combinations prone to bimetallic corrosion are common. Thus, it doesn't take much corrosion to destroy a conductor and only slightly corrosive conditions to cause electrochemical migration or bimetallic corrosion. For components used in electronics, sub μm corrosion loss is detrimental, for example, in a conductor, and even corrosion films below 0.1 μm thickness may result in contact disturbances.

Overall size of electronic equipment has been decreasing presently at a very fast rate. The size of the ICs has been decreased by a factor of 10 over the last couple of years, which means that the spacing between the IC components is ~200 nm. For components on a PCB, the spacing is around 5 microns, while in mid-90's it was 100 microns. The reduction in size and distance between components makes the system more susceptible to corrosion problems. The reduced spacing between components on a PCB due to miniaturization of device is a factor that has made easy for interaction of components in corrosive environments. A fault in the conducting path of such constructions may be generated by a material loss of the order of picograms (10^{-12} g) [62].

5.9.1.2 Multiplicity of Materials

In electronic components a variety of materials is used simultaneously and a set of material depends on type of electronic device/component, such as ICs, PCB, electronic connectors, switches, magnetic recording media (Hard disc), packaging and shielding parts, and so on.

For example, an *IC* is basically made of silicon, although other metals such as gold, silver, copper, zinc, aluminum, or their alloys are used for various purposes such as connecting leads, bumps, and so on. The bumps on the silicon wafer are made of gold and they are connected to the lead frames with a bonding process for electrical connection using a gold or aluminum wire. The end of the lead, which needs to be bonded to the wafer is selectively treated with 99.9% gold or 99.9% silver.

Typical *PCB* consists of copper connecting path integrated in a fiber glass reinforced epoxy polymer. PCB wiring use *ENIG*-process (EN-Gold). The ENIG consists of a few micron thick Ni–P (EN) coating on the base copper on PCB, followed by a thin gold layer on EN [63]. Thickness of the gold layer varies from 50 to 100 nm.

A large variety of materials are used for *electronic connectors*. The classical materials used for LV electronics are the copper and copper alloys such as CuSn6, CuBe2, CuNi10Sn2, or authentic stainless steel X12CrNi 177. These basic materials are usually electroplated with a noble metal such as gold (hard gold plating), palladium (Pd), or PdNi alloy with hard gold flash [61].

Computer *hard disk drive* consists of several components. The round disc platter is the magnetic recording media for hard disc. The platter is usually made of aluminum electroplated with nickel, and an over layer of 50 nm thick magnetically active cobalt alloy using PVD.

Electronic industries are on a transition path to *lead free solders*. The candidate alloy components involve Sn as the base element, Ag, Bi, Cu, and Zn as the major alloying elements, and some other minor additions such as In and Sb. However, based on the market developments, it appears that the ternary system Sn–Ag–Cu emerges as the primary choice for replacement. Alloys within the composition range Sn–[3.4–4.1] Ag–[0.45–0.9] Cu are generally recommended.

5.9.2 CAUSES OF CORROSION IN ELECTRONIC SYSTEMS

There are several environment-related factors that could accelerate the corrosion process, which needs to be controlled in order to reduce the corrosion effects in electronics systems. Sources of contamination include flux residues, cleaners, processing gases, coatings, and many circuit handling steps where salts, acids, and other ionic elements have access to the board and components. Among the environmental issues, a significant problem is *the residues* found on the PCBs. There are two types of residues: *process-related* residues (1) and *service-related* residues (2).

Process-related residues are the contamination on the surface due to the remains of the chemicals used for the manufacturing process. This could be the leftover of original chemicals or decomposed fractions of a compound formed during the production cycle. These are the fluxing agents, etching medium, plating bath residues, or additives from the polymeric materials. Although the solder fluxes (especially the no-clean variety) are designed to give essentially no residue after the soldering process, depending on the temperature cycles and applications, small amount of original compound or decomposed products could remain on the surface. In practice, tiny fractions of these chemicals are enough to accelerate the corrosion process. Use of some fluxes resulted in Ag dendrites due to electrolytic migration.

Specification ISA 71.04 "Environmental Conditions for Process Measurement and Control Systems: Airborne Contaminants" [64] covers four different species of airborne contaminates, including liquids, solids, gases, and biological influences. Although all four species are important, airborne contamination (gases) can contaminate and corrode electronic devices. These gases are hydrogen sulfide, active sulfur gases, inorganic chlorine, sulfur oxides, SO_2 and SO_3, nitrogen oxides, NO_x, hydrogen fluoride, ammonia and derivatives, ozone and other strong oxidants (bleach) most of which are catalysts for sulfur corrosion.

The service-related residues are the residues introduced during exposure to service environments. They can be *aggressive ions* like chlorides, SO_2 (gas), NO_2 (gas), or other types of chemically aggressive ions. The presence of such substances triggers corrosion to a large extent under humid conditions.

There are three types of *gases* that can be considered as prime candidates in the corrosion of data center electronics: acidic gases such as hydrogen sulfide, sulfur and nitrogen oxides, chlorine, and hydrogen fluoride; caustic gases, such as ammonia; and oxidizing gases, such as ozone. Of these, the acidic gases are of particular concern. For instance, it takes only 10 ppb of chlorine to inflict the same amount of damage as 25,000 ppb of ammonia.

Each site may have different combinations and concentration levels of corrosive gaseous contaminants. Performance degradation can occur rapidly or over many years, depending on the specific conditions at a site. Common sources of corrosive gases are shown in Table 5.3, based on information from Ref. [64].

TABLE 5.3

Sources of Reactive Environmental Contaminants

Sources of Contamination	Category of Contaminant	Chemical Name	Chemical Symbol
Aerosol content, oceanic processes, ore processing	Liquid	Chloride ions	Cl
Chlorine manufacture, aluminum manufacture, paper mills, refuse decomposition, cleaning products	Gas	Chlorine, chlorine dioxide	Cl_2, ClO_2
Automobile emissions, animal waste vegetable combustion, sewage, wood pulping	Gas	Active nitrogen	N_2
Automobile emissions, fossil fuel combustion, microbes, chemical industry	Gas	Nitrogen oxides	NO_x
Microbes, sewage, fertilizer manufacture, geothermal steam, refrigeration equipment, cleaning products, reproduction (blueprint) machines	Gas	Ammonia	NH_3
Incomplete combustion (aerosol constituent), foundry	Solid	Carbon	C
Automobile emissions, combustion, microbes, trees, wood pulping	Gas	Carbone monoxide	CO
Semiconductor manufacturing, wood and wood products, photo developing	Gas	Acetic acid	CH_3COOH
Fruit, vegetable, cut flower storage and transportation	Gas	Ethylene	C_2H_4
Semiconductor manufacturing	Gas	Arsine	AsH_3
Wood products, floor and wall covering, adhesives, sealants, photo developing, tobacco smoke	Gas	Formaldehyde	HCHO
Automotive emissions	Gas	Halogen compounds	HBr, HI
Automotive emissions, fossil fuel processing, tobacco smoke, water treatment, microbes, paper mills	Gas	Hydrocarbons (alcohols, aldehydes, ketones, organic acids)	HC, THC
Automotive emissions, combustion, oceanic processes, polymer combustion	Gas	Hydrogen chloride	HCl
Fertilizer manufacture, aluminum manufacture, ceramic manufacture, steel manufacture, electronic device manufacture, fossil fuel	Gas	Hydrogen fluoride	HF
Geothermal emission, microbiological activities, fossil fuel processing and combustion, wood pulping, sewage treatment, auto emissions, ore smelting, sulfuric acid manufacture	Gas	Hydrogen sulfide	H_2S
Foundries, sulfur manufacture	Gas	Mercaptans	S_8, R–SH
Fossil fuel combustion, automobile emissions, ore smelting, sulfuric acid manufacture, tobacco smoke	Gas	Sulfur oxides	SO_2, SO_3
Atmospheric photochemical processes, automotive emissions, tobacco smoke, electrostatic filters	Gas	Ozone	O_3
Crystal rock, rock, and ore processing, combustion, blowing sand, industrial sources	Solid	Inorganic dust	

Source: Adapted from ANSI/ISA Standard 71.04–1985–Environmental Conditions for Process Measurement and Control Systems: Airborne Contaminants, Research Triangle Park: International Society for Automation.

Dust is another component atmospheric contamination. Dust is everywhere. Even with the best filtration efforts, dust will be present in a data center and will settle on electronic hardware. Particularly, massive dust presence is found in industrial facilities which contain electrical and electronic equipment. Though most dust is benign, under specific circumstances dust will degrade electronic hardware [65]. See more about dust corrosive effect in data center in Section 7.9.5.

One mechanism by which dust degrades the reliability of PCBs involves the absorption of moisture from the environment by the settled dust. The ionic contamination in the wet dust degrades the surface IR of the PCB and, in the worst-case scenario, leads to electrical short circuiting of closely spaced features via ion migration.

When the dust acts as a moisture trapping agent, then formation of *water layer* is easier on a dusted surface compared with the clean one. Water layer or a tiny water droplet sitting on the surface of a PCB can generate micro galvanic cell by electrically connecting the two metallic parts. Under humid conditions, the condensed water layer at the surface of a PCB generates conducting path (electrolyte) for corrosion. It is like a tiny corrosion cell with adjacent components on a PCB acting as electrodes. In dusty conditions, corrosion is possible even at relatively low humidity such as 50%–70%.

If the surface of the electronic components is not contaminated, corrosion will not be a large issue. However, in practice significant levels of contamination could be detected on PCBs and components.

5.9.3 Major Forms of Corrosion Observed in Electronic Systems

Electronic systems could experience various types of corrosion, although the underlying mechanism in all cases is electrochemical in nature. The most prominent corrosion problems found for electronic systems are gas phase corrosion (1), anodic corrosion and electrolytic metal migration (2), cathodic corrosion (3), galvanic corrosion (4), fretting corrosion (5), stray current corrosion (6), and creep corrosion (7) [62].

5.9.3.1 Gas Phase Corrosion

The presence of low levels of hydrogen sulfide could create severe problems in electronics. The most susceptible material to H_2S in the electronic system is silver due to the formation of silver sulfide crystals. The damage to the silver–palladium thick film due to its exposure to H_2S [63] is presented in Ref. [63].

Devices where silver sulfide crystals have been found were used in steel plant, adjacent to blast furnaces that are certain to raise ambient levels of hydrogen sulfide, although no local measurements of H_2S concentration were carried out. The devices were likely to have seen several months of levels above 100 $\mu g/m^3$ and transient peaks of several milligrams per cubic meter. The problem is well known in environments even with low concentrations of H_2S (under 50 ppb), which is below the detection limit of H_2S by smell.

5.9.3.2 Anodic Corrosion and Electrolytic Metal Migration

Electrolytic migration occurs due to the presence of a potential gradient between two conductors connected by a thin layer of solution. Electrolytic metal migration is a very common form of corrosion observed for electronic systems containing metals such as Pb, Sn, Cu, Au, Ag, and so on. Due to electrolytic migration, dendrites grow from cathode to anode filling the gap finally leading to electric short and system failure [66]. A typical example of the dendrite formation between silver coatings on a PCB due to electrolytic metal migration presented in Ref. [61], as well as another example of corrosion with the pitting formed between copper conducting lines. Similar behavior could be observed for the corrosion of aluminum connecting lines on an IC chip. Failure from dendrite growth can occur in less than a day under the right environmental conditions and circuit layout [67].

5.9.3.3 Galvanic Corrosion

Galvanic corrosion, distinct from electrolytic corrosion, occurs when two dissimilar metals come in contact with one another in the presence of water. This can occur with solder joints, on plated conductors, and can be particularly problematic on conductors used in switches and plug contacts.

In electronic systems, *galvanic corrosion* manifest in many ways and is considered as dreadful for connectors and switches. For connectors and switches, the corrosion products formed by the galvanic corrosion induce high resistance in the circuit. Usually, the connectors are made of multilayer metallic coatings as described before. The metallic layers have distinctly different electrochemical properties. A typical example is the ENIG parts on the PCB.

A mobile phone key-pad made by the ENIG process corrosion is presented in Ref. [61]. The corrosion process is due to galvanic coupling between layers. The immersion gold (IM Au) layer is porous that exposes part of the EN layer. The large difference in electrochemical potential between EN and IM Au cause corrosion of EN, while the Au layer acts as a powerful cathode. As the corrosion proceeds, pitting of EN layer exposes Cu at deep pit areas.

5.9.3.4 Creep Corrosion

Creeping corrosion has been documented by Battelle labs, Texas Instruments, Rockwell Automation, Alcatel-Lucent, Hewlett Packard, Dell Computer, and the University of Maryland among others. In this extreme corrosion mechanism, the surface finish is corroded, exposing the underlying copper. Then the corrosive ions form copper salts. Liquid solvents, such as water, will carry the salts across the circuitry in an electrolyte solution. With sufficient liquid and surface tension, the electrolyte can bridge the gap between circuitry and solder mask.

When heated, the electrolyte may dry and deposit crystallized corrosion salts. If more liquid comes in contact with the circuitry, the cycle repeats, with the formation of rings of crystalline deposits. When analyzed, the material found to creep is copper with its corrosion product anion, usually sulfide.

Referred to as creep or creeping corrosion, this phenomenon is characterized by growth of copper sulfide crystals from copper features on the outer surfaces of the PCB or PWB [68,69].

The degree of corrosion may be to the extent that electrical failures can occur. Creep corrosion is not to be confused with electrochemical migration found with thick electrolytic plated silver. Electrochemical migration failure mechanism is characterized by long dendrites that grow in one direction, typically from one electrical source to another until a bridge is formed and a short occurs.

Creep corrosion is formed when unprotected copper reacts with another metal in an acidic medium such as moisture containing sulfur. The copper sulfide crystals grow in all directions equally [70].

As shown in Figure 5.22, creep corrosion is formed from copper sulfide oxidation products on the surface of a PCB, which caused field failure in high sulfur environment by bridging vias (via is an electrical connection between layers in a physical electronic circuit that goes through the plane of one or more adjacent layers). Creep corrosion product formed on hard disk drive is shown in Figure 5.23. The oxide layer can form on top of the solder mask and is typically seen in environments with elevated sulfur levels.

Immersion silver (ImAg) surface finish is often used on PCBs in high-sulfur to mitigate creep corrosion. However, a high rate of failure was recently found in computers with ImAg finished boards in an environment containing high levels of elemental sulfur.

Many industries release elemental sulfur into the air. For example, the source of sulfur could be the clays used in the prototype modeling of vehicles, at paper mills sulfur is used in the bleaching process, and power plants release sulfur where geothermal sources are used to turn steam turbines. Corrosion was found mostly inside *via* holes because copper remained exposed inside via holes due to incomplete silver finish covering. The corrosion products crept over via holes and caused shorts in the circuits [71].

In Ref. [72], the mechanism of creep corrosion in PCBs with immersion silver finish has been studied. It was found that under a high sulfur and high humidity environment ImAg finish surface

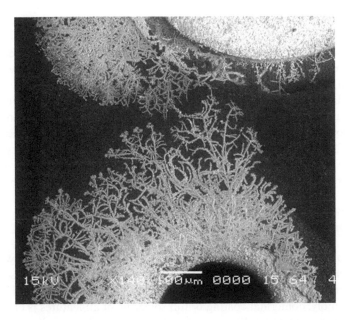

FIGURE 5.23 Creep corrosion on hard disk drive (HDD) with immerging silver (ImAg) surface finish: Cu_2S precipitates out of solution in a dendritic structure. (From Schueller R. Creep corrosion on lead-free printed circuit boards in high sulfur environments, *SMTA 2007*, Online. Available: http://www.dfrsolutions.com/wp-content/uploads/2012/06/2007_08_creep_corrosion_on_lead-free_pcb_in_high_sulfur_environments.pdf. With permission.)

was sulfurized. The dendrite-shaped creep corrosion products were formed by galvanic corrosion on ImAg finished pad edges.

It was found that the root reason behind the creep corrosion in PCBs finished with ImAg was that copper was not fully covered by Ag. Increasing the thickness of ImAg finish should reduce the probability and the areas of exposed copper in order to decrease the amount of creep corrosion.

5.9.3.5 Metallic Whiskers

This type of corrosion typically refers to electrically conductive, crystalline structures of tin that sometimes grow from surfaces where tin (especially electroplated tin) is used as a final finish. Tin whiskers have been observed to grow to lengths of several millimeters (mm) and in rare instances to lengths up to 10 mm. Short circuits caused by tin whiskers that bridge closely spaced circuit elements maintained at different electrical potentials.

"Whisker growth" can be also described as corrosion in which microscopic metal crystals grow out of the surface of the conductive metal. It is caused by the presence of sulfide molecules, for example, silver sulfide on a silver surface, which can migrate freely over the metallic surface and collect at dendrite boundaries where nucleation takes place and sulfide crystals grow out of the surface of the metal. These whiskers can be of size long enough to connect portions on a board, chip, and so on and cause short circuits. By some accounts, whisker growth accounts for approximately 10% of documented failures (see Chapter 6 on various failures in electronics caused by metallic whiskers).

5.9.4 Strategies and Means to Protect Electronic Components from Corrosion

The best coating, handling, shipping, and storage methods will not prevent corrosion of electronics used in heavily polluted environments. As electronics became widely used, computer workstations were placed directly on the production line. With worldwide industrialization, these factories were replicated in countries without strong environmental policy. Then, with the transition to lead-free technology, the computers became more vulnerable, as thick HASL coatings were replaced with thin organic and metallic finishes. Lastly, PCB is used in nonsoldered surface features. Previously, all surface areas were soldered. Now, many features are left unsoldered to allow test point, surface contact connection, and ground shielding functionality [69].

Controlling both the RH and rate of change in RH are very important factors in preventing corrosion failures. The amount of moisture in the air (RH) is the most important atmospheric parameter for electronic corrosion mechanisms. A few parts per billion of atmosphere contaminants combined with a humidity of greater than 60%, for example, micro level condensate, is very corrosive to electronic equipment. On the other hand, excessively low humidity (less than 40%) may cause electrostatic discharge problems. Therefore, the following is recommended [73]:

- Maintain humidity levels between 40% and 60%
- Control humidity fluctuations to less than 6% per hour rate of change

For very difficult environments, a corrosion prevention strategy can be viewed in three steps: *reinforcement, remediation, relocation*. The corrosion resistance of the PCB assembly can be reinforced with more robust protection. Protection can be at the PCB level, such as improved finishes. At component assembly, conformal coating or full soldering of all features is effective.

Very difficult environments can be improved with *filtering, air conditioning, humidity control, and general housekeeping*. Sensitive devices can be installed even in difficult industrial environments, with use of positive airflow, double doors, and interlocks.

If there is repeated failure of a device when introduced into an industrial environment, even when other preventative measures have been taken, *relocation* is the last resort. The computer workstation may have to be relocated to a less corrosive environment within the factory. For some devices, such as programmable controls on robotic manufacturing equipment, this may not be an option. More information on means to protect electronics from corrosion is presented in Section 7.11.

5.9.5 FAILURES OF ELECTRONICS IN DATA CENTERS

5.9.5.1 Failure Modes of Data Centers' Electronics

It was found that sulfur-bearing gases, even in the absence of moisture, attack silver-forming silver sulfide corrosion products that undermine the integrity of the package. The package with its integrity breached exposes the underlying silver to further corrosive attack until all the silver in the section is consumed, leading to an electrical open. The silver sulfide corrosion product on the field failed hardware is often visible as needles or nodules, under a low-power microscope [63].

Failure modes due to airborne dust include mechanical effects (obstruction of cooling airflow, interference with moving parts, interconnect interference, etc.), chemical effects (component corrosion, electrical short circuiting of closely spaced features, etc.), and electrical effects (impedance changes and electronic circuit conductor bridging).

Harmful dust in data centers is generally high in ionic content such as chlorine-bearing salts. The source of this harmful dust is mainly outdoor dust in the size range of 2.5–15 μm for coarse dust and 0.1–2.5 μm for fine dust. The coarse dust particles have mineral and biological origin, are formed mostly by wind-induced abrasion and can remain airborne for a few days. The fine dust particles are generally the result of fossil fuel burning and volcanic activity and can remain airborne for years. Large bodies of salt water are also a major source of airborne dust contamination in data centers. Sea salt can be carried 10 km (~6 miles) inland or farther by high winds present in coastal areas and can damage electronic devices at this range [63].

New industry-accepted specifications include particulate contamination limits that identify the quantity and RH of deliquescent dust. Additionally, research by *ASHRAE's Technical Committee 9.9 for Mission Critical Facilities, Technology Spaces, and Electronic Equipment* led to the publication of a white paper on contamination guideline for data centers [65] and the formulation of new gaseous contamination limits used to update International Society of Automation (ISA) Standard 71.04-2013 [64]. This research also led to the publication of an iNEMI position paper [74] and efforts to update the Chinese data center design guide GB 50174-2008 [75].

The observed increase in the rate of hardware failures in data centers resulted in recommendations to monitor and control dust and gaseous contamination in addition to temperature–humidity control. These additional environmental measures are necessary to reduce the two most common recent failure modes of copper creep corrosion on circuit boards and the corrosion of silver metallization in miniature surface-mounted components.

5.9.5.2 Control Process in Data Centers

With the changes to IT and datacom equipment mandated by various RoHS directives, data center owners, managers, and operators should include an environmental contamination monitoring and control section as part of an overall site planning, risk management, mitigation, and improvement plan [76].

The three parts of this plan should comprise:

1. Considerations for the *assessment of the outdoor air and indoor environment* with regards to corrosion potential: ISA Standard 71.04 can be used to provide site-specific data about the types and levels of gaseous contamination in the amount of corrosion being formed. Corrosion classification coupons (CCCs) can be used as a survey tool to establish baseline data necessary to determine whether environmental controls are needed and, if so, which ones.
2. A specific *contamination control strategy*: Corrosion in an indoor environment is most often caused by a short list of chemical contaminants or combination of contaminants. The contaminants present in a specific area are highly dependent on the controls put in place to mitigate them. Most of this would involve the selection and application of the appropriate chemical filtration systems to clean both the outdoor air being used for pressurization and/ or ventilation as well as any recirculation air.
3. A real-time *environmental monitoring program* based on the severity levels established in ISA Standard 71.04: Real-time atmospheric corrosion monitors can provide accurate and timely data on the performance of the chemical filtration systems as well as the room air quality.

ISA Standard 71.04 has been updated to include silver corrosion monitoring as a requirement in determining environmental severity levels. Many manufacturers of datacom and IT equipment currently reference this standard in their site planning/preparation guidelines as well as their terms and conditions for warranty compliance. The addition of silver corrosion rates as a required metric serves to bridge the gap between ambient environmental conditions and the reliability of RoHS-compliant (lead-free) electronic equipment.

Often the relationship between corrosion levels and hardware failures in data centers is overlooked or unknown. However, Advanced Micro Devices (AMDs), Cisco, Cray, Dell, EMC, Huawei, Hitachi, HP, IBM, Intel, Oracle, Seagate, Silicon Graphics International (SGI), and others are working hard to increase awareness of the problem and it solutions. These manufacturers are also working to develop successful corrosion monitoring and control programs.

5.10 CORROSIVITY OF SMOKE AND ITS EFFECT ON ELECTRICAL AND ELECTRONIC EQUIPMENT

Electrical and electronic equipment damage is often caused by so-called nonthermal damage produced by smoke and corrosive products generated from the combustion of materials, particularly plastics. Generally, nonthermal damage can be divided into two types: corrosive damage and noncorrosive damage [77–80].

5.10.1 Noncorrosive Damage from Smoke

Noncorrosive damage, from soot and smoke, can cause a loss of product, cosmetic repair costs, and damage to electrical components by conductivity (i.e., shorting of circuits). Noncorrosive damage includes the effects of relatively cool smoke and soot that condense on electronic equipment. The type of smoke and soot produced depends on the type of fire.

Smoldering fires are typically low temperature fires that produce particles heavily laden with nonconductive organic compounds. These particles, while not corrosive, condense and may cause insulating films that impair the contacts of electrical equipment.

High-temperature fires produce particles or so-called soot. There are different types of soot defined by the type of particles it contains [81]. *Conductive Soot* (graphitic)—may cause shorting of electrically insulated components. *Semi-conductive Soot* (carbonaceous)—may cause an increase in thermal insulation. *Insulating Soot* (organic)—may cause electrical isolation of conductors, connectors, and relays. *Metallic Soot* (vaporized metal or oxides)—may cause environmental health and safety issues and may be conductive, causing shorting of electrically insulated components. Additionally, soot may cause discoloration of surfaces and seizure of mechanical.

5.10.2 Corrosive Damage from Smoke

Electrical circuits and electronic components are at the greatest risk from this type of corrosive damage. Corrosive damage results when the by-products of combustion combine with moisture, condense, and form acidic compounds on the surface of equipment. The frequency and severity of losses due to nonthermal fire damage have increased in recent years, due to the increased use of thermoplastics in building materials and furnishing. Modern electronic equipment is sensitive to corrosive smoke produced by some thermoplastics.

PVC (and similar materials) is a thermoplastic compound that softens when heated. In its solid form, PVC can be difficult to ignite and generally will self-extinguish when the heat source is removed. However, when PVC burns, it releases hydrogen chloride that is both toxic and corrosive. When hydrogen chloride combines with moisture, it can form hydrochloric acid (HCI). There are a number of other materials that when burned produce potentially corrosive gases, including Teflon (PTFE). However, these materials typically produce hydrogen fluoride (HF), not HCI. HCI has been found to cause corrosive damage to electrical equipment at relatively low concentrations. HF and other potentially corrosive gases do not corrode electronics until much higher concentrations are reached. The specific concentrations vary based on the RH, type of circuit exposed, and the concentration of corrosive material present.

Typically, for corrosion to occur, the RH must be greater than 30% at 68°F (20°C). Corrosive gases (i.e., chlorine ions) require water or another electrolyte at the metal's surface before corrosion can occur. Rapid corrosion occurs at between 70%–80% RH. Since corrosive hydrochloric gas at lower concentrations does not have a significant smell or optical density, the presence of HCI may not be noticed until after substantial equipment damage has occurred. The amount of these corrosive gases produced and the spread of contamination will be proportional to the volume of plastic-based products present and the intensity of the fire. The greater the volume of plastic materials involved in combustion, the higher the volumetric concentration of corrosive gases produced.

5.10.3 Effect of Smoke on Electronic Equipment

Fires in telecommunication facilities, computer centers and other research, production or office facilities where large numbers of computers and electronic equipment are present can be very costly, even if the equipment is not directly exposed to the heat, but only to smoke. Exposure of the equipment to smoke gases and soot particulates can result in malfunctions, leading to extensive cleaning or replacement in order for a business to continue operating [79].

The problem was more serious when relay switches were common in electronic equipment, and especially in telecommunications centers. With the conversion to electronic switches, electronic equipment has become more resistant to corrosive gases and soot, however, the problem has not totally disappeared.

It is important to remember that corrosion does not stop after the flame exposure ceases, due to deposits of corrosive gases being present on the surface. Deposited soot also undergoes structural changes, becoming more compact with the passage of time and more difficult to remove by regular cleaning procedures. Soot in the particulate fraction of smoke efficiently adsorbs corrosive compounds and corrodes the surfaces by deposition. The presence of high molecular weight, sticky, organic compounds in the nonparticulate fraction of smoke increases the efficiency of deposition of soot with corrosive compounds on surfaces to enhance corrosion and leakage of current.

5.11 MEANS OF ENVIRONMENTAL CONTROL FOR CORROSION PROTECTION

5.11.1 ASSESSMENT OF ELECTRICAL AND ELECTRONIC EQUIPMENT EXPOSURE TO CORROSIVE ENVIRONMENT

Electrical equipment is exposed to the corrosive environment of many industrial facilities, including pulp and paper mills, chemical plants, oil refineries, and so on. In electrical power systems, the build up of corrosion products can cause overheating, leading to reduced equipment life or even to explosions of CBs and motor starters. While the cost of the systems themselves are high, the production losses from moderately sized operations can quickly add up to exceed the costs of repairing or replacing the affected systems.

Corrosive failure of electronic equipment is particularly common in the pulp and paper industries. Low concentrations, in the ppb range, of gaseous contaminants such as H_2S, SO_2, NO_x, Cl_2, NH_3, or mercaptans in air coming in contact with electrical switching circuits will quickly react with most metals, forming nonconducting layers of metal salts. As a result, substantial resistance to current flow develops, variations in voltages occur, and the entire circuit may become useless due to changes in component value.

There are many different kinds of failures in electronic process control equipment that are induced by corrosion. The failure could be due to the braking of corroded copper wires, caused by the acid condensates trapped under the insulation.

Electrical contacts could fail due to corrosion product creep following galvanic corrosion between precious metal plating and substrates, for example, between gold and copper or nickel. Gold, which does not react with the environment, is used as plating on base metal substrates, such as copper and nickel, which react with the environment.

Corrosion can occur where the gold plating is porous, and the rate of attack is accelerated when the humidity is high. Corrosion products creep over the gold-plated surface, resulting in high electrical CR and a functional loss of electrical contact.

Corrosion product creep is a rapid process, and in a simulated sulfur-containing environment, gold-plated surfaces were substantially covered with corrosion products in a few weeks [82]. Chemical substances found in paper mill environments, which are corrosive to electronic equipment and materials, may be gaseous or solid. The principal gaseous corrosives are the following: hydrogen sulfide, sulfur dioxide, the acid chlorine gases, and water vapor. Water vapor, sulfur dioxide, and hydrogen sulfide influence the rate of both copper and silver corrosion.

The results of experiments conducted at 86 control rooms in a single bleached Kraft mill to predict failure rates based on the atmospheric corrosivity are presented in Refs. [83,84]. The atmospheric corrosivity in each of the 86 electrical rooms in the mill was measured during a 90-day period, using the coupon monitoring procedure [55]. Corrosion-related failure histories of control equipment in these rooms were obtained from the electrical maintenance staff. According to Refs. [83,84], to eliminate premature corrosion failures in control rooms containing only electrical equipment (and no electronic equipment), the atmosphere must produce <0.5 μm/year of corrosion product on the copper coupons. Electrical contacts and wires will fail in about 3 years if corrosion deposits are thicker than 0.5 μm/year.

To eliminate premature failures of electronic equipment, the control room atmosphere must produce much less than 0.2 μm/year of corrosion product on the copper coupons. The circuit boards may fail in 3–4 years, but they may fail much sooner (in 2 years) if the layer of deposit forming on copper is thicker than 0.76 μm/year, and in 1 year if the corrosion layer is >1.3 μm/year. A limit of 0.1 μm/year was chosen as a practical maximum thickness of corrosion product to safeguard electronic equipment.

In addition to corrosive gases, particulates in the environment may contribute to the corrosion. Hygroscopic dirt, such as sodium sulfate dust in a pulp mill chemical recovery area, attracts moisture and forms a corrosive solution. This can pose problems if the hygroscopic dirt becomes lodged between two conductors so as to attack one of them or cause a short circuit.

5.11.2 Air Quality Monitoring

Most of the world's leading manufacturers of computer systems recognize the severity of the corrosion problem at industrial facilities. In many countries, the industrial standards define the acceptable levels of airborne contaminants and provide achievable and effective guidelines to protect electronics and electrical equipment from the damaging effects of corrosive gases.

The first line of defense against corrosion is an air monitoring program in the industrial facility, when a choice between an active and a passive air condition monitoring (CM) and between direct versus indirect monitoring techniques should be made.

For environmental corrosion assessment, the ideal situation would be continuous, online, multiple gas monitoring; however, this is usually an expensive solution. To continuously measure the ambient concentrations, which are typically quite low, and to monitor different combinations of gases, which produce varying reactivity levels, many different diagnostic methods have been developed. Reactivity monitoring can characterize the destructive potential of an environment. The growth of various corrosion films on specially prepared copper and silver coupons provides an excellent indication of the corrosion types.

Air monitoring is central to any environmental control program for achieving and maintaining air quality standards based on the presence (or absence) of gaseous air pollutants. Such a monitoring can also provide the short-term data required to manage and mitigate contaminant-specific episodes.

Several characteristics of any measurement technique must be evaluated to determine its suitability for use in air quality monitoring [85]. Among the more important characteristics are sensitivity, cost, and complexity. Sensitivity is a particularly demanding parameter for environments where near-ambient levels of many pollutants may be encountered and control levels are in the low ppb range. When deciding on a measurement technique, particularly in large surveys, the cost may be quite important. It is also important to consider the complexity of the technique and the degree of skill and training required to obtain quality results.

Other factors deserving consideration are selectivity and portability. Most measurement techniques are not optimized for all these parameters, and it is very important to evaluate the various characteristics and techniques of the monitoring system in order to achieve best results in monitoring the air quality at the industrial facility.

5.11.3 Direct Gas Monitoring

Electronic devices designed for real-time gas monitoring respond to changes in the measured variable very quickly. They are capable of detecting pollutant levels in the low ppb range, and are available for a wide range of pollutants. Individually, chemical pollutants may be monitored using

TABLE 5.4
Available Real-Time Chemical/Gas Monitors

Pollutant	Concentration Range (ppb)	Lower Detection Limit (ppb)	Response Time (s)	Selectivity	Susceptibility to Interferences
Ammonia	0–200	1.0	900	Medium	Low
Formaldehyde	0–1000	0.2	300	High	Low
Hydrochloric acid	0–200	1.0	900	Medium	High
Hydrogen sulfide	0–200	1.0	120	Medium	Low
Nitrogen oxides	0–200	0.1	90	NO-High NO$_2$-Low	Low
Ozone	0–1500	1.0	50	High	Low
Sulfur dioxide	0–200	0.1	120	High	Low
TVOC	0–20,000	20.0	120	Low	Low

Source: From Christopher O. Muller, Control of Corrosive Gases to Avoid Electrical Equipment Failure, Online. Available: http://www.purafil.com/Literature/Control_of_Corrosive_Gases_to_Avoid_Electrical_Equipment_Failure.pdf. With permission.

various analytical techniques to provide both the sensitivity and selectivity required for accurate low-level real-time monitoring.

The major disadvantage of the use of real-time gas monitors is the relatively high cost when compared to other techniques. Table 5.4 lists a number of different pollutants and the levels which can be monitored with real-time monitors [85].

5.11.4 CORROSION CONTROL TECHNOLOGY

The three methods of gaseous contaminant control most commonly employed in heating, ventilation, and air conditioning (HVAC) systems are source control, ventilation control, and removal control. Source control should always be the first strategy examined. Removing the sources of contaminants prevents them from becoming a problem in the first place.

However, in the industrial environments, the source of gaseous contaminants is the manufacturing processes themselves and cannot be easily removed. With many process technologies significantly improved, the emissions have been reduced over the last several years, but they are still a long way from levels that would be considered safe for electronic equipment. In some situations, the source control is not an option.

Then ventilation control should be the next option. Ventilation control involves the introduction of clean dilution air into the affected space, which allows reducing contaminant levels below acceptable threshold levels. Unfortunately, in some industries, such as pulp and paper mills, the outside air does not meet the required criteria with regard to gaseous contaminants as well. Therefore, if this air would be used for ventilation, it will simply increase the total level of contaminants at the facility.

It is clear that when neither source nor ventilation control is available to control the levels of gaseous contaminants in the affected space, removal control should be employed. Removal control may be provided by special gas-phase air filtration systems employing one or more dry-scrubbing media as an integral part of the HVAC system, which can effectively reduce gaseous contaminants to well below standard levels.

Three types of airborne contaminants need to be controlled [64,86]: liquids, solids, and gases. To deal with low-level airborne liquids and solid contamination, particle removal filtration, such as mechanical filters, is used. The most common technology available for removing gaseous contaminants is gas-phase, or dry-scrubbing, filtration.

Air filtration can effectively remove essentially all of the gaseous contaminants that could negatively impact electrical systems when proper considerations are made in choosing the gas-phase (chemical) filters. A typical gas-phase air filtration device has three major sections. The first section would consist of a one- or two-stage particulate prefilter to remove dust and other particulate matter that could "coat" the chemical filter media. The next section consists of 2–3 stages of the appropriate chemical filtration media, which would be followed by a final filter section.

5.11.5   Chemical and Particulate Filtration

Electronic process control equipment can be protected from corrosion in several ways. One way is to clean the air in the control room of corrosive gases or to keep corrosive gases out of the environment of the control room. The second option is to place the control equipment in an airtight cabinet and use vapor corrosion inhibitors (VCI). The third way is to use more resistant but substantially more expensive materials for building the control equipment.

The first option might be most practical and affordable. To keep corrosive gases out of the control room environments, the use of a modular filtration system using a deep bed of specialty activated carbon may be a valuable solution for the corrosive problem. This system works on the principle of supplying a purified air stream to pressurize the enclosed area to minimize the possibility of corrosive gas intrusion into the control room.

Air conditioning could be much more effective if fitted with activated carbon filters, which are housed in a separate box approximately the size of one of the air conditioning units. This method may effectively remove corrosive gases such as hydrogen sulfide, sulfur oxides, chlorine, and chlorine compounds. Solid particulates, primarily sodium sulfate and sodium carbonate, are removed by appropriate mechanical filters.

In order to meet removal requirements and to minimize installation costs, a skid-mounted, modular filtration system may be used. A carbon adsorber using a deep carbon bed of a specialty carbon may help to achieve a high degree of make-up air purity for a protection period of at least 1 year for most expected mill conditions.

5.11.6   Temperature Control

In many cases, corrosion is accelerated in the environment with elevated temperature; therefore temperature control is essential for corrosion control. There are two major factors causing overheating. For example, when silver is exposed to a sulfuric environment, it forms a nonconductive silver sulfide deposit, resulting in a highly resistant contact surface. With the increase in resistance, more heat is generated, speeding up the chemical reaction of silver with H_2S, which finally leads to thermal connection failure.

The second factor is the deterioration of the mechanical integrity of the components. It is very important to timely maintain the mechanism to keep the contact pressure at critical points, such as between the stab and the fingers, at the desired level. Otherwise as the pressure decreases, the resistance will increase and, depending on the current, more heat will be generated.

As follows from the literature review and observations (Section 7.5), an elevated temperature of the electrical equipment is one of the factors that accelerate the process of silver corrosion and whisker growth in particular. If the temperature of the electrical parts could be maintained at a relatively low level (below the standard maximum operating temperature of ~105°C) in the facilities contaminated with H_2S, the rate of silver corrosion and silver whiskers growth may decrease significantly.

To keep the temperature at a safe level, it is helpful to use one of available means of monitoring the temperature of the energized equipment. The techniques for such monitoring online will be described in Chapter 17.

5.12 CORROSION GLOSSARY

A

Acid: A substance which releases hydrogen ions when dissolved in water. Most acids will dissolve the common metals and will react with a base to form a neutral salt and water.

Acid rain: Atmospheric precipitation with a pH below 3.6–5.7. Burning of fossil fuels for heat and power is a major factor in the generation of oxides of nitrogen and sulfur, which are converted into nitric and sulfuric acids washed down in the rain. See also atmospheric corrosion.

Acidity: The quantitative capacity of water or a water solution to neutralize an alkali or base. It is usually measured by titration with a standard solution of sodium hydroxide and expressed in terms of its calcium carbonate equivalent.

Activity: Measure of the chemical potential of a substance, where chemical potential is not equal to concentration, which allows mathematical relations equivalent to those for ideal systems to be used to correlate changes in an experimentally measured quantity with changes in chemical potential.

Adhesion: A binding force that holds together molecules of substances whose surfaces are in contact or near proximity. The ability of dry paint to attach to and remain fixed on the surface without blistering, flaking, cracking, or being removed by tape.

Adsorb: To take in on the surface.

Adsorbent: A material, usually solid, capable of holding gases, liquids, and/or suspended matter at its surface and in exposed pores. Activated carbon is a common adsorbent used in water treatment.

Adsorption: The physical process occurring when liquids, gases, or suspended matter adhere to the surfaces of, or in the pores of, an adsorbent medium. Adsorption is a physical process which occurs without chemical reaction.

Aerosol: Aerosols are tiny particles suspended in the air. Some occur naturally, originating from volcanoes, dust storms, forest and grassland fires, living vegetation, and sea spray. Human activities, such as the burning of fossil fuels and the alteration of natural surface cover, also generate aerosols.

Alloy steel: Steel with modified properties made by combining iron with one or more elements in addition to carbon. Alloys change the properties of the steel making it, for example, harder, more formable, or more corrosion resistant, depending on the combination and amounts of alloys used.

Antipitting agent: An addition agent for electroplating solutions to prevent the formation of pits or large pores in the electrodeposits.

Aqueous: Pertaining to water; an aqueous solution is made by using water as a solvent.

Ash: The incombustible inorganic matter in the fuel.

ASTM: American Society for Testing of Materials.

Atmospheric air: Air under the prevailing atmospheric conditions.

Atmospheric corrosion: The gradual degradation or alteration of a material by contact with substances present in the atmosphere, such as oxygen, carbon dioxide, water vapor, and sulfur and chlorine compounds.

B

Bar: A shape of steel that is available in different forms such as rounds, squares, hexagons, and rectangles.

Base material (metal): Substrate.

Blister: A raised area, often dome shaped, resulting from either loss of adhesion between a coating or deposit and the base metal or delamination under the pressure of expanding gas trapped in a metal in a near-subsurface zone.

Blistering: Formation of dome-shaped projections in paints or varnish films resulting from local loss of adhesion and lifting of the film from the underlying surface.

Breakdown potential: The least noble potential where pitting or crevice corrosion, or both, will initiate and propagate.

Brittle fracture: Separation of a solid accompanied by little or no macroscopic plastic deformation. Typically, brittle fracture occurs by rapid crack propagation with less expenditure of energy than for ductile fracture.

C

Carbon steel: Steel that has properties made up mostly of the element carbon and which relies on the carbon content for structure. Most of the steel produced in the world is carbon steel.

Cathodic corrosion: Corrosion of a metal when it is a cathode (it usually happens to metals because of a rise in pH at the cathode or as a result of the formation of hydrides).

Cathodic inhibitor: An inhibitor that reduces the corrosion rate by acting on the cathodic (reduction) reaction.

Cathodic protection: A corrosion control system in which the metal to be protected is made to serve as a cathode, either by the deliberate establishment of a galvanic cell or by impressed current.

Caustic: Any substance capable of burning or destroying animal flesh or tissue. The term is usually applied to strong bases.

Caustic cracking: Stress-corrosion cracking of metals in caustic solutions.

Caustic soda: The common name of sodium hydroxide.

Chemical conversion coating: A protective or decorative nonmetallic coating produced *in situ* by chemical reaction of a metal with a chosen environment. It is often used to prepare the surface prior to the application of an organic coating.

Chemical stability: Resistance to attach by chemical action.

Chlorine: A widely used gas used in the disinfection of water and an oxidizing agent for organic matter.

Chromate treatment: A treatment of metal in a solution of a hexavalent chromium (Cr VI) compound to produce a conversion coating consisting of trivalent (Cr III) and hexavalent chromium compounds.

Chromium: A steel-gray, lustrous, hard, and brittle metallic element that takes its name from the Greek word for color (chrome) because of the brilliant colors of its compounds. It is found primarily in chromite. Resistant to tarnish and corrosion, it is a primary component of stainless steel and is used to harden steel alloys.

Coating: A paint, varnish, lacquer, or other finish used to create a protective and/or decorative layer. Generally used to refer to paints and coatings applied in an industrial setting as part of the OEM process.

Coating strength: (a) A measure of the cohesive bond within a coating, as opposed to coating-to-substrate bond (adhesive strength); (b) the tensile strength of a coating, usually expressed in kpa (psi).

Coating stress: The stresses in a coating resulting from rapid cooling of molten material or semi-molten particles as they impact the substrate. Coating stresses are a combination of body and textural stresses.

Cold rolling (CR): Rolling steel without first reheating it. This process reduces thickness of the steel, produces a smoother surface, and makes it easier to machine.

Conversion coating: A coating consisting of a compound of the surface metal produced by chemical or electrochemical treatments of the metal. Examples include chromate coatings on zinc, cadmium, magnesium, and aluminum and oxide and phosphate coatings on steel. See also chromate treatment and phosphating.

Copper plating: The electrolytic deposition of copper to provide either a corrosion barrier (often as an undercoat for hard chrome plate) or for reclamation of worn parts.

Copper strip corrosion: A standard for the evaluation of an oil's tendency to corrode copper or copper alloys (see ASTM D130). Test results are based on the matching of corrosion stains. Noncorrosiveness is not to be confused with rust inhibiting, which deals with the protection of a surface from some contaminant, such as water, rather than the oil itself.

Corrosion: The chemical or electrochemical reaction between a material, usually a metal, and its environment that produces a deterioration of the material and its properties.

Corrosion fatigue: The process in which a metal fractures prematurely under conditions of simultaneous corrosion and repeated cyclic loading at lower stress levels or fewer cycles than would be required in the absence of the corrosive environment.

Corrosion fatigue strength: The maximum repeated stress that can be endured by a metal without failure under definite conditions of corrosion and fatigue and for a specific number of stress cycles and a specified period of time.

Corrosion inhibitive: A type of metal paint or primer that prevents rust by preventing moisture from reaching the metal. Zinc phosphate, barium metaborate, and strontium chromate (all pigments) are common ingredients in corrosion-inhibitive coatings. These pigments absorb any moisture that enters the paint film.

Corrosion potential: The potential of a corroding surface in an electrolyte relative to a reference electrode measured under open-circuit conditions.

Corrosion product: Substance formed as a result of corrosion.

Corrosion rate: The amount of corrosion occurring in unit time (e.g., mass change per unit area per unit time; penetration per unit time).

Corrosion resistance: Ability of a metal to withstand corrosion in a given corrosion system.

Corrosive wear: Wear in which the chemical or electrochemical reaction with the environment is significant.

Corrosivity: Tendency of an environment to cause corrosion in a given corrosion system.

Creep: Time-dependent strain occurring under stress. The creep strain occurring at almost constant rate, secondary creep; and that occurring at an accelerating rate, tertiary creep.

Crevice corrosion: Localized corrosion of a metal surface at, or immediately adjacent to, an area that is shielded from full exposure to the environment because of close proximity between the metal and the surface of another material.

D

Deactivation: The process of prior removal of the active corrosive constituents, usually oxygen, from a corrosive liquid by controlled corrosion of expendable metal or by other chemical means, thereby making the liquid less corrosive.

Defect: A discontinuity or discontinuities that by nature or accumulated effect (e.g., total crack length) render a part or product unable to meet the minimum applicable acceptance standards or specifications.

Deposit: Foreign substance which comes from the environment, adhering to the surface of a material.

Deposit corrosion: Localized corrosion under or around a deposit or collection of material on a metal surface (see also crevice corrosion).

Dew point: Temperature at which moisture will condense from humid vapors into a liquid state.

Dezincification: Corrosion in which zinc is selectively leached from zinc-containing alloys. Most commonly found in copper–zinc alloys containing <83% copper after extended service in water containing dissolved oxygen; the parting of zinc from an alloy (in some brasses, zinc is lost, leaving a weak, brittle, porous, copper-rich residue behind).

Diffusion: Spreading of a constituent in a gas, liquid, or solid, tending to make the composition of all the parts uniform.

Diffusion coating: Any process whereby a base metal or alloy is either (a) coated with another metal or alloy and heated to a sufficient temperature in a suitable environment or (b) exposed to a gaseous or liquid medium containing the other metal or alloy, thus causing diffusion of the coating or of the other metal or alloy into the base metal with resultant changes in the composition and properties of its surface.

Discontinuity: Any interruption in the normal physical structure or configuration of a part, such as cracks, laps, seams, inclusions, or porosity. A discontinuity may or may not affect the usefulness of the part.

Dissolved solids: The weight of matter in true solution in a stated volume of water; includes both inorganic and organic matter; usually determined by weighing the residue after evaporation of the water.

Dry corrosion: See gaseous or hot corrosion.

E

Electrical steel: Steel that includes silicon. The silicon content allows the steel to minimize energy loss during electrical applications.

Electroless nickel: The autocatalytic deposition of nickel/phosphorus and nickel/boron has many useful corrosion and corrosion wear applications. Unlike the electrolytic processes, they produce a deposit with completely uniform coverage. In the case of nickel/phosphorus, deposits around 25–50 μm thick with a hardness of about 500 Hv are obtained, but thermal aging at temperatures around 400°C can develop hardness values in excess of 1000 Hv.

Electroplating: Electrodepositing a metal or alloy in an adherent form on an object serving as a cathode.

Embrittlement: The severe loss of ductility or toughness or both, of a material, usually a metal or alloy.

Environmental cracking: Brittle fracture of a normally ductile material in which the corrosive effect of the environment is a causative factor. Environmental cracking is a general term that includes corrosion fatigue, high-temperature hydrogen attack, hydrogen blistering, hydrogen embrittlement, liquid metal embrittlement, solid metal embrittlement, stress-corrosion cracking, and sulfide stress cracking.

Erosion: The progressive loss of material from a solid surface due to mechanical interaction between that surface and a fluid, a multicomponent fluid, or solid particles carried with the fluid.

Erosion–corrosion: A conjoint action involving corrosion and erosion in the presence of a moving corrosive fluid. Leads to the accelerated loss of material.

Exfoliation: Corrosion that proceeds laterally from the sites of initiation along planes parallel to the surface, generally at grain boundaries, forming corrosion products that force metal away from the body of the material, giving rise to a layered appearance.

F

Failure: An item of equipment is said to have suffered a failure when it is no longer capable of fulfilling one or more of its intended functions. Note that an item does not need to be completely unable to function to have suffered a failure.

Failure analysis: The systematic investigation of a component failure with the objectives of determining why the component failed and the corrective actions needed for preventing future failures.

Failure mode: The basic material behavior that resulted in the failure. Examples of failure mode include ductile fracture, brittle fracture, fatigue fracture, corrosion, erosion, wear, and distortion.

Filiform corrosion: Corrosion that occurs under some coatings in the form of randomly distributed thread-like filaments (a special form of crevice corrosion).

Film: A thin, not necessarily visible, layer of material.

Film build: Thickness produced in a paint application.

Filter: Porous material through which fluids or fluid and solid mixtures are passed to separate matter held in suspension.

Fines: The portion of a powder composed of particles which are smaller than the specified size.

Flakes: Short, discontinuous internal fissures in wrought metals attributed to stresses produced by localized transformation and decreased solubility of hydrogen during cooling after hot working. In a fracture surface, flakes appear as bright silvery areas; on an etched surface, they appear as short, discontinuous cracks.

Free corrosion potential: Corrosion potential in the absence of net electrical current flowing to or from the metal surface.

Fretting: Surface damage resulting from relative motion between surfaces in contact under pressure.

Fretting corrosion: The deterioration at the interface between contacting surfaces as the result of corrosion and slight oscillatory slip between the two surfaces.

Fuel: A substance containing combustible used for generating heat.

G

Galvalume: A proprietary zinc alloy coating containing 55% aluminum with superior corrosion resistance.

Galvanizing: The process by which steel is coated with a layer of zinc. The zinc coating provides the steel with greater corrosion resistance.

Galvanic corrosion: Accelerated corrosion of a metal because of an electrical contact with a more noble metal or nonmetallic conductor in a corrosive electrolyte.

Galvanic couple: A pair of dissimilar conductors commonly metals in electrical contact (see galvanic corrosion).

Galvanic current: The electric current between metals or conductive nonmetals in a galvanic couple.

Galvanic series: A list of metals and allots arranged according to their relative corrosion potentials in a given environment.

Galvanizing: To coat a metal surface with zinc using any of the various processes.

Galvanneal: To produce a zinc–iron alloy coating on iron or steel by keeping the coating molten after hot dip galvanizing until the zinc alloys completely with the base metal.

Gaseous corrosion: Corrosion with gas as the only corrosive agent and without any aqueous phase on the surface of the metal, also called dry corrosion.

General corrosion: A form of deterioration that is distributed more or less uniformly over a surface; see uniform corrosion.

Graphitic corrosion: A form of selective leaching specific to the deterioration of metallic constituents in gray cast iron, which leaves the graphitic particles intact (the term "graphitization" is commonly used to identify this form of corrosion but is not recommended because of its use in metallurgy for the decomposition of carbide to graphite).

H

Hard water: Water which contains calcium or magnesium in an amount which requires an excessive amount of soap to form lather.

Hexavalent chromium: Chromium (VI) or Cr VI.

Hot corrosion: An accelerated corrosion of metal surfaces that results from the combined effect of oxidation and reactions with sulfur compounds and other contaminants, such as chlorides, to form a molten salt on a metal surface which fluxes, destroys, or disrupts the normal protective oxide.

Hot cracking: Also called solidification cracking. Hot cracking of weldments is caused by the segregation at grain boundaries of low-melting constituents in the weld metal. Hot cracking can be minimized by the use of low-impurity welding materials and proper joint design.

Hot-dip coating: A metallic coating obtained by dipping the base metal or substrate into a molten metal.

Hot-rolling, HR: Rolling steel slabs into flat-rolled steel after it has been reheated.

Humidity test: A corrosion test involving exposure of specimens at controlled levels of humidity and temperature.

Hydrogen blistering: The formation of blisters on or below a metal surface from excessive internal hydrogen pressure (hydrogen may be formed during cleaning, plating, corrosion, etc.).

Hydrogen damage: A general term for the embrittlement, cracking, blistering, and hydride formation that can occur when hydrogen is present in some metals.

Hydrogen embrittlement: Hydrogen-induced cracking or severe loss of ductility caused by the presence of hydrogen in the metal. Hydrogen absorption may occur during electroplating, pickling, or other processes that favor the production of nascent or elemental hydrogen.

I

Impingement corrosion: A form of erosion–corrosion generally associated with the local impingement of a high-velocity, flowing fluid against a solid surface.

Inclusions: Particles of a foreign material in a metallic matrix. The particles are usually compounds (such as oxides, sulfides, or silicates), but may be of any substance that is foreign to (and essentially insoluble in) the matrix.

Incubation period: A period prior to the detection of corrosion while the metal is in contact with a corrosive agent.

Industrial atmosphere: An atmosphere in an area of heavy industry with soot, fly ash, and sulfur compounds as the principal constituents.

Inhibitor: A chemical substance or combination of substances that, when present in the proper concentration and forms in the environment, prevents or reduces corrosion.

Inorganic zinc-rich paint: Coating containing a zinc powder pigment in an inorganic vehicle.

Interconnected porosity: A network of pores in and extending to the surface of a coating.

Intercrystalline corrosion: See intergranular corrosion.

Intergranular: Between crystals or grains.

Intergranular corrosion: Preferential corrosion at or adjacent to the grain boundaries of a metal or alloy.

Intergranular cracking: Cracking or fracturing that occurs between the grains or crystals in a polycrystalline aggregate. Also called intercrystalline cracking.

Intergranular fracture: Brittle fracture of a metal in which the fracture is between the grains, or crystals that form the metal. Also called intercrystalline fracture. Contrast with transgranular fracture.

Internal oxidation: The formation of isolated particles of corrosion products beneath the metal surface. This occurs as the result of preferential oxidation of certain alloy constituents by inward diffusion of oxygen, nitrogen, sulfur, and so on.

Isocorrosion diagram: A graph or chart that shows constant corrosion behavior with changing solution (environment) composition and temperature.

L

Lamellar corrosion: A form of corrosion in which the expanding corrosion products stack up as layers. Similar to exfoliation of high strength aluminum alloys.

Liquid impingement erosion: Progressive loss of material from a solid surface due to continue exposure to impacts by liquid drops or jets.

Liquid metal embrittlement: Catastrophic brittle failure of a normally ductile metal when in contact with a liquid metal and subsequently stressed in tension.

Liter: The basic metric unit of volume; 3.785 L is equal to 1 U.S. gallon; 1 L of water.

Local action corrosion: Corrosion caused by local corrosion cells on a metal surface.

Local cell: A galvanic cell resulting from inhomogeneities between areas on a metal surface in an electrolyte. The inhomogeneities may be of physical or chemical nature in either the metal or its environment.

Local corrosion cell: An electrochemical cell created on a metal surface because of a difference in potential between adjacent areas on that surface.

Localized corrosion: Corrosion at discrete sites, for example, pining, crevice corrosion, and stress corrosion cracking.

Lubricant: Any substance interposed between two surfaces for the purpose of reducing the friction or wear between them.

M–N–O

Maximum contaminant level (MCL): The maximum allowable concentration of a contaminant in water as established in the U.S. EPA Drinking Water Regulations.

Mechanical filter: A filter primarily designed for the removal of suspended solid particles, as opposed to filters with additional capabilities.

Mesa corrosion: Mesa corrosion is one of the common types of corrosion experienced in service involving exposure of carbon or low-alloy steels to flowing wet carbon dioxide conditions at slightly elevated temperatures. An iron carbonate surface scale will often form in this type of environment which can be protective, rendering a very low corrosion. However, under the surface shear forces produced by flowing media, this scale can become damaged in a localized attack that produces mesa-like features.

Microbiologically influenced corrosion: Sometimes called microbial corrosion or bio corrosion, it refers to corrosion that is affected by the action of microorganisms in the environment.

Micrometer (μm): Formally known as micron. A linear measure equal to one millionth of a meter or 0.00003937 in. The symbol for the micrometer is μm.

Moisture: Water in the liquid or vapor phase.

Nickel plating: The electrolytic deposition of nickel to form a corrosion barrier or to reclaim a worn part. Can also include hard ceramic particles to from a wear-resistant composite coating.

Noble metal: A metal with a standard E.P that is more noble (positive) than that of hydrogen.

NO$_x$: Abbreviation for all of the family of oxides of nitrogen.

Oxidation: Loss of electrons by a constituent of a chemical reaction (also refers to the corrosion of a metal that is exposed to an oxidizing gas at elevated temperatures).

Oxidized surface (on steel): Surface having a thin, tightly adhering, oxidized skin (from straw to blue in color), extending in from the edge of a coil or sheet.

Oxidizing agent: A compound that causes oxidation, thereby itself being reduced.

Oxidizing atmosphere: An atmosphere which tends to promote the oxidation of materials.

Oxygen: Gas used to support combustion of fuel gases in combustion thermal spray processes. Achieves much higher flame temperatures than using air.

Oxygen attack: Corrosion or pitting in a boiler caused by oxygen.

P

Pack rust: This particular form of corrosion is often used in relation to bridge inspection to describe built-up members of steel bridges which are already showing signs of rust packing between steel plates.

Particle size: (a) A measure of dust size, expressed in microns or percent passing through a standard mesh screen; (b) the size of a particle suspended in water as determined by its smallest dimension.

Parting: The selective corrosion of one or more components of a solid solution alloy.

Parts per billion (ppb): A measure of proportion by weight, equivalent to one unit weight of a material per billion (10^9) unit weights of compound.

Parts per million (ppm): A measure of proportion by weight, equivalent to one unit weight of a material per million (10^6) unit weights of compound.

Passivation: The process in metal corrosion by which metals become passive. A type of inhibitor which appreciably changes the potential of a metal to a more noble (positive) value.

Passive: The state of the metal surface characterized by low corrosion rates in a potential region that is strongly oxidizing for the metal.

Passivity: A condition in which a piece of metal, because of an impervious covering of oxide or other compound, has a potential much more positive than that at the metal in the active state.

Patina: The coating, usually green, that forms on the surface of metals such as copper and copper alloys exposed to the atmosphere. Also used to describe the appearance of a weathered surface of any metal.

Percent solids: The percentage mass of nonliquid components in paint.

Permeability: A property measured as a rate of passage of a liquid or gas through a coating.

pH: A measure of the acidity or alkalinity of a solution; the negative logarithm of the hydrogen ion activity; it denotes the degree of acidity or basicity of a solution. At 25°C, 7.0 is the neutral value. Decreasing values below 7.0 indicate increasing acidity; increasing values above 7.0, increasing basicity.

Phosphatizing (phosphating or phosphate conversion coating): Phosphatizing is a metal pre-treatment primarily used to prepare steel for paint or coatings and to prevent corrosion. Phosphatizing is termed a conversion coating because, unlike paint or traditional coatings, it does not lie on the surface of the metal, but rather the surface of the metal is chemically changed to a new substance. The coating will, therefore, not chip- or scratch-off exposing the base material to corrosion and wear.

Pitting: Corrosion of a metal surface, confined to a point or small area that takes the form of cavities.

Pitting factor: Ratio of the depth of the deepest pit resulting from corrosion divided by the average penetration as calculated from weight loss.

Pitting resistance equivalent number: An empirical relationship to predict the pitting resistance of austenitic and duplex stainless steels.

Poultice corrosion: A term used in the automotive industry to describe the corrosion of vehicle body parts due to the collection of road salts and debris on ledges and in pockets that are kept moist by weather and washing. Also called deposit corrosion or attack.

Powder coating: A polymeric coating deposited via electrostatic attraction and applied to the surface as a dry, finely ground powder and then heated above its melting point so the powder particles flow together or cure.

Precious metal: One of the relatively scarce and valuable metals: gold, silver, and the platinum group metals. Also called noble metal(s).

Precipitate: To separate materials from a solution by the formation of insoluble matter by chemical reaction.

Products of combustion: The gases, vapors, and solids resulting from the combustion of fuel.

Protection potential: The most noble potential where pitting and crevice corrosion will not propagate.

Protective potential: The threshold value of the corrosion potential that has to be reached to enter a protective potential range. The term used in cathodic protection to refer to the minimum potential required to control corrosion.

Protective potential range: A range of corrosion potential values in which unacceptable corrosion resistance is achieved for a particular purpose.

Purge: To introduce air into a vessel flue passages in such volume and manner as to completely replace the air or gas–air mixture contained therein.

R

Reactive metal: A metal that readily combines with oxygen at elevated temperatures to form very stable oxides, for example, titanium, zirconium, and beryllium.

Red water: Water which has a reddish or brownish appearance due to the presence of precipitated iron and/or iron bacteria.

Reducing agent: A compound that causes reduction, thereby itself becoming oxidized.

Reducing atmosphere: An atmosphere which tends to (1) promote the removal of oxygen from a chemical compound; (2) promote the reduction of immersed materials.

Reduction: The gain of electrons by a constituent of a chemical reaction.

RH: The ratio, expressed as a percentage, of the amount of water vapor present in a given volume of air at a given temperature to the amount required to saturate the air at that temperature.

Ringworm corrosion: Localized corrosion frequently observed in oil well tubing in which a circumferential attack is observed near a region of metal "upset."

Rust: A corrosion product consisting primarily of hydrated iron oxide. A term properly applied only to ferrous alloys.

S

Sacrificial coating: A coating that provides corrosion protection wherein the coating material corrodes in preference to the substrate, thereby protecting the latter from corrosion.

Saline water: Water containing an excessive amount of dissolved salts, usually over 5000 mg/L.

Salinity: The total proportion of salts in seawater, often estimated empirically as chlorinity ×1.80655, also expressed in parts per thousand (or permille), which is approximately grams of salt per kilogram of solution.

Salt: In chemistry, the term is applied to a class of chemical compounds which can be formed by the neutralization of an acid with a base; the common name for the specific chemical compound sodium chloride used in the regeneration of ion exchange water softeners.

Salt fog: ASTMB-117 test procedure that attempts to simulate the corrosive environment caused by road salt and marine spray.

Saturated air: Air which contains the maximum amount of water vapor that it can hold at its temperature and pressure.

Scale: A deposit of mineral solids on the interior surfaces of water lines and containers often formed when water containing the carbonates or bicarbonates of calcium and magnesium is heated. Oxide of iron that forms on the surface of steel after heating.

Scanning electron microscope (SEM): An electron microscope in which the image is formed by a beam synchronized with an electron probe scanning the object. The intensity of the image forming beam is proportional to the scattering or secondary emission of the specimen where the probe strikes it.

Scoring: A severe form of wear characterized by the formation of extensive grooves and scratches in the direction of sliding.

Sealant, sealer: A preparation of resin or wax-type materials for sealing the porosity in coatings.

Sealing: A process which, by absorption of a sealer into thermal spray coatings, seals porosity and increases resistance to corrosion of the underlying substrate material.

Season cracking: See stress-corrosion cracking.

Shelf life: The length of time any unopened container (e.g., of paint) can be stored at the supplier-recommended storage temperature and still retain the properties in both the unmixed and mixed states as required by the specification or advertised in the product data sheets.

Silver plating: The electrodeposition of silver for electrical, decorative, or antifretting properties.

Sodium chloride: The chemical name for common salt.

Soft water: Water which contains little or no calcium or magnesium salts, or water from which scale-forming impurities have been removed or reduced.

Solution: A homogeneous dispersion of two or more kinds of molecular or ionic species. Solutions may be composed of any combination of liquids, solids, or gases, but they always consist of a single phase.

Soot: Unburned particles of carbon derived from hydrocarbons.

Sour water: Waste waters containing fetid materials, usually sulfur compounds.

Specialty steel: Steels such as electrical, alloy, or stainless steels. These generally are produced in smaller volumes to meet the specific needs of customers.

Stress-corrosion cracking: A cracking induced from the combined influence of tensile stress and a corrosive environment.

Substrate: The parent or base material to which a coating is applied.

Subsurface corrosion: See internal oxidation.

Sulfidation: The reaction of a metal or alloy with a sulfur-containing species to produce a sulfur compound that forms on or beneath the surface of the metal or alloy.

T

Thermogalvanic corrosion: The corrosive effect resulting from the galvanic cell caused by a thermal gradient across the metal surface.

Tinplate: Thin steel sheet with a very thin coating of metallic tin. Used primarily in can making.

Topcoat: Usually the final paint film applied to a surface.

Total acidity: The total of all forms of acidity, including mineral acidity, carbon dioxide, and acid salts. Total acidity is usually determined by titration with a standard base solution to the phenolphthalein end point (pH 8.3).

Total alkalinity: The alkalinity of a water as determined by titration with standard acid solution to the methyl orange end point (pH approximately 4.5); total alkalinity includes many alkalinity components, such as hydroxides, carbonates, and bicarbonates.

Total chlorine: The total concentration of chlorine in water, including combined and free chlorine.

Total dissolved solids (TDS): The weight of solids per unit volume of water which are in true solution, usually determined by the evaporation of a measured volume of filtered water, and determination of the residue weight.

Trace: A very small concentration of a material, high enough to be detected but too low to be measured by standard analytical methods.

Transgranular cracking: Cracking or fracturing that occurs through or across a crystal or grain. Also called transcrystalline cracking.

Transition metal: A metal in which the available electron energy levels are occupied in such a way that the d-band contains less than its maximum number of 10 electrons per atom, for example, iron, cobalt, nickel, and tungsten. The distinctive properties of the transition metals result from the incompletely filled d-levels.

Trap: A receptacle for the collection of undesirable material.

U–V–W–Z

Uniform corrosion: Corrosion that proceeds at about the same rate over a metal surface.

Varnish (clear coat): An unpigmented binder solvent solution applied to protect or decorate a surface.

VCI: Vapor-phase corrosion inhibitor.

Viscosity: The resistance of fluids to flow, due to internal forces and friction between molecules, which increases as temperature decreases.

Volatile: Capable of vaporization at a relatively low temperature.

Volatile matter: Those products given off by a material as gas or vapor, determined by definite prescribed methods.

Volatile organic compound (VOC): Organic chemicals and petrochemicals that emit vapors while evaporating. In paints, VOC generally refers to the solvent portion of the paint which, when it evaporates, results in the formation of paint film on the substrate to which it was applied.

Volatile solids: Matter which remains as a residue after evaporation at 105 or 180°C, but which is lost after ignition at 600°C. Includes most forms of organic matter.

Weak base load faction: The sum of the chloride, sulfate, and nitrate in a given water.

WR: Zinc oxide, the powdery product of corrosion of zinc or zinc-coated surfaces.

Zinc plating: The electrodeposition of zinc or zinc alloys (e.g., Zn/Ni, Zn/Sn) to provide galvanic corrosion protection.

REFERENCES

1. Leygraf C. and Graedel T. E. Atmospheric corrosion, *ECS Corrosion Monograph Series*, John Wiley & Sons Interscience Publishing, New York, 2000, 354pp.
2. Atmospheric Corrosion, Types of Corrosive Atmospheres, Online. Available: http://corrosion-doctors. org/Corrosion-Atmospheric/Types-of-atmospheres.htm.
3. Gaseous and Particulate Contamination Guidelines for Data Centers. Whitepaper prepared by ASHRAE Technical Committee (TC) 9.9 Mission Critical Facilities, Technology Spaces, and Electronic Equipment, 2009 (Update in 2011).
4. Brusse J. and Sampson M. NASA, Zinc whisker: Hidden cause of equipment failure, *IT Professional*, 6(6), 43–46, 2004.
5. Lahtinen R. and Gustafsson T. *The Driving Force Behind Whisker Growth*, Metal Finishing (Elsevier Inc.), 2008.
6. Miller S. K. Whiskers in the data center, *Processor*, 29(30), 24, July 27, 2007 in print issue, Online. Available: http://www.processor.com/editorial/article.asp?article=articles/P2930/31p30/31p30.asp.
7. Corrosion Resistance, Online. Available: www.corrosion-doctors.org/CorrosionAtmospheric/ Corrosion-resistance.htm.
8. Kirk W. W. *Atmospheric Corrosion*, Vol. 1239 of ASTM special technical publication, ASTM International, West Conshohocken, PA, 280pp.
9. Townsend H. E. *Outdoor Atmospheric Corrosion*, Vol. 1421 of ASTM special technical publication, ASTM International, West Conshohocken, PA, 385pp.
10. Dean S. W. and Rhea E. C. Atmospheric corrosion of metals: A symposium, *ASTM Committee G-1 on Corrosion of Metals*, Vol. 767 of ASTM special technical publication, 414pp., 1982.
11. Leygraf C. New fundamental aspects of atmospheric corrosion, *Proc. of NACE International Corrosion 2009 Conference and Expo,* Atlanta, Georgia, USA, March 22–26, 2009.
12. Chudnovsky B. Corrosion of electrical conductors in pulp and paper industrial applications, *IEEE Transactions on Industrial Applications*, 44(3), 932–939, 2008.
13. Silver Tarnishing by Paper and Paper Board, TAPPI Standard T444, 1985.
14. Chudnovsky B. Degradation of power contacts in industrial atmosphere: Silver corrosion and whiskers, *Proc. 48th Annual Holm Conference on Electrical Contacts*, Orlando, FL, pp. 140–147, October 2002, Online. Available: http://nepp.nasa.gov/whisker/reference/tech_papers/chudnovsky2002-paper-silver-corrosion-whiskers.pdf.
15. Dunn B. D., de Rooij A., and Collins D. S. Corrosion of silver-plated copper conductors, *ESA Journal*, 8, 307–325, 1984, Online. Available: https://escies.org/public/materials/Ag-Cu.pdf.
16. Investigation of Silver Plated Conductor Corrosion (Red Plague), SAE Standard AIR4487, June 1992.
17. Cooke R. W. Red Plague Control plan (RPCP), NASA Technical Report Server.
18. Under Coating Corrosion of Silver Plated Copper in a Circuit Breaker, Online. Available: http://www. corrosionlab.com/Failure-Analysis-Studies/20034.under-coating.silver-plated-copper.htm.
19. Bersirova O., Królikowski A., and Kublanovsky V. Deposition conditions and corrosion behavior of silver coatings from dicyanoargentate complexes, *Third Baltic Electrochemical Conference*, Gdansk, Poland, April 2003.
20. Imrell T. The importance of the thickness of silver coating in the corrosion behavior of copper contact, *Proc. 37th Annual Holm Conference on Electrical Contacts*, Chicago, IL, pp. 237–243, 1991.
21. Rudolphi A. K., Bjorkman C., Imrell T., and Jacobson S. Conduction through corrosion films on silver plated copper in power contacts, *Proc. 41st IEEE Holm Conference on Electrical Contacts*, Montreal, Canada, pp. 124–134, 1995.

22. Bauer R. Sulfide corrosion of silver contacts during satellite storage, *J. Spacecraft Rockets*, 25(6), 439–440, 1988.
23. Abbott W. H. Contact corrosion, in *Electrical Contacts*, edited by Slade P. G., Marcel Dekker, Inc., New York, Chapter 3, pp. 113–154, 1999.
24. Crossland W. A. and Knight E. The Tarnishing of Silver-Palladium surfaces and its effect on CR in low energy circuits. *Proc. 19th Annual Holm Conference on Electrical Contacts*, Pittsburgh, PA, pp. 248–264, 1973.
25. Tamai T. Ellipsometric analysis for growth of Ag2S film and effect of oil film on corrosion resistance of Ag contact surfaces, *IEEE Transactions on Components Hybrids and Manufacturing Technology*, 12(1), 43–47, 1989.
26. Simko S. J., Lee A., Gaarenstroom S. W., Dow A. A., and Wong C. A. Film formation on silver-based switching contacts, *Proc. 34th Annual Holm Conference on Electrical Contacts*, Chicago, IL, pp. 167–176, 1989.
27. Hawthorne J. G. and Smith C. S. *Theophilis on Diverse Arts*. Dover Publications, New York, Book 3, Chapter 80, p. 158, 1979.
28. Ankersmit B. The protection of silver from tarnishing: The use of silver tokens (SILPROT). *Presentation at 1999 Indoor Air Pollution Meeting*. Instituut Collectie Nederland, Amsterdam, Nederland, 26–27 August 1999, Online. Available: http://www.iaq.dk/iap/iap1999/1999_07.htm.
29. Wharton G., Maish S. L., and Ginell W. S. A comparative study of silver cleaning abrasives, *Journal of the American Institute of Conservation (JAIC)*, 29(1), 13–31, 1990, Online. Available: http://aic.stanford.edu/jaic/articles/jaic29-01-002.html.
30. Keil A., Meyer C.-L. Crystal growth in silver exposed to sulfur and the decay of silver sulfide Kristallwachstum bei Schwefel Einwirkung auf Silber und beim Zerfall von Silbersulfid, *Zeitschrift für Metallkunde*, 51, 253–255, 1960.
31. Keil A., Meyer C.-L. On the Origin of Hail-like crystals on metallic surfaces. Über die Entstehung haarförmiger Kristalle auf metallischen Oberflächen, *Elektrotechnische Zeitschrift*, ETZ, Ausgabe B, v. 14. Jahrgang 1962.
32. Van de Ven N., Ouëndag R., Pronk L., Vlutters H. "Whiskers" the root cause of spontaneous short circuits, *Proc. 3rd Petroleum and Chemical Industry Conference (PCIC) Europe Electrical and Instrumentation Applications*, pp. 119–123, Amsterdam, The Netherlands, June 7–9, 2006.
33. Muniesa J. The growth of whiskers on Ag surfaces. *Proc. International Conference on Electric Contact Phenomena*, Pub. VDE-Verlag GmbH, Berlin, pp. 56–59, June 1982.
34. Riddle K. A. Tin plating versus silver plating and the use of silver alloys on electrical equipment for the paper industry, *Conference Record of 1990 Annual Pulp and Paper Industry Technical Conference*, Seattle, WA, pp. 32–36, June 1990.
35. Walker E. Whisker Growth on Silver Tipped Contacts. Report on investigation of an electrical accident at Coking plant in 1979. UK Health and Safety Executive (HSE), UK British Standard Institute (BSI), 1999.
36. Key P. L. Surface morphology of whisker crystals of tin, zinc and cadmium, *IEEE Electronic Components Conference*, Arlington, VA, pp. 155–157, May 1977.
37. Devaney J. R. Corrosion and dendrite growth and other metallurgical phenomena in microelectronic packages, Electronic Packaging and Corrosion in Microelectronics, *Proc. of ASM's Third Conference on Electronic Packaging: Material and Processes and Corrosion in Microelectronics*, Minneapolis, MN, pp. 287–293, April 1987.
38. Tamai T. Ellipsometric analysis for growth of Ag2S film and effect of oil film on corrosion resistance of Ag contact surfaces, *Proc. 34th Annual Holm Conference on Electrical Contacts*, San Francisco, CA, pp. 281–287, 1988.
39. Muller O. Multiple contaminant gas effects on electronic equipment corrosion, *Corrosion*, 47(2), 146–151, 1991.
40. Blanton T., Misture S., Dontula N., Zdzieszynski S. In situ high-temperature X-ray diffraction characterization of silver sulfide Ag_2S, JCPDS-International Centre for Diffraction Data 2011 ISSN 1097-0002, pp. 110–117, *Powder Diffraction*, 26(2), SI (JUN), 114–118, 2011, Online. Available: www.icdd.com/resources/axa/vol54/V54_14.pdf.
41. De vorming en gevaren van silver whiskers (The formation and dangers of silver whiskers), Online. Available: http://schakelenverdeel.cobouw.nl/cases/cases-eaton/de-vorming-en-gevaren-van-silver-whiskers.
42. Warwick M. E. *Atmospheric Corrosion of Tin and Tin Alloys*, International Tin Research Institute, ITRA Publication No. 602.

43. Yasuda K., Umemura S., and Aoki T. Degradation mechanisms in tin- and gold-plated connector contacts, *IEEE Transactions on Components, Hybrids, and Manufacturing Technology*, 10(3), 456–462, 1987.

44. Cast Nonferrous: Corrosion of Zinc, Key to Metals Database, Online. Available: http://www.keytometals.com/Article40.htm.

45. Rahrig P. G. Zinc coatings on handrail tubing: A comparative analysis, *Materials Performance*, 41, 25–26, July 2002.

46. Robinson J. White rust on zinc coating—Causes, effects and remedies, *Corrosion Management*, 13(2), 15–18, 2006, Online. Available: http://www.scribd.com/doc/57812306/White-Rust-Coating.

47. White Rust, ProFence, Australia, Online, Available: www.profence.com.au/uploads/products/WHITE_RUST.pdf.

48. White Rust on Galvanized Steel, SAB-profiel, Netherland, Online. Available: http://www.sabprofiel.com/index.cfm/site/sabprofiel_engels/pageid/E310CFE7-E081-2F5B-428AEA7C92CD7CD5/index.cfm and https://www.sabprofiel.nl/assets/user/Diversen/White_rust_on_galvanized_and_galvanized_pre-painted_steel.pdf.

49. White Rust: An Industry Update and Guide Paper 2002, Association of Water Technologies (AWT), Online. Available: www.awt.org/IndustryResources/white_rust_2002.pdf.

50. Langill T. J. and Rahrig P. G. Predicting the service life of galvanized steel, May 29, 2003, Online. Available: http://www.thefabricator.com/article/metalsmaterials/predicting-the-service-life-of-galvanized-steel.

51. Gay P. A., Percot P., and Pagetti J. The protection of silver against atmospheric attack, *Plating and Surface Finishing*, 26(6), 71–73, 2005.

52. Whitley J. The tin commandments, *Plating and Surface Finishing*, 68(10), 38–39, 1981.

53. Antler M. Materials, coatings and platings, in *Electrical Contacts*, edited by Slade P. G., Marcel Dekker, Inc., New York, pp. 403–432, 1999.

54. Restriction of the Use of Certain Hazardous Substances in Electrical and Electronic Equipment (RoHS), Directive 2002/95/EC of the European Parliament and of the council of 27 January 2003, Official Journal of European Union, 46[L37], pp. 19–23, 02/13/2003, Online. Available: http://europa.eu.int/eurlex/pri/en/oj/dat/2003/l_037/l_03720030213en00190023.pdf.

55. Martin P. J., Johnson W. B., and Miksic B. A. Corrosion inhibitors of electronic metals using vapor phase inhibitors. *Proc. NACE Conference*, New Orleans, LA, paper No. 36, April 1984, Online. Available: http://www.cortecvci.com/Publications/Papers/VCIProducts/CTP-6.PDF.

56. Tarvin M. E. and Micsic B. A. Volatile corrosion inhibitors for protection of electronics, *1989 Conference of National Association of Corrosion Engineers (NACE)*, Paper #344, Wichita, KS, p. 9, Online. Available: http://www.cortecvci.com/Publications/Papers/VCIProdicts/CTP-19.pdf.

57. Miksic B. A., Jaeger P., and Chandler C. VCI's for mitigating electronic corrosion, *1996 Conference of National Association of Corrosion Engineers (NACE)*, Paper #631, Seattle, WA, p. 9, Online. Available: http://www.cortecvci.com/Publications/Papers/VCIProdicts/CTP-25.pdf.

58. Micsic B. A., Boyle R., and Wuertz B. F. N. Speller award lecture: Efficacy of vapor phase corrosion inhibitor technology in manufacturing, *The Journal of Science and Engineering: Corrosion Science*, 60(6), 515–522, 2004.

59. Chudnovsky B. H. Lubrication of electrical contacts, *Proc. 51th IEEE Holm Conference on Electrical Contacts*, Chicago, IL, pp. 107–114, 2005.

60. Sjögren L. Corrosion of electronics. *Corrosion News*, no. 5, October 2013, p. 5, Swerea KIMAB and Institut de la Corrosion, Online. Available: http://www.swerea.se/sites/default/files/corr_news_5_2013.pdf.

61. Ambat R. A review of Corrosion and environmental effects on electronics, 16pp., August 2013, Online. Available: http://www.smtnet.com/library/files/upload/A-review-of-Corrosion-and-environmental-effects-on-electronics.pdf.

62. Yunovich M. Appendix Z – Electronics, p. Z1-Z7, Online. Available: www.corrosioncost.com/pdf/electronics.pdf.

63. Impact of gaseous sulfides on electronic reliability, *ERA Technology*, Online. Available: http://www.edifgroup.com/case-studies/impact-of-gaseous-sulphides-on-electronic-reliability.

64. ANSI/ISA Standard S71.04-2013 *Environmental Conditions for Process Measurement and Control Systems: Airborne Contaminants*, International Society for Automation, ISA, Research Triangle Park, 30pp., 2013.

65. *2011 Gaseous and Particulate Contamination Guidelines For Data Centers*, American Society of Heating, Refrigerating and Air-Conditioning Engineers, Inc. (ASHRAE), 22 pp., 2011.

66. Frankel G. S. *Corrosion Mechanisms in Theory and Practice*, edited by Marcus P. and Oudar J., Marcel Dekker Inc., New York, 1995.

67. Protection from the Elements Part III: Corrosion of Electronics, *Lighting Global Technical Notes*, Issue 14, September 2013, Online. Available: https://www.lightingglobal.org/wp-content/uploads/bsk-pdf-manager/63_issue14_part-iii_corrosion_technote_final.pdf.

68. Smith J. More benefits of nickel barriers in soldering, *Electronic Manufacturing Science (EMS)*, Feb 2011, Online. Available: http://www.emsciences.com/blog/more-benefits-of-nickel-barriers-in-soldering/.

69. Cullen D. Preventing corrosion of PCB assemblies, *OnBoard Technology*, October 2008, 52-57, Online. Available: http://www.onboard-technology.com/pdf_ottobre2008/100808.pdf.

70. Schueller R. Creep corrosion on lead-free printed circuit boards in high sulfur environments, *SMTA 2007*, Online. Available: http://www.dfrsolutions.com/wp-content/uploads/2012/06/2007_08_creep_corrosion_on_lead-free_pcb_in_high_sulfur_environments.pdf.

71. Mazurkiewicz P. Accelerated corrosion of printed circuit boards due to high levels of reduced sulfur gasses in industrial environments, *Proc. 32nd International Symposium for Testing and Failure Analysis*, Austin, TX, pp. 469–473, Nov. 12–16, 2006, Online. Available: http://chavant.com/new_site/files/pdf/HP_ImAg_Corrosion_Paper_ISTFA_20061.pdf.

72. Zhou Y., Pecht M. Reliability assessment of immersion silver finished circuit board assemblies using clay tests, *8th International Conference on Reliability, Maintainability and Safety (ICRMS)*, pp. 1212–1216, July 2009.

73. Bowman J., Zajko R. A. The Environment in Control and Equipment Rooms: how important is it, and what to look for? *Honeywell White Paper*, Aug 2013, Online. Available: https://www.honeywellprocess.com/library/marketing/whitepapers/Control-Systems-Environment-WP.pdf.

74. iNEMI Position Statement on the Limits of Temperature, Humidity and Gaseous Contamination in Data Centers and Telecommunication Rooms to Avoid Creep Corrosion on Printed Circuit Boards, iNEMI 2012, Online. Available: http://thor.inemi.org/webdownload/projects/iNEMI_Position-Creep_Corrosion2012.pdf.

75. China National Standard GB 50174-2008: Code for Design of Electronic Information System, 2008.

76. Muller C., Singh P., White G. H. et al. Solving air contaminant problems in data centers, *Uptime Journal*, Online. Available: http://journal.uptimeinstitute.com/solving-air-contaminant-problems-data-centers/.

77. A Literature Review of the Effects of Smoke from a Fire on Electrical Equipment, United States Nuclear Regulatory Commission (US NRC), NUREG/CR-7123, prepared by Peacock Richard D., Cleary Thomas G., Reneke Paul A., et al, NIST, pp. 105, July 2012, Online. Available: pbadupws.nrc.gov/docs/ML1218/ML12185A086.pdf.

78. Electrical and Electronic Equipment—Non-Thermal Fire Damage, Hanover Risk Solutions, Online. Available: www.hanover.com/risksolutions/pdf/171-0932.pdf.

79. Levchik S., Hirschler M., Weil E. *Practical Guide to Smoke and Combustion Products from Burning Polymers - Generation, Assessment and Control*, Smithers Rapra, 247p., 2011, Online. Available: www.gbv.de/dms/tib-ub-hannover/635361043.pdf http://info.smithersrapra.com/downloads/chapters/PG%20to%20Smoke.pdf.

80. Shafer M. Effects of smoke corrosion to equipment and electronics, *Technical Bulletin*, Online. Available: http://www.er-emergency.com/technical-bulletin-effects-of-smoke-corrosion-to-equipment-and-electronics.

81. Krzyzanowski M. E. *Fire Event Electronic and Mechanical System Damage Mechanisms*, Equipment Damage Consultants, LLC, 2008, Online. Available: www.eqdamcon.com/images/EDC_Fire_White_Paper.pdf.

82. Huggins Jr. R. L. *Protecting Electrical/Electronic Systems in a Corrosive Treatment Plant Environment*, Charleston Commissioners of Public Works.

83. Sharp W. B. A. Protection of control equipment from atmospheric corrosion, *Proceedings of Corrosion 1990 Conference*, NACE International, Las Vegas, NV, pp. 1–12, Dec. 31, 1990.

84. Sharp W. B. A., Falat L., and Krasowski J. A. Corrosion prevention in electrical control rooms, *Tappi Journal*, 72(10), 143–145, 1989.

85. Muller C. O. Control of corrosive gases to avoid electrical equipment failure, *Paper Technology*, April 2000.

86. Muller C., Ruede D., Crosley G. ISA Standard 71.04: Changes required for protection of today's process control equipment, *Analysis*, 2012, Online. Available: https://www.isa.org/store/isa-standard-7104-changes-required-for-protection-of-today%E2%80%99s-process-control-equipment-analysis-2012/122344.

6 Lubrication of Distribution Electrical Equipment

6.1 LUBRICATION PRIMER

6.1.1 PURPOSE OF LUBRICATION

The primary purpose of lubrication is to reduce friction between moving metal surfaces. To keep metal surfaces separated, lubricating grease should wet the metal and resist being displaced by the pressures it encounters. A lubricant can also serve as a coolant, prevent corrosion, and block the entry of contaminants. Satisfactory grease will also resist separation, be reasonably stable, and not significantly change its nature within the temperature range for which it was designed [1–8].

Lubricants are widely used in electrical connectors, switches, and CBs. Lubricants improve the performance of electrical apparatus in many ways, such as *reducing mating forces, extending plating durability, and enhancing corrosion protection* (fretting, galvanic, and harsh environments). Good grease will flow into bearings when applied under pressure and remain in contact with moving surfaces. It will neither leak out from gravitational or centrifugal action nor will it stiffen in cold temperature where it unduly resists motion [9,10].

Lubricants *prevent environmental and galvanic corrosion* on electrical contacts. Airborne contaminants attack metals, causing oxides to gradually build up in pores until they reach the surface, where they impede current flow. Contact surfaces and switches made of dissimilar metals are especially susceptible to moisture, oxygen, and aggressive gases. Even noble metal plating is at risk if it is worn or porous. The purpose of lubricant is to prevent dust, water, dirt, and other contaminants from entering the part being lubricated.

Lubricants *reduce wear and heat* between contacting surfaces in relative motion. While wear and heat cannot be completely eliminated, they can be reduced to negligible or acceptable levels, especially on sliding electrical contacts which experience repetitive cycling or arc damage—two common causes of failures. Although evidence suggests that lubricants change or reduce arc patterns, the lubricant's real job on sliding contacts is to separate the surfaces during operation and keep debris out of the contact area. Otherwise, the microscopic wear particles oxidize quickly, turning into insulators.

Buildup of oxide particles also accelerates wear. In general, hydrocarbon lubricants work best at wear prevention because their molecular structure is more rigid than other base oils. Proper lubricants strike a balance between preventing wear and maintaining electrical continuity.

Because heat and wear are associated with friction, both effects can be minimized by reducing the coefficient of friction between the contacting surfaces. Lubricants *reduce the friction* between mechanical components, thus reducing the amount of force needed to activate a swift motion. Lubricants usually ensure a coefficient of friction of 0.1 or less, which means that it takes little force to operate a device with a high preload. This can be important in switches and CBs where high normal forces ensure low contact resistance (CR) and a stable signal or power path. Lubrication is also mechanically important because it gives the end user smooth, uniform operation.

Care must be taken to select the proper lubricant to achieve the desired results. In choosing grease, the three primary elements that go into the making of grease should be considered: the liquid lubricant (petroleum or synthetic fluid), the thickening agent, and additives. The chosen grease must be heavy enough to provide the right film strength, yet be light enough to flow well in cold temperatures.

The person selecting the grease should investigate which solid or liquid additives, such as the extreme pressure (EP) type, will do the job well. For example, to lubricate open gears, grease containing a tacky additive must be used to keep the grease in place. The grease should be compatible with other materials used with the parts being lubricated. Manufacturers may specify the type of grease to be used in their equipment.

6.1.2 Lubrication Terminology

Various lubrication terms have been accepted and widely used in the industry to describe the physical properties of lubricating materials [2–4,11].

Viscosity is a measure of flowability. It is the resistance to flow caused by internal friction between the lubricant molecules. This characteristic helps determine the load-carrying capacity, thickness of the lubricating film, and operating temperature. In selecting a lubricant for a particular application, definition of the required viscosity level at start-ups and during operation conditions is critically important to ensure optimum lubricant performance. Viscosity can be expressed as the number of seconds required for a measured volume of oil to flow through a specified orifice in standard conditions. In the United States, viscosity is usually measured and specified between 100°F and 210°F (~40–100°C).

Apparent viscosity is introduced to distinguish between the viscosity of oil and grease; the viscosity of grease is referred to as "apparent viscosity." Apparent viscosity is the viscosity of a grease that holds only for the shear rate and temperature at which the viscosity is determined. At start-up, grease has resistance to motion, implying high viscosity. However, as grease is sheared between wearing surfaces and moves faster, its resistance to flow reduces. Its viscosity decreases as the rate of shear increases. By contrast, oil at a constant temperature would have the same viscosity at start-up as what it has when moving.

Viscosity index (VI) indicates how viscosity varies with temperature. The property can be an important consideration in applications where operating temperatures vary widely, particularly when low temperatures are encountered.

Pour point is the lowest temperature at which oil flows and is most critical in low-temperature applications. Formation of wax crystals causes flow failure in paraffinic oils.

Dropping point is an indicator of the heat resistance of grease. As grease temperature rises, penetration increases until the grease liquefies and the desired consistency is lost. The dropping point is the temperature at which grease becomes fluid enough to drip. It indicates the upper temperature limit at which grease retains its structure, not the maximum temperature at which grease may be used. A few greases have the ability to regain their original structure after cooling down from the dropping point.

Flash point is the temperature at which oil gives off ignitable vapors. The flash point is not necessarily a safe upper limit for oil because some decomposition takes place below the flash point.

Fire point is the temperature at which oil will burn if ignited.

Oxidation stability is the ability of grease to resist a chemical union with oxygen. The reaction of grease with oxygen produces insoluble gum, sludges, and lacquer-like deposits that cause sluggish operation, increased wear, and reduction of clearances.

Prolonged high-temperature exposure accelerates oxidation in greases.

Pumpability is the ability of a grease to be pumped or pushed through a system. More practically, pumpability is the ease with which a pressurized grease can flow through lines, nozzles, and fittings of grease-dispensing systems.

Slumpability, or feedability, is its ability to be drawn into (sucked into) a pump. Fibrous greases tend to have good feedability but poor pumpability. Buttery-textured greases tend to have good pumpability but poor feedability.

Shear stability: Grease consistency may change as it is mechanically worked or sheared between wearing surfaces. A grease's ability to maintain its consistency when it is worked is its shear stability or mechanical stability. A grease that softens as it is worked is called thixotropic. Greases that harden when they are worked are called rheopectic.

Reversibility defines an ability of grease to return to its original consistency after it is exposed to conditions of high temperature or high shock loading for a short period of time. When a grease encounters abnormally high temperatures for short periods of time and then returns to normal operating temperatures, or encounters high shock loading conditions, bleeding of the base oil from the grease may occur. The grease with high reversibility must have the ability to recapture its bases in order to return to its original consistency.

6.1.3 TYPES OF LUBRICATING MATERIALS

There are three major types of the lubricants: oils (fluid lubricants), greases (soft substances), and solid lubricants [2–8].

6.1.3.1 Oil

Oil covers a broad class of fluid lubricants, each of which has particular physical properties and characteristics. Basically, petroleum lubrication oils (mineral oils) are made from naphthenic or paraffinic crude oil. Naphthenic oils contain little wax and their low pour point makes them good lubricants for most applications. Paraffinic oils, on the other hand, are very waxy. Lubricants made from them are used mainly in hydraulic equipment and other machinery.

6.1.3.2 Synthetic Oils

Synthetic oils are being used for instrument bearings, hydraulics, air compressors, gas and steam turbines, and many other applications. Silicones are not true oils; their principal advantages are excellent viscosity–temperature characteristics, good resistance to oxidation, and a wide operating temperature range.

6.1.3.3 Grease

Grease consists of a thickening agent, oil or synthetic fluid, and additives. Various thickeners provide different properties of grease [12–14]. Greases are usually ranked by their consistency on a scale set up by the National Lubricating Grease Institute (NLGI). Consistency or grade numbers range from 000 to 6, corresponding to specified ranges of penetration numbers (see Section 8.1.5.1).

6.1.3.4 Synthetic Lubricants

Base stocks for synthetic lubricant are esters, synthetic hydrocarbons (SHCs), polyglycols, silicones, and so on. Synthetic greases provide a cost-effective, lifetime lubrication for bearings and other moving parts, for gaskets, and seals. High thermal stability and chemical inertness make it useful for aerospace, electrical, automotive, and other high-tech or industrial applications. Synthetic lubricants usually keep their lubrication characteristics at temperatures in the range of −73–290°C (−100–550°F), maintain good viscosity over a wide temperature range, and are nonflammable with no autogeneous ignition, flash, or fire point up to 650°C (1200°F).

The label "synthetic lubricants" covers a broad category of fluids and pastes having varied properties. They are available in a variety of forms, from light oils to thick greases. Synthetic lubes [6,7,15] have a longer product life, are more inert than petroleum materials, and generate less waste. They have a capability for a wider range of temperatures, from extremely low to high; certain classifications are friendlier to elastomers, seals, and O-rings that come in contact with the lubricant.

6.1.3.5 Solid Lubricants

Solid lubricants are usually fine powders, such as molybdenum disulfide, graphite, and Teflon® (polytetrafluoroethylene (PTFE)). They can be used as additives in greases or dispersions, as dry film-bonded lubricants, or alone. Lubricating solids can provide longer term lubrication than unfortified oils and greases because of their ability to form burnished films on surfaces and stay longer (see Section 8.5).

6.1.3.6 Silicones

Silicones are very stable and very inert lubricants, which provide wider operating temperature ranges than nonsilicone synthetic lubes. *Fluorosilicones* have some of the advantages of other silicones, such as dimethyl and phenylmethyl silicones, plus higher resistance to harsh environments and the ability to carry heavy bearing loads.

6.1.3.7 Fluid Lubricants

Fluid lubrication has played a major role throughout history in the development of new technologies involving contacting interfaces. These technologies have been investigated for lubrication of micrometer-scale contacts in microelectromechanical systems (MEMSs), though effective lubrication strategies for micrometer and smaller scale contacts are still being developed [16].

6.2 GREASE COMPOSITION: PROPERTIES AND TESTING

6.2.1 GREASE COMPOSITION

Grease composition may be described by a simple formula: *Lubricating grease = Fluids + Thickener + Additives* [11–14]. Various types of fluids, thickeners, and additives are presented in Tables 6.1 through 6.3. The grease properties depend on how much and what type of each component is included in the grease composition.

6.2.2 GREASE PROPERTIES AND TESTING

6.2.2.1 Consistency

Consistency is defined as the degree to which a plastic material resists deformation under the application of force. In the case of lubricating greases, this is a measure of the relative hardness or softness and has some relation to flow and dispensing properties. Grease consistency is classified

TABLE 6.1
Grease Composition: Fluids (75%–95%)

Fluid Group	Fluid Type
Mineral oils	Paraffinic oil
	Naphthenic oil
	Phosphate esters
SHC	Diesters
	Organic esters
	Polyglycols
	Polyalphaolefin
	Polyphenyl ether
	Silicate ether or disiloxane
Silicones	Chlorinated phenyl methyl silicones
	Methyl silicones
	Phenyl methyl silicones
Fluorocarbons	PFPE

Abbreviation: SHC, synthetic hydrocarbon.

TABLE 6.2
Grease Composition: Thickeners (5%–20%)

Thickener Group	Thickener Type
Soaps	Lithium (Li) complex
	Sodium (Na) complex
	Calcium (Ca) (conventional, anhydrous, complex), Ca sulfonate
	Aluminum (Al) complex
	Strontium (Sr)
	Barium (Ba) (conventional, anhydrous, complex)
Nonsoap (inorganic)	Microgel
	Bentonite (clay)
	Carbon black
	Silica gel
	Fumed silica
Nonsoap (organic)	Urea compounds (polyurea)
	Polytetrafluoroethylene (PFTE, Teflon)
	Fluorine compounds
	Terephthalate, organic dyes

TABLE 6.3
Grease Composition: Additives (0%–15%)

Oxidation Inhibitors	Emulsifiers
Rust/corrosion inhibitors	Demulsifiers
Antiwear agents	Pour point depressants
EP additives	Tackiness/adhesive agents
Acid neutralizers	Viscosity modifiers
Antifoam additives	Oiliness enhancers
Detergents	Solid additives (MoS_2, graphite)
Dispersants	Perfumes, dyes

according to a scale developed by the NLGI. Table 6.4 shows the NLGI number relation to the consistency of the grease.

Grease consistency depends on the type and amount of thickener used and the viscosity of its base oil. Consistency of grease is its resistance to deformation by an applied force. The measure of consistency is called penetration. Penetration depends on whether the consistency has been altered by handling or working. ASTM D217 and D1403 methods measure penetration of unworked and worked greases.

The softest greases are rated at 000 (liquid), with higher numbers indicating harder grease. Most greases fall in NLGI range from 1 to 2 (soft) but some are hard which NLGI is in the range 5–6 (Table 6.4). Consistency is measured by ASTM D217, Cone Penetration of Lubricating Grease. NLGI is based on the degree of penetration achieved by allowing a standard cone to sink into the grease, which has been worked for 60 strokes in a grease worker, at a temperature of 25°C (77°F) for a period of 5 s. The depth of cone penetration is measured on a scale of tenths of mm and the softer greases allow the cone to penetrate further into the grease, hence the higher penetration number. The test is conducted in accordance to ISO 2137.

The choice of a certain consistency for a certain application depends on many operating conditions such as temperature, speed, shaft alignment, pumpability, and so on. For example, vertical

TABLE 6.4

NLGI Grease Classification

NLGI Number	ASTM Worked Penetration 0.1 mm $(3.28 \times 10^{-4}$ ft) at 25°C (77°F)	Consistency
000	445–475	Semifluid
00	300–430	Semifluid
0	355–385	Very soft
1	310–340	Soft
2	265–295	Common grease
3	220–250	Semihard
4	175–205	Hard
5	130–160	Very hard
6	85–115	Solid

Source: Lubrication of Power Plant Equipment, Facilities Instructions, Standards, and Techniques (FIST), Vol. 2–4, United States Department of the Interior Bureau of Reclamation (USBR), Springfield, VA, 33 pp., 1991, Online. Available: http://www.usbr. gov/power/data/fist/fist2_4/vol2–4.pdf.

Abbreviation: NLGI, National Lubricating Grease Institute.

shaft arrangements calls for stiff greases, but low operating temperatures calls for low consistency greases in general and so on. Once a certain consistency has been chosen for a certain application, it should not change drastically during the advised relubrication interval or storage time. This is related to grease mechanical stability. These are some possible causes for a consistency change whether it is softening or hardening of the grease [17,18].

6.2.2.2 Softening and Hardening

Softening (lower NLGI value) of the grease can be caused by excessive temperature for the grease used in application, by the presence of water in grease, and by the mixing of incompatible greases. Grease may soften if its shelf life exceeded. In some mechanical vibrating application, the grease with too soft consistency or poor mechanical stability may soften even further. In bearings, grease may soften if the bearing housing filled too much for the speed used which creates churning and excessive grease shearing. Grease softening in a bearing may eventually cause grease to leak out from the housing, requiring more maintenance and frequent grease replenishment to avoid premature failure resulting from lack of lubricant on the rolling elements.

Hardening (higher NLGI value) of the grease can be caused by mixing of incompatible greases, if large amount of solid contaminants (carbonized particles for instance) are present in grease, and if shelf life exceeded. The grease may harden if it has lost base oil (excessive oil bleeding) due to exceeded relubrication maintenance interval or because of evaporation due to continuous use at high temperature. If the grease is of poor quality and temperature in application changes often and fast, it may also harden.

6.2.2.3 Grease Shear and Structural Stability

Grease needs to maintain its consistency under high shear conditions. The *shear stability* test measures the softening of grease when sheared for 10,000 or 100,000 double strokes with a grease worker. Loss of less than one NLGI grease grade signifies a stable thickener under high shear conditions [19].

Grease *structural and mechanical stability* is an essential performance characteristic of lubricating grease as it is a measure of how the grease consistency will change in service when it is subjected to mechanical stress (shear) resulting from the churning action caused by moving elements or vibrations generated by, or external to, the application. In order to have good mechanical stability, greases are developed through careful selection of the thickener composition and optimization of the manufacturing process.

Mechanical stability is often measured using the ASTM D217 prolonged worker test (e.g., 100,000 double strokes) or the ASTM D1831 Roll Stability test. ASTM D1831 subjects the grease to shearing by rotating a cylinder containing a 5-kg roller at 165 rpm for 2 h. The change in penetration at the end of the tests is a measure of the mechanical stability. This test produces low shearing forces approximately equal to those found in the grease worker used for ASTM D217.

In application and use, ingress of environmental contaminants is unfortunately a common reality that often adversely affects the mechanical stability of the grease. It is important that greases not only be developed to provide excellent structural stability in a pristine state but also in the presence of environmental contaminants such as water, process fluids, or other contaminants. This can be assessed by means of laboratory bench tests operating in a variety of conditions with the presence of water.

6.2.2.4 Dropping Point

The dropping point of grease is the temperature at which the thickener loses its ability to maintain the base oil within the thickener matrix. This may be due to the thickener melting or the oil becoming so thin that the surface tension and capillary action become insufficient to hold the oil within the thickener matrix.

ASTM D2265 is the standard method used to determine the dropping point of grease. A small grease sample is placed in a cup and heated in a controlled manner in an oven-like device. When the first drop of oil falls from the lower opening of the cup, the temperature is recorded to determine the dropping point. Dropping point is a function of the thickener type. High drop points, typically above 240°C (465°F), are commonly observed for lithium complex, calcium complex, aluminum complex, polyurea, and clay greases, while much lower dropping points are typical of conventional lithium (180°C/355°F), calcium (180°C/355°F), and sodium (120°C/250°F) soaps.

The dropping point is one of the determinations that characterize the grease's thermal stability. However it is NOT an accurate prediction of the grease's upper operating temperature limit which is a function of many variables such as base oil oxidation stability, additive degradation, thickener shearing, oil separation, and so forth. A high dropping point, while not a predictor of upper operating temperature, is an indicator of the maximum peak temperature that the grease may be subjected to for a short duration while not releasing oil excessively and therefore drastically reducing the life of the grease and potentially damaging the application in the long run.

All grease properties should be tested according to ASTM standards and reported by the manufacturers [19]. These ASTM standards and corresponding test methods to define typical grease properties are summarized in Table 6.5.

6.2.2.5 Effect of Cold Temperatures on Lubricant Properties

Outdoor electrical equipment in the northern parts of the United States, Europe, and Canada are vulnerable to harsh outdoor conditions during winter months. At extremely low temperatures, such as those experienced outside on a cold winter morning in these parts of the world, the lubricants (oils and greases) within the housing can reach a point where they actually congeal, harden, and will no longer perform properly. Components will then starve of lubrication, which leads to an early failure.

Another problem the lubricant must endure from the cold is oil separation. At extreme low temperatures, blended base oils can begin to separate into different phases. This separation process is called *stratification*. The additives are also susceptible to becoming insoluble at colder temperatures. When they become insoluble, additives tend to gravitationally separate from the base oil and form deposits at the bottom of the lubricated device. If the equipment requires these additives and they are in the form of a sludge or deposit at the bottom, the lubricant's performance will be hindered and the machine could be damaged [20].

TABLE 6.5

ASTM Tests to Define Typical Grease Properties

ASTM #	Test Method	Parameter, Unit	Property
D217	Cone penetration unworked and 60 double strokes; worked penetration 10,000 and 100,000 double strokes	mm/10	Grease consistency, NLGI grade
D566	Determine temperature at which the grease passes from a semisolid to a liquid state	Temperature, °C(°F)	Dropping point
D97	Determine temperature at which grease becomes fluid enough to drip	Temperature, °C(°F)	Pour point
D1743	Corrosion of grease-lubricated tapered roller bearings stored under wet conditions	Pass/fail	Corrosion prevention
D1742	Determination of the tendency of a lubricating grease to separate oil during storage	Oil separated, %	Oil separation
D1264	Evaluation of the resistance of a lubricating grease to washout by water from a bearing	Grease washed out, %	Water washout
D4049	Evaluate the ability of a grease to adhere to a metal surface when subjected to direct water spray	Grease sprayed off, %	Water spray-off resistance
D2509	Measurement of load-carrying capacity of lubricating grease, Timken OK Load	Maximum weight, Kg (Lbs)	Performance of EP additives in lubricating grease
D2783	Measurement of extreme pressure properties of lubricating fluids (Four-Ball Method)	Weld point, Kg Load wear index, a number	EP properties, load-carrying properties of lubricating fluids
D2266	Wear preventive characteristics of lubricating grease (Four-Ball Method)	Scar diameter wear reading, mm	Relative wear-preventing properties of greases

Source: Adapted from Thibault R. *Maintenance Technology Magazine*, July 2009, Online. Available: http://www.mt-online.com/julyaugust2009/grease-basics.

Abbreviations: EP, extreme pressure; NLGI, National Lubricating Grease Institute.

With most mineral-based industrial oils (designated as turbine, hydraulic, industrial, and machine oils), pour point measured by test method according to ASTM D77 Standard Test Method for Pour Point of Petroleum Products corresponds to the temperature that freezes the paraffin molecules of the oil into a white crystalline wax that will eventually immobilize the overall oil. With their low paraffinic content, wax-free synthetic and naphthenic mineral oils can be further cooled to a lower pour point. At this point, the viscosity becomes so high (usually about 100,000 centistokes, cSt) it will eliminate any visible oil flow in the pour point test [21].

Most base oils and greases are able to withstand moderate temperature dips to 0°C (32°F) and many to −10°C (14°F) without much decrease in performance. However, at −20°C (−4°F) and beyond, some lubricants become unsuitable and begin to reach their pour point. The pour point is dictated by the base oil quality as well as the presence of certain additives. A good rule of thumb is to always select a lubricant with a pour point that is at least 10°C lower than lowest expected service temperature.

At temperatures lower than −20°C (−4°F), simple mineral base oils will no longer perform sufficiently, so alternatives must be found. Polyalphaolefin (PAO) synthetic oils are among the lubricants used at cold temperatures of −20°C (−4°F) and lower. PAO oils do not contain the wax and this allows for excellent flow, even at low temperatures. Some PAOs have pour points as low as −50°C (−58°F) [20].

TABLE 6.6
LT Limits for Greases

Base Oil Type	Grease Type	LT Limit, °C (°F)
Mineral oil	Conventional industrial greases	−34 (−30)
	LT greases	−45.6 (−50)
Silicones	High-temperature greases	−34 (−30)
	Special LT greases	−79 (−110)
PAO hydrocarbons	Synthetic greases	−54 (−65)
Diester	Synthetic greases	−62 (−80)

Abbreviations: LT, low temperature; PAO, polyalphaolefin.

Specific greases should be considered for outdoor electrical machines lubrication in cold climates [21]. Table 6.6 shows low-temperature limits for various types of greases. The limits in Table 6.6 are defined by the viscosity of the oil component that makes up 80%–90% of the grease composition. For example, conventional industrial greases made with mineral oils with viscosity ~100 cSt at 40°C (104°F) can be used at cold temperatures down to −34°C (−30°F), where viscosity of the oil phase reaches about 100,000 cSt.

6.2.3 LUBRICANT THICKENER: ROLE IN GREASE PROPERTIES

The thickener is a material that will, in combination with the selected lubricant, produce the solid to semifluid structure. A thickener's efficiency depends on how much of it is needed to make a given grade (stiffness) of grease. The primary lubricating component is oil, so it is beneficial to have as much oil in the formulation as possible. But a lubricant should not be too viscous, otherwise it causes hydroplaning (an open-circuit condition), especially at low temperature.

There are many different classes of materials used as thickeners in grease manufacturing: simple and complex metal soaps, and inorganic and organic nonsoaps; the role of a thickener is to give grease its characteristic consistency and the thickener is sometimes thought of as a "three-dimensional fibrous network" or a "sponge" that holds the oil in place. Lubrication function has been compared to that of a sponge (thickener) gradually releasing its liquid over a period of time. The major difference is in manufacturing process of thickeners [14,17,22,23]. The summary of different grease properties provided by various thickeners is shown in Tables 6.7 and 6.8.

A thickener's ability to resist water is also an important consideration. Even in devices that are generally protected from the environment, humidity can condense inside them and displace the grease or it can become entrained and accelerate corrosion. Lithium soap greases have good freshwater resistance, but poor saltwater resistance. Clay and PTFE generally perform well in wet applications. PTFE also lowers friction, especially on plastic components [22].

6.2.3.1 Soap Thickeners

The primary type of thickener used in grease is metallic soaps. These soaps include lithium, aluminum, clay, polyurea, sodium, and calcium (Table 6.2). Soap-type greases are produced using so-called saponification process similar to a soap manufacturing technique. Typical lithium-based greases (the most common) are made from a fatty acid, usually 12-hydroxystearic acid, and a lithium base to produce a simple soap which acts as the grease thickener.

Calcium sulfonate greases are made by converting a fluid detergent that contains amorphous calcium carbonate to a grease containing calcite particles. Because of lubricating properties of the calcite particles, performance additives containing sulfur, phosphorus, or zinc may not be needed. This is why some calcium sulfonate-based greases are attractive to the food industry.

TABLE 6.7

Grease Properties by Thickener Types

Grease Thickener	Appearance	Shear Stability	Pumpability	Protection Against Rust	Heat Resistance	Water Resistance
Calcium soap	Buttery	Good	Fair	Poor to fair	Fair	Excellent
Calcium complex	Buttery to grainy	Good	Fair	Fair	Good	Good to excellent
Calcium sulfonate	Buttery to grainy	Good	Good	Excellent	Excellent	Excellent
Sodium soap	Fibrous	Fair	Poor		Good to excellent	Poor
Barium soap	Fibrous	Good	Poor	Fair to good	Excellent	Excellent
Lithium soap	Buttery	Excellent	Good to excellent	Poor to fair	Good to excellent	Excellent
Lithium complex	Buttery	Good	Good to excellent	Poor to fair	Excellent	Excellent
Aluminum complex	Buttery to grainy	Good to excellent	Good	Fair to good	Excellent	Excellent
Clay (bentonite)	Buttery	Good	Good	Poor to fair	Excellent	Excellent
Polyurea	Buttery	Good	Good	Poor to fair	Excellent	Excellent

TABLE 6.8

Physical Properties of the Greases with Different Thickeners

Grease Thickener	Dropping Point, °C (°F)	Maximum Usable Temperature, °C (°F)	Other Properties
Calcium soap	96 (205)	93 (195)	EP grades
Calcium complex	232 (450)	177 (350)	EP grades
Calcium sulfonate	300 (572)	177 (350)	EP grades, heavier than water
Calcium 12-hydroxystearate	135 (275)	110 (230)	EP grades
Sodium soap	120 (250)	130 (266)	EP grades
Barium soap	188 (370)	120 (250)	EP grades
Barium complex soap	193 (380)	141 (285)	EP grades
Lithium soap	177–204 (350–400)	135 (275)	EP grades
Lithium complex	260 (500)	149–177 (300–350)	EP grades
Lithium 12-hydroxystearate	177 (350)	121 (250)	EP grades
Aluminum soap	110	79	EP grades
Aluminum complex	204 (400)	177 (350)	EP grades
Clay (bentonite)*	260+ (500+)	177–377 (350–700)	
Polyurea	238 (460)	177 (350)	EP grades

Abbreviation: EP, extreme pressure.

The most used in industrial and manufacturing environments are lithium and lithium *complex greases*, which have strong properties in many categories. These greases have high dropping points and excellent load-carrying abilities, excellent long-term work stability, strong high-temperature characteristics, and have acceptable wash- and corrosion-resistance capabilities.

With additive enhancement, the wash and corrosion resistance can be improved. These greases also have good low-temperature shear performance, making them suitable for extremely low-temperature applications.

Complex greases are made by combining the conventional metallic soap with a complexing agent. The most widely used complex grease is lithium based. These are made with a combination of the conventional lithium soap and a low-molecular-weight organic acid as the complexing agent. Each soap type provides different performance properties [24].

6.2.3.2 Nonsoap thickeners

Nonsoap thickeners are made with a variety of products and processes, and deliver range of performance results. The clay-based (bentonite) products silica aerogel and polyurea products represent nonmetallic thickeners. Bentonite and silica aerogel are two examples of thickeners that do not melt at high temperatures.

Polyurea thickeners are a type of polymer formed by a reaction between amines and isocyanates, which occurs during the grease formation process. Polyurea greases are widely used in ball bearing applications. Polyurea greases contain little to no heavy metals and have favorable high-temperature performance resulting in very good oxidation resistance. Polyureas provide good work stability and wash and corrosion resistance. Some polyureas have a low level of compatibility with other soap and nonsoap greases, including other polyureas. Nonetheless, there are individual products being manufactured that demonstrate strengths in all of these categories, including the important issue of compatibility.

Bentonite products are created by direct addition of the thickener to the base and additive mixture. These products require significant milling to assure uniformity. Bentonite is a type of clay. Bentonite greases are incompatible with most other grease types.

Greases with clay bentonite thickener were the original nonmelting grease. Some clay-thickened (bentonite) greases may have very high melting points, with dropping points noted on the product data sheets as 500°C or greater, some data sheets say that there is no dropping point. For these nonmelting products, the lubricating oil burns off at high temperatures, leaving behind hydrocarbon and thickener residues. Other base oils may evaporate before the clay material becomes hot enough to melt. This is both a strength and weakness. When used for extended periods of time at elevated temperatures, bentonite grease residues may cause a filling of the housing that can make long-term relubrication difficult.

The maximum usable temperature is dependent upon the quality and type of base oil used in the formulation of the grease and its relubrication cycle.

6.2.4 LUBRICANT ADDITIVES: THE ROLE IN GREASE PROPERTIES

6.2.4.1 Types of Grease Additives

The additives are used in the greases to enhance desirable properties and suppress undesirable ones, as well as add new properties to a lubrication product [14]. Various additives are listed in Table 6.3 and their functions include surface protection, grease performance enhancement, and lubricant preservation. These functions are explained below.

Oxidation inhibitors (antioxidants) reduce oxidation. Their various types are oxidation inhibitors, retarders, anticatalyst metal deactivators, and metal passivators. Oxidation inhibitors, or antioxidants, lengthen a lubricant's service or storage life by increasing its oxidation resistance either through binding free oxygen in the oil or through neutralizing the catalytic effect of metals.

Rust/corrosion inhibitors reduce the rusting of ferrous surfaces swept by oil as well the corrosion of cuprous and other metals. Rust inhibitors protect ferrous (iron or steel) parts by forming a film on the part that resists attack from water. Corrosion inhibitors act in a similar way to protect nonferrous parts and also act to neutralize acids with a basic compound such as calcium carbonate.

Antiwear agents reduce wear and prevent scuffing or rubbing surfaces under steady load operating conditions. Antiwear additives work by coating the metal surface. If light metal-to-metal contact is made, the heat from the friction melts the additives, forming a liquid layer between the surfaces. This molten additive layer, being softer than the metal, acts as a lubricant, preventing wear of the metal surfaces.

EP additives prevent scuffing or rubbing surfaces under severe operating conditions, for example, heavy shock load, by the formation of a mainly inorganic surface film. EP additives work by reacting with a metal to form a compound that acts as a protective layer on the metal's surface. Because this layer is softer than the metal itself, under EP conditions, the compound layer wears away first, protecting the metal. As this layer is removed, the EP additive acts to form another layer. In contrast to the action of antiwear additives, EP additives control wear instead of preventing it. Some EP additives, because of their reactive nature, can be corrosive to brass or copper-containing alloys. To prevent excessive corrosion, most EP additives are activated by the heat of friction created during EP conditions but do not react at room temperature.

Acid neutralizers neutralize contaminating strong acids that are formed, for example, by combustion of high-sulfur fuels or by decomposition of active EP additives.

Antifoam additives reduce surface foam. Foam inhibitors prevent lubricant foaming by decreasing the surface tension of air bubbles, allowing them to combine into large ones, which break more rapidly.

Detergents reduce or prevent deposits formed at high temperatures (e.g., in internal combustion engines).

Dispersants prevent deposition of sludge by dispersing a finely divided suspension of the insoluble material formed at low temperature. Detergents and dispersants are primarily used in engine oils to keep the surfaces free of deposits and to keep contaminants dispersed in the lubricant.

Demulsifiers promote the separation of oil and water in lubricants exposed to water.

Emulsifiers form emulsions, either water-in-oil or oil-in-water, according to type. An emulsifier promotes the rapid mixing of oil and water to form a stable emulsion. Emulsifiers are used in motor oils to allow water, formed by combustion of fuel, to be kept in emulsion until engine heat can evaporate it. Emulsifiers are also used in soluble oils used in some metal working operations and in fire-resistant hydraulic fluids. Emulsification is usually not a desirable property in most hydraulic fluids or turbine oils.

Pour point depressants reduce the pour point of paraffinic oils and enable lubricants to flow at low temperature.

Tackiness agents reduce the loss of oil by gravity, for example, from vertical sliding surfaces, or by the centrifugal force. Tackiness agents act to increase the adhesiveness of an oil or grease.

Viscosity modifiers (index improvers) reduce the decrease in viscosity caused by an increase in temperature. VI improvers lower the rate of change in viscosity with temperature and are used to produce multigrade motor oils.

Oiliness enhancers/lubricity additives reduce friction under boundary lubrication conditions; they increase the load-carrying capacity where it is limited by TR by the formation of mainly organic surface films. Lubricity, also referred to as oiliness, with respect to lubricating oil, is defined as the ability of an oil to reduce friction between moving surfaces. Lubricity additives, usually vegetable or animal fats, enhance lubricity by tenaciously adhering to the metal's surface, forming an adsorbed film of high lubricating value.

Scavengers react with undesirable contaminants such as acids or sulfur to render them less harmful.

Friction modifiers reduce the friction between moving parts by surface adsorption.

Metal deactivators inhibit the metals contacting the lubricant from catalyzing oxidation of the lubricant.

Seal swell agents assist elastomer seals and gaskets to perform their function.

6.2.4.2   Lubricant Additive Chemistry

There are two basic forms of additive protection available to surfaces that are either temporarily or permanently in intimate contact, *physisorped* and *chemically reacted* [25].

The physical (physisorped) form is associated with the migration of polar molecules to the active metal sites on the surface. This form typically occurs at contact temperatures in the range of 70–150°C and gives rise to the surface model called "hairy molecule."

The chemically reacted layers are associated with actual chemical reaction between additives and the active metal sites on the surface. These are frequently EP (or antiscuffing) additives and are usually activated at temperature above 170°C. With conventional lubricant additive packages, these layers frequently fail at contact temperatures above 240°C.

Changes in running conditions, for example, changing load or frequency of events, will usually cause a temporary change in the consistency of these protective layers or films, indicating some kind of limited dynamic stability. It was postulated (Cameron and others) that these films were partially destroyed and reformed each cycle. In other words, each specimen pass knocked the top off a number of surface asperities creating new active metal sites. The dwell time between passes allowed time for the chemistry to work so that the film, chemical or physical, reformed before the next specimen pass.

There is no way of telling how many asperities could be scraped off in this way, although it must be related to wear. The rate of formation of the chemically reacted films in particular is considered to be a direct function of the contact temperature, following the Arrhenius equation, and most importantly the repetition rate. In other words, one needs temperature for activation and for controlling the rate of reaction plus a finite time for the chemistry to take effect.

The higher the temperature, the faster a given chemical reaction will proceed described by an exponential relationship with reaction rates accelerating with increasing temperature. Evidence of the process of continuous regeneration of the chemically reacted films and the time dependence issue has been reported in respect of gear contacts.

At very high speeds, there has been some suggestion that insufficient time elapses between the successive engagements of matching teeth for the chemistry to work, thus leaving the surfaces unprotected, giving rise to scuffing failure. However, it may be that the actual cause of this type of film failure is desorption of the reacted film caused by excessive contact temperature.

6.2.4.3 Lubricant Deterioration Caused by Decomposition of Additives

6.2.4.3.1 Effect of Electrical Current on Additives

This study is related to electrical equipment such as motors and alternators. It was established that in these electrical machines under certain operating conditions and depending on resistivity of the lubricant, electrical current may flow though the bearing. The effect of electrical current on bearing life and deterioration of lubricants used in noninsulated bearing was studied in [26–30].

Electrical current passing through bearing may cause the lubricant deterioration and failure. For example, it was found that the zinc additive, namely, zinc dithiophosphate (ZDTP), used as multifunctional additive in grease, protects the rubbing metal surfaces under rolling friction and contributes to friction wear reduction, which depends partly on the amount of additive on these surfaces. Decomposition of ZDTP in lithium-based greases under the influence of electric current leads to the formation of lithium zinc silicate ($Li_{3.6} Zn_{0.2} SiO_2$) in the presence of high relative percentage of free Li and silica impurity in the grease under high temperature in the asperity contacts, along with the formation of gamma lithium iron oxide (γ-$LiFeO_2$) [28]. During this process, lithium hydroxide is also formed which corrodes the bearing surfaces. Original structure of lithium stearate changes to lithium palmitate.

On the contrary, these changes are not detected under rolling friction without the influence of electric current. Thus, decomposition of the grease and the gradual formation of corrugation on the bearing surfaces due to the current passage lead to deterioration and failure of bearings [30].

6.2.4.3.2 Effect of Magnetic Field on Oil Dispersed with the Additives

This study is related to the change of the friction coefficient for sliding lubricated steel surfaces when they are exposed to magnetic field. The lubricants in the study was paraffin oil containing various additives such as zinc dialkyldithiophosphates (ZDDPs), molybdenum disulphide (MoS_2), heteropolar organic based additive (CMOC), graphite (C), detergent additive (DA; calcium sulphonate), PTFE, and polymethyl methacrylate (PMMA) [31].

It was found that the magnetic field decreased friction coefficient when the oil was dispersed by CMOC by means of increasing the adherence of CMOC particles into the sliding surfaces due to their electronic properties. Besides, it was observed that magnetic field much affected the performance of oil dispersed by additives of specific electrical properties.

For example, CMOC, DA, and PTFE particles dispersed in the oil were strongly influenced by the magnetic field leading to friction decrease. The same trend of friction decrease was observed for PMMA particles dispersed in oil. Significant friction decrease was observed by dispersing oil by ZDDP additive. However, the performance of ZDDP additive was not affected by the application of magnetic field. Significant friction increase was observed in the presence of graphite (C) in lubricating oil and application of magnetic field caused slight friction reduction.

6.2.5 Role of Oil in Lubricant Properties and Performance

6.2.5.1 Effect of Oil Bleeding on Grease Properties

Base oil makes up 60%–95% of grease. Grease dries out as base oil in grease continuously bleeds out, although the rate of bleeding gets slower and slower with time. This process is called aging. The time scale for this process depends on a number of factors such as operating temperature [32]. Base oil in greases has a certain kinematic viscosity expressed in mm²/s or cst. High temperature could promote its oxidation and thus increase its kinematic viscosity.

The amount and viscosity of the base oil should not change drastically within the advised relubrication interval. A change in oil bleeding properties may proceed both ways: become slower or faster with time compared to fresh grease.

A slower bleeding in used grease might be due to loss of base oil, which is generally accompanied with an increase in consistency. In high-temperature application, lower oil bleeding may be caused by base oil oxidation resulting in base oil viscosity increase at high temperature. Slower bleeding may be caused by amount of hard particles or the mix of greases.

A higher bleeding might be due to intensive shearing or vibrations (especially true for sheared polyurea greases) when grease cannot keep base oil in its structure. Mixing greases or oil contamination from neighboring systems may also cause higher oil bleeding. Grease with poor mechanical stability may also harden faster.

There are a number of different tests that can measure grease bleeding and oil separation characteristics. These tests can be categorized into two groups: static and dynamic bleed tests. Two kinds of the most common tests used to evaluate oil separation and bleeding are static and dynamic [33]. Static tests are defined in *ASTM D1742 Oil Separation from Lubricating Grease During Storage*. This test predicts the tendency of grease to separate oil during storage when stored at room temperature. Another standard for static test is *ASTM D6184 Test Method for Oil Separation from Lubricating Grease (Conical Sieve Method)* which determines the tendency of the oil in a lubricating grease to separate at elevated temperatures. Dynamic test is specified in *U.S. Steel Pressure Oil Separation Test*. This test is used to measure the oil separating and caking characteristics of grease under fixed conditions that indicate the stability of a grease under high pressures and small clearances in a centralized grease pumping system. The standard *ASTM D4425 Oil Separation from Grease by Centrifuge* evaluates the oil separation tendency of a grease when subjected to high centrifugal forces. Another dynamic test is presented in *Trabon Method 905A* used to predict the tendency of grease to separate oil while under pressure in a centralized lubrication system.

6.2.5.2 Degradation of Lubrication Oil

Degradation of oil serving as a lubricant and as a grease component is responsible for many kinds of equipment failures. A lubricant in service is subjected to a wide range of conditions which can degrade its base oil and additive system. Such factors include heat, entrained air, incompatible gases, moisture, internal or external contamination, process constituents, radiation, and inadvertent mixing of a different fluid. There are several degradation mechanisms, such as oxidation, thermal breakdown, microdieseling, additive depletion, electrostatic spark discharge, and contamination. The relationship between oil analysis and degradation mechanisms is examined in Ref. [34].

Oxidation: The reaction of materials with oxygen (oxidation) can be responsible for viscosity increase, varnish formation, sludge and sediment formation, additive depletion, base oil breakdown, filter plugging, loss in foam properties, acid number increase, rust, and corrosion. Controlling oxidation is a significant challenge in trying to extend the lubricant's life.

Thermal breakdown: The temperature of the lubricant is a primary concern in lubricant breakdown. In addition to separating the moving parts of the machinery, the lubricant must also dissipate heat. If the lubricant is heated above its recommended stable temperature, it can cause the light components of the lubricant to vaporize or the lubricant to decompose. Certain additives evaporate from the system without performing their job. The viscosity of the lubricant may increase. At temperatures greatly exceeding the thermal stability point of the lubricant, larger molecules will break apart into smaller molecules. This thermal cracking, often referred to as thermal breakdown, can initiate side reactions, induce polymerization, produce gaseous by-products, destroy additives, and generate insoluble by-products. In some cases, thermal degradation will cause a decrease in viscosity.

Microdieseling: This phenomenon is known as pressure-induced thermal degradation, and it is a process in which an air bubble moves from a low-pressure to a high-pressure zone, resulting in adiabatic compression. This may produce localized temperatures in excess of 1000°C, resulting in the formation of carbonaceous by-products and accelerated oil degradation.

Additive depletion: Most additive systems are designed to be sacrificial. Monitoring additive levels is important not only to assess the health of the lubricant but it also may provide clues related to specific degradation mechanisms. Monitoring additive depletion can be complex depending upon the chemistry of the additive component.

Electrostatic spark discharge: When clean, dry oil rapidly flows through tight clearances, internal friction on molecular level within the oil can generate static electricity and may accumulate to the point where it produces a sudden discharge or spark. These sparks are estimated to be between 10,000°C and 20,000°C and typically occur in mechanical filters.

Contamination: Foreign substances can greatly influence the type and rate of lubricant degradation. Metals such as copper and iron are catalysts to the degradation process. Water and air can provide a large source of oxygen to react with the oil. Therefore, a contaminant-free lubricant is ideal and monitoring a fluid's contamination levels provides significant insight to the machine's health.

6.3 INCOMPATIBILITY OF LUBRICANTS

6.3.1 DEFINITION OF INCOMPATIBILITY

Greases are considered incompatible when the physical or performance characteristics of the mixed greases fall below original specifications. The NLGI Lubricating Grease Guide [35] defines grease incompatibility as the lessening of the performance capability and physical properties when two greases are mixed. Grease incompatibility will be revealed by any of the several areas, such as: (1) lower heat resistance, (2) change in consistency, usually softening, or (3) decrease in shear stability.

Mixtures that show none of these changes are considered compatible. Since each of the greases in a mixture is a combination of thickener, base oil, and additives, incompatibility is not always due to the thickener alone. Sometimes, the thickener of one grease is incompatible with the oil or additives present in the other grease. Incompatibility is not predictable and is best determined in service test or in service-related tests.

ASTM International Committee D02.G developed the *ASTM D6185* Standard Practice for Evaluating Compatibility of Binary Mixtures of Lubricating Greases in 1997. This document defines the procedure for evaluating the basic compatibility of greases, which is determined by measuring the dropping point, the mechanical stability, and the change in consistency of the mixture upon heating.

A mixture of two greases is considered to be incompatible if the test proves that the mixture is significantly softer; less shear stable, or less heat resistant than the original greases. Certain thickener combinations are generally recognized to be incompatible, for example, lithium and sodium grease and organoclay and most soap greases. However, there is no practical rule one can apply to all mixtures of different greases to determine the compatibility properties [36].

6.3.2 CAUSES OF INCOMPATIBILITY

6.3.2.1 Thickeners

An important component of grease is a thickening agent whose properties define the grease consistency [14,37]. Detailed information on the study of various thickener systems' compatibility is given in Refs. [38,39] and shown in Table 6.9, which presents the latest and most complete and updated grease compatibility chart.

When greases made with incompatible thickeners are mixed, a substantial softening may occur. This is especially dangerous since softened grease can run out, particularly in vertical applications. From the other hand, lithium grease hardens in some mixtures, as did some clay mixtures. Barium grease blends look like grease on the bottom and oil on the top; mixing may cause the second grease to liquefy, while the barium remains intact. Not all polyurea greases are mutually compatible. The final determination of compatibility comes with proper testing of greases in the application for key performance properties [38–41].

6.3.2.2 Base Oils

Lubricating oil or grease is a mixture of oil (synthetic or mineral) and different additives. Compatibility problems may occur when different types of oils from one or more suppliers are being mixed. Insoluble reaction products are formed in the presence of water in a mixture of different oils (e.g., hydraulic and trunk piston engine oils). Such mixtures should always be avoided. Mixtures of various synthetic lubricants should also be avoided, as the different types of synthetics are not always compatible.

It is important to know the compatibility of some common types of synthetic lubricants with mineral oil-based lubricants. It was found that the mixture of mineral oil and synthetic fluids made from alkylbenzenes (ABs) and PAOs is safe; these fluids are compatible [42]; and the compatibility of mineral oil with diesters is considered to be good, but with polyglycols it is poor, and it is fair with phosphate esters and polyolesters (POEs) [43]. Compatibility chart of some base mineral and synthetic oils is presented in Table 6.10.

6.3.2.3 Additives

Very often the cause of incompatibility is a chemical reaction between different types of additives, and grease may have up to a dozen different chemicals added to the base oil to maintain the properties of grease and improve its lubricating abilities [44].

As an example, oils incompatibility may occur if one oil contains an acidic additive and another oil has an alkaline additive, which will neutralize each other if these two oils are mixed. For some oils, such a reaction occurs in the presence of water. Raising the temperature speeds up the reaction. Different fatty acids and/or additive packages affect the compatibility.

Combining greases of different base oils can produce a fluid component that will not provide a continuous lubrication film. Additives can be diluted when greases with different additives are mixed. Mixed greases may become less resistant to heat or have lower shear stability. If a new brand of grease must be introduced, then the component part should be disassembled and thoroughly cleaned to remove all of the old grease. If this is not practical, the new grease should be injected until all traces of the prior product are flushed out.

TABLE 6.9

Grease Incompatibility Chart

	Aluminum Complex	Barium Complex	Calcium Stearate	Calcium 12-Hydroxy	Calcium Complex	Calcium Sulfonate Complex	Lithium Stearate	Lithium 12-Hydroxy	Lithium Complex	Sodium	Clay (Nonsoap)	Polyurea	Polyurea Shear (Stable)
Aluminum Complex	—	I	I	C	I	B	I	I	B	I	I	I	C
Barium Complex	I	—	I	C	I	C	I	I	I	I	I	I	B
Calcium Stearate	I	I	—	C	I	C	C	B	C	I	C	I	C
Calcium 12-Hydroxy	C	C	C	—	B	B	C	C	C	I	C	I	C
Calcium Complex	I	I	I	B	—	I	I	I	C	I	I	C	C
Calcium Sulfonate Complex	B	C	C	B	I	—	B	B	C	I	I	I	C
Lithium Stearate	I	I	C	C	I	B	—	C	C	I	I	I	C
Lithium 12-Hydroxy	I	I	B	C	I	B	C	—	C	I	I	I	C
Lithium Complex	B	I	C	C	C	C	C	C	—	B	I	I	C
Sodium	I	I	I	I	I	I	I	I	B	—	I	I	I
Clay (Nonsoap)	I	I	C	C	I	I	I	I	I	I	—	I	B
Polyurea	I	I	I	C	C	C	C	I	I	I	B	—	C
Polyurea Shear (Stable)	C	B	C	C	C	C	C	C	C	I	B	C	—

Note: Relative Compatibility Rating: **C** = Compatible; **I** = Incompatible; **B** = Borderline.

TABLE 6.10

Base Oil Compatibility Chart

Base Oils	Mineral	Ester	PAG[a]	Methyl Silicone	Phenyl Silicone	PPE[a]	PFAE[a]
Mineral	X	C	I	I	B	I	I
Ester	C	X	C	I	C	C	I
PAG[a]	I	C	X	I	I	I	I
Methyl silicone	I	I	I	X	B	I	I
Phenyl silicone	B	C	I	B	X	C	I
PPE[a]	I	C	I	I	C	X	I
PFAE[a]	I	I	I	I	I	I	X

Source: Adapted from Grease Compatibility Information, Online. Available: http://www.petroliance. com/sites/default/files/PDF/Grease/Grease%20Compatibility%20Information.pdf.

[a] PAG, polyglycol; PPE, polyphenyl esters; PFAE, perfluorinated aliphatic ester.

6.3.3 Symptoms of Lubricants Incompatibility

In practice, different lubricants could be mixed in various binary concentrations from time to time. This occurs for a variety of reasons. For example, there was practical inability to drain out all remnants of a previous lubricant when switching lubricant types to another lubricant which the original one is incompatible with. The maintenance team could follow bad advice regarding compatibility offered by lubricant suppliers or unwise use of generic cross-reference guidelines. Accidental mixing of lubricants may be caused by untrained personnel or packaging/labeling error, and so on. Even the best maintenance organizations may face lubricant incompatibility issues occasionally.

Once the threat has been recognized, damage control and remediation may be the only practical alternative. This should begin with vigilant monitoring and inspection, looking for symptoms of incompatibility, side effects, and/or impaired performance.

There are different degrees of incompatibility ranging from harsh rejections leading to catastrophic machine failure to a change barely noticeable in a single lubricant's performance parameter. To resolve the incompatibility issue in timely manner long before it causes failure, it is important to be able to recognize the problem.

The following are examples of potential symptoms of lubricant incompatibility [45].

Increase in wear metal production, heat generation, vibration, or acoustic emission (AE), a symptom of *impaired film strength.*

Emulsions may begin to form where previously water would settle rapidly, which signals about *water demulsibility.*

Some changes in lubricant color and opacity are normal as a lubricant is exposed to operating temperature and the service environment. In other cases, a pronounced and *unusual change in color* or *turbidity* may be a sign of incompatibility from cross-contamination.

Impaired air release and foam suppression is a potential symptom of incompatibility.

Possible side effects of incompatible additives and/or base oils may result in formation of *soft insoluble,* such as additive floc, additive dropout, varnish, sludge, deposits, bathtub rings, premature filter plugging, and bottom sediment.

There are many symptoms of *premature base oil oxidation,* including a sudden increase in acid number and viscosity excursions combined with sludge and a rancid odor.

An unexplainable *increase in particle count* may be additive precipitation or other insolubles resulting from additive and/or base oil interactions.

Oil samples may exhibit *separation of fluid phases,* color, consistency, and/or refractive indices (e.g., streamers).

Some lubrication mixtures can cause *thick gels* to form on filters, reservoirs, and sight glasses. This can lead to lubricant starvation and even a drop in oil pressure.

Leakage from several points in the system may be associated with viscosity change and/or interfacial tension change, both possibilities caused by lubricant mixing.

6.3.4 GREASE COMPATIBILITY TESTING

It is necessary to perform compatibility testing on combinations of the greases to determine whether the greases can be mixed in service. It is incumbent upon the user of the greases to verify compatibility when making a change from one product to another. Most grease suppliers have data on certain grease combinations or are willing to perform the required testing for their customers [46].

The necessity of testing lubricating materials (fluids, oils, and greases) for compatibility was specially addresses in the United States by the Department of Defense in 1986 [47]. Individual military specifications include requirements for assuring compatibility of like or similar fluids/oils. Some of these requirements involve mixing different fluids and subjecting them to heating up to 135°C (275°F) for 2 or more hours, followed by cooling down to −54°C (−65°F) for an additional 2 or more hours. Visual separation to any degree is considered evidence of an incompatibility and ground for rejection of the fluid. These requirements can be found in military specifications for most fluids, including hydraulic fluids, general preservative oils, automotive engine and gear oils, and aviation engine and gear oils. Many of the procedures for determining fluid or lubricating oil compatibility are available and published in the Federal Standard 791C titled "Lubricants, Liquid Fuels, and Related Products; Methods of Testing."

ASTM D6185 "Standard Practice for Evaluating Compatibility of Binary Mixtures of Lubricating Greases" details the procedure for evaluating the basic compatibility of greases, which is determined by measuring their most important grease properties: the dropping point, the mechanical stability, and the change in consistency of the mixture upon heating.

The standard defines a protocol to evaluate the compatibility of binary mixtures of 10:90, 40:40, and 90:10 of lubricating greases by comparing their properties or performance relative to those of the original greases comprising the mixture. The principle of the test is to blend and shear under controlled and identical conditions the two greases at various ratios, determining after a short period of rest at room temperature any change in structural stability compared to the fresh greases' stability. Three properties are evaluated in a primary testing protocol using standard test methods: (1) dropping point, (2) shear stability by 100,000-stroke worked penetration, and (3) storage stability at elevated temperature via change in 60-stroke penetration after storage. The overall assessment of the test results determines if greases are compatible (all changes within the repeatability of the least performing grease), borderline compatible (change beyond the repeatability but still within the test reproducibility of the least performing grease), or incompatible (change beyond the reproducibility of the least performing grease) [48].

Greases are considered to be compatible if the following conditions are met:

- The dropping point of the mixture is not significantly lower than that of the individual greases.
- The mechanical stability of the mixture is within the range of consistency of the individual greases.
- The change in consistency of the mixture following elevated temperature storage is within the range of the change in consistency of the individual greases following elevated temperature storage.

Once two greases are determined to be compatible in the above three areas, further testing to determine the impact on other performance parameters of the products may be warranted.

Any test that is designed for measuring grease performance may be used on a mixture of greases to determine the effect on that parameter when the greases are mixed. The tests that are conducted should be agreed upon between the user and grease supplier to assure that the properties that are critical to the proper function of the product in service are covered.

6.4 GREASE CONTAMINATION

Grease should be free of contaminants. Contaminated grease will reduce bearing performance and lifetime can be reduced to a large extent. Contamination may be coming from the outside, introduced by poor sealing, dirty grease guns, poor bearing mounting methods. Contaminants can be found in grease used for lubrication of any moving parts of electrical equipment, particularly in mechanical parts.

Grease contamination can be caused by various natural sources such as sand, water, dust, fibers, steam flow, and so on. Lubricant contamination may be a result of oil leaking from neighboring systems or introduction of wrong or incompatible grease. If the grease reaches end of life, it deteriorates with formation of carbonized particles that are formed and stick to surfaces promoting friction. The grease used in bearing is contaminated with bearing wear material [49].

More than 20 different elements can be found in used grease. Information about the elemental composition of the used grease may help to evaluate degree of wear, contamination, condition of the additives, and thickeners. These elements include wear metals such as iron, chromium, tin, copper, lead, nickel, aluminum, molybdenum, and zinc. Contamination from environment may include such elements as silicon, calcium, sodium, potassium, and aluminum. Decomposition of additives or thickeners may add magnesium, calcium, phosphorus, zinc, barium, silicon, aluminum, molybdenum, and boron to the grease composition [50].

Besides solid contaminants, which can be identified by the presence of silicon, calcium, or aluminum, water is a type of contamination that is often the cause of corrosion. Depending on the grease type and application, the water content in the grease should not exceed the recommended values. Too much water in grease can produce a variety of adverse effects, including corrosion on bearing metals, increased oxidation of the base oil, softening of the grease, and water washout of the grease.

6.4.1 GREASE CONTAMINATION TESTING TECHNIQUES

For diagnostics of a bearing or grease condition, the amount of iron and chromium, which are present as wear particles from the bearing material, is very important. To determine type of wear (corrosive or abrasive) of the bearing cage, one might look for nonferrous materials like copper, lead, and tin, which is significant. The information about the presence of silicon or calcium from dust or sodium, potassium, or magnesium from sea water may help to determine what causes the bearing metal to wear. Other components such as metallic soap elements present in grease or a comparison of the additive content in fresh and used grease can also reveal whether the correct, recommended by OEM grease was used in the application.

Size, amount, shape, and nature of the contaminants give an indication on the proper functioning of the bearing and other mechanical parts. Contamination test is usually performed after the consistency test by inspecting the grease samples under microscope or by more precise techniques such as *optical emission spectroscopy (OES)* providing elemental analysis of the grease, *particle quantifier (PQ)* method, which characterizes and iron particles present in the sample independent of the particle size, it gives the total content of magnetic wear particles.

Particle count method defined in *ISO 11171, ISO 11500,* and *NAS 1638* is a test designed to count the number of particles present greater than a given micron size per unit volume of fluid. The results reflect the insoluble contaminants present within that size range and are applied to assess fluid cleanliness and filtration efficiency.

Cleanliness levels are also represented by the *ISO 4406 Solid Contamination Code*, which is the system to classify the particles larger than 4, 6, and 14 μm per milliliter of fluid.

Spectrochemical analysis (ASTM D6595, D5185) is using elemental spectroscopy which is an analytical method to measure and monitor specific trace metals for wear and corrosion levels, airborne or internally generated contaminants, and certain additives. Particles detected are typically 8 μm or less and the results are reported in parts per million (ppm) by weight.

Another technique *Fourier transform infrared (FTIR) spectroscopy (ASTM E2412)* identifies the type of base oil and thickener of the used grease. Using this method, additive depletion or contamination by another grease type can be determined by comparing the unused fresh grease reference to the used grease sample. The FTIR method can also show whether synthetic or mineral base oils are used. If a mineral oil is used as the base oil, FTIR can indicate what caused oxidation of the base oil: too much time passed without regreasing or because the temperature was too high. If the grease contains EP additives with zinc and phosphorus, FTIR method could determine if the additives degraded. The water content in the grease may also be provided by FTIR.

More traditional method to determine water contamination in oil and grease is *Karl Fischer titration method*. For water determination according to the Karl Fischer method, a small grease quantity (approximately 0.3 g) is placed into a glass vial and sealed with a septic cap. In a small oven, the sample is heated to ~120°C. The steamed out water is transferred by nitrogen into a titration vessel in which an electrochemical reaction between the water and a Karl Fischer reagent takes place. A titration curve is recorded and the water content is defined precisely. If the result for water content according to the Karl Fischer method is compared to the elemental analysis by OES, it can be determined whether the water in the sample is "hard" or sea water, which contains minerals like sodium or potassium, or if it is soft water like condensate or rain water. If sodium, potassium, calcium, and magnesium are found in the used grease but are not in the fresh grease, the presence of "hard" water is the likely reason. Comparing these two methods, Karl Fischer titration and OES, can also indicate whether the water was already present in the fresh grease as part of the production process.

There are also other tests which are in use for more complex analysis, particularly if failure analysis is required. Among them are oil bleeding test, penetration test for grease consistency, dropping point test, copper corrosion test, and so on. Some of these existing analysis tools include wear particle and contamination analysis methods such as *ferrographic analysis, gravimetric filtration, atomic absorption spectroscopy*, and *SEM spectroscopy*—all of which are aimed at understanding the material makeup and characteristics of the contamination. Particle sizing and counting techniques employ both manual microscopic methods and automatic direct counting through equipment using light scattering methods to determine particle size distributions and concentration levels.

There are several more tests that are used for determination of lubricant degradation degree: *Membrane Patch Colorimetry (MPC)* or *Quantitative Spectrophotometric Analysis (QSA^{SM})*. This is a laboratory method of extracting the insoluble contaminants from a used oil sample, followed by spectral analysis of the separated material. With QSA, a direct correlation is made from the color and intensity of the insolubles to oil degradation. The test is designed to identify soft contaminants (those directly associated with oil degradation) and is not strongly influenced by larger hard contaminants unrelated to oil degradation. This test is considered to be highly sensitive and reliable for detecting subtle changes in insoluble levels.

Test called *ultracentrifuge (UC) sedimentation rating* is a method to isolate all insolubles in a sample. It is accomplished by spinning the sample at 20,000 rpm in a centrifuge for 30 min. The insolubles are separated from the fluid centrifugally, allowing a visual sediment rating scale to be used. The minimum value of one represents low levels of insolubles. The maximum value of eight represents a critical level. Limitations of the test include the inability to differentiate between oil degradation by-products and other insoluble contaminants (dirt). Unfortunately, the centrifugation process may also remove additives (VI improvers, dispersants, and sulfonates) and may be laborious to run.

Linear sweep voltammetry (LSV): ASTM D6971 is a test designed to detect the oxidative health of a lubricant by measuring the primary antioxidants in the fluid. It is performed using an instrument called the RULER and is an essential condition monitoring (CM) tool for lubricant health. The level of remaining additive, and thus remaining useful life of the lubricant, is determined by comparison to original levels. The results of LSV can be correlated to fluid degradation, provided a significant amount of data is available from that particular fluid type.

Rotating pressure vessel oxidation test (RPVOT) according to ASTM D2272 is a controlled, accelerated oxidation test of a lubricant used to measure the performance of remaining antioxidant additives. Results are then evaluated and compared to new oil levels. This method has limited value as a primary RCA test because fluid degradation may take place on isolated segments of the lubricant, causing insolubles to be created without meaningful drops in RPVOT values. It is not uncommon for sludge and varnish problems to occur in oils even with high RPVOT values. In fact, several lubricants with high RPVOT values have additive chemistries that are more prone to produce deposits such as phenyl-alpha-naphthylamine (PANA).

Another option is a lubrication analysis in which the physical properties of the lubrication, contamination levels, and wear debris are monitored over time. If samples of oil, or in some cases grease, are taken on a regular basis, physical and chemical testing will help determine how well the system is running [51].

In summary, grease analysis has proven to be a useful tool to evaluate grease and bearing condition. Different situations and influencing factors for wear, contamination, and grease condition have shown complex coherences between the grease analysis results and their practical meaning. This leads to the conclusion that observing and interpreting these factors with expert knowledge can enable proactive maintenance strategies to be applied in a reasonable way for grease-lubricated components.

6.4.2 Lubricant Particle Contamination and Its Role in Mechanism Wear

Particle contamination in the lubricant causes abrasion, erosion, and fatigue of mechanical parts of machinery, which in turn are predominant wear mechanisms in wear-related failures of lubricated machinery. These wear mechanisms cause 82% of mechanical wear. It was found in the study performed at the National Research Council of Canada in conjunction with the Society of Tribologists and Lubrication Engineers. The study examined 3722 failures across several industries, including pulp and paper, mining, forestry, transportation, and power generation. The results of this study indicate that the major cause of machine wear is lubricant contamination.

For example, abrasive wear is usually a result of three-body cutting wear caused by dust contamination of the lubricant. Dust, which is much harder than steel, gets trapped at a nip point between two moving surfaces. The trapped particles tend to embed in the relatively softer material and then cut grooves in the harder metal such as the process by which sandpaper cuts steel. These wear mechanisms are aggravated by the presence of water contamination in a lubricant. Therefore, most failures in lubricated equipment, whether they were catastrophic and sudden or just premature replacements, are caused by particle and moisture contamination. Using a lubricant that is either incorrect for the application or has degraded beyond the point of being suitable for use may also cause a premature failure [52].

The *ISO 4406 Solid Contamination Code* is the most widely used method for characterizing particle counts in oils, measuring number of particles in milliliter of the oil. The latest version of the code employs a three-number system, which represents range of the number of particles corresponding to certain sizes. Control of particle contamination in the lubricant should be employed to remove particle debris from the system while minimizing dust ingression through air breather ports, seals, and incoming lubricants.

It is essential to control particle contamination by establishing target cleanliness levels based on particle counting measurements. These measurements are specified in ASTM D7416, D7647, and D7596 (wear particle detection and classification). The technique to analyze the particles in the grease is wear particle analysis (WRA), as guided by ASTM D7684 and explained in ASTM D7416 and D7690 [53].

6.5 SOLID LUBRICANTS

Various solid materials protect and lubricate interacting surfaces in extreme conditions such as high and low shaft speeds, high and low temperatures, high pressures, concentrated atmospheric and process contaminants, and inaccessibility.

Solid film lubricants offer protection beyond the normal properties of most mineral and synthetic oil-based fluid lubricants. Conditions that warrant the use of these agents in a pure form, or as an additive, include extremes of temperature, pressure, and chemical and environmental contamination. Some agents have a strong affinity for metallic surfaces and will adhere to those surfaces through loose covalent forces. These may be applied directly as a topical coating or indirectly in the form of an additive to a fluid lubricant. Some agents have no natural attractiveness to metallic surfaces, and therefore must be bonded to the surface through specialized treatment.

The more common types of materials include the following: molybdenum disulfide (MoS_2)— also known as moly; PTFE—also known as Teflon; graphite, and others. These materials are characterized as dry film or solid film lubricants. Moly, graphite, and Teflon are the most commonly recognized by practitioners of machinery lubrication. Molybdenum and graphite are agents that are extracted from mined ore. Teflon was created by DuPont™ Chemical Company and is manufactured by various companies for many purposes.

The solid lubricating materials tend to have upper temperature ranges well above the surface-protecting capabilities of mineral and most synthetic base stocks. Fluorinated hydrocarbons are stable in liquid or solid form to roughly 600°F, but will begin to degrade and may produce noxious fumes at that temperature. Graphite and molybdenum can operate in a similar temperature range, and molybdenum disulfide can also function in a vacuum without losing its slippery property.

6.5.1 MOLYBDENUM DISULFIDE

MoS_2 is one of the best known solid lubricants, and although it originally gained popularity in aerospace and military applications, it is now commonly found in a variety of lubrication applications. It is widely used in greases and specialized grease-like products known as pastes, in fluid lubricants such as automotive and industrial gear oils, in solid film lubricants including but not limited to burnished (rubbed-on) films, sputtered coatings, resin bonded, and impingement coatings [54–56].

The electrical properties of MoS_2 are very important, particularly for implementation in electrical brushes and spacecraft. Until now there is no clear agreement about electrical conductivity of MoS_2. However, the general view is that molybdenum disulfide could be classified as a "p"-type semiconductor. Greases containing molybdenum disulfide are implemented in a wide variety of applications mostly in mechanical applications where greases are used and in virtually all types of grease thickener systems, including bentonite clay, lithium, lithium 12-hydroxysterate, lithium complex, aluminum complex, and polyurea.

Greases containing MoS_2 are used as OEM lubricants by most electrical equipment manufacturers and are also supplied to the aftermarket. Applications include kingpins, ball joints, pivot pins, and spherical frame pivot bearings. These greases are used to enhance the running-in process of new components as well as protecting load-bearing surfaces operating under difficult conditions [54].

Grease-like products containing a high percentage (50%–70%) of MoS_2 and other solid lubricants are referred to as *pastes*. These products are typically used as antiseize lubricants for threaded connections. Military applications for MoS_2-containing greases are similar to nonmilitary ones. For example, Mil G-23549 grease is a general-purpose clay-thickened grease containing 5% MoS_2 that is used to lubricate Navy catapults on aircraft carriers and also in military automotive vehicles. Mil G-21164 is a lithium complex grease containing 5% MoS_2 in synthetic base oil recommended for heavily loaded sliding surfaces.

Solid film of MoS_2 can provide lubrication to sliding surfaces without the presence of a fluid film (i.e., a grease or oil film) being a material with inherent lubricating properties, which is firmly bonded to the surface of a substrate by some method. One of the properties of MoS_2 is that it can be physically rubbed on most metal substrates establishing a lubricant film about 1–5 μm in thickness (for the longest life, the optimum thickness is 3–5 μm). Solid films with MoS_2 can be applied by several techniques, including burnishing, impingement, sputtering, and resin bonding.

The burnished coating is frequently used as a lubricant in applications such as metal forming dies, threaded parts, sleeve bearings, liquid oxygen valves, and electrical contacts in relays and switches. In these applications, the MoS_2 burnished coating can be produced from loose powder, dispersions, wire brushing, or even "sandblasted" onto the metal surface to form a thin adherent film [55]. Solid MoS_2 burnished film provides high load-carrying capacity, very low friction and wear rate. Also, it is one of the best techniques for lubrication bearings with very small clearances in air.

Another type of solid MoS_2 coating is *sputtered coating* of MoS_2 formed by bombarding a target material of compacted MoS_2 with a charged gas (e.g., argon) which releases atoms in the target that coat the nearby substrate. The process takes place in a magnetron vacuum chamber under low pressure, producing extremely adherent coatings of MoS_2 that are typically less than 1-μm thick which is much thinner, more adherent, and shows greater endurance compared to other types of solid lubricant coatings. Also, MoS_2 sputtering can be performed on a wide variety of substrate materials such as Mo, Fe, Ni, Co, W, Ta, Al, plastics, glass, ceramics, and oxidized surfaces (before the sputtering process) of Cu and Ag [55].

Specific applications where sputtered MoS_2 is widely implemented include the most critical aerospace applications such as spacecraft and satellite moving mechanical assemblies, solar array drives, antenna control systems, despin mechanisms, as well as similar mechanisms on earth that operate in a vacuum or inert environment [56].

In some applications, MoS_2 is used in CB mechanisms in mixture with oils with the concentration of MoS_2 between 0.1% and 60%.

6.5.2 GRAPHITE

Graphite as a lubricant dates to ancient times; it was first referenced in the mid-1500 s as being used as pencils. Graphite is a soft, crystalline form of carbon. It is gray to black, opaque, has a metallic luster, and is flexible but not elastic. Graphite occurs naturally in metamorphic rocks such as marble, schist, and gneiss.

It exhibits the properties of a metal and a nonmetal, which makes it suitable for many industrial applications. The metallic properties include thermal and electrical conductivity. The nonmetallic properties include inertness, high thermal resistance, and lubricity. Some of the major end uses of graphite are in high-temperature lubricants, brushes for electrical motors, friction materials, and battery and fuel cells.

Graphite is the other most frequently utilized particle in lubricating greases [57]. It has low friction, low chemical reactivity, and low abrasiveness, high thermal and electrical conductivity, and high thermal stability. Just like MoS_2, it has a lamellar structure with strong covalent bonds within each layer and weaker bonds between the layers. The incorporation of graphite, as well as other particles, is less problematic in greases compared to other lubricants since the inherent high viscosity prevents particle settling. Since graphite requires adsorbed vapors for its lubricity, it works better in regular atmospheric conditions rather than in vacuum. Its synergistic effect with MoS_2 is well documented for normal conditions. Under humid conditions, however, where MoS_2 is susceptible to cause corrosion, or at high temperatures, graphite is more effective on its own.

The useful temperature range for graphite is up to 450°C (842°F) in an oxidizing atmosphere compared to 350°C (662°F) for MoS_2. The level of oxidation stability depends quite simply on the quality of the graphite. It is important to understand under which conditions the more high purity and thus more expensive graphite will be needed for particular application. However, it shows that a more expensive solid does not provide improved properties to the grease under all conditions. One type of graphite can cost four times more than another type but, in terms of performance in the four-ball wear test, there is virtually no difference.

6.5.3 OTHER SOLID LUBRICANTS

Polymers are used as thin films, as self-lubricating materials, and as binders for lamellar solids. Films are produced by a process combining spraying and sintering. Alternatively, a coating can be produced by bonding the polymer with a resin. Sputtering can also be used to produce films. The most common polymer used for solid lubrication is PTFE, a polymer which can be used for its ability to lubricate even once frictional heat increases in the contact, due to its high softening point. It has one of the lowest static and dynamic friction coefficients of solid lubricants.

The main advantages of PTFE are wide application range of −200 to 250°C (−328 to 418°F) and lack of chemical reactivity. Disadvantages include lower load-carrying capacity and endurance limits than other alternatives. Common applications include antistick coatings and self-lubricating composites. Since PTFE particles can build networks within grease, they also serve as thickening agents.

Clays and fumed silicas are particles that have been extensively used as thickeners in greases, in particular for greases intended for high-temperature applications. In grease thickened with bentonite clay, for instance, other filler particles may act as a "reinforcement" creating a stronger network and providing higher shear strength. For soap-thickened greases, the interaction between the particles and the soap molecules is even more complex and the order of adding the components during production can influence the rheological characteristics of the grease [57].

Many *soft metals* such as lead, gold, silver, copper, and zinc possess low shear strengths and can be used as lubricants by depositing them as thin films on hard substrates. Deposition methods include electroplating, evaporating, sputtering, and ion plating. These films are most useful for high-temperature applications up to 1000°C (1832°F) and roller bearing applications where sliding is minimal [58].

6.6 LUBRICANT WORKING TEMPERATURE LIMITS

6.6.1 LUBRICANT WORKING TEMPERATURE

Lubricants are used for the protection of electrical and mechanical parts from corrosion, because they reduce the ingress of corrosive gases on the contact metals or through pores to react with the substrate metal. Application of a lubricant reduces the possibility of electric contact failure due to insulated particulates such as dust or wear fragments of tarnished films, which may deposit on the contact area.

On the other hand, grease is also exposed to many environmental factors that cause contamination and oxidation of the grease. When the grease oxidizes, it usually darkens; there is a buildup of acidic oxidation products. These products can have a destructive effect on the thickener, causing softening, oil bleeding, and leakage. Because grease does not conduct heat easily, serious oxidation can begin at a hot point and spread slowly through the grease. This produces carbonization and progressive hardening or crust formation [59]. Lubricants may degrade through oxidation or polymerization, forming insulating films when time in service increases and temperature changes.

6.6.1.1 Temperature Limits

Thermal limitations, both high and low, represent the major factors of many lubricants' durability. At low temperatures, many lubricants appear to solidify, developing high shear strength films, and leading to high CR. Some lubricants are susceptible to cracking due to long-term exposure to high temperature [60]. The temperature limits for the use of greases are determined by the drop point, oxidation, and stiffening at low temperatures.

6.6.1.2 Maximum Temperature

Grease has a maximum temperature up to which it can be safely used. The higher the temperature to which grease is exposed, the higher is the rate of oxidation. As was observed for most greases, the grease undergoes a gradual change in consistency until a certain critical temperature is reached. At this point, the gel structure breaks down and the whole grease turns liquid. This critical temperature is called the *dropping point* (see Section 8.2.2.4 and Table 6.8). When grease is heated above its dropping point and then allowed to cool, it usually does not fully regain its grease-like consistency and its subsequent performance will be unsatisfactory. At no time should the drop point be exceeded.

6.6.1.3 Minimum Temperature

This is the point where the grease becomes too hard for the bearings or other greased components to be used. Again, the base oil of the grease determines the minimum temperature (see Section 8.2.2.5). Obviously, the base oil of the grease for low-temperature service must be made from oils having a low viscosity at that temperature. While choosing the best grease for a specific application, all thermal factors should be considered, including the environment, application conditions, and working thermal range of the grease.

6.7   LUBRICANT STORAGE CONDITIONS AND SHELF LIFE

The foundation of effective lubrication, sometimes called "the four Rs of lubrication," which means *right place, right time, right amount, and right condition.* One should distinguish service life and shelf life. Service life is the life performance of a lubricant in the operating conditions; shelf life is based on the oxidative stability of the lubricant.

On the shelf, standard lubricants generally remain motionless for extended period of time, they can be exposed to cyclic variations in temperature and other environmental conditions, such as vibration, or potentially can absorb contaminants from the environment and react with atmospheric oxygen and become less effective. If this happens, the appearance of the grease may remain unchanged, but it will not perform as well. In general, the service life of grease will be shorter than its shelf life.

Under ideal conditions, with cool operating temperatures, no contamination, no oil seepage past the seals (perfect seals), and no bearing wear (perfect lubrication), the service life will be equivalent to the shelf life. However, service life is unpredictable since there are so many variables that can shorten it.

6.7.1 Factors Affecting Shelf Life

Oxidation occurs in all oils and greases that are in contact with air, including stored lubricants. The base oil and additive combination affects the rate of oxidation, and the presence of the thickener in grease can increase the degradation rate. But the environmental and storage conditions have the greatest influence on the rate at which the lubricant degrades [61]. The storage environment greatly affects the shelf life of lubricants and greases [62]. There are several important storage conditions that may greatly accelerate lubricant deterioration.

Temperature: Both high heat (>45°C or 113°F) and extreme cold (<−20°C or −4°F) can affect the lubricant stability. Heating increases the rate of oil oxidation, which may lead to the formation of deposits and an increase of viscosity. Cold can result in wax and possible sediment formation. In addition, alternating exposures to heat and cold may result in air being drawn into drums, which may result in moisture contamination. A temperature range of −20°C to 45°C (−4°F to 113°F) is acceptable for the storage of most of the lubricating oils and greases. Ideally, the temperature range for storage should be 0–25°C (32–77°F).

Light: Light may change the color and appearance of lubricants. Lubricants should be kept in their original metal or plastic containers.

Water: Water may react with some lubricant additives, sometimes forming an insoluble matter. Water can also promote microbial growth at the oil/water interface. Lubricants should be stored in a dry location, preferably inside.

Particulate contamination: Drums and pails should not be stored in areas where there is high level of airborne particles. This is especially important when a partially used container is stored.

Atmospheric contamination: Oxygen and carbon dioxide can react with lubricants and affect their viscosity and consistency. Keeping lubricant containers sealed until the product is needed is the best protection.

Oil separation: Oil will naturally separate from most greases. Temperatures in excess of 45°C (113°F) can accelerate oil separation. If grease is removed from a drum or pail, the surface of the remaining grease should be smoothed to prevent oil separation into the cavity.

Storage condition affecting greases: Grease properties can change during storage depending on the type of thickener, its concentration, the base fluids, and the additives used.

Lubricant shelf life: Limitations are not formalized into a standard yet. But there is some guidance coming from major lubricant manufacturers with general recommendation for lubricant shelf life. As an example, some grease manufacturers [62,63] estimated the shelf life of base and lubricating oils and greases in unopened containers to be 5 years for base oils, mineral or synthetic lubricating oils, and greases. A limited shelf life (2 years) is listed for rust preventives and open gear lubricants. However, for some advanced greases, the shelf life is longer (6 years in any container). If oil is observed on top of the grease, it should be remixed back into the top inch (2.5 cm) of the grease prior to use. Horizontally stored grease cartridges may bleed minor amounts of oil. This may make the container cosmetically unpleasing, but the grease remains suitable for use.

6.7.2 Products Exceeding the Estimated Shelf Life

A product in an unopened container, which is beyond the estimated shelf life, may still be suitable for service. Exceeding OEM shelf life of the lubricants may cause the product to be useless or severely worsen its performance.

So-called FIFO method (the first-in, first-out) is one to use when choosing which lubricant on the storage shelf to use. This method simply requires the maintenance professional to use the oldest lubricants that were put into the storage system first and the newest lubricants put into the storage system last. This will help ensure lubricants do not accidentally exceed their recommended shelf life. The product should be tested and evaluated against the original product specifications. To collect a representative and uniform sample for testing, the product should be thoroughly mixed. If the product's test results fall within the original specifications of the grease, it should be still suitable for use. Following testing, if the product is not consumed within a year, the product should be marked for reclamation or disposal. The user should validate the product's performance claims against the equipment manufacturer's recent specifications.

6.7.3 Shelf Life Estimated by OEMs

Shelf life imitations are not formalized into a standard yet, however there is some guidance coming from major lubricant manufacturers with general recommendation for lubricant [62,63].

TABLE 6.11

Estimated Shelf Life of Base Oils, Lubricating Oils, and Greases from Chevron

Products	Type	Shelf Life, Years	Products	Type	Shelf Life, Years
Base oils		5+	Greases	Mineral	3
Lubricating	Mineral	5		Synthetic	3
Oils	Synthetic	5	Known exceptions		
Coolants	General	5	Rust preventatives		2
			Open gear lubricants		2

Source: Adapted from Lubricant storage, stability, and estimated shelf life, *Chevron USA Information Bulletin 1*, 2011, Online. Available: http://www.chevronmarineproducts.com/docs/Chevron_ InfoBulletin01_ShelfLife_v1013.pdf.

Equipment designs and specifications can change over time, making an old product obsolete for use with new equipment. The manufacturers of the lubricants have recommendations on their products in regards of shelf life; one of these recommendations adopted from Chevron [62] is shown in Table 6.11. However, recommendations from other grease manufacturer may vary significantly, for example, Amsoil recommends 2-year limits storage life for their greases [64].

ExxonMobil also comes up with the shelf life recommendation, specifying the shelf life for each specific product, namely, aviation oils and greases [65], many of which are commonly in use in electrical industry. Among them are Mobilgrease 28 and Mobilgrease SHC 100, with recommended 6 years of shelf life in any container. But it is necessary to notice that if oil is observed on top of the grease, it should be mixed back into the top inch (2.5 cm) of grease prior to use. Horizontally stored grease cartridges may bleed minor amounts of oil. This may make the container cosmetically unpleasing, but the grease remains suitable for use.

Very often there are no recommendations from lubricant suppliers about the storage environment and shelf life limits. Manufacturers of the lubricants usually develop technical recommendations on the shelf life of their products and their recommendations greatly depend on the makeup and properties of the greases.

The manufacturers producing advanced modern greases, such as DuPont, declare that their products Krytox® PFPE grease and oil lubricants with no additives have an *indefinite shelf life* if left unopened and stored in a clean dry location [66]. Unopened containers of these products have shown no change in properties after 20 or more years of storage at ambient temperatures. It was determined that retained samples of DuPont Krytox grease that are 40 years old have been tested and still perform well as they did when they were originally made. Krytox oils are perfluoropolyethers (PFPE), inert to oxygen and most chemicals. They do not oxidize or degrade while sitting in storage or in use. They are stable under conditions up to their decomposition temperature, which is above 350°C. The most common thickener is PTFE which is also inert and nonoxidizing, so it would not degrade with age.

Furthermore, some Krytox greases contain additives for enhanced performance. Though there is no long-term storage history on Krytox grease with additives, it is not expected to be significantly different than grease without additives. Although some of the additives may degrade and become less effective over time, Krytox greases are designed to handle these minor changes, and neither their performance nor their characteristics will be affected.

The other less advanced DuPont products such as Krytox XP series lubricants can develop an odor and a slight amber color over time; as a result, they have been given a 3-year shelf life. Testing has shown that these products retain their antirust properties and lubricate and perform properly over their expected life, even if used near the end of the shelf life of *3 years*.

TABLE 6.12

Grease Components or Conditions that Influence the Storage Life of New Lubricants

Grease Component	Decrease Storage Life	Increase Storage Life
Base oil	Lower grade mineral oils, inorganic esters	Highly refined mineral oils, synthetic hydrocarbons, inert synthetic
Additive	EP additives	Rust and oxidation inhibitors
Thickener	Yes	No
Storage condition		
Temperature	High	Low
Temperature variability	High	Low
Humidity	High	Low
Agitation	High	Low
Container type	Metal drums, especially poorly conditioned ones	Plastic container or liners
Outdoor storage	Yes	No

Source: Adapted from Troyer D. and Kucera J. *Machinery Lubrication Magazine*, May 2001, Online. Available: http://www.machinerylubrication.com/Read/172/lubricant-storage-life.

However, according to the study performed by the editorial staff of *Machinery Lubrication magazine*, the issue of storage life of the lubricants is more complicated. Based on the polls of several lubricant suppliers, large and small, and lubrication consultants, it was found that the stored lubricants degrade for a number of reasons, not discussed earlier [61]. Grease components and conditions of grease storage which influence storage life of the lubricants are summarized in Table 6.12.

Oxidation occurs in all oils that are in contact with air, including stored lubricants. The base oil and additive combination affects the rate of oxidation, and the presence of the thickener in grease can increase the degradation rate. But the environmental and storage conditions have the greatest influence on the rate at which the lubricant degrades. Increasing the temperature at which the lubricant is stored by 10°C (18°F) doubles the oxidation rate, which cuts the usable life of the oil in half. The presence of water, usually introduced as a result of temperature variations, increases the rate of oxidation.

Frequent agitation of the lubricant incorporates air into the oil. This increases the surface area contact between air and the oil, increasing the rate of oxidation. Agitation also serves to emulsify water into the oil, increasing its catalytic effect on the oxidation process. The storage container itself can affect the rate of oxidation. A poorly prepared steel drum can expose the oil to iron, which catalyzes the oxidation process. Of course, the use of nonreactive (plastic) containers or drum liners eliminates the metal catalyst effect on oxidation. It is suggested that lubricants are used within a short period of time from 3 to 12 months, depending upon the lubricant type, for example, lithium greases should be used within 12 months and calcium complex greases in 6 months [65].

However, there is no consensus between recommendations from the lubrication professionals. Though the study [67] was not based on scientific study, the authors were able to summarize the responses from lubrication professionals, suppliers, and consultants on the subject of recommended length of storage time for 10 different lubricating materials from Mineral R&O 32 to PAO-based EP polyurea grease in various conditions: indoors at 20°C (68°F), outdoors in a warm climate at 30°C (85°F), and outdoors in a cool climate at −13°C (9°F). The results are varied for indoor storage of mineral lithium complex grease or PAO-based polyurea grease at 20°C (68°F) from 1 year to 15 years.

Some independent grease manufacturer suggested that shelf life for these greases is infinite. In outdoor storage in warm climate (30°C or 85°F), for the same greases, these limits are said could be from less than 1 year to 2–3 years. In outdoor storage in cold climate (–13°C or 9°F), these limits are thought to be from 3 months to 3–4 years.

The absence of consensus in the issue of storage life of the lubricants shows that Industry needs best practices and guidelines to follow. The lack of consensus in the polls [64] suggests that end users receive little guidance about lubricant storage and shelf life limits, and the information which is provided may be inconsistent. However, no industry standard was yet developed and adapted so far.

6.7.4 Practical Advice on Lubricant Storage

In practice, it is important to keep grease storage in good conditions to prevent shortening of grease life. Open grease tubes and drums are attracting airborne contaminants such as lint and dust. Securing used grease tubes that will be reused in sealable, washable containers is considered the best practice. The containers will hold one tube of grease and allow for great contaminant exclusion. Used drums of grease are at an even higher risk of contamination. These drums are often opened and used over a greater period of time, leading to more and more opportunities for contaminants to enter. If not using a sealed air style grease dispensing unit for drums of grease to fill grease guns, some of the best methods for contaminant exclusion are to use Velcro style covers or snap-on caps. Using these types of contaminant exclusion devices will keep the grease cleaner and prolong its life [67].

NLGI recommends the following steps for best practice of grease storage:

1. Store the grease in a cool, dry indoor area where airborne debris is at a minimum.
2. Use the oldest container first.
3. Keep containers tightly covered.
4. Wipe off the edges of a container before opening it to avoid intrusion of dirt.
5. Where necessary, grease should be brought to a satisfactory dispensing temperature just before being put into service.
6. Clean grease-handling tools (such as spatulas, drum pumps, etc.).
7. When a container has been partially emptied and the remainder will not be immediately used, all void spaces within the remaining grease should be filled with grease, and the surface leveled and smoothed.
8. Store grease cartridges (tubes) vertically with the removable cap up.

6.8 LUBRICATION PRINCIPLES AND CHOICE OF LUBRICANTS FOR ELECTRICAL CONTACTS

Application of lubricants is an effective way to protect an electrical connection from the factors that cause an increase of electrical resistance. In electrical and electronic devices, there are benefits from applying lubricants to the electrical contacts. The type and function of the contact define the lubricant required. By definition, an electrical contact is a combination of two surfaces held together by force with the passage of current, voltage, or signals [68].

Electrical contacts in very high-power circuits with currents up to mega-amperes do not have much in common with electronic circuits, with currents as small as microamperes [69]. Power dissipation ranges from dozens of watts to a tenth of a watt [70]. The force holding the contact surfaces together may vary from low values ($P < 0.2$ N) to high (several tens of N) [71]. Many factors can lead to the degradation of contact surfaces.

Sliding and rolling contacts constantly move against each other and the surfaces degrade due to friction and wear debris. Static or wiping contacts experience vibration while the current is passing through, which destroys the plated coating (fretting). Nonconductive deposits on the contact surfaces formed due to various types of corrosion (atmospheric and galvanic) lead to a rise in CR.

A properly chosen lubricant may slow down these processes without interfering with the electrical resistance of the contact.

6.8.1 Principles of Contact Lubrication

The general principle of lubrication is to use a substance that protects an electrical connection from degradation and increase in electrical resistance. The positive effect of lubrication can be achieved only if the product is carefully chosen based on conformity between lubricant properties and the conditions of application, such as the contact design, load, heat dissipation, and environment. The long-term effect could be negative if the lubricant cannot withstand the service conditions and degrades itself, inducing an additional cause of contact deterioration.

Lubricant function depends on application. In separable connectors, it reduces friction during installation, minimizes mechanical wear during connector service, and slows down the destructive effect of fretting corrosion. The lubricant in a permanent electrical connection blocks the access of corrosive gases or particulate material to the interfaces of the conductors [70]. An important factor in choosing the right lubrication product is the environment in which a particular electronic or electrical apparatus is to function. Various environmental factors and their combinations are aggressive toward traditional metals used in electrical contacts (copper, aluminum, and various alloys) and their coatings (silver, tin, nickel, gold, etc.), but they may also be harmful to lubrication materials.

The most important and general roles of the lubrication of electrical contacts are protection from environmental and galvanic corrosion, wear/friction, and fretting corrosion. All known types of lubricants (among them mineral and synthetic oils and greases, solids, and dispersions) [72–74] have been tested and applied to electrical contacts. Compared with other lubricants, polymeric ethers such as polyphenyl ether (PPE) and perfluorinated polyether (PFPE) have better characteristics: stability at high temperatures and a very low vapor pressure. Above all, PPEs demonstrate an ability to remain at the contact point without migrating. Polymeric ethers also provide a high degree of protection from atmospheric contaminants for connectors [75,76].

6.8.2 Choice of Lubricants Based on Design and Contact/Plating Materials

A summary of the last 20 years of lubricants testing for their application to various types of electrical contacts made of different types of materials is presented in Tables 6.13 through 6.15 [60]. In these tables, the following abbreviations have been used: PPE—polyphenyl ether; PFPE—perfluoropolyether; PAO—polyalphaolefin; POE—polyolester; PAG—polyalkylene glycol; FA—fatty acid; MCW—microcrystalline wax; PTFE—polytetrafluoroethylene; PFAE—perfluoropolyalkylether; FxHy—fluorinated thiol; FxOHy—fluorinated ether thiol; SH-PFPE—PFPE with sulfur end groups.

Lubricants evaluation was performed on the basis of their effect on the performance and stability of the contacts, protection from atmospheric and fretting corrosion, reduction of friction and wear, and their ability to withstand degradation under the contact working conditions. For example, many lubricants have been tested on a bolted joint under various current cycling conditions and fretting to determine the spreading tendency, stability to thermal degradation, stability to ultraviolet (UV) radiation and field conditions, as well their ability to protect the contact against fretting [77].

In a series of tests [78] on tin- and silver-plated stationary parts, it was shown that lubricants generally reduced the severity of wear and tendency for fretting corrosion (thereby reducing the CR), and gave no increase of CR or any other negative effect.

It was also demonstrated in modeling experiments that lubrication might be effective in restoring a degraded system [79]. However, it was confirmed that lubrication delays but does not totally prevent fretting corrosion [80,81]. After prolonged exposure to fretting, the beneficial effect of lubrication may be lost. Further conduction is impeded by the formation of a layer composed of nonconducting wear debris and lubricant [92].

TABLE 6.13
Lubricants for Electronic Connections: Separable, Low Load, Low Power, Made of Copper

Base Metal/Plating/ Underplating	Mating Part Base Metal/Plating/ Underplating	Lubricants Tested	Test	Refs.
Cu/Sn	Same	PPE, POE, esters, PFAE, PAG, silicones, hydrocarbon oils (mineral and synthetic), chlorofluorocarbons	Volatility, fretting, spreading, wetting, gum formation	82
Cu/Sn	Same	Chlorophenyl silicone (best), phenylmethyl silicone, and methyl alkyl silicone systems	Fretting, high T, oxidation, reactivity	83
Cu/Ag	None	Stearic acid	Corrosion	84
Cu/Au	Same	Synthetic paraffin mixture, wax mixture, synthetic paraffin with PPE, PPE	Resistivity, friction, wear, corrosion, dust	85
Cu/Au	Same	Liquid lubricants, wax lubricants[b]	Dust	86
Cu/Au or Ag or Sn	Same	Liquid lubricants: paraffin oil with FA ester, five-ring PPE	Corrosion, dust	87
Cu/Sn	Cu/Sn	PAO	Fretting, oxidation	88
Cu/Au or 60Pd40Au or Ni inlay	Brass/Au/Pd/Ni[a] or Au/Ni[a]	PPE, mineral oil, POE	Friction, wear	89
Cu Au flash/Pd/Ni[a]	Phosphor bronze/ Au/Ni[a]	POE, five-ring PPE, six-ring PPE, six-ring PPE with MCW, different additives	Thermal aging, corrosion	90
Cu/Au	Pd/none or Au	Mineral oil, PPE	Wear, fretting, resistance	91

Abbreviations: PAG, polyglycol; PAO, polyalphaolefin; PFAE, perfluorinated aliphatic ester; POE, polyolester; PPE, polyphenyl esters.

[a] Underplating.
[b] Lubricants performed unsatisfactory.

Fretting tests showed that at low loads none of the lubricants could protect the contact zone against fretting, but increasing the contact loads significantly improved the lubricant's protective ability. This improvement was observed in copper-to-copper and copper-to-tin plated copper contacts, but not in the case of aluminum fretting against aluminum, since none of the lubricants tested had the ability to suppress the effect of fretting.

The results of recent cycling tests showed that all lubricants had a beneficial effect on the performance of bolted joints as manifested by stable operation [93]. Having a lubricated contact surface reduces the friction force and the wear of gold plating during sliding or micromotion. In contactor applications, it appears that the insertion and withdrawal force can be reduced to as low as 20% of the nonlubricated surface [87]. Detailed information on the requirements, methods of application, and examples of widely used fluids, greases, and solid lubricants may be found in Ref. [94].

6.8.3 LUBRICATION AS PROTECTION FROM FRETTING CORROSION, MECHANICAL WEAR, AND FRICTION

The practice of contact lubrication is as old as connectors themselves [71]. Early testing demonstrated that lubrication is essential to achieve low and repeatable coefficients of *friction*, whether

TABLE 6.14

Lubricants for Electronic Connections: Separable, Low Load, Low Power, Made of Copper Alloys

Base Metal/Plating/ Underplating	Mating Part Base Metal/Plating/ Underplating	Lubricants Tested	Type of Testing	Refs.
Cu alloys/Au	Brass/Au/Pd/Ni[a] or Au/Ni[a]	PPE, mineral oil, POE	Wear, friction	91
Cu alloys/Au-Ag-Pt or Au/ Pd-Ag inlays	Phosphor bronze/Ni or Au	Five-ring PPE	Corrosion	95
Brass/bare, Au, Au/Ni[a], Pd/Ni[a], Sn	Same	PAO with FA, liquid paraffin with FA ester, paraffin wax with FA, alkyl mercaptan with wax, and five-ring PPE	Friction, resistance	79
Brass/none or Au/Ni[a] or Sn-Pb	Same	Six-ring PPE, petroleum jelly	Resistance, wiping	80
Brass C26000/Sn-Pb/Cu[a]	Same	Five-ring PPE, PAG (liquid and grease)	Fretting	96
N/A (socket)/Au/Ni[a]	Ne-Fe alloy/Sn/Cu[a]	Mil-L-87177 grease[b]	Fretting	94
Cu-Ni-Sn alloy CA725/ Au flash/Pd/Ni[a] or Au/Ni[a]	Same	Six-ring PPE, 6-ring PPE mixed with MCW	Corrosion, fretting	97
Cu-Ni-Sn alloy CA725 Au flash/Pd/Ni[a] or Au/Ni[a]	Clad Au-Pd-Ag alloy/none	Six-ring PPE	Friction, wear	98
Phosphor bronze/none or Sn-Pb	Same	Six-ring PPE, petroleum jelly	Resistance, wiping	80
Phosphor bronze/Au/Ni[a]	Same	PPE (thermal aging)	Thermal aging	99
Phosphor bronze/Au/Ni[a]	Same	Water-soluble lubricant, paraffin oil with FA ester, five-ring PPE	Corrosion, wear, fretting	100
Phosphor bronze/Au/Ni[a]	Same	FxHy, FxOHy, SH-PFPE	Corrosion	101
Copper alloys/Au/Ni[a] or Diffuse Au alloy or Sn or Ag/Ni[a]	Same	PFPE silicone grease, Stabilant-22[b]	Resistance, oxidation	102
CuSn4 bronze/Sn	same	PFPE linear low viscosity and branched low and high viscosity	Friction, wear, fretting	103 104
Au-Ni clad/none, Au flash	Same	Six-ring PPE	Wear	92

Abbreviations: PAG, polyglycol; PAO, polyalphaolefin; POE, polyolester; PPE, polyphenyl esters.

[a] Underplating.

[b] Lubricants performed unsatisfactory.

using inlay or electroplated material. The effectiveness of lubricants in stabilizing the CR, in wear reduction, and in retarding the rate of oxide formation by shielding the surface from air appears to be not strongly dependent on their composition and viscosity, and both fluids and greases are of value [91].

Some lubricants (PPEs) appeared to be more effective than others in preventing adhesive wear, particularly when they are thick [89,91]. Mineral oil and POE are not very effective in reducing the coefficient of friction, probably due to their lower viscosity [89]. The results of the multiple *fretting* tests (Tables 6.13 through 6.16) showed that many contact compounds have beneficial effects on the CR behavior. Application of a lubricant helps to decrease CR fluctuations and to lower the CR.

TABLE 6.15

Lubricants for Permanent High-Load Electrical Connections Made of Al, Cu, and Ag Alloys

Base Metal/Plating/ Underplating	Mating Part Base Metal, Plating/ Underplating	Lubricants Tested	Test	Refrences
Bare Al	Same	Molykote G-n[b], silicone vacuum grease, Penetrox[b], petroleum jelly, Aluma Shield[b], Kearnalex[b] inhibitor, Contactal[b], Fargolene[b], Alcan[b] jointing compound	High T, consistency, resistance, fretting	77,105
Al/Ni	Same	Petroleum jelly	Fretting	106
Bare Al	Cu	Silicone, Penetrox[b], petroleum jelly, NO-OX-ID[b], Kopr-Shield[b], Nikkei[b], Contactal[b]	High T, spreading, fretting	107
Bare Al	Same	Stearic acid[c]	Friction, wear	108
Bare Al Bare Cu	Same	MoS$_2$ grease[c]	Fretting	93
Bare Al	Bare Cu or Cu/Sn	Mobil grease[b], Belray[b], Convoy[b], Shell Darina[b], Fomblin[b], Penetrox[b], Felpro[b] Ni antiseize, GEC M111[b], Kopr-Shield[b]	High T, UV, spreading, fretting	109
Bare Al Bare Cu	Bare Al Bare Cu or Cu/Sn	PFPE greases, Superlube[b] grease (Teflon), DuPont Krytox[b], PFPE grease with Cu additive	High T, resistance, fretting	78
Bare Al	Same	Teflon, PFPE/PTFE, PFAE/ PFPE	High T, UV	110
Cu/Ag or Sn or Ag/Ni[a]	Same	Li synthetic grease	Resistance, adhesive and fretting wear, friction, cold welding, oxidation	111–113
Cu/Sn	Al-Mg alloy/bare or Ni Al/Ni[a] and bare Cu	Commercial grease TSIATIM-222[b] with or without Ni powder	High T, oxidation	114
Cu/N/A	Same	Li soap grease, metalless grease, silicate-based grease[c]	Wiping and arcing effects	115
Bare Cu	Same	Li stearate PAO grease with and without corrosion inhibitor	High T, oxidation	116
Cu (OLTC contacts)/ Ag	Same	Silver iodide	Friction, wear	117
Cu (sliding contacts)/ none or Sn or Ag	Same	Naphthenic mineral oil	High T, friction, wear	118
Ag-CdO/AgP or AgCd brazing alloys	Same	Graphite-S, silicone paste, MoS$_2$ grease, Na soap mineral grease, mineral oil	Contamination	119
Ag-CdO rivet/none	Cu blade/none	Silicone[c]	Contamination	120

Abbreviations: OLTC, on-load tap changer; PAO, polyalphaolefin; PFAE, perfluorinated aliphatic ester; PTFE, polytetrafluoroethylene; UV, ultraviolet.

[a] Underplating.

[b] Proprietary products.

[c] Lubricants performed unsatisfactory.

TABLE 6.16
Failure Events with RTBs

Year	# of Units	Description of Event	Piece Part	Cause
1982	4	Breakers failed to trip on undervoltage	UV trip assembly	Inadequate lubrication, excessive PM interval
1983	3	Same	Same	Same
1984	N/A	Undervoltage trip response time was found out of specifications	Mechanical assembly	Lack of proper amount of lubricant on the bearings of the breakers' frame assemblies
1984	2	Undervoltage trip response time was found out of specifications high	Mechanical assembly	Same
1985	2	Undervoltage trip response time was found to be out of the allowable tolerance	Mechanical assembly	Lack of lubrication and dirt accumulation on from frame assembly
1992	1	Breaker failed to close	Latch assembly	Inadequate lubrication of the breaker operating mechanism latch
1994	2	Breakers would not close from the hand switch in the main control room, the latch stuck in mid travel	Latch assembly	Insufficient lubrication of the inertia latch

Source: Modified from *Common-Cause Failure Event Insights Circuit Breakers*, United Stated Nuclear Regulatory Commission, NUREG/CR-6819, Vol. 4 "Circuit Breakers", INEEL/EXT-99–00613, May 2003, Online. Available: http://www.nrc.gov/reading-rm/doc-collections/nuregs/contract/cr6819/.

Abbreviations: PM, periodic maintenance; RTBs, reactor trip breakers.

The efficacy of a compound in reducing the fretting damage depends on the ability of a particular compound to flow back over freshly exposed metal pushed aside by the contact slip [77]. This ability depends on the lubricant's initial consistency and the change in consistency [121].

6.8.4 LUBRICATION AS PROTECTION FROM CORROSION

Corrosive factors in the environment are all those that are chemically aggressive toward contact materials and plating, and produce a nonconductive deposit on the contacts (corrosive gases and vapors) or the environmental factors that accelerate corrosion (temperature extremes and humidity). An additional contaminating factor is dust. Another phenomenon that leads to contact degradation is galvanic corrosion between dissimilar metals.

The benefits of applying lubricant for corrosion protection are not as well understood as its application in the reduction of friction and wear. For many years, the benefits of using lubricant for corrosion inhibition were overlooked because prospective users were skeptical that a nonconductive substance could be applied onto contacts without interfering with the conduction [122].

It is important to consider all the factors that may result in the negative effect of lubrication if the chosen product deteriorates while exposed to a high-temperature environment, migrates or leaks out, or collects dust [123]. The choice of lubricant for corrosion protection should be based on thorough qualification of the product for survival in long-term use and the ability of such products to provide friction and wear reduction. Very few lubricants have been identified that provide exceptional long-term corrosion protection and yet do not produce any adverse effect on connector surfaces.

Field testing showed that lubricated surfaces collect and retain dust. Sliding electric contact experiments show that, in dusty environments, liquid lubricants appear to perform better than wax lubricants. Experimental and theoretical analyses indicated that the high permittivity of the lubricants plays an important role in attracting dust [124].

The mixture of dust particles and wax could be very harmful to the contact, especially under low normal force. It was found that liquid lubricants with lower permittivity provide much better contact behavior in dusty environments, either for sliding electric contacts or static contact applications [86]. There was little indication of any adverse electrical effects for the best lubricants. It appears that while dust may be retained by the lubricant, it is also effectively dispersed away from the mating interface during wipe or micromotion events. The tests found that the beneficial effects of corrosion inhibition by the lubricant overshadow any negative effects due to dust [87,124].

An example of a lubricant that proved useful for corrosion protection for connectors made or plated with precious metals is a six-ring PPE or its mixture with microcrystalline wax, although the wax content should be controlled [81,92,97]. Multiple products, such as synthetic soap greases and other commercial greases, have been tested and proven useful for bolted contact application [77,78,107,109,114].

Galvanic or bimetallic corrosion is another phenomenon that is harmful to the contact surfaces. Whenever dissimilar metals are in the presence of an electrolyte, a difference in electric potential develops. When these metals are in contact, an electrical current flows, as in the case of any short-circuited electric cell. This electrolytic action causes an attack on the anodic metal, leaving the cathodic metal unharmed. The extent of the attack depends on the relative position of two metals in contact in the electrolytic potential series.

Galvanic corrosion is common with aluminum-to-copper connections, since copper and aluminum are quite far apart in the series, copper being cathodic and aluminum being anodic. Hence, when aluminum and copper are in contact in an electrolyte, the aluminum may be expected to be severely attacked.

Some products, such as petrolatum-type compounds containing zinc dust, effectively protect contacts made of dissimilar metals from galvanic corrosion [125]. However, some lubricants may induce galvanic corrosion; therefore, the products for this application should be chosen very carefully. For example, graphite, which has a noble potential, may lead to severe galvanic corrosion of copper alloys in a saline environment [40].

Multiple studies showed that lubrication reduces the ingress of corrosive gases on the contact metals or through pores (which is particularly important for thin gold plating) to react with the substrate metal. Application of lubricant reduces the possibility of electric contact failure from insulated particulates such as dust or wear fragments of tarnished films, which may deposit on the contact area.

However, application conditions of connectors (such as electric current, normal force, the number of contact pairs within one housing, sliding, micromotion, etc.) and environmental conditions (such as humidity, temperature, corrosive gases, dust, etc.) are very complicated. That is the reason why it is almost impossible to choose just one kind of lubricant to fulfill all the requirements and to match all the above conditions. Unfortunately, there is no such thing as a "universal lubricant."

6.8.5 Durability of Lubricants

The purpose of lubrication is to improve electrical contact functionality and prolong service life. Therefore, it is important to ensure that none of the lubrication product properties would work against this purpose. Lubrication products should not deteriorate or contaminate the contact surfaces for a predetermined period of time within the range of application conditions. A lubricant should remain stable in sufficient quantities at the contact interface to perform its intended function throughout life.

Lubricants applied to the contact surfaces are exposed to four major temperature-sensitive aging mechanisms: evaporation, surface migration, polymerization, and degradation [81,82,123,126]. With time and temperature, lubricants may degrade by oxidation or polymerization, forming insulating films or gums [82].

Due to evaporation and mechanical loss, a thin coating of a contact lubricant can disappear from surfaces. Mechanical loss is its physical removal as a result of sliding, such as by wear debris that carries it away. Evaporative loss will occur if the lubricant has a significant vapor pressure and is important at elevated temperatures, particularly with the fluid lubricants [76,81].

When a lubricant disappears from the contact interface by evaporating or spreading, the contact is exposed to failure by wear or by corrosion. Several material properties, such as viscosity, surface tension, vapor pressure, and thermal stability, are important in selecting the proper lubricant for contact application. Thermal limitations, both high and low, represent major factors in the durability of many lubricants. At low temperatures, many lubricants appear to solidify, developing high-shear-strength films, and leading to high CR [127]. Some lubricants are susceptible to cracking due to long-term exposure to high temperature [78].

Lubricants may degrade due to chemical reaction with the atmosphere and may also polymerize when heated in the presence of a copper-based alloy [128]. Copper is known to degrade polymers through catalytic-enhanced oxidation, so grease would undergo changes in contact with copper at higher application temperatures, which in turn induces copper corrosion. Applying grease with a copper corrosion inhibitor to copper contacts at high temperature may significantly decrease lubricant-induced copper corrosion [116].

Because of their composition, some lubricants, such as greases with silicate thickener, have undesirable properties when applied to electrical contacts [115,120]. Silicate as a high-temperature grease thickener may pose problems if entrapped under the contact. Near an arc, splattering effects can occur, coating the surface with a nonconductive layer. Particles may also be fused together by the arc. Even with lower voltages, arcing will occur if the inductance in the circuit is significant. Silicon compounds, both inorganic and organic, are detrimental contaminants for arcing electrical contacts [120]. Any refractory material in the area of arcing can present similar problems.

6.8.6 Dry Lubricant in Specific Electrical Contact Applications

The need to operate in extreme environmental conditions (ultrahigh vacuum, high temperatures, aerospatial, etc.) and the miniaturization toward microelectromechanical system is demanding new materials in the field of low-level electrical contacts lubrication. A dry and chemically immobilized layer avoiding the traditional wet lubricating fluids would have many advantages.

In Ref. [129], the first results are presented on films obtained by the electrochemical reduction of different *diazonium salts* and their use as protective coatings able to lubricate the metallic surfaces of an electrical contact while preserving at the same time electrical conduction. Gold surfaces coated with different diazonium salts have been in regards to an application as protection for electrical contacts. After being characterized by techniques such as attenuated total reflection-infrared (ATR-IR) and XPS, macroscopic friction tests have shown outstanding wear resistance with low CRs.

Using of graphite for lubrication of monolithic silver brushes required a special attention because of high wear rates in high-power electrical motors. A new solution was suggested in Ref. [130] by loading a sacrificial solid lubricant, *graphite*, against the rotor to apply a thin, conductive transfer film to the interface of the sliding electrical contact *in situ* to reduce wear and friction while retaining high electrical efficiency. These films may help in improving the operational life and efficiency of brush/rotor sliding electrical contacts. Decoupled *in situ* lubrication of monolithic silver brushes with graphite has demonstrated the ability to eliminate mechanical wear at low sliding velocities (1.6 m/s) and low current densities. High current densities, however, still present the problem of electrically induced wear of the brush contacts and significant Ohmic heating due to the thickness of the graphite transfer layer.

6.9 ELECTRICAL CONNECTOR LUBRICATION

6.9.1 Failure Mechanisms of Connectors

The most common cause of failure in portable systems such as cellular telephones, computers, instruments, and so on is a defect at the metallic interface inside a connector. The metallic interface inside a connector is typically tin or tin–lead, but may be plated with a very thin layer of a noble metal such as gold. Whether a pin is tin or gold, the failure mechanism begins with wear due to friction at the metallic interface.

In cellular telephones, wear is usually the result of micromotion, which may involve displacement between the pin and the socket of only a few microns. Micromotion in turn is caused by vibration from handling of the telephone and temperature change. The base metals at the point of contact and the plastics used in the connector have different coefficients of thermal expansion, and a thermal change of only 10°C can cause linear micromotion of 5 microns.

Micromotion exposes less noble sublayers below the less corrosive tin or gold layers. These sublayers are quickly oxidized and are sensitive to contact with moisture and other airborne contaminants. The moisture and airborne contaminants (sulfuric acid, nitric acid, and many more) create electrolytic cells that begin the corrosion of the pin and promote rapid failure. Failure actually occurs when the corrosion products grow and eventually separate the metallic surfaces at the point of contact.

Contact corrosion is as primary failure mechanism. Airborne contaminants are present everywhere, including dozens of various organic and inorganic contaminants in form of suspended aerosols or fine particulate solids, which attack not only connectors but also plastic-packaged integrated circuits (ICs). In plastic ICs, contaminants and moisture may enter along the external leads or they may settle on the package surface and then migrate as ions through the epoxy.

In cellular telephones, oxygen, moisture, and contaminants penetrate to the connector along the surface of the pin, but they also may migrate into the plastic housing around a connector and may eventually reach the metallic surfaces of the connector.

Another mechanism known as fretting corrosion leads to abrasion of contact surfaces. When metallic surface is plated with gold, which is usually very thin—fraction of a micron, plating may be removed; if the pin is plated with tin–lead, the surface is scored, which also may lead to the connector failure.

Fretting and atmospheric corrosion are two mechanisms that lead to the failure of connectors in other than portable instruments. An example of connector corrosion in electronics used in automotive industry is shown in Ref. [131].

6.9.2 Protection of Connectors from Failure with the Lubricants

Lubricants are basically insulators—nevertheless they have a positive effect on the performance of electrical contacts. Lubricants often have to be used on electrical contacts if the CR and the actuating forces are to remain constantly low for as long as possible under following operating requirements:

- A high number of plug or contact cycles is required (e.g., smart card connectors and plug-in connectors in automation technology); in these cases, the focus is on *wear reduction*.
- Low plug and unplug forces are required (back planes in telecommunications, multipin plug-in connectors for data lines); in these cases, the focus is on *friction reduction*.
- Where long service life and contact reliability are required also when operating under vibration and frequent temperature cycles (automobiles, automation technology); in these cases, the focus is on the reduction of *fretting corrosion*.
- Contact erosion is to be reduced due to switching arcs.

Without a lubricant, electrical contacts are unlikely to meet the ever increasing requirements of today's applications. Even electroplating or chemical coating of the contacts does not always have the desired or required effect. Additionally, the use of coatings can become prohibitively expensive if layers of a certain thickness are required. Furthermore, ambient and operating conditions are often such that the contact surfaces get coated by layers of foreign matter or change chemically. The resistance is increased by the foreign layers that are frequently found on the contact surfaces (e.g., metal oxide layers or plastic deposits). It takes a sufficiently high contact force or heat generation due to the power loss to penetrate these layers.

The lubricant covers the open contact (during storage), thus preventing the formation of a detrimental layer of foreign matter (e.g., oxide) on its surface. When the electrical contact is closed, the lubricant forms a separating film due to the relative motion of the two contacts, despite the contact force pressing the contacts together. The lubricant film reduces the friction coefficient and wear (abrasion) notably. Due to the contact force, the lubricant is displaced from between the roughness peaks while the contact is stationary, so there is direct metal-to-metal contact between the surfaces. The result is a low electrical CR [132].

Another benefit of using contact lubricant is that electric arcing during the switching operation is impeded by the dielectric strength of the lubricant, which is higher than that of air. In lubricated contacts, electric arcing occurs only when the distance between the contacts is already very short. When opening the contact, the electric arcs will cease much sooner. Shorter electric arcing times mean a general reduction of contact erosion and hence longer contact life.

6.9.3 FORMS, TYPES, AND ROLES OF CONNECTOR LUBRICANTS

Connector lubricants are available in three forms [131]:

Oils: Most often applied as a field fix to correct a problem. They are occasionally used during production where they are atomized and sprayed onto terminals as they are stamped

Greases: They are injected into the female part of the connector. Greases can be applied both before and after a connection is made, and they can contain a variety of additives to solve specific problems.

Dispersions: Consist of greases dissolved in a solvent to make them more liquid. They are easy to apply in the production environment by spraying or dipping. After a dispersion is applied, the solvent evaporates to leave a thin film of lubricant on the contact.

Connector lubricants are formulated from a variety of chemicals:

1. *PAO* is the most common synthetic lubricant and is typically known as an antifretting lubricant. It is available in a range of grease thickener systems and provides good protection at temperatures to 125°C or higher.
2. *Perfluoropolyether (PFPE)* is used in high-temperature applications and provides excellent protection at temperatures up to 250°C. It provides good insertion force reduction.
3. *PPE* is typically used on gold contacts. It provides unique film strength capability that prevents galling of gold when the contact is made. PPE is the most expensive lubricant used.
4. *Specialty silicones* are used occasionally for contacts that require high-temperature resistance above 250°C (482°F).

Factors determining choice of connector lubricant:

1. *Temperature range*: Below 135°C (275°F) use PAO; above 135°C use PFPE.
2. *Elastomer compatibility*: It is virtually impossible to make accurate, all-inclusive compatibility recommendations without actually testing a lubricant on a material. This is because a single elastomer category (e.g., nitrile) can have as many as 100 possible formulations, each with different compatibility issues. When testing compatibility, materials should be exposed to lubricants under various temperatures and loads. The most accurate compatibility testing is done at the expected operating extremes. For further assurance, test materials of nearby components in case of oil migration or outgassing and condensation.
3. *Insertion force reduction*: PFPE/PTFE provides the greatest reduction.
4. *Cost*: PAO provides the lowest cost where it can be used.

Contact lubricants play three important roles in protecting electrical circuits [133]. First, they prevent damage to the contact surface from the nearly imperceptible movements that occur during operation. Second, lubricants seal the connection to stop moisture from entering and corroding the contacts. This is especially important for connections exposed to the elements. Finally, the lubricants reduce insertion forces to speed assembly and prevent workforce fatigue and injury.

Experience shows that connector lubricants keep connectors, sensors, and switches functioning long after warranties expire. Types of lubricants include

1. Specially formulated synthetic lubricants that prevent wear, corrosion, and fretting caused by vibration and thermal changes in the connector housing.
2. Unique synthetic fluids that reduce friction to prevent galling, scratching, and deformation of gold-plated contacts (e.g., in airbags).
3. Fluoroether-based lubricants that survive temperatures in excess of 200°C while limiting insertion forces, especially for multipin connectors, reducing the potential for repetitive motion injuries.
4. Oils that resist oxidation at high temperatures. The contact lubricant is an often overlooked component in keeping auto electronics operating for the life of the car. These oils, greases, and dispersions can actually pay for themselves by reducing both warranty claims and worker injuries.

A properly selected lubricant lowers insertion force by decreasing the coefficient of friction between mating surfaces. It reduces mechanical wear by placing a film of oil between the mating surfaces. Coating the contacts with an antifretting lubricant reduces mechanical wear, provides an oxygen barrier, and helps keep oxide debris away from the contact area.

6.9.4 POLYPHENYL ETHERS (PPE) AS CONNECTOR LUBRICANTS

It was found that connector's metallic interface can be protected from both abrasion and corrosion by a specialized class of lubricants called PPEs, which have the characteristics that practically no other lubricants have: stability at high temperatures, very low vapor pressure, and above all, their ability to remain at the contact point without migrating. They remain stable at temperatures higher than those commonly encountered in cellular phones or other electronic systems. Chemically, they are relatively inert, and when used on connector pins would not react with nearby metal and plastic elements even if they could migrate to them. They also have a very low vapor pressure. A very thin film of PPE deposited onto a tin–lead connector pin will, at normal temperatures and pressures, evaporate only after 40–50 years.

It was determined in Ref. [133] that PPE provides a high degree of protection of connectors from atmospheric contaminants, though the mechanism of this protection is not yet fully understood. It was determined that applied to connector pins in a very thin layer, PPE can extend the service life of connectors by a factor of 1000 or more. Increasing the service life of connectors—the most frequent failure site in cellular telephones—greatly increases the reliability of the telephone. A very interesting test was conducted by coating a connector pin that has already become corroded with PPE. The performance of the pin immediately improved because the droplets of the lubricant have absorbed the corrosion products that were preventing contact between the pin and the socket. Studies have shown that PPEs provided the best performance on connectors; however, PPEs are expensive and have limited low temperature operability, solidifying at 20°C (68°F).

Another important field where connector failure is not acceptable—medical equipment, such as dialysis machines or in pulmonary monitors, which has to exhibit not merely long-term reliability but uncompromising reliability [134]. However, when a connector lubricant is used, it nearly always is applied only once—during electronic assembly of the system. Relatively few connectors in installed systems are readily accessible for easy lubrication, and few repair technicians carry connector lubricants with them. When connector failure does occur, in most cases it necessitates costly equipment disassembly for the purpose of replacing the corroded component.

Whichever type of connector lubricant is used in a particular medical system, the material must have several distinctive properties. Probably most important, it has to remain where it has been applied. The lubricant is useless if it migrates away from the connector, or, more specifically, from the two in-contact surfaces.

Most general lubricants have little or no positional stability and work briefly or not at all on electronic connectors. The substance must lubricate the surfaces against micromotion and also against larger excursions. And it must continue to provide effective lubrication over a fairly wide temperature range. Because the lubricant must be counted on to protect reliability during the life span of the equipment, which may be several years, it must have a low vapor pressure. That is, it must evaporate only very slowly. Other physical and chemical requirements for the material are that it should be thermally and oxidatively stable.

The lubricant must operate well both on noble metals, including gold and palladium, and on such nonnoble, or base, metals as tin. Finally, it must be chemically inert, that is, nonreactive to the connector and the connector housings. PPEs meet all of these requirements. An example of the superiority of performance offered by PPEs is that they remain stable and continue to act as effective lubricants at temperatures as high as 453°C (847°F), while petroleum-based and other synthetic lubricants decompose at around 200°C (392°F). PPEs are also extremely resistant to ionizing radiation, a property that has led to their widespread use in nuclear facilities and on earth-orbiting satellites.

Not all electronic connectors require the top-level lubricating performance of PPEs to achieve high reliability. A wide performance gap existed between low-end connector lubricants and PPEs, and an assembler who needed superior performance had only PPEs to choose from. This gap has now been filled, however, through the development and introduction of advanced phenyl ether (APE) lubricants. The single most important performance property of connector lubricants, as mentioned, is positional stability: the lubricant cannot protect the connector if it has migrated to a different location. APE lubricants have a surface tension that is lower than that of PPEs but significantly higher than the surface tension of ordinary connector lubricants. When applied to the connectors in, for example, an x-ray system or a ventilator, an APE lubricant ensures reliable equipment function by remaining dependably in place at the connector interface.

6.9.5 Effect of Lubricant Losing Weight on Contact Performance

Lubricants applied to electrical contact provide protection of plating from friction, wear, and corrosive gases. Because the lubricant is usually a dielectric material, it is important that the lubricant can be easily displaced by relatively low normal forces to allow good metal-to-metal electrical connection between two mating contacts.

Modern lubricants used in electrical industry for lubrication of electrical contacts are often a complex mixture of solid, liquid, and polymeric ingredients. It also includes multiple additives improving various lubricant properties, such as thermal stability and corrosion resistance. In using such complex dielectric materials on conductive surface may cause a serious problem affecting performance of the contact.

One of these problems is a possibility of the liquid component of the lubricant to evaporate over time in service, which results in drier lubricant and the formation of a solid-like film. Such film is harder to displace under low normal forces. The risk is that a loss of weight from the lubricant could create an impenetrable film on the contact and interfere with the electrical connection across a contact interface. This problem was studied in Ref. [135] by conducting and analyzing the results of the tests using coupons plated with 15 microinches (0.38 μm) of gold plating and a proprietary TE Connectivity (TE) contact lubricant.

The first test was conducted to evaluate the lubricant's continued ability to prevent porosity corrosion after losing 25% of its weight. Figure 6.1 shows unlubricated and lubricated test coupons after 10 days exposure to the Bellcore Class IIA MFG test. The unlubricated sample (a) shows

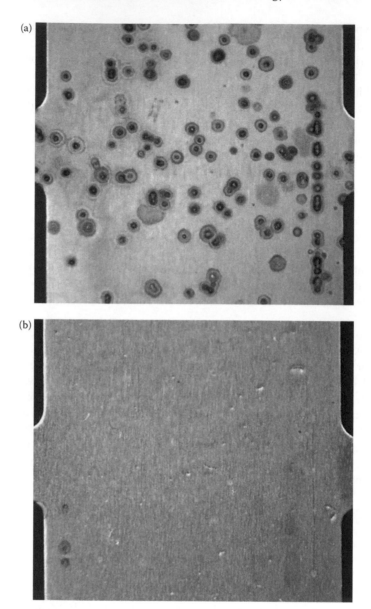

FIGURE 6.1 MFG porosity on unlubricated (a) and lubricated (b) gold plated coupons. (From Pawlikowski G. T. Tyco electronics HM-15 lubricant stability projected weight loss, Technical Paper, Tyco Electronics Corporation (a TE Connectivity ltd. Company), November 2006, Online. Available: http://application-notes. digchip.com/166/166-47954.pdf. With permission.)

porosity and creep corrosion. The sample lubricated with the TE contact lubricant (b) shows that the lubricant remains effective in preventing porosity corrosion despite the loss of 25% of its weight.

Another critical requirement of a contact lubricant is that it does not interfere with good electrical connections. The second test conducted in Ref. [135] was to determine the *low-level CR* (LLCR) across a lubricated interface at various normal forces. By increasing the force of a conductive probe on a conductive substrate while measuring resistance between the two interfaces allows to measure a relation between applied force and CR. The test of the lubricant's ability to resist the formation of an impenetrable film showed that there is little difference in the median LLCR of the as applied

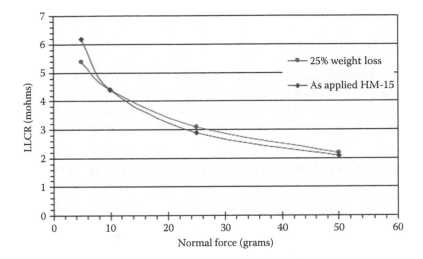

FIGURE 6.2   Electrical resistance across an interface containing HM-15 lubricant. (Pawlikowski G. T. Tyco electronics HM-15 lubricant stability projected weight loss, Technical Paper, Tyco Electronics Corporation (a TE Connectivity ltd. Company), November 2006, Online. Available: http://application-notes.digchip. com/166/166-47954.pdf. With permission.)

lubricant compared to the lubricant that lost 25% of its weight through high-temperature exposure as shown in Figure 6.2.

6.9.6   USE OF CORROSION INHIBITIVE LUBRICANTS (CILS) ON CONNECTORS IN ELECTRONICS

Today's avionics systems assume a major responsibility for the performance, safety, and success of commercial and general aviation. These avionics systems control the operation of flight-critical and flight-essential equipment, including navigation, communications, power distribution, flight and engine controls, displays, and wiring. The reliability of these complex and often interrelated systems in any environment is critical for safe operation [136].

The wiring on the aircraft can act as conduits for water condensed by changes in temperature and altitude during flight or on the ground. Once the condensation forms within the wire bundles, it starts to travel to the lowest point in the harness—usually the line replaceable unit (LRU). These units are the modularized avionics systems equipment units (black boxes) that support communication, navigation, autoflight, in-flight entertainment, or other systems. If the connectors are not properly sealed, water will eventually enter the LRU through the connectors, resulting in premature failure or corrosion problems. The November 2004 DoD Corrosion Prevention and Mitigation Strategic Plan references one of the United States Air Force (USAF) sustainment priority projects as follows: "Improved avionics reliability through the use of corrosion-inhibiting lubricants" [137].

In the process of corrosion, thin and often invisible insulating films can form on the surfaces of electrical connectors by the reactions of natural environments with the elements used in the manufacture of commercial and military electrical connectors. Such films may represent a significant source of faults such as Cannot Duplicate (CND) and Retest OK (RETOK) faults, the terms used in military avionics. USAF studies for the last 20 years have shown that application of certain CILs or *corrosion preventative compounds* (CPCs) can substantially impact CND and RETOK rates, thus improving avionics reliability [138,139].

Corrosion between contacts may have materially contributed to several F-16 crashes. After the cause of the crash has been determined, corrosion inhibiting lubricant spray, MIL-L-87177A Grade B (the primary component of MIL-L-87177A is PAO, which is not flammable or hazardous), has been identified and is used annually in a preventative maintenance context. Treatment of F-16 electrical connectors with the MIL-L-87177A Grade B was remarkably effective. Conductivity of the tin-plated pins was fully restored and the CPC prevented continued corrosion, so much so that in a test at one base, the aircraft so treated demonstrated a 16% improved mission-capable (MC) rate.

In addition, millions of dollars saved by cost avoidances were documented by treating the aircraft and aircraft ground equipment (AGE) connectors [140]. The U.S. Air Force has been using and testing MIL-L-87177A Grade B aerosol spray for corrosion control in electrical connectors for more than 10 years. As of March 2002, MIL-L-87177 Grade B has been added to the Air Force Tech Order/Manual 1-1-689, Avionics Cleaning and Corrosion Prevention/Control.

6.10   LUBRICANTS FOR HIGH-TEMPERATURE ELECTRICAL APPLICATIONS

6.10.1   CHOICE OF HIGH-TEMPERATURE LUBRICANTS

There are many applications when electrical connections are exposed to high temperatures and should be lubricated to avoid fretting and atmospheric corrosion. With respect to temperature, the major contributors are viscosity, thermal degradation, and oxidation. If a lubricant performance at high temperature is important, it should provide the protection at high temperatures (very high VI), that would not thermally degrade or "cook" onto hot machine surfaces and leave deposits, and that would not oxidize at the elevated temperatures.

When lubricant is heated above its recommended stable temperature, it can cause the light components of the lubricant to vaporize or the lubricant itself to decompose. This can cause certain additives to be removed from the system without performing their job, or the viscosity of the lubricant may increase.

At temperatures greatly exceeding the thermal stability point of the lubricant, larger oil molecules will break apart into smaller molecules. This thermal cracking, often referred to as thermal breakdown, can initiate side reactions, induce polymerization, produce gaseous by-products, destroy additives, and generate insoluble by-products. In some cases, thermal degradation will cause a decrease in viscosity [44]. The thermal decomposition temperature is measured according to the ASTM Standard D2879.

A hydrocarbon-based lubricant would not be an option if the temperature in application is more than 400°C (750°F), as well as SHCs, which practical temperature limits are less than 200°C. Their initial thermal degradation temperatures are higher than this, but they oxidize very rapidly at these elevated temperatures, effectively making their lifespan only a few hours.

As lubricants applied to live electrical contacts must withstand very high temperatures over a very long period of time, the base oils should show a very low tendency to evaporate and a high resistance to oxidation. If temperatures become excessively high for a short time, the lubricant should evaporate or burn away without residues such that no foreign matter (coking) remains, which would interfere with the function of the contact later. When faced with requirements of this kind, PFPE perform much better than hydrocarbon-based lubricants.

Effective lubrication is only available at temperatures below 170°C. Anything above this temperature will almost always require a compromise to be made. These compromises come in the form of reduced life, lower load-carrying capacity, slower speed, higher levels of friction, difficulties in application, compatibility issues, and so on.

The upper temperature limit of grease also depends on the thickener. The dropping point of grease is the temperature at which the thickener loses its ability to maintain the base oil within the thickener matrix. This may be due to the thickener melting or the oil becoming so thin that the surface tension and capillary action become insufficient to hold the oil within the thickener matrix.

Dropping point is a function of the thickener type. High dropping points, typically above 240°C (465°F), are commonly observed for lithium complex, calcium complex, aluminum complex, polyurea, and clay greases. For example, dropping point of polyurea and bentonite is above 250°C. Much lower dropping points are typical of conventional simple soaps: lithium and calcium 180°C (355°F), sodium 120°C (250°F), and aluminum 90°C (194°F). The dropping point is one of the determinations that characterize the grease's thermal stability; however, it is not an accurate prediction of the grease's upper operating temperature limit. Working temperature limit of a grease is a function of many variables such as base oil oxidation stability, additive degradation, thickener shearing, oil separation, and so forth. A high dropping point, while not a predictor of upper operating temperature, is an *indicator* of the maximum peak temperature that the grease may be subjected to for a short duration while not releasing oil excessively, and therefore drastically reducing the life of the grease and potentially damaging the application in the long run [141]. ASTM D2265 (preferred over the older and less precise standard ASTM D566) is the standard method used to determine the dropping point of grease [142].

Simple lithium soaps are often used in low-cost general-purpose greases and perform relatively well in most performance categories at moderate temperatures. Complex greases such as lithium complex provide improved performance particularly at higher operating temperatures. A common upper operating temperature limit for lithium grease might be 120°C (250°F), while that for lithium complex grease might be 180°C (355°F). Another thickener type that is becoming more popular is polyurea. Like lithium complex, polyurea has good high-temperature performance as well as high oxidation stability and bleed resistance. For temperatures above 400°C, only a small handful of lubricants are available, and the only lubricants with long life at these temperatures are liquid metals, liquid oxides, glasses, and a few solid lubricants like molybdenum disulfide [143].

From the other hand, there is another point of view of determination of high-temperature performance of the grease [144]. Dropping point was historically designed as manufacturing quality control test to confirm proper thickener formation, which indicates the temperature at which the grease thickener loses the capacity to retain oil under test conditions. In real life, it may not be related to high-temperature grease performance.

Factors limiting grease high-temperature performance include degradation resulting from thickener and base oil oxidation, and the loss of base oil due to grease bleed and evaporation. There are several bearing tests that allow evaluating grease high-temperature limits, and these limits may be lower than the limits based on dropping point (Figure 6.3). These tests are ASTM method D3336, known as the "Spindle Life" or "Pope" test, the SKF R0F+ Test, and the DIN 51821 (or FE9) test and they are conducted under accelerated operating conditions to promote grease aging processes.

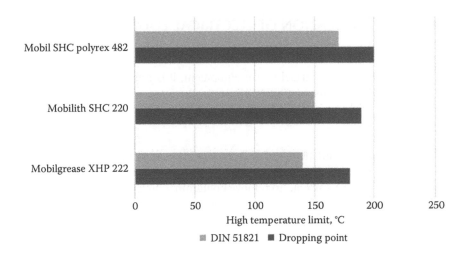

FIGURE 6.3 High temperature operating limits based on dropping point and on bearing tests, providing more realistic operational temperature guidelines. (Adapted from Grease high temperature performance, Mobil Technical Topic, 2009, Online. Available: http://www.chemcorp.co.uk/creo_files/upload/related-items/grease_high_temp_performance.pdf.)

6.10.2 Lubrication of High-Temperature Terminals in Automotive Industry

The number of electrical systems in a typical passenger vehicle continues to grow, powering everything from headlights and DVD players to body impact sensors and global positioning systems. And with each new system comes additional connectors. Luxury cars, for example, now have more than 400 connectors with 3000 individual terminals, which means there are 3000 potential trouble spots.

Connectors have to be protected against water, changing temperatures, road grit, and vibrations, all of which speed oxidation and fretting corrosion on connectors. Corrosion generates resistive oxides on the connectors, which causes intermittent faults and electrical failure. So the challenge for connector manufacturers is to extend the operating life of their products as more are used on each car and car companies continue to extend their warranties. It is not hard to see that lubricants have an important role to play in connector performance, including corrosion prevention and reducing costs. This is especially true for LV connectors (0.1–0.5 W), which now constitute about 75% of connectors in passenger cars.

The right lubricant lowers the insertion force needed to assemble connectors, reduces mechanical wear by placing a film of oil between mating surfaces, and, if the lubricant contains the proper additives, it minimizes corrosion. Such lubricants can be based on a variety of chemistries, but SHCs and ethers currently dominate the market.

PFPE-based greases, for example, are among the most frequently chosen synthetic lubes. They are usually selected for high-temperature applications (up to 250°C or 480°F) and thermo-oxidative stability (i.e., it does not oxidize at higher temperatures). To make a lubricant with low coefficient of friction, in the 1980s, lubricant engineers developed greases that combined the stability of PFPEs with PTFE with exceptionally low coefficient of friction [145].

However, in automotive application, the high normal force and heat, combined with the action of putting it together and taking it apart several times, burnishes PTFE into the surface of the contact, insulating the contact asperities that actually carry current. This leads to intermittent and catastrophic connector failures. Another potential problem of PFPE used as a base oil due to its high-temperature performance and resistance to oxidation is that inert chemistry and high specific gravity of PFPE make it difficult to thicken the grease. The solution was found in the grease where urea replaces PTFE to avoid burnishing that increases CR. Urea is a soluble, slightly basic nitrogen compound, $CO(NH_2)_2$. And when urea is a part of the mix, the grease separates at high temperatures due to density differences between PFPE and urea.

6.11 PRACTICAL LUBRICATION OF ELECTRICAL EQUIPMENT

6.11.1 Periodic Lubrication Maintenance of Electrical Power Equipment

During periodic lubrication of electrical power equipment, it is important to follow the directions usually given in manufacturers' user manuals. However, there are several rules that are the same no matter which type of equipment is maintained.

6.11.1.1 Cleaning

Before periodical lubrication, all traces of the old lubricant must be removed from surfaces by the use of commercial cleaners: kerosene, mineral spirits, and so on. To loosen the old lubricant, soak the disconnected parts in solvent. To remove contamination from the part, the solvent can be agitated or flushed through the part. If the part cannot be removed, adding oil could be helpful. It is important to use the same type of oil that was used as a base in grease to be removed. To speed the process of cleaning, soft bristled brushes may be used. After removal from the solvent, parts should be carefully dried and relubricated as soon as possible.

6.11.1.2 Penetrating Oil

Penetrating oil should not be used as a lubricant in electrical equipment. Penetrating oil is not designed for lubrication; it always contains solvent(s). Penetrating oil works only briefly, is contaminated easily, and may change into a viscous mess. In comparison with grease, penetrating oil has much lower viscosity (flows easily), very low boiling temperature, and high vapor pressure at ambient temperature. It will leak out under gravity or centrifugal action, leaving the lubricated parts dry. Penetrating oil will attack, dissolve, and wash out factory-installed lubricants and hasten failure. Most penetrating oils or aerosols are flammable and should not be applied in areas where sparks or arcing may occur. Penetrating oils are recommended only for rust removal and ease of part disconnection.

6.11.1.3   Lubrication in Field

For lubrication in field when disassembling is not supposed to be done, low- or medium-viscosity oil is recommended. In some cases, lubrication characteristics of petroleum oils may be improved by adding a stable dispersion of molybdenum disulfide in a premium mineral oil, which is available from some major manufacturers. These materials may extend lubricant and gears life, reduce metal-to-metal contact, lower friction, and so on, and may also be used as lubricant concentrates.

6.11.1.4   Troubleshooting Lubrication

A procedure for lubrication troubleshooting of electrical equipment such as CBs is usually recommended by the manufacturer. If the manufacturer's troubleshooting procedure is unavailable or ineffective, one can consider using diagnostic instruments, which can measure the trip time, force, resistance, vibration, and so on. Certain profiles will indicate lubrication problems. If this type of diagnostic is not available, but slow operation is encountered, it may be caused by inadequate lubrication.

When analyzing the lubrication problem, it is important to identify a symptom that is specific to the lubrication fault: excessive leakage of the lubricant, overheating, wear and scoring, and so on. The problem may be due to manifold reasons: it could be that incorrect lubrication product was used or too little of the lubricant was applied. The presence of water in the system could dramatically change the lubricant's properties.

To take corrective actions, it is important to examine what kind of product was used in the previous maintenance and what the properties of the used product are. The final step would be to define a proper product and relubricate the apparatus accordingly to the OEM recommendations.

6.11.2   General Lubrication Recommendations for Electrical Equipment

6.11.2.1   Choice of Lubricants

Depending on application, manufacturers may specify various greases, which have been thoroughly tested for a specific application. Usually, each manufacturer lists all greases which should be applied at specific points and the greases are different for the mechanical and electrical parts. These greases are applied during assembly of the original equipment and it is always recommended to use the same or a similar lubrication product for relubrication, which will guarantee that it will perform properly in service.

6.11.2.2   OEM Specifications

Lubrication should be performed according to the manufacturer if an alternative lubrication practice is not implemented. If either the lubrication product is obsolete or the electrical manufacturer is no longer available in the market, then OEM's lubrication specification cannot be followed. In such a case, the lubricant to use could be chosen strictly based on the application and physical properties of the recommended greases.

6.11.2.3   Change of Lubrication Product

Any change of lubrication product often involves testing and must be performed under the supervision of engineering staff. The engineering staff has to specify what type and brand of grease or oil should be applied during maintenance or overhaul of a particular piece of equipment. Any failure of equipment which might be caused by insufficient lubrication must be analyzed. Any experience has to be used to improve the lubrication technique and lubricant choice.

Lubricants availability: Any lubricant specified or chosen for lubrication must be readily available locally and nationwide. The user should keep a sufficient supply of greases in properly organized lubrication storage together with the products' material safety data sheet (MSDS) and technical data sheets.

6.11.2.4   Lubrication of Electrical Contacts

Electrical contacts should not be lubricated with metal-filled lubricants unless tested and proved to be effective in the long term. Many can accelerate corrosion, create conductive paths, and eventually cause failure. The general rule is to avoid graphite, molybdenum disulfide, or PTFE (Teflon) lubricants for electrical contacts because they could cause a rise in resistance after multiple operations [146,147]. For most switches and the ones that operate infrequently, to keep the contact just clean and dry with no lubricant might be a viable option. Main and arcing contacts should never be lubricated. Proper lubrication practices for various types of LV and MV power distribution equipment are presented in multiple papers [6,7,148–154] and OEM's Data Bulletins [155,156] available for customers.

6.11.2.5   Application of Lubricants

The rules that are listed below are general and can be applied to all maintenance lubrication procedures.

1. Use the factory-recommended lubricants whenever it is possible.
2. Do not use petroleum lubricants in areas with extreme cold temperatures.
3. Petroleum oil greases get thick and hard and do not function properly below −20°C (0°F).
4. Do not apply metal filled lubricants to electrical contacts. They can accelerate corrosion and create conductive paths and eventually failure.
5. Do not apply silicone lubricants to silicone rubber (SIR), fluorosilicone lubricants to fluorosilicone rubber, and petroleum lubricants to rubber. They are incompatible.
6. Do not use penetrating oils for lubrication of CBs; they work only briefly and tend to become thick and sticky with age.
7. Do not use flammable penetrating oils or aerosols in areas where arcing or sparks may occur.
8. Do not over-lubricate.
9. Avoid mixing lubricants.

6.12   LUBRICATION FAILURE MODES

6.12.1   Causes of Lubrication Failure

The term "lubrication failure" includes different issues, such as inadequate amount of lubricant applied or no lubricant at all, improper product chosen for lubrication, contamination and aging of the lubricant because of ambient conditions, and so on.

It is found that as many as 60%–80% of all bearing failures are lubrication related, whether it be poor lubricant selection, poor application, lubricant contamination, or lubricant degradation. Many components are failing early because lubrication best practices have not been established [157].

When failure happens, a thorough examination of the lubricant's properties is required in many cases, as well as analysis of the amount of lubricant applied, where it was applied, and the operating conditions of the mechanical or electrical parts. All these factors contribute to the damage caused by "lubrication failure," which led to excessive wear and corrosion of parts and the failure of the electromechanical unit to perform properly [146,158].

The lubricant plays a very important role. Lubricants separate the working surfaces preventing metal-to-metal contact, and reduce friction and wear. Lubricants protect against corrosion, exclude contaminants, and dissipate heat. Reducing friction and corrosion through proper lubrication is equally important for electromechanical equipment with moving parts and for nonrotating apparatus.

A power switching device is an example of such apparatus. Its operation depends upon friction-free movement of many different mechanical components, usually through only a short distance or slight angle, often exposed to moisture and environmental contamination, as well as to long periods of idleness during which the lubricant deteriorates and materials corrode. CBs, which must operate quickly and reliably during unpredictable emergencies, are especially vulnerable [159].

In this chapter, different lubrication failure modes will be summarized. Lubrication failure may affect the functionality and performance of various parts of electromechanical apparatus. The role of the proper choice of the lubrication products, terms, and procedures is discussed in relation to operational and environmental parameters. Lubrication failure mode describes the specific causes of failure associated with a component or functionality of a process. The failure may be caused by lubrication procedures, terms, or products. For example, the mode for CB (component) failure to open (functionality of a process) is lack of lubrication.

A number of different factors may affect lubricant performance. It is important to choose the correct product for each particular application, which means that many factors should be considered, such as operational and thermal temperature ranges, mechanical load (pressure) on the areas where the lubricant is applied, presence of moisture, possible contamination (particulate matter, dust, corrosive gases), and so on. The lubricant properties (working temperature range, moisture content, and presence of corrosive chemical components) should comply with the application parameters and materials to avoid adversely affecting the units and environment.

Knowledge of the composition of the lubricant is particularly important for its application to electrical contacts, since some additives may affect electrical resistance of conductors. It is also important to apply the right amount of lubricant because the lack of it or an excessive amount may increase the friction and wear of the parts or lead to contamination of other parts of the unit with lubricants.

6.12.2 Wrong Lubricant for Application

The lubricant choice for a specific application is determined by matching the machinery design and operating conditions with the desired lubricant characteristics [160]. Grease, and not oil, is generally used for the lubrication of CBs, equipment that do not run continuously and whose parts can stay inactive for long periods of time. One reason for the preference of grease is because CBs are not easily accessible for frequent lubrication, and high-quality greases can lubricate the components for extended periods of time without replenishing. Grease, and not oil, should be chosen for lubrication of most parts of CBs, since they can be exposed to high temperatures, shock loads, or high speeds under heavy load. In the application to CBs, choosing the right lubricant is a very important step.

6.12.3 Thermal Limitations

Thermal limitations, both high and low, represent major factors in the durability of many lubricants (see Section 8.7.5). At low temperatures, many lubricants appear to solidify, developing high-shear-strength films, and leading to high CR. Some lubricants are susceptible to cracking due to long-term exposure to high temperature.

Grease has a maximum temperature at which it can safely be used. This critical temperature is called the *drop point* at which the gel structure breaks down and the whole grease becomes liquid. When grease is heated above its drop point and then allowed to cool, it usually fails to fully regain its grease-like consistency and its subsequent performance will be unsatisfactory. Accordingly, if the temperature of CB-lubricated components in operation is higher than the upper limit of the lubricant working temperature range, it could leak out at high temperatures, which will affect the mechanical performance of the parts or expose them to corrosive environment.

Elevated operating temperature shortens the life of greases. In fact, if a switchgear operates at temperatures above 70°C (158°F), it cuts grease life by a factor of 1.5 for each 10°C rise [161]. High temperature promotes oxidation and increases oil evaporation rates and oil loss by creep, which accelerate grease drying and shorten grease life. Grease life can be estimated for operating temperatures above 70°C (158°F) with moderate loads and no contamination. This approach is applicable to fresh industrial greases of Grade 2 consistency with thickeners such as lithium, complex metal soaps, and polyureas.

Other ways to estimate lubricant lifecycles are based on operating conditions and require the application of factors to account for real-life conditions, such as solid contamination, moisture, air, catalytic effect of wear debris, temperature variations in a circulating system, and so on. As the temperature increases, the rate of grease degradation increases. The Arrhenius rule suggests doubling the lubricating oil degradation rate for each 10°C increase in temperature. Synthetic lubricants generally last longer at elevated temperatures than their mineral oil counterparts.

The base oil of the grease determines the minimum working temperature for the grease. The base oil of the grease for low-temperature service must be made from oils having a low viscosity at that temperature. For example, if the operational temperature of the unit in operations is lower than the lower limit of the lubricant working temperature range, then the lubricant may stiffen at cold temperatures, which could happen in the CBs' outdoor installations [21].

6.12.4 LUBRICANT COMPOSITION AND WRONG AMOUNT OF LUBRICANT

If the lubricant containing nonconductive solid additives is applied to electrical contact surfaces, it will increase electrical resistance, leading to overheating failures. To avoid such problems, it is important to use both the type and quantity of lubricant specified by the manufacturer, not only at the time of installation but also during maintenance procedures.

A common misconception among maintenance personnel is that it is better to over-lubricate than to under-lubricate bearings and matching parts. Both methods are undesirable. Under-lubrication risks metal-to-metal contact. Over-lubrication pushing excessive grease into the cavity causes heat buildup and friction as the moving elements continuously try to push extra grease out of the way. To ensure that moving parts are not over- or under-lubricated, follow the manufacturer's instructions. This recommendation applies both to grease and oil lubrication. In electrical applications, over-greasing may lead to nonconductive parts' contamination with grease oil, which will adversely affect dielectric properties of insulating materials.

6.12.5 CONTAMINANTS OR CORROSIVES IN THE LUBRICANT

Although contaminants are sometimes difficult to detect, they often cause lubrication failure in CBs. Dirt, sand, and water are the most common contaminants, but acid and other corrosives also can deteriorate the lubricant (grease and oil). They can dilute the oil film, reducing viscosity, and they can corrode metal surfaces, disrupting the lubrication film and causing erosion, creating thousands of abrasive particles. Moisture contamination increases chemical wear by rusting iron and steel surface; it also increases the corrosive strength of resident acids attacking copper and lead surface [162].

Solid particulate contaminants introduce most damage from the moment they come into contact with the surfaces of rotating parts (bearings). Depending on particle size, they may be seen or felt as grittiness in the lubricant. Particle contamination primarily affects the lubricant's ability to control friction and wear. Obstructing the separation of moving components and interfering with chemical oil films provided by antiwear and EP additives, particle contamination can substantially increase the rate of abrasion, adhesion, and surface fatigue. The effects of particle contamination are slow and imperceptible.

In fact, the loss of the lubricant's functional ability to reduce wear and friction is often overlooked due to the slowness with which particle contamination affects the system [162]. The best safeguard against contaminants is a clean, dry CB operating environment. If operating circumstances do not permit this environment, select sealed enclosures or shields to keep contaminants out. If humidity is a problem, consider selecting a lubricant with a good rust inhibitor. Harsh environments can sometimes be overcome by increasing the frequency of lubrication maintenance.

6.12.6 Environmental Factors Causing Grease Deterioration

Grease is exposed to many environmental factors that cause contamination and oxidation of the grease. These factors are atmospheric oxygen, dust and dirt, high and cold temperatures, and light. Exposure of lubricant to atmospheric oxygen forms gums, resins, and acidic products with viscosity increase. When the grease oxidizes, it usually darkens; there is a buildup of acidic oxidation products. These products can have a destructive effect on the thickener, causing softening, oil bleeding, and leakage. Because grease does not conduct heat easily, severe oxidation can begin at a hot point and spread slowly through the grease. This produces carbonization and progressive hardening or crust formation [159]. To protect the grease from contact with oxygen, it is recommended that lubricants containing antioxidant additive be used.

Keep the grease in sealed containers [162]. Dirt and dust in the grease increase the rate of wear between bearing surfaces, which also necessitates keeping the lubricants covered or in containers.

Exposure of the grease to high temperatures increases the rate of deterioration, forms gums, resins, and acidic products, and increases oil separation from greases, while the exposure of water-containing materials (e.g., cutting oils and certain fire resisting hydraulic oils) to cold causes water to separate out. Therefore, it is recommended to keep the storage temperature not higher than 20°C (68°F) and above the freezing point. Light promotes the formation of gums, resins, and acids in the lubricants; to protect the materials from exposure to light, they should be stored in metal or opaque containers.

With increasing time in service and temperature fluctuations, lubricants may degrade by oxidation or polymerization, forming insulating films. Oxidation occurs in all oils and greases that are in contact with air, including stored lubricants. The base oil and additive combination affects the rate of oxidation, and the presence of a thickener in grease can increase the degradation rate.

If the lubricant was badly degraded or reacted chemically to environmental contaminants, deposits in the form of gum, resin, varnish, sludge, or other deposits could be found in places where the lubricant was applied. Elemental analysis of these residues may help to identify the sources of deterioration and to determine if they were degraded grease additives or thickeners, or environmental contaminants.

Environmental and storage conditions have the strongest influence on the rate at which the lubricant degrades [159]. The storage environment strongly affects the shelf life of lubricants and greases [158]. The characteristics of some greases may change in storage. Grease may bleed, change consistency, or pick up contaminants during storage. Storage conditions that may accelerate lubricant deterioration are given in Section 8.3.

6.12.7 Lubricants Incompatibility

Incompatibility occurs when a mixture of two greases shows physical or performance characteristics significantly inferior to those of either grease before mixing.

An incompatibility issue could be reflected in reduced lubricating performance due to the modified composition of the fluids and additives from intermixing. Usually, problems are not obvious until the lubricated apparatus is in use. However, if two incompatible greases are mixed, lubrication failure is inevitable. Changes in physical properties would be reflected by softening or hardening of grease consistency, or decreased shear stability upon mechanical working and even reduction in thermal stability identified by a reduction in dropping point. Soft or runny grease is an indication of the problem. A grease that is much thicker than its original consistency may indicate incompatible mixing [14].

Grease color should also be observed. For example, if the original lubricant was green but has turned brown by the time of inspection, incompatible greases may have been mixed. Incompatible greases are a factor in many failures. Polyurea and lithium-based greases, for example, break down rapidly when mixed. It is best to know in advance which types of greases can be used together and which should not. Some greases cannot be mixed with others even when both types meet specifications. Unless this incompatibility is understood and accounted for, a switch to different grease can have serious consequences. The causes of the incompatibility of the lubricants are given in Section 8.3.

6.13   LUBRICATION FAILURES OF ELECTRICAL EQUIPMENT: CASE STUDIES

6.13.1   CB Failures Caused by Lubrication at U.S. Commercial Nuclear Power Plants

The data shown in this section are derived from the report [163] prepared for the Division of Risk Analysis and Applications of Office of Nuclear Regulatory Research, U.S. NRC. This report documents a study performed on the set of common cause failures (CCFs) of CBs from 1980 to 2000. The data studied here were derived from the NRC CCF database, which is based on U.S. commercial nuclear power plant event data (Tables 6.16 through 6.18).

Only the failures caused by improper lubrication are presented in the tables. It was found that improper lubrication determined on about 50 units caused 30% (24 out of 80) of all LV and MV CB failure events at nuclear power plants in 1980–2000. The events were classified as either fail-to-open or fail-to-close. The failure mode for the majority of CB events is fail-to-close (55%). The fail-to-open failure mode accounted for the remaining 45% of the events.

As followed from data presented in this report, either lack of lubrication of different mechanical parts or a hardened lubricant was the major cause of the failure. It was concluded that improper

TABLE 6.17
Failure Events with MV CBs

Year	# of Units	Description of Event	Piece Part	Cause
1986	4	4160 Vacuum CBs would nut close	Spring chaining motor	Lack of lubrication, dirty breakers, loose connections
1987	2	4160 Vacuum CBs would not charge	Same	Lack of lubrication, dirty contacts, and dirty closing mechanism
1991	N/A	One 4160 vacuum CB failed to open and several more degraded	Latch assembly	Hardened crease and lack of lubrication, maintenance incomplete
1995	1	4 kV supply CB closed during testing but failed to instantly recharge	Mechanical assembly	Aging of the latch monitor pivot bearing lubrication
1996	1	A failure of a roll pin securing a spring for a latch pawl on a 4 kV CB	Latch assembly	Lubricant was inadvertently removed from breaker parts
1997	5	Five safety-related CBs failed to open on demand during separate evolution	Mechanical assembly	The main and auxiliary contacts were not lubricated as recommended by OEM operating mechanism latch

Source: Modified from *Common-Cause Failure Event Insights Circuit Breakers*, United Stated Nuclear Regulatory Commission, NUREG/CR-6819, Vol. 4 "Circuit Breakers", INEEL/EXT-99–00613, May 2003, Online. Available: http://www.nrc.gov/reading-rm/doc-collections/nuregs/contract/cr6819/.

Abbreviations: CBs, circuit breakers; MV, medium voltage; OEM, original equipment manufacturer.

TABLE 6.18

Failure Events with 480 VAC CBs

Year	# of Units	Description of Event	Piece Part	Cause
1984	5	One CB failed to trip when the undervoltage device was de-energized and two CBs failed to trip within specified time limit	Mechanical assembly	Dirt and lack of lubrication
1985	4	Feeder breakers failed to close on demand	Mechanical assembly	Dirty and dry lubricant on the trip latch adjustment parts
1986	2	Feeder breakers were binding	Mechanical assembly	Dirty hardened grease
1987	2	Breakers found not operational, CBs would not trip	Mechanical assembly	Lack of lubrication of the internal moving parts
1988	1	The normal feeder breaker from a transformer would not close when transferring from alternate to normal power	Relay	Lack of lubrication
1989	1	The alternate breaker failed to close when the normal breakers were tripped	Mechanical assembly	Lack of proper lubrication of the trip rod bearings
1991	2	Breakers failed to close due to mechanical binding	Mechanical assembly	Dried out, hardened lubricant
1992	2	The normal supply breaker failed to close on demand. The alternate feed breaker failed to close during a hot transfer	Mechanical assembly	Lack or hardening of lubrication
1997	3	Breakers failed to close on demand	Mechanical assembly	Hardened grease on the stop and main drive link rollers, the rollers were not cleaned and lubricated during PM
1997	N/A	Several breakers failed to trip	Mechanical assembly	Lack of lubrication, lubricant was removed during refurbishment and not reinstalled
1997	N/A	Several breakers on all three phases did not trip in the required time	Mechanical assembly	Aging and degraded lubricants resulting from ineffective maintenance program

Source: Modified from *Common-Cause Failure Event Insights Circuit Breakers*, United Stated Nuclear Regulatory Commission, NUREG/CR-6819, Vol. 4 "Circuit Breakers", INEEL/EXT-99–00613, May 2003, Online. Available: http://www.nrc.gov/reading-rm/doc-collections/nuregs/contract/cr6819/.

Abbreviations: CBs, circuit breakers; PM, periodic maintenance.

lubrication was a result of an ineffective maintenance program, which was either incomplete or not performed according to OEM's recommended schedule. Review of the data given in the Nuclear Regulatory Commission (NRS) report confirms the importance of properly provided lubrication, which in most cases is given in detail in OEM Instruction Manual for each particular power device. It usually includes a list of the points/parts to lubricate, the lubrication products to use, and the recommended time interval between maintenance lubrications. Very often an instruction manual suggests shortening the lubrication interval if the unit might be exposed to a corrosive environment.

6.13.2 OVERHEATING OF THE MV SWITCH

Electrical equipment overheating is often caused by the increased CR. The cause of high resistance may be related to improper maintenance, including insufficient lubrication. An example of ineffective maintenance is based on the case of overheating failure of the MV switch.

FIGURE 6.4 The blades of the switch (This picture and associated copyrights is proprietary to, and used with the permission of, Schneider Electric.)

After the failure, the switch was disassembled and thoroughly inspected. The contact parts (blades and jaws) have been inspected to determine the cause of overheating [164]. During the examination, it was found that the contact parts on different phases look different. The surfaces of two blades are discolored across the whole area of the blades, which is a signature of continuous overheating (Figure 6.4). The surface of the blades on the third phase still had a metallic luster of the original tin plating and did not show signs of overheating.

The material of the blades is tin-plated copper. It is well known that tin, when it is heated in the presence of air, oxidizes with the formation of highly resistant tin oxide (SnO_2), which has off-white color. It is possible that the color of presumably overheated blades (frosty and dull) is the color of oxidized tin plating.

An examination of the jaws showed that the surfaces also look different: two phases have dark oxidized surfaces and the third phase has a metallic luster (Figure 6.5). The areas where the blades come in contact with the jaws on two phases with signs of overheating also have signs of arcing, but no arcing fingerprints are found on the third phase.

An additional difference between phases was the brown deposit found on the blade and the jaw surfaces of one phase, which did not show signs of overheating. No deposit was found on or around the contact area of the blades and the contact area on jaws on the other two phases where arcing occurred. Finding out why the parts look so different may help to identify the cause of overheating.

The brown deposit is most likely composed of the residues of the dried out grease applied to the surfaces of the blades around the contact area. The appearance and locations of the deposit allow inferring that the grease was heated above the lubricant maximum working temperature, causing the grease to melt and leak down toward pivoted points at the bottom of the blade.

To determine the cause of the overheating, the plant maintenance records have been reviewed. It was found that the switch was installed 13 years before the overheating happened. OEM Instruction Manual for this type of switch defines the maintenance terms, procedure, and lubrication products.

However, the annual maintenance log showed that the maintenance tasks did not include relubrication of the disconnect assembly. Therefore, during the 13-year period of the switch being in service, the switch contacts have not been properly lubricated and a new lubricant has never been added to the switch. It is well known that no lubricant can survive 13 years in service. After 13 years in service, most of the lubricant presumably applied on the contacts during the assembly of the switch disappeared with some dry grease still visible on one of the phases.

FIGURE 6.5 The jaws of the switch (This picture and associated copyrights is proprietary to, and used with the permission of, Schneider Electric.)

With the lack of lubrication, three different deterioration processes could have caused the rise in electrical resistance on the switch contacts. First, on the blades on two phases with the least amount of lubricant, tin plating was exposed to atmospheric oxygen and thus not protected from *oxidation*. Due to heavy oxidation, a thick layer of highly resistant gray tin oxide covered the contact areas, which resulted in high resistance on the switch electrical contacts.

Second, the role of the lubricants is very important in protection of the contact from fretting, another damaging type of corrosion to which tin plating is specifically subjected [165]. Therefore, a lack of lubricants on tin-plated surfaces could lead to surface damage by *fretting corrosion*, resulting in an additional resistance growth (see Section 6.5 of Chapter 6).

There is another possible cause of the growing resistance. While oxidation and fretting corrosion led to growing CR, they in turn caused more heat generation on the energized contacts and a temperature increase on the contacts during the service. An extended exposure of the contacts made of tin-plated copper to elevated temperatures usually induces the formation of *IMC* on the interface between plating (tin) and base metal (copper) [166]. Intermetallic compounds (IMCs) have a very high electrical resistance. This factor added to further growth of contacts resistance, which in turn further increased the amount of heat generated.

In the scenario of the switch overheating suggested in Ref. [164], a major cause of the overheating was an improper maintenance procedure and lack of lubrication, which resulted in severe atmospheric and fretting corrosion of the contacts, and the formation of copper–tin IMC. All these deteriorating factors together led to the growth of contact electrical resistance. Abnormally resistant contact areas produced a continuous flow of extreme heat. Continuous overheating led to decreased jaw spring strength and thus to a decreased contact pressure, which all together resulted in arcing.

6.13.3 Lubricant Contamination in Electrical Connector

Lubricant may cause formation of corrosive residues. It may happen if environment contains various chemicals, such as cleaners/disinfectants, soaps, detergents, which is typical for hospital environment. Some of these chemicals are corrosive to medical devices that function in the hospital environments. Corrosion of electrical devices include discoloration or appearance of a new material (like a solid or liquid), or the production of a gas (that may be sensed through smell). An example of

FIGURE 6.6 Corroded electrical connector. (Fom Jaiyeola G. Failure analysis of a corrosive residue on a medical device, *Case Study*, Online. Available: http://www.element.com/information/resources/articles-index/case-study-failure-analysis-of-a-corrosive-residue-on-a-medical-device.)

such failure is presented in Ref. [167]. The electrical connector (composed of copper and steel materials) of a hospital bed device corroded as a result of contamination from a lubricant and a cleaner/disinfectant that the hospital was using on and near the bed device. The corrosion of the connector resulted in the performance failure of the bed device. The corrosion of the connector showed up as a heavy residue (Figure 6.6).

Detail analysis of the residue on corroded connector showed that the main cause of the corrosion of the electrical connectors in the hospital bed medical device was due to the contamination by the lubricant that the hospital was applying to the device to reduce friction. It was also determined that the cleaner/disinfectant used at the hospital added to the corrosive environment that surrounded the connectors and that they were exposed to. The cleaner/disinfectant contained ammonium compounds, which are corrosive to copper, a key component of the electrical connectors. The use of silicone grease as a lubricant and nonammonium containing cleaners/disinfectants was recommended, both of which are compatible with steel and copper.

6.13.4 Causes of Lubricant Failure in Bearings

Multiple types of bearing failures are caused by lubricants. According to a major ball bearing company, 54% of bearing failures are lubrication related. In a study by MIT, it was estimated that approximately $240 billion is lost annually (across U.S. industries) due to downtime and repairs to manufacturing equipment damaged by poor lubrication. Improper bearing lubrication or relubrication accounts for up to 40%–50% of machine failures [168] which altogether lead to 40%–50% of such failures. There are numerous causes for lubricant failure [169], including *insufficient lubricant quantity or viscosity* and deterioration due to *prolonged service without replenishment*. The lubricants fail when exposed to excessive temperatures or being contaminated with foreign matter.

Lubrication with grease when conditions dictate the use of static or circulating oil also may lead to bearing failure. Oil is the preferred lubricant when speed or operating conditions preclude the use of grease or where heat must be transferred from the bearing. Often oil is used to meet the operating requirements of other components such as seals and gears.

If *incorrect grease base* for a particular application is chosen, the bearing may fail. Grease selection should vary with the application. Factors to consider include hardness (consistency), stability (ability to retain consistency), and water resistance (emulsification). However, grease is oil suspended in a base or carrier, and when these bases are exposed to moisture or heat, they can turn into soap or carbon ash. Therefore, it may be necessary to use synthetic additives to prevent deterioration of the base.

Adequate bearing failure requires a correct choice of lubricant viscosity, which is just as important as oil quantity. Required viscosity depends on operating temperature. Inadequate lubricant viscosity appears as a highly glazed or glossy surface. As damage progresses, the surface appears frosty and eventually spalls. This type of spalling is comparatively fine grained compared to the more coarsely grained pattern produced by fatigue failure [169].

Bearing may fail if it is *over-lubricated* [170]. Over-lubrication may cause a rapid rise in temperature, particularly at high speeds because the rolling elements have to push the grease out of the way. This leads to churning in the grease, which produces heat. This churning action will eventually bleed the base oil from the grease and all that will be left to lubricate the bearing is a thickener system with little or no lubricating properties. The heat generated from the churning and insufficient lubricating oil will begin to harden the grease.

This will prevent any new grease added to the bearing from reaching the rolling elements. The end result is bearing failure and equipment downtime. Adding more grease only worsens the problem, adding the risk of blowing out a seal. Ironically, an attempt to sufficiently lubricate a bearing by giving it several extra pumps from a grease gun will eventually result in its failure due to under-lubrication [170].

6.13.4.1 Hardened and Oxidized Lubricant in Bearings

In ball bearings, the continuous presence of a very thin—millionths of an inch—film of lubricant is required between balls and races, and between the cage, bearing rings, and balls. Failures are typically caused by restricted lubricant flow or excessive temperatures that degrade the lubricant's properties. Discolored (blue/brown) ball tracks and balls are symptoms of lubricant failure [171]. Excessive wear of balls, rings, and cages will follow, resulting in overheating and subsequent catastrophic failure. In rolling and ball bearings, improper lubrication may result in bearing fracture and seizure, formation of cracks, and wear on outer/inner rings and cages; if grease is too high viscosity, it may cause scratches on raceway surface [172].

Lubricant failures can be detected by the presence of discolored (blue/brown) raceways and rolling elements. Excessive wear on rolling elements, rings, and cages follows, resulting in overheating and subsequent catastrophic failure. In addition, if a bearing has insufficient lubrication, or if the lubricant has lost its lubricating properties, an oil film with sufficient load-carrying capacity cannot form. The result is metal-to-metal contact between rolling elements and raceways, leading to adhesive wear. If the grease is stiff or caked and changed in color, it indicates lubrication failure. The original color will usually turn to a dark shade or jet black. The grease will have an odor of burnt petroleum oil. Lubricity will be lost as a result of lack of oil. In cases of lithium base greases, the residue appears like a glossy, brittle varnish which will shatter when probed with a sharp instrument. Examples of grease failures in the bearings are shown in Figures 6.7 and 6.8.

One can observe strong oxidation and hardening of the grease that occur following high-temperature stress, which is produced through electrical grounding (arcing). Loss of lubricant health produces mixed friction and wear in the roller contact area. The fact that a bearing cannot be easily relubricated from the outside plays a crucial role in eventual element failure. If the hardened and oxidized lubricant already presents in the bearing, the newly added grease cannot displace it. It makes an exchange of grease impossible even with normal relubrication intervals; bearing failure is inevitable [173].

6.13.4.2 Lubricant Water Contamination in Bearings

Moisture is known to enter lubricated bearing systems in several different ways resulting in dissolved, suspended, or free water. Both dissolved and suspended water can promote rapid oxidation of the lubricant's additives and base stock resulting in diminished lubricant performance.

FIGURE 6.7 Bearing lubrication failures: Aged grease between cage and inner ring (a); Hardened and oxidized lubricant in roller bearing (b). (From Weigand M. Lubrication of rolling bearings—Technical solutions for critical running conditions, *Machinery Lubrication*, no.1, 2006, Online. Available: http://www.machinerylubrication.com/Read/844/lubrication-rolling-bearings. With permission.)

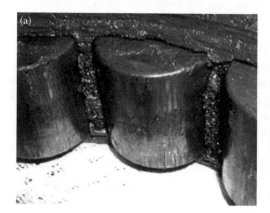 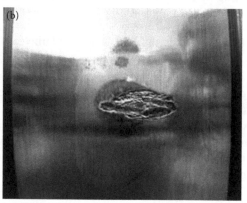

FIGURE 6.8 Bearing lubrication failures: Mixed lubrication in a *cylindrical roller bearing* (a); Roller damage from bad lubrication (b). (From Weigand M. Lubrication of rolling bearings—technical solutions for critical running conditions, *Machinery Lubrication*, No. 1, 2006, Online. Available: http://www.machinerylubrication.com/Read/844/lubrication-rolling-bearings. With permission.)

Rolling element bearings may experience reduced fatigue life due to hydrogen embrittlement caused by water penetrated bearing surfaces. Many other moisture-induced wear and corrosion processes are common in both rolling element and journal bearings. As little as 1% water in grease can have a significant impact on bearing life. A tapered roller bearing cone and rollers and a ball bearing outer race and balls show rusting with pitting and corrosion from moisture/water exposure. This condition is referred to as *etching* [174]. Etching is most often caused by condensate collected in the bearing housing from temperature changes. Water etching is a common type of corrosion occurring on bearing surfaces and their raceways. This aqueous corrosion is caused primarily by the generation of hydrogen sulfide and sulfuric acid from water-induced lubricant degradation. This occurs as a result of the liberation of free sulfur during hydrolysis reactions between the lubricant and suspended water.

Moisture or water can get in through damaged, worn, or inadequate seals. Improperly washing and drying bearings when they are removed for inspection also can cause considerable damage. Moreover, moisture can contaminate the lubricant and condense on the gear surfaces forming sludge, corrosion, and micropitting.

The best defense against moisture contamination is a three-step, proactive maintenance strategy called Target–Exclusion–Detection (TED). Only when lower moisture levels are consistently stabilized can the life extension of lubricants and bearings be effectively achieved.

6.14 RECOMMENDATIONS FROM MANUFACTURERS OF THE PRODUCTS FOR ELECTRICAL INDUSTRY

Lubricant manufacturers often publish the results of detail and thorough studies, focused on the lubrication properties of the products they manufacture. It is very helpful to learn what they recommend for specific applications. In this book, their studies and comments on specifics of electrical apparatus lubrications in different circumstances and environments are summarized.

6.14.1 Dow Corning® Corporation

6.14.1.1 Lubrication of Outdoor Electrical Equipment

Lubricants to use and their application is a very important part or component of modern mechanical and electrical equipment. Wrong lubrication technique, interval, or choice of lubricant can result in equipment failure. In many applications in CBs or any other electromechanical device requiring lubrication, conventional or older technology lubricants continue to be used.

However, highly engineered lubricants would deliver much better performance. Lack of understanding of the features and benefits of engineered lubricants can lead to decreased reliability of electrical apparatus. Modern engineered lubricants have been demonstrated to provide long service life and represent one possible opportunity to increase reliable equipment operation particularly HV switchgear and CBs. Modern lubricants are able to outperform conventional lubricants and excel at temperature extremes, resist drying out, and help prevent wear and corrosion that may lead to equipment seizure.

Proper lubricant selection for any given application may be aided by paying attention to application details and consideration of goals and objectives. Factors influencing lubricant selection include component type (gear, bearing, etc.), kinds of surfaces in contact (metal-to-metal, metal-to-plastic, etc.), their relative motion, and temperature. Goals or objectives to be met may include extended lubrication intervals, increased reliability, elimination of corrosion, eliminating slow trip time, and many others. Once application details are understood and goals are prioritized, knowledge of modern lubricant technology can be put to best use to achieve one's objectives. Key application details can be summarized by the acronym "L.E.T.S.," or *Load, Environment, Temperature, Speed* [175].

After summarizing application L.E.T.S., end users and designers can begin to make decisions on how to properly lubricate any particular application. Applications that move frequently and operate under light-to-moderate loads outside of temperature extremes can be successfully lubricated with liquid-based lubricants. As speed or frequency of motion decline and loads or vibration increase and temperatures enter extremes, one may wish to explore lubricants where lubricating solids are present. Examples may include grease fortified with molybdenum disulfide or graphite or other lubricant forms such as lubricating paste.

Reliability of outdoor electrical equipment may in some cases be increased through careful lubricant selection. Two common causes of poor CB and disconnect reliability are seizure and degraded lubricant. Two common causes of seizure are rust and galvanic corrosion. Metal surfaces exposed to oxygen, humidity, fly ash, or other metal surfaces could undergo corrosion leading to seizure. Metal-to-metal contact in equipment exposed to the elements and static conditions for long periods of time may be best served by applying lubricating paste or dry film lubricants rather than grease. The reason is due to the ability of solids to provide a more robust, continuous protective film that lasts longer than films provided by liquid-based lubricants. Applications such as yoke adjustments, ball and socket connections, pins, bushings, some gears, and cams may be able to operate more freely over time if lubricated with solid-based lubricants [175].

Lubricant degradation can occur in a number of ways. Common to outdoor electrical equipment is grease degradation via evaporation and oil separation.

Evaporation is a cooling process that can be accelerated by increasing temperature, surface area, or introducing an air current to the environment. Molecules of less mass tend to evaporate faster than those of greater mass.

Loss of base oil to evaporation leads to increasingly stiff grease that may in time impact trip times. Mineral oil-based grease tends to be more prone to evaporation than synthetic greases. Fluorosilicone-based grease offers excellent resistance to evaporation.

Static conditions may lead to *oil separation* and loss due to the effect of gravity. Thickeners have different densities than base oils. Gravity acts to separate them from each other. Thickeners are designed to release oil over time and they do not chemically bind with oil molecules. Most thickeners hold oil through capillary action or physical entanglement. The pull of gravity may overcome those things over time. Oxidation can occur but this form of degradation is more likely to occur in industrial applications where operating temperatures exceed 70°C (160°F) on a continuous basis. Solvent washout can occur when penetrating oils are used to treat equipment. Most lubricants are hydrocarbon based and solvents in penetrating oils are hydrocarbon based. Chemically speaking, similar things tend to dissolve each other quite well. Fluorosilicone-based grease is resistant to hydrocarbon-based solvents and resists washout by penetrating oils.

Equipment manufacturers and end users can affect equipment reliability by the choice of lubricant and selected or suggested lubrication interval. Given operating conditions and resources available to maintain breakers and disconnects, lubricant choice should prefer products that stand the test of time by resisting evaporation and offer resistance to corrosion. Increased reliability may come from use of lubricants that resist degradation over time and that are properly applied to components they are supposed to protect from wear and corrosion.

6.14.1.2   Molykote® and Dow Corning Brand Lubricants for Power Equipment

In electric power generation and distribution, HV CBs and disconnect switches (DS) are expected to remain in service for up to 20 years without major service. The temperature inside a CB may range from −40 to 60°C (−40 to 140°F) over the course of a year with wide daily fluctuations based on night and day during the same season. When compounded with environmental extremes such as severe cold, extreme heat, wind, and water and airborne contamination, conventional mineral oil-based lubricants are often unable to measure up.

In this industry, synthetic lubricants such as silicone- and PAO-based lubricants provide longer service life due to their inherent ability to resist changes in viscosity and drying out. For example, in needle bearings used in trip latch mechanisms, fluorosilicone grease has been demonstrated to provide many years of successful service without drying out.

Mineral oil-based lubricants are more prone to drying out when exposed to high temperatures due to their increased volatility compared to the more thermally stable synthetics. Mineral oils also tend to have low VIs. They tend to thicken quickly as temperature decreases and thin quickly when temperature increases. In the case of a trip coil bearing, use of the wrong lubricant in colder conditions may mean the difference between an acceptable trip versus a delayed or failed trip. Over time, mineral oil-based grease may also dry out enough to prevent a successful trip.

The engineered properties of synthetics stand up in a wide range of applications to eliminate seasonal lubricant change-outs. Synthetics are widely used by OEMs as life lubricants. Most HV equipment today is constructed without grease ports, making it very difficult to relubricate unless it is brought into the shop and disassembled. Because of this, maintenance personnel should use synthetic lubricants to take advantage of longer service life and better performance at temperature extremes [176,177].

According to Ref. [177], Molykote and Dow Corning brand lubricants can provide proven long-term, reliable lubrication over a wide service temperature range and in a variety of environments.

Fluorosilicones in particular resist solvents such as penetrating oil, and they do not dry out and harden, even after many years of exposure to the elements and static conditions. Formulated to address specific requirements, Molykote and Dow Corning lubricants can help reduce seizure and increase CB performance and reliability. They may allow extended maintenance intervals and reduced lifecycle cost. The lubricants may also generate increased customer satisfaction by reducing equipment failures.

An example of such products is Molykote 3451 Chemical Resistant Bearing Grease, which provides exceptional lubrication on needle bearings in CBs. This is fluorosilicone grease providing superior resistance to most chemicals and can be used at high temperatures and with heavy loads. This grease combines fluorosilicone oil with PTFE thickener, its service temperature range is −40 to 232°C (−40 to 450°F). It may be used in rollers, needles and sleeve bearings, silicone O-rings and gaskets, and so on. This grease reduces wear and corrosion and resists drying out; it is solvent resistant and chemically inert.

Another Molykote product is 33 Extreme Low Temperature Bearing Grease, which delivers exceptional performance in cold environments. It lubricates trip latch and coil bearings, antifriction bearings, and plastic and rubber parts under light-to-moderate loads. Molykote 33 grease is resistant to oxidation and moisture, and it is compatible with many plastics and elastomers. It is based on phenylmethyl silicone oil and lithium soap thickener, its service temperature range is −73 to 180°C (−100 to 356°F), it may be used on gaskets in the presence of SF_6, on flange surfaces, it resists drying out, it is solvent resistant and chemically inert, and may enable longer service life.

Original electrical manufacturer may apply different greases, including Dow Corning grease; however, a thorough testing of the equipment is always required for each specific lubrication product application to be accepted as OEM recommendation and included into OEM equipment manuals.

Some of the most useful lubricants for electrical applications manufactured and tested by Dow Corning are presented in the table in Ref. [177]. This table lists composition, service temperature range, potential applications, and benefits and features of Molykote brand lubricants.

6.14.2 NYE Lubricants Products for Electrical Industry

NYE designed and manufactured many lubricants for separable electrical connectors. Lubricant reduces friction and eases mating. A thin film of lubricant can reduce mating force by as much as 80%, an important factor in connector assembly. For electronic connectors with dozens or even hundreds of pins, or for automotive connectors that are in hard-to-reach places, a low insertion force makes assembly more efficient and ensures solid connections. For gold-plated connectors, an effective lubricant reduces the potential for noble metal wear during mating and separation.

Lubricant applied to connector guards against corrosion and the effects of harsh environments. With gold-plated connectors, this means protection against substrate corrosion. Thin gold plating can be microscopically porous, and a thin film of lubricant can seal the pores, prevent substrate attack, and assure low CR. Lubricant also prevents fretting corrosion to which tin/lead connectors are also subjected to "fretting corrosion." Fretting corrosion is the result of low amplitude vibration caused by thermal expansion and contraction or nearby motion, as from fans, motors, or merely opening and closing a cabinet door. Fretting corrosion continually exposes fresh layers of metal surface to oxidation. A lubricant film minimizes metal-to-metal contact during vibration, protecting the connector from metal wear.

NYE's connector lubricants can be divided into two general classes: lubricants for noble metal connectors and tin/lead connectors. For noble metals, fluoroethers are the lubricants of choice. They withstand extreme temperatures and resist aggressive chemicals and solvents. Five- and six-ring PPEs are also extremely stable in thin film, and offer an excellent track record on gold-plated connectors. For tin/lead connectors, SHCs provide excellent film strength, broad temperature serviceability, and protection against fretting corrosion. Popular NYE lubricants are used for stationary separable electrical connectors [178]. Some of them are presented in Table 6.19. All

TABLE 6.19

NYE Lubricants for Stationary Separable Electrical Connectors

Lubricants	Temp Range, °C (°F)	Tin/Lead	Silver	Noble Metal	High Current	Plastic Compatible
Fluorocarbon gel 813-1	−70 to 200 (−94 to 392)	■		■	■	■
NyeTact® 502J-20-UV	−40 to 120 (−40 to 248)	■				■
NyeTact® 523V-2-UV	+20 to 250 (68 to 482)			■		
NyeTact® 561J-20-UV	−40 to 175 (−40 to 347)	■		■	■	■
NyeTact® 561J-35-AG	−40 to 175 (−40 to 347)	■	■	■	■	■
NyeTact® 570H-2	−20 to 225 (−4 to 437)			■		■
NyeTact® 570H-25-UV	−20 to 225 (−4 to 437)			■		■
NyoGel® 760G	−40 to 135 (−40 to 275)	■		■	■	■
NyoSil M25	−70 to 200 (−94 to 392)	■		■	■	■
Rheotemp® 761AG	−40 to 175 (−40 to 347)	■	■	■	■	■
Rheotemp® 761G	−40 to 175 (−40 to 347)	■		■	■	■
UniFlor™ 8511	−50 to 225 (−58 to 437)	■		■	■	■
UniFlor™ 8917	−70 to 225 (−94 to 437)	■		■	■	■

Source: Adapted from Lubricants for Stationary Separable Electrical Connectors, NYE LubeNotes, Online. Available: http://www.nyelubricants.com/stuff/contentmgr/files/0/24d7a2dc2442a71906571ee688ad6bf3/en/lubenote_stationary_separable_electrical_connectors.pdf.

lubricants from this table are applicable to low-current electrical connectors. Some of the lubricants contain so-called UV tracer, which is a fluorescent dye added to the lubricant. The system is then inspected with a high-intensity UV or blue light lamp. Since the dye collects at all leak sites, the lamp will show the precise location of every leak with a bright glow. The dye can remain in the system indefinitely and does not affect system components or performance.

Recently NYE Lubricants bought the company Anderol®, well known for manufacturing many lubricants accepted by electrical equipment manufacturers as the products for use in CBs and switches. Since 2013, NYE Lubricants is the official manufacturer of six of the Anderol lubricants which have had their place in electrical industry for many years. NYE is offering the Anderol 732, Anderol 752 Grade 1, Anderol 752 Grade 2, Anderol 757, Anderol 786, and Anderol 793 products. NYE is manufacturing these products according to the original process and control specifications as provided by Anderol [179]. The new Anderol greases manufactured by NYE Lubricants are presented in Table 6.20.

6.14.3 CONTRALUBE: UK MANUFACTURER OF A LUBRICANT FOR HV CONTACTS

It is a new name among lubricants' manufacturers and is a UK branch. The Contralube brand consists of state-of-the art, synthetic gels that are primarily used for the protection of electrical, electronic, data connections, and contact areas/surfaces. Composition of the lubricants is not disclosed. One of these products (Contralube 880) is a green gel that is primarily used as a contact surface lubricant and protector on HV/arcing contacts, CBs, and disconnects [180].

TABLE 6.20

NYE's Family of Anderol® Replacement Lubricants

Composition, Properties	AND-732	AND-752-1	AND-757	AND-786	AND-793
Base oil	Diester	PAO	Diester	Diester	Diester
Thickener	Li soap	Li complex	Li soap	Clay	Li soap
Color	Amber	Tan	Amber	Black	Light brown
Texture	Tacky	Smooth	Tacky	Tacky	Smooth
NLGI grade	0	1	1.5	1.5	2
Working temperature range, °C (°F)	−40 to 150 (−40 to 302)	−60 to 150 (−76 to 302)	−40 to 150 (−40 to 302)	−20 to 150 (−4 to 302)	−60 to 150 (−76 to 302)
Dropping point, °C (°F)	181 (358)	296 (565)	192 (378)	>300 (>572)	192 (378)
Penetration worked (unworked), mm/10	359 (397)	315 (307)	295 (289)	303 (301)	289 (261)
Evaporation,%	0.66	0.20	1.10	1.60	0.60
Oil separation, %	15	3.7	1.4	0.6	4.8
Performance	Friction and wear reduction, rust protection	Antiwear, EP	Antiwear, adhesion to metal	Corrosion/rust inhibited, load carrying at high speed	High-temperature stability
Applications	Chain, gears, slides, bearings	Gears, slides, bearings	General-purpose electrical contacts and switches	Gears, slides, bearings, pulleys, and general maintenance	Fine precision instruments

Source: Adapted from Nye's Family of Anderol® Replacement Lubricants, NYE LubeNotes, February 2013, Online. Available: http://www.nyelubricants.com/stuff/contentmgr/files/0/2ab16cba30c7287dd39ff65a9fbf9a04/en/lubenote_anderol.pdf.

Abbreviations: EP, extreme pressure; NLGI, National Lubricating Grease Institute; PAO, polyalphaolefin.

Technically, HV contact grease faces a number of challenges. They must not only cope with the extreme temperatures that occur under arcing conditions but also minimize friction, withstand broad ambient temperatures without oxidizing, evaporating, or causing resistive failures. A high-quality grease will also inhibit wear and corrosion plus provide environmental sealing and control free motion.

Typical applications for Contralube 880 gel include power distribution/switchboards, CBs and disconnects, HV electrical substation contacts, power infrastructure, and distribution equipment. Contralube gel 880 is used to prevent contact wear, corrosion, and the buildup of carbon deposits from arcing that can form a resistive layer which leads to overheating and a possible fire hazard. When arcing occurs across electrical contacts, the temperatures reached are sufficient to degrade any organic molecule and so a lubricant's ability to burn cleanly is a definite advantage. Contralube 880 is designed with a specialist formulation that burns away cleanly under arcing conditions, leaving no carbon deposits on the contact surface. It is a synthetic product, therefore cannot evaporate, it is water and UV resistant, and it does not contain silicone.

6.14.4 KLUBER LUBRICATION ON ELECTRICAL SWITCHES AND CONTACTS

Kluber lubricant are designed and tested for application in electrical switches and contacts [132,181]. Important issue to resolve in applying a lubricant to electrical switches and contacts is ensuring its compatibility with the contact materials and any surrounding materials. Electrical contacts are normally held by plastic parts. Occasionally, rubber-elastic materials—also referred to as elastomers—can also be found in the vicinity of the contacts. The selection criteria for a lubricant depend on the parameters that are most important to be optimized. Some of these parameters are reduction of plug and unplug forces and resistance to fretting corrosion.

For lubricant selection, it is also of major importance to consider what kind of metal surfaces are moved against one another with what contact force. The adhesion of the lubricant depends not only on its chemical composition and consistency but also on the contact material, the surface roughness, and the orientation of the roughness. As lubricants applied to live electrical contacts must withstand very high temperatures over a very long period of time, the base oils should show a very low tendency to evaporate and a high resistance to oxidation. If temperatures become excessively high for a short time, the lubricant should evaporate or burn away without residues such that no foreign matter (coking) remains, which would interfere with the function of the contact later. When faced with requirements of this kind, PFPEs perform much better than hydrocarbon-based lubricants.

6.14.5 SILVER-BASED GREASES FOR ELECTRICAL CONTACTS

6.14.5.1 Mineral Oil-Based Conductive Grease from Cool-Amp

U.S. Company Cool-Amp has been family owned and operated by three generations since 1944. The company manufactures Cool-Amp, a silver plating powder, which is applied by hand and recommended for use on copper, brass, or bronze stationary parts. Cool-Amp can be used on busbars, cable terminals, current transformer (CT) terminals, clamps and fittings, ham radios, printed circuit boards (PCBs), welders, or where conductivity is needed on bolted copper, brass, or bronze parts. The powder reduces power loss and overheating, maximizes conductivity, and assures long life and low maintenance; it prevents oxidation and corrosion [182]. It is an alternative to silver plating/dipping of damaged electrical parts. The silver dipping process requires parts to be shipped for offsite dipping and shipped back before they can be used.

The company produces Conducto-Lube, a silver conductive lubricant for high-amperage connections. There is no voltage limits in application. The grease is not soluble in water; however, it will gradually wear off if exposed to water for long periods of time. It does not change conductivity, but will conduct electricity, and resistance does not change appreciably with temperature. Working temperature range is from −31 to 210°C (−25 to 410°F). Conducto-Lube contains 65%–75% of silver and 25%–35% of mineral oil. The silver used has particles that are 3–4 microns in size.

Conducto-Lube provides electrical and thermal conductivity, lubrication, and protection. The lubricant eliminates hot spots, increases conductivity, and prevents galling or pitting that will freeze joints. Conducto-Lube is successful in reducing resistance in hinge joint and knife blade switches due to arcing because the solids in the lubricant eventually imbed themselves into the minute imperfections of the contact surface resulting in smooth surface.

6.14.5.2 Silicone-Based, Silver-Filled Conductive Grease from Chemtronics

Another silver-based grease utilizes an advanced silicone lubricant that is compatible with metal, rubber, and plastic; the grease is manufactured by ITW Chemtronics® [183]. It features high electrical conductivity, excellent thermal conductivity, and provides protection against wear, moisture, and corrosion. Because it is based on silicone oil, it remains stable in a wide temperature range from −57°C to 252°C (−70 to 485°F). The grease consistency is smooth paste; it has a very low viscosity versus temperature change. This silver conductive grease may be used for high- and low-power applications, including lubrication of substation switches or CBs and low- or medium-speed sliding contacts. Other usages could be for static grounding on seals or O-rings [183].

6.14.5.3 Sanchem Grease for Corrosion Protection of Electrical Contacts: NO-OX-ID

Electrically conductive grease NO-OX-ID has been used in the power industry for more than 65 years to prevent corrosion in electrical connectors from low-micropower electronics to HV switchgear. NO-OX-ID "A-SPECIAL" is the electrical contact grease of choice for new electrical installations and maintenance. This grease keeps metals free from rust and corrosion; prevents the formation of oxides, sulfides, and other corrosion deposits on copper, aluminum, and steel surfaces and conductors; and lubricates the connection for easier maintenance. NO-OX-ID "A-Special" electrical grease prevents corrosion attack on all metal surfaces. Attack can come from battery acid, salt, moisture, and various industrial chemical vapors in the environment. When this conductive paste is used on aluminum connectors in joints, NO-OX-ID "A-Special" prevents the reformation of oxide films, which cause high resistance and subsequent failures.

NO-OX-ID "A-Special" conductive grease is recommended by connector manufacturers for trouble-free joint connections. When nuts, mounting bolts, and cotter keys are coated with NO-OX-ID "A-Special," they will never rust or freeze assuring easy, trouble-free removal. NO-OX-ID "A-Special" should be used wherever the formation of a corrosive product will affect the proper functioning of the metal surface. This electrical contact grease is easily applied, easily removed, and gives long-lasting reliable performance even on dissimilar metals. Some of the applications of this protective grease are bolted connections, brackets, busbar systems, cable and clamp connectors, contact points in CB and switches [184].

6.14.5.4 Contact Lubricants from Electrolube (United Kingdom)

Electrolube, a division of the international HK Wentworth Ltd., have been the leading supplier of contact lubricants since their invention by the founder in the 1950s. The lubricants increase the reliability and lifetime of all current carrying metal interfaces, including switches, connectors, and busbars [185]. Some lubricants and their properties are shown in Table 6.21.

Electrolube has earned an unsurpassed reputation for the manufacture and supply of special lubricants to the automotive, military, aerospace, industrial, and domestic switch manufacturing sectors. The range has been developed over the years to accommodate many advances in such rapidly advancing industries, combining excellent electrical properties and lubricity, to improve movement

TABLE 6.21
Contact Lubricants from Electrolube (UK)

Property	SGA	SGB	CG53A	CG60	EGF
Base oil	Complex ester	Poly alkylene glycol	Poly alkylene glycol	PAO/complex ester	PFPE
Thickener	Clay	Clay	Li complex soap	Li complex soap	Fumed silica
Temperature range, °C (°F)	−40 to +125 (−40 to 257)	−35 to +130 (−31 to 266)	−35 to +130 (−31 to 266)	−45 to +130 (−49 to 266)	−25 to +300 (−13 to 572)
Dropping point, °C (°F)	267 (512)	250 (482)	200 (392)	200 (392)	>250 (482)
NLGI	1	1	1	1	2
Applications	Electrical fixed or moving contacts of contactors, busbars, knife switches, switchgear in heavy arcing conditions, and corrosive environments	All types of electrical contacts and with most types of thermoplastics	Switches and connectors, high-voltage contacts in corrosive environments	Column switches, rocker switches, and push-push switches in automotive and high-quality domestic switch industries	Printed circuit edge connectors, plug connectors, rotary and sliding switches, gold contacts in aggressive environments

Source: Adapted from Benefits of contact lubricants, Electrolube, Online. Available: http://www.contact-lubricants.com/the-benefits-of-contact-lubricants.html.

Abbreviations: NLGI, National Lubricating Grease Institute; PAO, polyalphaolefin.

and "feel' characteristics, with plastics compatibility. Electrolube products are electrically insulating in thick films, preventing tracking. In ultrathin films, that is, between closed metal contacts, they allow the current flow; they also exhibit a neutral pH, thereby avoiding surface corrosion.

6.14.5.5 Electrical Joint Protection from Galvanic and Atmospheric Corrosion from Tyco Electronics (TE)

Electric joints may be damaged by corrosion either atmospheric or galvanic (see Chapter 7 on Corrosion). For atmospheric action, there must be moisture and oxygen. Galvanic action results in corrosion when two dissimilar metals in the electrolytic series, for example, aluminum and copper, are in physical contact. In this case, moisture acts as an electrolyte. In such an instance, the copper becomes a cathode and receives a positive charge; the aluminum becomes the anode and receives a negative charge. The resultant current flow attacks the aluminum leaving the copper unharmed. Both factors are influenced by environmental conditions—the chemical attack of airborne pollutants. Corrosion damage of electrical joints is the strongest in heavy industry locations such as steelworks, chemical plants, refineries, and so on, and also may occur in urban and rural areas.

Galvanic problems are not present for aluminum-to-aluminum connectors; however, the oxide film forms rapidly on the surface of freshly cleaned aluminum exposed to air. This oxide film is an insulator and must be removed with a scratch brush in order to achieve a satisfactory and reliable electrical joint. The problem with aluminum is that the freshly cleaned surface will quickly reoxidize; hence, it is important to coat the surface with a *contact sealant*.

For Al-to-Cu connections, it is a good practice to use contact sealant on the aluminum connector body and brushed into the strands of the aluminum conductor. Wherever possible, the aluminum conductor should be installed above the copper to prevent pitting from the galvanic action of copper salts washing over the aluminum connector and conductor when in a lower position.

Various sealant formulations have been developed to provide improved electrical and mechanical performance as well as environmental protection to the contact area [186]. The use of sealants is recommended for aluminum-to-aluminum or aluminum-to-copper connections. Sealants are also recommended for copper-to-copper joints which are subject to severe corrosive environments. Nongritted sealants are recommended for flat connections and as a groove sealant in bolted connectors such as parallel groove clamps. Gritted sealant is primarily used in compression connectors. The sharp metallic grit particles provide multicontact current carrying bridges through remaining oxide films to ensure superior electrical conductivity. The joint sealants are mineral oil-based corrosion inhibitor with added chemicals to dissolve aluminum oxide.

The sealants with added *fluoride* (dropping point 65.6°C) are recommended for palm-to-palm joints (Al-to-Al and Al-to-Cu) and for use on aluminum surface-to-surface bolted joints, such as busbar joints and terminal lugs. The sealant assists in breaking up the oxide film by chemically etching the connecting surfaces to ensure a low-resistance joint.

Sealants with added *lithium* are recommended for bolted connections (Al-to-Al, Al-to-Cu, Cu-to-Cu) and on palm-to-palm joints (Cu-to-Cu). This sealant has higher drop point (180°C) than the one with fluoride. Such sealant seals the exposed surface to prevent reoxidation and permanently excludes the future ingress of air and moisture. It is extremely adhesive, resistant to water, and has high-temperature resistance to ensure continuous operation under all situations.

Joint compounds with added *zinc particles* have highest drop point, 188°C, provide excellent outdoor protection, and are recommended for compressing joints and bolted connections (Al-to-Al and Al-to-Cu). Conductive zinc granules are suspended in a viscous petroleum oil base. Under pressure, these granules make high-pressure contact points with the parent metal to provide a sound electrical connection, while the base material seals the joint to prevent further corrosion. Zinc-containing paste used many years in electrical industry is the product called Penetrox, an oxide-inhibiting compound producing low initial CR, sealing out air and moisture, and thus preventing oxidation or corrosion. This paste is usable over wide temperature ranges and provides a high conductivity, so-called gas-tight-joint (GTJ). All Penetrox compounds contain homogeneously suspended particles. The particles assist in the penetration of thin oxide films, act as electrical "bridges" between conductor strands, aid in gripping conductor, improve electrical conductivity, and enhance integrity of the connection. The specially formulated Penetrox compounds are for use with compression and bolted connectors providing an improved service life for both Cu and Al connections. Additionally, the nontoxic compounds are an excellent lubricant for threaded applications reducing galling and seizing.

6.15 INFORMATION SOURCES FOR LUBRICANTS

There are many valuable information resources on the subject of lubrication [40].

1. *Operations and maintenance manuals*: The primary information sources are the manufacturer's installation, operation, and maintenance manuals. The information contained in these manuals applies specifically to the equipment requiring servicing. The operation and maintenance (O&M) manual will usually outline the required characteristics of the lubricants as well as a recommended schedule for replacement or filtering. If the maintenance manual is not available, or is vague in its recommendations, lubricant manufacturers and distributors are other sources of information.

2. *Lubricant manufacturers* are probably the most valuable sources of information and should be consulted for specific application situations, surveys, or questions [187]. The details of the companies are easily available via the Internet and phone services. When choosing a lubricant for a particular piece of equipment, all the pertinent information on the equipment, such as operating speed, frequency of operation, operating temperature, and any other special or unusual conditions should be provided to the lubricant manufacturer or distributor so that a lubricant with the proper characteristics can be chosen. Some discretion should be used when dealing with a lubricant salesperson in order to prevent purchasing an expensive lubricant with capabilities in excess of what is required.

3. *Industry standards* organizations such as ANSI, ASTM, American Gear Manufacturers Association (AGMA), and Institute of Electrical and Electronics Engineers (IEEE) publish standard specifications for lubricants and lubricating standards for various types of equipment.

4. *Engineering and trade publications and journals* such as *Lubrication, Lubrication Engineering, Machinery Lubrication,* and *Wear* specialize in the area of lubrication and tribology. Articles featured in these publications are generally technical in nature and describe the results of current research. Occasionally, research results are translated into practical information that can be readily applied. General trade publications and magazines such as *Power, Power Engineering, Hydraulics and Pneumatics, Machine Design, Pump and Systems,* and *Plant Engineering Magazine* frequently contain practical articles pertaining to lubrication of bearings, gears, and other plant equipment.

5. *Lubrication charts*: *Plant Engineering Magazine* is publishing "Exclusive Guide to Synthetic Lubricants" and "Interchangeable Lubricants Guide" [15,188]. These charts cross-reference lubricants by the application and the company producing the product. Lubrication product names are provided by the manufacturers, and the publishing of the data does not reflect the quality of the lubricant, imply the performance expected under particular operating conditions, or serve as an endorsement.

   As an example of the information contained in the interchangeable lubricant chart, the 2004 chart identifies available products from 118 lubricant companies in nine categories. Fluid products in each category are listed within viscosity ranges. Greases comprise NLGI 2 only. Included is a chart titled "Viscosity/Grade Comparison Chart" that tabulates viscosity equivalents for ISO viscosity grade, kinematic viscosity (CSt), Saybolt viscosity (Saybolt universal seconds (SUS)), gear lubricant (AGMA) specifications, EP gear lubricant, and worm gear lubricant (Comp). *Plant Engineering Magazine* notes that the synthetic lubricant products presented in each category are not necessarily interchangeable or compatible. Interchangeability and compatibility depend on a variety of interrelated factors, and each application requires an individual analysis.

6. *The Internet* offers access to a large amount of information, including lubrication theory, product data, and application information. The Internet also provides a means of communicating and sharing information with personnel at other facilities. Problems, causes, and solutions are frequently described in great detail, although caution should be used when evaluating information obtained through the Internet. The amount of information located depends on the user's skills at applying the most pertinent keywords on any of the search engines. Hyperlinks are usually available and lead to other information sources. Users should note that broad search categories, such as "lubrication," will provide the greatest returns but will undoubtedly include much extraneous data. Generally, inserting too many words in the search field narrows the scope of the search and may produce little or no useful information. The search field must be adjusted until the desired information is obtained or the search is abandoned for another reference source.

Libraries: In a manner similar to the Internet searches, librarians can also help locate information within their collections or outside their collections by conducting book and literature searches. Unlike the Internet, literature searches rely on large databases that require password entry not available to the general public. Therefore, these searches are usually conducted by a reference librarian. The search process is a very simple method used for locating books on a specific subject or specific articles that have been included in technical publications. Again, the amount of information located depends on using the proper search keywords. Searches can be expanded or contracted until the desired information is obtained.

6.16 LUBRICATION GLOSSARY

A–B

Additive: A compound that enhances some property of, or imparts some new property to, the base fluid. In some hydraulic fluid formulations, the additive volume may constitute as much as 20% of the final composition. The more important types of additives include antioxidants, antiwear additives, corrosion inhibitors, VI improvers, and foam suppressants.

Additive stability: The ability of additives in the fluid to resist changes in their performance during storage or use.

Adhesion: The property of a lubricant that causes it to cling or adhere to a solid surface.

ANSI: American National Standards Institute.

Antifoam agent: One of two types of additives used to reduce foaming in petroleum products: silicone oil to break up large surface bubbles and various kinds of polymers that decrease the number of small bubbles entrained in the oils.

Antioxidants: Prolong the induction period of base oil in the presence of oxidizing conditions and catalyst metals at elevated temperatures. The additive is consumed and degradation products increase not only with increasing and sustained temperature but also with increases in mechanical agitation or turbulence and contamination (water, metallic particles, and dust).

Antiwear additives: Improve the service life of tribological elements operating in the boundary lubrication regime. Antiwear compounds (e.g., ZDDP and 1,2,3-trichloropropane [TCP]) start decomposing at 90–100°C and even at a lower temperature if water (25–50 ppm) is present.

ASTM: American Society for Testing Materials.

Base stock: The base fluid, usually a refined petroleum fraction or a selected synthetic material, into which additives are blended to produce finished lubricants.

Boundary lubrication: Form of lubrication between two rubbing surfaces without development of a full-fluid lubricating film. Boundary lubrication can be made more effective by including additives in the lubricating oil that provide a stronger oil film, thus preventing excessive friction and possible scoring. Antiwear additives are commonly used in more severe boundary lubrication applications. The more severe cases of boundary lubrication are defined as EP conditions; they are met with lubricants containing EP additives that prevent sliding surfaces from fusing together at high local temperatures and pressures.

C

Centipoise (cp): A unit of absolute viscosity. 1 centipoise = 0.01 poise.

Centistoke (cst): A unit of kinematic viscosity. 1 centistoke = 0.01 stoke.

Centralized lubrication surfaces: A system of lubrication in which a metered amount of lubricant or lubricants for the bearing of a machine or group of machines is supplied from a central location.

Chemical stability: The tendency of a substance or mixture to resist chemical change.

Cloud point: The temperature at which waxy crystals in an oil or fuel form a cloudy appearance.

Compound: (1) Chemically speaking, a distinct substance formed by the combination of two or more elements in definite proportions by weight and possessing physical and chemical properties different from those of the combining elements. (2) Compounds are fatty oils and similar materials added to lubricants to impart special properties during petroleum oil manufacturing.

Compounded oil: Petroleum oil to which has been added other chemical substances.

Consistency: The degree to which a semisolid material such as grease resists deformation (See ASTM designation D217). Sometimes used qualitatively to denote viscosity of liquids.

Contaminant: Any foreign or unwanted substance that can have a negative effect on system operation, life, or reliability.

Corrosion: The decay and loss of a metal due to a chemical reaction and its environment. It is a transformation process in which the metal passes from its elemental form to a combined (or compound) form.

Corrosion inhibitor: Additive for protecting lubricated metal surfaces against chemical attack by water or other contaminants. There are several types of corrosion inhibitors. Polar compounds wet the metal surface preferentially, protecting it with a film of oil. Other compounds may absorb water by incorporating it in a water-in-oil emulsion so that only the oil touches the metal surface. Another type of corrosion inhibitor combines chemically with the metal to present a nonreactive surface.

D–E

Degradation: The progressive failure of a machine or lubricant.

Detergent: In lubrication, a detergent is either an additive or a compounded lubricant having the property of keeping insoluble matter in suspension, thus preventing its deposition where it would be harmful. A detergent may also redisperse deposits already formed.

Dielectric strength: A measure of the ability of an insulating material to withstand electric stress (voltage) without failure. Fluids with high dielectric strength (usually expressed in volts or kilovolts) are good electrical insulators (ASTM Designation D877).

Emulsibility: The ability of a nonwater-soluble fluid to form an emulsion with water.

Emulsifier: An additive that promotes the formation of a stable mixture, or emulsion, of oil and water. Common emulsifiers are metallic soaps, certain animal and vegetable oils, and various polar compounds.

Emulsion: Intimate mixture of oil and water, generally of a milky or cloudy appearance. Emulsions may be of two types: oil-in-water (where water is the continuous phase) and water-in-oil (where water is the discontinuous phase).

Environmental: All material and energy present in and around an operating contaminant system, such as dust, air moisture, chemicals, and thermal energy.

EP additive: Lubricant EP additive that prevents sliding metal surfaces from seizing under conditions of EP. At the high local temperatures associated with metal-to-metal contact, an EP additive combines chemically with the metal to form a surface film that prevents the welding of opposing asperities, and the consequent scoring that is destructive to sliding surfaces under high loads. Reactive compounds of sulfur, chlorine, or phosphorus are used to form these inorganic films.

EP lubricants: Extreme pressure (EP) lubricants that impart to rubbing surfaces the ability to carry appreciably greater loads than would be possible with ordinary lubricants without excessive wear or damage.

F–G

Film strength: Property of a lubricant that acts to prevent scuffing or scoring of metal parts.

Fire point: Temperature to which a combustible liquid must be heated so that the released vapor will burn continuously when ignited under specified conditions.

Fire-resistant fluid: Lubricant used especially in high-temperature or hazardous hydraulic applications. Three common types of fire-resistant fluids are (1) water–petroleum oil emulsions, in which the water prevents burning of the petroleum constituent; (2) water–glycol fluids; and (3) nonaqueous fluids of low volatility, such as phosphate esters, silicones, and halogenated hydrocarbon-type fluids.

Flash point: Temperature to which a combustible liquid must be heated to give off sufficient vapor to form a flammable mixture with air when a small flame is applied under specified conditions (ASTM Designation D92).

Force feed: A system of lubrication in which the lubricant is supplied to lubrication the bearing surface under pressure.

Friction: Resisting force encountered at the common boundary between two bodies when, under the action of an external force, one body moves or tends to move relative to the surface of the other.

Graphite: A crystalline form of carbon having a laminar structure, which is used as a lubricant. It may be of natural or synthetic origin.

Grease: A lubricant composed of oil or oils thickened with a soap or other thickeners to a semisolid or solid consistency.

H

Hydraulic fluid or oil: Fluid serving as the power transmission medium in a hydraulic system. The most commonly used fluids are petroleum oils, synthetic lubricants, oil–water emulsions, and water–glycol mixtures. The principal requirements of a premium hydraulic fluid are proper viscosity, high VI, antiwear protection (if needed), good oxidation stability, adequate pour point, good demulsibility, rust inhibition, resistance to foaming, and compatibility with seal materials. Antiwear oils are frequently used in compact, high-pressure, and capacity pumps that require extra lubrication protection.

Hydrocarbons: Compounds containing only carbon and hydrogen. Petroleum consists mostly of hydrocarbons.

Immiscible: Incapable of being mixed without separation of phases. Water and petroleum oil are immiscible under most conditions, although they can be made miscible with the addition of an emulsifier.

Inhibitor: Any substance that slows down or prevents such chemical reactions as corrosion or oxidation.

Insolubles: Particles of carbon or agglomerates of carbon and other materials. Indicates deposition or dispersant dropout in an engine. Not serious in a compressor or gearbox unless there has been a rapid increase in these particles.

ISO: International Standards Organization, sets viscosity reference scales.

ISO viscosity: A number indicating the nominal viscosity of an industrial fluid grade lubricant at 40°C/104°F as defined by ASTM Standard Viscosity System for Industrial Fluid Lubricants D2422 (identical to ISO Standard 3448).

L–M–N

Liquid: Any substance that flows readily or changes in response to the smallest influence. More generally, any substance in which the force required to produce a deformation depends on the rate of deformation rather than on the magnitude of the deformation.

Load carrying: Property of a lubricant to form a film on the lubricated surface, which resists rupture under given load capacity conditions. Expressed as the maximum load, the lubricated system can support without failure or excessive wear.

Lubricant: Any substance interposed between two surfaces in relative motion for the purpose of reducing the friction and/or the wear between them.

Lubricity: Ability of an oil or grease to lubricate; also called film strength.

Mineral oil: Oil derived from a mineral source, such as petroleum, as opposed to oils derived from plants and animals.

Miscible: Capable of being mixed in any concentration without separation of phases, for example, water and ethyl alcohol are miscible.

Moly: Molybdenum disulfide (MoS_2), a solid lubricant and friction reducer, colloidally dispersed in some oils and greases.

Multigrade oil: An oil meeting the requirements of more than one Society of Automotive Engineers (SAE) viscosity grade classification, and may therefore be suitable for use over a wider temperature range than a single-grade oil.

Naphthenic: A type of petroleum fluid derived from naphthenic crude oil, containing a high proportion of closed ring methylene groups.

O

Oil: A greasy, unctuous liquid of vegetable, animal, mineral, or synthetic origin.

Oil ring: A loose ring, the inner surface of which rides a shaft or journal and dips into a reservoir of lubricant from which it carries the lubricant to the top of a bearing by its rotation with the shaft.

Oiliness: That property of a lubricant that produces low friction under conditions of boundary lubrication. The lower the friction, the greater is the oiliness.

Oxidation: Occurs when oxygen attacks petroleum fluids. The process is accelerated by heat, light, metal catalysts, and the presence of water, acids, or solid contaminants. It leads to increased viscosity and deposit formation.

Oxidation inhibitor: Substance added in small quantities to a petroleum product to increase its oxidation resistance, thereby lengthening its service or storage life, also called antioxidant. An oxidation inhibitor may work in one of these ways: (1) by combining with and modifying peroxides (initial oxidation products) to render them harmless, (2) by decomposing the peroxides, or (3) by rendering an oxidation catalyst inert.

Oxidation stability: Ability of a lubricant to resist natural degradation upon contact with oxygen.

P–R

Paraffinic: A type of petroleum fluid derived from paraffinic crude oil and containing a high proportion of straight-chain, saturated hydrocarbons. Often susceptible to cold flow problems.

Poise: A measure of absolute viscosity numerically equal to the force required to move a plane surface of 1 cm²/s when the surfaces are separated by a layer of fluid 1 cm in thickness. It is the ratio of the shearing stress to the shear rate of a fluid and is expressed in dyne seconds per square centimeter (dyne s/cm²); 1 centipoise is equal to 0.01 poise.

Polymerization: The chemical combination of similar-type molecules to form larger molecules.

Pour point: The lowest temperature at which an oil or distillate fuel is observed to flow, when cooled under conditions prescribed by the test method ASTM D97. The pour point is 3°C (5°F) above the temperature at which the oil in a test vessel shows no movement when the container is held horizontally for 5 s.

Pumpability: The low temperature, low shear stress–shear rate viscosity characteristics of an oil that permit satisfactory flow to and from the engine oil pump and subsequent lubrication of moving components.

Refining: A process of reclaiming used lubricant oils and restoring them to a condition similar to that of virgin stocks by filtration, clay adsorption, or more elaborate methods.

Ring lubrication: A system of lubrication in which the lubricant is supplied to the bearing by an oil ring.

Rings: Circular metallic elements that ride in the grooves of a piston and provide compression sealing during combustion. Also used to spread oil for lubrication.

S

SAE: Society of Automotive Engineers, serving the automotive industry.

Saybolt universal viscosity (SUV): The time in seconds (SUV or SUS) required for 60 cm^3 of a fluid to flow through the orifice of the standard Saybolt universal viscometer at a given temperature under specified conditions (ASTM Designation D88).

Semisolid: Any substance having the attributes of both a solid and a liquid. Similar to semiliquid but more closely related to a solid than a liquid. More generally, any substance in which the force required to produce a deformation depends both on the magnitude and on the rate of the deformation.

Sludge: Insoluble material formed as a result either of deterioration reactions in oil or of contamination of oil, or both.

Solid: Any substance having a definite shape which it does not readily relinquish. More generally, any substance in which the force required to produce a deformation depends upon the magnitude of the deformation rather than upon the rate of deformation.

Specific gravity: The ratio of the weight of a given volume of material/liquid to the weight of an equal volume of water.

SUS: SUS, a unit of measure used to indicate viscosity at 100°F.

Stoke (St): Kinematic measurement of a fluid's resistance to flow defined by the ratio of the fluid's dynamic viscosity to its density.

Surfactant: May increase the oil's affinity for metals and other materials.

Synthetic hydrocarbon: Oil molecule with superior oxidation quality tailored primarily out of paraffinic materials.

Synthetic lubricant: A lubricant produced by chemical synthesis rather than by extraction or refinement of petroleum to produce a compound with planned and predictable properties.

T–V–W–Z

Thermal conductivity: Measure of the ability of a solid or liquid to transfer heat.

Thermal stability: Ability of a fuel or lubricant to resist oxidation under high-temperature operating conditions.

Thin film lubrication: A condition of lubrication in which the film thickness of the lubricant is such that the friction between the surfaces is determined by the properties of the surfaces as well as by the viscosity of the lubricant.

Tribology: The science and technology of interacting surfaces in relative motion, including the study of lubrication, friction, and wear. Tribological wear is wear that occurs as a result of relative motion at the surface.

Vapor pressure: Pressure of a confined vapor in equilibrium with its liquid at specified temperature, thus a measure of a liquid's volatility.

Varnish: When applied to lubrication, a thin, insoluble, nonwipeable film deposit occurring on interior parts, resulting from the oxidation and polymerization of fuels and lubricants. Can cause sticking and malfunction of close clearance moving parts. Similar to, but softer, than lacquer.

Viscometer or viscometer: An apparatus for determining the viscosity of a fluid.

Viscosity: Measurement of a fluid's resistance to flow. The common metric unit of absolute viscosity is the poise, which is defined as the force in dynes required to move a surface 1 cm^2 in area past a parallel surface at a speed of 1 cm/s, with the surfaces separated by a fluid film 1-cm thick. In addition to kinematic viscosity, there are other methods for determining viscosity, including the SUV, Saybolt Furol viscosity, Engler viscosity, and Redwood viscosity. Since viscosity varies inversely with temperature, its value is meaningless until the temperature at which it is determined is reported.

Viscosity grade: Any number of systems which characterize lubricants according to viscosity for particular applications, such as industrial oils, gear oils, automotive engine oils, automotive gear oils, and aircraft piston engine oils.

Viscosity index (VI): A commonly used measure of a fluid's change of viscosity with temperature. The higher the VI, the smaller is the relative change in viscosity with temperature.

Viscosity index improvers: Additives that increase the viscosity of the fluid throughout its useful temperature range. Such additives are polymers that possess thickening power as a result of their high molecular weight and are necessary for the formulation of multigrade engine oils.

Viscosity modifier: Lubricant additive, usually a high-molecular-weight polymer that reduces the tendency of an A148 viscosity to change with temperature.

Viscosity, absolute: The ratio of the shearing stress to the shear rate of a fluid. It is usually expressed in centipoise.

Viscosity, kinematic: The time required for a fixed amount of oil to flow through a capillary tube under the force of gravity. The unit of kinematic viscosity is the stoke or centistoke (1/100 of a stoke). Kinematic viscosity may be defined as the quotient of the absolute viscosity in centipoises divided by the specific gravity of a fluid, both at the same temperature.

Viscosity, SUS: SUS, which is the time in seconds for 60 mL of oil to flow through a standard orifice at a given temperature (ASTM Designation D88-56).

Viscous: Possessing viscosity. Frequently used to imply high viscosity.

Volatility: This property describes the degree and rate at which a liquid will vaporize under given conditions of temperature and pressure.

Wear: The attrition or rubbing away of the surface of a material as a result of mechanical action.

ZDDP: Zinc dialkyldithiophosphate, an antiwear additive found in many types of hydraulic and lubricating fluids.

REFERENCES

1. Runyon W. Pretty sticky, *Equipment Management*, 18(9), 30–35, September 1990.
2. Bailey C. A. and Aarons J. S. (Eds). *The Lubrication Engineers Manual*, United States Steel, Philadelphia, Pennsylvania, p. 460, 1971.
3. Lansdown A. R. In *Lubrication and Lubricant Selection: A Practical Guide*. UK Mechanical Engineering Publications, Pergamon Press, Oxford, UK, Section 1, p. 285, 1996.
4. Drozda T. J. and Wick C. Fundamentals of lubrication. *Tool and Manufacturing Engineers Handbook* (TMEH Series), Society of Manufacturing Engineers, Dearborn, MI, Vol. 1, Machining, Chapter 4, pp. 4-35–4-60, 1983.
5. Blau P. J. (Ed). Lubricants and lubrication. *ASM Handbook*, Vol. 18, ASM International, Materials Park, OH, pp. 79–171, 1992.
6. Richard D. M. Application of lubricants in operating equipment, *Minutes of the 61st Annual International Conference of Doble Clients*, Boston, MA, pp. 5-4.1–5-4.8, 1994.

7. Richard D. M. Lubrication products and practices in circuit-breaker maintenance, *Power Supply World*, 69–72, 1996.
8. Richard D. M. Getting a grip on specialty lubricants, *Machinery and Equipment MRO*, 26–28 December, 1997.
9. Lubrication, S. K. F. *Bearing Maintenance Handbook*, Chapter 7, SKF Group Publications, 38 pp., November 2010.
10. Stevens C. Lubricant selection vital to maintenance solutions, *Plant Engineering*, 49(10), 64–68, August 1995.
11. *Engineering and Design—Lubricants and Hydraulic Fluids*, U.S. Army Corps of Engineers (USACE), Manual No. 1110–2–1424, February 1999.
12. *Lubrication Guide, Timken*, 2003, Online. Available: http://www.timken.com/industries/torrington/catalog/pdf/general/form640.pdf.
13. Ball and roller bearings, In *Lubrication*, Publications of NTN Corporation, Japan, Chapter 11, pp. A72–A79, 2002.
14. *Lubrication of Power Plant Equipment*, Facilities Instructions, Standards, and Techniques (FIST), Vol. 2–4, United States Department of the Interior Bureau of Reclamation(USBR), Springfield, VA, 33pp., 1991, Online. Available: http://www.usbr.gov/power/data/fist/fist2_4/vol2–4.pdf.
15. Holzhauer R. Exclusive guide to synthetic lubricants, *Plant Engineering Magazine*, Online. Available: http://www.plantengineering.com/industry-news/top-stories/single-article/plant-engineering-magazine-s-exclusive-guide-to-synthetic-lubricants/1428b70360.html.
16. Brenner D. W., Irving D. L., Kingon A. I., and Krim J. Multiscale analysis of liquid lubrication trends from industrial machines to micro-electrical-mechanical systems, *Langmuir*, 23(18), pp. 9253–9257, 2007.
17. Grease— Its components and characteristics, Mobil, Technical Topics, 2000, Online. Available: http://www.mobilindustrial.com/IND/English/Files/tt-components-and-characteristics-of-grease.pdf.
18. SKF Grease Test Kit TKGT 1, SKF Group, Online. Available: http://www.skf.com/binary/12-35956/MP5366E.pdf.
19. Thibault R. Grease basics, *Maintenance Technology Magazine*, July 2009, Online. Available: http://www.mt-online.com/julyaugust2009/grease-basics.
20. Wright J. How cold temperatures affect your lubricants, *Machinery Lubrication*, February 2015, Online. Available: http://www.machinerylubrication.com/Read/30093/cold-temperatures-lubricants.
21. Khonsari M. and Booser E. R. Low temperature and viscosity limits, *Machinery Lubrication*, March 2007, Online. Available: http://www.machinerylubrication.com/Read/1014/low-temperature-viscosity-limits.
22. Akin K. D. Lubricating electrical switches, *Machine Design*, October 2001, Online. Available: http://machinedesign.com/archive/lubricating-electrical-switches.
23. Johnson M. Understanding grease construction and function. *Tribology and Lubrication Technology*, 32–38, June 2008.
24. Schaeffers Grease Seminar, Presentation, 83pp, Online. Available: http://schaefferoil.de/sheets/praesentationen/greasebergbau_en.pdf.
25. Plint G. *Guidance Notes on Lubricated Wear Testing*, pp. 1–13, 2005, Online. Available: http://www.phoenix-tribology.com/cat/at2/index/lubricated%20testing.pdf.
26. Rudnick L. R. *Lubricant Additives: Chemistry and Applications*, CRC Press, Boca Raton, FL, 790 pp., 2009.
27. Komatsuzaki S., Uematsu T., and Nakano F. Bearing damage by electrical wear and its effect on deterioration of lubricating grease, *Lubric. Eng.*, 43, 25–30, 1987.
28. Prashad H. Diagnosis of deterioration of lithium greases used in rolling-element bearings by X-ray diffractometry, *Trib. Trans.*, 32(2), 205–214, 1989.
29. Mas R. and Magnin A. Rheological and physical studies of lubrication greases before and after use in bearings, *ASME J. Tribol.*, 118, 681–686, 1996.
30. Prashad H. Diagnosis and cause analysis of rolling-element bearing failure in electrical power equipment due to current passage, *Proceedings of ASME/STLE Tribology Conference*, Toronto, Ontario, Canada, October 26–28, 1998.
31. Abdel-Jaber G. T., Mohamed M. K., Al-Osaimy A. S., and Ali W. Y. Effect of magnetic field on the performance of lubricant additives, *J. Appl. Ind. Sci.*, 1(1), 25–31, April 2013.
32. Grease static oil bleed, *Mobil Technical Topics*, Online. Available: http://www.mobil.com/IND/English/files/tt-grease-static-oil-bleed.pdf.

33. Ludwig L. G. Storing grease to avoid bleed and separation, *Machinery Lubrication*, February 2012, Online. Available: http://www.machinerylubrication.com/Read/28761/storing-grease-to-avoid-bleed-separation-.

34. Livingstone G., Wooton D., and Brian T. Finding the root causes of oil degradation, *Proceedings of Lubrication Excellence Conference*, 2006, Online. Available: http://www.machinerylubrication.com/Read/989/fluid-degradation-causes.

35. *NLGI Lubricating Grease Guide*, The National Lubricating Grease Institute (NLGI), Kansas City, MO, 148 pp., 2006.

36. *Grease Compatibility and Grease Lubrication Practices*, Chevron Technical Bulletin, February 2012, 2 pp., Online. Available: http://www.michiganpetroleum.com/images/application_category_documents/13.pdf.

37. *Incompatibility of Greases,* Online. Available: http://www.ca.nsk.com/page.asp?ID=444.

38. Kurosky J. Maximizing grease performance through optimal compatibility—An overview of compatibility testing, *Machinery Lubrication*, July 2003. Online. Available: http://www.machinerylubrication.com/Read/514/grease-performance-compatibility.

39. Gebarin S. Recommendations for mixing greases, *Machinery Lubrication*, 23–24, May 2006. Online. Available: http://www.machinerylubrication.com/Read/882/mixing-greases.

40. *Engineering and Design—Lubricants and Hydraulic Fluids*, U.S. Army Corps of Engineers (USACE), Manual No. 1110–2–1424, February 1999.

41. *Grease Compatibilty Chart,* Online. Available: http://www.mpc-home.com/technicaldata/GreaseCompatibility.pdf.

42. *Compatibility of Lubricating Oils*, Marine Lubricants Information, Online. Available: http://www.fammllc.com/famm/publications/lubes_bulletins_07.pdf.

43. *Grease Compatibility Information*, Online. Available: http://www.petroliance.com/sites/default/files/PDF/Grease/Grease%20Compatibility%20Information.pdf.

44. Shah R. Additives keep the wheels of industry turning, *Plant Services*, May 2006, Online. Available: http://www.plantservices.com/articles/2006/138/.

45. Fitch J. Recognizing the symptoms of lubricant incompatibility, *Practicing Oil Analysis*, September 2006, Online. Available: http://www.machinerylubrication.com/Read/911/lubricant-incompatibility.

46. Turner D. The skinny on grease compatibility, *Machinery Lubrication*, January, 2009, Online. Available: http://www.machinerylubrication.com/Read/1865/grease-compatibility.

47. Le Pera M. E. Testing lubricants for compatibility, *Practicing Oil Analysis*, January 2002, Online. Available: http://www.machinerylubrication.com/Read/283/testing-lubricants-compatibility.

48. Grease compatibility—to be or not to be!, *Mobil Technical Topic*, Exxon Mobil Corporation, 2009, Online. Available: http://www.mobilindustrial.com/ind/english/files/tt-grease-compatibility.pdf.

49. Johnson M. Planning for destruction: The problem of grease contamination, *Lubrication Excellence*, 7 pp., 2004, Online. Available: http://www.amrri.com/PDF/Cleanliness_Grease54.aspx.

50. Steffen B. Grease analysis: Early warning system for failures and proactive maintenance tool, *Machinery Lubrication*, No 2, February 2013, Online. Available: http://www.machinerylubrication.com/Read/29284/grease-analysis-system.

51. Springer T. E. *Improving Reliability Through a Clearer Understanding of Contamination*, The Timken Company, September 2006, Online. Available: http://www.mt-online.com/september2006/improving-reliability-through-a-clearer-understanding-of-contamination.

52. Potteiger J. Understanding lubrication failure, *Uptime Magazine*, 30–31, August/September 2011. Online. Available: http://reliabilityweb.com/articles/entry/Understanding_lubrication_failures/.

53. Ray C. Identifying root causes of machinery damage with condition monitoring, *Machinery Lubrication*, November–December 1–4, 2012.

54. Epshteyn Y. and Risdon T. J. Molybdenum disulfide in lubricant applications—A review, *Presented at the 12 Lubricating Grease Conference*, NLGI-India Chapter, 12pp., January 2010, Online. Available: http://www.nlgi-india.org/images/PDF/Yakov%20Ephsteyn.pdf.

55. Lansdown A. R. *Molybdenum Disulphide Lubrication, Tribology Series*, 35, Elsevier, Amsterdam, Netherlands, 380 pp., May 1999.

56. Spalvins T. Lubrication with sputtered MoS2 films principles, operation, limitations, *NASA Technical Memorandum* 105292, 1991.

57. Persson K., Fathi-Najafi M. Grease containing solid particles, White Paper, 08–08, Axel Christiernsson AB, pp. 1–11, Online. Available: http://axelamericas.com/pdf/White_Paper_09.pdf.

58. Non-fluid lubrication. *Engineering and Design—Lubricants and Hydraulic Fluids,* Chapter 6, 10 pp., EM 1110-2-1424, February 28, 1999, Online. https://www.cedengineering.com/userfiles/Lubricants%20 and%20Hydraulic%20Fluids.pdf.

59. Hughes R. L. *Understanding the Basics of Grease,* Online. Available: http://www.reliability.com/ industry/articles/article66.pdf.

60. Chudnovsky B. Lubrication of electrical contacts, *Proceedings of the 51st Annual Holm Conference on Electrical Contacts*, Chicago, IL, pp. 107–114, 2005.

61. Troyer D. and Kucera J. Lubricant storage life limits—Industry needs a standard, *Machinery Lubrication Magazine*, May 2001, Online. Available: http://www.machinerylubrication.com/Read/172/ lubricant-storage-life.

62. Lubricant storage, stability, and estimated shelf life, *Chevron USA Information Bulletin 1*, 2011, Online. Available: http://www.chevronmarineproducts.com/docs/Chevron_InfoBulletin01_ShelfLife_v1013.pdf.

63. *Exxon Mobile Lubricants and Specialties, Aviation Lubes Tech Topics, Shelf Life,* Online. Available: https://lubes.exxonmobil.com/aviation/learningandresources_shelf-life-bulletin.aspx.

64. Lubricant shelf life and conditions affecting duration, *Amsoil Technical Service Bulletin*, 11/21/05, Revision: 7/26/2012, Online. Available: http://www.amsoil.com/techservicesbulletin/other/tsb%20 ot-2005-11-21%20storage%20and%20handling.pdf.

65. Aviation lubricant shelf life, *ExxonMobil Technical Bulletin*, 2015, Online, Available: https://lubes.exxonmobil.com/aviation/Files/learningandresources_shelf-life-bulletin_2015.pdf.

66. Shelf life profile of DuPont™ Krytox® lubricants, *DuPont Technical Bulletin K22117_1*, Online, Available: http://www2.dupont.com/Lubricants/en_US/assets/downloads/DuPont_Krytox_Shelf_Life_ Profile_K-22117-2.pdf.

67. Sumerlin S. 10 Ways to improve lubricant storage and handling, *Machinery Lubrication*, March 2011, Online. Available: http://www.machinerylubrication.com/Read/28429/improve-lubricant-storage.

68. Glossenbrenner E. W. Sliding contacts for instrumentation and control, in *Electrical Contacts*, edited by Slade P. G., Marcel Dekker, Inc., New York, pp. 885–942, 1999.

69. Slade P. G. Introduction to contact tarnishing and corrosion, in *Electrical Contacts*, edited by Slade P. G., Marcel Dekker, Inc., New York, Chapter 2, pp. 89–112, 1999.

70. Timsit R. S. *Electrical Connector Lubricants,* Online. Available: http://www.timron-inc.com/newsletter. htm.

71. Holm R. In *Electric Contacts: Theory and Application*, 4th edition, Springer, New York, 483 pp., 1999.

72. Bailey C. A. and Aarons J. S. (Eds). *The Lubrication Engineers Manual*, United States Steel, Pittsburgh, Pennsylvania, pp. 460, 1971.

73. Grease, in *Lubricants and Hydraulic Fluids*, U.S. Army Corps of Engineers, Engineering Manual, Chapter 5, EM 1110-2-1424, February 1999, Online. Available: http://www.scribd.com/doc/47468785/ Lubricants-and-Hydraulic-Fluids.

74. Braunovic M., Konchits V. V., and Myshkin N. K. Lubricated electrical contacts, in *Electrical Contacts, Fundamentals, Applications and Technology*, Section 9.3, Taylor & Francis Group/CRC Press, Boca Raton, FL, pp. 414–454, 2006.

75. Stone D. S. and Joaquim M. E. The many roles of polyphenyl ethers, *Sensors*, 19(11), 2002, Online. Available: http://archives.sensorsmag.com/articles/1102/34/main.shtml.

76. Aukland N. and Joaquim M. E. Lubricants extend the life of sensor connectors, *Sensors*, 17(5), May 2000, Online. Available: http://archives.sensorsmag.com/articles/0500/78/.

77. Braunovic M. Evaluation of different types of contact aid compounds for aluminum to aluminum connectors and conductors, *Proceedings of the International Conference on Electric Contacts Phenomena and 30th Annual Holm Conference on Electrical Contacts*, Chicago, IL, pp. 97–104, 1984.

78. Gagnon D. and Braunovic M. High temperature lubricants for power connectors operating at extreme conditions, *Proceedings of the 48th Annual Holm Conference on Electrical Contacts*, Orlando, FL, pp. 273–282, 2002.

79. Sugimura K. and Nacae A. Lubricants for some plated contacts, *Proceedings of the 36th Annual Holm Conference on Electrical Contacts*, Montreal, Canada, pp. 417–424, 1990.

80. van Dijk P. Some effects of lubricants and corrosion inhibitors on electrical contacts, *AMP Journal of Technology*, 2, 56–62, November 1992.

81. Antler M. Tribology of electronic connectors: Contact sliding wear, fretting, and lubrication, in *Electrical Contacts*, edited by Slade P. G., Marcel Dekker, Inc., pp. 403–432, 1999.

82. Freitag W. O. Lubricants for separable contacts, *Proceedings of the Holm Seminar on Electrical Contacts*, Illinois Institute of Technology, Chicago, IL, pp. 57–63, 1976.

83. Wang S. S., Lee A., and Mamrick M. S. An electrochemical technique for evaluating electrical contact lubricants, *Proceedings of the 33rd Annual Holm Conference on Electrical Contacts*, Chicago, IL, pp. 9–14, 1987.
84. Tamai T. Ellipsometric analysis for growth of Ag2S film and effect of oil film on corrosion resistance of Ag contact surface, *Proceedings of the 34th Annual Holm Conference on Electrical Contacts*, San Francisco, CA, pp. 281–287, 1988.
85. Zhang J. G. and Zhu M.-D. Lubricant testing for electrical connectors, *Proceedings of the 33rd Annual Holm Conference on Electrical Contacts*, Chicago, IL, pp. 22–33, 1987.
86. Zhang J. G. and Chen W. Wipe on various lubricated and non-lubricated electric contacts in dusty environments, *Proceedings of the 36th Annual Holm Conference on Electrical Contacts*, Montreal, Canada, pp. 410–416, 1990.
87. Zhang J. G. The application and mechanism of lubricants on electrical contacts, *Proceedings of the 40th Annual Holm Conference on Electrical Contacts*, Chicago, IL, pp. 145–154, 1994.
88. Shao C. B. and Zhang J. G. Electric contact behavior of Cu–Sn intermetallic compounds formed in tin plating, *Proceedings of the 44th Annual Holm Conference on Electrical Contacts*, Washington, DC, pp. 26–33, 1998.
89. Capp P. O. and D. Williams W. M. Evaluation of friction and wear of new Palladium alloy inlays and other electrical contact surfaces, *Proceedings of the International Conference on Electric Contacts Phenomena and 30th Annual Holm Conference on Electrical Contacts*, Chicago, IL, pp. 410–416, 1984.
90. Abbott W. H. and Antler M. Connector contacts: Corrosion inhibiting surface treatments for gold-plated finishes, *Proceedings of the 41st Annual Holm Conference on Electrical Contacts*, Montreal, Canada, pp. 97–123, 1995.
91. Antler M. Survey of contact fretting in electrical connectors, *Proceedings of the International Conference on Electric Contacts Phenomena and 30th Annual Holm Conference on Electrical Contacts*, Chicago, IL, pp. 3–22, 1984.
92. Aukland N., Hardee H., and Lees P. Sliding wear experiments on clad gold-nickel material systems lubricated with 6-Ring polyphenyl ether, *Proceedings of the 46th Annual Holm Conference on Electrical Contacts*, Chicago, IL, pp. 27–35, 2000.
93. Fournier D. Aging of defective electrical joints in underground power distribution systems, *Proceedings of the 44th Annual Holm Conference on Electrical Contacts*, Washington, DC, pp. 179–192, 1998.
94. Abbott W. H. Performance of the gold-tin connector interface in a flight environment, *Proceedings of the 44th Annual Holm Conference on Electrical Contacts*, Washington, DC, pp. 141–150, 1998.
95. Williams D. W. M. The effect of test-environment on the creep of base metal surface films over precious metal, *Proceedings of the 33rd Annual Holm Conference on Electrical Contacts*, Chicago, IL, pp. 79–85, 1987.
96. Antler M., Aukland N., Hardee H., and Wehr A. Recovery of severely degraded tin-lead plated connector contacts due fretting corrosion, *Proceedings of the 43rd Annual Holm Conference on Electrical Contacts*, Philadelphia, PA, pp. 20–32, 1997.
97. Antler M. Corrosion control and lubrication of plated noble metal connector contacts, *Proceedings of the 41st Annual Holm Conference on Electrical Contacts*, Montreal, Canada, pp. 83–96, 1995.
98. Antler M. Sliding wear and friction of electroplated and clad connector contact materials: Effect of surface roughness, *Proceedings of the 42nd Annual Holm Conference on Electrical Contacts*, Chicago, IL, pp. 363–374, 1996.
99. Rosen G., Coghill A., and Tunca N. Prediction of connectors long term performance from accelerated thermal aging tests, *Proceedings of the 38th Annual Holm Conference on Electrical Contacts*, Seattle, WA, pp. 257–263, 1992.
100. Zhang J. G., Zhou Y. L., Sugimura K. et al. Comprehensive experiment of water soluble lubricant covered on gold plated surface, *Proceedings of the 42nd Annual Holm Conference on Electrical Contacts*, Chicago, IL, pp. 444–454, 1996.
101. Noël S., Lécaudé N., Alamarguy D. et al. A new mixed organic layer for enhanced corrosion protection of electric contacts, *Proceedings of the 22nd International Conference on Electrical Contacts and 50th IEEE Holm Conference*, Montreal, Canada, pp. 274–280, September 2004.
102. Leung C. H. and Lee A. Thermal cycling induced wiping wear of connector contacts at 150°C, *Proceedings of the 43rd Annual Holm Conference on Electrical Contacts*, Philadelphia, PA, pp. 132–137, 1997.
103. Noël S., Lécaudé N., Bodin C. et al. Electrical and tribological properties of hot-dipped tin separable contacts with fluorinated lubricant layers, *Proceedings of the 45th Annual Holm Conference on Electrical Contacts*, Pittsburgh, PA, pp. 225–235, 1999.

104. Noël S., Lécaudé N., Alamarguy D. et al. Lubrication mechanisms of hot-dipped tin separable electrical contacts, *Proceedings of the 47th Annual Holm Conference on Electrical Contacts*, Montreal, Canada, pp. 198–202, 2001.

105. Braunovic M. Effect of contact aid compounds on the performance of bolted aluminum-to-aluminum joints under current cycling conditions, *Proceedings of the 31st Annual Holm Conference on Electrical Contacts*, Chicago, IL, pp. 19–27, 1985.

106. Braunovic M. Fretting in nickel-coated aluminum conductors, *Proceedings of the 36th Annual Holm Conference on Electrical Contacts*, Montreal, Canada, pp. 464–471, 1990.

107. Braunovic M. Evaluation of different contact-aid compounds for aluminum-to-Copper connections, *Proceedings of the 36th Annual Holm Conference on Electrical Contacts*, Montreal, Canada, pp. 509–517, 1990.

108. Timsit R. S., Bock E. M., and Corman N. E. Effect of surface reactivity of lubricants on the properties of aluminum electrical contacts, *Proc. 43rd Annual Holm Conference on Electrical Contacts*, Philadelphia, PA, pp. 57–66, 1997.

109. Braunovic M. Further studies of different contact aid compounds for aluminum-to-copper connections, *Proceedings of the 45th Annual Holm Conference on Electrical Contacts*, Pittsburgh, PA, pp. 53–62, 1999.

110. Johnson B., Ladin D., Gagnon D., and Braunovic M. Evaluation of high-temperature fluorinated ethers (PFPE) for power connectors operating at extreme service conditions, *17th International Conference on Electricity Distribution*, Barcelona, May 2003.

111. Kassman R. A., Imrell T., Hogmark S., and Jacobson S. Deteriorating mechanisms of silver and tin plated copper connectors subjected to an oscillating movement, *Proceedings of the 36th Annual Holm Conference on Electrical Contacts*, Montreal, Canada, pp. 395–401, 1990.

112. Kassman R. A. and Jacobson S. The role of cross plastic fretting in serviceability and deterioration of power contacts, *Proceedings of the 42nd Annual Holm Conference on Electrical Contacts*, Chicago, IL, pp. 352–362, 1996.

113. Kassman R. A. and Jacobson S. The contact resistance of rolling silver coated copper contacts, *Proceedings of the 43rd Annual Holm Conference on Electrical Contacts*, Philadelphia, PA, pp. 33–40, 1997.

114. Dzekster N. N. and Izmailov V. V. Some methods for improving aluminum contacts, *Proceedings of the 36th Annual Holm Conference on Electrical Contacts*, Montreal, Canada, pp. 518–520, 1990.

115. Klungtvedt K. A study of effects of silicate thickened lubricants on the performance of electrical contacts, *Proceedings of the 42nd Annual Holm Conference on Electrical Contacts*, Chicago, IL, pp. 262–268, 1996.

116. McCarthy S. L., Carter R. O., and Weber W. H. Lubricant—induced corrosion in copper electrical contacts, *Proceedings of the 43rd Annual Holm Conference on Electrical Contacts*, Philadelphia, PA, pp. 115–120, 1997.

117. Arnell S. and Andersson G. Silver iodide as a solid lubricant for power contacts, *Proceedings of the 47th Annual Holm Conference on Electrical Contacts*, Montreal, Canada, pp. 239–244, 2001.

118. Nakagawa H. and Matsukawa K. Friction and wear properties of tin plated sliding contacts under oil lubricated condition, *Proceedings of the 48th Annual Holm Conference on Electrical Contacts*, Orlando, FL, pp. 151–155, 2002.

119. Braumann P. and Schroder K.-H. Changes in switching behavior due to surface contamination for contact materials in low voltage power applications, *Proceedings 32nd Annual Holm Conference on Electrical Contacts*, Boston, MA, pp. 85–90, 1986.

120. Witter G. J. and Leiper R. A. A study of contamination levels measurement techniques, testing methods, and switching results for silicon compounds on silver arcing contacts, *Proceedings of the 38th Annual Holm Conference on Electrical Contacts*, Pittsburgh, PA, pp. 173–180, 1992.

121. Fletcher R. G. *Lubricants for Aluminum Connectors*, B. C. Hydro Research Report No. 31145–78, October 1979.

122. Abbott W. H. Contact corrosion, In *Electrical Contacts*, edited by Slade P. G., Marcel Dekker, Inc., pp. 113–154, 1999.

123. Antler M. Electronic connector contact lubricants: The polyether fluids, *Proceedings of the 32nd Annual Holm Conference on Electrical Contacts*, Boston, MA, pp. 35–44, 1986.

124. Zhang J. G., Mei C. H., and Wen X. M. Dust effects on various lubricated sliding contacts, *Proceedings of the 35th Annual Holm Conference on Electrical Contacts*, Chicago, IL, pp. 35–42, 1989.

125. Basic connection principles, in *Burndy Reference,* pp. O2–O5, Online. Available: http://portal.fciconnect.com/res/en/pdffiles/brochures/MC02_Section_O-Reference.pdf.

126. Kulwanoski G., Gaynes M., Smith A., and Darrow R. Electrical contact failure mechanism relevant to electronic packages, *Proceedings of the 37th Annual Holm Conference on Electrical Contacts*, Chicago, IL, pp. 184–192, 1991.

127. Abbott W. H. Field and laboratory studies of corrosion inhibiting lubricants for gold plated connectors, *Proceedings of the 42nd Annual Holm Conference on Electrical Contacts*, Chicago, IL, pp. 414–428, 1996.

128. Steenstrup R. V., Fiacco V. M., and Schultz L. K. A Comparative study of inhibited lubricants for dry circuit, sliding switches, *Proceedings of the 28th Holm Conference*, IIT, Chicago, IL, pp. 59–68, 1982.

129. Alamarguy D., Benedetto A., Balog M. et al. Tribological and electrical study of fluorinated diazonium films as dry lubricants for electrical contacts, *Surf. Interface Anal.,* 40, 802–805, 2008. Published online in Wiley Interscience: February 6, 2008.

130. Bares J. A., Argibay N., Dickrell P. L. et al. In situ graphite lubrication of metallic sliding electrical contacts, *Wear*, 264, 1–8, 2009.

131. Synthetic lubes protect electrical connections, reduce warranty claims, *NYE Synthetic Lubricants*, 9 pp., Online. Available: http://www.nyelubricants.com/connectors.

132. Keeping contact: Specialty lubricants for electrical contacts. *Kluber Lubrication*. Bulletin B053001002, Edition 11-2012, pp. 1–11, München, Germany, Online. Available: http://pdf.directindustry.com/pdf/kluber-lubrication/keeping-contact-lubricant/12008-367115.html.

133. Joaquim M. E. and Tom A. Stationary lubricants increase connector reliability, *Santovac Fluids Inc.*, Technical Paper, Online. Available: http://www.chemassociates.com/products/findett/PPEs_Swedish_Cell.pdf.

134. Hamid S. Using lubricants to avoid failures in medical electronic connectors, *Medical Electronic Design*, October 2006, Online. Available: http://www.medicalelectronicsdesign.com/article/using-lubricants-avoid-failures-medical-electronic-connectors.

135. Pawlikowski G. T. Tyco electronics HM-15 lubricant stability projected weight loss, Technical Paper, *Tyco Electronics Corporation (TE Connectivity Ltd. Company)*, November 2006, Online. Available: http://application-notes.digchip.com/166/166-47954.pdf.

136. Sabatini N. A. Inspection, prevention, control, and repair of corrosion on avionic equipment, *US D.O.T F.A.A Advisory Circular*, May 30, 2001.

137. Undersecretary of defense AT&L, *Report to Congress, Department of Defense Status Update on Efforts to Reduce Corrosion and the Effects of Corrosion on the Military Equipment and Infrastructure of the Department of Defense*, May 2005, Online. Available: http://www.nstcenter.com/docs/PDFs/TechResourcesReportToCongress.pdf.

138. Abbott W. H., Hernandez H., Kinzie R, Siejke S. Effects of corrosion inhibiting lubricants on electronics reliability, *Proceedings of the 21st Digital Avionics Systems Conference*, Irvine, CA, 2002, vol. 2, pp. 12D2-1–12D2-13.

139. Dotson S. L. *Effects of Corrosion Inhibitive Lubricants on Electronics Reliability*, pp. 1–14, Online. Available: http://correxllc.com/resources/pdf/Corrosion_inhibitive_lubricants_Scott_Dobson.pdf.

140. Horne D. H. *Catastrophic Uncommanded Closures of Engine Feedline Fuel Valve from Corroded Electrical Connectors Paper# 00719*, NACE, April 2000, Online. Available: http://www.lektrotech.com.

141. Grease—its components and characteristics, *Technical Topic, ExxonMobil*, Online. Available: http://www.mobilindustrial.com/ind/english/files/tt-components-and-characteristics-of-grease.pdf.

142. ASTM Standard D2265-00 Standard Test Method for Dropping Point of Lubricating Grease Over Wide Temperature Range, Current edition approved December 10, 2000. Published January 2001. Originally published as D 2265 – 64. Last previous edition D 2265 – 94a, ASTM International, West Conshohocken, PA 19428-2959, United States., Online. Available: http://www.shxf17.com/pdf/ASTMD2265-00.pdf.

143. Momah L. *Choosing a High-temperature Lubricant*, May 2013, Online. Available: http://lubepoint.wordpress.com/2013/05/15/choosing-a-high-temprature-lubricant/.

144. Grease high temperature performance, *Mobil Technical Topic*, 2009, Online. Available: http://www.chemcorp.co.uk/creo_files/upload/related-items/grease_high_temp_performance.pdf.

145. Mraz S. Giving electrical connectors the slip, *Machine Design*, February 2004, Online. Available: http://machinedesign.com/archive/giving-electrical-connectors-slip.

146. Chudnovsky B. Degradation of power contacts in industrial atmosphere: Silver corrosion and whiskers, *Proceedings of the 48th Annual Holm Conference on Electrical Contacts*, Orlando, FL, pp. 140–147, 2002, Online. Available: https://nepp.nasa.gov/whisker/reference/tech_papers/chudnovsky2002-paper-silver-corrosion-whiskers.pdf.

147. *Lubrication Manual: Lubrication Instructions for Major Manufacturers' Medium and Low Voltage Circuit Breakers*, 6th Edition, Schneider Electric/Square D, p. 84, 2008.

148. Crino A. D. The effects of exposed contact lubricants on the temperature rise of outdoor disconnect switches. *Proceedings of the 63rd Annual International Conference of Doble Clients*, Boston, MA, pp. 4-2.1–4-2.5, 1996.

149. Chudnovsky B. Lubrication practices in the electrical Industry: The key to reliability of circuit breakers, *NETA World*, pp. 55–59, Winter 2001–2002.

150. *Lubrication Guide of the Doble Circuit Breaker Committee*, Doble Engineering Company, Watertown, MA, Rev C, p. 27, 1995.

151. Stevens C. Practical pointers for grease and antiseize selection, *Plant Engineering*, 52(5), 67–69, May 1998.

152. Chudnovsky B. Lubrication practices can make or break circuit breaker reliability, *Lubrication and Fluid Power*, 4(4), 23–26, November/December 2003.

153. Chudnovsky B. Ensuring protection and reliability of load interrupter switches, *Plant Safety and Maintenance*, 4(2), 31, 2004.

154. Chudnovsky B. Smooth breakers. *Plant Services*, November 2008, Online. Available: http://www.plantservices.com/articles/2008/246.html.

155. *Lubrication and Circuit Breakers*, Square D Data Bulletin #0600DB0109, September 2001, Online. Available: http://static.schneider-electric.us/docs/Circuit%20Protection/0600DB0109.pdf.

156. *Preventative Maintenance for HVLTM Load Interrupter Switches*, Square D Data Bulletin #6040DB0201, November 2002, Online. Available: http://static.schneider-electric.us/docs/Electrical%20Distribution/Medium%20Voltage%20Switchgear/Metal-Enclosed%20Air%20Insulated%20Switchgear/HVL/6040DB0201.pdf.

157. Mark B. What exactly is a lubrication failure? *Machinery Lubrication Magazine*, January 2007, Online. Available: http://www.machinerylubrication.com/article_detail.asp?articleid=967&relatedbookgroup=Lubrication.

158. Snyder D. R. Unearthing root causes, *MRO Today Magazine*, December 2005/January 2006.

159. Nailen R. L. Lubrication: Important for circuit breakers, too, *Electrical Apparatus*, March 2004, Online. Available: http://www.findarticles.com/p/articles/mi_qa3726/is_200403/ai_n9361544.

160. Wright J. Grease basics, *Machinery Lubrication*, pp. 50–52, May–June 2008, Online. Available: http://www.machinerylubrication.com/Read/1352/grease-basics.

161. Khonsari M. and Booser E. R. Predicting lube life—Heat and contaminants are the biggest enemies of bearing grease and oil, *Machinery Lubrication Magazine*, September 2003, Online. Available: http://www.machinerylubrication.com/Read/537/predict-oil-life.

162. Neale M. J. *Lubrication: A Tribology Handbook*, Society of Automotive Engineers, Butterworth-Heinemann Ltd, Oxford, England, 640 pp., 1995.

163. *Common-Cause Failure Event Insights Circuit Breakers*, United Stated Nuclear Regulatory Commission, NUREG/CR-6819, Vol. 4 "Circuit Breakers," INEEL/EXT-99–00613, May 2003, Online. Available: http://www.nrc.gov/reading-rm/doc-collections/nuregs/contract/cr6819/.

164. Overheating of the MV switch, Section 4.7.2, *Electrical Power Transmission and Distribution: Aging and Life Extension Techniques*, 1st Edition, CRC Press, Boca Raton, FL, 2012.

165. Braunovic M. Fretting damage in tin-plated aluminum and copper connectors, *IEEE Trans. Compon. Hybrids Manuf. Technol.*, 12(2), 215–223, June 1989.

166. Braunovic M. Effect of intermetallic phases on the performance of tin-plated copper connections and conductors, *Proceedings of the 49th Annual Holm Conference on Electrical Contacts*, Chicago, IL, pp. 124–131, September 2000.

167. Jaiyeola G. Failure analysis of a corrosive residue on a medical device, *Case Study*, Online. Available: http://www.element.com/information/resources/articles-index/case-study-failure-analysis-of-a-corrosive-residue-on-a-medical-device.

168. Mowry M. *The True Cost of Bearing Lubrication*, Spring 2011, Online. Available: http://www.igus.com/_Product_Files/Download/pdf/igus_White-Paper_True-Cost-of-Lubrication_June2011.pdf

169. Lubricant Failure = Bearing Failure, *Machinery Lubrication*, January/February 2009, Online. Available: http://www.applied.com/site.cfm/Lubricant_Failure=Bearing_Failure.cfm.

170. *Bearing Failure Due to Over Lubrication*, Online. Available: http://www.belray.com/bearing-failure-due-over-lubrication.

171. Bearing failure: Causes and cures, *Barden Precision Bearings*, Online. Available: http://www.schaeffler.com/remotemedien/media/_shared_media/08_media_library/01_publications/barden/brochure_2/downloads_24/barden_bearing_failures_us_en.pdf.

172. *Ball & Roller Bearings: Failures, Causes and Countermeasures*, JTEKT Corporation, Online. Available: http://donoupoglou.gr/wp/wp-content/uploads/2015/03/catb3001e.pdf

173. Weigand M. Lubrication of rolling bearings—Technical solutions for critical running conditions, *Machinery Lubrication,* No.1, 2006, Online. Available: http://www.machinerylubrication.com/Read/844/lubrication-rolling-bearings.

174. Inadequate grease lubrication in bearings: Water contamination and debris contamination, *Timken Automotive TechTips*, 3(5), Part 3 of a 3-Part Series, 2009, Online. Available: http://www.timken.com/en-us/solutions/automotive/aftermarket/lightduty/TechTips/Documents/Vol3Iss5_Inadaquate_Grease_Lubrication_Part3of3.pdf.

175. Finner G. *Modern Lubrication Practices for Improved Outdoor Electrical Equipment Reliability*, Dow Corning Corporation, Online. Available: http://www.dowcorning.com/content/publishedlit/80-3503.pdf.

176. Wagman D. Lubrication: Industry leaders look to fix the varnish issue, *Power Engineering*, February, 2008, Online. Available: http://www.power-eng.com/articles/print/volume-112/issue-2/features/lubrication-industry-leaders-look-to-fix-the-varnish-issue.html.

177. Dow Corning Power & Utilities Solutions: There's a lot on the line, Online, Available: http://www.dowcorning.com/content/publishedlit/80-3459.pdf.

178. *Lubricants for Stationary Separable Electrical Connectors*, NYE LubeNotes, Online. Available: http://www.nyelubricants.com/stuff/contentmgr/files/0/24d7a2dc2442a71906571ee688ad6bf3/en/lubenote_stationary_separable_electrical_connectors.pdf.

179. Nye's family of Anderol® replacement lubricants, *NYE LubeNotes*, February 2013, Online. Available: http://www.nyelubricants.com/stuff/contentmgr/files/0/2ab16cba30c7287dd39ff65a9fbf9a04/en/lubenote_anderol.pdf.

180. *High Voltage Contact Grease*, Contralube, UK, Online. Available: http://highvoltagecontactgrease.com/.

181. Klüber lubricants for electrical switches and contacts, Bulletin B053001002, Edition May 2008, *Klüber Lubrication*, München KG, Germany, Online. Available: http://www.industrialbearings.com.au/uploads/catalogs/klueber_elektrische_kontakte-en_1351825422.pdf.

182. *Manufacturer & Distributor of Two Silver Based Products*, Online. Available: http://www.cool-amp.com/index.html.

183. *Silver Filled Conductive Grease*, Online. Available: http://www.chemtronics.com/products/americas/TDS/Cw7100tds.pdf.

184. *Conductive Grease and Electrical Contact Lubricant NO-OX-ID A-Special Electrical Grade*, Online. Available: http://www.sanchem.com/aSpecialE.html.

185. *Benefits of Contact Lubricants, Electrolube*, Online. Available: http://www.contact-lubricants.com/the-benefits-of-contact-lubricants.html.

186. *Electrical Jointing*, Tyco Electronics, Online. Available: http://www.gvk.com.au/pdf/electrical_jointing.pdf.

187. EPRI/NMAC Lube Notes, *Nuclear Maintenance Applications Center (NMAC)*, 2008–2011.

188. Foszcz J. L. Interchangeable lubricant guide, *Plant Engineering Magazine*, September 2004, Online. Available: http://www.plantengineering.com/industry-news/top-stories/single-article/interchangeable-lubricant-guide/dee266c482.html.

7 Insulation, Coatings, and Adhesives in Transmission and Distribution Electrical Equipment

7.1 INSULATING MATERIALS IN POWER EQUIPMENT

A dielectric material that resists the flow of electric charge is called an insulator. In an electrical machine, the electrical insulation ensures that current flows only along the conductors and not between individual conductors or between the conductor and the ground. Insulating materials are used in electrical equipment not only for separation of current-carrying parts from each other and from the environment, but also for support of electrical conductors. Insulating materials can serve as practical and safe insulators for low to high voltages (up to thousands of volts).

In electrical systems, insulators are commonly used as a flexible coating on electric wire and cable to avoid wires from touching each other and be touched, and therefore preventing electrocution and fire hazards. In coaxial cables, the center conductor must be supported exactly in the middle of the hollow shield in order to prevent electromagnetic wave reflections. An insulated wire or cable has a voltage rating and a maximum conductor temperature rating. It may not have an ampacity (current-carrying capacity) rating, since this depends on the surrounding environment (e.g., ambient temperature).

7.1.1 INSULATING MATERIALS USED IN THE ELECTRICAL INDUSTRY

Dielectric materials do not conduct electricity; they have a very low electrical conductivity. Three classes of solid dielectric materials are used for insulation: inorganic, organic, and composites [1]. Inorganic materials are *ceramics*, such as porcelain, alumina, and so on, *glass*, and *cements* or *minerals* (mica). *Polymer* insulators, or nonceramic insulators (NCI), were first developed in the 1960s and installed in the 1970s, and developed into mature products in the 1980s. Polymers made of organic materials or *composites* such as Kevlar, carbon, and fiberglass are used for insulation. Lower overall cost, easy installation, and light weight are among the major reasons for using polymer insulators, although there are some concerns, such as reduced life expectancy and lack of knowledge about polymers' aging. There are also many nonsolid dielectric materials that are used in the electrical industry, for example, oil, gas, and vacuum. These insulating media will be discussed in Section 7.6 in more detail.

Solid organic insulating materials may be divided into two subclasses: thermoplastic and thermosetting (Table 7.1). A thermoplastic material is a polymer that softens or melts on heating, and becomes rigid again on cooling. Thermosetting insulation made of plastic that, when cured by heat or other means, such as radiation, catalysts, and so on, changes into a substantially infusible and insoluble product hardened by heating.

In electronic systems, PCBs are made from epoxy plastic and fiberglass. The nonconductive boards support layers of copper foil conductors. In electronic devices, the tiny and delicate active components are embedded within nonconductive epoxy or phenolic plastics, or within baked glass or ceramic coatings. In microelectronic components such as transistors and ICs, the silicone

TABLE 7.1

Organic Insulating Materials

Type of Solid Insulation	Material	Chemical Variety, Abbreviated Name or Trademark
Thermoplastic	Rubber	Natural, butyl, and silicone
	Polyamide	PA, nylon
	Polyester	PE, mylar
	Polypropylene	PP
	Polysterene	PS
	Polyvinyl chloride	PVC
	Polymethylmetacrylate	PMMA
	Polycarbonate	PC
	PTFE	PTFE, teflon
Thermosetting	PE	PE, LDPE, MDPE, HDPE, XLPE
	Ethylene–propylene	EPR
	Polyimide	PI
	Polyetheretherketone	PEEK
	Epoxy	Phenolic, silicone, and polyester resins

material is normally a conductor because of doping, but it can easily be selectively transformed into a good insulator by the application of heat and oxygen. Oxidized silicone is quartz, that is, crystalline silicone dioxide.

In HV systems containing transformers and capacitors, using oil as a liquid insulator is the typical method of preventing arcs. The oil replaces the air in any spaces, which must support a significant voltage without electrical breakdown. Other methods of insulating HV systems are by using ceramic or glass wire holders, gas, or vacuum, and simply placing the wires with a large separation, using air as the insulation.

The safety and longevity of electrical power equipment strongly depend on the technical performance of multiple insulating materials used in the design of electromechanical machines. It is important to choose and apply insulating materials that perform well and do not fail in various environments. Insulating materials should withstand environmental parameters such as humidity, dust, temperature, and contamination for practical applications of electrical apparatus indoor and outdoor.

Insulating materials should endure working and environmental conditions such as mechanical stresses, electrical fields, moisture, ozone, UV light, surface arcing, and contamination. In the following sections, the aging issues of some insulating materials used in power distribution and transmission will be discussed.

In electric power transmission, suspended wires are bare except where they enter buildings, and they are insulated by the surrounding air. Insulators are required at points where they are supported by utility poles or pylons. Insulators are also required where the wire enters buildings or electrical devices, such as transformers or CBs, to insulate the wire from the case. These hollow insulators with a conductor inside them are called bushings.

Insulators used for HV power transmission are made from glass, porcelain, or composite polymer materials (Figure 7.1). Porcelain insulators are made from clay, quartz, or alumina and feldspar, and are covered with a smooth glaze to shed water. Insulators made from porcelain rich in alumina are used where high mechanical strength is a criterion. Porcelain has a dielectric strength of about 4–10 kV/mm [2]. The advantages of porcelain insulators include superior electrical properties, good mechanical properties (especially tensile strength), good creep resistance at room temperature, high corrosion resistance, minimal leakage problems, and less adverse effects from changing temperatures.

FIGURE 7.1   Insulators made of dielectric composite material (silicone). (Piedmont Bushings & Insulators, Online. Available: http://www.pbi.com/insulator_pbi.html. With permission.)

Glass has a higher dielectric strength, but it attracts condensation and the thick irregular shapes needed for insulators are difficult to cast without internal strains [3]. Some insulator manufacturers stopped making glass insulators in the late 1960s, switching to ceramic materials.

Recently, some electric utilities have begun switching to *polymer composite* materials for some types of insulators. These are typically composed of a central rod made of *fiber-reinforced plastic* and an outer weather shed made of *SIR* or *ethylene propylene diene monomer (EPDM) rubber*. EPDM is a type of synthetic rubber, an elastomer used in a wide range of applications. Composite materials have been utilized for several years to replace porcelain as insulation media.

There are proven advantages of the composites such as lower weight, pollution resistance, and weather and explosion proof. An additional advantage is that composite insulation allows for a more compact and reliable design of lines and substations for higher ratings. Composite material may have different properties depending on the mixture of substances, manufacturing process, surface and mechanical design, and so on. Composite insulators are less costly, are lighter in weight, and have excellent hydrophobic capability. This combination makes the composites ideal for service in polluted areas. However, these materials do not yet have the long-term proven service life of glass and porcelain.

To confirm better performance of composite insulators, a thorough testing including dielectric and long-term "field test" is required [4]. The composite insulation property of "hydrophobicity" (an ability to repel water) significantly improves the insulation's pollution resistance behavior, but this should also be confirmed by accelerated pollution and stress aging tests for composite insulation in comparison with porcelain.

Insulators made of SIR have been tested to find out whether the insulators could permanently work when electrically stressed beyond the recommended limits in polluted and clean tropical environments. The tested insulators included eight types of SIR composite insulators, one type of hybrid silicone ceramic insulator and one semiconducting glazed porcelain insulator, while ordinary (a)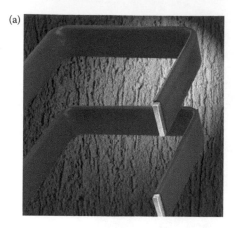

(b)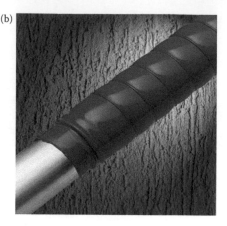

FIGURE 7.2 A 5–35 kV bus with heat shrinkable tubing (a) and tape (b). Users should independently evaluate the suitability of the insulating material pictured above for their particular use. (From TE Connectivity, Online. Available: http://energy.te.com/index.asp?act=page&pag_id=2&prl_id=5&pls_id=29&prf_id=135&prr_parent=968. Photographs reprinted with permission.)

porcelain and glass insulators were used as the Ref. [5]. The tests showed that there is a possibility to increase the electric stress on SIR insulators to levels higher than those used today on glass and porcelain insulators.

Insulation of electrical conductors (busbars) in transformers and CBs is often composed of SIR or epoxy resin, while the supporting insulator and bushing are made of glass, porcelain, ceramic, hybrid silicone ceramic, and semiconducting glazed porcelain (or their combinations).

For busbar insulation, heat shrinkable tubing (Figure 7.2a), sleeves, tapes (Figure 7.2b), and sheets are used. This insulating material is made of specially formulated, cross-linked, flame-retardant, track-resistant polyolefin that will shrink to fit rectangular, square, or round busbars. It will also cover and insulate inline bolted connections of regular busbars. Other heat-shrinkable materials are made of fluoropolymers and elastomers.

Often busbars and CBs in switchgear may be insulated with glass-reinforced plastic insulation, treated to have low flame spread and to prevent tracking of current across the material. Thermosetting *epoxy* powders are specially formulated for electrical insulation applications. The epoxy is applied to a preheated metal surface by the fluidized-bed process. A dry epoxy powder melts, cures, and produces a uniform film of insulating coating with excellent adhesion, corrosion, and chemical resistance, and a good combination of mechanical, thermal, and electrical properties.

Epoxy powder-coated busbar insulation allows for closer component location in a system, which is ideal for single conductors or multiple conductor assemblies that may have numerous forms (Figure 7.3). Epoxy powder coating is also well suited for insulating thick conductors, in addition to conductors with multiple electrical contact points. The material has a high dielectric strength and is a durable insulator, impervious to most elements. Epoxy powder coating's high dielectric strength can be varied based on the application process and component preparation.

Insulation of busbar shrouds and boots is usually made of polyvinyl chloride (PVC) material, which is formulated specially for electrical applications at HV. These insulators are flexible and reusable electrical insulating covers for bus and switchgear connections with special buttons making it easy to install, remove, or replace them during maintenance.

7.1.2 THERMAL LIMITATION FOR ELECTRICAL INSULATION

Power electrical equipment when in service is a source of thermal energy. Its conductors may be heated up to very high temperatures, particularly in systems with high currents flowing through the

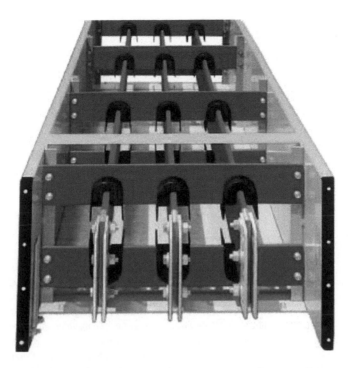

FIGURE 7.3 Epoxy powder insulated MV bus. (From Park Electric Company Busway, Online. Available: http://www.parkdetroit.com/What-We-Do/Busways. With permission.)

conductors. On the other hand, every electrical insulation material has its own working temperature range beyond which it may melt or become flammable. Either event would cause a disastrous failure. Even at minor overheating of insulation above the working temperature maximum, the aging of insulation would accelerate and create defects and a decrease of insulating properties [6].

Institute of Electrical and Electronics Engineers (IEEE) Standard 1-1986, "IEEE Standard General Principles for Temperature Limits in the Rating of Electric Equipment and for the Evaluation of Electrical Insulation" [7], was developed to guide in the preparation of IEEE and other standards that deal with the selection of temperature limits and the measurement of temperature for specific types of electric equipment. Principles are included for the development of test procedures for thermal evaluation of electrical insulating materials (EIMs), insulation systems, and thermal classification for use in the rating of electric equipment.

IEEE Standard 1-2001, "IEEE Recommended Practice—General Principles for Temperature Limits in the Rating of Electrical Equipment and for the Evaluation of Electrical Insulation" [8], is intended to help in the preparation of standards that are principally concerned with the thermal endurance of EIM and simple combinations of such materials, with the establishment of limiting temperatures of electrical insulation system (EIS), and with the provision of general principles for thermal classification of EIS.

According to these standards and to ensure the safety and long life of electrical power equipment, a system of choosing a suitable insulating material in the electrical design has been developed. This system classified insulating materials based on the material thermal properties and the temperature working range and introduced the so-called temperature classes of insulation.

In the past, the temperature classes of electric insulators were simply divided into natural organics, inorganics, and their combinations. In the late 1930s, various types of synthetic resins have been developed and their application range gradually expanded to insulation. International Electrotechnical Commission (IEC) issued a recommendation on the temperature classes of electric insulators, followed by IEC Publication 85 (1957), which established the classification system

TABLE 7.2

Thermal Class of Insulation Based on IEEE and IEC Standards

	Thermal Classification	
Thermal Class IEEE	IEEE	IEC
90	None	Y
105	A	A
120	E	E
130	B	B
155	F	F
180	H	H
200	N	200
220	R	220
240	S	240
>240	C	–

Source: Modified from IEEE Standard 1–2001. *Recommended Practice—General Principles for Temperature Limits in the Rating of Electrical Equipment and for the Evaluation of Electrical Insulation.* IEEE Publishing, New York © 2001, IEEE; IEC Publication 85. *Recommendations for Classification of Materials for the Insulation of Electrical Machinery and Apparatus in Relation to their Thermal Stability in Service*, Publication of International Electrotechnical Commission (IEC), Geneva, Switzerland, 1957; IEC standards 60085. *Electrical Insulation— Thermal Evaluation and Destination.* Publication of International Electrotechnical Commission (IEC), Geneva, Switzerland, 2007.

that is currently in use [9]. Later, when silicone resin became a popular insulator material, the new class (H) of insulators was added to the temperature classes (Table 7.2), and it became increasingly necessary to find applications for heat-resistant materials in electric instruments. The IEC standards [9,10] established several insulation classes (Y, A, E, B, F, and H).

According to these standards, insulation in each class cannot be exposed to temperatures higher than the *maximum allowed operation temperature.* In other standards, such as the National Electrical Manufacturers Association (NEMA) standard, classifications also rated insulation systems according to maximum allowable operating temperatures for motor insulation, where this classification is named "temperature tolerance classes" (A, B, F, and H). The testing techniques for the high-temperature life of new materials for temperature classification were investigated intensively in the United States [11].

Low-temperature Class-Y insulation (not classified in the United States) is typically made of cotton, silk, or paper; and its maximum temperature is 90°C, while high-end temperature (Class H) insulation is made of inorganic material glued with silicone resin or adhesives of equivalent performance with a maximum operating temperature of 180°C. Class C insulation is made of 100% inorganic material and its temperature limit is over 180°C.

International Standard IEC 60034-1 [12] classifies the TR limits of insulation materials based on the maximum permissible temperatures that the various classes of insulation materials could withstand. The TR of the electrical machine is the increase in temperature that is permissible above this maximum ambient temperature. IEC standard [12] specifies a maximum ambient temperature of 40°C. A safe temperature is the sum of the maximum specified ambient temperature and the permitted TR due to the mechanical load. For example, for Class E insulation with a maximum permissible temperature of 120°C, the maximum allowed TR is 80°C.

The importance of applying a proper insulation in the design of a specific electrical apparatus is hard to overestimate. The limit of temperature which the insulation is allowed to reach during service is also a limit to which a conductor under this insulation or touching this insulation may be heated. The whole design of electrical apparatus would be regulated and determined by this limit to not allow the temperature of the conductors surrounded by insulation to rise above this limit under any circumstances.

7.1.3 Thermal Degradation of Insulators

The weakest and the fastest degrading component of any electrical equipment is the insulation. In cables, transformers, reactors, trip coils of CBs, operating coils of contactors, motors, and capacitors, the type of insulation used is solid synthetic and paper. The degradation of this insulation depends strongly upon the maximum temperature to which it is subjected. Various types of insulation are deteriorating at different rates while being exposed to elevated temperature, and an aging process is going on even if the exposure temperature is below the maximum temperatures defined by standards.

The longer the insulation material is under elevated temperature, the shorter the useful life of the insulation. In the electrical industry, there are some widely accepted terms that define the thermal performance of a material based on an accepted test method for electrical insulation materials.

Thermal endurance of insulation is defined as the relationship between temperature and the time spent at that temperature required for producing such degradation of an electrical insulation that it fails under specified conditions of stress, electric or mechanical, in service or under test (IEEE Standard Dictionary of Electrical and Electronic Terms [13]). The point of failure, also referred to as the "Thermal Life," is typically considered the time at which the measured property falls below 50% of its original untreated value.

The so-called relative temperature index (RTI) characterizes an insulation system [14] and is obtained when a candidate insulation system is simultaneously tested with a known insulation system using identical conditions (e.g., the same temperatures, time intervals, and oven). The *temperature index* according to ASTM Standard D2304 [15] is a number that permits a comparison of the temperature/time characteristics of an electrical insulation material, or a simple combination of materials, based on the temperature in degrees Celsius that is obtained by extrapolating the Arrhenius plot of life versus temperature to a specified time, usually 20,000 h.

Any analysis of the thermal performance of electrical insulation material should be based on the established test methods covered in IEEE, UL, NEMA, and ASTM standards. Table 7.2 shows the thermal class designations of insulation based on IEEE and IEC standards [8–10].

IEC Standard 60216, "Electrical Insulating Materials—Thermal Endurance properties" [16], defines thermal aging processes and specifies the testing techniques for determining thermal endurance properties of electrical insulation materials. Temperature limits for paper insulation in liquid-immersed distribution, power, and regulating transformers have been established in IEEE Standard C57.12.00-2010 [17].

Test programs and methods for the determination of thermal performance of insulation materials were developed based on the assumption that heat is the chief cause of insulation degradation. Other factors being equal, thermal degradation is accelerated as the temperatures increases, and the other mechanical and electrical properties deteriorate with increasing temperature.

According to ASTM D2304, "Thermal degradation is often a major factor affecting the life of insulating materials and the equipment in which they are used." The Standard also states that "Experience has shown that the thermal life is approximately halved for a 10°C increase in exposure temperature."

Losing insulating properties under the influence of increased temperature is a major cause of insulation failure. For example, certain glasses and ceramics exhibit large increases in dielectric loss with increasing temperature. To determine thermal failure of solid EIM, various test procedure have been developed and standardized [18]. ASTM [19] established a method to determine thermal degradation of solid EIM.

TABLE 7.3

Thermal Limits for Insulating Materials in MC Switchgear Assemblies

Class of Insulating Material	Limit of Hottest-Spot Total Temperature (°C)	Limit of Hottest-Spot Temperature Rise (°C)
90	90	50
105 (A)	105	65
130 (B)	130	90
155 (F)	155	115
180 (H)	180	140
220 (R)	220	180

Source: Modified from ANSI/IEEE Standard C37.20.2–1999, *IEEE Standard for Metal-Clad Switchgear.* © IEEE.

7.1.4 TEMPERATURE LIMITATIONS FOR SWITCHGEAR ASSEMBLY BASED ON INSULATION CLASS

An example of temperature limitations for distribution power equipment determined by applied insulation can be learned from ANSI/IEEE Standard C37.20.2-1999, "IEEE Standard for Metal-Clad Switchgear" [20]. The metal-clad (MC) MV switchgear contains drawout electrically operated CBs; it is compartmentalized to isolate all the components such as instrumentation, main bus, and both incoming and outgoing connections with grounded metal barriers. In this Standard, temperature limitations and classification of the insulating materials used in switchgear are specified (Table 7.3).

The limiting temperature for the MC switchgear is the maximum temperature permitted for: (a) any component, such as insulation, buses, IT, and switching and interrupting devices; (b) air in cable termination compartments; (c) any noncurrent-carrying structural parts; and (d) the air surrounding the devices. The total temperature to which insulating materials are subjected shall not exceed the values defined for the various classes of insulating materials, based on the maximum ambient temperature 40°C.

7.2 AGING OF INSULATING MATERIALS DUE TO ELECTRICAL STRESS

Life of insulating materials strongly depends on the ability to resist deterioration under the influence of environmental factors, such as humidity, pollution (corrosive gases, particles, and dust), temperature changes, and others. In the case of electrical insulation, it is important to note that when in service, the insulation is exposed to multiple stresses: electric, thermal, and mechanical stresses (bending, vibration, etc.).

All the stresses the insulation is exposed to irreversibly change the material properties with the onset of time and thus reduce progressively the ability of the insulation to endure the stress itself. This process is called aging and ends when the insulation is no more able to withstand the applied stress. In the following sections, several particularly damaging effects caused by the presence of electromagnetic field are described that shorten the life of the insulation. Various methods for monitoring the presence of harmful phenomena, maintenance techniques, and corrective action to extend the life of insulation will be presented in the following parts of the book.

7.2.1 ELECTRICAL BREAKDOWN IN INSULATION

When the electric field is high enough, a dielectric material may suddenly lose its property of nonconduction, permanently or temporarily, showing an electrical breakdown. In other words, electrical breakdown is the result of a rapid reduction in the resistance of an electrical insulator that can lead to a spark jumping around or through the insulator. It may be a momentary event (as in an electrostatic discharge), or may lead to a continuous arc discharge if protective devices fail to interrupt the current in a high-power circuit.

The breakdown of the insulation of an electrical wire or other electrical components usually results in a short circuit or a blown fuse. This occurs at the breakdown voltage. Actual insulation breakdown is more generally found in HV applications, where it sometimes causes the opening of a protective CB. Electrical breakdown is often associated with the failure of solid or liquid insulating materials used inside HV transformers or capacitors in the electricity distribution grid. Electrical breakdown can also occur across the insulators that suspend overhead power lines, within underground power cables, or lines arcing to nearby branches of trees.

Under a sufficient electrical stress, electrical breakdown can occur within solids, liquids, gases, or vacuum. However, the specific breakdown mechanisms are significantly different for each kind of dielectric medium. All this leads to catastrophic failure of power equipment. During electrical "flashover"—arcing between conductors—the insulation is penetrated and usually destroyed by the particles of conducting material boiled off the conductors during the passage of current.

7.2.2 CORONA

7.2.2.1 Destructive Nature of Corona

This phenomenon received its name because it is visible to the unaided eye under certain conditions. A partial breakdown of the air occurs as a corona discharge on HV conductors at those points with highest electrical stress. Conductors whose configuration includes sharp points, or balls with small radii, are more prone to dielectric breakdown.

Corona is sometimes seen as a glow around HV wires and heard as a sizzling sound along HV power lines. Corona also generates RF noise that can also be heard as "static" or buzzing on radio receivers. A true corona involves only dielectric breakdown—ionization—of the air adjacent to a conductor. The heat that is generated during corona is too little to thermally break down the insulation material or to melt the conductors. No dielectric failure occurs within the insulation.

However, corona is destructive [21]. During corona, ionization of air converts oxygen to ozone, a highly reactive gas that chemically attacks insulation materials. In addition, physical bombardment of insulation surfaces by ionized air particles wears away the solid material as by sandblasting. This deterioration has several consequences. In coil insulation, prior to complete winding breakdown, a whitish residue, like mildew, may appear on end turn surfaces. This "powdering" consists of fine particles of insulation eroded away from coil surfaces. Coils later removed from the stator can show tape or wrapper sections eaten away as metal would be by corrosion. That is particularly true for polyesters. Corona discharges on rubber-based insulators, tape, and insulation board leave a white, fibrous power residue or dust (Figure 7.4) [22]. This dust is the result of physical breakdown of the material. The white power left by corona is conductive and it can support arcing conditions. Inside electrical components such as transformers, capacitors, electric motors, and generators corona progressively damages the insulation, leading to premature equipment failure. One form of attack is ozone cracking of elastomer items such as like O-rings.

7.2.2.2 Corona Tracking

The term "surface discharge" is also used for the sparking associated with "tracking" or current flow along the contaminated surfaces or end windings. Such tracking quickly leads to carbonization of insulation through local overheating. It will cause insulation damage much sooner than the mechanical or chemical erosion caused by air ionization (Figure 7.5a). Practically, once corona becomes active, it leaves behind a conductive "tracking" path on surfaces and also creates a very

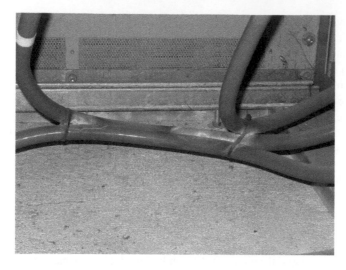

FIGURE 7.4 White powder/dust residue formed on 13 kV power cables that are tie-wrapped together, forming tight air spaces between each other. (From Brady J. Corona and tracking conditions in *MC Switchgear Case Studies*, Online. Available: http://www.irinfo.org/articles/article_8_1_2006_Brady.pdf. With permission.)

conductive cloud of air around itself. A flash-over can occur once a tracking pathway is completed from phase to phase or phase to ground [22]. It can also occur from the conductive cloud of the surrounding air once it finds a path to the ground. Three major causes for corona development should be considered: geometric factors, spatial factors, and contamination [22,23].

7.2.2.3 Corona in Switchgear

Geometric factors causing corona in switchgear include sharp edges on conductors, connections, and switchgear cabinet components. There are multiple locations within switchgear where corona may develop, such as sharp or squared tape wraps in conductor terminations, tag ends on conductors, and corners and points on cabinet bracing and support shelves. Spatial factors in switchgear include small air spaces between conductors, insulation board, and switchgear cabinet components. Lack of space may result from conductors being tie wrapped together as well as from conductors

(a)

(b)

FIGURE 7.5 Advanced stage of carbon tracks on insulation board and insulated tape wrap on 4160 V bus (a); severely deteriorated power feed cable in 4160 V switchgear cabinet displaying white power and carbon tracks (b). (From Brady J. *Corona and Tracking Conditions in MC Switchgear Case Studies*, Online. Available: http://www.irinfo.org/articles/article_8_1_2006_Brady.pdf. With permission.)

FIGURE 7.6   Unusual weathering pattern on 13 kV copper bus under attack by corona produced nitric acid. (From Brady J. *Corona and Tracking Conditions in MC Switchgear Case Studies*, Online. Available: http://www.irinfo.org/articles/article_8_1_2006_Brady.pdf. With permission.)

touching insulators, conduit, and edges of cabinets. Corona may be initiated near nonshielded cables in contact with grounded surfaces or on busbars in close proximity to fiber–resin supports and insulator material. Contamination such as dust, oils/fluids, and other particulates on conductors and insulators will also create corona.

Humid and wet conditions are very favorable for developing corona. Humid and wet conditions inside switchgear cabinets will allow nitric acid to form, which attacks the copper surface, leaving unusual weathering patterns. Cabinets lacking heaters, cabinets with poor weather seals, and those next to wet industrial processes are especially vulnerable to corona conditions.

As the corona condition worsens, carbon tracks develop on conductors and insulators. The distance between the phase and the ground will determine the time to reach a flashover. In worst case scenarios, cables will be severely deteriorated (Figure 7.5b). Other indicators include discoloration and pitting on cable insulation. Usually, dull finishes and microcrack stains on cable insulation will be noticed. Unusual weathering patterns on copper bus and conductors are also good indicators of corona (Figure 7.6).

7.2.3   PARTIAL DISCHARGE

In electrical engineering, *partial discharge* (PD) is a localized dielectric breakdown of a small portion of a solid or fluid electrical insulating system (EIS) under HV stress, which does not bridge the space between two conductors. While a corona discharge is usually revealed by a relatively steady glow or brush discharge in air, PDs within solid insulation system are not visible. PD can occur in a gaseous, liquid, or solid insulating medium. It often starts within gas voids, such as voids in solid epoxy insulation or bubbles in transformer oil. Protracted PD can erode solid insulation and eventually lead to the breakdown of insulation.

Therefore, the existence of the defects in insulating material, particularly the presence of voids and pores, usually leads to PD, which accelerates the process of aging of the insulator. PD usually begins not only within voids, cracks, or inclusions in a solid dielectric, but also at conductor–dielectric interfaces within solid dielectrics (Figure 7.7) [24,25]. PD can also occur along the boundary between different insulating materials and along the surface of solid insulating materials if the surface tangential electric field is high enough to cause a breakdown along the insulator surface.

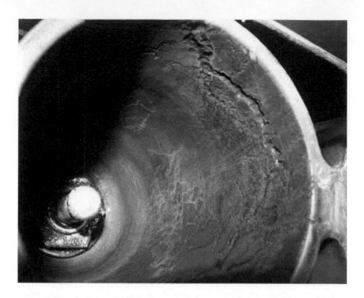

FIGURE 7.7 PD damage on a cast resin circuit breaker spout. (From N. Davies. *Why Partial Discharge Measurement Can Pay Dividends*, Online. Available: http://www.powerengineeringint.com/articles/print/ volume-18/issue-3/features/why-partial-discharge-measurement-can-pay-dividends.html. With permission.)

This phenomenon commonly manifests itself on overhead line insulators, particularly on contaminated insulators during days of high humidity. Overhead line insulators use air as their insulation medium. Once begun, PD causes progressive deterioration of insulating materials, ultimately leading to electrical breakdown. The effects of PD within HV cables and equipment can be very serious, ultimately leading to complete failure. Partial discharge within solid dielectrics is the formation of numerous, branching, and partially conducting discharge channels, a process called treeing.

Repetitive discharge events cause irreversible mechanical and chemical deterioration of the insulating material. The chemical transformation of the dielectric also tends to increase the electrical conductivity of the dielectric material surrounding the voids. This increases the electrical stress in the as yet unaffected gap region, accelerating the breakdown process. Damage is caused by the energy dissipated by high-energy electrons or ions, UV light from the discharges, ozone attacking the void walls, and cracking as the chemical breakdown processes liberate gases at high pressure.

The first solid insulation used for electrical applications was porcelain. A number of inorganic dielectrics, including glass, porcelain, and mica materials are very good insulators; they are also basically immune to electrical degradation, being significantly more resistant to PD damage than organic and polymer dielectrics. Nowadays, solid insulation is often replaced by polymeric materials, which generally offer good electrical insulation properties. However, these materials are subject to surface aging and degradation processes. Any electrical activity on the surface of such materials causes tracking damage.

In practice, PD in HV/MV insulation exists in two forms, surface PD and internal PD. When surface PD is present, tracking occurs across the surface of the insulation, which is intensified by airborne contamination and moisture, leading to erosion of the insulation.

Surface degradation occurs as a result of the charring of the insulating material by electrical discharge according to the following scenario. The degradation is promoted by the presence of contaminants and/or moisture on the surface forming a conducting film on insulating parts, which allows a leakage current to permeate through the film. When this current is interrupted at a narrow point in the film, it results in a small discharge arc. High localized heating of the surface at the point of the arc produces a small spot of charred material. Further contamination of the surface allows the process to repeat until the spot becomes extended in lines in the direction of the field. Multiple permanent carbonized paths or lines caused by the degradation of the insulating material are formed on the surface of an insulator, resulting in a phase to earth fault.

Internal PD occurs within the bulk of insulation materials and is caused by age, poor materials, or poor-quality manufacturing processes. If allowed to continue unchecked, either mechanism of PD will lead to the failure of the insulation system, potentially resulting in catastrophic failure of the equipment.

7.2.3.1 Partial Discharge in Switchgear

PD activity has long been accepted as a major cause of failure of HV/MV switchgear [24,25], where during a temporary overvoltage, during an HV test, or under transient voltage conditions during operation PDs may occur on nonself-restoring insulation, which includes gas, liquid, and solid materials.

IEEE Standard 493-1997, "IEEE Recommended Practice for Design of Reliable Industrial and Commercial Power Systems—Gold Book" [26], concludes that "If these partial discharges are sustained due to poor materials, design, and/or foreign inclusions in the insulation, degradation and possible failure of the insulation structure may occur." It has been found [27,28] that environmental conditions play a significant role in the inception and development of PD in switchgear.

High levels of RH lead to an almost instantaneous increase in the levels of recorded surface PD, particularly when the moisture is condensed on the surfaces of the CB. It was observed that the PD activity could cease during dry periods, even during the advanced stages of damage to the cast resin insulation. It was also observed that the levels of RH correlate with the presence and magnitude of PD activity.

7.2.3.2 Partial Discharge in Paper-Insulated HV Cables

PD in HV cables begins as small pinholes penetrating the paper windings that are adjacent to the EC or outer sheath. As PD activity progresses, the repetitive discharges eventually cause permanent chemical changes within the affected paper layers and impregnating dielectric fluid [29]. Over time, partially conducting carbonized trees are formed. This places greater stress on the remaining insulation, leading to further growth of the damaged region, resistive heating along the tree, and further charring (also called *tracking*; Figure 7.8). This eventually culminates in the complete dielectric failure of the cable and, typically, an electrical explosion.

PD generally dissipates energy in the form of heat. Sometimes, however, the energy is lost in the form of sound and light, such as the hissing and dim glowing from the overhead line insulators. Heat energy dissipation may cause thermal degradation of the insulation, although the level is generally low. For HV equipment, the integrity of the insulation can be confirmed by monitoring the PD activities that occur through the equipment's life.

To ensure supply reliability and long-term operational sustainability, PD in HV electrical equipment should be monitored closely with early warning signals for inspection and maintenance. PD can be prevented through careful design and material selection. In critical HV equipment, the integrity of the insulation is confirmed using PD detection equipment during the manufacturing stage as well as periodically through the equipment's useful life. PD prevention and detection are essential to ensure reliable, long-term operation of HV equipment used by electric power utilities.

There are several national and international standards that define the techniques of PD evaluation and testing: (1) EC 60270:2000/BS EN 60270:2001, "High-Voltage Test Techniques—Partial Discharge Measurements"; (2) IEEE 400-2001, "IEEE Guide for Field Testing and Evaluation of the Insulation of Shielded Power Cable Systems"; (3) IEEE 1434-2000, "IEEE Trial-Use Guide to the Measurement of Partial Discharges in Rotating Machinery."

(a)

(b)

FIGURE 7.8 Partial discharge tracking on CB insulation (a and b). (From Davies N. and Goldthorpe S. Testing distribution switchgear for partial discharge in the laboratory and the field, *Proceedings of 20th International Conference on Electricity Distribution*, CIRED 2009, Prague, 8–11 June 2009, paper 0804, Online. Available: http://www.cired.be/CIRED09/pdfs/CIRED2009_0804_Paper.pdf. With permission.)

7.3 ENVIRONMENTAL AGING OF INSULATING MATERIALS

The polymer insulator is constructed from two basic components, a core and weather shed. The core consists of a rod made of glass-fiber reinforced resin (silicone resin—SID) coupled to metal fittings at both ends. The rod is protected against external agents by polymeric sheds, which also provide the shape and leakage distance required for the rod to withstand electrical stresses.

The working and environmental conditions such as mechanical stresses, electrical fields, moisture, ozone, UV light, surface arcing, and contamination affect the performance of the polymeric insulators.

7.3.1 INSULATION DETERIORATION UNDER ENVIRONMENTAL CONDITIONS

Longevity of insulators depends on ability of the material to withstand detrimental effects of the environment, such as high humidity, pollution (corrosive gases, particles, and dust), cold and hot temperature, and their fast changes and other naturally existing factors. UV radiation and acid rain are the other sources of the aging caused by the surface chemical and structural changes derived from these stresses. Particulates deposited on the insulator surfaces can induce corona discharge and local TR may speed up insulator aging [30].

Different areas where electrical equipment is installed may differ significantly in what kind of pollution may be found on the surfaces of insulators, which cause destructive electrical discharges.

These areas are usually divided into several pollution categories: natural, industrial, and mixed. Natural surroundings are also different. In the inland environment, the source of pollution is mostly soil dust, which is not very conductive and adhesive and usually can be easily removed by natural and artificial washing. In the desert, sand may deposit on the surfaces and might have a high conductivity; some sands contain >20% of soluble material. In coastal areas where the wind carries sea water, salt crystals may form on the insulating surfaces. This deposit is not very adhesive and could be easily removed by natural and artificial washing.

In industrial areas, pollution is much worse. If electrical equipment is installed near or at steel mills, coke plants, cement factories, chemical plants, paper mills, generating stations, quarries, and so on, the deposits coming from the industrial atmosphere may be highly conductive and often combine with insoluble materials. When an industrial facility is located close to the sea coast or in the desert, these pollution factors combine and may produce very adhesive and very conductive deposits.

In the study of aging of silicon rubber (SIR) [5], it was shown that oxidation and formation of by-products of oxidation are the most dominant chemical reactions during the aging of SIR.

7.3.2 BIOLOGICAL CONTAMINATION AND CORROSION OF INSULATORS

In the tropical condition, biofilms may grow on SIR insulators. The earlier experience was with finding a biological growth [31] on SIR insulators in Sri Lanka, but a similar contamination is frequent in other parts of the world [32,33]. Intensive biological growth of algae and fungus was found on energized insulators. The contaminants are usually present on the upper sides of the sheds, except for the top ones. In most cases, the contaminant covered the shed area almost completely.

In the marine environment, sea spray, mist, or fog can carry sodium chloride several miles from the coast, which makes the installations located a substantial distance from the ocean vulnerable. It is found that exposure to high humidity and intensive salt contamination leads to corrosion of metallic connections between the tower and the insulator end fittings as well as insulator end fittings [31], although the salt deposits are not often visible on the insulators because it is easily removed by rains.

7.3.3 ENVIRONMENTAL AGING OF INSULATORS IN TRANSMISSION LINES

Overhead transmission or distribution lines are widely used in the present power system to transmit electric power from generation stations to customer points. Their proper function largely depends on the insulation system with the supporting structures [34]. The performance of outdoor insulators, as the main insulating material, is influenced by many environmental parameters. Two of these parameters are pollution and humidity.

Transmission lines sometimes pass through many polluted or emitted areas, such as coastal, industrial, and even geothermal areas. In the particular case of geothermal areas, transmission lines are used to transmit electric energy from a geothermal power plant (GTPP) to a switchyard of distribution. Geothermal power stations create emission as chemical materials. Most chemical compounds are present in gaseous form, especially CO_2 (94%–98%) and H_2S (1%) [33].

Geothermal liquids (steam or hot water) usually contain gases, such as CO_2, H_2S, NH_3, CH_4, and trace amounts of other gases, as well as dissolved chemicals whose concentrations usually increase with temperature. For examples, sodium chloride (NaCl), boron (B), arsenic (As), and mercury (Hg) are sources of pollution if discharged into the environment [35].

GTPP could release the following pollutants: CO_2 (98%), H_2S (1.5%), methane (0.4%), and hydrogen (0.1%). The oxidation of H_2S to SO_2 and its subsequent reaction with sulfate ions produces aerosols representing a major component of acid rain. GTPP could also release trace amounts (0.001%) of ammonia, radon, boron, arsenic, cadmium, and antimony. Ammonia and mercury may also enter local waters from geothermal steam condensation [35].

All these pollutants dissolved in water could easily deposit on the surface of insulators and eventually destroy it. An example of the composition of sticky pollutants found on the surface of insulators in Kamojang switchyard (in Thailand) is given in Table 7.4 [35], which shows that pollution contains many elements (metallic and nonmetallic), whose compounds could be very chemically active and harmful.

TABLE 7.4
The Composition of Pollution Deposit on Insulator Surface

Element	Amount (%)	Element	Amount (%)	Element	Amount (%)
C	61.23	Cu	5.82	S	2.40
Si	11.92	Fe	5.63	K	1.33
Al	6.81	Ca	4.32	V	0.54

Source: Modified from Waluyo, Ngapuli I. S., Suwarno, and Djauhari M. A. *ECTI Trans. Electrl. Eng. Electron. Commun.* 8(1), 126–145, 2010.

7.3.4 SCC IN COMPOSITE INSULATORS

NCI, also referred to as composite or polymeric insulators, are used in overhead transmission lines (OHTL) with line voltages in the range of 69–735 kV. Despite the many benefits that NCI offer in comparison with their porcelain counterparts, they can fail mechanically in service by rod fracture. One of the mechanical failure modes of the insulators is a failure process called brittle fracture, which is caused by the SCC of the GRP rods.

It was found [36] based on experimental evidence that brittle fracture failures of composite (non-ceramic) HV insulators could be caused by water and mechanical stresses. However, in Refs. [37,38], it has been clearly shown that only water, in the absence of electrical field, cannot cause SCC of unidirectional E-glass/polymer composites and thus brittle fracture of composite NCI. It was rather seen as a cumulative effect of the formation of nitric acid in service through corona discharges, ozone, and moisture.

The process is catastrophic and unpredictable, leading to the drop in energized transmission lines. The most important characteristic of the brittle fracture process is that it can occasionally affect HV nonceramic transmission line insulators, leading to their catastrophic in-service failures [39,40].

To prevent brittle fracture, water or nitric acid must never be allowed inside the insulators; however, it is difficult to effectively provide for long periods of time considering the harsh nature of the environment in the vicinity of the energized ends of the insulators (nitric acid, pollution, acid rain, ozone, corona discharges, etc.). It can be expected that even the best shielding against moisture ingress will sooner or later fail to protect the insulators. Some manufacturers are making every effort to design such a protection in the best way they can [39]. However, it might be a better solution to develop an insulator with a high resistance to brittle fracture.

7.4 HV BUSHINGS IN TRANSFORMERS AND CBs

Bushing is a very important component in the design of switchgear, CBs, and transformers. It is a hollow insulator, allowing a conductor to pass along its center and connect at both ends to other conductive elements. Bushings are often made of wet-process fired porcelain, and may be coated with a semiconducting glaze to assist in equalizing the electrical stress along the length of the bushing. The inside of the bushing may contain paper insulation and the bushing is often filled with oil to provide additional insulation.

Bushings for MV and LV apparatus may be made of resins reinforced with paper. The use of polymer bushings for HV applications is becoming more common. The largest HV bushings made are usually associated with HV DC converters.

Some of the HV types are called capacitor bushings because they form a low-value capacitor between the conductor and the wall. They are made of layers of conductive paper, film, ink, or aluminum foil with an insulating medium. This is done to reduce the electrical field stress that would otherwise occur and cause breakdown.

7.4.1 TYPES OF BUSHINGS

HV bushings manufactured for use on transformers and breakers are of several types [41]:

1. *Composite bushing*—A bushing in which insulation consists of two or more coaxial layers of different insulating materials.
2. *Compound-filled bushing*—A bushing in which the space between the major insulation (or the conductor where no major insulation is used) and the inside surface of a protective weather casing (usually porcelain) is filled with a compound having insulating properties.
3. *Condenser bushing*—A bushing in which cylindrical conducting layers are arranged coaxially with the conductor within the insulating material. The length and diameter of the cylinders are designed to control the distribution of the electric field in and over the outer surface of the bushing. Condenser bushings may be one of several types: resin-bonded paper insulation; oil-impregnated paper insulation; or others.
4. *Dry or unfilled-type bushing*—Consists of porcelain tube with no filler in the space between the shell and conductor. These bushing are usually available from 38 kV and below (Figure 7.9).
5. *Oil-filled bushing*—A bushing in which the space between the major insulation (or the conductor where no major insulation is used) and the inside surface of a protective weather casing (usually porcelain) is filled with insulating oil.
6. *Oil-immersed bushing*—A bushing composed of a system of major insulations totally immersed in a bath of insulating oil.
7. *Oil-impregnated paper-insulated bushing*—A bushing in which the internal structure is made of cellulose material impregnated with oil.

FIGURE 7.9 Porcelain dry-type bushings in ratings from 15 to 38 kV. (From Piedmont Bushings & Insulators, Online. Available: http://www.pb-i.com/images/4bushings_1.jpg. With permission.)

TABLE 7.5

Bushing Defects and Possible Causes of the Failure

Bushing Defects	Possible Causes of the Failure
Cracks in porcelain, deterioration of cemented joints	Moisture inside bushing, oil, gas, or filler leaks
Gasket leaks, moisture in insulation, solder seal leak	Moisture inside bushing, oil, gas, or filler leaks
Broken connection between ground sleeve and flange	Sparking in apparatus tank or within bushing. Discolored oil
Voids in compound	Internal corona
Oil migration	Filler contamination
No oil	Oil leaks out. Moisture enters
Displaced grading shield	Internal sparking discolors oil
Electrical flashover, lightning	Cracked or broken porcelain
Corona	Internal breakdown. Radio interference. Treeing along surface of paper or internal surfaces
Short-circuited condenser sections	Increased capacitance. Reduced voltage at capacitance tap terminal. Internal stress to insulation
Darkened oil	Radio interference, poor test results

Source: From Facilities instructions, standards, and techniques, in *Testing and Maintenance of High-Voltage Bushings*, Vol. 3–2, US Department of the Interior Bureau of Reclamation, 1991, Online. Available: www.usbr.gov/power/data/fist/fist3_2/vol3-2.pdf. With permission.

8. *Resin-bonded, paper-insulated bushing*—A bushing in which the major insulation is provided by cellulose material bonded with resin.

7.4.2 BUSHINGS: POSSIBLE CAUSES OF FAILURES

About 90% of all preventable bushing failures are caused by moisture entering the bushing through leaky gaskets or other openings according to operating records [41]. Table 7.5 shows the defects found in the bushings and common causes of bushing failures. Bushing failures may cause very costly outages. HV bushings, if allowed to deteriorate, may explode with considerable violence and cause extensive damages to adjacent equipment.

After long exposure to all stresses, multiple chips, cracks, and contamination may be found on the surface of porcelain bushings. Minor chips may be painted with an insulating varnish to obtain a glossy finish which will shed dirt and moisture. Bushings with major chips or cracks can no longer stay in service since these defects significantly decrease the creepage distance.

Deposits of dirt on the bushings, particularly in areas where there are contaminants such as salts or conducting dusts in the air, may cause flashovers. These deposits should be removed by periodic cleaning of the surface of the porcelain to remove dirt, oil, and other deposits that may reduce the flashover value. Multiple tests are recommended to determine the physical condition of the bushings [41,42].

7.5 POWER CABLE INSULATION

The power cable is an essential part of any electrical system. Power cables may be installed as permanent wiring within buildings, buried in the ground, made to run overhead, or exposed. Cables consist of three major components: conductors, insulation, and a protective jacket.

The construction and material of individual cables vary according to the application and are determined by several major factors: working voltage, current-carrying capacity, and environmental conditions such as temperature, water, chemical, or sunlight exposure, and mechanical impact. The thickness of the insulation is determined by working voltage, while the cross-sectional size of the conductor(s) is determined based on the current-carrying capacity. The form and composition of the outer cable jacket is usually chosen based on the environmental conditions where the cable would be installed.

Power cables use stranded copper or aluminum conductors, although small power cables may use solid conductors. Some power cables for outdoor overhead use may have no overall sheath. Other cables may have a plastic or metal sheath enclosing all the conductors. The materials for the sheath will be selected for resistance to water, oil, sunlight, underground conditions, chemical vapors, impact, or high temperatures.

In nuclear industry applications, the cable may have to meet special requirements for ionizing radiation resistance. It may be specified that the cable materials should not produce large amounts of smoke if burned. Cables intended for underground use or direct burial in earth will have heavy plastic or metal, most often lead sheaths, or may require special direct-buried construction. When cables must run where they are exposed to mechanical impact damage, they may be protected with flexible steel tape or wire armor, which may also be covered by a water-resistant jacket.

There are many types of cables depending on voltage and application. MV cables may be used for underground residential distribution (URD; 15–35 kV), as cable in conduit (CIC), as a single or multiconductor (5–69 kV), as armored submarine cable, mining cable, and so on. A paper-insulated underground distribution cable carries power ranging from 5 to 46 kV, and a paper-insulated underground transmission cable carries power ranging from 69 to 345 kV.

7.5.1 Cable Insulation Types

Electrical insulation materials are used over the metallic conductors of underground cables at all voltage ratings. *Paper* insulation was used first in the power industry, and was later replaced in LV and MV applications with polymeric materials, but the nature of the polymer may vary with the voltage class. After synthetic polymer development, *PE* and *cross-linked polyethylene* (XLPE) have been used as insulation material; the latter version is used in most countries.

XLPE is considered to be the material of choice due to its ease of processing and handling. A version of XLPE insulation is Tree-Retardant Cross-Linked Polyethylene (TR-XLPE). Another insulation synthetic material used for insulation is *ethylene propylene rubber* (EPR).

Cables for transmission are operating above 46 kV; they have traditionally used paper and oil systems as the insulation. The paper is applied as a thin film wound over the cable core. A variation of paper insulation was developed as a *laminate of paper with polypropylene* (PPP or PPLP), and paper-insulated lead-covered (PILC) cables are paper insulated with lead-sheathed cables. MV power cables operate in the voltage range of 5–45 kV and the following types of cables are used more often than others: XLPE-insulated power cables; paper-insulated power cables; EPR trailing cables for mines; and bundle conductors MV XLPE types.

Cables with paper-impregnated insulation have served the consumers and industry extremely well; they are reliable and continue operating for up to 40 years. With the introduction of the cables with XLPE insulation, the sales pitch was that these cables could be laid into an unprepared trench and would survive.

The first generation of XLPE cables (before 1980) were badly constructed with tape or loose core screens. The latest PVC or XLPE cables can be as reliable as cables with paper-impregnated insulation, provided they are correctly installed and maintained. Solid dielectric cables will eventually replace the cables with paper-impregnated insulation.

7.5.2 Aging of Cable Insulating Materials

Many cables as a part of electric power systems have been manufactured and installed over many decades. The oldest oil-paper insulated cables still in use were installed in the 1940s. Installation of XLPE cables started in the middle of the 1970s. Some of the oil-paper insulated cables are rather old and the first XLPE cables are old enough to show possible degradation results. The insulation systems of power cables and their accessories are subject to different kinds of stresses during their service life and thus suffer degradation. This can lead to a reduction of life, which in turn can lower the reliability of electrical power systems.

Aging of insulation in MV and HV cables is caused by the same factors as solid insulation: electrical and mechanical stresses and environmental deterioration. One of the most serious electrical stresses leading to significant deterioration of insulation in power cables is PD, which eventually results in failure.

7.5.2.1 XLPE Cable Insulation Degradation

The degradation processes of XLPE insulation can be categorized into two main groups: (1) due to voids, contaminants, physical imperfections, or poorly dispersed components; and (2) due to physical or chemical changes or trapped charges in the cables [43–46], many of them leading to *thermal* degradation of the insulation. During normal operation, cable XLPE insulation temperatures should be below 90°C, while during fault conditions cables should withstand temperatures of up to 120°C. At temperatures between 150°C and 225°C, chemical changes are activated in the polymer, which decrease the mechanical strength and the density of insulation. At temperatures higher than 225°C, chemical deterioration is detected, which further damages the insulating properties.

Residual internal *mechanical* stresses from manufacturing have a significant effect on the breakdown strength of cable insulation [44].

The most dangerous defects in polymeric insulation are caused by *electrical* degradation. PDs, electrical trees, and water trees are the most important mechanisms. Defects in insulation structure increase the effect of electrical degradation and the risk of electrical breakdown. Polymeric insulation is sensitive even to small PDs [45]. Small cavities or gas bubbles can form inside the insulation or in the surface between insulation layers during the manufacturing or installation of accessories on the cable. When the electric field inside a cavity or bubble is high enough, partial breakdown will occur.

PDs in one location within the cable insulation will lead to an electrical degradation that affects the insulation locally and randomly. Electrical degradation does not affect the whole cable as thermal degradation does. The defect leading to final breakdown will usually be a local phenomenon. Low electric field intensity and long development time are common for electrical degradation mechanisms in MV cables.

7.5.2.2 Electrical and Water Treeing

An electrical tree is a network of fine conductive channels that spreads relatively quickly through the insulation to cause failure. *Electrical trees* can "grow" from the eroded surface of the void or in the micro cavities of the polymer. They develop further from the expanded void due to the PDs within the branches. The tree growth rate depends on the applied electrical stress, the temperature, and the environmental and mechanical stresses [45,46].

The majority of XLPE insulated cables installed in the ground are exposed to moisture. The result is another degradation process called *water trees*. Water trees are generally found to initiate and grow in XLPE insulation exposed to an alternating electric field and humidity. Impurities inside the insulation material increase the risk of water tree formation. Water can penetrate the insulation from the outside of the cable if water blocking barriers are not used. Water can also enter cable insulation from termination or joint faults. Moisture inside the insulation will start to propagate in the direction of the electric field in a bush or tree form.

Growth of the water trees depends on the presence of water, the intensity and frequency of the electric field, the insulation material, temperature, and mechanical stresses. Total breakdown of the insulation can happen when a branch of a water tree bridges the electrodes. The underground cables installed in moist environments should be protected against moisture penetration into the insulation.

7.6   OTHER INSULATING MEDIA

7.6.1   INSULATING OIL

The dielectric oil functions are to insulate, suppress corona and arcing, and to serve as a coolant. Therefore, insulating oil should remain stable at high temperatures for an extended period of time. Insulating oil is usually a highly refined mineral, vegetable, or synthetic oil that is stable at high temperatures and has excellent electrical dielectric properties. It is used in oil-filled transformers, some types of HV capacitors, and some types of HV switches and CBs.

7.6.1.1   Transformer Oil

The purpose of using oil in a transformer is to cool it. To improve the cooling of large power transformers, the oil-filled tank may have external radiators through which the oil circulates by natural convection. Very large or high-power transformers (with the capacities of thousands of KVA) may also have cooling fans, oil pumps, and even oil-to-water heat exchangers. To ensure that the transformer is completely free of water vapor before the cooling oil is introduced, large HV transformers undergo prolonged drying processes, using electrical self-heating, the application of a vacuum, or both. This helps prevent corona formation and subsequent electrical breakdown under load. The oil also provides electrical insulation between internal live parts.

Multiple gases are produced in the oil as a side effect of corona or an electric arc in the windings. Oil-filled transformers with a conservator (an oil tank above the transformer) tend to be equipped with safety devices (Buchholz relays) that detect the build-up of gases (such as acetylene) inside the transformer and switch off the transformer. Transformers without conservators are usually equipped with sudden pressure relays, which perform a similar function as the Buchholz relay.

7.6.1.2   Oil Switches and CBs

Mineral transformer oil is used for interrupting and cooling an electrical arc. Oil circuit breakers (OCBs) are CBs that have their contacts immersed in oil. Current interruption takes place in oil, which cools the arc developing during interruption, and thereby oil quenches the arc. Because of the high temperature of the arc, the oil is evaporated rapidly and oil vapors are partially decomposed, producing ethylene, methane, and other gases. A gas bubble is formed in the arcing zone; the pressure in the bubble is very high. The arc is then extinguished, both because of its elongation upon the parting of contacts and because of intensive cooling by the gases and oil vapor.

There are two types of oil-filled switches and CBs: (1) tank-type oil and (2) oil-minimum or low-oil-capacity. In the first type, the main contacts and the arc-quenching devices are located in a grounded metal tank; in the second type, they are in an insulating or ungrounded metal enclosure filled with oil. Tank-type OCBs are inferior to other types of HV breakers in many respects. However, their low cost and high reliability have led to their continued use.

7.6.1.3   Aging of Transformer Oil

Where oil is used as insulating media, there are multiple mechanisms of oil aging and deterioration, such as ingress of moisture, ingress of air, and oversaturation with gas. Aging oil displays thermal instability and is often contaminated with metal-containing colloids and particles. All these oil-deteriorating processes affect the performance of the electrical machine. One of the most important aging mechanisms is decomposition of the oil itself during arc interruption and overheating, which leads to deterioration of its dielectric properties. The second important aging problem is the deposition of conductive and polar particles accumulating in oil during the service. These particles form a deposit on the solid insulation of the parts, leading to insulation surface contamination, which causes a distortion of electrical field and a reduction in the electrical strength of the insulation system [47,48]. For more about transformer oil aging, see Section 17.3.

There are several transformer oil tests that should be periodically performed to determine the condition of transformer oil.

7.6.1.3.1 Dissolved Gas Analysis

Determines the types and concentrations of the gases dissolved in transformer oils. The gases are generated in the oil due to the breakdown of the paper or oil under stress or degradation. With regular dissolved gas analysis (DGA) testing, multiple problems can be detected months in advance.

7.6.1.3.2 Furanic Analysis

Determines the concentration of furans in transformer oils. Furans are degradation products of the insulation paper found in transformers. Analysis of furans is important in predicting the degradation of the insulating paper.

7.6.1.3.3 Polychlorinated Biphenyl Analysis

Determines the amount of Polychlorinated biphenyl in transformer oils. Polychlorinated biphenyls are toxic compounds; their quantity in transformer oils is usually regulated and should not be exceeded.

7.6.1.3.4 Corrosive Sulfur Analysis

Determines the amount of sulfur in the oil. Sulfur reacts with copper conductors and silver contacts to form metal sulfides that contaminate the insulating paper; sulfur also forms acidic conditions in transformers. If corrosive sulfur is found, oil will need to be replaced.

7.6.1.3.5 Moisture Content

Determining the moisture content in transformer oils allows finding out whether there are any leaks in the transformer, and the increase in moisture indicates insulating paper degradation.

7.6.1.3.6 Total Acid Number

Total acid number (TAN) is the quantity of base (mg of KOH) that is required for neutralizing acid constituents in 1 g of a sample. An increase in acid indicates that sludge formation is beginning to occur or is occurring.

7.6.1.3.7 Dielectric Strength

Dielectric strength is the voltage at which breakdown of the oil occurs. Insulating power decreases as the amount of contaminants in the transformer oil increases, so the insulating quality of the oil can be predicted.

7.6.1.4 Thermal and Electrical Faults of Transformer Oil

At high temperatures, a transformer oil ages rapidly; however, there are multiple factors that accelerate oil aging, such as moisture, copper, paint, varnish, and oxygen. There are limits for the oil temperature of transformers and bushings that are defined by the standards.

ANSI/IEEE Standard C57.12.00-2010, "General Requirements for Liquid-Immersed Distribution, Power, and Regulating Transformers" [17], states that the ambient temperature shall average 30°C, and the average winding TR shall not exceed 65°C. This would mean that the top transformer oil will average 95°C in a 24-h period. Another transformer standard ANSI/IEEE C57.12.10-1997 determines top-liquid temperature-range limits. It states that the transformer shall be suitable for operation over a range of top-liquid temperatures from −20°C to +105°C, provided that the liquid level has been properly adjusted to the indicated 25°C level. The bushing standard ANSI/IEEE C57.19.00-2004, "General Requirements and Test Procedure for Power Apparatus Bushings," states that the temperature of the ambient air should not be above 40°C or below −30°C, and the temperature of the transformer insulating oil in which the lower end of the bushing is immersed and the bushing mounting surface does not exceed 95°C averaged over a 24-h period. This means that exceeding the maximum standard temperature 95°C of transformer oil for prolonged periods of time would lead to accelerated aging of the oil.

The principal mechanism of transformer oil aging is oxidation, which results in the formation of acids and other polar compounds. A transformer oil, when subjected to thermal and electrical stresses in an oxidizing atmosphere, gradually loses its stability and becomes decomposed and oxidized; its acidity increases and finally it begins to produce mud [49].

At oil temperatures from 150°C to 300°C, relatively large quantities of the low molecular weight gases, such as hydrogen (H_2) and methane (CH_4), and trace quantities of the higher molecular weight gases ethylene (C_2H_4) and ethane (C_2H_6) are produced by poor cooling or stray losses in winding or leads; between core laminations or in core, tank, or supporting structures.

As the temperature of mineral oil increases from 300°C to 700°C, the hydrogen concentration exceeds that of methane and is accompanied by significant quantities of higher molecular weight gases, first ethane and then ethylene. Beyond 700°C (the upper end of the thermal fault range), increasing quantities of hydrogen and ethylene and traces of acetylene (C_2H_2) may be produced. In general, nonelectrical thermal faults produce temperatures below 700°C; however, welding on oil-filled equipment or oily surfaces produces acetylene due to the very high temperature [50].

Under the conditions of normal transformer functioning, aging can occur at a slow pace; however, this process is accelerating under continuous electrical and thermal faults. Aging transformer oil contains various soluble hydrocarbon gases, such as methane, ethane, ethylene, acetylene, and propane, along with carbon monoxide, carbon dioxide, hydrogen, oxygen, and nitrogen [49]. Concentration and the rate of gas generation usually depend on fault characteristics, such as type and intensity.

The most important dissolved gases in transformer oil to watch are combustible gases such as hydrogen, acetylene, carbon monoxide, and carbon dioxide. An increase in the generation of dissolved combustible gases is also caused by high temperatures of transformer oil, which may happen if the cooling system failed or is not operating automatically.

Electrical faults range in energy and temperature from intermittent low-energy PDs to steady discharges of high energy (arcing). As the discharge progresses from low energy to higher energy, the acetylene and ethylene concentrations rise significantly. PDs at low temperature (150–300°C) mainly produce hydrogen, with lesser quantities of methane and trace quantities of acetylene. Some thermal decomposition of oil and cellulose is observed in this temperature range.

Arcing and continuing discharge at low and high energy generate temperatures ranging from 700°C to 1800°C, at which acetylene is produced. National and international standards are developed to define the factors of oil deterioration and the means of testing and controlling the degree of oil aging [50,51]. Typical defects and failure mechanisms of oil-filled electrical equipment and various means of testing will be presented in the following parts of the book.

7.6.2 Sulfur Hexaflouride (SF$_6$) as Insulating and Cooling Media

SF_6 is used as an insulating gas in substations, as an insulating and cooling medium in transformers, and as an insulating and arc-quenching medium in switchgear for HV and MV applications. These are all closed systems which are extremely safe and unlikely to leak.

The gas-insulated substations (GIS) use a superior dielectric gas, sulfur hexaflouride (SF_6), at moderate pressure for phase-to-phase and phase-to-ground insulation. Because of much better dielectric properties of SF_6 gas, GIS can be much smaller than air-insulated substations (AIS). A GIS is mostly used where space is expensive or not available. In a GIS, the active parts are protected from the deterioration caused by exposure to atmospheric air, moisture, contamination, and so on. As a result, a GIS is more reliable and requires less maintenance than AIS.

In the breakers filled with SF_6, the gas provides electrical insulation and effectively controls arcing. The HV conductors, CB interrupters, switches, current transformers, and voltage transformers (VTs) are enveloped in SF_6 gas inside grounded metal enclosures.

7.6.2.1 Insulating Properties and Decomposition of SF_6

SF_6 is an inert, nontoxic, colorless, odorless, tasteless, and nonflammable gas consisting of a sulfur atom surrounded by and tightly bonded to six fluorine atoms. It is about five times as dense as air. SF_6 has a dielectric withstand capability 2.5 times better than air. Usually the gas is used at 3–5 times atmospheric pressure and then the dielectric properties are 10 times better than for air. The pressure is chosen so that the SF_6 will not condense into a liquid at the lowest temperatures the equipment experiences. It is the universally used interrupting medium for HV CBs, replacing the older media of oil and air [52,53].

SF_6 gas is very stable, but it will partly decompose in association with electric discharges and arcs, for example, in a breaker. Some reactive decomposition by-products are formed because of the trace presence of moisture, air, and other contaminants. Then, gaseous and solid decomposition products are produced. Normally, the level of gaseous decomposition products is kept low through the use of absorbers built into the switchgear. In large concentrations, the decomposition products are corrosive and poisonous. Therefore, there are established routines for service personnel when opening SF_6-filled equipment for maintenance or scrapping.

The solid decomposition products are mainly metal fluorides in the form of a fine gray powder. The powder only appears where arcing has occurred, for instance, in used CBs. The powder can be easily taken care of as separate waste. The decomposition products are reactive, which means that they will decompose quickly and disappear without any long-term effect on the environment.

In modern GIS, SF_6 gas is contained for the entire lifetime of the equipment; there is no need to open the equipment for servicing and no gas can escape. The quantities of decomposition products are very small. Molecular sieve absorbents inside the GIS enclosure eliminate these reactive by-products. Small conducting particles significantly reduce the dielectric strength of SF_6 gas. The particles are moved by the electric field, possibly to the higher-field regions inside the equipment or deposited along the surface of the solid epoxy support insulators, leading to dielectric breakdown at operating voltage levels. Cleanliness in assembly is therefore very important for GIS [54].

7.6.2.2 SF_6 as a Greenhouse Gas

SF_6 is a strong greenhouse gas that could contribute to global warming. At an International Treaty Conference in Kyoto in 1997, SF_6 was listed as one of the six greenhouse gases whose emissions should be reduced. SF_6 is a very minor contributor to the total amount of greenhouse gases due to human activity, but it has a very long life in the atmosphere (the half-life is estimated at 3200 years), so the effect of SF_6 released to the atmosphere is effectively cumulative and permanent [55].

The major use of SF_6 is in electrical power equipment. However, SF_6 gas is contained in closed systems—in many cases, sealed for life. Major manufacturing companies produce the products with guaranteed leakage rates below 0.5%/year and the next generation should lose <0.1%/year. Accidental escapes of SF_6—due to mistakes during manufacturing, installation, maintenance, and decommissioning—are a bigger problem than leaks. In GIS, SF_6 is contained and can be recycled. According to international guidelines on the use of SF_6 in electrical equipment, the contribution of SF_6 to global warming can be kept at <0.1% over a 100-year horizon [56].

The emission rate of SF_6 due to the use in electrical equipment has been reduced over the last several years. Most of this reduction has been simply due to the adoption of better handling and recycling practices. Standards now require GIS to leak <1%/year. The leakage rate is normally much lower. Field checks of GIS in service for many years indicate that the leak rate objective can be as low as 0.1%/year when GIS standards are revised.

7.6.3 Air and Vacuum as Insulating Media

The atmospheric *air* is used in air-insulated CB and substations (AIS) as insulating media. Miniature LV CBs use the air alone to extinguish the arc. Larger ratings will have metal plates or nonmetallic arc chutes to divide and cool the arc. Magnetic blowout coils or permanent magnets deflect the arc into the arc chute. Air CBs use compressed air (puff) to blow out the arc, or alternatively, the contacts are rapidly swung into a small sealed chamber escaping displaced air by blowing out the arc (air-magnetic and air-blast CBs).

The development of medium and high CBs using *vacuum* as arc interrupting media began in the United States and England in the 1950s. The benefits of using vacuum in circuit interrupters are all based on the unusual dielectric properties of vacuum. Vacuum has a minuscule number of possible charge carriers (atoms and molecules); therefore, there is practically no capability to have an electrical breakdown or lower vacuum ability of quenching an arc between electrical contacts when they separate. Vacuum has a very high dielectric strength and a high natural interruption capability [57].

The principle of interrupting of AC currents by vacuum arcs is mainly used in LV and MV CBs at voltage rating up to 40.5 kV; application of vacuum at very HV above 72 kV is rare [58]. When the contacts separate in vacuum circuit breakers (VCB), the current flowing through electrodes initiates a metallic vapor discharge in the contact called the vacuum arc. The arc is then extinguished and the conductive metal vapor condenses on the metal surfaces. Another benefit of using vacuum in CBs is that there is no risk to the environment compared with SF_6 CBs [59].

The worst cause of failure of vacuum interrupters is the leak that allows air to get inside the vacuum interrupter, thus increasing the pressure inside and creating the means for electrical breakdown. Failures related to the CB interrupter chamber losing vacuum are very rare, and these devices are very reliable and safe. However, vacuum arc creates metallic vapor, which deposits on the surfaces of insulation in vacuum interrupters, thus significantly decreasing the dielectric performance of the ceramic envelope usually made of high-purity alumina (Al_2O_3), This deposition may eventually lead to the failure of the insulation; however, the density of the vapor in the arc is extremely low [60] and various designs of vacuum interrupters may confer protection of the ceramic surface against metal vapor condensation [61].

7.7   POWDER COATING AND PAINT FOR ELECTRICAL ENCLOSURES

7.7.1   Electrical Enclosures: Types and Materials

Electrical equipment should be protected from the adverse influence of the environment, particularly when they are exposed to a polluted industrial atmosphere or installed outdoor. Therefore, electrical machines are locked in boxes or enclosures that are made from many different materials, such as steel, aluminum, polyester, polycarbonate, and so on [62]. The NEMA and UL established a rating system that guarantees that an enclosure meets certain minimum conditions of corrosion resistance [63]. In Tables 7.6 and 7.7, different types of outdoor and indoor enclosures are listed together with the degree of protection from specific conditions.

Some types of electrical enclosures are not listed in Table 7.6. In hazardous locations, Type 7 and 10 enclosures are designed to contain an internal explosion without causing an external hazard. Type 8 enclosures are designed to prevent combustion through the use of oil-immersed equipment. Type 9 enclosures are designed to prevent the ignition of combustible dust [63].

The NEMA Standard 250-2003 for Enclosures for Electrical Equipment does test for environmental conditions such as corrosion, rust, icing, oil, and coolants. IEC Standard 60529 [64] also provides a system for specifying the enclosures of electrical equipment, using IP code on the basis of the degree of protection provided by the enclosure. However, it does not specify degrees of protection against mechanical damage of equipment, risk of explosions, or conditions such as moisture (produced, e.g., by condensation), corrosive vapors, fungus, or vermin. For this reason, and because the test and evaluations for other characteristics are not identical, the IEC Enclosure Classification Designations cannot be exactly equated with the enclosure Type numbers in NEMA Standard. In 2004, IEC Standard was adopted by ANSI and a comparison of the two rating systems has been provided [64,65].

The various materials used for building enclosures are listed in Table 7.8.

TABLE 7.6

Comparison of Specific Applications of Enclosures for Indoor Nonhazardous Locations

Provides a Degree of Protection against the Following Conditions	Type of Enclosure									
	1[a]	2[a]	4	4X	5	6	6P	12	12K	13
Access to hazardous parts	X	X	X	X	X	X	X	X	X	X
Ingress of solid foreign objects (falling dirt)	X	X	X	X	X	X	X	X	X	X
Ingress of water (dripping and light splashing)	...	X	X	X	X	X	X	X	X	X
Ingress of solid foreign objects (circulating dust, lint, fibers, and flyings[b])	...	...	X	X	...	X	X	X	X	X
Ingress of solid foreign objects (settling airborne dust, lint, fibers, and flyings[b])	...	...	X	X	X	X	X	X	X	X
Ingress of water (hosedown and splashing water)	...	...	X	X	...	X	X	...	...	...
Oil and coolant seepage	...	...	...	..	...	...	...	X	X	X
Oil or coolant spraying and splashing	...	...	...	...	...	...	...	...	...	X
Corrosive agents	...	...	...	X	...	...	X	...	...	...
Ingress of water (occasional temporary submersion)	...	...	...	...	...	X	X	...	...	...
Ingress of water (occasional prolonged submersion)	...	...	...	...	...	...	X	...	...	...

Source: NEMA Standards Publication 250-2003. *Enclosures for Electrical Equipment (1000 Volts Maximum)*, Online. Available: http://www.global.ihs.com. With permission.

[a] These enclosures may be ventilated.

[b] These fibers and flyings are nonhazardous materials and are not considered Class III type ignitable fibers or combustible flyings. For Class III type ignitable fibers or combustible flyings see the National Electrical Code, Article 500.

TABLE 7.7

Comparison of Specific Applications of Enclosures for Outdoor Nonhazardous Locations

Enclosure Provides a Degree of Protection against the Following Conditions	Type of Enclosure									
	3	3X	3R[a]	3RX[a]	3S	3SX	4	4X	6	6P
Access to hazardous parts	X	X	X	X	X	X	X	X	X	X
Ingress of water (rain, snow, and sleet[b])	X	X	X	X	X	X	X	X	X	X
Sleet[c]	...	...	...	...	X	X	...	...	...	...
Ingress of solid foreign objects (windblown dust, lint, fibers, and flyings)	X	X	...	...	X	X	X	X	X	X
Ingress of water (hosedown)	...	...	...	...	...	...	X	X	X	X
Corrosive agents	...	X	...	X	...	X	...	X	...	X
Ingress of water (occasional temporary submersion)	...	...	...	...	...	...	...	...	X	X
Ingress of water (occasional prolonged submersion)	...	...	...	...	...	...	...	...	...	X

Source: NEMA Standards Publication 250-2003. *Enclosures for Electrical Equipment (1000 Volts Maximum)*, Online. Available: http://www.global.ihs.com. With permission.

[a] These enclosures may be ventilated.

[b] External operating mechanisms are not required to be operable when the enclosure is ice covered.

[c] External operating mechanisms are operable when the enclosure is ice covered.

TABLE 7.8

Materials Used for Enclosures and Applications

Material	Applications
Hot-rolled pickled and oiled steel (ASTM A569)	General indoor and outdoor use after a suitable finish has been applied to protect against corrosion
Cold-rolled steel (ASTM A366)	General indoor and outdoor use after a suitable finish has been applied to protect against corrosion
Galvanized steel (ASTM A653)	Indoor or outdoor use in neutral pH. Resists oil, gas, glycerine, dichromates, borates, and silicates. The most frequent application is outdoor including seacoast atmospheres. Additional finishes may be applied to improve corrosion resistance
18–8 Stainless steel (Type 302–304)	Indoor and outdoor use. Ideally suited for use in food processing areas, dairies, breweries, or any wet area. Also works well in areas where caustic elements or alkalies are present
316 or 316 L Stainless steel	Indoor or outdoor use in almost any environment. Provides improved resistance to salt, some acids, and high temperature. Resistance to sulfates and chlorine is less than Type 304. These stainless steels are the most corrosion resistant metal materials for enclosures. A superior material choice for marine environments
5052 Aluminum	Indoor or outdoor use, particularly in marine environments. Also an excellent choice for enclosures exposed to solvents, petrochemicals, some acids, most sulfates and nitrates
Monel	Frequently specified for marine and chemical plant applications. Monel hardware is used on many of nonmetallic enclosures. High cost
Galvannealed steel (ASTM A653)	Indoor or outdoor use where a painted finish is important. Used in the automotive industry. Corrosion resistant, excellent paintability, easy weldability. The combined paint/galvannealed coating is resistant to pealing and/or blistering with properly selected paints and primers

Source: Modified from Enclosure Metal Specifications Technical Data, Online. Available: http://www.hubbell-wiegmann. com/catalog/sectionL.pdf.

7.7.2 POWDER COATING/PAINT USED FOR ENCLOSURES

Enclosures made of aluminum and stainless steel usually do not require additional protection. The choice of coating for steel depends on what kind of environmental chemicals the equipment might be exposed [62]. To protect steel enclosures from corrosive solvents, fumes and gases, various powder coatings (polyester, epoxy), polyurethane paint, or galvanized steel is recommended.

7.7.2.1 Criteria for Paint Type Selection

Various criteria and their order of importance need to be established when selecting the best coating for enclosures, which basically performs two functions: protection and decoration. It is important to know the exposure conditions, which are subdivided into three categories:

- *Severe* conditions including total immersion in chemicals or salt water, coastal situations, areas of high industrial pollution, and areas close to chemical processing plants.
- *Moderate* conditions are found in outdoor exposure to heavy rainfall and continuous humidity in light industrial and urban conditions, rural and inland areas more than two miles from industrial districts, and coastal areas.
- *Mild* conditions are experienced in interiors subject to condensation, as well as interiors where there is a source of pollution. Very mild conditions are found in warm, dry interiors not subject to condensation or pollution.

TABLE 7.9

Types and Properties of Different Powder Coatings and Paints for Enclosures

Coating Type	Properties	Max Service Temperature	Use	Applications
Polyester urethane	Good adhesion, flexibility, corrosion and weather resistance	120°C/250°F	Indoor or outdoor	Mild steel, aluminum, or stainless steel. UL approved
Polyester TGIC	Good adhesion and flexibility, corrosion and weather resistance	80°C/180°F	Indoor or outdoor	Mild steel, aluminum, or stainless steel. UL approved
Epoxy	Good adhesion and flexibility, corrosion and weather resistance	150°C/300°F	Indoor	Mild steel, aluminum, or stainless steel
Hybrids (blend of polyester and epoxy resins)	Excellent mechanical strength and corrosion resistance, chemically resistant to many acids, oils, and solvents		Indoor	Mild steel, aluminum, or stainless steel
Polyurethane paint	Excellent corrosion resistance and good outdoor performance	80°C/180°F	Indoor or outdoor	Meets UL Type 4[a] requirements
Epoxy enamel coating	Excellent protection against corrosion and chemicals	120°C/250°F	Indoor	Meets UL Type 4[a] requirements
Polyamide epoxy	Excellent protection against corrosion and chemicals	95°C/200°F	Indoor or outdoor	Meets UL Type 4[a] requirements

[a] NEMA Type 4 Enclosures constructed for either indoor or outdoor use to provide a degree of protection to personnel against incidental contact with the equipment enclosed within; to provide a degree of protection against falling dirt, rain, sleet, snow, windblown dust, splashing, and hose-directed water; and that will not be damaged by the formation of ice on the exterior of the enclosure (see Tables 7.5 and 7.6).

Additionally, the coating choice also depends on whether or not the enclosure will be exposed to bright sunlight and high temperatures. According to the application and service conditions, the paint and powder coatings usually recommended are presented in Table 7.9.

7.7.2.2 Techniques of Applying a Powder Coating/Paint to Metal Panels

To improve corrosion resistance of a powder coating, primer coating is usually used. For example, either zinc phosphate (for outdoor applications) or iron phosphate (for indoor applications) is used as a primer under polyester power coating to improve the corrosion resistance of the paint.

Zinc-rich primers are used to protect steel surfaces from corrosion. Unlike regular paints or epoxies, which resist corrosion by forming an impermeable barrier between the metal and atmospheric moisture, zinc-rich primers provide corrosion protection by electrical means. The zinc and steel form a tiny electrical cell that protects the steel at the expense of the zinc, forming the so-called sacrificial layer. A zinc primer under a top coat of powder coating provides a backup or secondary method of protecting the steel from corrosion.

After applying primer, either through automatic or manual electrostatic application, a dry coating powder is charged as it passes through an ionized field and is attracted by the grounded part. After application, the paint is cured; this thermal curing causes cross-linking that produces a hard and durable finish with above average physical properties.

There are a number of different powder coating specifications that vary depending on the manufacturer of electrical equipment and the type of equipment. For example, pad-mounted equipment

TABLE 7.10
Powder Coating Specifications for Pad-Mounted Equipment

Equipment Type	Base Steel	Primer Type, Thickness	Paint, Thickness	Salt Fog Test (h)
MV 3-phase capacitor bank	Hot-roll 12 gauge	Zinc rich epoxy coating, 50 µm (2 mils)[a]	Polyester TGIC, 50 µm (2 mils)	N/A
MV capacitor bank	Hot-roll 12 gauge, stainless steel optional	Primer, 25 µm (1 mil)	Two finish coats, 50 µm (2 mils)	N/A
MV gear PMH	Hot-roll 12 gauge, stainless steel optional	N/A	Super Durabake II finish[b]	1500[c]
MV switchgear	A-60 galvanneal steel	N/A	N/A	4000
MV capacitor bank	11-gauge A-60 galvanneal steel	N/A	N/A	4000
Equipment substation transformers	N/A	Electro-deposited epoxy[d]	Two-component urethane paint, 50–100 µm (2–4 mils)	N/A

[a] Prepaint procedure—Iron phosphate conversion, plus polymeric coating.
[b] Durabake—spay polyurethane coat, applied to firearms, marine surfaces, and so on.
[c] 3000 h on stainless steel.
[d] Prepaint procedure—Iron phosphate conversion coating.

is designed for mounting on a concrete pad with underground connecting cables, with the enclosures usually made of galvanized, galvannealed, or stainless steel. To provide superior corrosion resistance of these enclosures primers are usually used, such as zinc-rich epoxy, high-solid epoxy, and electrodeposited epoxy. Also two, and not one, finish coats are applied. Some examples of OEM coating specifications are given in Table 7.10 for outdoor pad-mounted equipment with the enclosures made of different types of steel.

7.7.3 Defects and Failures of Powder Coatings and Paints

Signs of paint and coating failure may differ depending on the type of coating and especially on the type of environment the coating is exposed to. The coating may be metallic as in galvanized and galvannealed steel. The failure of the metallic coating may be defined as an excessive corrosion of the coating and steel under the coating. The failure may be caused by the inappropriate metallic coating type and its thickness, the type of pretreatment and its compatibility with the metallic coating, the type of primer, and its thickness.

The coating may be either organic or inorganic, thermoset or thermoplastic material in powder coating or paint. Different mechanical defects (blisters, pinholes, tearing) could be found on the powder coating or paint, such as pinholes, which appear as small spots without any coating [66]. Blisters in the coating are caused by entrapped air, gases, or moisture during the coating process. Pinhole in a powder coating is caused by entrapped moisture from the pretreatment which evaporate during the curing process. Another defect—cissing or nonuniform wetting of the surfaces—appears as areas or spots without any coating.

The coating may also have too low thickness, which results in poor surface finish. If the coating has too high thickness, the profile surface may get an orange peel appearance. A rough surface finish may be caused by coarse powder particles which have been cured before melting and the formation of a smooth coating layer. All these defects in coating may result in its premature failure.

If corrosion of painted steel is observed, the failure may be explained by the wrongly chosen type of paint and its thickness, or defects in the paint caused by improper painting techniques.

(a)    (b)

FIGURE 7.10   Corrosion of steel panels (a) and brackets and bolts (b) of CB enclosure caused by coating failure.

The specific environmental conditions must also be taken into account. An example of a failed powder coating which resulted in corrosion of a steel enclosure is shown in Figure 7.10.

ASTM Standard D1654 [67] defines a means of evaluating and comparing the basic corrosion performance of paint, including the substrate, pretreatment, or coating system, or combinations thereof, after exposure to corrosive environments. The test method covers the treatment of previously painted or coated specimens for accelerated and atmospheric exposure tests and their subsequent evaluation with respect to corrosion, blistering associated with corrosion, loss of adhesion at a scribe mark, or other film failures. SS testing is usually used for determining corrosion protection properties of the coating [68].

7.7.4   HV RTV Coating

In 1970, Dow Corning was approached by major U.S. utilities to develop and supply a permanent and longer-term solution to protect HV glass and porcelain insulators from pollution that causes severe flashovers, resulting in power outages and losses together with reduced reliability of power lines. The ideal coating would be the one that cannot be easily contaminated. The concept of using a room temperature vulcanization silicone (RTV) as a permanent coating was developed by Dow Corning in the form of high-voltage insulator coatings (HVIC) [69].

RTV coating is used to protect outdoor porcelain and glass HV insulators from environmental contamination, which has been a long-term source of severe flashover problems that have resulted in power outages and losses together with reduced reliability. HVIC was extensively tested by major U.S. utilities between 1974 and 1984. Global applications of RTV coating included the Americas, Europe, and Asia and with voltages ranging from 11 to 500 kV DC.

IEEE developed a guide [70] that summarized various important aspects that are needed for satisfactory long-term performance of HVIC. Various possible application scenarios, maintenance issues on coated insulators, factors affecting long-term performance, the question of aging, laboratory accelerated tests, and functional outdoor evaluation are described in this guide.

RTV coating when applied to the insulator modifies the surface properties of the glass or porcelain insulator by creating a hydrophobic material surface that interrupts the flashover mechanism at an early stage. RTV prevents the formation of a conductive water film, thereby limiting and successfully managing surface leakage currents. Arc track resistant additives are used to control surface temperature and avoid surface damage under high electrical activity.

RTV coating has been mostly applied as a retrofit solution to existing substation insulators which are in highly polluted environments. However, lately, a number of new OHTL have also been RTV coated prior to installation; these have been mostly in the Middle and Far East.

7.7.4.1   Aging of RTV Insulation

Since its introduction in the eighties of the last century, RTV coatings have been modified and manufactured by many companies, and have also been tested in the field and the laboratory to determine the moisture absorption properties of the coating and aging factors [71–73]. It was found [72] that RTV silicone coatings applied to the porcelain outdoor insulators darkened in color and the cracks had developed in the RTV coating on all energized insulators. The cracks extended to the surface of the porcelain insulator after the coating had been subjected to 345 kV system voltage and exposure to weather conditions for a period of 7 years. Most of the original mechanical strength and elasticity of the RTV silicone coatings had been lost. It was suggested that the aging process was accelerated by acid rain, which is not rare along coast lines.

7.7.4.2   The Role of Fillers in RTV Coatings

The factors having a significant influence on the aging performance of RTV coatings include the composition of the polymer and inorganic material (filler), which is an essential part of the coating affecting its dielectric and hydrophobic properties [74]. Inorganic fillers in SIR dielectrics enhance the properties of thermal conductivity, relative permittivity, and electrical conductivity, making them useful in outdoor HV insulation applications. The addition of alumina trihydrate (ATH) or silica (SiO_2) fillers to silicone polymers produces binary composites with enhanced thermal conductivity.

The filler type, particle size, shape, and concentration also play an important role in improving the coating performance. The enhanced relative permittivity of silicone dielectrics is obtained through additional fillers, such as barium titanate powder, which can be further increased with the addition of aluminum powder, also improving the dielectric properties of RTV. The use of antimony-doped tin oxide filler binary composites when the coating is applied to outdoor bushings greatly enhances the pollution performance of the coating.

The way the filler is bonded to the polymer matrix and the existence of other fillers can improve the performance of RTVs in severe ambient conditions. The study [75] indicated that the initial loss of hydrophobicity can be correlated to the surface discharges that develop on the material surface due to the droplets formed. The hydrophobicity loss was observed for silica-filled coating, which also have the lowest value of heat conductivity. Lower levels of surface activity and material deterioration were observed for the ATH-filled coatings in comparison with the SiO_2–filled ones.

7.8   ELECTRICAL INSULATION STANDARDS AND GLOSSARY

7.8.1   National and International Standards and Regulations on Insulation

The following national and international standards specify the requirements for various insulation materials and types of insulators, and they are only a partial list of the standards on insulation. Other standards describe the procedures for testing insulation.

For example, in ANSI C29.1-1988 (R2002), "American National Standard for Electrical Power Insulators—Test Methods," procedures for conducting tests to determine the characteristics of the insulators used on electric power systems are given.

European Standard IEC 60156 Ed. 2.0 b:1995, "Insulating liquids—Determination of the Breakdown Voltage at Power Frequency (PF)—Test Method," describes an empirical test procedure intended to indicate the presence of contaminants such as water and solid suspended matter, and the advisability of carrying out drying and filtration treatment. Standardized testing procedures and equipment are essential for the unambiguous interpretation of test results.

7.8.1.1 North American Standards for Solid Insulation

ANSI/IEEE Standard C29.1-1988 (R2002), "American National Standard for Electrical Power Insulators—Test Methods"

ANSI/IEEE Standard C29.2-1992 (R1999), "Wet-Process Porcelain and Toughened Glass Suspension Type"

ANSI/IEEE Standard C29.5-1984 (R2002), "Wet-Process Porcelain Insulators—Low and Medium Voltage Types"

ANSI/IEEE Standard C29.7-1996, "Wet Process Porcelain Insulators, High-Voltage Line-Post Type"

ANSI/IEEE Standard C29.8-1985, "Wet Process Porcelain Insulators, Apparatus, Cap and Pin Type"

ANSI/IEEE Standard C29.9-1983, "Wet Process Porcelain Insulators, Apparatus, Post-Type"

ANSI/IEEE Standard C29.10-1989 (R2002), "Wet Process Porcelain Insulators, Indoor Apparatus Type"

ANSI/IEEE Standard C29.11-1989 (R1996), "Composite Suspension Insulators for Overhead Transmission Lines, Tests"

ANSI/IEEE Standard C29.12-1997 (R2002), "Insulators Composite, Suspension Type"

ANSI/IEEE Standard C29.13-2000, "Insulators Composite, Distribution Dead End Type"

ANSI/IEEE Standard C29.17-2002, "Insulators Composite, Line Post Type"

ANSI/IEEE Standard C29.18-2003, "Insulators Composite, Distribution Line Post Type"

ANSI/IEEE Standard C57.19.00-2004, "General Requirements and Test Procedure for Power Apparatus Bushings"

ANSI/IEEE Standard 1572 (2004), "Guide for Application of Composite Line Post Insulators"

ANSI/IEEE Standard 95-2002, "IEEE Recommended Practice for Insulation Testing of AC Electric Machinery (2300 V and Above) with High Direct Voltage"

ANSI/IEEE Standard 1313.2-1999, "IEEE Guide for Application of Insulation Coordination"

ANSI/IEEE Standard C57.12.00-2010, "General Requirements for Liquid-Immersed Distribution, Power, and Regulating Transformers"

IEEE 400-2001, "IEEE Guide for Field Testing and Evaluation of the Insulation of Shielded Power Cable Systems"

IEEE 1434-2000, "IEEE Trial-Use Guide to the Measurement of Partial Discharges in Rotating Machinery"

NEMA Standards Publication No. HV 2-1991 (R1996, R2002), "Application Guide for Ceramic Suspension Insulators"

UL 1446, "Standard for Safety of Systems of Insulating Materials—General"

CAN411.1-10, "AC Suspension Insulators"

CAN411.4, "Composite Insulators for A.C. Overhead Lines"

CAN411.6, "Line Post Composite Insulators for Overhead Distribution Lines"

CAN411.7, "Composite Insulator for Guy Wires."

7.8.1.2 International Standards for Solid Insulation

European Standard EN 50151, "Fixed installations—Electric traction—Special Requirements for Composite Insulators"

IEC 120 third edition (1984), "Dimensions of Ball and Socket Couplings of String Insulator Units"

IEC 273 third edition (1990-02), "Characteristics of Indoor and Outdoor Post Insulators for Systems with Nominal Voltages Greater than 1000 V"

IEC 305 fourth edition (1995-12), "Insulators for Overhead Lines with A Nominal Voltage Above 1000 V. Ceramic or Glass Insulator Units for A.C. Systems. Characteristics of Insulator Units of the Cap and Pin Type"

IEC 720 first edition (1981), "Characteristics of Line Post Insulators"

IEC 60085, "Electrical Insulation—Thermal Evaluation and Designation"

IEC 60137 fifth edition (2003-08), "Insulated Bushings for Alternating Voltages Above 1000 V"

IEC 60216–1 ed5.0 (2001-07), "Electrical Insulating Materials—Properties of Thermal Endurance—Part 1: Ageing Procedures and Evaluation of Test Results"

IEC 60216–2 ed4.0 (2005-08), "Electrical Insulating Materials—Thermal Endurance Properties—Part 2: Determination of Thermal Endurance Properties of Electrical Insulating Materials—Choice of Test Criteria"

IEC 60270:2000/BS EN 60270:2001, "High-Voltage Test Techniques—Partial Discharge Measurements"

IEC 60433 third edition (1998-08), "Insulators for Overhead Lines with A Nominal Voltage Above 1000 V. Ceramic Insulators for AC Systems. Characteristics of Insulator Units of the Long Rod Type"

IEC 60471 second edition (1977-01), "Dimensions of Clevis and Tongue Couplings of String Insulator Units"

IEC 60660 second edition (1999-10), "Tests on Indoor Post Insulators of Organic Material for Systems with Nominal Voltages Greater than 1000 V Up to But not Including 300 kV"

IEC 60672-1, IEC 60672-2 and IEC 60672-3, "Specification for Ceramic and Glass Insulating Materials"

Part 1 second edition (1995-07), "Definitions and Classification"

Part 2 second edition (1999-12), "Methods of Tests"

Part 3 second edition (1997-10), "Specifications for Individual Materials"

IEC 60815, "Selection and Dimensioning of High-Voltage Insulators Intended for Use in Polluted Conditions"

IEC 61109 second edition (2008-05), "Composite Suspension and Tension Insulators for AC Systems with A Nominal Voltage Greater than 1000 V. Definitions, Test Methods and Acceptance Criteria"

IEC 61466-1 and IEC 61466-2, "Composite String Insulator Units for Overhead Lines with A Nominal Voltage Greater than 1000 V"

Part 1 first edition (1997-02), "Standard Strength Classes and End Fittings"

Part 2 edition 1.1 (2002-02), "Dimensional and Electrical Characteristics"

IEC 61857, "Electrical Insulation Systems—Procedures for Thermal Evaluation—Part 1: General Requirements—Low Voltage"

IEC 61858, "Electrical Insulation Systems—Thermal Evaluation of Modification to an Established Wire-Wound EIS"

IEC 61952 second edition (2008-05), "Insulators for Overhead Lines. Composite Line Post Insulators for Alternative Current with A Nominal Voltage Greater than 1000 V—Definitions, Test Methods and Acceptance Criteria"

IEC TS 62073 Technical Specification first edition (2003-06), "Guidance on the Measurement of Wettability of Insulator Surfaces"

IEC 62217 second edition (2005-10), "Polymeric Insulators for Indoor and Outdoor Use with A Nominal Voltage Greater than 1000 V—General Definitions, Test Methods and Acceptance Criteria"

IEC 62231 Ed. 1.0, "Composite Station Post Insulators for Substations with A.C. Voltages Greater than 1000 V Up to 245 kV—Definitions, Test Methods and Acceptance Criteria"

DIN VDE0302, "German Standard for Testing and Requirements of Electrical Insulating Materials in General"

DIN VDE0441, Part 2 and Part 2, "German Standard for Testing & Material Requirements of Composite Insulators."

7.8.1.3 National and International Standards for Transformer Oil

ANSI/IEEE Standard C57.12.00-1993, "General Requirements for Liquid-Immersed Distribution, Power, and Regulating Transformers"

ANSI/IEEE Standard C57.104-1991, "IEEE Guide for the Interpretation of Gases Generated in Oil-Immersed Transformers"

ANSI/IEEE Standard C57.106.2002, "IEEE Guide to Acceptance and Maintenance of Insulation Oil in Equipment"

ANSI/IEEE Standard PC57.127/D10.0, 2007, "Draft Guide for the Detection and Location of Acoustic Emissions from Partial Discharges in Oil-Immersed Power Transformers and Reactors"

ASTM D877-02(2007), "Standard Test Method for Dielectric Breakdown Voltage of Insulating Liquids Using Disk Electrodes"

ASTM D1533-00(2005), "Standard Test Method for Water in Insulating Liquids by Coulometric Karl Fischer Titration"

ASTM D1816-04, "Standard Test Method for Dielectric Breakdown Voltage of Insulating Oils of Petroleum Origin Using VDE Electrodes"

ASTM Standard D3612-02(2009), "Standard Test Method for Analysis of Gases Dissolved in Electrical Insulating Oil by Gas Chromatography"

IEC 60156 Ed. 2.0 b: 1995, "Insulating liquids—Determination of the breakdown voltage at PF—Test method"

IEC 60296, "Mineral Insulating oils for transformers & switchgear"

AS 1767.1—1999, "Australian Standard for insulating liquids. Part 1: Specification for unused mineral insulating oils for transformers and switchgear."

7.8.1.4 National Standards for Paints and Coatings for Steel

ASTM D610, "Standard Test Method for Evaluating Degree of Rusting on Painted Steel Surfaces"

ASTM D714, "Standard Test Method for Evaluating Degree of Blistering of Paints"

ASTM D1654, "Standard Test Method for Evaluation of Painted or Coated Specimens Subjected to Corrosive Environments"

ASTM D4228, "Standard Practice for Qualification of Coating Applicators for Application of Coatings to Steel Surfaces"

IEEE Standard 1523-2002, "Guide for the Application, Maintenance, and Evaluation of Room Temperature Vulcanizing (RTV) Silicone Rubber Coatings for Outdoor Ceramic Insulators."

7.8.2 Insulation Glossary

7.8.2.1 Solid Insulation Glossary

C–D

Carbon black: A black pigment. It imparts useful UV protective properties, and hence is frequently suspended into plastic and elastomeric compounds intended for outside weather exposure.

Chlorinated polyethylene (CPE): A synthetic rubber jacketing compound.

Chlorosulfonated polyethylene (CSPE): A synthetic rubber jacketing compound manufactured by Du Pont under trade name of Hypalon.

Chlorinated polyethylene (CPE): Jacketing material.

Dielectric breakdown: The voltage at which a dielectric material is punctured; which is divisible by thickness to give dielectric strength.

Dielectric constant: A physical property of an insulating material which is the ratio of the parallel capacitance (C) of a given configuration of electrodes with the material as the dielectric, to the capacitance of the same electrode configuration with vacuum as the dielectric.

Dielectric strength: The voltage which an insulating material can withstand before breakdown occurs, usually expressed as a voltage gradient (such as volts per mil).

Dielectric tests: (1) Tests which consist of the application of a voltage higher than the rated voltage for a specified time for the purpose of determining the adequacy against breakdown of insulating materials and spacings under normal conditions. (2) The testing of insulating materials by the application of constantly increasing voltage until failure occurs.

Double insulation: Insulation comprising both basic insulation and supplementary insulation.

Durabake: Spray polyurethane coat, applied to firearms, marine surfaces, and so on.

E

EIS: Electrical insulation systems.

EIM: Electrical insulating materials.

Epoxy: Also known as *polyepoxide*, epoxy is a thermosetting polymer formed from the reaction of an epoxide "resin" with a polyamine "hardener." Epoxy has a wide range of applications, including fiber-reinforced plastic materials and general-purpose adhesives. Two part epoxy coatings were developed for heavy-duty service on metal substrates; they use less energy than heat-cured powder coatings.

ETFE: Modified ethylene tetrafluoroethylene compound used for insulation and jackets.

Ethylene propylene rubber (EPR): Synthetic rubber insulation based upon ethylene propylene hydrocarbon.

F–H

Fluorinated ethylene propylene (FEP): Insulation and jacket compound having specific electrical characteristics.

Functional insulation: Insulation needed for correct operation of the equipment.

Fibrous filler: A material used to fill interstices in cables made from fibers such as jute, polypropylene, cotton, glass, and so on.

Filler: Any material used in multiconductor cables to occupy interstices between insulated conductors or form a core into a desired shape (usually circular). Also, any substance, often inert, added to a plastic or elastomer to improve its properties or decrease its cost.

Flame resistance: The ability of a burning material to extinguish its own flame, once its flame-initiating heat source is removed.

Flame retardance: Ability of a material to prevent the spread of combustion by a low rate of travel so the flame will not be conveyed.

Hygroscopic: Attracting or absorbing moisture from the ambient atmosphere.

Hypalon: Du Pont trademark for CSPE synthetic rubber.

I–J

Insulation level-100%: Cable for use on grounded systems or where the system is provided with relay protection such that grounds faults will be cleared as rapidly as possible but in any case within 1 min.

Insulation level-133%: Cable for use on ungrounded or grounded systems or where the faulted section will be de-energized in a time not exceeding 1 h.

Insulation resistance: The amount of leakage current that will flow through when a voltage is applied across a layer of insulation. It is determined by the voltage divided by the insulation resistance.

Jacket: A material covering over a wire insulation or an assembly of components, usually an extruded plastic or elastomer.

M–N–O

Metal-clad cable: NEC type designation for power and control cables enclosed in a smooth metallic sheath (Okoclad), welded and corrugated metallic sheath (C-L-X), or an interlocking tape armor (Loxarmor). (NEC Article 330.)

Medium-voltage cable: NEC designation for single and multiple conductor insulated cables rated 2001–35,000 V. (NEC Article 328.)

Migration: The loss of plasticizer from a plastic, usually due to heat or aging. It is undesirable since it will make the plastic hard and brittle. It is also called leaching.

Mining cable: A flame retardant cable especially constructed to withstand rough handling and exposure to moisture for underground use in the environment of a mine or tunnel, or surface use where it is exposed to sunlight and extremes of temperature.

Moisture absorption: The amount of water that an insulation or jacket, which is initially dry, will absorb under specified conditions. It is expressed as the percentage ratio of the absorbed water's weight to the weight of the jacket or insulation alone.

Neoprene: Trade name for polychloroprene, used for jacketing (see polychloroprene).

Nitrile rubber: A rubbery copolymer of butadiene and acrylonitrile. It is usually compounded and vulcanized.

Nordel: Du Pont trademark for EPDM synthetic rubber.

Oxygen bomb test: A test to determine the ability of conductors and insulations to withstand physical and electrical change when immersed in pure oxygen gas of a specified temperature and pressure for a specified time.

P

Plastic: Any solid material employing organic matter of a high molecular weight as a principal constituent, which can be shaped by heat and pressure during manufacturing or processing into a finished article.

Plasticizer: A substance incorporated into a material to increase its workability or flexibility.

Polychloroprene: Chemical name for neoprene A, a rubber-like compound used for jacketing where the wire and cable will be subjected to rough usage, moisture, oil, greases, solvents, and chemicals.

Polyester: A resin generally used as a thin film in tape form. Depending on the chemical structure, polyester can be a thermoplastic or thermoset; however, the most common polyesters are thermoplastics.

Polyester, thermoset: A family of resins formed by the reaction of a dibasic organic acid and a polyhydric alcohol. Notable for fast cure, the unusual variety of processing methods available, and its combination of excellent physical, electrical, heat, and chemical properties at reasonable cost. Particularly compatible with reinforcing glass fibers.

Polyethylene: A thermoplastic material composed of polymers of ethylene.

Polymer: A material formed by the chemical combination of monomers having either the same or different chemical compositions.

Polypropylene: A thermoplastic polymer of propylene.

Polyvinyl chloride (PVC): A thermoplastic material composed of polymers of vinyl chloride, which may be rigid or elastomeric, depending on specific formulation. *PVC* is used as insulating and jacketing material which is usually flame-retardant and resistant to many chemicals.

Power-limited circuit: A circuit that is either inherently limited requiring no over-current protection or limited by a combination of a power source and over-current protection.

Primer: A preparatory coating put on materials before painting. Priming ensures better adhesion of paint to the surface, increases paint durability, and provides additional protection for the material being painted.

R

Reinforced insulation: A single insulation system which provides a degree of protection against electric shock equivalent to double insulation under the conditions specified in IRC 60950 Standard.

RTV: Room temperature vulcanization silicone (RTV).

Rubber: A material that is capable of recovering from large deformations quickly and forcibly and that can be, or already is, modified to a state in which it is essentially insoluble (but can swell) in boiling solvent.

S

Salt spray (fog) test: Standard test method for evaluation of painted or coated specimens subjected to corrosive environments (ASTM Standard D1654-08). An SS test is an accelerated corrosion test that produces a corrosive attack to the coated samples in order to predict its suitability for use as a protective finish.

Screen: A semiconducting nonmetallic layer used under and over the insulation of power cables rated over 2 kV to reduce electrical stresses and corona.

Secondary insulation: Any extremely high-resistance material which is placed over the primary insulation to protect it from abrasion.

Semiconducting: An extruded layer or tape of such a resistance that, when applied between two elements of a cable, the adjacent surfaces of the two elements will maintain almost the same potential.

Specific dielectric strength: The dielectric strength per millimeter of thickness of an insulating material.

Specific inductance capacitance: That property of a dielectric material which determines how much electrostatic energy can be stored per unit volume when unit voltage is applied.

Stabilizer: Any ingredient added to plastics to preserve their physical and chemical properties.

Supplementary insulation: Independent insulation applied in addition to basic insulation in order to ensure protection against electric shock in the event of failure of the basic insulation.

T

Tear strength: The force required to initiate or continue a rip in a jacket or other insulation under specified conditions.

Temperature rating: The maximum temperature at which a given insulation or jacket may be safely maintained during continuous use, without incurring any thermally induced deterioration.

Thermocouple cable: A cable consisting of two dissimilar metals or alloys that, when electrically joined at one end, can be used to measure the temperature. These cables have no voltage rating.

Thermoplastic: A plastic that can be formed or melted repeatedly when heated enough. The change with temperature is physical rather than chemical. Some examples are nylon, polycarbonate, and polyethylene.

Thermoset: A plastic that undergoes a nonreversible chemical reaction when it is cured. Subsequent heating will not melt it. Examples are polyester, phenolic, melamine, epoxy, and silicone.

Thermoplastic polyolefin (TPPO): A thermoplastic jacket material with low smoke characteristics and free of halogens.

TPR: A trade name of Uniroyal Inc. for their thermoplastic rubber.

V–W–X

Vulcanization: An irreversible process during which a rubber compound through a change in its chemical structure (e.g., cross-linking), becomes less plastic and more resistant to swelling by organic liquids and the elastic properties are conferred, improved, or extended over a greater range of temperature.

Water absorption: The ratio of the weight of water absorbed by a given material under specified conditions, to the weight of that material when dry. It is generally expressed as a percentage. Materials such as electrical insulation that absorbs little or no water will also be more dimensionally stable and have a smaller reduction of electrical properties when wet.

XLPE: Cross-linked polyethylene, an insulating compound.

XLPO: Cross-linked polyolefin, a thermoset jacket material with low-smoke characteristics and free of halogens.

7.8.2.2 Insulating Oil Glossary

D–F

Degasification: Degasification removes harmful gases and water from the oil. This is an essential aspect of transformer maintenance since water decreases the dielectric strength of the oil, and dissolved gases cause arcing and corrosion, resulting in overheated connections. The amount of water held in suspension changes in relation to the temperature of the oil. Degasification is routinely performed in conjunction with all other oil reclamation services.

Dissipation factor: This test measures the dielectric losses in insulating oils due to heat dissipation in an electric field. High values indicate the presence of contaminants, oxidation products, and metal particles.

Fuller's earth systems: Fuller's earth removes both acid and particulates from the oil, including sludge and other contaminants. Sludging is a direct result of high acid content, and over time, sludge can harden over internal parts of the transformer, preventing the oil from circulating properly. Reclamation of oil with a high acid content requires Fuller's earth filtration and degasification. If reclamation is done in the early stages of acid build-up before sludging occurs, the oil will retain its properties longer under normal operating conditions.

H–I

Hot oil dryout: Once the used oil is removed from the transformer, the core and coil assembly are rinsed with clean oil to remove sludge or sediment caused by oxidation. The transformer is then partially filled with hot oil which is circulated to remove water and contaminants from the insulation. By cleaning the transformer in this manner, the cause of the oil contamination is addressed. Following this, the replacement oil will retain its properties longer under normal operating conditions.

Inhibitor: A number of organic chemical compounds are known to slow down the oxidation when added to lubricating or insulating oils. Such additives are known as antioxidants or oxidation inhibitors. Unfortunately, all known oxidation inhibitors become depleted with time. When all of the inhibitor has been used up, the oil starts to deteriorate again, and deterioration proceeds in the same way at the same (or greater) rate as in the uninhibited oil. The useful life of the oil is extended by the amount of time the inhibitor remains effective, before it becomes depleted.

Insulating oil (dielectric oil, transformer oil): Usually a highly refined mineral (petroleum-based) oil that is stable at high temperatures and has excellent electrical insulating properties. It is used in oil-filled transformers, some types of HV capacitors, fluorescent lamp ballasts, and some types of HV switches and CBs. Its functions are to insulate, suppress corona and arcing, and to serve as a coolant.

O

Oil circuit breaker: A CB in which the contacts open and close in insulating oil.

Oil condition tests: A series of oil tests and analysis required to determine the condition of the insulating oil or as specified in the maintenance standards.

Oil leakage sump: A receptacle which is intended to receive the oil of a transformer or other oil-filled equipment in case of leakage.

Oil pass: The quantity of oil indicated on the transformer or tap-changer nameplate which is circulated through the oil treatment plant.

Oil reclaiming: An oil treatment process which removes soluble and insoluble contaminants. Heated oil is vacuum-dehydrated and then percolated through Fuller's earth and filtered. Additionally, the vacuum process will degas the oil.

Oil reconditioning: An oil treatment process where water and solids are removed by filtration, centrifuging, and vacuum dehydration. Additionally, the vacuum process will degas the oil and remove some of the more volatile acids.

Oil regeneration: A proprietary process which uses refining and blending techniques to produce insulating oil with essentially new oil characteristics.

Oil spray nozzle: A device that sprays oil into a transformer to degasify the oil while under vacuum and to uniformly distribute the hot oil over the core and windings.

Oil treatment: A term used in this standard to include the processes of reconditioning oil, reclaiming oil, and the desludging of transformers.

Oil water valve: A valve that automatically closes when an effluent with a quantity of oil flows through it. The oil may be sensed by a change in the specific gravity of the effluent or by an oil-in-water detector.

Oil-immersed-type transformer: A transformer of which the magnetic circuit and windings are immersed in oil [IEC 76-1].

Oily effluent: A mixture of oil and water.

Oily tighten: Tightening of oiled windings and major insulation after dry out and oil impregnation.

P–R

Polychlorinated biphenyl: Polychlorinated biphenyls are very toxic and persistent in the environment. Polychlorinated biphenyls are widely used in electrical equipment, including transformers. Many transformers will still contain traces of Polychlorinated biphenyl-contaminated oil even after the oil has been changed several times.

PPM: Parts per million.

Retro filling: Involves removing the used oil from the transformer and replacing it with reclaimed oil.

T–V–W

TAN: Total acid number is the amount of potassium hydroxide (KOH) in milligrams that is needed to neutralize the acids in 1 g of oil. When transformer oils oxidize, they begin to produce acids. These acids can adversely affect the insulation and inner components of the transformer. When significant amounts of acid are produced, they will polymerize with each other or react with the metallic surfaces inside the transformer to form sludge. A low TAN (<0.3 mg KOH/g oil) is necessary to maximize the life of the insulating system and prevent sludge.

Vacuum filling: Vacuum filling is a critical step in ensuring that the new quality oil does not become contaminated during retro filling of the transformer.

Water content: Excessive water in the oil destroys the lubricant's ability to separate moving parts, allowing severe wear to occur. In addition, water increases corrosion rates and accelerates oil degradation. Degradation of oil results in loss of lubrication quality. The presence of water leads to premature plugging of filters, minimizing the effect of additives and supporting the growth of bacteria. A procedure that extracts water from an oil sample and determines the amount of water is called the Karl–Fisher method.

REFERENCES

1. Tommasini D. In *Dielectric Insulation & High Voltage Issues*, CERN Lectures, European Organization for Nuclear Research (CERN), Bruges, Belgium, pp. 16–25, June 2009.

2. Wadhwa C. L. Overhead line insulators, in *Electrical Power Systems*, Chapter 8, New Age International Publishers, New Delhi, India, p. 904, 2005.
3. Cotton H. and Barber H. High voltage insulators, in *The Transmission and Distribution of Electrical Energy*, Chapter 7, English University Press, London, UK, 1958.
4. Wikström D. and Öhlen C. Composite insulation for reliability centered design of compact HVAC & HVDC, *Publication of Swedish Transmission Research Institute (STRI)*, Ludvika, Sweden, pp. 1–6, 2004.
5. Fernando M. A. R. M. and Gubanski S. M. Ageing of silicone rubber insulators in coastal and inland tropical environment, *IEEE Transactions on Dielectrics and Electrical Insulation* 17(2), 326–333, April 2010.
6. Bruetsch R., Tari M., Froehlich K., Weiers T., and Vogelsang R. Insulation failure mechanisms of power generators, *IEEE Electrical Insulation Magazine* 24(4), 17–25, 2008.
7. IEEE Standard 1-1986. *General Principles for Temperature Limits in the Rating of Electric Equipment and for the Evaluation of Electrical Insulation*, IEEE Publishing, New York, 1992.
8. IEEE Standard 1-2001. *Recommended Practice—General Principles for Temperature Limits in the Rating of Electrical Equipment and for the Evaluation of Electrical Insulation*, IEEE Publishing, New York, 2001.
9. IEC Publication 85. *Recommendations for Classification of Materials for the Insulation of Electrical Machinery and Apparatus in Relation to Their Thermal Stability in Service*, Publication of International Electrotechnical Commission (IEC), Geneva, Switzerland, 1957.
10. IEC Standard 60085. *Electrical Insulation—Thermal Evaluation and Designation*, Publication of International Electrotechnical Commission (IEC), Geneva, Switzerland, 2007.
11. *Polymeric Materials; Long Term Property Evaluations*, UL Standard 746B, January 27, 2011.
12. IEC Standard 60034-1. *Rotating Electrical Machines, Part 1: Rating and Performance*, Publication of International Electrotechnical Commission (IEC), Geneva, Switzerland, 2004.
13. IEEE Standard dictionary of electrical and electronics terms. In *Software, Standards Information Network*, John Wiley & Sons, Inc., Hoboken, NJ, January 2008.
14. *Relative Thermal Life and Temperature Index for Insulated Electric Wire*, SAE Standard Product Code AS4851, March 6, 2005.
15. *Standard Test Method for Thermal Endurance of Rigid Electrical Insulating Materials*, ASTM Standard D2304 – 10, 2010.
16. IEC Standard 60216. *Electrical Insulating Materials—Properties of Thermal Endurance—Part 1: Ageing Procedures and Evaluation of Test Results*, ed. 5.0 and Part 2: *Determination of Thermal Endurance Properties of Electrical Insulating Materials—Choice of Test Criteria*, ed. 4.0, Publications of International Electrotechnical Commission (IEC), Geneva, Switzerland, Part 1—79pp., July 2001; Part 2—5pp., August 2005.
17. IEEE Standard C57.12.00-2010. *General Requirements for Liquid-Immersed Distribution, Power, and Regulating Transformers*, IEEE Publishing, New York, pp. 1–70, 2010.
18. *Standard Test Method for Thermal Failure of Solid Electrical Insulating Materials under Electric Stress*, ASTM Standard D3151, August 26, 1988.
19. ASTM Standard D 3850-94. In *Standard Test Method for Rapid Thermal Degradation of Solid Electrical Insulating Materials by Thermogravimetric Method (TGA)*, 2006.
20. *IEEE Standard for Metal-Clad Switchgear*, Online. Available: http://www.standards.ieee.org/findstds/standard/C37.20.2-1999.html.
21. Nailen R. L. Corona: What it does; How to detect it, *Electrical Apparatus*, Online. Available: http://www.findarticles.com/p/articles/mi_qa3726/is_199901/ai_n8835205/ElectricalApparatus/Jan1999
22. Brady J. Corona and tracking conditions in *Metal-Clad Switchgear Case Studies*, Online. Available: http://www.irinfo.org/articles/article_8_1_2006_Brady.pdf.
23. Lautenschlager M. *Corona—Is There Anything Good About It?* NETA World, Fall 1998.
24. *Partial Discharge Measurement*, Online. Available: http://www.hoestarinsp.com.sg/services-partial-discharge.html.
25. *Partial Discharge Surveys—Level II*, Online. Available: http://www.eatechnologyusa.com/fieldservices/partialdischargeservices/pdlevel2.
26. IEEE Gold Book™ Standard 493™-2007, *IEEE Recommended Practice for the Design of Reliable Industrial & Commercial Power Systems*, IEEE Publications, New York, 383pp., 2007.
27. Lowsley C. J., Davies N., and Miller D. M. Effective condition assessment of MV switchgear, *Proceedings of the Twenty First AMEU Technical Convention*, South Africa, Paper 32, 2006, Online. Available: http://www.maintenanceonline.co.uk/maintenanceonline/content_images/Pages%2046,%2047,%2048,%2049,%2050,%2051.pdf.

28. Davies N. and Goldthorpe S. Testing distribution switchgear for partial discharge in the laboratory and the field, *Proceedings of the twentieth International Conference on Electricity Distribution*, CIRED 2009, Prague, pp. 8–11, June 2009.

29. Liu Z., Phung B. T., Blackburn T. R., and James R. E. *The Propagation of Patrial Discharge Pulses in a High Voltage Cable*, Online. Available: http://citeseerx.ist.psu.edu/viewdoc/download?doi=10.1.1.136. 5192&rep=rep1&type=pdf.

30. Tao L., Jian X., and Baoshan Z. How to eliminate ESP insulator thermal breakdown and insulation aging, *Proceedings of the International Conference on Electrostatic Precipitation (ICESP)*, Pretoria, South Africa, Paper A19, 11pp., May 2004, Online. Available: www.isesp.org/ICESP%20IX%20 PAPERS/ICESP%2009%20A19.pdf.

31. Gubanski S. M., Karlsson S., and Fernando M. A. R. M. Performance of biologically contaminated high voltage insulators, *IEEE International Conference on Industrial and Information Systems (ICIIS), University of Peradeniya*, Sri Lanka, pp. 30–35, 2006.

32. Kumagai S. Influence of algal fouling on hydrophobicity and leakage current on silicone rubber, *IEEE Trans. Dielctr. Electr. Insul* 14, 1201–1206, 2007.

33. Gubanski S. M., Dernfalk A., Wallström S., and Karlsson S. Performance and diagnostics of biologically contaminated insulators, *IEEE 8th International Conference on Properties and Applications of Dielectric Materials*, Denpasar, Indonesia, pp. 23–30, 2006.

34. Fernando M. A. R. M. Performance for non-ceramic insulators in tropical environments, *PhD dissertation, Department of Electric Power Engineering*, Chalmers University of Technology, Goteborg, Sweden, pp. 1–2, 70–91, 1999.

35. Waluyo, Ngapuli I. S., Suwarno, and Djauhari M. A. Leakage current and pollutant properties of porcelain insulators from the geothermal area, *ECTI Trans. Electr. Eng. Electron. Commun.* 8(1), 126–145, 2010.

36. Kumosa M., Kumosa L., and Armentrout D. Can water cause brittle fracture failures of composite non-ceramic insulators in the absence of electric fields? *IEEE Trans. Dielectr. Electr. Insul.* 11(3), 523–533, June 2004.

37. Kumosa M., Kumosa L., and Armentrout D. Failure analyses of nonceramic insulators: Part 1—Brittle fracture characteristics, *IEEE Electr. Insul. Mag.* 21(3), 14–27, May/June 2005.

38. Kumosa M., Kumosa L., and Armentrout D. Failure analyses of nonceramic insulators: Part II—The brittle fracture model and failure prevention, *IEEE Electr. Insul. Mag.* 21(4), 28–41, July/August 2005.

39. *Case Study–Acid Induced Stress Corrosion Cracking*, Online. Available: http://www.rapra.net/ consultancy/case-studies-acid-induced-stressed-erosion-cracking.asp.

40. Burnham J. T. et al. IEEE task force report: Brittle fracture in nonceramic insulators, *IEEE Trans. Power Del.* 17(3), 848–856, July 2002.

41. Facilities instructions, standards, and techniques. In *Testing and Maintenance of High-Voltage Bushings*, Vol. 3–2, United States Department of the Interior Bureau of Reclamation (USBR), Springfield, VA, 8pp., 1991, Online. Available: www.usbr.gov/power/data/fist/fist3_2/vol3-2.pdf.

42. *Methods of Measurement of Radio Influence Voltage (RIV) of High-Voltage Apparatus*, National Electrical Manufacturers, NEMA Publication No. 107-1987 (R1993), 1993.

43. Densley J., Bartnikas R., and Bernstein B. S. Multi-stress ageing of extruded insulation system for transmission cables, *IEEE Electr. Insul. Mag.* 9(1), 15–17, January/February 1993.

44. Amyot N., David E., Lee S. Y., and Lee I. H. Influence of post-manufacturing residual mechanical stress and crosslinking by-products on dielectric strength of HV extruded cables, *IEEE Trans. Dielectr. Electr. Insul.* 9(3), 458–466, June 2002.

45. Harlin A., Danikas M. G., and Hyvonen P. Polyolefin insulation degradation in electrical field below critical inception voltages, *J. Electr. Eng.* 56(5–6), 135–140, 2005.

46. Hyvönen P. Prediction of insulation degradation of distribution power cables based on chemical analysis and electrical measurements, In *Doctoral Dissertation*, Helsinki University of Technology, Helsinki, Finland, 83pp., 2008, Online. Available: lib.tkk.fi/Diss/2008/isbn9789512294039/isbn9789512294039.pdf.

47. Lokhanin A. K., Shneider G. Y., Sokolov V. V., Chornogotsky V. M., and Morozova T. I. Internal insulation failure mechanisms of HV equipment under service condition, *CIGRE General Session Paris*, France August 25–30, 2002.

48. El-Refaie., El-Sayed M. M., Salem M. R., and Ahmed W. A. Prediction of the characteristics of transformer oil under different operation conditions, *World Acad. Sci., Eng. Technol.* 53, 764–768, 2009.

49. Aragon P. J. and Tenbohlen S. *Improved Monitoring of Dissolved Transformer Gases on the Basis of a Natural Internal Standard*, NIS Standard CIGRE23, 2007.

50. *Guide for the Interpretation of Gases Generated in Oil—Immersed Transformers*, IEEE Publishing, New York, 2008.

51. IEC Publication No. 60599. *Mineral Oil-Impregnated Equipment in Service—Guide to the Interpretation of Dissolved and Free Gases Analysis*, Publications of International Electrotechnical Commission (IEC), Geneva, Switzerland, 2007.

52. IEEE Standard C37.122.1. *Standard for Gas-Insulated Substation*, IEEE Publishing, New York, 1993.

53. IEC Standard 1634–(1995). *High-Voltage Switchgear and Controlgear—Use and Handling of Sulphur Hexafluoride (SF$_6$) in High-Voltage Switchgear and Controlgear*, IEC Standard 1634, Publications of International Electrotechnical Commission (IEC), Geneva, Switzerland, 1995.

54. Cigre W. G. Handling of SF$_6$ and its decomposition products in gas-insulated switchgear (GIS), *Electra*, 136 and 137, 69–101, 1991, SF$_6$ recycling guide, *Electra* 173, 43–69, 1997.

55. Cigre W. G. SF$_6$ and the global atmosphere, *Electra*,164, 121–131, October 23, 1996.

56. O'Connel P. et al. SF$_6$ in the electric industry, *Electra*, 200, 16–25, February 2002.

57. Slade P. *The Vacuum Interrupter: Theory, Design, and Application*, CRC Press, Taylor and Francis Group, Boca Raton, FL, Chapter 1, pp. 11–49, 2007.

58. Budde M., Horn A., Körner F., Kurrat M., and Steinke K. Dielectric behaviour of vacuum circuit-breakers, *International Symposium on High Voltage Engineering (ISH)*, Beijing, China, Paper I-04, p. 536, August 2005.

59. Iturregi A., Torres E., Zamora I., and Abarrategui O. High voltage circuit breakers: SF$_6$ vs. Vacuum, *Proc. of International Conference on Renewable Energies and Power Quality (ICREPQ)*, Valencia, Spain, 6pp., April 2009, Online. Available: http://icrepq.com/ICREPQ'09/abstracts/458-iturregi-abstract.pdf.

60. Suzuki S., Matsuo T., Sakuma R., and Yanabu S. Diffusion of metal vapor from electrodes to the vacuum interrupter shield and its dependence on electrode construction, *IEEE Trans. Plasma Sci.* 34(2), 485–489, April 2006.

61. Fink H., Gentsch D., and Heimbach M. Condensed metal vapor on alumina ceramic in vacuum interrupters, *IEEE Trans. Dielectr. Electr. Insul.* 9(2), 201–206, April 2002.

62. *Materials and Paint Finishes: Chemical Resistance*, Online. Available: http://www.hoffmanonline.com/stream_document.aspx?rRID=16157&pRID=16154.

63. NEMA Standards Publication 250-2003. *Enclosures for Electrical Equipment (1000 Volts Maximum)*, Online. Available: http://www.global.ihs.com.

64. ANSI/IEC Publication 60529-2004. *Classification of Degrees of Protection Provided by Enclosures*, Online. Available: http://www.nema.org/stds/complimentary-docs/upload/ANSI_IEC%2060529.pdf.

65. NEMA Publication. *A Brief Comparison of NEMA 250 and IEC 60529*, Online. Available: http://www.nema.org/stds/briefcomparison.cfm.

66. Powder coating defects. *Troubleshooting Powder Coating Problems*, Online. Available: http://www.powder-coater.com.

67. ASTM D1654-08. *Standard Test Method for Evaluation of Painted or Coated Specimens Subjected to Corrosive Environments*, Online. Available: http://www.astm.org/Standards/D1654.htm.

68. ASTM B117–09. *Standard Practice for Operating Salt Spray (Fog) Apparatus*, Online. Available: http://www.astm.org/Standards/B117.htm.

69. Goudie J. L. and Collins T. P. Development and evaluation of an improved RTV coating for outdoor insulation, Conference Record of the *IEEE International Symposium on Electrical Insulation*, Indianapolis, IN, pp. 475–479, September 19–22, 2004.

70. *Maintenance, and Evaluation of Room Temperature Vulcanizing (RTV) Silicone Rubber Coatings for Outdoor Ceramic Insulators*, IEEE Guide for the Application 1523–2002, 2003.

71. Homma H., Mirley C. L., Ronzello J., and Boggs S. A. Field and laboratory aging of RTV silicone insulator coatings, *IEEE Trans. Power Deliv.* 15(4), 1298–1303, October 2000.

72. Mirley C. L., Ronzello J., Homma H., Boggs S. A., and Eldridge K. Aging of field-installed RTV silicone insulator coatings. *Presented at the 1995 Doble Clients Conference*, Boston, pp. 1–8, March 1995.

73. Jahromi A. N., Cherney E. A., Jayaram S. H., and Simon, L. C. Aging characteristics of RTV silicone rubber insulator coatings, *IEEE Trans. Dielectr. Electr. Insul.*, 15(2), 444–452, April 2008.

74. Cherney E. A. Silicone rubber dielectrics modified by inorganic fillers for outdoor high voltage insulation applications, *IEEE Trans. Dielectr. Electr. Insul.* 12(6), 1108–1115, December 2005.

75. Siderakis K., Agoris K. D., and Gubanski S. Salt fog evaluation of RTV SIR coatings with different fillers, *IEEE Trans. Power Deliv.*, 23(4), 2270–2277, October 2008.

76. Davies N. *Why Partial Discharge Measurement Can Pay Dividends*, Online. Available: http://www.powerengineeringint.com/articles/print/volume-18/issue-3/features/why-partial-discharge-measurement-can-pay-dividends.html.

8 Electrical Equipment Life Expectancy, Aging, and Failures

8.1 LIFE EXPECTANCY FOR DISTRIBUTION AND TRANSMISSION EQUIPMENT

8.1.1 ESTIMATION OF ELECTRICAL EQUIPMENT LIFETIME

A power system comprises a number of components, such as lines, cables, transformers, CBs, disconnectors, and so on. Every component in a power system has an inherent risk of failure. Multiple outside factors influence the possibility of component failure; for example, the current loading of the component, damage by third parties (human or animal), and trees. Atmospheric conditions such as temperature, humidity, pollution, wind, rain, snow, ice, lightning, and solar effect may play a significant role in the component failure [1].

It is often assumed that the lifetime of installed electrical equipment is about 35–40 years [2].

However, when trying to estimate the life of equipment it is necessary to consider multiple factors, such as the range of extreme operating conditions and environments, and the different levels of past maintenance.

For example, if the electrical equipment is located in a high-rise, modern office building, there is a good chance that the environment is very clean and temperatures are moderate. For the same equipment working in a paper mill or in a dust-filled environment, or in a hot and humid climate, the life expectancy cannot be the same [3]. Therefore, the typical life expectancy of the equipment may vary dramatically depending on the conditions of service and the environment [4].

Generally, the useful life of such power system components depends upon the level of care given to them and their duty cycles. For example, a CB on mainly switching duty can last 40–50 years. The majority of transformers at the utilities serve about 40 years if none of the catastrophic events such as lightening happen. On the other hand, on- LTCs in HV transformers are prone to failure.

HV bushings are accounted for as one of the most significant single causes of failure in HV substations. The failure mechanisms tend to develop to a critical level at a midlife point for the surrounding assets and such mechanisms generally result in a sudden and catastrophic failure of an explosive nature, thus significantly shortening the life span of HV substations [5].

Models used for the estimation of lifetime are based on stress tests (electrical, mechanical, and thermal) specified in the standards [6–8] and on the mathematical model of the mean time to first fail [9,10].

8.1.2 OVERLOADING AND ESTIMATED LIFE OF ELECTRICAL EQUIPMENT

Electric power components are typically assigned a capacity rating and an estimated life. The two parameters are related, and it is assumed that if the component does not experience overloading (an excess of its rating), it will stay in service in the stated useful life, whereas if it experiences excessive loading, its useful life will be decreased [11].

8.1.2.1 Circuit Breakers

A CB's function is to operate properly to interrupt fault currents. When a CB detects a fault, it acts to open its contacts, across which an electrical arc will develop. The arc either extinguishes itself, as in the case of an oil-filled breaker, or, as in the case of a gas-filled breaker, a blast of air or other gases (e.g., SF_6) is directed at the arc to extinguish it. Failure to interrupt the current within the required time will not only destroy the CB, but will probably also severely damage the equipment that it is designed to protect (transmission line, generator, etc.). Even if the fault currents are within the design parameters of the breaker, each breaker operation results in some mechanical erosion of the contacts and in the production of contaminants in the oil or gas insulation. The higher the fault current, the greater the degree to which the contacts are eroded and contaminants are produced and the sooner the breaker will fail or must be overhauled. Therefore, the higher the fault current level, the lesser the number of operations the CB is expected to perform before failure occurs [12,13].

8.1.2.2 Transformers

The negative effect on a transformer is mainly produced by the excess heat caused by overloading [14–16]. A transformer consists of an iron core around which are wrapped various coils of insulated wires. The core is positioned inside a tank filled with insulating oil, along with connectors, bushings, and various other small components.

Excess heat produced by overloading on a transformer causes degradation of the Kraft paper insulation around the wires, leading to internal failures of the coils. Another negative effect of the transformer overloading is in generating an excessive tank pressure and degradation of the insulating oil, either of which can cause catastrophic failures, even explosions, and leaking gaskets and seals. Overloading also generates harmonics, resulting in thermal cycling and mechanical vibration of the transformer, which cause physical damage by loosening electrical connections.

Any faults near the transformer, when they occur, will be greater than normal at overload, leading to an increased chance of damage to the transformer from fault currents. These include coil failures, bushing flashovers, blown gaskets and seals, connector failures, oil explosions and fires, and physical displacement of internal components due to electromechanical torques.

8.1.2.3 Conductors

The principal component of transmission and distribution lines is the conductor: a wire or bundle of wires made of aluminum, copper, or steel. If the overloading is severe enough and protection systems do not act properly to disconnect the line, only a few minutes or hours of overloading could lead to severe overheating. During these periods the conductor temperature will increase due to the heat generated by the resistive losses.

Too much high-temperature operation will cause annealing of the wires when they lose their elasticity and strength. The conductors will sag and violate clearance limits, which will require premature replacement with new conductors. Sometimes, splices or connectors in the line will overheat faster than the conductor itself, causing the connectors to fail, in which case the conductor may fall to the ground [17,18].

8.1.2.4 Underground Transmission

Transmission cables for underground circuits consist of copper conductors surrounded by paper or PE insulation, and jacketed by layers of PVC or metal cladding. High-temperature operation will cause generally the same type of insulation degradation as in transformers. Exposure to high temperature will also result in connector failures and the leakage of fluid or gas. In underground transmission the conductors are not under tension; therefore annealing is not a concern.

8.1.3 Temperature and Estimated Life of Electrical Equipment

In the previous section, it was explained that overloading conditions may lead to elevated temperature of all components. Elevated temperature is one of the most serious deteriorating factors for any component of transmission and distribution equipment and decreases the useful life of the electrical apparatus.

The majority of electrical distribution equipment are designed to operate properly and achieve normal life expectancy under ambient conditions, if the temperature does not go below 0°C (32°F) and does not rise above 40°C (104°F) at RH levels from 0% to 99%.

Unless serious extremes of temperature are anticipated, heating requirements are limited to maintaining equipment surfaces above the dew point to avoid condensation and providing an acceptable work environment for service. It can be accomplished using electrical resistance heaters installed inside the enclosures, making it possible to install even MV (over 600 V and less than 72 kV) equipment in unheated spaces or outdoors.

The effect of exceeding the 40°C (104°F) ambient temperature varies with the type of equipment. For example, uninterruptible power supply (UPS) systems may shut down if the entering air temperature or internal temperature exceeds certain limits. Electronic devices such as relays and meters that are intended to be used in electrical switchgear may have ambient temperature specifications up to 70°C (158°F). In some cases, elevated temperatures require derating of the equipment.

Protective devices such as fuses and thermal-magnetic CBs detect overloads based on the heat produced by the current flowing through them. These devices will trip at lower-than-rated currents at elevated ambient temperature. Less sophisticated equipment, such as motors, switchgear bus, and transformers, will continue to operate properly at elevated ambient temperature, but will develop higher internal temperatures that can cause insulation deterioration and reduce life expectancy [19]. As a rule of thumb, the life of the insulation will decrease by 50% for every 10°C rise in temperature. It is evident that the higher the temperature, the shorter the life expectancy of the insulation [20,21].

Another cause of the temperature increase in an electrical machine is because of a rise in the electrical resistance of different components of the current-carrying path, which may be of many different origins [13]. The primary circuit resistance rise may be caused by the loosening of an electrical connection in the primary circuit as well as by a reduction of spring pressure within the fixed main contacts assembly of the interrupter. The resistance rise may be caused by chemical contamination, deposits of particles of dirt or grit, or oxidation (corrosion) of the surfaces of contact surfaces. Mechanical damage of the surfaces of main contacts may also lead to a significant electrical resistance rise of the current-carrying path. All these deteriorating processes in an electrical machine lead to the TR, shortening its useful life.

8.2 SIGNATURES OF AGING OF ELECTRICAL EQUIPMENT IN NUCLEAR, INDUSTRIAL, AND RESIDENTIAL ENVIRONMENTS

8.2.1 AGING FACTORS

During its useful life, electrical equipment experiences multiple stresses that affect its performance. These factors are electrical stresses (voltage, current, on–off cycling, overloading), electrostatic discharges (partial discharge and corona), mechanical stresses (vibrations, shocks), extreme temperatures and temperature changes, humidity, corrosive environment, dust, and so on.

The effect of all these stresses gradually build up, leading to mechanical damage to the parts, corrosion of conductive parts, deterioration of insulation and eventually to failure of one or several components of the structure. The most significant aging is found on connectors when they are exposed to oxidation and contamination, high humidity, or fretting corrosion. Cables may be mechanically damaged and chemically attacked. The insulation may degrade due to moisture, high temperature, or radiation. Electromechanical components may get corroded, the contacts may be oxidized or contaminated, and the coils may burn out.

The longer the electrical equipment is subjected to all these stresses without due attention and maintenance, the shorter is its useful life. However, aging of power equipment seldom leads to a sudden, catastrophic failure. Instead, the effect of all stress factors is cumulative; it is a slow but inevitable worsening each year.

To keep aging power equipment in good condition, to extend its useful life as much as possible, and to prevent a catastrophic failure, it is important to know how to timely identify the different modes, types, and outcomes of deterioration of electrical equipment.

Material fatigue and other losses of mechanical strength, losses of electrical properties like insulation strength or conductivity, increased friction, and binding of moving mechanical parts—these are some electrical equipment deterioration modes. It is essential to identify the causes of deterioration, either internal (overloading, lack of lubrication, mechanical stresses, overheating) or external (environmental factors such as contamination, dust, corrosive gases, humidity, etc.). Depending on the type of industry where the electrical equipment is serving, there may be significant differences in the causes and signs of deterioration.

8.2.2 Aging Equipment in an Industrial Environment

8.2.2.1 Nuclear Facilities

The necessity of maintaining the electrical equipment in a nuclear power plant in safe and proper conditions is intensified by an elevated risk of any even minor failures at such facilities to the environment and the population. At a nuclear power plant, a special program called Aging Management Guidelines (AMGs) provides guidance on how to perform detailed evaluations of aging mechanisms and aging management strategies applicable to critical groups of equipment, among which are motor control centers (MCC), transformers, switchgear, cables, battery chargers, inverters, and UPS [22]. In Table 8.1, the aging factors for electrical equipment at nuclear power plants are summarized.

8.2.2.2 Aviation

Aging electrical equipment in the aviation industry poses an actual danger to aircraft, since even minor defects in the electrical system may threaten the lives of the passengers and crew. The Federal Aviation Administration (FAA) developed a special research program for investigating aging electrical systems [23]. It was specified that the severity of an electrical system failure is related to both the loss of functionality and the potential for fire or other physical damage to the aircraft. The most often found aging mechanisms in aviation are arc faults, cable degradation, some inconsistent mechanical connections (loose screws), and degradation of connections at high-temperature conditions.

8.2.2.3 Chemical and Oil Refining Industries

Degradation of electrical equipment at chemical plants proceeds much faster owing to the presence of higher amounts of various corrosive fumes and gases and very often a higher density

TABLE 8.1
Aging Factors of Electrical Equipment at Nuclear Facilities

Electrical Equipment	Document	Aging Factors
Motor control centers	June 1994 (SAND93-7069)	"Harsh" environments and/or subject to a high number of operating cycles
Electrical switchgear	July 1993 (SAND93-7027)	Lubrication practices, "harsh" environments and/or subject to a high number of operating cycles
Power and distribution transformers	May 1994 (SAN093-7068)	Insulation, insulating fluid, bushing contamination, "harsh" environments and/or subject to a high number of operating cycles

Source: Data from Nakos J. T. and Rosinski S. T. Research on U.S. Nuclear Power Plant Major Equipment Aging, SAND94-2400C, DOE's. *Office of Scientific and Technical Information (OSTI)*, 1994, Online. Available: http://www.osti.gov/bridge/servlets/purl/10190825-J7vdeV/webviewable/10190825.pdf.

of corrosive particles and dust than at other types of industrial facilities. It was found that the electrical equipment at a chemical plant usually has a shorter life than other kinds of equipment of the plant. Modern electrical equipment contains many electronic components, which are very sensitive to their environmental conditions, to chemically and mechanically active substances in particular. The environmental factors included in the specific designs of ventilation and air conditioning systems for control rooms at chemical industrial facilities are climatic and biological conditions and chemically and mechanically active substances. The major aging mechanism of electrical machines in the chemical and oil processing industries is corrosive deterioration of the current-carrying paths, accelerated by the elevated temperature and humidity in the environment of industrial facilities.

8.2.3 Aging Equipment in Power Generation and Transmission and Distribution

8.2.3.1 Overhead Power Transmission

Overhead transmission lines (OHTL) structures include many components, but not all of them may be defined as actual electrical equipment. However, all these components should be considered when evaluating the degree of aging. The performance and safety of transmission lines strongly depend on the aging of supporting structures, such as wood poles and concrete poles with steel enforcement. Wood poles deteriorate fast because of humidity, losing their mechanical strength. Concrete supports are degraded by corrosion of the steel reinforcement. When steel is corroded, rust forms on steel beam surfaces and triggers the splitting of concrete cover because it is more bulky than steel, which decreases the pole mechanical strength.

Conductors (transmission cables) are exposed to harsh weather conditions, such as high temperatures and their variations, rains, sleet, wind, lightning, and so on. Therefore, the climatic condition is a parameter that defines not only the mechanical strength of the transmission network but also causes its degradation. These degradation factors may differ according to the geographic zones.

The deterioration modes of transmission cables include creep, the nonelastic nonlinear behavior of conductors, as well as broken strands caused by sunshine, lightning impacts, and vibration. The cable fittings wear due to the insulator sets swinging caused by wind action [24].

8.2.3.2 Power Plant

At a power plant, there are a number of critical components whose failure could endanger plant safety, could cause an extended forced outage, a long lead time in maintenance and repair, and lost revenues. In a power production plant, the cables and station main transformers are critical components. Power cables and electrical equipment such as motors and transformers must operate as long as possible within reliability and safety standards, which require timely recognition of signs of equipment aging.

The other equipment is called the influence component. This is the equipment whose failure would probably not result in an extended outage, would not endanger plant safety, and is unit specific.

Power cables are used for supplying power to plant equipment, such as large motors, auxiliary transformers, precipitators, and backup diesel generators. These cables are rated from 2 to 15 kV and are single-conductor or three-conductor cables, of shielded or unshielded construction. Before the year 1970, power cables were insulated with extruded dielectrics, such as XLPE, and butyl rubber. In some cases, the paper-insulated lead-covered (PILC) cable was also used. The majority of cable failures in an extruded cable system are related to water treeing, which then progresses to electrical trees. Once a water tree progresses to an electrical tree, the time to failure is normally very short because the initiated electrical tree propagates rapidly through the already weakened dielectric material (Section 7.5).

In PILC cables, failures are commonly associated with moisture ingress. Moisture in PILC cables increases the dielectric losses, resulting in localized heat generation that thermally degrades the paper insulation and normally leads rapidly to a cable failure. PD may only be present at advanced stages of such degradation. Cable accessories such as splices and terminations are most likely to fail because of PD that causes degradation [25].

In a *power transformer*, several operating and physical conditions, such as high temperature and high load currents above name plate rating, degrade the transformer cellulose and oil insulation system. Other aging and damaging factors are moisture inside the transformer, oxidation of cellulose and oil; voltage surges, lightning; and physical damage due to moving and relocating. As cellulose and oil age, their insulating properties degrade. Degraded insulation produces gases that can form bubbles in the insulating oil (Section 7.6). These bubbles are where PD may occur. Aging can progress to a level that will lead to failure of the transformer, most commonly in the coil structure.

In addition to insulation aging, the laminated core of a transformer can degrade. Discharge may take place between laminations at locations where the lamination insulation (usually a metal oxide layer) has broken down. Core discharge can take place over a long time period, but usually does not lead to catastrophic failure of the transformer. Transformers typically fail in the core structure.

8.2.4   Aging Power Equipment in a Residential Environment

Examples of distribution electrical equipment include transformers, fuses, CBs, and other overcurrent protection devices, fixed wiring, switches, outlets, and receptacles, cords and plugs, meters and meter boxes, and so on.

Factors affecting the service life of industrial and residential electrical equipment [26] may be divided into several groups, such as electrical and mechanical stresses and environmental factors. Electrical stresses are excessive loads, overvoltage, harmonics, electrical surges, glowing contacts, and arcing faults. Environmental factors include water and humidity damage, corrosion, heat-related damage.

Important factors in aging of distribution industrial and residential equipment are age-related deterioration, insulation deterioration, and physical damage of wiring or devices. Added would be other defects caused by improper original installations, improper additions and modifications, loose connections and poor workmanship.

In residential areas, aging of residential wiring causes great concerns for fatalities, personal injuries, and loss of property [27]. In 2002, in response to questions regarding the influence of aging and installation quality on the fire safety of electrical systems, the Fire Protection Research Foundation initiated a collaborative research and development project designed to address the issue through a comprehensive survey of the conditions of representative samples of residential electrical components installed in different time periods and the U.S. locations. A total of 30 homes were selected from across the United States ranging in age from 30 to 110 years. The report [28] presents a summary of the results of that study; it provides critical information to code writers, especially for NFPA 73, "Standard for Electrical Inspections for Existing Dwellings" [29], and The National Electrical Code (NEC) [30].

It was reported [28] that the disproportionately high incidence of fire in the electrical systems of older homes can be attributed to one or more of the following factors:

- Inadequate and overburdened electrical systems
- Thermally reinsulated walls and ceilings burying wiring
- Defeated or compromised overcurrent protection
- Misuse of extension cords and makeshift circuit extensions
- Worn-out wiring devices not being replaced
- Poorly done electrical repairs

Apart from the natural effects that age can have on wire insulation and electrical equipment over time, residential electrical systems are seldom inspected after their original installation. In addition to the above, the quality of the original installation may also be a factor, as well as inappropriate upgrades or additions that may have been done by unqualified homeowners or others throughout the years. Examples of aged residential equipment are shown in Figure 8.1.

(a)   (b)

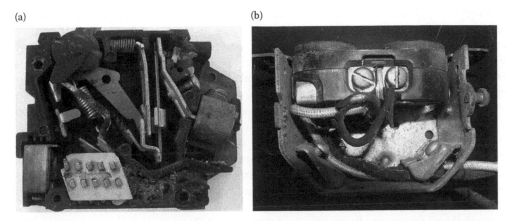

FIGURE 8.1 Examples of aged residential electrical equipment: corrosion damage of the CB (a); brittle rubber insulation (b). (From *Residential Electrical System Aging Research Project, Technical Report,* National Fire Protection Association, Quincy, MA July 1, 2008, Online. Available:http://www.nfpa.org/assets/files/PDF/Research/RESAReport.pdf. With permission.)

8.2.4.1 Aging of Conductors

Degradation of conductors and insulation increases with age. The following aging factors for conductors and connections in residential electrical equipment should be considered:

- Increased resistance of conductors and at connections, caused by oxidation/corrosion films
- Embrittlement of conductors, caused by excess heat from *overlamped* light fixtures, which happen in old houses when a light fixture has a bulb with a higher wattage than what the fixture is designed for
- Loose or disconnected conductor connections in electrical outlets, switches, and light fixtures
- Conductors deterioration following expansion and contraction due to temperature changes seasonally and daily, as well as by vibrations from trains, sonic booms from jet aircraft, other noisy vehicles, high-decibel-level music/radio equipment, and thunder
- Conductors aging caused by wind and air pressure changes

8.2.4.2 Aging of Insulation

Electrical equipment aging is directly related to deterioration of insulation caused by many factors specific to residential areas [28]:

- Heat from overlamped light fixtures and other installations heavily loaded electrically causes insulation aging
- Continuous exposure to light causes insulation deterioration by accelerating loss of plasticizer in insulation
- Dry locations lead to loss of the plasticizer in PVC insulation, which depends on the length of time and high-temperature factors as well
- Water absorption into insulation in wet locations reduces its dielectric strength
- Rodents and other creatures can damage insulation over the years

8.2.5 Aging Electrical Equipment in Rural/Agricultural Applications

In rural areas there are multiple aging factors that are different from those in industrial and residential areas, which required establishing a set of codes and standards specific to rural areas [31].

These include applications to grain bins, storage structures, irrigation pump motors, oil pumps, other motor applications, animal confinement buildings, and wet or corrosive installations. These documents standardize methods of grounding and wiring, as well as set up the rules of equipment selection. One of the major concerns in rural areas is aging transmission lines that are exposed to the rural environment much longer than in industrial areas without proper maintenance and replacement [32].

Agricultural buildings that house livestock require special care when selecting wiring materials, wiring methods, and electrical equipment because of corrosive dust, gases, and moisture. Corrosion of metallic conduit, boxes, and fixtures frequently leads to electrical system failure. Boxes and fixtures made of a nonmetallic material or corrosion-resistant stainless steel, that is, nonmagnetic, are recommended for all agricultural buildings and are required in any buildings that house livestock or contain corrosive dust. Accelerated corrosion due to condensation occurs on electrical panels that have not been properly designed [33].

Electrical enclosures are often made of galvanized steel. Galvanized structures corrode when the protective zinc coating has been corroded away by acid or salty environments (or by strong alkali). In rural environments, galvanized sheet may well last more than 30 years, but life in a severely polluted farming areas may be only 5–10 years. Galvanized steel does not satisfactorily withstand aggressive soils (e.g., peats or marshes containing sulfates and chlorides) or acid conditions that may occur at ground level in some buildings, nor can it withstand the severe condensation that is not uncommon in some farm buildings [34].

Electrical devices may experience very severe corrosion attack at or near piggeries and poultry houses. These agricultural facilities usually have an internal atmosphere which can contain high concentration of ammonia. This gas is damaging for most of the materials that electrical apparatus is built of. In poorly ventilated enclosures, acid vapors emitted by timber as part of the natural seasoning process can corrode metals that are nearby but not actually in the contact. Excessive corrosion can result in a variety of defects such as seizure of moving parts, distortion of metering equipment, blocked outlets, and so on, which can seriously affect the functioning of electrical machinery. Most corrosion damage occurs in poorly drained areas that remain damp and dirty, for example, in overlapped joints, on spot-welds and in crevices generally, and on moving parts, for example, hinges.

Some of the major harmful effects of corrosion can be summarized as the loss of technically important surface properties of any metallic component. These could include frictional and bearing properties, and electrical conductivity of electrical contacts [35].

8.3   FAILURE MODES AND FAILURE RATES OF AGING ELECTRICAL EQUIPMENT

8.3.1   DEFINITIONS OF FAILURE, FAILURE MODE, AND FAILURE RATE OF ELECTRICAL EQUIPMENT

Failure is any trouble with a power system component that causes any of the following to occur: (1) partial or complete plant shutdown, or below-standard plant operation; (2) unacceptable performance of the user's equipment; (3) operation of the electrical protective relaying or emergency operation of the plant electrical system; and (4) de-energization of any electrical circuit or equipment. This definition of failure is given in IEEE Gold Book Standard 493™-2007, "IEEE Recommended Practice for the Design of Reliable Industrial & Commercial Power Systems" [36].

A failure on a public utility supply system may cause the user to have either a power interruption or a loss of service or a deviation from the normal voltage or a frequency outside the normal utility profile. A failure of an in-plant component causes a forced outage of the component; that is, the component is unable to perform its intended function until it is repaired or replaced. The terms "failure" or "forced outage," which is an outage that cannot be deferred, are often used synonymously.

Failure rate is defined as the mean number (arithmetical average) of failures of a component per unit exposure time. Usually, exposure time is expressed in years and failure rate is given in failures per year.

Distribution system components, such as lines, transformers, and CBs, are subject to a variety of failure modes. The failure modes have different impact on system reliability performance. It is sometimes useful to categorize system components as switching devices (such as CBs) or non-switching devices (such as lines and transformers) [36].

Failure modes for nonswitching components are the events that make the component unable to fulfill its current-carrying function, generally due to a fault and the subsequent isolation of the fault component by a protective device. For switching devices such as CBs or protection systems, which have both a continuously required current-carrying function and response functions, failure modes would make them unable to perform both functions. An inability to perform a current-carrying function will immediately impact the whole electrical system performance, while an inability to perform a response function, such as tripping open on command, will be manifested only when the response is required.

According to the IEEE Gold Book [36], a common cause of electrical failure is dust and dirt accumulation and the presence of moisture. This can be in the form of lint, chemical dust, day-to-day accumulation of oil mist and dirt particles, and so on. These deposits on the insulation, combined with oil and moisture, become conductors and are responsible for tracking and flashovers. Deposits of dirt can cause excessive heating and wear, and decrease apparatus life. Moisture condensation in electrical apparatus can cause copper or aluminum oxidation and connection failure.

Loose connections are another cause of electrical failures. Creep or cold flow is a major cause of joint failure. Friction can affect the freedom of movement of devices and can result in serious failure or difficulty. Dirt on moving parts can cause sluggishness and improper electrical equipment operations such as arcing and burning.

8.3.2 BATH TUB CURVE, THE HYPOTHETICAL FAILURE RATE VERSUS TIME

The statistical failure rates rise over the years in service according to the well-known bath tub curve (Figure 8.2). The bath tub curve consists of three periods: (1) an infant mortality period with a decreasing failure rate followed by (2) a normal life period (also known as "useful life") with a low, relatively constant failure rate. The curve is concluded with a wear-out period (3) that exhibits an increasing failure rate [37]. The bath tub curve displayed in Figure 8.2 describes the relative failure rate of an entire population of products over time.

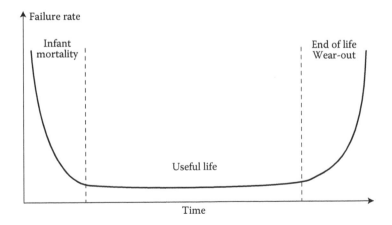

FIGURE 8.2 The bath tub curve: hypothetical failure rate versus time.

The bath tub curve is typically used as a visual model to illustrate the three key periods of product failure and not calibrated to depict a graph of the expected behavior for a particular product family. Some individual units will fail relatively early (*infant mortality* failures), others will last until wear out and end of life, and some will fail during the relatively long period typically called *normal (or useful) life* [37]. Failures during infant mortality are highly undesirable and are always caused by defects and blunders: material defects, design mistakes, errors in assembly, and so on. Normal life failures are normally considered to be random cases of "stress exceeding strength."

However, many failures often considered normal life failures are actually infant mortality failures. Wear-out is a fact of life due to fatigue or depletion of materials. A product's useful life is limited by its shortest-lived component. A product manufacturer must ensure that all specified materials are adequate to function through the intended product life.

8.3.3 FAILURE CAUSES OF CBS

8.3.3.1 LV and MV CB Failure Causes

The U.S. NRC collected and summarized the data on CCF of LV and MV CBs from 1980 to 2000, which are presented in Ref. [38]. The data studied in the report were derived from the NRC CCF database, which is based on the U.S. commercial nuclear power plant event data. The report states that in a CB, each component fails because of its susceptibility to the conditions created by the root cause.

By the definition given in Ref. [38], a *proximate cause* of a failure event is the condition that is readily identifiable as leading to the failure. The proximate cause can be regarded as a symptom of the failure cause, and it does not in itself necessarily provide a full understanding of what led to that condition. As such, it may not be the most useful characterization of failure events for the purposes of identifying appropriate corrective actions.

The proximate cause classification consists of six major groups or classes:

- Design/construction/installation/manufacture inadequacy
- Operational/human error
- Internal to the component, including hardware-related causes and internal environmental causes
- External environmental causes
- Other causes
- Unknown causes

The causal chain can be long and, without applying a criterion, identifying an event in the chain as a "root cause" is often arbitrary. Identifying root causes in relation to the implementation of defenses is a useful alternative. The root cause is therefore the most basic reason or reasons for the component failure, which if corrected, would prevent recurrence.

The leading proximate cause of the report [38] was "internal to component" and accounted for about 61% of the total events. In the cause group "Design/construction/installation/manufacture inadequacy" faults accounted for 18% of the total events. Human error accounted for 13% of the total events. To a lesser degree, the "external environment and the other causes" proximate cause categories were assigned to the CB component.

8.3.3.2 Failures of CBs According to the IEEE Gold Book

Failure modes of CBs are summarized in IEEE Gold Book Standard 493-2007, "Recommended Practice for the Design of Reliable Industrial & Commercial Power Systems" [36].

The largest percentage of CB failures (i.e., 42%) were of the type "opened when it should not," which is a primary concern for industrial plants and makes this the number one failure mode of CBs. This type of CB failure can significantly affect plant processes and may result in total plant shutdown.

Also, a large percentage (i.e., 32%) of the CBs "failed while in service (not while opening or closing)," which is the second most common failure mode of CBs. During this time, the CB could not have performed its designed primary function, which is protection against overcurrent. The above information is given for all types of CBs.

For CBs up to 600 V types, it was found that the highest percentage of failures of MC draw-out CBs (71%) occur when the breaker "opened when it should not," while only 5% of fixed-type breakers (including molded case circuit breakers (MCCB)) experience this type of failure.

The alarming statistic for molded case breakers is that the "failed while in-service" is 77%. This shows the lack of ability of the MCCB to perform its primary function, which is protection against overcurrents. This is not surprising since typically the MCCB does not have maintenance performed on it.

It was found [36] that 43% of CB failures were caused by flashovers or arcing of all types, 19% of breakers failed because of other electrical defects, and mechanical defects caused 11% of CBs failures. After failure, damaged insulation of all types was found in 17% of the breakers, mechanical damage (bearings, other moving parts, and other mechanical parts) was found in 18% of the breakers, and in 28% of failed breaker protective devices have been damaged.

8.3.4 Failure Causes and Failure Rates of Power Transformers

8.3.4.1 MV and LV Power Transformers

On the basis of the IEEE survey [39,40], failure rates for liquid-filled power transformers for all subclasses from 300 to 10,000 kVa were approximately equal for three age groups (1–10 years old, 11–25 years old, and older than 25 years). The slightly higher failure rates for transformer units aged 1–10 years and for units older than 25 years may be attributed to "infant mortality" and to units approaching the end of their life, respectively.

The major *failure (MF)-initiating* causes for *power transformers* leading to the highest percentage of failures are found to be the following: transient overvoltage disturbance (switching surges, arcing ground fault, etc.) resulting in 16.4% of failures, winding insulation breakdown (29.1%), and insulation bushing breakdown (13.6%). Other failure causes are mechanical breaking, cracking, loosening, abrading, or deforming of static or structural parts (7.3%), loose connection or termination (7.3%), and so on.

For *rectifier transformers*, about 40% of all failures are caused by breakdown of winding, bushings, and other insulation breakdown, and 20% of all failures caused by mechanical breaking, cracking, loosening, abrading, or deforming of static or structural parts. Other failure causes are transient overvoltage disturbance (13.3% of failures) and mechanical burnout, friction, or seizing of moving parts (13.3% of all failures). Overheating of rectifiers transformers caused 6.7% of all failures versus 2.7% of failures of power transformers.

Normal deterioration from age and cooling medium deficiencies were reported to have contributed to a large number of both power transformers (13.3%) and rectifier transformer (28.6%) failures.

It is important also to know what the most important *failure-contributing* causes for all power transformers are. It was found [5] that exposure to aggressive chemicals, solvents, dusts, moisture, or other contaminants contributed to 14.4% of failures. Also environmental conditions such as abnormal temperature (contributed to 5.5% of failures) and severe wind, rain, snow, sleet, or other weather conditions (4.4% of failures) play an important role in power transformer failures. Loss, deficiency, contamination, or degradation of oil or other cooling media contributed to 10.0% of all failures of power transformers.

8.3.4.2 HV Power Transformers

In several studies and surveys, it was found that problems with bushings are a major source of HV transformer failures [5]. One study [41] of 106 aged (23–39 years old) 110–500 kV transformers noted bushing defects in 70% of the transformers surveyed. The same survey also concluded that transformer life is limited by the deterioration of accessories, of which were listed just three: bushings, LTC, and the cooling system. Over 46% of transformer defects were found to be attributable to these items [41,42].

Another study [43] from the 63 members of the Edison Electric Institute lists bushings among the three most common transformer failure modes reported. Australian and New Zealand reliability statistics [44] in a survey on 2096 transformers in 1970–1995 found similarly that bushings were second only to tap changers as the component initially involved in failure and were among the top three contributors to costly transformer failures.

Reported findings from European statistics [45], indicate that bushing failures cause 30% of transformer failures but that 80% of bushing failures occur between 12 and 20 years in service or at midlife of the transformer population. It was also confirmed that a bushing or HV CT failure may often be followed by a catastrophic event such as fire, tank rupture, violent explosion of the bushing propelling large broken shards of porcelain some hundreds of meters at velocities enough to imbed the material in concrete walls [5].

8.3.5 Failure Causes of MV Switchgear

Multiple defects may be listed as causes of switchgear failures [46]. Loose and faulty connections may cause complete thermal failure of the connection due to the heat created by increased resistance. It was estimated that approximately 25% of all electrical failures occur due to loose connections. An MV insulation is exposed to voltage stresses that may lead to a failure.

The cables are connecting switches to transformers, potential transformers, control power transformers, and surge arresters to the bus; and connecting transformer coils and taps together. If these cables come in contact with the ground, other phases, or even with other types of insulation used for supporting them, they will likely develop stresses at those points that will lead to a future failure. Cable terminations can also create localized partial insulation breakdown or PD that will lead to failure.

Bus insulation barriers for separating compartments and supporting the bus often have small air gaps between the bus and the insulation. The localized voltage stress is great enough to cause partial breakdown or PD that may lead to failure.

An MV switchgear that is exposed to high-humidity conditions will absorb moisture, and voltage stresses will attack the hydrophobic insulation surfaces which are designed to inhibit moisture absorption. Water inside the switchgear can create instant short circuits, long-term insulation damage, and long-term metallic component corrosion, among other complications that may lead to failure.

It is sometimes difficult to obtain definitive information on the causes of faults in MV switchgear; however, some information is always available and it can be used as a guide to the typical problems faced.

On the basis of the faults in MV vacuum switchgear operating on electricity distribution networks within the United Kingdom, the distribution of the failure causes has been determined [47]. The fault information included defects identified during operation or maintenance as well as disruptive failure of the switchgear. The biggest contributors to faults are mechanical problems that caused 30% of the faults and maloperations that caused 8% of the failures.

The second largest fault cause was associated with PD activity (26%). It should also be noted that faults reported with cable termination boxes (6%), voltage transformers (VTs; 9%), and CTs (3%) are often associated with PD activity and therefore the actual figure for PD-related problems actually lie somewhere between 26% and 44%.

The general construction of SF_6 switchgear is comparable with vacuum switchgear; and it would be reasonable to suggest that a similar fault breakdown will also exist for MV SF_6 switchgear. The situation is slightly different when it comes to oil-filled switchgear. Much research has been carried out into how bulk oil-filled switchgear ages and degrades; and it has been established that the condition of the oil is a key indicator of the overall condition of the switchgear [37,38].

8.3.6 FAILURE CAUSES OF OTHER MV AND LV POWER ELECTRICAL EQUIPMENT

In the case of electrical equipment other than MV CBs and transformers, analysis of failures [36] shows that most of the motor starters, DSs, bus ducts, wires, and cables "failed in service" (Table 8.2).

Based on the survey, it appears that a majority of failures were caused by flashovers or arcing of all types in bus ducts (100%), cables (74%), cable joints (79%), and cable terminations (59%) [36]. A large percentage of insulation has been found damaged during failure of bus duct and cables and their accessories (joints and terminations), while mechanical parts (bearings, other moving parts, and other mechanical parts) were found mostly damaged in the motor starters and DS (Table 8.3).

There are multiple suspected failure-initiating and failure-initiating causes for all types of electrical equipment. In Tables 8.4 and 8.5 these causes are listed with the percentage of failures.

TABLE 8.2
Dominant Failure Characteristics of Electrical Equipment

Type of Equipment	Failed in Service (%)	Failed during Testing (%)	Damage Discovered during Testing or Maintenance (%)
Motor starters	37	6	36
Disconnect switches	72	3	18
Bus duct	90	5	0
Open wire	68	2	1
Cable	92	2	2
Cable joint	96	4	0
Cable terminations	80	2	12

Source: IEEE Gold Book™ Standard 493™-2007, *IEEE Recommended Practice for the Design of Reliable Industrial & Commercial Power Systems*, IEEE Publications, New York, 2007, 383pp. © 2007, IEEE.

TABLE 8.3
Damaged Parts during the Failure of Electrical Equipment

Type of Equipment	Insulation— Winding (%)	Insulation— Bushings (%)	Insulation— Other (%)	Mechanical Parts (%)
Motor starters	5	0	10	18
Disconnect switches	0	1	14	39
Bus duct	15	10	65	0
Open wire	0	1	6	4
Cable	5	0	83	4
Cable joint	0	0	91	0
Cable terminations	0	12	74	4

Source: IEEE Gold Book™ Standard 493™-2007, *IEEE Recommended Practice for the Design of Reliable Industrial and Commercial Power Systems*, IEEE Publications, New York, 2007, 383pp. © 2007, IEEE.

TABLE 8.4
Major Failure-Initiating Causes of Electrical Equipment Failures

Type of Equipment	Manufacturer Defective Component (%)	Improper Application (%)	Inadequate Installation and Testing Prior to Startup (%)	Inadequate Operating Procedures (%)
CBs	23	4	3	6
Motor starters	18	51	0	3
Disconnect switches	29	6	4	39
Bus duct	26	16	5	0
Open wire	0	2	9	2
Cable	16	8	14	3
Cable joint	0	0	50	0
Cable terminations	0	18	38	0

Source: IEEE Gold Book™ Standard 493™-2007, *IEEE Recommended Practice for the Design of Reliable Industrial and Commercial Power Systems*, IEEE Publications, New York, 2007, 383pp. © 2007, IEEE.

TABLE 8.5
Major Failure-Contributing Causes of Electrical Equipment

Type of Equipment	Exposure to Aggressive Chemicals (%)	Exposure to Abnormal Moisture or Water (%)	Severe Weather Conditions (%)	Normal Deterioration from Age (%)
CBs	2	3	1	17
Motor starters	0	0	0	40
Disconnect switches	0	4	0	5
Bus duct	0	17	11	49
Open wire	28	1	30	3
Cable	14	8	16	30
Cable joint	13	22	2	29
Cable terminations	10	12	16	24

Source: IEEE Gold Book™ Standard 493™-2007, *IEEE Recommended Practice for the Design of Reliable Industrial and Commercial Power Systems*, IEEE Publications, New York, 2007, 383pp. © 2007, IEEE.

It is important to note that about a quarter of the CBs, motor starters, DS, and bus ducts failed because of some damaged components installed by manufacturers. Improper application of the apparatus caused half of the motor starters failures; and the majority of cable joints and terminations failed due to inadequate installation and testing prior to startup.

8.3.7 FAILURE CAUSES OF POWER CONNECTORS

There are multiple failure causes of power connectors: extreme environments, the use of incompatible materials, inappropriate use of materials, insufficient maintenance, as well as manufacturing flaws.

8.3.7.1 Aluminum Connectors

Aluminum is a widely used material in power connectors, and this metal has several physical properties that can ultimately lead to failure. An insulating oxide layer building on an aluminum surface can reduce the metal's ability to form a strong connection to another aluminum component.

In a connector, this can be especially harmful, as the device depends on the primary connection between two (often aluminum) wires. Additionally, aluminum is prone to thermal expansion, corrosion, fretting, and stress relaxation, all of which can further compromise the connector's ability to transmit electricity via a strong aluminum-to-aluminum contact.

8.3.7.2 Corrosion

Corrosion is a common concern in connectors that are exposed to extreme conditions or outdoor environments, which first begins with the development of a thin film on the connector's contacts. As the film builds, the contacts' ability to conduct electricity between one another can be lost. Different metals react to surface films in different ways. Some metals are extremely corrosion resistant, but may be more prone to other issues, such as fretting.

8.3.7.3 Contact Fretting

Fretting corrosion is a form of accelerated atmospheric oxidation which occurs at the interface of contacting materials undergoing slight cyclic relative motion. In electrical contacts involving non-noble metals, fretting action can cause rapid increases in contact resistance, proceeding to virtual open circuits in a matter of minutes in the worst cases. Fretting corrosion is a build-up of insulating, oxidized wear debris that can form when there is small-amplitude fretting motion between electrical contacts. More about fretting corrosion is given in Section 4.5. When two different metals are used, such as nonlubricated aluminum and copper, heat can cause one metal to expand at a different rate than the other. In the case of nonlubricated aluminum and copper, aluminum expands faster than copper, compromising their connection and leading to fretting. To avoid contact fretting, simply lubricating the contact material can make a significant difference.

8.3.7.4 Stress Relaxation

When a connector is under a high amount of pressure over an extended period of time, stress relaxation can occur. Because of the amount of pressure, the metallic contact undergoes a shift in structure (and not in dimension) causing the connection to weaken and the pressure between the contacts to slacken. If a connector is exposed to high temperature, stress relaxation intensifies and the results are more severe. With aluminum in particular, increased temperature aggravates stress relaxation.

8.3.8 Inadequate Maintenance and Maintenance Quality as a Cause of Failure

Maintenance quality has a significant effect on the percentage of all failures blamed on "inadequate maintenance," as was found in the survey [4]. Of the 1469 failures reported from all causes, "inadequate maintenance" was blamed for 240, or 16.4% of all the failures. Twelve main classes of electrical equipment have been surveyed and presented in the IEEE Gold Book [36]. The results are based on information received from 30 companies, covering 68 plants in nine industries in the United States and Canada.

The percentage of electrical equipment failed when improper maintenance was suspected as a failure-initiating cause [36] is shown in Table 8.6. This information might be outdated, since the survey was performed in 1972, and with the continuous improvement of maintenance strategies and enforcement of preventive maintenance implementation in the majority of industrial and power facilities, the situation would improve in the twenty-first century.

The IEEE data [39] also showed that "months since maintenance" is an important parameter when analyzing the failure data of electrical equipment. Analysis of the number of failures caused by "inadequate maintenance" for CBs, motors, open wire, transformers, and all equipment classes combined showed that the percentage of failures attributed to "inadequate maintenance" is in close correlation with the number of months ago that the equipment was maintained. Thus, 77.8% of CBs, 44.4% of motors, and 36.4% of transformers, in which the failures were caused by "inadequate maintenance," failed after being maintained more than 24 months ago.

TABLE 8.6

Inadequate Maintenance as a Failure-Initiating Cause of Electrical Equipment

Type of Equipment	Inadequate Maintenance (%)
CBs	23
Motor starters	8
Disconnect switches	13
Bus duct	16
Open wire	30
Cable	10
Cable joint	18
Cable terminations	32

Source: IEEE Gold Book™ Standard 493™-2007, *IEEE Recommended Practice for the Design of Reliable Industrial and Commercial Power Systems*, IEEE Publications, New York, 2007, 383pp. © 2007, IEEE.

It is important to realize that maintenance and testing are needed due to the fact that nearly 1/4 of all CB failures were caused by a manufacturer defective component and nearly another 1/4 of all CB failures were due to inadequate maintenance. Thus, if proper maintenance and testing are performed, potentially 50% of the failures could be eliminated or identified before a problem occurs. The results of the 1996 IEEE study show that better technology has improved the failure rate of LV power CBs and could potentially be cut by almost half, but maintenance and testing are still needed.

NFPA 70B, "Recommended Practice for Electrical Equipment Maintenance" [48], makes a very clear statement about an effective Electrical Preventive Maintenance (EPM) program: "Electrical equipment deterioration is normal, but equipment failure is not inevitable. As soon as new equipment is installed, a process of normal deterioration begins. Unchecked, the deterioration process can cause malfunction or an electrical failure. Deterioration can be accelerated by factors such as a hostile environment, overload, or severe duty cycle. An effective EPM program identifies and recognizes these factors and provides measures for coping with them." Further development of preventive maintenance techniques and its influence on the performance and longevity of electrical equipment will be discussed in Chapter 19.

8.4 FAILURE CAUSES AND RATES OF ELECTRICAL EQUIPMENT BASED ON CIGRÉ SURVEY

CIGRÉ (International Council on Large Electric Systems) is one of the leading worldwide organizations on electric power systems, covering their technical, economic, environmental, organizational, and regulatory aspects. Electrical networks worldwide have an increasingly aged population of electrical equipment. Before the latest CIGRÉ survey started in 2004, the results of two other surveys have been published.

The first CIGRÉ survey covered the time interval 1974–1977, which included all types of CBs with service voltage above 63 kV (published in Electra No. 79, 1981). The second CIGRÉ survey, which covered the time interval 1988–1991, was published in CIGRÉ Technical Report 83, 1994. It included single pressure SF_6 CBs (live tank, dead tank, and the breakers at GIS), placed in service in 1978–1991; the lowest rated voltage was 72.5 kV. There was no age limitation to equipment population.

During 4 years (January 2004–December 2007) CIGRÉ was carrying out a worldwide survey of failures in service on HV equipment rated for voltages greater than or equal to 60 kV. The survey based on the information from utilities about their equipment populations covered SF6 CB; DS, earthing switches (ES), IT, and GIS.

When the data acquisition was completed, the information was subjected to extensive statistical analyses with the purpose of obtaining trustworthy information on reliability and failures for the considered component types. The survey's application helps the owners and operators of these networks to understand the impact of aged equipment on network performance and helps mitigate the effects through proper operation, effective maintenance, monitoring, asset refurbishment and asset replacement.

Basically, two different failure rates may be estimated based on survey data: absolute failure rates and failure rate with age [49].

Absolute failure rates: The data can be used to derive a failure rate, typically the number of failures per 100 years, for a particular family of HV equipment. This failure rate can then be used, for example, at the initial life of the asset to ensure that the application of the HV equipment is consistent with the expected failure rate. The absolute failure rate has a number of uses during the operating phase of the asset and, for example, can assist in determining the optimum number of spares.

Failure rate with age: The survey includes questions regarding the age of the failed items and the age profile of the population. This allows determining the failure rate for different ages of equipment. It should be noted that as design and maintenance practices change over time, it is not possible to assume that any definitive age profile or bath tub curve can be found. However, with care such information should enable maintenance to be more targeted and should help inform the difficult decision on when to refurbish or replace an item of HV equipment.

Types of the failures: Two different types of failure may be considered: major and minor [50]. By the definition of CIGRÉ, *failure* is a lack of performance by an item of its required function or functions. The occurrence of a failure does not necessarily imply the presence of a defect if the stress or the stresses are beyond those specified.

MF: MF is a failure of a switchgear or control gear which causes the cessation of one or more of its fundamental functions. An MF will result in an immediate change in the system operating conditions, for example, the backup protective equipment being required to remove the fault, or will result in mandatory removal from service within 30 min for unscheduled maintenance, or will result in unavailability for the required service.

Minor failure (mf): Minor failure (mf) is a failure of equipment other than an MF or any failure, even complete, of a constructional element or a subassembly which does not cause an MF of the equipment.

8.4.1 Results of the Older CIGRÉ Surveys of HV CB Failures

The results of two past CIGRÉ surveys (1974–1977 and 1988–1991) have been compared in Ref. [50].

8.4.1.1 Main Results of the First Survey

- 70%—MF of mechanical origin
- 19%—MF of electrical origin concerning the auxiliary and control circuits
- 11%—MF of electrical origin concerning the main circuit
- 48%—MF classified as "Does not open or close on command"

The *operating mechanism* was the part of the CB responsible for the highest number of failures (37% of MF). Different types of operating mechanism have about the same MF rate. Most of the minor failures of operating mechanisms are either hydraulic oil or air leakages. The minor failure rate of spring, pneumatic, and hydraulic drive systems is 1:2:7, respectively.

The percentage of failures in *auxiliary interrupters or resistors* of CBs with rated voltage 500 kV and above was of the same order as the percentage of failures in *making and breaking units.*

Compared with the first survey, in the second survey it was determined that the MF rate for single pressure SF_6 CBs is about 40% of the value in the first survey (for CBs of all technology), and the minor failure rate is about 30% higher than in the first survey. This might be explained by SF_6 leakage problems and more complex control and auxiliary systems of modern CBs. The MF rate increases rapidly with voltage, but in comparison with the first survey the improvement in the higher voltage ranges is much greater, due to improved design.

In reviewing the survey results, the maintenance and mechanical aspects should be considered.

8.4.1.2 Maintenance Aspects

- The average interval between scheduled overhaul is 8.3 years, which in many cases could be extended.
- The number of failures caused by incorrect maintenance has decreased compared to the first enquiry (85% decrease for MFs and 26% for minor failures), but there is still room for improvement.
- About a quarter of the failures are caused by inadequate instructions and incorrect erection, operation, and maintenance.

8.4.1.3 Mechanical Aspects

- As with the first enquiry, a majority of the MFs have mechanical origin.
- The operating mechanism and the electrical control and auxiliary circuits are the components responsible for the majority of both major and minor failures.
- The dominant MF modes are "Does not open or close on command" and "Locked in open or closed position." These modes add up to almost 70% of the MFs.

8.4.2 Failure Causes of GIS

8.4.2.1 Older CIGRÉ Surveys

The first CIGRÉ survey on GIS was published in 1992–1994; it covered the experience up to 1990 and was based on information from both manufacturers and users. The second survey was published in the final form in 2000 CIGRÉ Technical Brochure 150 and covered the experience up to 1995; it was based on information from users only. Both surveys covered the whole population of GIS, including apparatus, and all failures without age limitations. At this time the GIS was still a relatively new technology, average age 9 years.

Comparison of the results from the first and second CIGRÉ GIS surveys showed an increasing reliability of GIS in service. Total downtime due to failure was 13 days on average. There are three components most frequently involved in MFs: CB or switch, busbar or bus duct, and disconnector.

About 50% of all failures are dielectric breakdowns, which occur during normal service. About 50% of all failures are claimed to be caused by inadequate design and manufacture. SF_6 tightness is reported to be better than standard requirements; the majority report 0.5% leakage per year.

8.4.2.2 Major GIS Failure Modes

In the first decade of the twenty-first century (January 2004 to December 2007) CIGRÉ was carrying out a survey based on the information from utilities about their equipment population of SF_6 CBs and GIS. During the previous CIGRÉ surveys, GIS was still a relatively new technology; however, in the latest survey, 55 utilities from 24 countries participated in the 2004–2007 survey. The data of this survey [51,52] determined the following major GIS failure modes:

- Failing to perform requested operation—63%
- Dielectric breakdown—23%

- Loss of mechanical integrity—4%
- Loss of electrical connections integrity in primary and secondary circuits—1%
- Other modes—9%

8.4.2.3 Age of CIS and MF Mode Distribution

For older breakers (manufactured before 1993), the MF mode is still "Failing to perform requested operation," but at higher rate, ~80%. However, the dielectric breakdown failure mode (in normal service without switching operations) was determined much less often in older GIS (45% and 12% for GIS made before 1884 and in 1984–1993, respectively). In new CIS built after 1994, the dielectric breakdown failure mode plays a major role in the failures, about 43%. Loss of mechanical integrity (3%–6% in older GIS) was not found to be responsible in any MFs of new GIS.

8.4.2.4 Location, Origin, and Environmental Contribution in GIS MF

An important observation was location distribution of failures: about 80% of all MFs happened in outdoor GIS, with the rest (20%) MFs occurring in indoor GIS. It was found that 18% of all GIS MFs originated from mechanical parts other than the operating mechanism; 24% originated from the operating mechanism, while electrical main and secondary circuits were specified as the origin of failure in 11% and 38%, respectively. The environment was not found to be contributing into 88% of all failures.

8.4.2.5 Component and Voltage Class of CIS

The frequency of GIS MFs depends on voltage classes and type of GIS equipment: CBs, IT, DS, ES, and other components (busbars and bus ducts). For example, it was found that for GIS >700 kV, the majority of failures involved CBs (100%). However, for GIS of lower voltage (60–100 kV) CBs caused about 27% of failures, with the "weak link" being DS/ES (60% of all failures). CBs have been found to cause failures in about 50% of all GIS failures for GIS classes from 300 to 700 kV. For all voltage classes the ITs' role in GIS failures was about 5%, while CBs caused about 34% of all failures. Switches caused almost half of the GIS failures of all voltage classes [52].

The data clearly demonstrated the correlation of MF rate and age of the GIS. MF rate tripled for the GIS manufactured before 1984 compared with the newest GIS built in 1994–2005. MF rate was more than twice higher for older GIS than for those manufactured in 1984–1993. Various voltage classes of GIS of different ages experience different rates of MF. The oldest GIS MF rate was highest for the substation of the class 300–500 kV and lowest for the class 200–300 kV [51].

8.4.2.6 Age of GIS Components

Depending on GIS age, different components failed. For example, in older GIS manufactured before 1993, the distribution of failed components is about the same: CBs failed in about 45% MF, switches (DS and ES) failed also in about 45% of the cases. The rest of the failures were caused by the failure of busbars/bus ducts and ITs. But in new GIS manufactured in 1994–2005, the roles of the components in MFs changed; it was found that the failures of busbars/bus ducts and ITs grew significantly and caused about 33% and 7%, respectively, of all failures. The failures of CBs and switches of all types play a lesser role (~22% and ~37%, respectively) in MF events of new GIS than of older GIS [51].

8.4.2.7 Service Conditions of MF Discovery

Analysis of survey data shows that the majority of GIS failures were discovered during switching operation either using CBs or switches (51%); the second largest discovery was during normal service (31%); 10% of all MF was found during testing and maintenance, and the rest was found in de-energized condition or during operation (each about 4%).

8.4.2.8   Time of MF Cause Introduced

The new survey also determined when the cause of the GIS MF was introduced, considering two important periods: during a period of putting GIS in service or during service.

Multiple conditions during the service are included in survey, such as current in excess of rating, mechanical stress in excess of rating, various environmental stresses, corrosion, wear and aging, incorrect operation, incorrect monitoring, and so on. Electrical and mechanical failures of adjacent equipment, human error, incorrect maintenance, external damage caused by animals, humans, and other factors may also play an important role in introducing failures.

It was found that in older GIS, the cause of most GIS failures (64%), was introduced before GIS was put into service, while in only 28% of GIS MFs the cause was introduced during service. In new GIS built after 1994, this relation was reversed: in 61% of all GIS MFs, the cause was introduced during service, and only in 17% of all GIS failures, the cause was introduced during the period before GIS was put into service [51].

8.4.2.9   Age of the CIS and Primary Cause of the Failure

In the older CIS, the primary cause of failure was attributed to wear and aging (50%–60%), while in new GIS these cause only about 12%, which is reasonable for recently built substations. The role of incorrect maintenance was decreasing with age from about 10% for the oldest GIS to 3% for the newest GIS. And the reverse relation to age was found for the role of manufacturing faults, which increased 10 times from 5% for the oldest substations to 20% for MFs of GIS.

8.4.2.10   Failure Rates of GIS Components

According to Ref. [53], the role of various components of GIS failures is distributed almost evenly between three major groups. Among all components, CBs or switches are most often involved in MF of 300–500 kV GIS (27.3%), the next components most often involved in failure are bus ducts and interconnecting parts (24.2%), and the third type of components most often involved in failure are the disconnectors (20.5%). SF_6/air bushings (9.3%), busbars (6.2%), voltage, and CTs (4.3% and 1.9%, respectively) play a lesser role in major HV GIS. Much less often transformer interfaces are involved in MF (2.5%) and surge arresters (1.9%).

8.4.3   Failure Causes of SF_6 CBs

Latest CIGRÉ survey data on CBs failure causes were reviewed in Ref. [54]. It was found that in most failures (93%) environmental circumstances did not play any role; however, in the rest (7%) of all MFs, the environment contributed to the failure. Among the components of CBs, the operating mechanism plays a leading role in MFs (82% of all failures), and electrical control and auxiliary circuits caused 67% of all MFs.

The role of the operating mechanism in MFs was found to depend on the age of CBs; thus, a lesser number of failures were caused in older CBs built before 1984 than in CBs built in 1994–2003. As to the failures caused by electrical control and auxiliary circuits, the finding is the reverse: a greater number of failures were caused by these circuits in older breakers than in newly built CBs.

MF rate if found to be directly related to the age of CBs, going from 87% for the breakers built before 1984% to 78% for breakers manufactured in 1985–1993, and down to 68% for new breakers built in 1993–2003.

Based on the latest survey of SF_6 CBs reliability, the main conclusions are

- The majority of the CBs are used at service voltages between 60 and 200 kV.
- The majority of the CBs are installed outdoors.
- 54% of the CBs are used for overhead line switching.

- The most widely used type of operating mechanism has changed from hydraulic to spring design.
- Most of the failures seem to happen during normal service.
- Leakage of SF_6 or oil seems to remain a problem.
- Operating mechanisms are still the most often reported components responsible for MFs.

8.4.4 FAILURE CAUSES OF DISCONNECTORS AND EARTHING SWITCHES

First survey on reliability of Disconnectors and Earthing switches (DES) was based on the review of responses from a total of 21 countries that participated both in gas insulated and in air insulated substations (GIS and AIS) surveys of almost a quarter million disconnectors and ES [55].

Six different types of disconnectors have been reviewed: center-break, double-break, knee-type, vertical break, semipantograph, and pantograph disconnectors [53]. It was found that most MFs occurred with center-break-type DES (43%), 21% of MFs happened with double-break DES, and knee-type and vertical break DES together had 10% of all MFs (3% and 7%, respectively). Low-speed ES had 18% MFs, while high-speed switches experienced no failures.

Failure cases have been sorted by type of substation. It was found that MFs of DES happened more often at AIS (more than 60%) than at GIS (20% at one-phase GIS and 32% at three-phase GIS) compared with minor failures of the switches. Most cases of DES MFs were found with the electric motor operating mechanism (55%), compared with 39% with the pneumatic operating mechanism and 6% with the manual one. The important factor in DES MFs was determined to be the service condition in which the failure happens. Most failures occurred during "normal service operation demanded" (61%), much less during "normal service—no operation" (24%), and 8% of all MFs happened in the de-energized condition.

8.5 FAILURE CASES OF HIGH-VOLTAGE ELECTRICAL EQUIPMENT

In this section, several failure cases of different types of HV electrical equipment will be presented as they have been described in original papers and reports. These cases have been selected because they represent those events when the equipment failure originated from the failure causes repeatedly mentioned by most industries utilizing electrical equipment (Section 8.3).

8.5.1 FAILURES OF HV BUSHINGS

An impressive account of HV bushings' failures is presented in IEEE PES "Progress Report on Failures of High Voltage Bushings with Draw Leads" [56]. The report states that the failure rate of 230 kV bushings increased significantly in 2007 with eight failures, an additional related failure in 2009, and two failures in 2010. Failed bushing age ranged between 4 and 13 years.

Bushing failure points in sealed-type transformer tanks are found below the flange for gas space transformers, with greatest damage in the lower insulator, in bushing tube and in the lead. For conservator-type tanks the failure point is typically above the flange for conservator transformers, which destroys the upper porcelain, leaving the lower insulator intact.

A detailed root cause investigation has been performed to reveal the cause(s) of multiple HV bushings failure. The evidence collected points to the arcing issue: arc marks confirmed by laboratory analysis, with the worst affect found in the dry portion of the bushing tube.

It was proved that noninsulated leads are able to flash over and make contact, and that random contact between the cable and the tube contributes to the problem as an unintended current path is formed. These facts led to a conclusion that the common cause of HV bushings' failure was *arcing* between the bushing tube and the draw lead cable, with the *flashovers* as the initial cause of failures.

8.5.2 Failures of HV Transformers

8.5.2.1 Case: Failure of Winding Insulation and Bushing

The case of a violent failure of an HV three-phase autotransformer (220, 132, and 11 kV, 50 MVA) at a substation in India in 2006 is presented in Ref. [57]. This transformer was in operation for 22 years since 1984 and failed with openings at welded joints on top of the tank. It was found that the lower part of R-phase of HV (220 kV) bushing had been damaged, with the cellulose insulation exposed, the upper part of the R-phase bushing had popped up. Y-phase and B-phase of HV (220 kV) bushings were found to be displaced from the bottom part. The gaskets were in a very bad condition and continuous oil leakage was found from the bottom of Y-phase of HV bushing. Both the inspection window and the main tank of the transformer were found lifted up from the welded joint portion under pressure and many other kinds of damage were also discovered. No previous failure or major repair of the transformer was reported.

Analysis of the failure showed that the transformer was quite old and was in operation for more than two decades. The average loading on the transformer has rarely exceeded 50% of its rated capacity and has never been overloaded beyond its maximum capacity during its service life. Even on the day of failure, loading on the transformer was only 18 MW. Therefore, the cellulose insulation of the transformer was most likely to have passed through the normal aging process, and fast aging due to overloading was not expected.

The oil condition according to the data of the test performed a year before the failure showed the measured parameters of oil within the desired limits. But the DGA conducted also during the previous year showed that the content of oxygen (O_2) was 6640 ppm, which is considered dangerous for the transformer. The concentration of oxygen might have increased further over a period of about 1 year until the date of failure in 2006.

Both the moisture and oxygen content of the insulation system have a decisive influence on the aging rate of the power transformer. Moisture, especially in the presence of oxygen, is extremely hazardous to transformer insulation. Oxygen together with a water content of 2% in paper insulation can raise the rate of aging by a factor of 20. Oxygen with concentration above 2000 ppm dissolved in transformer oil is very destructive for paper insulation. Additional moisture, especially in the presence of oxygen, is extremely hazardous to transformer insulation. This might have accelerated the aging process of paper insulation. Low-energy arcing and aged paper insulation might have resulted in insulation failure.

Arcing might have taken place inside the transformer due to either a failure of winding insulation/HV bushing and/or bad contact in the tap changer, which has resulted in the formation of gases and development of gas pressure. A continuous increase in content of oxygen and nitrogen and insulation failure might have led to the high-energy arcing, accumulation of other fault gases and rise in gas pressure.

High gas pressure development inside the transformer and nonoperational condition of the pressure relief device (PRD) possibly due to blockage have resulted in the creation of openings at vulnerable regions and weak points of the transformer tank and some welded junctions to vent out the gases. It was concluded that the failure of the transformer was caused by the *failure of winding insulation and bushing*.

8.5.3 Failure Mechanisms of HV Transformers and Bushings

In several studies and surveys, it was found that problems with bushings are a major source of HV transformer failures. To develop a means of preventing violent, dangerous, and costly failure events, it is important to clearly identify the possible failure mechanisms of HV bushing. Defects in a bushing may originate from one or more main parts of the bushing construction: core, core surface, oil, the porcelain inner surface, and so on [5]. Core defects include ingress of moisture or air, or high losses of impregnated liquid, all of which might result from leaking or deteriorated gaskets. These defects may be residual moisture, poor impregnation of oil into paper, migration of conductive graphite ink used in some bushings instead of foils, shorted layers, overstressing, or dielectric heating. Wrinkles and delamination in papers may also cause defects.

- *Core surface defects* may arise from surface moisture, contamination from oil aging products, and deposits of metal or carbon particles on the surface. These deposits may give rise to PD on the core surface, reduced surface resistance, and increased dielectric losses. Failure modes resulting from core problems include ionization, gassing, thermal runaway, puncture, flashover, and explosion.
- *Aging defects of the bushing oil* develop as the oil deteriorates from the effects of high temperature from core heating, moisture, electrical field, and bushing solid materials. The oil may be contaminated with colloid-type inclusions containing copper, aluminum, zinc, and so on, and lose its dielectric strength. Since the oil/paper ratio is small in the bushing, even small quantities of oil loss or moisture ingress from the atmosphere or core may significantly degrade the insulation qualities inside the bushing.
- *Aging of the internal porcelain surface* may be caused by deposits of carbon, and semiconductive sediments from oil breakdown and even from particles of iron from pump bearing wear. Impurities in the porcelain may give rise to PD, and catastrophic failures from this mechanism have certainly been noted in epoxy bushing construction. Failure modes from oil and porcelain surface problems include PD, surface discharge, and gassing.
- *Conductor defects* are caused by overheated connections at the top and bottom of the bushing and may be worsening by the use of dissimilar metals.
- *External defects* such as cracks, contamination, and surface discharge may occur and result in a flashover. Improper storage of spare bushings has been known to promote failures. Problems with seals may also cause major defects from loss of oil or moisture ingress.
- *Bushing failure scenarios.* Faults in the bushing core [58] result in the development of destructive ionization at the points of overstressing, overheating, and thermal instability of dielectric materials. These processes create a defective area with excessive conductance appearing between two or more core layers, which results in a burning through between papers and the occurrence of a short circuit between two of the several layers. As core layers are shorted, this failure mechanism leads to further heating and finally to a state of thermal runaway: an explosion is practically inevitable from this condition. In another failure scenario occurring in an inner capacitive layer of the bushing, the overheating causes the paper to dry and then burn outward. As the paper burns, hydrocarbons are created that cause gases to be generated in the bushing. Eventually these gases build up pressure, resulting in explosion and fire.

8.5.4 FAILURES OF HV CBS

8.5.4.1 Case 1: Failure of Mechanical Linkage

The case was reported in Ref. [57] about an HV SF_6 CB failure that occurred at a substation in India in 2006. The interrupter chambers of the Y pole had exploded and porcelain housing had shattered into pieces. The root cause investigation found that the flashover of the insulator string of this phase at the tower location caused ground fault. The flashover had taken place on a normal day without any rain. The fault on the line initiated the tripping of the three-phase gang-operated CB. During the opening of the CB, the R and B poles of the breaker successfully opened.

Unfortunately, the Y pole of the gang-operated breaker failed to break the fault current. The fault current continued to flow through the Y pole of the CB for quite some time. The flow of high current through the breaker increased SF_6 gas pressure in the interrupting chamber, leading to blasting of the chamber to release excess gas pressure. The CB has served only for about 5 years. The visible holes on the metallic part of the main contact shows severe arcing marks.

The Y pole of the CB got disengaged from the gang/common operating rod connected to a common motor. The blasted Y pole of the breaker being disengaged from the common operating rod fell over the nearby isolator of the same bay, causing bus fault and outage of both 132 and 66 kV systems for about 31 min. It was concluded that the HV CB had failed due to failure of the mechanical linkage of the Y pole of the breaker with a common driving assembly while breaking the fault current.

8.5.4.2   Case 2: Trapped Water in Internal Bolt Holes

In April 2004, the failure of a 115 kV SF_6 gas CB was reported [59]. Although there was no visible external damage, relay and protection analyzed the relay information and all indications led to an internal failure of the K98 CB. After inspection of the disassembled breaker, the flashover was noted at the bushing assembly, with evidence indicating that the arc struck from the bushing conductor to the bushing adapter plate. The quantity of the arc by-product in the center phase tank was moderate, indicating that the flashover was not severe. Investigation into the cause of flashover indicated that randomly occurring water was the likely cause of the flashover. Water was detected using an ultrasonic inspection technique at different levels, in three of the six bushing nozzles, in bolt holes.

The manufacturer disclosed their knowledge of three other similar failures of the same type of breaker and admitted to a known process control that was identified and corrected in 2001, affecting breakers manufactured between 1999 and 2001. The manufacturing process began with pressure washing the cast breaker tanks with water. The tanks were set outside to dry and the failure was attributed to trapped water in internal bolt holes. This water then eventually migrated into the tank, providing a point of discharge, ultimately faulting the breaker internally at the point of water droplets on the tank wall. The manufacturer did not issue a service advisory to all their customers notifying them of this problem.

8.5.4.3   Case 3: Contact Jamming or Mechanism Failure

This failure case of SF_6 CBs was reported by an electrical transmission and distribution company operating in Zambia [60]. In August 2008, a 66 kV SF_6 CB failed when the feeder, on which it is installed, was de-energized to allow a high load to pass under the line. This CB was less than 6 years old. The blue phase interrupting column completely shattered approximately 30 s after de-energizing the concerned transmission line. There was a flashover, resulting in the snapping of the conductor and catastrophic failure of the CB.

The consequences of the failure were extremely damaging. The flying porcelain debris from the CB damaged the post insulators for the 66 kV line isolator, the line main and reserve busbar insulators and the conductors to transformer 66 kV main busbar isolator. The flying porcelain debris from the CB also damaged insulator discs for the busbar flyovers on two 66 kV line bays. In addition, the red phase VT on 66 kV line had oil leak and cracked porcelain from the flying debris. The blue phase VT on an adjacent line bay was also affected and had an oil leak. SF_6 CB-interrupting columns and support insulators were completely shattered. The blue phase assembly components (fixed and moving contacts) were found on the ground. CB and line earth switch control cables were damaged as a result of the explosion. The explosion on the blue phase was vented in all directions and also downwards through the CB-operating mechanism and control cabinet.

On the basis of a root cause investigation, it was suggested that the failure of SF_6 CBs could happen due to a lack of proper moving contact separation in the blue phase because of contact jamming or mechanism failure. The following scenario could explain how the failure had developed. At first, breaker contacts in the blue phase-interrupting column did not open, or did not fully open; therefore, when an abnormal load was crossing the line, it made contact with the blue phase conductor, thus generating a fault current. The fault current generated in the blue phase, with the contacts not fully separated, led to arcing, heating, and decomposition of SF_6 gas in the CB blue phase column, which resulted in high pressure in the column and eventual explosion.

It has also been suggested that decomposition of SF_6 gas could play an important role in failure. SF_6 gas properties may have been compromised after arcing across the contacts during fault clearing operations. Products of SF_6 gas decomposition include hydrogen fluoride, carbon tetra-fluoride, and hydrolysable fluorides, all of which are corrosive in nature and can lead to loss of insulating properties and arcing across open CB contacts. These corrosive by-products can also degrade insulating materials that are present in the CB-interrupting columns.

8.6 FAILURE CASES OF LV AND MV ELECTRICAL EQUIPMENT

8.6.1 Bushing Failures in MV Switchgear

Violent bushing failures in MV switchgear, if they happen, may cause catastrophic damage to the surrounding buildings and adjacent plants. A significant number of such catastrophic failures are presented in a report from the City of Cape Town in South Africa [61]. Multiple failures began occurring after the manufacturers replaced previously used Bakelite paper bushings with resin cast bushings. The replacement was related to the expansion of SF_6 and vacuum technology with the supposed advantages of resin cast technology. Resin cast insulators are considered have greater suitability for mass production, superior fault toleration, resistance to scratch, and mechanical damage resistance. Some manufacturers would redesign existing breakers to SF_6 and resin cast technology to fit them in the same panel as the previous generation oil breaker. South African utilities used three models of SF_6 MV switchgear from the same manufacturer, which are all interchangeable with previous generation oil breakers.

However, contrary to the expectations, the City of Cape Town experience showed that bakelite paper bushings outperform resin cast bushings by far. A large number of bushing failures, mostly due to PD activity across the surface of bushings, were experienced on at least two models of the same panel and manufacturer, while severe PD activity but no catastrophic failures were experienced on a third model of the same manufacturer.

Some of the first recorded failures in the City of Cape Town were at the substation in 2004. Staff who entered the substation for switching operations experienced a strong smell of chlorine in the substation and, on further investigation, they found the breaker to be with severe discharge degradation on bushing insulation and cluster contacts [61]. Severe corrosion, tracking, and cluster degradation were also found at the substations (Figures 8.3 and 8.4).

During inspection, other defects have been found, such as pores beneath the skin of the bushing resin (manufacturing defects), as well as misalignment of shutter boxes (signs of poor quality control during manufacturing).

(a)   (b)

FIGURE 8.3 Degradation of CB clusters and bushings. (From van Heerden C. and Rogerson A. Bushing failures in medium voltage switchgear, *2009 Convention*, Online. Available: http://www.ameu.co.za/Portals/16/Conventions/Convention%202009/Papers/Bushing%20failures%20in%20medium%20voltage%20switchgear%20-%20Coetzee%20van%20Heerden.pdf. With permission.)

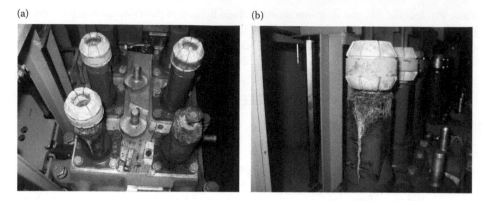

FIGURE 8.4 Severe corrosion degradation of clusters (a) and tracking on the bushings (b). (From van Heerden C. and Rogerson A. Bushing failures in medium voltage switchgear, *2009 Convention*, Online. Available: http://www.ameu.co.za/Portals/16/Conventions/Convention%202009/Papers/Bushing%20failures%20in%20 medium%20voltage%20switchgear%20-%20Coetzee%20van%20Heerden.pdf. With permission.)

Based on inspection results and root cause investigation, severe *PD* activity was determined as the primary cause of failures. The probability of PD activity was increased due to multiple local regions of electrical stress created within voids, pores, and bubbles in resin moldings.

Highly elevated humidity in the area where substations are located should be added to the failure causes. Humid air and, in particular, droplets of *moisture* accelerate PD activity in the bushing assembly. Moisture carried in air modifies the dielectric properties of air, which in modern switchgear is a significant part of the insulation system. The reduced breakdown strength of the air increases the probability of PD. However, while most of the new switchgear panels are fitted with heaters, some manufacturers specify panel heaters as optional.

Misalignment of the breaker may play a significant role in increased PD activity. If the breaker bushings do not align with the respective orifice bushings, this results in an inconsistent electrical field around the bushing due to irregular positioning of the shutter box apertures. Misaligned shutter boxes may have the same effect. Misaligned breakers may also cause the orifice (female) bushing to be damaged and tracking may occur. These breakers may be dangerous to operate.

8.6.2 Case Studies of MV Switchgear Failures

The MV switchgear failures can be attributed to inadequate design. However, more often the failures could result from inadequacy of the maintenance practices, which fail to prevent and detect mechanical and electrical problems that lead to failures. Increasing load densities also put a lot of pressure on the existing aging MV switchgear infrastructure. Solutions often try to accommodate the conflicting requirements of maintaining high reliability and availability while keeping the costs low and providing the highest levels of safety.

Given the space constraints in existing receiving stations, a feasible solution for enhancing the life of the existing switchgear is often resolved by retrofitting and refurbishment actions. However, refurbishment actions require providing operational flexibility and rating, and involve giving special attention to safety aspects and integration. Errors in design, defects of the components, and human mistakes, and a combination of all these, may cause severe MV switchgear failure [62].

8.6.2.1 Case 1: Component Defect

The following failure happened with a feeder breaker, which was taken out for performing CM tests. The wires for the timing test were connected to the breaker arms and the breaker was racked in the "test" position for performing the operations. The breaker flashed over while inserting in "test" position. It was found that flashover was caused by the failure of the mechanical interlock.

The metallic clamp of the test equipment connected to the breaker came into arcing distance of the live busbar, causing the flashover.

8.6.2.2 Case 2: Arcing, Design Errors

The MV breaker failed during fault clearance, causing extensive damage to doors and windows of the switchgear room. Root cause investigation led to the conclusion that during breaker failure, the arc produced the pressure wave that could not find a way to escape and caused damage to the door and windows. This damage could be prevented if the outlets for the pressure wave have been installed at various locations in the switchgear room in the form of hinged louvers. These louvers remain closed normally and do not allow the outside dust to enter the room. In case a pressure wave is caused due to the electrical arc, the pressure is released through these louvers, thus preventing damage to the switchgear room and also injury to the operating personnel present in the room.

8.6.2.3 Case 3: Flashover, Water Condensation

A flashover occurred in the breaker compartment of a feeder which was charged and was carrying no current. It was found that the space heater in the breaker compartment was defective (manufacturing defect). The failure could have been avoided if it had been mandatory for the space heater to be in service even when the breaker carries a very low percentage of rated current (<30%). This would help avoid condensation in the breaker compartment. As the load current was zero in the case of failure and the heater was off, there was water condensation in the compartment, which resulted in flashover.

8.6.2.4 Case 4: Overheating

An MV vacuum CB (22 kV, 2500 A) failed during service. On analysis of the failure, it was suspected that the current-carrying part of the breaker was getting overheated while carrying the nominal load current. The overheating was proved by carrying out a Heat Run test, which showed that the TR was exceeding allowable limits. To prevent overheating with the breaker inside the cubicle, a derating factor of 20% had to be applied, especially for feeders that carry more than 70%–75% of the breaker/switchgear-rated current.

8.6.3 Metal-Clad Switchgear Failures

Two failure cases of MC switchgear are presented in NRC Notice 2002–01 [63], one at the U.S. Nuclear Generating Station (NGS) and another at a foreign nuclear power station (NPS). The most interesting aspect in both events was propagation of damage from an electrical fault in one breaker cubicle to other breakers and buswork in the same enclosure. These electrical events provide us insight into possible collateral damage and cascading failures resulting from a single electrical failure and the consequent challenges to plant operation.

8.6.3.1 Case 1: Failure of the 25-Year-Old CB, and Lack of Maintenance

While shifting loads to the unit auxiliary transformers, a fault on a 4.16 kV supply CB from the unit auxiliary transformer caused a fire and the loss of offsite power. The fault was caused by the failure of the 4.16 kV breaker's C phase main contacts to fully close. This resulted in arcing and the production of a thick, dark ionized smoke. The breaker was a three-pole, MV AC power CB rated for 3000 A (continuous) and 350 MVA (interrupting). The breaker was approximately 25 years old and had its last preventive maintenance performed 4 years prior to failure. Due to the extensive fire damage, the cause of the breaker's failure to close could not be definitively established.

Offsite power was lost when ionized smoke (which is conductive) diffused through holes (through which wires passed) and conduits between adjacent cubicles. This shorted the energized incoming terminals of the offsite power supply from the reserve auxiliary transformer. The fault blew open the cubicle door of the offsite supply CB and blew off an insulating boot that covered the A phase busbar. The HV supply breakers upstream of the reserve auxiliary transformer opened to clear the fault. This interrupted the nonvital offsite power to the unit. The loss of offsite power resulted in the need for DC backup power to operate the non-safety-related turbine lube oil pump. The failure of the DC supply breaker for the lube oil pump resulted in significant damage to the main turbine.

8.6.3.2 Case 2: Insulator Failure

At a foreign NPS, a fault in a 4.16-kV load center caused a fire and the loss of offsite power, while the reactor was shut down but with significant decay heat. This, combined with a subsequent independent failure in the onsite standby power supply, resulted in a station blackout (i.e., a loss of AC power to both redundant safety systems). Recovery from the event was further complicated by smoke and the dependence on AC-powered emergency lighting and ventilation.

For several days prior to the event, foggy, misty weather had caused salt deposition on the insulators on the 345-kV transmission system and consequent power fluctuations and interruptions. On the day prior to the failure, the 345-kV transmission system had been interrupted, resulting in an automatic reactor shutdown and transfer to the backup 161-kV offsite source. On the day of failure, with the reactor shut down and with offsite power being supplied to the unit from the 161-kV backup source, the 345-kV source was recovered and the circuit into the plant was reenergized.

After the switchyard 345-kV CB was closed (energizing the 345/4.16-kV transformer and the 4.16-kV circuits into the vital load centers, while the 4.16-kV supply breakers were still open), a fault occurred in the A train 4.16 kV load center. The fault was caused by insulator failure on one phase of the A train 4.16 kV safety-related switchgear on the supply side of the supply breaker from the 345/4.16-kV transformer. The cause of the insulator failure is unknown. The fault produced voluminous smoke and ionized gas. Migration of the smoke through the A train 4.16-kV switchgear enclosure resulted in multiple arcing faults. Smoke from the event spread out, causing multiple failures in the switchgear, loss of both the sources of offsite power to both trains, and total loss of power to one safety train. An unrelated failure of the other source of onsite power led to total loss of power to both trains. In addition, the loss of offsite power disabled the backup station blackout power supply.

8.6.4 FAILURE OF MV POWER CABLES

There are several factors causing the failure of MV power cables [64]. The types of insulation of cables that are most widely used lately are PE (XLPE, tree-retardant cross-linked polyethylene [TR-XLPE], and high molecular weight polyethylene [HMWPE]) and ethylene propylene rubber (EPR). Several causes of underground power cable failures have been defined [65,66], such as excessive pulling tension, water treeing, and corrosion. HV surges may blow holes in the jacket and damage the shield, and water may enter through the holes and cause corrosion.

MV power cables are often located in inaccessible locations such as conduits, cable trenches and troughs, duct banks, underground vaults, or in other directly buried installations; under such conditions, they can fail due to insulation degradation. Cable failures are one of the most important concerns for the nuclear industry, because the cables supply power to several loads at a nuclear power plant and the failure event may be very dangerous, causing loss of power to safety buses, service water and emergency service water [67]. An extensive account of cable failures can be found in the NRC report [68].

Very often, PD activity in power cables causes the failures. PD is accelerated by various defects, such as voids, shield protrusions, contaminants, advanced stage of water trees, and so on. Partial discharges will gradually degrade and erode the dielectric materials, eventually leading to the final breakdown [69].

8.6.5 LV SWITCHBOARD FAILURE

This switchboard failure occurred in a community infrastructure-related facility [70]. The damaged switchboard (Figure 8.5) had a supply bus rating of 2000 A, a section bus rating of 2000 A, and a neutral bus rating of 1000 A. The switchboard short circuit rating was 50,000 A. The system was

FIGURE 8.5   Failed LV switchboard. (From Ross B. Anatomy of a Meltdown, Electrical Construction & Maintenance, September 1, 2002, Online. Available: http://ecmweb.com/news/electric_anatomy_meltdown/. With permission.)

three-phase, four-wire, 480/277 V. A physical inspection covered the switchboard, the main breaker, the bus, the metering compartment, and another bus nearby. The arcing damage was found primarily in the main breaker cubicle, vertical bus, and the metering compartment above the breaker.

The examination revealed evidence of arcing (vaporized and missing metal) on the control wiring, on the load side vertical bus, on the line side vertical bus and cabinet metal, and on the main breaker finger clusters and stabs. Both the damaged and undamaged busbars showed considerable corrosion and flaking. The inspection revealed corrosion within the failed breakers and their connections, as well as on the nearby breakers. It was found that hydrogen sulfide was present in the air of the breaker's electrical room. Over time, the hydrogen sulfide caused corrosion and flaking of the finger cluster surfaces. This gas also attacked the mating surfaces of the breaker cubicle bus connections (stabs), so they too corroded and flaked.

The following chain of events was suggested to cause the failure. About 20 days prior to the failure, electricians racked in the main breaker. Because the contact surfaces of its finger clusters and stabs were irregular and corroded, electrical connection had higher than normal electrical resistance. Load current passing through these connections produced extra heat, which accelerated further oxidation and deterioration of the connection surfaces. That, in turn, worsened the heating problem. The heating will likely be most intense on the center (B) phase, and it was enough to melt the B phase connection surfaces. The normal load current produced an arc that bridged across the small gap where the copper surfaces had melted.

Since the load current continued to flow through the arc, it melted and vaporized more of the contact surface material, producing extreme heat and ionized (conductive) gases. Intense heat and ionized gases entered the upper compartment and produced a short circuit of the 480 V (potential transformer) control wiring. Melted wiring insulation may have allowed separate phases to make contact, or the melted fuse block may have similarly allowed separate phases to make contact. The control wiring short circuit was on the line side of the breaker and the line side of the control wiring fuses, and consequently did not cause the main breaker to open or the fuses to blow.

REFERENCES

1. Lehtonen M. and Bertling L. Asset management in power systems, a PhD course in co-operation between KTH and TKK, 2005.
2. Kundur P. *Power System Stability and Control*, Electric Power Research Institute (EPRI) Power System Engineering, McGraw-Hill, New York, 1994.
3. Olsen T. W. *Expected Life of Electrical Equipment*, Tec Topics #15, Siemens, April 2001, Online. Available: http://www.energy.siemens.com/us/pool/us/services/power-transmission-distribution/medium-voltage-product-services/tech-topics-application-notes/techtopics15rev0.pdf.

4. EPA Publication, *Asset Management: A Handbook for Small Water Systems*, EPA 816-R-03-016 September 2003, Online. Available: http://www.cdph.ca.gov/certlic/drinkingwater/Documents/TMFplanningandreports/Typical_life.pdf.

5. Lord T. and Hodge G. *On-Line Monitoring Technology Applied to HV Bushings*, Publication of LORD Consulting, Bondi Junction, Australia, June 2008, Online. Available: http://www.avo.co.nz/technical-papers/transformers/102-on-line-monitoring-technology-applied-to-hv-bushings.

6. IEEE Std C57.115-1991, *IEEE Guide for Loading Mineral-Oil-Immersed Power Transformers Rated in Excess of 100 MVA (65°C Winding Rise)*, March 1991, IEEE Publishing, New York.

7. ANSI/IEEE C57.91-1981, *Guide for Loading Mineral-Oil-Immersed Overhead and Pad-Mounted Distribution Transformers Rated 500 KVA and Less with 65°C or 55°C Average Winding Rise*, IEEE Publishing, New York, 1981.

8. IEEE Std C57.91-1995, IEEE guide for loading mineral-oil-immersed transformers, *Transformers Committee of the IEEE Power Engineering Society*, New York, June 14, 1995.

9. Montanari G. C. and Simoni L. Ageing phenomenology and modeling, *IEEE Transactions on Electrical Insulation*, 28(5), October 1993, 455–776.

10. Mamdouh Abd El A. M., Khalil I. D., and Kamel H. A. Estimation of the lifetime of electrical components in distribution networks, *The Online Journal on Electronics and Electrical Engineering (OJEEE)*, 2(3), Reference Number: W10-0002, 269–273.

11. The effects of overloading on electrical system component life, Appendix O in *Comparing Conventional, Modular and Transportable Electric Transmission and Distribution Capacity Alternatives Using Risk-adjusted Cost*, Sandia Report SAND2014-18540, September 2014, Online. Available: http://www.sandia.gov/ess/publications/SAND2014-18540.pdf.

12. Stoving P. N. and Baranowski J. F. Interruption life of vacuum circuit breakers, *Proceedings of the Nineteenth International Symposium on Discharges and Electrical Insulation in Vacuum* (ISDEIV), Xi'an, China, vol. 2, 18–22 September 2000.

13. Sprague M. J. Service-life evaluations of low-voltage power circuit breakers and molded-case circuit breakers, *IEEE Transactions on Industry Applications*, 37(1), January/February 2001, 145–152.

14. Morgan V. T. Effects of frequency, temperature, compression, and air pressure on the dielectric properties of a multilayer stack of dry kraft paper, *IEEE Transactions on Dielectrics and Electrical Insulation*, 5(1), February 1998, 125–131.

15. Sen P. K. and Pansuwan S. Overloading and loss-of-life assessment guidelines of oil cooled transformers, *Rural Electric Power Conference*, April 29–May 1, 2001.

16. IEEE Standard C57.91-2011 (Revision of IEEE C57.91-1995, *IEEE Guide for Loading Mineral-Oil-Immersed Transformers and Step-Voltage Regulators,* IEEE Publications, New York, 2012, 1–123.

17. Morgan V. T. Effect of elevated temperature operation on the tensile strength of overhead conductors, *IEEE Transactions on Power Delivery*, 11(1), January 1996, B4/1–B4/8.

18. Parise G., Rubino G., and Ricci M. Life loss of insulated power cables: An integrated criterium to improve the ANSI/IEEE and the CENELEC/IEC method for overload protection, *Conference Record of the Thirty-First Industrial Applications Society Annual Meeting*, San Diego, CA, October 1996.

19. Coyle T. Keeping your cool: HVAC for electrical equipment spaces, *Current Affairs, Engineered Systems*, August, 2002, Online. Available: http://findarticles.com/p/articles/mi_m0BPR/is_8_19/ai_90511127.

20. Senn F., Muhr M., Ladstätter W., and Grubelnik W. Complexity of determining factors for the thermal evaluation of high voltage insulation systems on the example of rotating machines, *XVII International Conference on Dielectric and Insulating Systems in Electrical Engineering*, DISEE2008, September 2008, Demänovská Dolina, Slovakia, Online. Available: http://www.isovolta.com/images/GetDownload.aspx?miid=161978bb-ec72-4484-846f-466e16539eb9.

21. IEC 60085: Electrical insulation—Thermal evaluation and designation, 2007–11.

22. Nakos J. T. and Rosinski S. T. Research on U.S. Nuclear Power Plant Major Equipment Aging, SAND94-2400C, DOE's Office of Scientific and Technical Information (OSTI), 1994, Online. Available: http://www.osti.gov/bridge/servlets/purl/10190825-J7vdeV/webviewable/10190825.pdf.

23. Pappas R. A. *FAA Aging Electrical Systems Research Program Update*, Jan 2003, Online. Available: https://www.caasd.org/atsrac/meeting_minutes/2003/January-2003-FAA_Research&Development_Update.pdf.

24. Minaud A., Pezard J., and Priam J-P. Ageing of medium voltage overhead lines, CIRED, *20th International Conference on Electricity Distribution*, Paper 0647, Prague, June 2009, Online. Available: http://www.cired.net/publications/cired2009/pdfs/CIRED2009_0647_Paper.pdf.

25. Srinivas N. and Morel O. Condition assessment of electrical equipment in power plants, *Electricity Today*, March 2009, pp. 28–35, Online. Available: http://www.vanchanvishesh.com/documents/CONDITION%20 ASSESSMENT%20OF%20ELECTRICAL%20EQUIPMENT%20IN%20POWER%20PLANTS.pdf.

26. Wafer J. A. *Keynote address, Proceedings of the Aged Electrical Systems Research Application Symposium*, The Fire Protection Research Foundation (NFPA), Chicago, IL, October 18, 2006.

27. Rasdall J. *Aging Residential Wiring Issues: Concerns for Fatalities, Personal Injuries, and Loss of Property*, Online. Available: http://www.kngelectricalservices.com/Aging-Residential-Wiring-Issues.pdf.

28. *Residential Electrical System Aging Research Project*, Technical Report, prepared by Dini D.A, National Fire Protection Association, Quincy, MA July 1, 2008, 71pp., Online. Available: http://ewh. ieee.org/cmte/pses/ffat/support/RESAReport.pdf.

29. NFPA 73: *Standard for Electrical Inspections for Existing Dwellings*, The Fire Protection Research Foundation (NFPA), Quincy, MA, 2011.

30. NFPA 70®: *National Electrical Code® (NEC®) Handbook*, 2011 Edition.

31. Stetson L. E. Electrical codes and standards in rural applications, *IEEE Transactions on Industry Applications*, 34(6), November/December 1998, 1419–1429.

32. *An Overview of Transmission System Studies*, United States Department of Agriculture, Rural Electrification Administration, REA BULLETIN 1724E-202, February 1993, Online. Available: http:// www.usda.gov/rus/electric/pubs/a/1724e202.pdf.

33. *Corrosion Cost*, Appendix X "Agricultural Production," CC Technologies, Online. Available: http:// corrosionda.com/pdf/agriculture.pdf.

34. *Corrosion control of Agricultural Equipment and Buildings*, Online. Available: http://www.npl.co.uk/ upload/pdf/corrosion_control_of_agricultural_equipment_and_buildings.pdf.

35. Eker B. and Yuksel E. Solutions to corrosion caused by agricultural chemicals, *Trakia Journal of Sciences*, 3(7), 1–6, 2005, Online. Available: http://www.uni-sz.bg/tsj/vol3No7_1_files/Eker_2_1_[1].pdf.

36. IEEE Gold Book™ Standard 493™-2007, *IEEE Recommended Practice for the Design of Reliable Industrial & Commercial Power Systems*, IEEE Publications, New York, 2007, 383pp.

37. Wilkins D. J. The bathtub curve and product failure behavior, Part I—The bathtub curve, infant mortality and burn-in, reliability, *Hot Wire*, 21, November 2002, Online. Available: http://www.weibull.com/ hotwire/issue21/hottopics21.htm.

38. *Common-Cause Failure Event Insights Circuit Breakers*, NUREG/CR-6819, vol. 4, INEEL/EXT-99-00613, May 2003, Online. Available: http://pbadupws.nrc.gov/docs/ML0317/ML031710861.pdf.

39. IEEE Committee Report. Report on Reliability Survey of Industrial Plants, part 6, *IEEE Transactions on Industry Applications*, IA-10, July/August and September/October 1974, 456–476, 681.

40. IEEE Committee Report. Reliability of electric utility supplies to industrial plants, *IEEE-ICPS Technical Conference Record*, 75-CH0947-1-1A, May 1980, Toronto, Canada, May 5–8, 1979, pp. 70–75.

41. Sokolov V. Transformer Life Management, *II Workshop on Power Transformers-Deregulation and Transformers Technical, Economic, and Strategical Issues*, Salvador, Brazil, 29–31 August 2001.

42. Sokolov V. Changing World Perspectives – A Report from CIGRE, *Fourth AVO New Zealand International Technical Conference*, Methven, New Zealand, April 2005.

43. Janick M. Operations-extreme loading, *Electrical World T&D Magazine*, Nov/Dec 2001.

44. *Australia/New Zealand Transformer Reliability Survey 1996 Report*, Western Power Transmission Projects Branch.

45. Stead J. Bushing failure rates/mechanisms etc., *Wiedmann 2002 LV Conference Presentation*.

46. Genutis D. A. Top five switchgear failure causes and how to avoid them, *NETA World Journal*, 1-3, Summer 2010, Online. Available: http://www.netaworld.org/sites/default/files/public/neta-journals/ NWsu10-NoOutage-Genutis.pdf.

47. Lowsley C., Davies N. and Miller D. Effective condition assessment of MV switchgear, *ME--- Maintenance & Asset Management*, 27(4), 46-51, July/Aug. 2012, Online. Available: http://www. maintenanceonline.co.uk/maintenanceonline/content_images/Pages%2046,%2047,%2048,%2049,%20 50,%2051.pdf

48. NFPA 70B, *Recommended Practice for Electrical Equipment Maintenance*, The Fire Protection Research Foundation (NFPA), Quincy, MA, 2002.

49. Waite F., Kopejtkova D., Mestrovic K., Skog J. E., and Solver C. E. Use of data from CIGRE high voltage equipment. reliability survey, *CIGRE 2009 6th Southern Africa Regional Conference*, Paper C207, Cape Town, South Africa.

50. Sölver C. E. Past CIGRÉ surveys on reliability of HV equipment, *CIGRE Session 2004*, Paris, Online. Available: http://www.mtec2000.com/cigre_a3_06/Rio/past.pdf.

51. Kopejtkova D., Kudoke M., and Furuta H. Gas Insulated Substations (GIS), *Preliminary Results from Present CIGRE Survey*, CIGRE WG A3-06 "HV Equipment Reliability," October 2008, Seoul, Online. Available: http://www.mtec2000.com/cigre_a3_06/Seoul/GIS.pdf.

52. *Gas Insulated Substations (GIS) Reliability Results*, CIGRE WG A3-06 "HV Equipment Reliability," 2010, Online. Available: http://www.mtec2000.com/cigre_a3_06/SCB3_GIS_2010.pdf.

53. Coventry P. Asset Policy, Gas-insulated switchgear (GIS), presented at US National Grid (USNG), GIS Meeting, June 2008, Online. Available: https://www.scribd.com/document/309910867/GISmeeting10June2008Presentationv3.

54. Makareinis D., Sölver C. E., Hyrczak A., Mestrovic K., Lopez C. M., Carvalho A., and Sweeney B. Reliability of high voltage equipment. intermediate results: Circuit-breakers, on-line. *CIGRE WG A3-06 "HV Equipment Reliability,"* October 2008, Seoul, Online. Available: http://www.mtec2000.com/cigre_a3_06/Seoul/Circuit_Breakers.pdf.

55. Krone J. C., Hyrczak A., Park Kyong-Yop, Desanti S. S., and Kuntze T. Preliminary results from present CIGRE survey on "disconnectors and earthing switches," *CIGRE Meeting*, June 2006, Rio de Janeiro, Online. Available: http://www.mtec2000.com/cigre_a3_06/Rio/Disconnectors.pdf.

56. Hopkinson P., Del Rio A., Wagenaar L. et al. Progress report on failures of high voltage bushings with draw leads, *2010 Power and Energy Society General Meeting*, 25–29. July 2010, pp. 1–16, Online. Available: http://www.electric-connectionco.com/assets/documents/Progress%20Report%20of%20HV%20Draw%20Lead%20Bushing%20Failures.pdf.

57. *Report on Failure of Circuit Breaker and Transformator at Ranasan and Limdi Substations of Gujrat Electric Transmission Company (GETCO)*, Central Electricity Authority Ministry of Power Government of India, 15 pp., New Delhi, September 2006, Online. Available: http://cea.nic.in/reports/others/ps/pspa2/failure_ranasan_limdi_ss.pdf.

58. *Guidelines for Life Management Techniques for Power Transformers* Prepared by CIGRE WG 12.18 "Life management for transformers," January 20, 2003, Online. Available: http://www.buenomak.com.br.

59. Wright J. M. *Mitsubishi SF6 Gas Circuit Breaker Type 100 SFMT-40E, Breaker Failure Report*, Online. Available: http://www.weidmann-solutions.cn/huiyi/Seminar%202004%20Sacramento/Wright Presentation2004.pdf.

60. Silweya J. O. and Katepa E. S. Copper belt Energy Corporation Experience on To Failures of Merlin Gerin Type SB6-72 Circuit Breakers, Online. Available: http://www.ameu.co.za/library/convention2008./21.%20John%20Silweya.doc.pdf.

61. van Heerden C. and Rogerson A. Bushing failures in medium voltage switchgear, *2009 Convention*, Online. Available: http://www.ameu.co.za/Portals/16/Conventions/Convention%202009/Papers/Bushing%20failures%20in%20medium%20voltage%20switchgear%20-%20Coetzee%20van%20Heerden.pdf.

62. Jain P. K., Murugan P., and Raina D. Case studies on MV switchgear, *Fifteenth National Power Systems Conference (NPSC)*, Indian Institute of Technology (IIT Bombay), Mumbai, India, December 2008, pp. 446–449.

63. NRC Information Notice 2002-01: Metalclad switchgear failures and consequent losses of offsite power, United States Nuclear Regulatory Commission, Office of Nuclear Reactor Regulation, Washington, DC, January 8, 2002, Online. Available: http://www.nrc.gov/reading-rm/doc-collections/gen-comm/info-notices/2002/in02001.html.

64. IEEE Standard 1511.1-2010 *Guide for Investigating and Analyzing Shielded Power Cable Failures on Systems Rated 5 kV Through 46 kV*, Issue date: March 25, 2011, pp. 30.

65. Bristol R. Underground cable failure analysis, *Norwest Public Power Association Engineering and Operations Conference*, April 23, 2009, Spokane, WA.

66. Marion A. Underground medium voltage cable failures and status of testing, Online. Available: http://www.nrc.gov/public-involve/conference-symposia/ric/past/2007/slides/alexmarionslides.pdf.

67. Calvo J., Jenkins R., and Koshi T. Briefing on potential common-mode failures of medium voltage underground cables, *NRC Public Meeting with Nuclear Energy Institute*, June 2, 2004, Online. Available: http://pbadupws.nrc.gov/docs/ML0612/ML061220137.pdf.

68. Nuclear Power Plant Generic Aging Lessons Learned (GALL). http://pbadupws.nrc.gov/docs/ML9936/ML993620048.pdf.

69. Liu Z., Phung B. T., Blackburn T. R., and James R. E. The propagation of partial discharges pulses in a high voltage cable, *Proceedings of the AUPEC'99, Northern Territory University*, September 26–29, 1999, pp. 287–292, Online. Available: http://citeseerx.ist.psu.edu/viewdoc/download?doi=10.1.1.136.5192&rep=rep1&type=pdf.

70. Ross B. *Anatomy of a Meltdown, Electrical Construction & Maintenance*, September 1, 2002, Online. Available: http://ecmweb.com/news/electric_anatomy_meltdown/.

Section II

Renewable Energy Equipment Challenges

9 Solar Energy

9.1 RENEWABLE GREEN ENERGY SOURCES

There are five renewable sources and these are solar, wind, geothermal, water, and biomass.

Solar energy is the energy received from the sun.

Wind energy is the energy produced from the windblow.

Water energy: There are several ways that water can produce or can act as a source of energy. There are three basic forms of "water" energy:

- *Wave energy* is produced by the waves in the sea which are caused by winds.
- *Tidal energy* is caused by the gravitational pool of the moon on the seas.
- *Hydroelectric energy* is produced by falling water.

Geothermal energy is the energy received from the earth's natural heat. This kind of energy is produced and stored in the earth.

Biomass energy is the energy given by either heating or fermenting biomass. All organic materials dead or alive are considered biomass. Biomass energy sources are wood and municipal solid waste, landfill gas, biodiesel, and ethanol.

In this chapter, we will discuss the challenges of the solar renewable energy equipment types.

9.1.1 SOLAR ENERGY

Solar panel using renewable energy from the sun is an array of photovoltaic (PV) cells. Solar cells are arranged in a grid-like pattern on the surface of solar panel. These cells collect sunlight during daylight hours and convert it into electricity. One of the most popular solar energy technologies involves PV systems that consist of a PV module array and other electrical components needed to convert solar energy into electricity.

PV systems are often perceived to be quintessentially "green," since virtually no pollution, greenhouse gas emissions, or noise are produced during their use phase. However, they experienced a slow and somewhat sluggish start, suffering from well-known limits, which include low conversion efficiency, high capital costs, low social acceptability because of poor aesthetics, and concerns of extensive land occupation.

Today, PV provides less than 1% of the electricity that is produced in the world. But the PV sector has been growing steadily for the last two decades, and is now rapidly gaining momentum; estimates indicate that the global cumulative installed capacity is likely to keep doubling roughly every 2 years [1].

Energy generation systems and supporting structures are installed in the areas known for extreme conditions such as deserts and saltwater seas. Solar equipment is being installed in areas that are light intensive, which are in the desert, along coastlines, on top of buildings if they are smaller, and so on. Therefore, this equipment exposed to a corrosive environment become vulnerable to natural processes.

9.1.1.1 Construction of Typical Solar Panel

Basic structure of solar module is shown in Figure 9.1 based on design of Mitsubishi model [2].

The following materials are used in solar module construction [3]:

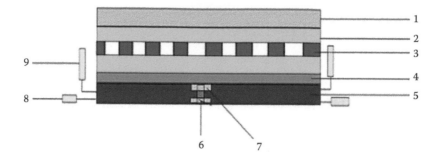

FIGURE 9.1 Typical solar panel (based on Mitsubishi's solar module design): 1—glass; 2—encapsulation; 3—Si strip; 4—back film; 5—frame; 6—protection bars; 7—solder pad; 8—junction box; 9—connectors. (Modified from Kaisare A. Reliability challenges for solar modules, Tutorial 13-9, *ASME 2011 Pacific Rim Technical Conference and Exhibition*, InterPACK 2011, Portland, Oregon, July 2011.)

Glass: It is a tempered and textured glass of 4-mm thickness with low iron content.

Encapsulation: Made of polymers such as ethylene vinyl acetate (EVA), the most dominant material. It is a thermosensitive transparent resin. Four properties make the EVA encapsulation often a choice material:

- High electrical resistivity which makes it as a good electric insulator
- Low fusion and polymerization temperature
- Very low water absorption ratio
- Good optical transmission

The EVA protects the sensitive cells from weather influences such as moisture and UV radiation. There are other materials used for encapsulation around the world.

Solar cells in PV modules are made of different materials, such as crystalline silicon (c-Si), either monocrystalline or polycrystalline; thin film modules, including amorphous silicon (a-Si), copper indium gallium arsenide (CIGS) and cadmium telluride (CdTe); and photoelectrochemical cells that include dye-sensitized cells and polymer or plastic cells.

Insulation: Mylar or polyethylene terephthalate film (PET) is used to electrically isolate the output connections from the back of the cells. PET is hard and strong, is a dimensionally stable material that absorbs very little water, and has good gas barrier properties and good chemical resistance.

9.1.1.2 Off-Grid Solar Power

Solar panels have mostly been used as stand-alone systems for energy. These systems are now being deployed throughout the industrialized and developing world on a commercial scale. Today the global market demand for PV exceeds 5 billion USD annually. The market for PV has developed in both industrialized countries and in the developing countries where off-grid and hybrid village grid electrical services are now becoming available to thousands of remote villages. Such rural populations of developing countries without the benefits of grid connections can enjoy an electrical supply from stand-alone PV systems with their inherent advantages of modularity and independence from imported fuels.

9.1.1.3 Grid-Connected Solar Power

It is now technically possible to connect solar panels to the electricity grid, meaning those who own them could sell excess energy back to their power company. Three developments show how important this branch is becoming:

- The world's largest solar PV power plant, a 10-MW facility in Bavaria, Germany, became fully operational at the beginning of 2005.

- The world's largest rooftop PV installation, a 5-MW roof-integrated design, is now operating in South Hessen, Germany.
- Also during 2005, a leading American manufacturer started marketing a 3-kW grid-tie solar inverter for home use.

PV grid-connected systems are rapidly increasing in numbers supported by government-sponsored programs in Australia, Europe, Japan, and the United States. Most of these systems are located on residences and public/commercial/industrial applications. Installations of large-scale centralized PV power stations, typically owned by utilities, continue at a very slow rate.

9.2 DETERIORATION OF SOLAR POWER EQUIPMENT

9.2.1 EXPOSURE OF SOLAR PANELS TO ENVIRONMENT

Solar panel mounts are available in three primary categories:

1. *Flush mounts*, which are the cheapest and simplest and are used for small solar panels
2. *Roof/ground (or universal) mounts*, which are used for large solar panel system, they are more expensive than flush mounts
3. *Pole mounts*, which are divided into three subcategories: top of pole mounts, side of pole mounts, and poll tracking mounts, which maximizes operating efficiency of the solar panel unit [2].

Since PV panels are installed outdoor, they are exposed to widest range of extreme environmental conditions, such as temperatures from −20 to 11°C (−4 to 230°F) for 365 days during ~20 years of thermal cycles, and from 0 to 100% of air humidity, including snow, rain, and hail [3]. Temperature conditions include

- Maximum ambient temperature: 58°C (137°F)
- Minimum ambient temperature: −67°C (153°F)
- Maximum temperature change $\Delta T/24$ h = 103°C, $\Delta T/1$min = 20°C
- Highest average temperature change $\Delta T/24$ h = 30°C

Maximum yearly average UV (280–400 nm) exposure the panel is receiving on horizontal surface is about 185 kWh/m². Possible environmental conditions on the planet differ in different areas within very wide ranges. The PV panel must survive atmospheric pressures up to 10 kPa static and 4 kPa dynamic. Hail might bring up to 200-g ice deposit. RH may rise up to 95% RH at 35°C. Near salted roads, farms, and seas, pH is about 11 (basic ambient). Near industrial complexes and highways, pH is about 3 (acidic ambient). In these conditions, the PV module must deliver electric energy with minimal performance losses [4,5].

Due to extreme environmental conditions, many types of solar failures occur [2]. Types of solar panel failures are presented in Figure 9.2.

As shown in Figure 9.2, more than 40% of solar cell failures are caused by damp heat. Solar cell equipment or PV cells exposition to damp heat may cause corrosion at cell interconnects, solder joints, and various metallic parts.

The installation of PV modules in coastal areas or on agricultural buildings may lead to additional environmental stress. Besides corrosion of metallic parts, power loss and affected adhesives were identified in the field. Corrosion of silicone-based adhesive sealing may result in consequential loss of insulation properties as well as adhesive strength and can be hazardous to human beings and animals or the infrastructure. Corroded contacts may ultimately cause DC arcing and hence potentially result in fire [5]. Corrosion of PV solar panel [6] is shown in Figure 9.3.

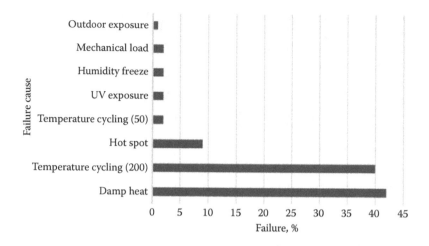

FIGURE 9.2 Solar panel failures. (Modified from Kaisare A. Reliability challenges for solar modules, Tutorial 13-9, *ASME 2011 Pacific Rim Technical Conference and Exhibition*, InterPACK 2011, Portland, Oregon, July 2011.)

FIGURE 9.3 Corroded PV module. (From Mathiak G. et al. PV Module Corrosion by ammonia and salt mist. Experimental study with full-size modules, *Proceedings of 27th European Photovoltaic Solar Energy Conference and Exhibition*, 2012, Online. Available: http://www.tuv.com/media/germany/10_industrialservices/downloadsi06/poster_pvsec/PV_Module_corrosion_by_ammonia.pdf. With permission.)

Corrosion failures of solar panels often occur on installations close to the coast, and in maritime applications (e.g., on buoys or ships) and may be caused by the salt particles deposited from salt-containing sea atmosphere. These failures also take place on installations close to highways (e.g., noise barriers) due to deicing salt present in atmosphere. Another concern is corrosion of the supporting structures that are going into the ground and holding the panels up. Due to corrosion, these posts may degrade and fall over.

9.2.2 Galvanic Corrosion in Solar Panels

PV modules are designed for an operational life span of more than 25 years in the field. The design of the entire installation, not just these expensive components, should target a similar life span.

FIGURE 9.4 Effect of water and corrosion in a solar panel. (From www.booki.cc/tech-cahier/solarpower2/)

There is a strong potential impact of galvanic corrosion on array structures and grounding techniques for solar module frames [6,7]. Atmospheric contaminants, such as chlorides (in marine environments) and sulfur dioxide and nitrous oxides (in industrial locations), are deposited on array structures. Once deposited, the contaminants react with oxygen and water and typically increase corrosion rates by releasing electrons from the metal surface (Figure 9.4). However, corrosion rates can be effectively decreased in areas of high rainfall, as contaminants are regularly washed away from the structural materials [6].

9.2.2.1 Corrosion of Grounding Connection in Solar Panels

PV modules are typically installed on aluminum or galvanized, painted, or stainless-steel frame structures. Stainless-steel grounding components in solar panels are often pressed into the base aluminum, which is creating a dissimilar metals connection, where galvanic corrosion may develop in the presence of electrolyte. Galvanic corrosion could erode the aluminum frame.

Failures from corrosion can have many causes. Galvanic corrosion, which occurs when dissimilar metals like aluminum and copper are in direct contact, has been a common cause of equipment grounding system failures in PV arrays. However, corrosion can also occur as a result of long-term exposure of components to cycling leakage current, which produces an electrolysis process.

Failures due to corrosion can be attributed to the following general causes [8]:

- Improper selection of materials for the bonded connection. Copper and aluminum bonds are the most common and have dramatic results, but other less obvious combinations may break down over time.
- Dissimilar metals in close proximity, which—depending on the electrolyte involved—causes corrosion when exposed to water, soil, or other conductive debris elements.
- Insufficient barriers between dissimilar metals, such as undersized or badly installed stainless-steel washers separating copper and aluminum.
- Good but inadequately protected connections after long-term exposure to leakage current, water, salt humidity, and/or other corrosive agents. An example is a tin-coated assembly joining a copper wire and aluminum frame where the coating is inadequate to serve as a sacrificial barrier over the long term.

9.2.2.2   Prevention of Galvanic Corrosion in Solar Panel Grounding

To prevent failures, one needs to carefully consider whether the methods and materials are suitable for the intended use and for the installation environment. The grounding issue requires a very detail and specified analysis in choosing a proper connector to avoid or minimize a galvanic corrosion issue for solar panel grounding. According to the Solar America Board for Codes and Standards (Solar ABCs)' interim grounding report [8], the following materials are generally compatible: nickel, tin, zinc, zinc–aluminum alloys, 5000 or 6000 series aluminum alloys (alloyed with magnesium and silicon), commercially pure aluminum, and stainless steel containing a minimum of 16% chromium. Copper is not on this list.

PV system designers and installers must provide reliable means of galvanic isolation—such as stainless-steel washers or components with suitable plating or coatings—when using copper conductors to bond dissimilar metals [7]. It might be beneficial to use a grounding lug connected with copper grounding wire to solar panel frames. This connector body is made with extruded serrations for breaking the anodized finish on PV modules and frames. The serrations in the wire binding break corrosion on the grounding wire, which means that no surface preparation is required for the PV module, frame, or ground wire.

9.2.2.3   Standards Regulating Module Frame Grounding Techniques

Standards that currently address module frame grounding include [8]

- UL 1703: Flat-Plate Photovoltaic Modules and Panels
- UL 467: Grounding and Bonding Equipment
- UL 1741: Inverters, Converters, Controllers, and Interconnection System Equipment for Use With Distributed Energy Resources
- UL 2703: Rack Mounting Systems and Clamping Devices for Flat-Plate Photovoltaic Modules and Panels
- UL 61730-1: Photovoltaic (PV) Module Safety Qualification—Part 1: Requirements for Construction

UL 1703 is currently the "primary" standard affecting module grounding and devices. Methods certified to UL 1703 and documented in module manufacturers' listed installation instructions are almost universally accepted by inspectors and authorities having jurisdiction (AHJs). The standard covers a range of safety and construction-related requirements for modules, with a few sections dedicated to frame bonding, grounding, and continuity. UL 1703 establishes requirements for the means of grounding as well as continuity requirements subject to applied current and environmental (accelerated life) testing.

The primary issues or concerns associated with UL 1703 as reported by the industry [8] are that it does not provide adequate assurance or guidance for long-term grounding reliability and it is too restrictive in its approach and process to facilitate certification of third-party devices. Many solar panels' manufacturers proposed to improve current standards to resolve remaining issues [8]. There is a range of proposals for clarifying and/or expanding on the use of appropriate metals, and avoiding inappropriate dissimilar metals.

9.2.3   Solar Panel Corrosion: Mitigation Techniques

Several corrosion mitigation techniques are suggested for solar panels installations [6].

9.2.3.1   Material Coatings

Corrosion is most often combated with paints, electroplating, or other coatings such as hot-dip galvanization. If intact and properly maintained, the coating prevails in the material interactions and limits potential galvanic corrosion. Aluminum used in module frames and array rails is anodized to increase the material's corrosion resistance. Aluminum naturally builds up an oxide layer when exposed to oxygen. The anodizing creates a thicker oxide layer, which acts as a barrier to corrosion. In some cases, one metal's corrosion potential is used to protect another metal.

For example, galvanized channel is protected by a zinc coating (sacrificial coating). The zinc in the galvanized coating reacts with the atmosphere to create layers of zinc oxide and zinc carbonate, protecting the steel underneath it. When used adjacent to other metals, the galvanized steel coating sacrifices itself to protect other, more structurally important metals. A thicker galvanization coating provides greater protection. The minimum coating for steel array structures is a G90 coating that is approximately 0.02-mm (0.75 mils) thick. The coating's service life varies considerably based on a project's location and environmental conditions, so thicker galvanization may be necessary.

Coatings should be repaired if damaged, especially if they are on an anodic material such as steel. When galvanized channel is cut in the field, always apply cold galvanizing paint to protect the steel. Since the occasional paint scratch is practically unavoidable during installation, repair or restore a chipped coating whenever possible. Unsightly rust can be reduced and the service life of the product extended. More than one product manufacturer recommends suitable paint for maintenance of a compromised coating.

9.2.3.2 Contact Surface Area

The metals' surface contact area is also an important consideration in determining the rate of corrosion, since it determines the ratio of cathode to anode. Keep in mind that the galvanic cell is in solution. By overwhelming the solution with anodic material—but limiting the cathodic component—the solution becomes saturated with electrons and corrosion is limited. In the reverse condition, as in steel nails used to fasten a copper material, the cathode dominates with its ability to absorb or dissolve any and all electrons offered by the anode. For example, if zinc-coated steel roofing nails are used to secure copper flashing, the nails quickly fail. It is best to avoid using a small amount of anode (steel) in contact with a large amount of cathode (copper).

9.2.3.3 Isolation Strategies

To minimize solar panel corrosion, it is necessary to physically and electrically separate potentially problematic metal combinations. Using rubber washers to isolate galvanized screws from painted steel sheet goods is common practice in the roofing industry. Stainless-steel washers with an ethylene propylene diene monomer (EPDM) gasket already adhered are commonly available at hardware supply houses.

Metal channel manufacturers have introduced products that isolate copper from the steel channel by way of plastic clamps. These are designed for plumbers who are clamping copper pipe to steel strut, whether electroplated with zinc or hot-dip galvanized. Cooper B-Line Iso-Pipe isolation wrap and isolating clamps such as Vibra-Clamps also provide material isolation and eliminate the metal-to-metal contact that can start the galvanic corrosion process.

EPDM rubber can be used to isolate dissimilar metals, whether built into the washer or inserted as a separate material sheet. Certain plastics may suffice if they are rated for outdoor conditions. Some installers have more confidence in EPDM roofing materials than in plastics, even if the plastic material selected is designed to sustain UV exposure.

9.2.3.4 Fastener Selection

Since many corrosion events take place at a bolted connection, the integrity of each connection is a legitimate concern. The fastener selected should not be anodic in relation to the structural members. In solar panel applications, stainless fasteners commonly available as 18-8 variety (stainless steel having approximately 18% chromium and 8% nickel) meet this requirement. The use of stainless-steel fasteners may prevent galvanic corrosion, but it is also true for galvanized or zinc-plated fasteners. However, some manufacturers recommend stainless-steel hardware only or stainless-steel hardware with aluminum channel.

While stainless steel has become commonplace in PV hardware assemblies, the new Dura-Con line of fasteners offered by Mudge Fasteners uses coatings that reportedly meet or exceed the performance of stainless steel. The new line of nuts, bolts, washers, and lag bolts offers a lower cost alternative to stainless-steel hardware and reduces the potential for galvanic corrosion between the fastener and the aluminum structure.

The cost of materials is always a factor. While metal prices change on a regular basis due to market factors, at the time of writing, aluminum strut channel costs twice as much as galvanized channel, and stainless-steel channel is four times the cost of galvanized. For hardware, stainless-steel fasteners are about four times more expensive than their zinc-plated counterparts.

9.2.3.5 Long-Term Durability

To provide long-term durability, one needs to pay close attention to mechanical connections that may be susceptible to galvanic corrosion and possible structural failures after 10 or 20 years in service. One should consult with equipment manufacturers regarding installation's best practices that take into account the climate where the array is located. It is also important to consult with material trade associations, such as the American Galvanizers Association, which can offer sound guidance and summaries of past research related to galvanic corrosion.

9.2.4 ROLE OF ENVIRONMENT IN POTENTIAL-INDUCED DEGRADATION MECHANISM OF SOLAR PV MODULES

Over the past decade, degradation and power loss have been observed in PV modules resulting from the stress exerted by system voltage bias. The extent of the voltage bias degradation is linked to the leakage current passed from the silicon active layer through the encapsulation and glass to the grounded module frame.

Evaluation of the environmental conditions that influence the system PID mechanisms is done in Refs. [9,10]. It was shown that the wet module associated with the morning dew or rains leads to an elevated leakage current as the system voltage rises with the sun. The leakage current decreases significantly when the module dries and the surface resistance increases. Despite that conductivity of glass and encapsulating material increases with temperature, the leakage current remains controlled by humidity. It is therefore concluded that a wet environment will activate system voltage degradation mechanisms more than a hot, dry environment based on the elevated leakage current.

Humidity is seen to be a key factor in the circuit that enables a leakage current. Conductivity of glass increases with RH up to 100% and PV modules regularly see surfaces at 100% RH (dew, rain, wet snow). The study in Ref. [9] was conducted in testing chamber with carefully chosen several environmental and stress factors. It was found that PV module leakage current is highest in the morning associated with dew and remnant precipitation from the night. Over the course of a day, the magnitude of the leakage current moves opposite to module temperature, despite the higher conductivity of glass and the encapsulating material, because the module dries out and leaves no surface humidity to complete the circuit to ground.

Modules in continuously wet climates are expected to show the greatest system voltage-related degradation considering the observation that wetness is the most important environmental factor for elevated leakage current and the relationship between module leakage current and power degradation. It was determined that the results of the study [9], in combination with those found in the literature, suggest that a constant stress with humidity and system voltage will be more damaging than that stress applied intermittently or with periods of recovery comprising hot and dry conditions or alternating bias in between [11].

Testing modules for susceptibility to PID is now important in any large-scale PV project, and is often needed to obtain financing. Testing will also determine if the mechanism causing PID is reversible, which determines whether mitigating measures are required and what measures are appropriate. A full list of testing agencies and the modules tested to date is provided in the November 2012 issue of *Photon International Magazine* [12]. Many testing organizations currently offer PID testing, including National Renewable Energy Laboratory (NREL), Fraunhofer, Intertek, CFV Solar Test Laboratory, PV Evolution Labs, TÜV Rheinland PTL, TÜV SÜD America, and others.

A test standard is being developed by a consortium of industry experts led by Dr. Peter Hacke of NREL. The team is expecting to have the Final Draft International Standard (FDIS) for IEC 62804 in 2014. The standard is expected to be adopted by testing organizations, and many are already using its preliminary provisions. The IEC 62804 standard will prescribe a very specific test procedure and the basic conditions for conducting the test, including

- Module-rated system voltage and polarities
- Chamber air temperature $60 \pm 2°C$
- Chamber RH $85 \pm 5\%$
- Test duration of 96 h at above stated temperature and RH with applied stated voltage

Under the IEC 62804 standard, modules will be deemed to be PID resistant if:

- Power loss is less than 5%.
- There is no evidence of any major defect as defined in IEC 61215 clauses 10.1, 10.2, 10.7, and 10.15.

9.2.5 DETERIORATION OF PV MODULES DUE TO CLIMATIC EFFECTS

In the general study of PV reliability, three areas of the United States are targeted for their climatic conditions: Colorado—cold and humid, Florida—hot and humid, and Arizona—hot and dry. The most common causes of failure in PV modules are known to be moisture penetration and temperature fluctuation [13–16]. Many deterioration effects have been found in PV modules, such as delamination, cracking of glass or backing sheets, solder debonding, and other slowly occurring degradation phenomena [13]. An example of PV module degradation of the antireflection coating of a solar cell caused by water vapor ingress is shown in Figure 9.5.

A common observation has been that delamination is more frequent and more severe in *hot and humid climates*, sometimes occurring after less than 5 years of exposure. Delamination first causes a performance loss due to optical decoupling of the encapsulant from the cells. Of greater concern

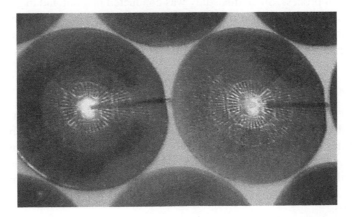

FIGURE 9.5 Degradation of the antireflection coating of a solar cell caused by water vapor ingress. (From Bremner S. Photovoltaic modules, ELEG620: Solar Electric Systems University of Delaware, ECE, Spring 2009, Online. Available: www.solar.udel.edu/ELEG620/13_PV_modules.pdf. With permission.)

from a module lifetime perspective is the likelihood that the void resulting from the delamination will provide a preferential location for moisture accumulation, greatly increasing the possibility of corrosion failures in metallic contacts [15].

Corrosion has been found to occur in some earlier generation a-Si:H PV modules that were grid connected in an array operating at 300-V DC in hot and humid environment at Florida. Moisture ingress caused severe degradation due to corrosion in these modules. Moisture ingress could be limited by using of improved PV module manufacturing technology, and hence the consequent corrosion and degradation would not result in complete destruction of the cell.

Chemical analyses of dozens of cell and encapsulant samples from field-aged modules have been conducted by Florida Solar Energy Center (FSEC) [16,17]. The primary analytical method used for the surface chemical analysis was Auger electron spectroscopy (AES). By comparing samples from unexposed modules to those from field-aged modules, these chemical analyses provided strong evidence of the dynamic chemical activities occurring in the module during field exposure. From these analyses, it was clear that sunlight, temperature, and moisture migration through the encapsulant provided the components required for a variety of chemical reactions, many of which may degrade the integrity of the encapsulant's adhesive bond to the cell. Phosphorous, titanium, oxygen, solder flux, encapsulant additives, and even sodium that has migrated through the encapsulant from the glass are typical reactive materials found at the cell–encapsulant interface after extended field exposure.

An extensive study of PV modules' degradation mechanism aged on extended *damp heat* exposure was performed in Ref. [18]. The Damp Heat (85°C/85% RH) protocol has been used to accelerate potential degradation mechanisms by increasing the temperature and humidity of the modules under test. This accelerated aging test was applied to PV modules with silicone and EVA encapsulation materials to compare their performance. These modules were aged in damp heat several times longer than the IEC 61215 standard of 1000 h. At 3000 h of exposure, a significant reduction in power output was noted in the EVA set of modules, while minimal power degradation was observed in the silicone set. These reductions were interpreted as corrosion taking place at the electrical interconnects caused by high concentrations of acetic acid from the reaction of EVA with water. No acetic acid was detected in the aged silicone PV modules.

In the study [19], moisture penetration was extremely low since the investigation was performed in Arizona *desert climate*. In this study, about 1900 PV modules aged in the field from 10 to 17 years have been analyzed to determine the occurrence and classify various failure modes. Difference in module construction appears to have an effect on module performance. After observing the rates of delamination, it became apparent that excessive heating caused by glass/glass construction accelerates the delamination process. Since this study was conducted in a hot and dry climate, it does not mean that the glass/glass module construction would not succeed elsewhere. It may be an excellent choice in an area that is cooler or more humid. The impermeability of the glass substrate makes it an excellent choice for a moisture barrier but limits the ambient temperature in which it will uphold.

9.2.6 "Snail Trails" in Solar Panels

In recent years, the solar industry has been facing a mysterious phenomenon which experts have dubbed snail trail due to its unusual appearance. After a period of time ranging from several months to several years after initial installation, solar modules show some discoloration on the cells, and crisscrossing narrow dark lines about the thickness of a finger begin to appear on the surface of the modules. Snail trail has become a widespread phenomenon, with more than 13 module makers from around the world facing a similar technical obstacle. An example of snail trails on solar module is shown in Figure 9.6.

"Snail trails" or "snail tracks" is a jargon name for localized discoloration of the contact grid on solar cells. Some modules are found completely covered with features looking like actual snail trails. This phenomenon was first discovered in Spanish solar panels installation in 2007. In a year,

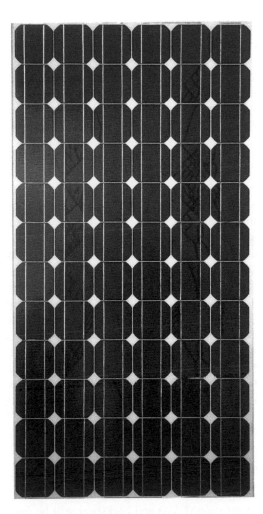

FIGURE 9.6 Snail trails on solar module. (From Köntges M. et al. Snail tracks (Schneckenspuren), worm marks and cell cracks, *IEA PVPS Task 13 Workshop at 27th EU PVSEC*, 2012. With permission.)

it was found that "snail trails" progressed from one side of the modules to the other. It was assumed that this indicates the presence of microcracks in the cells, which can have a negative impact on the output of the module [17,20]. It is also assumed that these discolorations of the contact grids are triggered by a chemical reaction which involves silver in the print paste, water, and several other components in the backsheet.

Three types of snail tracks can be distinguished: type I spreads back and forth across the solar cell along microcracks; type II is found on edges of the cells, and type III appears near the rim of the cell [20]. Some of the snail trails, which are located along cell edges are called "framing," those which start from cell interconnect along the finger are called "finger points." Cell cracks, snail trails, and framing appear after 3–5 months outdoor exposure, about 80% of cells have framing, and about 20% have snail trails, but after 8 months at nearly all cell cracks snail tracks appear [21].

Currently there is no clear understanding of the origin of snail tracks, but root cause analysis is conducted in several laboratories. In Ref. [22], the authors suggested that the origin of the snail trails is the formation of silver carbonate nanoparticles which discolor the silver grid. Micropore arrays on the silicon substrate within the snail trail region could accelerate the discoloration by offering reactive compounds via penetration and release. Moisture, oxygen, carbon dioxide, and other compounds could gain access through micropores, cracks, and cell edges or pass through the debonded areas between the encapsulation layer and the Si substrates to interact with and oxidize the silver grid. The potential mechanisms of PV module discoloration are proposed, yielding clues as to how snail trail formation can be mitigated by technical solutions. Aging tests on the discolored modules in Ref. [22] suggest no significant power degradation or discolored area enlargement after accelerated aging. Canadian Solar, one of the top five module manufacturers in the world, is also conducting an extensive research to identify the cause and proper steps to take to treat snail trails.

Canadian Solar's modules with snail trail have been tested both internally by the Canadian Solar Photovoltaic Testing Laboratory and externally by the Frauenhofer Institute for Solar Energy Systems, a third-party, Germany-based research organization, as well as in the field at a power plant. After extensive testing, the output performance reveals no significant drop in wattage, and the drop exhibited by a very small number of modules was due to cell breakages instead of snail trail. Acceleration aging tests also occurred internally and externally.

The reports show the snail trail does not spread after aging, and it does not affect the modules' long-term power output and reliability, echoing many other independent studies [23]. However, this study resulted in different hypothesis what causes snail trails. Canadian Solar's hypothesis of the chemical reactions are the following: the transition metal oxides in the glass frits of the silver paste, the peroxide cross-linker, the UV absorber, and the antioxidants are all involved in the chemical reaction. The transition metal oxides help break down the antioxidants in certain environments. The disappearing of antioxidants makes the residual peroxide cross-linker react with the UV absorber, generating dark-colored materials. To test the company's hypothesis, first there were changes to the concentration of additives in the EVA, and then there was experimentation of different levels of additives with cells of different types of silver pastes. The results supported Canadian Solar's hypothesis. Since metal oxide resides in the silver grid lines, the reaction happens only over the grid lines and does not spread over time [24].

9.2.7 Hot-Spot Failures

If a part of the solar cell is shaded, the cell can heat up to such extreme temperatures that the cell material as well as the encapsulation (EVA) and backsheet will be permanently damaged. Under normal operation condition, the cell generates current. In contrast, a shaded cell does not produce any electricity any more but uses the current from the other cell. The current from the other cells of the strings is driven through the darkened cell. The current flow is then converted into heat. When so-called hot spot develops, the temperature rise near a defect can vary from mild (up to 80°C) with low damage probability to extreme temperature rise (>200°C) leading to cell damage. Hot spots in solar cell may be seen with infrared (IR) camera [25] (Figure 9.7).

Detection of hot spots in completed modules can identify potential failures before the module is installed in the field. The use of an infrared measurement technique, IRIS™ presented in Ref. [26], may help to quickly identify and characterize the severity of module hot spots. Using this technique, hot spots may be conclusively identified before or during field installation with IRIS inspection machines capable of >25 modules per hour. The technique works in ambient light and directly measures the local heating due to defects.

To prevent the cells from hot spots, bypass diodes are used in all standard modules. If a cell is shaded, the bypass diodes get into operation and redirect the current for the full cell string via a bypass and prevent the cells from the hot-spot effect. Hot spots may still occur. For example, if the bypass diodes are faulty, or if only a very small part of the cell is shaded, the bypass diode is not enabled.

Other reasons for hot spots can be high contact resistance at the busbars of the cells (busbars are the silver-colored lines connecting the cells). Reasons for high contact resistance can be cracked solder joints on the busbars. The power loss of a module with a hot spot is often very low. Unless there

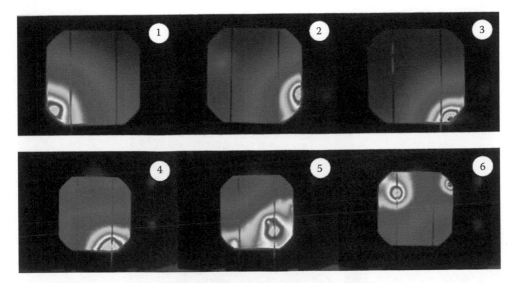

FIGURE 9.7 Hot spots in solar cell (from TUV). (Adapted from Wohlgemuth J. et al. *Failure Modes of Crystalline Si Modules, BP Solar, PV Module Reliability Workshop*, 2010, Online. Available: http://www1. eere.energy.gov/solar/pdfs/pvrw2010_wohlgemuth_silicon.pdf.)

are already big areas with hot spots. Nevertheless, these modules should be replaced, especially when not only the cell but also the surrounding encapsulation material is burned.

9.2.8 DELAMINATION AND DISCOLORATION OF ENCAPSULATION IN SOLAR CELLS

Encapsulants are typically polymers such as EVA, the most dominant material, polydimethylsiloxane (PDMS), polyvinyl butyral (PVB), polyethylene ionomers, polyolefins, and thermoplastic polyurethane (TPU).

Encapsulants for solar cells must have excellent adhesion to the cells, backsheet (insulation), and glass. They need to have good transmission of light over a long lifecycle and the ability to withstand weather extremes, protection from UV radiation, and they must provide a good moisture barrier.

Some of the drawbacks to EVA include less than ideal mechanical and thermal properties, high diffusivity for water, production of acetic acid, and poor electrical insulation. These drawbacks can translate into delamination, electric arcing, and hot spots. Several failure modes of solar module are related to encapsulant [27]: discoloration, delamination, and corrosion (Figure 9.8).

During manufacturing, the module composite is laminated under a precisely defined pressure and process temperature. It is important to keep the defined process temperature and time to ensure that the EVA cures correctly during the lamination process. Module delamination, resulting from loss of adhesion between the encapsulant and other module layers, is also a failure mechanism that needs to be addressed in order for manufacturers to achieve 30-year lifetime for PV modules. Delamination has occurred to varying degrees in a small percentage of modules from all manufacturers.

Wrong process parameters or cheap material can result in a delamination of the EVA later in lifetime. The layers of EVA dissolve and get a "milky" color. Example of the solar module failure caused by delamination around the busbars may be seen in Ref. [26]. Most of the delamination observed in the field has occurred at the interface between the encapsulant and the front surface of the solar cells in the module. Water and water vapor ingress cause various failures (delamination and deterioration) in solar cells [13].

Delaminated solar modules should also be replaced. Due to the delamination, moisture can get to the cells, which leads to cell corrosion and an ongoing performance loss. Further, the light transmission is reduced.

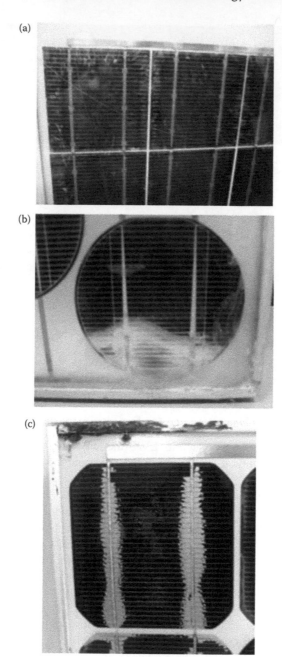

FIGURE 9.8   Failure modes of solar module related to encapsulant: discoloration (a), corrosion (b), and delamination (c). (Adapted from Shioda T. Delamination failures in long-term field-aged PV modules from point of view of encapsulant, *2013 NREL PV Module Reliability Workshop*, Denver, CO, February 2013, Online. Available: pvmrw13_openingsession_shioda_mitsui.pdf.)

Another defect is called yellowing or *browning*, which is a yellow or brown discoloration of the EVA (cell encapsulation) [25]. The EVA contains additives to improve the UV resistance and to prevent it from the browning effect. If low-quality EVA is used, if the EVA be stored under improper conditions or too long before manufacturing, the additive in the EVA partly disappears, and due to UV radiation and heat, the EVA starts to get yellow or brown color. As a result of browning cell bleaching, bubble formation at the EVA and backsheet and also cell corrosion can happen. The heat

FIGURE 9.9 Discoloration in encapsulant and adhesives in solar panel. (Adapted from Wohlgemuth J. et al. *Failure Modes of Crystalline Si Modules, BP Solar, PV Module Reliability Workshop*, 2010, Online. Available: http://www1.eere.energy.gov/solar/pdfs/pvrw2010_wohlgemuth_silicon.pdf.)

absorption increases with browning. Higher heat absorption enforces further browning. Modules showing browning should be replaced.

Discoloration may also be caused by heat and UV and by oxygen bleaching. It was found that discoloration (Figure 9.9) was caused not by encapsulating material (EVA), but by additives in the formulation [25].

9.2.9 ROLE OF INSULATION AND ADHESIVES IN PV PANELS FAILURES

Solar panel manufacturing, whether thin film or rigid silicon, is usually discussed in terms of efficiently converting energy from the sun into electric current and delivering that current as useable power. The manufacture of more efficient, higher performance solar panels can seem to be solely focused on ways to increase the conduction of electrical energy—with scarcely a mention of electrical insulation.

The materials that are used as insulation between panel components play an important role in panel performance, and it is critical for PV solar manufacturers to give the proper attention to insulation materials during the panel's design stage [28]. Electrical insulation, electrically conductive, thermal insulation, thermally conductive, and moisture barrier materials can be laminated together to block or transport current, moisture, or heat.

Panel failures such as arcing, overheating, and component corrosion can be contributed to using the wrong insulation materials. In many cases, more than one material must be laminated together to get the desired insulation effect.

Purity of the adhesive used in manufacturing of PV panels is crucial for any application that involves the inside of a solar panel. Some adhesives contain corrosive impurities in their makeup. Normally, these corrosive components are found in lower cost, pressure-sensitive adhesives (PSAs). Solar panels run at a high internal temperature, causing these PSAs to outgas corrosive impurities into the panel. Outgassing can lead to component corrosion throughout the panel.

9.2.10 The Other Types of Deterioration of Solar Modules

PV modules that are sold on the worldwide market today have to pass the relevant IEC tests for certification. These tests are only a mark for a certain quality level, not a reliability test. Nevertheless, manufacturers of PV modules give performance guaranties of 20–25 years on their products, some even more. Therefore, the question asked in Ref. [29] is how PV modules can survive 25+ years between the pole and the equator or why do PV modules fail. To answer this important question, general failure modes should be determined.

Extrinsic PV module failures can be caused by different climatic stress factors and by defective installations. In Refs. [29–31], these factors are classified. However, intrinsic failure reasons are to be taken into account. Some intrinsic failure can be explained on material level. An overview of major failures and their frequency detected during certification allow to determine why solar modules fail. Methods to detect and measure the relevant failure factors and to test new materials, components, and modules for future PV systems are being developed. There are at least 20 major failure modes discovered in solar panels and presented in Refs. [29,32].

The studies found failure causes over the lifespan such as cell breakage, corrosion of cell connections, delamination of the solar cell backsheet, discolored busbars, hot spots on the panel, microcracks, and oxidation on the busbar. Several kinds of failures show up during damp heat tests when humidity penetrates the cell structure and causes chemical reactions on various cell components. Similarly, troubles arise during hot spot tests, which simulate the reverse biasing of a cell that can happen when it is shaded, while other cells in its string are still in the sunshine.

Studies at Fraunhofer Institute for Solar Energy Systems noted numerous defects in the polymer cell coatings such as bubbles and holes. Among the defects Fraunhofer noted is a degradation of the antireflective coating on PV cells and bubbles that form in the polymer materials on the cells during hot spot tests that simulate conditions when a lone cell in a string is in a shady spot.

Various defects have been found in PV cell materials. In front and/or back cover glass, for example, there were instances of corrosion, abrasion, frosting, and breakage. Metal frames, connectors, cables, and other connections experienced distortion, stripping, corrosion, and disconnections. Module backsheets, cable insulation, the polymer front cover, and adhesives experienced problems that included delamination, blistering, cracks and embrittlement, and penetration of sealants from humidity [31,32].

9.2.11 PV Module Failures Due to External Causes

PV panels are subjected to deterioration caused by multiple external conditions, such as temperature (heat, frost, night–day cycles), irradiance (sun, sky), mechanical stress (wind, snow load, hail impact), humidity, moisture (rain, dew, frost), and surrounding atmosphere (salt mist, dust, sand, pollution) [33]. Some of those conditions and their effect on PV module are presented below.

9.2.11.1 Lightning and Snow

Important PV module failures may originate from external causes [34], such as defective bypass diode caused by a lightning strike. However, this effect has often been found and may cause subsequent safety failures, but the PV module is not the source of the failure. Typical induced defects caused by a lightning strike are open-circuit bypass diodes or a mechanically broken PV module directly hit by the lightning strike. Both defect types may cause hot spots as subsequent failures.

Many PV modules have been designed and applied for heavy snow load regions. To test and certify the PV modules for the heavy snow load regions, the snow load test of the IEC 61215 should be used. Regarding real snow load characteristics, the mechanical load test cannot apply extraordinary stress to the framing section at the lower part of a module at an inclined exposure. Snow loads creep downhill and intrude into the potential space between the frame edge and top surface. The ice

(a)   (b)   (c)

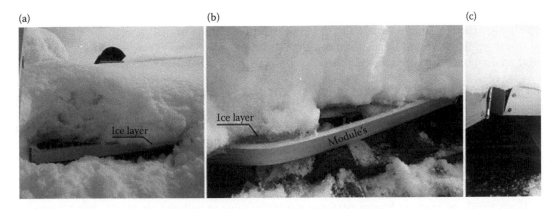

FIGURE 9.10 Damaged module frames after heavy snow load of 1.2 m, melted down to 35 cm, in winter 2012 in Alpine location: ice layer slides over module's edge (a); ice bends frame (b); failure of the corner screw joints (c). (From Köntges M. et al. *Review of Failures of Photovoltaic Modules, International Energy Agency, Photovoltaic Power System Programme (IEA PVSP)*, Report IEA-PVPS T13-01:2014, March 2014, Online. Available: http://www.isfh.de/institut_solarforschung/files/iea_t13_review_of_failures_of_pv_modules_final.pdf. With permission.)

formed by compression of the lower snow areas pushes against the exposed tip of the frame. Heavy snow load may damage module frames [34] (Figure 9.10).

However, for rooftop PV systems, the weight of the snow itself is not the only problem. As the snow melts and then freezes at night, additional forces from ice buildup occur on the panels, racking, and mounts. These thaw–freeze cycles may dislodge the panels, bend the racking, and/or loosen the mounts. Ice dams may develop under the panels and under the shingles as puddles of water from snow and ice melt and refreeze, backing up other meltwater. The most common result is that water seeps through the fasteners from the loosened mounting system, damaging the roof and rooms below [35].

9.2.11.2 Impact, Abrasion, and Breakage from Windborne Debris

Recent analysis [14] of module breakage in a roof-mounted array identified impact-induced failure mode that must be addressed by system designers. Field investigation, followed by fracture analysis, and then impact testing in the lab confirmed that small (2–4 g) stones propelled in storms to relatively low velocities (10–15 m/s) are capable of fracturing the tempered glass superstrates used on most PV modules. The significance of this finding for designers of roof-mounted systems is that stones (gravel) of this size are commonly used as the top covering on commercial and residential roofs.

9.2.11.3 Corrosion of PV Panel Caused by Ammonium Hydroxide

Aggressive ammonium hydroxide develops in livestock barns, which may lead to corrosion on solar panels. The pungent gas develops in livestock barns when animal excrements rot. The concentration is especially high in pigsties and chicken coops. There the ammonia gas burns the eyes, causes fits of coughing, and the smell stings in the nose. To prevent health damage, the barn air is extracted and transported off via the roof—where solar panels are installed. Then a creeping chemical reaction sets in on the panels caused by the ammonia: racks and frames corrode, the glued connections on the sockets become discolored, and glass turns dull. The barn air makes the panels age faster and may lead to damage that results in loss of performance and, in the worst case, arcing.

As a result, many farmers are uneasy about their planned investments in PV on their barn roofs. That is why the German Agricultural Society (DLG) developed a method at the end of 2009 to test the resistance of solar panels to ammonia. Module manufacturer Schott Solar assisted the society, which also tests food products and agricultural engineering. Various institutions in Germany, among them the German Agricultural Society (DLG) and TÜV Rheinland, have reacted and now offer an ammonia test for panels [35].

9.2.11.4　Clamping

Other external cause is clamping when failure in the field is glass breakage of frameless PV modules caused by the clamps. The origin of the failure is, on the one hand, at the planning and installation stage either because of (a) poor clamp geometry for the module, for example sharp edges; (b) too short and too narrow clamps; or (c) the positions of the clamps on the module not being chosen in accordance with the manufacturer's manual. The second origin, which induces glass breakage, could be excessively tightened screws during the mounting phase or badly positioned clamps.

Glass breakage leads to loss of performance in time due to cell and electrical circuit corrosion caused by the penetration of oxygen and water vapor into the PV module. Major problems caused by glass breakage are electrical safety issues. First, the insulation of the modules is no longer guaranteed, in particular in wet conditions. Second, glass breakage causes hot spots, which lead to overheating of the module.

9.2.11.5　Transport and Installation

Transport and installation are the first critical stages in a PV module's life [36]. The glass cover of some PV modules may break. In this case, it is easy to attribute the glass breakage to the transportation or installation. This is clearly no PV module failure. Another failure is breaking of the cells in the laminate due to vibrations and shocks. The cause of cell breakage is very difficult to determine. Visually it cannot be seen and in many cases it cannot be detected by a power rating of the PV module directly after occurrence of the cell breakage. The damage can be revealed only by an electroluminescence image or a lock-in thermography image.

9.2.12　Potential Problems with Rooftop Solar Systems

There are currently approximately 400,000 homes in the United States with rooftop solar arrays. By 2016, the number of solar homes will exceed one million, or roughly 5 gigawatts (GW) of solar capacity. And by 2024, rooftop solar panels are likely to be more common than satellite dishes.

It would be concerning for both consumers and contractors if these systems, consisting of solar panels, inverters, racking, roof penetrations, and rooftop wiring connections started experiencing problems, such as roof leaks, loose panels, or defective wiring, in significant numbers. The solar industry, standards organizations, and code officials have been working thoroughly to improve the safety and reliability of equipment, as well as the quality of the installations themselves.

To gain additional insight into the reliability of solar mounting systems, primary data were gathered on 20 rooftop systems in the San Francisco Bay area in California. The average age of these systems was 10 years. These systems were installed on a variety of residential roof types by experienced contractors, which passed applicable inspections, and are still operating properly.

It was found that the most serious problems include two areas: installation practices and component selection [36]. In installation practices, the following problems occur:

1. Leaks around mounts that attach the racking to the roof
2. Missed rafters when attaching mount
3. Improper grounding of panels or racking
4. Improper wiring
5. Improper attachment of panels to racks, or racks to mounts
6. Snow and ice damage

In component selection, the following corrosion issues can cause the following panel failures:

1. *Corrosion of rooftop mounting components*: Metal components on rooftops are exposed to rain as well as daily heating/cooling cycles. Components made of improper materials will corrode over time and can eventually fail, causing severe damage to the panels and rooftop. For expediency and cost issues, nonoutdoor-rated materials are sometimes used in rooftop installations. Ordinary steel, even when painted or galvanized, will rust and weaken over time in certain environments. Although commonly used, wood and plastic components are generally not durable enough to provide structural strength after long-term exposure to bright sunlight and heat. Almost all plated or coated fasteners will corrode and weaken on rooftops. Coastal environments make these problems even more severe because of additional corrosion from saltwater droplets in the air.

2. *Corrosion of grounding components*: The electrical industry has a wide variety of grounding components that are used to connect electrical equipment safely. Unfortunately, most of these components are designed for indoor applications, not for rooftop applications. On roofs, components are exposed to the elements, so the use of common indoor grounding components frequently leads to rusting corrosion after only a few years. More significantly, copper in contact with aluminum outdoors rapidly corrodes from a galvanic reaction between the metals, leading to a degraded grounding connection.

   Examples of corrosion caused by incorrect hardware and materials being used in solar panel are shown in Figure 9.11a is a braided copper grounding wire attached to the identified grounding lugs on a PV module; galvanic corrosion between the aluminum and copper caused both metals to degrade. Correct practice is to prevent copper and aluminum from coming into contact, typically using an appropriate barrier material (like stainless steel) between the adjacent aluminum and copper components. Shown in Figure 9.11b is a grounding lug that is intended to provide the necessary isolation between the copper wire and aluminum frame of the module. Although there was no corrosion of the aluminum and copper, the carbon steel fasteners used to attach the lug and wire rusted and lost some of their strength.

(a)   (b)

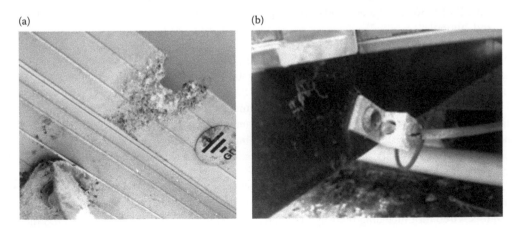

FIGURE 9.11 Examples of corrosion caused by incorrect hardware and materials being used in solar panel (a) and corroded screws securing a grounding lug (b). (From Solar Panels Last 25 Years–But will they stay safely attached to your roof? *The Importance of Reliable Solar Mounting Systems*, White Paper, 29pp., March 2014, Online. Available: http://cinnamonsolar.com/wp-content/uploads/2013/05/The-Importance-of-Reliable-Solar-Mounting-Systems-White-Paper_March-2014-copy.pdf. With permission.)

9.3   STANDARDS REGULATING SOLAR ENERGY INDUSTRY

9.3.1   NATIONAL STANDARDS FOR SOLAR POWER

Many new solar energy standards are introduced each year regulating testing energy conversion, reflectance or materials properties, fabricating arrays, integrating into smart grid, or ensuring workplace safety by national organizations, such as ASTM, IEC, ISO, American Society of Heating, Refrigerating and Air-Conditioning Engineers (ASHRAE), and Society of Automotive Engineers (SAE).

Standards regulating *testing and measurement* are

- ISO 9060:1990 Solar energy—Specification and classification of instruments for measuring hemispherical solar and direct solar radiation
- ASTM E424-71(2007) Standard Test Methods for Solar Energy Transmittance and Reflectance (Terrestrial) of Sheet Materials
- ISO 9553:1997 Solar energy—Methods of testing preformed rubber seals and sealing compounds used in collectors
- ANSI/ASHRAE 96-1980 (RA 1989) Methods of Testing to Determine the Thermal Performance of Unglazed Flat-Plate Liquid-Type Solar Collectors
- SAE J 1559-2011 (SAE J1559-2011) Measurement of Solar Heating Effect
- ASTM E1084-86(2009) Standard Test Method for Solar Transmittance (Terrestrial) of Sheet Materials Using Sunlight
- ISO 14438:2002 Glass in building—Determination of energy balance value—Calculation method
- ASTM E491-73(2010) Standard Practice for Solar Simulation for Thermal Balance Testing of Spacecraft
- ASTM F1192-11 Standard Guide for the Measurement of Single Event

Phenomena (SEP) Induced by Heavy Ion Irradiation of Semiconductor Devices
National standards that determine *terminology* in solar energy are

- ISO 9488:1999 Solar energy—Vocabulary
- ASTM E772-13 Standard Terminology of Solar Energy Conversion
- IEC/TS 61836 Ed. 2.0 en: 2007 "Solar photovoltaic energy systems—Terms, definitions and symbols"

Many other standards determine calibration and performance of solar systems and equipment, which is used in and with solar panels such as rubber seals, water heating systems, low-voltage fuses, automatic electrical controls, and so on [37].

Underwriter's Laboratory (UL) develops and manages the various standards that apply to solar equipment. UL also provides testing services to manufacturers to certify that products meet these standards. Several other testing laboratories, including extract, transform, load (ETL), compliance, safety, accountability (CSA), and TUV, provide these testing services and can certify that products meet relevant UL standards [36].

UL 1703 Solar Panels is the standard that applies to flat-plate PV modules and panels; this standard has historically focused on safety issues related to the solar panel itself and the way in which it is wired (not including the way in which it is mounted). A new version of UL 1703 includes additional fire testing requirements.

These requirements cover flat-plate PV modules and panels intended for installation on or integral with buildings, or to be freestanding (i.e., not attached to buildings), in accordance with the National Electrical Code, NFPA 70, and Model Building Codes. These requirements cover modules and panels intended for use in systems with a maximum system voltage of 1000 V or less (residential systems are not allowed to exceed 600 V per National Electric Code [NEC] requirements). These requirements also cover components intended to provide electrical connection to mounting facilities for flat-plate PV modules and panels.

UL 2703—Solar Panel Mounting Systems is a newer standard that addresses combinations of solar panels and mounting systems, including grounding. Revisions and improvements to this standard have been in progress since 2013. This standard requirements cover rack mounting systems, mounting grounding/bonding components, and clamping/retention devices for specific (manufacturer/model designation) flat-plate PV modules and panels that comply with the Standard for Flat-Plate PV Modules and Panels intended for installation on or integral with buildings, or to be freestanding (i.e., not attached to buildings), in accordance with the National Electrical Code, ANSI/NFPA 70, and Model Building Codes.

These requirements cover rack mounting systems and clamping devices intended for use with PV module systems with a maximum system voltage of 600 V. These requirements cover rack mounting systems and clamping, retention devices pertaining to ground/bonding paths, mechanical strength, and suitability of materials only. Revisions and improvements to this standard have been in progress since 2013. These revisions are nearing completion, and UL 2703 is anticipated to become an American National Standards Institute (ANSI)-accredited standard sometime in 2014.

9.3.2 INTERNATIONAL STANDARDS FOR SOLAR POWER

The IEC Technical Committee 82 (TC82) prepares international standards for systems of PV conversion of solar energy into electrical energy and for all the elements in the entire PV energy system. In this context, the concept "PV energy system" includes the entire field from light input to a solar cell to, and including, the interface with the electrical system(s) to which energy is supplied. IEC TC82 has prepared standards for terms and symbols, salt mist corrosion testing, design qualification and type approval of crystalline silicon and thin-film modules, and characteristic parameters of stand-alone systems, among others. In the future, TC 82 work will include

- System commissioning, maintenance, and disposal
- Characterization and measurement of new thin-film PV module technologies such as CdTe, Center for Internet Security (CIS), CuInSe2, and so forth
- New technology storage systems
- Applications with special site conditions, such as tropical zone, northern latitudes, and marine areas

TC 82 also expects to address several system and component safety issues including grid-connected systems on buildings and utility-connected inverters, as well as various aspects of environmental protection. This includes safeguarding the natural environment from such things as radio frequency and electromagnetic pollution, disposal of toxic PV materials, and atmospheric contamination from PV manufacturing processes, among other topics.

9.3.2.1 IEC Standards for Solar Energy

Shown below is only a portion of all standards developed by IEC related to solar modules [38]:

IEC 61836, 2007 Ed 3, IEC/TS 61836 Ed. 3.0, Solar photovoltaic energy systems—Terms, definitions, and symbols

IEC 60891, 2009 Ed 2, IEC 60891 Ed. 2.0, Photovoltaic devices—Procedures for temperature and irradiance corrections to measured I–V characteristics

IEC 60904-1, 2006 Ed 2, Photovoltaic devices—Part 1: Measurements of PV current–voltage characteristics

IEC 60904-2, 2007 Ed 2, Photovoltaic devices—Part 2: Requirements for reference solar devices

IEC 60904-3, 2008 Ed 2, Photovoltaic devices—Part 3: Measurement principles for terrestrial photovoltaic (PV) solar devices with reference spectral irradiance data

IEC 60904-4, 2009 Ed 1, Photovoltaic devices—Part 4: Reference solar devices—Procedures for establishing calibration traceability

IEC 60904-5, 2011 Ed 2.0, IEC 60904-5 Ed. 2.0, Photovoltaic devices—Part 5: Determination of the equivalent cell temperature (ECT) of photovoltaic (PV) devices by the open-circuit voltage method

IEC 60904-6, 1994 Ed. 1.0, Photovoltaic devices—Part 6: Requirements for reference solar modules

IEC 60904-7, 2008 Ed 3, Photovoltaic devices—Part 7: Computation of the spectral mismatch correction for measurements of photovoltaic devices

IEC 60904-8, 1998 Ed 3, IEC 60904-8 Ed. 3.0 Photovoltaic devices—Part 8: Measurement of spectral response of a photovoltaic (PV) device

IEC 60904-9, 2007 Ed 2, Photovoltaic devices—Part 9: Solar simulator performance requirements

IEC 60904-10, 2009 Ed 2, Photovoltaic devices—Part 10: Methods of linearity measurement

IEC 61215, 2005 Ed 2.0: Scope of the work in progress includes design qualification and type approval for crystalline silicon terrestrial PV module, Ed 3; publish 4Q 2011

IEC 61345, 1998 Ed 1, UV test for photovoltaic (PV) modules

IEC 61646, 2008 Ed 2, Thin-film terrestrial photovoltaic (PV) modules—Design qualification and type approval

IEC 61701:1995 Ed 1.0: Scope of the work in progress includes salt mist corrosion testing of PV modules for Edition 2; publish 4Q 2010

IEC 61730-1:2004 Ed 1.0: Scope of the work in progress includes PV module safety qualification requirements for construction for Part 1/Amendment 1; publish 4Q 2010

IEC 61730-2, 2004 Ed 1, Photovoltaic (PV) module safety qualification—Part 2: Requirements for testing. Amendment 1 in process; publish 2Q 2011

IEC 61829, 1995 Ed 1.0: Scope of the work in progress includes on-site measurement of I–V characteristics for crystalline silicon PV array, Ed 2; publish 1Q 2011

IEC 61853-1: Ed 1.0: Scope of the work in progress includes irradiance and temperature performance measurements and power rating for PV module performance testing and energy rating, Part 1; publish 1Q 2011

IEC 61853-2: Ed 1.0: Scope of work in progress includes spectral response, incidence angle, and module operating temperature measurements for PV module performance testing and energy rating, Part 2; publish 4Q 2010

IEC 62548 Ed. 1.0 Design requirements for photovoltaic (PV) arrays, 2013

IEC 62716, Ed 1, Ammonia corrosion testing of photovoltaic (PV) modules. Publish 2Q 2012

IEC/TS 62738 Ed. 1.0 Design guidelines and recommendations for photovoltaic power plants, 2012

IEC/TS 62748 Ed. 1.0 PV systems on buildings, 2012

IEC 62759-1 Ed. 1.0 Transportation testing of photovoltaic (PV) modules—Part 1: Transportation and shipping of PV module stacks, 2013

IEC 62775 Ed. 1.0 Cross-linking degree test method for Ethylene-Vinyl Acetate applied in photovoltaic modules—Differential Scanning Calorimetry (DSC), 2014

IEC 62782 Ed. 1.0 Dynamic mechanical load testing for photovoltaic (PV) modules

IEC 62788-1-2 Ed.1 Measurement procedures for materials used in photovoltaic modules—Part 1-2: Encapsulants—Measurement of resistivity of photovoltaic encapsulation and backsheet materials, 2015

IEC 62788-1-4 Ed.1 Measurement procedures for materials used in Photovoltaic Modules—Part 1-4: Encapsulants—Measurement of optical transmittance and calculation of the solar-weighted photon transmittance, yellowness index, and UV cut-off frequency, 2015

9.3.2.2 Solar Power Standards in Other Countries

Many countries which manufacture solar panels are also developing standards and regulations. In Australia, there is a system of Australian standards that comply with international standard testing procedures [39].

For example, the installation for solar panels is defined by Australian Standard AS/NZS 5033:2005, Installation of photovoltaic (PV) arrays. This standard indicates that components used to manufacture solar panels should be tested according to the following IEC standards:

- IEC 61215 Ed 2.0, Crystalline silicon terrestrial photovoltaic (PV) modules—Design qualification and type approval or I.S. EN 61215:2006, Crystalline silicon terrestrial photovoltaic (PV) modules—Design qualification and type approval (for PV Modules—Silicon PV modules)
- IEC 61646 Ed 2.0, Thin-film terrestrial photovoltaic (PV) modules—Design qualification and type approval (for PV Modules—Thin Material)
- IEC 61730, Photovoltaic (PV) module safety qualification Series and I.S. EN 61730 Series (for PV Modules—Safety)

9.4 SOLAR ENERGY GLOSSARY

The solar energy glossary contains definitions for technical terms related to solar power and PV technologies, including terms having to do with electricity, power generation, and concentrating solar power (CSP) [40].

A

Absorber: In a PV device, the material that readily absorbs photons to generate charge carriers (free electrons or holes).

Acceptor: A dopant material, such as boron, which has fewer outer shell electrons than required in an otherwise balanced crystal structure, providing a hole, which can accept a free electron.

Air mass (sometimes called air mass ratio): Equal to the cosine of the zenith angle—the angle from directly overhead to a line intersecting the sun. The air mass is an indication of the length of the path solar radiation travels through the atmosphere. An air mass of 1.0 means the sun is directly overhead and the radiation travels through one atmosphere (thickness).

Amorphous semiconductor: A noncrystalline semiconductor material that has no long range order.

Amorphous silicon: A thin-film, silicon PV cell having no crystalline structure. Manufactured by depositing layers of doped silicon on a substrate. *See also* single-crystal silicon, a polycrystalline silicon.

Angle of incidence: The angle that a ray of sun makes with a line perpendicular to the surface. For example, a surface that directly faces the sun has a solar angle of incidence of zero, but if the surface is parallel to the sun (e.g., sunrise striking a horizontal rooftop), the angle of incidence is 90°.

Annual solar savings: The annual solar savings of a solar building is the energy savings attributable to a solar feature relative to the energy requirements of a nonsolar building.

Antireflection coating: A thin coating of a material applied to a solar cell surface that reduces the light reflection and increases light transmission.

Array: *See* PV array.

Array current: The electrical current produced by a PV array when it is exposed to sunlight.

Array operating voltage: The voltage produced by a PV array when exposed to sunlight and connected to a load.

Availability: The quality or condition of a PV system being available to provide power to a load. Usually measured in hours per year. One minus availability equals downtime.

Azimuth angle: The angle between true south and the point on the horizon directly below the sun.

B

Balance of system: Represents all components and costs other than the PV modules/array. It includes design costs, land, site preparation, system installation, support structures, power conditioning, operation and maintenance costs, indirect storage, and related costs.

Bandgap: In a semiconductor, the energy difference between the highest valence band and the lowest conduction band.

Bandgap energy (Eg): The amount of energy (in electron volts) required to free an outer shell electron from its orbit about the nucleus to a free state, and thus promote it from the valence to the conduction level.

Barrier energy: The energy given up by an electron in penetrating the cell barrier; a measure of the electrostatic potential of the barrier.

BIPV: *See* building integrated PV.

Blocking diode: A semiconductor connected in series with a solar cell or cells and a storage battery to keep the battery from discharging through the cell when there is no output, or low output, from the solar cell. It can be thought of as a one-way valve that allows electrons to flow forward, but not backward.

Boron (B): The chemical element commonly used as the dopant in PV device or cell material.

Boule: A sausage-shaped, synthetic single-crystal mass grown in a special furnace, pulled and turned at a rate necessary to maintain the single-crystal structure during growth.

Building-integrated photovoltaics: A term for the design and integration of PV technology into the building envelope, typically replacing conventional building materials. This integration may be in vertical facades, replacing view glass, spandrel glass, or other facade material; into semitransparent skylight systems; into roofing systems, replacing traditional roofing materials; into shading "eyebrows" over windows; or other building envelope systems.

Bypass diode: A diode connected across one or more solar cells in a PV module such that the diode will conduct if the cell(s) become reverse biased. It protects these solar cells from thermal destruction in case of total or partial shading of individual solar cells, while other cells are exposed to full light.

C

Cadmium (Cd): A chemical element used in making certain types of solar cells and batteries.

Cadmium telluride (CdTe): A polycrystalline thin-film PV material.

Capacity (C): *See* battery capacity.

Capacity factor: The ratio of the average load on (or power output of) an electricity-generating unit or system to the capacity rating of the unit or system over a specified period of time.

Captive electrolyte battery: A battery having an immobilized electrolyte (gelled or absorbed in a material).

Cathode: The negative pole or electrode of an electrolytic cell, vacuum tube, and so on, where electrons enter (current leaves) the system; the opposite of an anode.

Cathodic protection: A method of preventing oxidation of the exposed metal in structures by imposing a small electrical voltage between the structure and the ground.

Cell (battery): A single unit of an electrochemical device capable of producing direct voltage by converting chemical energy into electrical energy. A battery usually consists of several cells electrically connected together to produce higher voltages (sometimes the terms cell and battery are used interchangeably). *See also* PV cell.

Cell barrier: A very thin region of static electric charge along the interface of the positive and negative layers in a PV cell. The barrier inhibits the movement of electrons from one layer to the other, so that higher energy electrons from one side diffuse preferentially through it in one direction, creating a current and thus a voltage across the cell. Also called depletion zone or space charge.

Cell junction: The area of immediate contact between two layers (positive and negative) of a PV cell. The junction lies at the center of the cell barrier or depletion zone.

Charge: The process of adding electrical energy to a battery.

Charge carrier: A free and mobile conduction electron or hole in a semiconductor.

Charge controller: A component of a PV system that controls the flow of current to and from the battery to protect it from overcharge and overdischarge. The charge controller may also indicate the system operational status.

Chemical vapor deposition (CVD): A method of depositing thin semiconductor films used to make certain types of PV devices. With this method, a substrate is exposed to one or more vaporized compounds, one or more of which contain desirable constituents. A chemical reaction is initiated, at or near the substrate surface, to produce the desired material that will condense on the substrate.

Cleavage of lateral epitaxial films for transfer (CLEFT): A process for making inexpensive gallium arsenide (GaAs) PV cells in which a thin film of GaAs is grown atop a thick, single-crystal GaAs (or other suitable material) substrate and then is cleaved from the substrate and incorporated into a cell, allowing the substrate to be reused to grow more thin-film GaAs.

Cloud enhancement: The increase in solar intensity caused by reflected irradiance from nearby clouds.

Combined collector: A PV device or module that provides useful heat energy in addition to electricity.

Concentrating photovoltaics (CPV): A solar technology that uses lenses or mirrors to concentrate sunlight onto high-efficiency solar cells.

Concentrating solar power (CSP): A solar technology that use mirrors to reflect and concentrate sunlight onto receivers that convert solar energy to heat. This thermal energy is then used to produce electricity with a steam turbine or heat engine driving a generator.

Concentrator: A PV module, which includes optical components such as lenses (Fresnel lens), to direct and concentrate sunlight onto a solar cell of smaller area. Most concentrator arrays must directly face or track the sun. They can increase the power flux of sunlight hundreds of times.

Conduction band (or conduction level): An energy band in a semiconductor in which electrons can move freely in a solid, producing a net transport of charge.

Contact resistance: The resistance between metallic contacts and the semiconductor.

Conversion efficiency: *See* PV (conversion) efficiency.

Converter: A unit that converts a DC voltage to another DC voltage.

Copper indium diselenide (CuInSe2 or CIS): A polycrystalline thin-film PV material (sometimes incorporating gallium (CIGS) and/or sulfur).

Copper zinc tin sulfide/selenide (CZTS): A polycrystalline thin-film PV material.

Crystalline silicon: A type of PV cell made from a slice of single-crystal silicon or polycrystalline silicon.

Czochralski process: A method of growing large-size, high-quality semiconductor crystal by slowly lifting a seed crystal from a molten bath of the material under careful cooling conditions.

D

DC-to-DC converter: Electronic circuit to convert DC voltages (e.g., PV module voltage) into other levels (e.g., load voltage). Can be part of a maximum power point (MPP) tracker.

Defect: *See* light-induced defects.

Dendrite: A slender thread-like spike of pure crystalline material, such as silicon.

Dendritic web technique: A method for making sheets of polycrystalline silicon in which silicon dendrites are slowly withdrawn from a melt of silicon, whereupon a web of silicon forms between the dendrites and solidifies as it rises from the melt and cools.

Design month: The month having the combination of insolation and load that requires the maximum energy from the PV array.

Diffuse insolation: Sunlight received indirectly as a result of scattering due to clouds, fog, haze, dust, or other obstructions in the atmosphere. Opposite of direct insolation.

Diffuse radiation: Radiation received from the sun after reflection and scattering by the atmosphere and ground.

Diffusion furnace: Furnace used to make junctions in semiconductors by diffusing dopant atoms into the surface of the material.

Direct beam radiation: Radiation received by direct solar rays. Measured by a pyrheliometer with a solar aperture of 5.7° to transcribe the solar disc.

Direct insolation: Sunlight falling directly upon a collector. Opposite of diffuse insolation.

Disconnect: Switch gear used to connect or disconnect components in a PV system.

Distributed systems: Systems that are installed at or near the location where the electricity is used, as opposed to central systems that supply electricity to grids. A residential PV system is a distributed system.

Donor: In a PV device, an n-type dopant, such as phosphorus, that puts an additional electron into an energy level very near the conduction band; this electron is easily exited into the conduction band where it increases the electrical conductivity more than of an undoped semiconductor.

Donor level: The level that donates conduction electrons to the system.

Dopant: A chemical element (impurity) added in small amounts to an otherwise pure semiconductor material to modify the electrical properties of the material. An n-dopant introduces more electrons. A p-dopant creates electron vacancies (holes).

Doping: The addition of dopants to a semiconductor.

Downtime: Time when the PV system cannot provide power for the load. Usually expressed in hours per year or that percentage.

Duty cycle: The ratio of active time to total time. Used to describe the operating regime of appliances or loads in PV systems.

Duty rating: The amount of time an inverter (power conditioning unit) can produce at full rated power.

E–F

Edge-defined, film-fed growth (EFG): A method for making sheets of polycrystalline silicon for PV devices in which molten silicon is drawn upward by capillary action through a mold.

Energy contribution potential: Recombination occurring in the emitter region of a PV cell.

Extrinsic semiconductor: The product of doping a pure semiconductor.

Fill factor: The ratio of a PV cell's actual power to its power if both current and voltage were at their maxima. A key characteristic in evaluating cell performance.

Fixed tilt array: A PV array set in at a fixed angle with respect to horizontal.

Flat-plate array: A PV array that consists of nonconcentrating PV modules.

Flat-plate module: An arrangement of PV cells or material mounted on a rigid flat surface with the cells exposed freely to incoming sunlight.

Flat-plate photovoltaics (PV): A PV array or module that consists of nonconcentrating elements. Flat-plate arrays and modules use direct and diffuse sunlight, but if the array is fixed in position, some portion of the direct sunlight is lost because of oblique sun angles in relation to the array.

Float-zone process: In reference to solar PV cell manufacture, a method of growing a large-size, high-quality crystal whereby coils heat a polycrystalline ingot placed atop a single-crystal seed. As the coils are slowly raised, the molten interface beneath the coils becomes single crystal.

Fresnel lens: An optical device that focuses light like a magnifying glass; concentric rings are faced at slightly different angles so that light falling on any ring is focused to the same point.

Full sun: The amount of power density in sunlight received at the earth's surface at noon on a clear day (about 1000 W/m²)

G–H

Gallium (Ga): A chemical element, metallic in nature, used in making certain kinds of solar cells and semiconductor devices.

Gallium arsenide (GaAs): A crystalline, high-efficiency compound used to make certain types of solar cells and semiconductor material.

High-voltage disconnect: The voltage at which a charge controller will disconnect the PV array from the batteries to prevent overcharging.

High-voltage disconnect hysteresis: The voltage difference between the high-voltage disconnect set point and the voltage at which the full PV array current will be reapplied.

Homojunction: The region between an n-layer and a p-layer in a single material, PV cell.

Hybrid system: A solar electric or PV system that includes other sources of electricity generation, such as wind or diesel generators.

Hydrogenated amorphous silicon: Amorphous silicon with a small amount of incorporated hydrogen. The hydrogen neutralizes dangling bonds in the amorphous silicon, allowing charge carriers to flow more freely.

I

Incident light: Light that shines onto the face of a solar cell or module.

Indium oxide: A wide bandgap semiconductor that can be heavily doped with tin to make a highly conductive, transparent thin film. Often used as a front contact or one component of a heterojunction solar cell.

Infrared radiation: Electromagnetic radiation whose wavelengths lie in the range from 0.75 to 1000 μm; invisible long wavelength radiation (heat) capable of producing a thermal or PV effect, though less effective than visible light.

Ingot: A casting of material, usually crystalline silicon, from which slices or wafers can be cut for use in a solar cell.

Input voltage: This is determined by the total power required by the alternating current loads and the voltage of any DC loads. Generally, the larger the load, the higher is the inverter input voltage. This keeps the current at levels where switches and other components are readily available.

Insolation: The solar power density incident on a surface of stated area and orientation, usually expressed as Watts per square meter or Btu per square foot per hour. *See also* diffuse insolation and direct insolation.

Interconnect: A conductor within a module or other means of connection that provides an electrical interconnection between the solar cells.

Internal quantum efficiency (internal QE or IQE): A type of quantum efficiency. Refers to the efficiency with which light not transmitted through or reflected away from the cell can generate charge carriers that can generate current.

Intrinsic layer: A layer of semiconductor material, used in a PV device, whose properties are essentially those of the pure, undoped material.

Intrinsic semiconductor: An undoped semiconductor.

Inverted metamorphic multijunction (IMM) cell: A PV cell that is a multijunction device whose layers of semiconductors are grown upside down. This special manufacturing process yields an ultralight and flexible cell that also converts solar energy with high efficiency.

Inverter: A device that converts DC electricity to alternating current either for stand-alone systems or to supply power to an electricity grid.

Ion: An electrically charged atom or group of atoms that has lost or gained electrons; a loss makes the resulting particle positively charged; a gain makes the particle negatively charged.

Irradiance: The direct, diffuse, and reflected solar radiation that strikes a surface. Usually expressed in kilowatts per square meter. Irradiance multiplied by time equals insolation.

ISPRA guidelines: Guidelines for the assessment of PV power plants, published by the Joint Research Centre of the Commission of the European Communities, Ispra, Italy.

I-type semiconductor: Semiconductor material that is left intrinsic or undoped so that the concentration of charge carriers is characteristic of the material itself rather than of added impurities.

I–V curve: A graphical presentation of the current versus the voltage from a PV device as the load is increased from the short-circuit (no load) condition to the open-circuit (maximum voltage) condition. The shape of the curve characterizes cell performance.

J–K

Junction: A region of transition between semiconductor layers, such as a p/n junction, which goes from a region that has a high concentration of acceptors (p-type) to one that has a high concentration of donors (n-type).

Junction box: A PV generator junction box is an enclosure on the module where PV strings are electrically connected and where protection devices can be located, if necessary.

Junction diode: A semiconductor device with a junction and a built-in potential that passes current better in one direction than the other. All solar cells are junction diodes.

Kerf: The width of a cut used to create wafers from silicon ingots, often resulting in the loss of semiconductor material.

L

Langley (L): Unit of solar irradiance. One gram calorie per square centimeter. $1\,L = 85.93\,kwh/m^2$.

Lattice: The regular periodic arrangement of atoms or molecules in a crystal of semiconductor material.

Lead–acid battery: A general category that includes batteries with plates made of pure lead, lead–antimony, or lead–calcium immersed in an acid electrolyte.

Levelized cost of energy (LCOE): The cost of energy of a solar system that is based on the system's installed price, its total lifetime cost, and its lifetime electricity production.

Life: The period during which a system is capable of operating above a specified performance level.

Life cycle cost: The estimated cost of owning and operating a PV system for the period of its useful life.

Light-induced defects: Defects, such as dangling bonds, induced in an amorphous silicon semiconductor upon initial exposure to light.

Light trapping: The trapping of light inside a semiconductor material by refracting and reflecting the light at critical angles; trapped light will travel further in the material, greatly increasing the probability of absorption and hence of producing charge carriers.

Line-commutated inverter: An inverter that is tied to a power grid or line. The commutation of power (conversion from DC to alternating current) is controlled by the power line, so that, if there is a failure in the power grid, the PV system cannot feed power into the line.

M

Majority carrier: Current carriers (either free electrons or holes) that are in excess in a specific layer of a semiconductor material (electrons in the n-layer, holes in the p-layer) of a cell.

Maximum power point (MPP): The point on the current–voltage (I–V) curve of a module under illumination, where the product of current and voltage is maximum. For a typical silicon cell, this is at about 0.45 V.

Maximum power point tracker (MPPT): Means of a power conditioning unit that automatically operates the PV generator at its MPP under all conditions.

Maximum power tracking: Operating a PV array at the peak power point of the array's I–V curve where maximum power is obtained. Also called peak power tracking.

Measurement and characterization: A field of research that involves assessing the characteristics of PV materials and devices.

Microgroove: A small groove scribed into the surface of a solar cell, which is filled with metal for contacts.

Minority carrier: A current carrier, either an electron or a hole that is in the minority in a specific layer of a semiconductor material; the diffusion of minority carriers under the action of the cell junction voltage is the current in a PV device.

Minority carrier lifetime: The average time a minority carrier exists before recombination.

Modified sine wave: A waveform that has at least three states (i.e., positive, off, and negative). Has less harmonic content than a square wave.

Modularity: The use of multiple inverters connected in parallel to service different loads.

Module: *See* PV module.

Module derate factor: A factor that lowers the PV module current to account for field operating conditions such as dirt accumulation on the module.

Monolithic: Fabricated as a single structure.

Movistor: Short for metal oxide varistor. Used to protect electronic circuits from surge Currents such as those produced by lightning.

Multicrystalline: A semiconductor (PV) material composed of variously oriented, small, individual crystals. Sometimes referred to as polycrystalline or semicrystalline.

Multijunction device: A high-efficiency PV device containing two or more cell junctions, each of which is optimized for a particular part of the solar spectrum.

Multistage controller: A charging controller unit that allows different charging currents as the battery nears full state of charge.

N–O

Normal operating cell temperature (NOCT): The estimated temperature of a PV module when operating under 800 W/m^2 irradiance, 20°C ambient temperature, and wind speed of 1 m/s. NOCT is used to estimate the nominal operating temperature of a module in its working environment.

N-type: Negative semiconductor material in which there are more electrons than holes; current is carried through it by the flow of electrons.

N-type semiconductor: A semiconductor produced by doping an intrinsic semiconductor with an electron donor impurity (e.g., phosphorus in silicon).

N-type silicon: Silicon material that has been doped with a material that has more electrons in its atomic structure than does silicon.

Open-circuit voltage (Voc): The maximum possible voltage across a PV cell; the voltage across the cell in sunlight when no current is flowing.

Operating point: The current and voltage that a PV module or array produces when connected to a load. The operating point is dependent on the load or the batteries connected to the output terminals of the array.

Orientation: Placement with respect to the cardinal directions, N, S, E, W; azimuth is the measure of orientation from north.

P

Packing factor: The ratio of array area to actual land area or building envelope area for a system; or the ratio of total solar cell area to the total module area, for a module.

Panel: *See* PV panel.

Parallel connection: A way of joining solar cells or PV modules by connecting positive leads together and negative leads together; such a configuration increases the current, but not the voltage.

Passivation: A chemical reaction that eliminates the detrimental effect of electrically reactive atoms on a solar cell's surface.

Peak power point: Operating point of the I–V curve for a solar cell or PV module where the product of the current value times the voltage value is a maximum.

Peak power tracking: *See* maximum power tracking.

Peak sun hours: The equivalent number of hours per day when solar irradiance averages 1000 W/m^2. For example, six peak sun hours means that the energy received during total daylight hours equals the energy that would have been received had the irradiance for 6 h been 1000 W/m^2.

Peak Watt: A unit used to rate the performance of solar cells, modules, or arrays; the maximum nominal output of a PV device, in Watts (Wp) under standardized test conditions, usually 1000 W/m^2 of sunlight with other conditions, such as temperature specified.

Phosphorus (P): A chemical element used as a dopant in making n-type semiconductor layers.

Photocurrent: An electric current induced by radiant energy.

Photoelectric cell: A device for measuring light intensity that works by converting light falling on, or reach it, to electricity, and then measuring the current; used in photometers.

Photoelectrochemical cell: A type of PV device in which the electricity induced in the cell is used immediately within the cell to produce a chemical, such as hydrogen, which can then be withdrawn for use.

Photon: A particle of light that acts as an individual unit of energy.

Photovoltaic(s) (PV): Pertaining to the direct conversion of light into electricity.

Photovoltaic (PV) array: An interconnected system of PV modules that function as a single electricity-producing unit. The modules are assembled as a discrete structure, with common support or mounting. In smaller systems, an array can consist of a single module.

Photovoltaic (PV) cell: The smallest semiconductor element within a PV module to perform the immediate conversion of light into electrical energy (DC voltage and current). Also called a solar cell.

Photovoltaic (PV) conversion efficiency: The ratio of the electric power produced by a PV device to the power of the sunlight incident on the device.

Photovoltaic (PV) device: A solid-state electrical device that converts light directly into DC electricity of V–I characteristics that are a function of the characteristics of the light source and the materials in and design of the device. Solar PV devices are made of various semiconductor materials, including silicon, cadmium sulfide, cadmium telluride, and gallium arsenide, and in single crystalline, multicrystalline, or amorphous forms.

Photovoltaic (PV) effect: The phenomenon that occurs when photons, the "particles" in a beam of light, knock electrons loose from the atoms they strike. When this property of light is combined with the properties of semiconductors, electrons flow in one direction across a junction, setting up a voltage. With the addition of circuitry, current will flow and electric power will be available.

Photovoltaic (PV) generator: The total of all PV strings of a PV power supply system, which are electrically interconnected.

Photovoltaic (PV) module: The smallest environmentally protected, essentially planar assembly of solar cells and ancillary parts, such as interconnections and terminals (and protective devices such as diodes), intended to generate DC power under unconcentrated sunlight. The structural (load carrying) member of a module can either be the top layer (superstrate) or the back layer (substrate).

Photovoltaic (PV) panel: Often used interchangeably with PV module (especially in one module systems), but more accurately used to refer to a physically connected collection of modules (i.e., a laminate string of modules used to achieve a required voltage and current).

Photovoltaic (PV) system: A complete set of components for converting sunlight into electricity by the PV process, including the array and balance of system components.

Photovoltaic thermal (PV/T) system: A PV system that, in addition to converting sunlight into electricity, collects the residual heat energy and delivers both heat and electricity in usable form. Also called a total energy system or solar thermal system.

Physical vapor deposition: A method of depositing thin semiconductor PV films. With this method, physical processes, such as thermal evaporation or bombardment of ions, are used to deposit elemental semiconductor material on a substrate.

P-I-N: A semiconductor PV device structure that layers an intrinsic semiconductor between a p-type semiconductor and an n-type semiconductor; this structure is most often used with amorphous silicon PV devices.

Plug-and-play PV system: A commercial, off-the-shelf PV system that is fully inclusive with little need for individual customization. The system can be installed without special training and using few tools. The homeowner plugs the system into a PV-ready circuit and an automatic PV discovery process initiates communication between the system and the utility. The system and grid are automatically configured for optimal operation.

P/N: A semiconductor PV device structure in which the junction is formed between a p-type layer and a n-type layer.

Pocket plate: A plate for a battery in which active materials are held in a perforated metal pocket.

Point-contact cell: A high-efficiency silicon PV concentrator cell that employs light trapping techniques and point-diffused contacts on the rear surface for current collection.

Polycrystalline: *See* multicrystalline.

Polycrystalline silicon: A material used to make PV cells, which consist of many crystals unlike single-crystal silicon.

Polycrystalline thin film: A thin film made of multicrystalline material.

Power conditioning: The process of modifying the characteristics of electrical power (e.g., inverting DC to alternating current).

Power conditioning equipment: Electrical equipment, or power electronics, used to convert power from a PV array into a form suitable for subsequent use. A collective term for inverter, converter, battery charge regulator, and blocking diode.

Power conversion efficiency: The ratio of output power to input power of the inverter.

Projected area: The net south-facing glazing area projected on a vertical plane.

P-type semiconductor: A semiconductor in which holes carry the current; produced by doping an intrinsic semiconductor with an electron acceptor impurity (e.g., boron in silicon).

Pulse width-modulated (PWM) wave inverter: A type of power inverter that produce a high-quality (nearly sinusoidal) voltage, at minimum current harmonics.

PV: *See* photovoltaic(s).

Pyranometer: An instrument used for measuring global solar irradiance.

Pyrheliometer: An instrument used for measuring direct beam solar irradiance. Uses an aperture of 5.7° to transcribe the solar disc.

Q–R

Qualification test: A procedure applied to a selected set of PV modules involving the application of defined electrical, mechanical, or thermal stress in a prescribed manner and amount. Test results are subject to a list of defined requirements.

Quantum efficiency (QE): The ratio of the number of charge carriers collected by a PV cell to the number of photons of a given energy shining on the cell. QE relates to the response of a solar cell to the different wavelengths in the spectrum of light shining on the cell. QE is given as a function of either wavelength or energy. Optimally, a solar cell should generate considerable electrical current for wavelengths that are most abundant in sunlight.

Ramp: A change in generation output.

Ramp rate: The ability of a generating unit to change its output over some unit of time, often measured in megawatt per minute.

Rankine cycle: A thermodynamic cycle used in steam turbines to convert heat energy into work. CSP plants often rely on the Rankine cycle. In CSP systems, mirrors focus sunlight on a heat-transfer fluid. This is used to create steam, which spins a turbine to generate electricity.

Rated battery capacity: The term used by battery manufacturers to indicate the maximum amount of energy that can be withdrawn from a battery under specified discharge rate and temperature. *See also* battery capacity.

Rated module current (A): The current output of a PV module measured at standard test conditions (STCs) of 1000 W/m² and 25°C cell temperature.

Rated power: Rated power of the inverter. However, some units cannot produce rated power continuously. *See also* duty rating.

Reactive power: The sine of the phase angle between the current and voltage waveforms in an alternating current system. *See also* power factor.

Recombination: The action of a free electron falling back into a hole. Recombination processes are either radiative, where the energy of recombination results in the emission of a photon, or nonradiative, where the energy of recombination is given to a second electron which then relaxes back to its original energy by emitting phonons. Recombination can take place in the bulk of the semiconductor, at the surfaces, in the junction region, at defects, or between interfaces.

Rectifier: A device that converts alternating current to DC. *See also* inverter.

Regulator: Prevents overcharging of batteries by controlling charge cycle usually adjustable to conform to specific battery needs.

Resistive voltage drop: The voltage developed across a cell by the current flow through the resistance of the cell.

Reverse current protection: Any method of preventing unwanted current flow from the battery to the PV array (usually at night). *See also* blocking diode.

Ribbon (photovoltaic) cells: A type of PV device made in a continuous process of pulling material from a molten bath of PV material, such as silicon, to form a thin sheet of material.

S

Sacrificial anode: A piece of metal buried near a structure that is to be protected from corrosion. The metal of the sacrificial anode is intended to corrode and reduce the corrosion of the protected structure.

Satellite power system (SPS): Concept for providing large amounts of electricity for use on the Earth from one or more satellites in geosynchronous Earth orbit. A very large array of solar cells on each satellite would provide electricity, which would be converted to microwave energy and beamed to a receiving antenna on the ground. There, it would be reconverted into electricity and distributed the same as any other centrally generated power through a grid.

Schottky barrier: A cell barrier established as the interface between a semiconductor, such as silicon, and a sheet of metal.

Scribing: The cutting of a grid pattern of grooves in a semiconductor material, generally for the purpose of making interconnections.

Semiconductor: Any material that has a limited capacity for conducting an electric current. Certain semiconductors, including silicon, gallium arsenide, copper indium diselenide, and cadmium telluride are uniquely suited to the PV conversion process.

Semicrystalline: *See* multicrystalline.

Series connection: A way of joining PV cells by connecting positive leads to negative leads; such a configuration increases the voltage.

Series controller: A charge controller that interrupts the charging current by open circuiting the PV array. The control element is in series with the PV array and battery.

Series regulator: Type of battery charge regulator where the charging current is controlled by a switch connected in series with the PV module or array.

Series resistance: Parasitic resistance to current flow in a cell due to mechanisms such as resistance from the bulk of the semiconductor material, metallic contacts, and interconnections.

Shunt regulator: Type of a battery charge regulator where the charging current is controlled by a switch connected in parallel with the PV generator. Shorting the PV generator prevents overcharging of the battery.

Siemens process: A commercial method of making purified silicon.

Silicon (Si): A semimetallic chemical element that makes an excellent semiconductor material for PV devices. It crystallizes in face-centered cubic lattice like a diamond. It is commonly found in sand and quartz (as the oxide).

Sine wave: A waveform corresponding to a single-frequency periodic oscillation that can be mathematically represented as a function of amplitude versus angle in which the value of the curve at any point is equal to the sine of that angle.

Sine wave inverter: An inverter that produces utility-quality, sine wave power forms.

Single-crystal material: A material that is composed of a single crystal or a few large crystals.

Single-crystal silicon: Material with a single crystalline formation. Many PV cells are made from single-crystal silicon.

Single-stage controller: A charge controller that redirects all charging current as the battery nears full state of charge.

Soft costs: Nonhardware costs related to PV systems, such as financing, permitting, interconnection, and inspection.

Solar cell: *See* PV cell.

Solar constant: The average amount of solar radiation that reaches the Earth's upper atmosphere on a surface perpendicular to the sun's rays; equal to 1353 W/m^2 or 492 Btu per square foot.

Solar cooling: The use of solar thermal energy or solar electricity to power a cooling appliance. PV systems can power evaporative coolers ("swamp" coolers), heat pumps, and air conditioners.

Solar energy: Electromagnetic energy transmitted from the sun (solar radiation). The amount that reaches the Earth is equal to one billionth of total solar energy generated, or the equivalent of about 420 trillion kilowatt-hours.

Solar-grade silicon: Intermediate-grade silicon used in the manufacture of solar cells. Less expensive than electronic grade silicon.

Solar insolation: *See* insolation.

Solar irradiance: *See* irradiance.

Solar noon: The time of the day, at a specific location, when the sun reaches its highest, apparent point in the sky.

Solar panel: *See* PV panel.

Solar resource: The amount of solar insolation a site receives, usually measured in kWh/m^2/day, which is equivalent to the number of peak sun hours.

Solar spectrum: The total distribution of electromagnetic radiation emanating from the sun. The different regions of the solar spectrum are described by their wavelength range. The visible region extends from about 390–780 nm (a nanometer is one billionth of one meter). About 99% of solar radiation is contained in a wavelength region from 300 nm (UV) to 3000 nm (near-infrared). The combined radiation in the wavelength region from 280 to 4000 nm is called the broadband, or total, solar radiation.

Solar thermal electric systems: Solar energy conversion technologies that convert solar energy to electricity, by heating a working fluid to power a turbine that drives a generator. Examples of these systems include central receiver systems, parabolic dish, and solar trough.

Space charge: *See* cell barrier.

Split-spectrum cell: A compound PV device in which sunlight is first divided into spectral regions by optical means. Each region is then directed to a different PV cell optimized for converting that portion of the spectrum into electricity. Such a device achieves significantly greater overall conversion of incident sunlight into electricity. *See also* mulitjunction device.

Sputtering: A process used to apply PV semiconductor material to a substrate by a physical vapor deposition process where high-energy ions are used to bombard elemental sources of semiconductor material, which eject vapors of atoms that are then deposited in thin layers on a substrate.

Square wave inverter: A type of inverter that produces square wave output. It consists of a DC source, four switches, and the load. The switches are power semiconductors that can carry a large current and withstand a high-voltage rating. The switches are turned on and off at a correct sequence, at a certain frequency.

Staebler–Wronski effect: The tendency of the sunlight to electricity conversion efficiency of amorphous silicon PV devices to degrade (drop) upon initial exposure to light.

Stand-alone system: An autonomous or hybrid PV system not connected to a grid. May or may not have storage, but most stand-alone systems require batteries or some other form of storage.

Standard reporting conditions (SRCs): A fixed set of conditions (including meteorological) to which the electrical performance data of a PV module are translated from the set of actual test conditions.

Standard test conditions (STCs): Conditions under which a module is typically tested in a laboratory.

Standby current: This is the amount of current (power) used by the inverter when no load is active (lost power). The efficiency of the inverter is lowest when the load demand is low.

Stand-off mounting: Technique for mounting a PV array on a sloped roof, which involves mounting the modules a short distance above the pitched roof and tilting them to the optimum angle.

String: A number of PV modules or panels interconnected electrically in series to produce the operating voltage required by the load.

Substrate: The physical material upon which a PV cell is applied.

Subsystem: Any one of several components in a PV system (i.e., array, controller, batteries, inverter, load).

Superconducting magnetic energy storage (SMES): SMES technology uses the superconducting characteristics of low-temperature materials to produce intense magnetic fields to store energy. It has been proposed as a storage option to support large-scale use of PV as a means to smooth out fluctuations in power generation.

Superstrate: The covering on the sunny side of a PV module, providing protection for the PV materials from impact and environmental degradation while allowing maximum transmission of the appropriate wavelengths of the solar spectrum.

System availability: The percentage of time (usually expressed in hours per year) when a PV system will be able to fully meet the load demand.

System operating voltage: The PV array output voltage under load. The system operating voltage is dependent on the load or batteries connected to the output terminals.

T

Temperature factors: It is common for three elements in PV system sizing to have distinct temperature corrections; a factor used to decrease battery capacity at cold temperatures; a factor used to decrease PV module voltage at high temperatures; and a factor used to decrease the current-carrying capability of wire at high temperatures.

Thermophotovoltaic (TPV) cell: A device where sunlight concentrated onto absorber heats it to a high temperature, and the thermal radiation emitted by the absorber is used as the energy source for a PV cell that is designed to maximize conversion efficiency at the wavelength of the thermal radiation.

Thick crystalline materials: Semiconductor material, typically measuring from 200- to 400-micron thick, that is cut from ingots or ribbons.

Thin film: A layer of semiconductor material, such as copper indium diselenide or gallium arsenide, a few microns or less in thickness, used to make PV cells.

Thin-film photovoltaic module: A PV module constructed with sequential layers of thin-film semiconductor materials. *See also* amorphous silicon.

Tilt angle: The angle at which a PV array is set to face the sun relative to a horizontal position. The tilt angle can be set or adjusted to maximize seasonal or annual energy collection.

Tin oxide: A wide bandgap semiconductor similar to indium oxide; used in heterojunction solar cells or to make a transparent conductive film, called NESA glass when deposited on glass.

Total AC load demand: The sum of the alternating current loads. This value is important when selecting an inverter.

Total internal reflection: The trapping of light by refraction and reflection at critical angles inside a semiconductor device so that it cannot escape the device and must be eventually absorbed by the semiconductor.

Tracking array: A PV array that follows the path of the sun to maximize the solar radiation incident on the PV surface. The two most common orientations are (1) one axis where the array tracks the sun east to west and (2) two-axis tracking where the array points directly at the sun at all times. Tracking arrays use both the direct and diffuse sunlight. Two-axis tracking arrays capture the maximum possible daily energy.

Transparent conducting oxide (TCO): A doped metal oxide used to coat and improve the performance of optoelectronic devices such as PV and flat panel displays. Most TCO films are fabricated with polycrystalline or amorphous microstructures and are deposited on glass. The current industry standard TCO is indium tin oxide. Indium is relatively rare and expensive, so research is ongoing to develop improved TCOs based on alternative materials.

Two-axis tracking: A PV array tracking system capable of rotating independently about two axes (e.g., vertical and horizontal).

U–V–W–Z

Ultraviolet: Electromagnetic radiation in the wavelength range of 4–400 nm.

Underground feeder (UF): May be used for PV array wiring if sunlight-resistant coating is specified; can be used for interconnecting balance-of-system components, but not recommended for use within battery enclosures.

Utility-interactive inverter: An inverter that can function only when tied to the utility grid, and uses the prevailing line-voltage frequency on the utility line as a control parameter to ensure that the PV system's output is fully synchronized with the utility power.

Vacuum evaporation: The deposition of thin films of semiconductor material by the evaporation of elemental sources in a vacuum.

Vertical multijunction (VMJ) cell: A compound cell made of different semiconductor materials in layers, one above the other. Sunlight entering the top passes through successive cell barriers, each of which converts a separate portion of the spectrum into electricity, thus achieving greater total conversion efficiency of the incident light. Also called a multiple junction cell. *See also* multijunction device and split-spectrum cell.

Voltage at maximum power (Vmp): The voltage at which maximum power is available from a PV module.

Voltage protection: Many inverters have sensing circuits that will disconnect the unit from the battery if input voltage limits are exceeded.

Wafer: A thin sheet of semiconductor (PV material) made by cutting it from a single crystal or ingot.

Window: A wide bandgap material chosen for its transparency to light. Generally used as the top layer of a PV device, the window allows almost all of the light to reach the semiconductor layers beneath.

Zenith angle: The angle between the direction of interest (e.g., of the sun) and the zenith (directly overhead).

REFERENCES

1. Kurtz S. Reliability challenges for solar energy, *IEEE International Reliability Physics Symposium (IRPS)*, March 27, Montreal, Canada, 2009, Online. Available: http://www.nrel.gov/docs/fy09osti/44970.pdf.

2. Kaisare A. Reliability challenges for solar modules, Tutorial 13-9, *ASME 2011 Pacific Rim Technical Conference and Exhibition*, InterPACK 2011, Portland, Oregon, July 2011.

3. Amarani A. El, Mahrane A., Moussa F. Y. et al. Solar module fabrication, *Int. J. Photoenergy*, 2007, Article ID 27610, 5pp., Hindawi Publishing Corporation, Cairo, 2007.

4. Alers G. B. Solar photovoltaic module failure analysis. In: *Microelectronics Failure Analysis: Desk Reference*, R. J. Ross (Ed.). 6th edition, October 2011, pp. 99–103. ASM International, Ohio. Online. Available: http://www.asminternational.org/documents/10192/3461432/09110Z_TOC.pdf/96fcf653-8d7f-462c-9ddd-997049f03001.

5. Sinicco I. Stress tests and failure modes of thin film silicon photovoltaic modules, *Durability of Thin Film Solar Cells - EMPA Academy*, Dubendorf ZH, April 2012, Online. Available: www.swissphotonics.net/libraries.files/Sinicco.pdf.

6. Mathiak G., Althaus J., Menzler S. et al. PV Module corrosion by ammonia and salt mist. Experimental study with full-size modules, *Proceedings of the 27th European Photovoltaic Solar Energy Conference and Exhibition*, 2012, Online. Available: http://www.tuv.com/media/germany/10_industrialservices/downloadsi06/poster_pvsec/PV_Module_corrosion_by_ammonia.pdf.

7. Weliczko E. Galvanic corrosion considerations for PV arrays, *Solar Professional*, (4.4), 1–3, June/July 2011, Online. Available: http://solarprofessional.com/articles/products-equipment/racking/galvanic-corrosion-considerations-for-pv-arrays.

8. Hren R. and Mehalic B. Grounding compendium for PV systems, *Solar Professional*, (6.5), 1–13, August/September 2013, Online. Available: http://solarprofessional.com/articles/design-installation/grounding-compendium-for-pv-systems.

9. Ball G. Grounding photovoltaic modules: The Lay of the Land, *Solar ABCs Interim Report*, Solar America Board for Codes and Standards Report, March 2011, Online. Available: http://www.solarabcs.org/about/publications/reports/module-grounding/pdfs/module-grounding_studyreport.pdf.

10. Hacke P., Terwilliger K., Smith R. et al. System voltage potential induced degradation mechanisms in PV modules and methods for Test, *Proceedings of the 37th IEEE Photovoltaic Specialists Conference (PVSC 37)*, Seattle, Washington, June 19–24, 2011, Online. Available: www.nrel.gov/docs/fy11osti/50716.pdf

11. Berghold J., Frank O., Hoehne H. et al. Potential induced degradation of solar cells and panels, 7 pp. Online. Available: http://www.solon.com/export/sites/default/solonse.com/_downloads/global/article-pid/Berghold_et_al_PID_of_Solar_Cells_and_Panels.pdf.

12. Understanding potential induced degradation, *Advanced Energy White Paper*, Online. Available: http://solarenergy.advanced-energy.com/upload/File/White_Papers/ENG-PID-270-01%20web.pdf.

13. Dhere N. G., Pandit M. B., Jahagirdar A. H. et al. Overview of PV module durability and long term exposure research at FSEC, *Proceedings of the NCPV Program Review Meeting*, October 15–17, Denver, CO, 2000, pp. 313–314, Online. Available: http://citeseerx.ist.psu.edu/viewdoc/download?doi=10.1.1.199.5494&rep=rep1&type=pdf.

14. Bremner S. Photovoltaic modules, *ELEG620: Solar Electric Systems University of Delaware*, ECE, Spring 2009, Online. Available: www.solar.udel.edu/ELEG620/13_PV_modules.pdf.

15. King D. L., Quintana M. A., Kratochvil J. A. et al. Photovoltaic module performance and durability following long-term field exposure, *Progress in Photovoltaic*, 8(2), 241–256, March/April 2000, Online. Available: http://www.cleanenergy.com.ph/projects/CBRED/TA%20RE%20Manufacturers%20SubContract/Compendium%20of%20References/Solar%20References/Collection%20of%20Solar%20Standards%20and%20Articles/C13%20PV%20module%20performance%20and%20durability.pdf.

16. Dhere N. G., Wollam M. E., and Gadre K. S. Correlation between surface carbon concentration and adhesive strength at the Si-cell/EVA interface in a PV module, *Proceedings of the 26th IEEE Photovoltaic Specialists Conference*, Anaheim, CA, pp. 1217–1220, 1997.

17. Ketola B. and Norris A. Degradation mechanism investigation of extended damp heat aged PV modules, *Proceedings of the 26th European Photovoltaic Solar Energy Conference and Exhibition (EU PVSEC)*, Hamburg, Germany, 6pp., 2011, Online. Available: http://www.dowcorning.com/content/publishedlit/06-1084.pdf.

18. Dhere N. G., Gadre K. S., and Deshpande A. M. Durability of photovoltaic modules, *Proceedings of the 14th European Photovoltaic Solar Energy Conference and Exhibition (EU PVSEC)*, Barcelona, Spain, pp. 256–259, 1997.

19. Suleske A. A. Performance degradation of grid-tied photovoltaic modules in a desert climatic condition, PhD thesis, November 2010, Online. Available: https://repository.asu.edu/attachments/56112/content/Suleske_asu_0010N_10107.pdf.

20. Rutschman I. and Matz M. D. Unlocking the secret of snail tracks, *Photon International*, pp. 82–89, January 2012, Online. Available: http://www.wrecked.net/Solar/Photon-Mag%202012-01%20SnailTrails.pdf.

21. Rutschmann I. Defying all findings, *Photon International*, pp. 76–80, June 2012.

22. Köntges M., Kunze I., Naumann V. et al. Snail tracks (Schneckenspuren), worm marks and cell cracks, *IEA PVPS Task 13 Workshop at 27th EU PVSEC*, 2012.

23. Peng P., Hu A., Zheng W. et al. Microscopy study of snail trail phenomenon on photovoltaic modules, *RSC Advances*, 2, 11359–11365, 2012, Online. Available: http://mme.uwaterloo.ca/~camj/pdf/2012/P%20Peng_RSC%20Adv_2012.pdf.

24. Xu A., Lee S. L., and Han R. Snail trail, *Today's Energy Solutions (TES)*, May 2012, Online. Available: http://www.onlinetes.com/Author.aspx?AuthorID=5891.

25. Wohlgemuth J., Cunningham D. W., Nguyen A. et al. *Failure Modes of Crystalline Si Modules, BP Solar, PV Module Reliability Workshop*, 2010, Online. Available: http://www1.eere.energy.gov/solar/pdfs/pvrw2010_wohlgemuth_silicon.pdf.

26. Gallon J., Horner G. S., Hudson J. E., Vasilyev L. A., and Lu K. PV Module hotspot detection, *2015 PV Module Reliability Workshop (PVMRW)*, Golden, CO, February 2015, Online. Available: http://www.nrel.gov/pv/performance_reliability/pdfs/2015_pvmrw_19_gallon.pdf.

27. Shioda T. Delamination failures in long-term field-aged PV modules from point of view of encapsulant, *2013 NREL PV Module Reliability Workshop*, Denver, CO, February 2013, Online. Available: pvmrw13_openingsession_shioda_mitsui.pdf.

28. Traver R. Insulation plays a critical role in solar panel manufacturing, *Global Solar Technology*, November/December 2013, Online. Available: www.globalsolartechnology.com.

29. Ferrara C. and Philipp D. Why do PV modules fail?, *Proceedings of the International Conference on Materials for Advanced Technologies (ICMAT)*, Singapore, June, 2011, *Energy Procedia*, 15, pp. 379–387, 2012, Online. Available: http://publica.fraunhofer.de/ise/2012.htm.

30. Jordan D. C. and Kurtz S. R. *Photovoltaic Degradation Rates— An Analytical Review, National Renewable Energy Laboratory* (NREL), Technical Report NREL/JA-5200-51664, 32 pp., June 2012, Online. Available: www.nrel.gov/docs/fy12osti/51664.pdf.

31. Packard C. E., Wohlgemuth J. H., and Kurtz S. R. *Development of a Visual Inspection Data Collection Tool for Evaluation of Fielded PV Module Condition, National Renewable Energy Laboratory (NREL)*, Technical Report NREL/TP-5200-56154, 52 pp., August 2012, Online. Available: www.nrel.gov/docs/fy12osti/56154.pdf.

32. Quality problems bring PV testing to the fore, *Power Electronics*, October 1, 2012, Online. Available: http://powerelectronics.com/content/quality-problems-bring-pv-testing-fore.

33. Jahn U. PV Module reliability issues including testing and certification, IEA International Energy Agency, Photovoltaic Power Systems Programme Task 13: Performance and Reliability of PV Systems, *Proceedings of the 27th European Photovoltaic Solar Energy Conference and Exhibition*, Frankfurt, Germany, September 2012.

34. Köntges M., Kurtz S., Packard C., Jahn U. et al. *Review of Failures of Photovoltaic Modules, International Energy Agency, Photovoltaic Power System Programme (IEA PVSP)*, Report IEA-PVPS T13-01:2014, March 2014, Online. Available: http://www.isfh.de/institut_solarforschung/files/iea_t13_review_of_failures_of_pv_modules_final.pdf.

35. Petzold K. Ammonium hydroxide attacks panels, *PV Magazine*, September 2011, Online. Available: http://www.pv-magazine.com/archive/articles/beitrag/ammonium-hydroxide-attacks-panels-_100004126/86/?tx_ttnews%5BbackCat%5D=174&cHash=9bee60fd1317f217f56e8b8c32c26133#ixzz2sZ2mZel1.

36. Solar panels last 25 years – But will they stay safely attached to your roof? *The Importance of Reliable Solar Mounting Systems*, White Paper, 29pp., March 2014, Online. Available: http://cinnamonsolar.com/wp-content/uploads/2013/05/The-Importance-of-Reliable-Solar-Mounting-Systems-White-Paper_March-2014-copy.pdf.

37. Solar Energy Standards, Online. Available: http://webstore.ansi.org/solar_energy/.

38. Solar ABCs: Codes and Standards, Solar America Board for Codes and Standards, International Electrotechnical Commission, Online. Available: solarabcs.org/codes-standards/IEC/index.html.

39. Guide to Standards – Solar Panels, SAI Global, Australia, November 2011, 11 pp.

40. Solar Energy Glossary. Online. Available: http://energy.gov/eere/sunshot/solar-energy-glossary.

10 Tidal and Wave Power

10.1 RENEWABLE TIDAL AND WAVE ENERGY SOURCES

Ocean energy has enormous potential for energy production in future. Ocean energy technologies are relatively new and applications are developing at very fast pace. As a result, classification, applications, and conversion concepts have yet to be defined. This section is devoted to presenting these issues based on current available literature and industrial trends. Ocean energy can be tapped in multiple forms such as energy from waves, kinetic energy from tidal and marine currents, potential energy from tides, and energy from salinity and thermal gradient [1]. The ocean energy can be classified on the basis of resources such as tides, currents, waves, salinity gradient, and ocean thermal energy conversion (OTEC) systems (Table 10.1).

10.1.1 TIDAL ENERGY

Energy from tides is mainly captured during the rise and fall of the sea level. Their rise and fall is due to the interaction of gravitational pull in earth, sun, and moon system. Tide is accompanied by vertical water movement (rise and fall) and horizontal water movement (tidal current). Tidal range and tidal current have often been confused in the literature in the past, difference between tidal range and tidal current is that the tidal range is the difference between the high and low tide (potential energy) whereas the tidal current is the horizontal water movement (kinetic energy).

Tidal energy is one of the oldest forms of energy. The tides are an enormous and consistent untapped resource of renewable energy. Tide mills, in use on the Spanish, French, and British coasts, date back to eighth century AD. The technology required to convert tidal energy into electricity is very similar to the technology used in traditional hydroelectric power plants—dam, gates, and turbines. Turbines, similar to wind turbines, can be anchored to the seabed to generate electricity from tidal currents. Not all tidal power companies are proposing seabed-mounted turbines, some are proposing floating designs.

Currents are generated not only by tides but also from wind, temperature, and salinity differences. The concept of tapping kinetic energy from ocean currents is the same as tidal currents. Marine and tidal current technologies share same principle of operation. Open center turbine utilizes the kinetic energy of currents to produce power.

10.1.1.1 Tidal Power System Classification

Tidal power allows many ways of electricity production and can be classified into two different main types [2]:

- *Tidal stream* systems make use of the kinetic energy of moving water to power turbines, in a similar way to windmills that use moving air. This method is gaining in popularity because of the lower cost and lower ecological impact compared to barrages.
- *Barrages or dams* across estuaries (or lagoons) to trap tidal waters. The "potential energy" of the water trapped at high tide is harnessed, and can be realized when the water is later released through turbines to generate electricity. Barrages make use of the potential energy in the difference in height (or head) between high and low tides. Barrages suffer from very high civil infrastructure costs, a worldwide shortage of viable sites, and environmental issues.

TABLE 10.1
Classification of Ocean Energy

Source of Ocean Energy	Type of Ocean Energy
Tides	Kinetic and potential
Currents	Kinetic
Waves	Kinetic and potential
Salinity gradient	Chemical
OTEC	Thermal

Tidal power stations are a variation of the classical hydroelectric power plants using the level difference of the water between low tide and flood, also called tidal range, for the production of electricity. For this purpose, dams can be established in estuaries or sea bays as a separation from the high seas. Such embankments get notches so that water will stream in when the tide is high and drain off when the tide is out, a process under the driving mechanism of the turbines. In both operation phases, electric current is produced. The big advantage of this technology is that low tide and flood continuously alternate and, hence, the tidal force displays a computable and dependable energy source.

The best known representative of this technology is the tidal power station in the mouth of the river La Rance near Saint-Malo at the French Atlantic coast. It was established already in 1966 and has a tidal range of 12–16 m with which a maximum power of 240 MW is reached.

Alternative to those tidal power stations are sea flow power stations, resembling the wind power equipment with the difference that the rotor turns underwater and that it is driven by the constant tidal change. A pilot plant was successfully installed in summer 2003 near the coast of North Devon in England under the project name "Seaflow." In another step, an equipment with a power of 1.2 MW was installed in April 2008 near Strangford Lough under the project name "SeaGen."

Tidal power plants are presently put forward as a cheap energy production technology [3–5]. Tidal-driven underwater turbines will operate in sea regions with high tidal potential. Giant tidal power scheme, built off the South Korean coast (300 units) and off the Welsh coast (8 units) made of turbines that are deep in the water to avoid any danger to ships [4]. Another U.K. project is underwater 10 megawatt tidal stream in the Sound of Islay between the Hebridean islands of Islay and Jura [3]. These projects will be challenged by eminent materials and corrosion problems. Economical solutions for materials and materials protection will need extensive R&D efforts.

Potential energy exploitation in the case of tidal energy has 80% efficiency which is considered to be high compared to other forms of renewable green energy sources. However, the cost of building a tidal powered electricity station is high and varies from 1.3 million USD per MW to 1.8 million USD per MW depending on the location and technology used. This high cost is one of the prohibitive factors of the expansion of tidal power.

On top of the generator, gearbox, and power electronics expertise, in order to get a prototype into the water needs shipbuilding and offshore technology. The associated expenses usually have been subsidized in many cases by grants from governments, including the United Kingdom and the European Union.

In 2008, the first experimental wave farm was opened in Portugal, at the Aguçadoura Wave Park. Currently, there are many functioning wave farms located in Portugal, the United Kingdom, Australia, and the United States. One of the latest developments of wave energy industry was a full-scale commercial near shore unit, *greenWAVE*, with a capacity of 1 MW installed off Port MacDonnell in South Australia in 2013.

Various parts of wave power equipment such as the piston, accelerator tubes, pumps and Pelton turbine, various check valves, manifolds, and other piping, are all exposed to the corrosive effects of seawater. There is an obvious need for using noncorrosive metals and coatings to protect wave power equipment from corrosion, the problem common for other green power equipment exposed to seawater (see Section 10.2 and Chapter 12).

10.1.1.2 Tidal Turbines Types

Tidal turbines work using similar techniques to the techniques that wind turbines use. Tidal turbines can be classified into four groups: horizontal axis, vertical axis, venture based and cross-flow turbines [6–9].

Horizontal axis turbines: These tidal turbines work in the same way as wind turbines do. These turbines are located on the seabed. The water stream is parallel to the seabed and also to the axis of the rotor of the turbine. The water flow moves the blades of the turbine producing energy by means of the generator.

One example of this kind of turbine is that developed by SeaGen (Figure 10.1). It was connected in 2008 and it generates 1.2 MW. Each turbine is mounted in a two arm-like extension with a rotor diameter of 16 meters (m). The maximum efficiency achieved is 89%.

Another type of horizontal axis tidal turbines DeltaStream has been developed by Tidal Energy Ltd. It works like a wind turbine, but it uses water flow instead of wind flow. The water flow turns the blades of the turbine and electricity is obtained by means of the generator located in the nacelle (Figure 10.2).

The electricity from tidal turbines is transmitted to shore through submarine cables. The communication between the tidal turbines and shore is via submarine cables as well.

FIGURE 10.1 The SeaGen tidal current turbine developed by Marine Current Turbines Ltd. (MCT), Bristol, the United Kingdom, is rated at 1.2 MW. (Adapted from Bearings for "underwater windmills" keep gearboxes operating, *Design World On-Line*, August 15, 2012.)

FIGURE 10.2   Delta Stream tidal turbines. (From DeltaStream Tidal Energy Solution, Tidal Energy Limited, Cardiff, Online. Available: http://www.tidalenergyltd.com/cms/wp-content/uploads/downloads/2012/10/DeltaStream_White_Paper_Aug12.pdf. With permission.)

The power unit of the DeltaStream turbine is composed of three turbines connected in a triangular frame 30 m wide and weighing 250 tons. They are located on the seafloor around 31.5 m below the surface of the sea. The minimum distance from the tip rotor to the sea surface is approximately 10 m at the lowest tide. Thereby, the turbine is underwater for high and low tides. The rotor of each turbine is 15 m in diameter and the nacelle of each turbine has a dimension of 2×9 m [10]. Each unit produces 1.2 MW of power. They are located in Ramsey Sound, Pembrokeshire in Wales. Condition monitoring system is installed on DeltaStream turbine (see Chapter 18).

Vertical axis: The axis of this turbine is orthogonal to the water flow and also to the seabed. The main advantage of these devices is that the blades can be enlarged horizontally, engaging with higher amounts of water flow. The turbine developed by Kobold is an example (Figure 10.3). It has three vertical blades and with a 6 m of diameter. At water flow of 2 m/s, the power produced is approximately 25–30 kW [7].

Venturi-based turbine: These turbines have a horizontal axis, but they present a *venturi*-shaped duct, whose function is to accelerate the water stream and consequently, increasing the power extracted for a determined value of the radius of the rotor (Figure 10.4). The *venturi effect* is the reduction in fluid pressure that results when a fluid flows through a constricted section (or choke) of a pipe. The venturi effect is named after an Italian physicist Giovanni Battista Venturi (1746–1822).

Lunar Energy Limited [8] has developed a technology known as the Lunar Tidal Turbine (LTT). The LTT turbine is a bidirectional horizontal axis turbine housed in a symmetrical venturi duct. The venturi draws the existing ocean currents into the LTT in order to capture and convert energy into electricity. The use of a submersible gravity foundation will allow the LTT to be deployed quickly and with little or no seabed preparation at depths in excess of 40 m. This gives the LTT a distinct advantage over most of its competitors and opens up a potential energy source that is five times the size of that available to companies using other installation techniques.

Cross-flow turbine: The axis of this turbine is parallel to the seabed, but it is orthogonal to the water flow. Ocean Renewable Power Company's (ORPC) developed TidGen™ Power System cross-flow turbine foils (Figure 10.5, top). In 2012, the TidGen became the first commercial, grid-connected, hydrokinetic tidal energy project in North America. This is the only ocean energy project, other than one using a dam, which delivers power to a utility grid anywhere in North, Central,

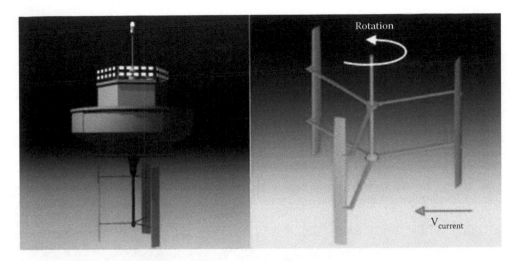

FIGURE 10.3 Kobold turbine (vertical axis). (From Calcagno G. and Moroso, A. The Kobold marine turbine: From the testing model to the full scale prototype. *Tidal* Energy Summit, November 28–29, 2007, London, *Tidal Today*, 2007, Online. Available: http://www.tidaltoday.com/tidal07/presentations/GuidoCalcagnoMoroso.pdf. With permission.)

and South America. Another ORPC system—a RivGen® Power System (Figure 10.5, middle) based on the same principle harvests energy from both river and tidal currents to generate electricity. RivGen Power System's core component, the turbine generator unit (TGU), is at the forefront of marine hydrokinetic technology, and utilizes innovative control systems to drive two advanced design cross-flow turbines that efficiently provide reliable energy even within highly turbulent flow environments.

German company Atlantisstrom has developed a tidal turbine, which may be used in rivers and offshore (Figure 10.5, bottom right and left) [9]. In June 2014, Atlantisstrom's first tidal power plant

FIGURE 10.4 Venturi-type tidal turbine. (Adapted from E.ON and Lunar Energy to Build 8-MW Tidal Power Project.)

FIGURE 10.5 Tidal power systems based on cross-flow turbine: TidGen Power System (top); RivGen Power System (middle) by Ocean Renewable Power Company's (ORPC); (Adapted from First Commercial, Grid-Connected, Hydrokinetic Tidal Energy Project in North America, US Department of Energy, Office of Science) and tidal turbine model (bottom left) and an actual turbine (bottom right) by Atlantisstrom GmbH & CoKG. (From Kai J. Atlantisstrom seeks finance take innovative design to next level, Atlantisstrom GmbH & CoKG, *Tidal Today*, 2015, Online. Available: www.atlantisstrom.de/files/Atlantisstrom_paper_Draft3.pdf. With permission.)

was installed and connected to the public electricity grid in the Faroe Islands for testing in the Vestmanna Sund about 400 m off the island of Vagar.

10.1.2 WAVE ENERGY

Another alternative of producing electricity is using *wave energy*. There are different ways in which wave energy technologies can be categorized, for example, by the way the wave energy is converted into mechanical energy or by the technology used.

10.1.2.1   Wave Energy Technologies

Wave power equipment is not widely used in as a green energy source yet. Currently, wave power generation is not a widely employed commercial technology, although there have been attempts to use it since the end of nineteenth century.

There are several types of wave energy technologies. One type uses *oscillating water column* (OWC) devices to generate electricity at the shore using the rise and fall of water within a cylindrical shaft (see Section 10.1.2.2). The rising water drives air out of the top of the shaft, powering an air-driven turbine. Third, a tapered channel, or overtopping device can be located either on or offshore. They concentrate waves and drive them into an elevated reservoir, where power is then generated using hydropower turbines as the water is released.

The vast majority of recently proposed wave energy projects would use *offshore floats, buoys, or pitching devices* to generate electricity using the rise and fall of ocean swells to drive hydraulic pumps. Energy output in wave power is determined by wave height, wave speed, wavelength, and water density. The floating wave power farm operates on the wave energy that is created when a float on a buoy flows with the natural up and down movement of the waves.

A prototype of a wave energy machine, AquaBuoy, has been tested in the Pacific Ocean, off Newport, Oregon, in the Northwestern United States [11]. Developed by a Canadian company, Finavera Renewables Inc. of Vancouver, AquaBuoy is a cylindrical device that floats vertically in the ocean and bobs up and down under the action of the waves (Figure 10.6).

Only the top portion of the AquaBuoy is visible above the surface. The current prototype is half scale; full-size units will be 13 m in diameter, 48 m long, and weigh about 65 tons. The longer, submerged portion of the largely empty cylinder has a piston that moves up and down with the waves, pushing water alternately up and down through "accelerator tubes" to two-stroke hose pumps that pressurize the water and direct it to a Pelton turbine, one of the most efficient types of water turbines in high-force, low-flow situations. The turbine in turn powers an electric generator in the upper portion of the device. Power then flows through an underwater cable to onshore users. Each AquaBuoy has the potential to generate up to 250 kW in 4- to 5-m waves [11]. Groups of them deployed offshore in various arrays could produce tens of megawatts of renewable electricity.

Another technology is developed in the United Kingdom by AWS Ocean Energy Ltd [12]. A multicell array of flexible membrane absorbers which converts wave power to pneumatic power through

FIGURE 10.6   Prototype of wave power system. (Adapted from The Wave of The Future? *Nickel*, 23(1), 4, December 2007, Online. Available: http://www.nickelinstitute.org/~/media/Files/Magazine/Volume23/Vol23-02Dec2007.ashx#Page=3.)

compression of air within each cell. The cells are interconnected, thus allowing interchange of air between cells in antiphase. Turbine-generator sets are provided to convert the pneumatic power to electricity.

A typical device will comprise an array of 9 cells, each measuring around 16 m wide by 8 m deep, arranged around a catamaran structure. Such a device is capable of producing an average of 2.4 MW from a rough sea while having a structural steel weight of less than 3500 tons. The AWS-III will be slack moored in water depths of around 100 m using standard mooring spreads. The AWS-III has a number of significant advantages for utility-scale offshore wave power. The flexible absorbers are highly efficient and are the only moving part of the power train exposed to the sea.

The use of air as a transmission medium removes end-stop issues while the air turbines employed are reliable proven technology. The large hulls provide a stable and safe working environment to allow onboard maintenance of the turbines, generators, and all ancillary systems to ensure high reliability. Transformers and switchgear are adequately housed onboard and again easily accessed for maintenance. The diaphragms are designed with maintenance in mind using a patented "cassette" system to allow change-out at sea.

10.1.2.2   Types of Wave Energy Converters

In a very broad categorization, there are three types of converters: OWCs, oscillating body converters, and overtopping converters.

Oscillating water columns are conversion devices with a semisubmerged chamber, keeping a trapped air pocket above a column of water. Waves cause the column to act like a piston, moving up and down and thereby forcing the air out of the chamber and back into it. This continuous movement generates a reversing stream of high-velocity air, which is channeled through rotor blades driving an air turbine-generator group to produce electricity.

Oscillating body converters are either floating (usually) or submerged (sometimes fixed to the bottom). They exploit the more powerful wave regimes that normally occur in deep waters where the depth is greater than 40 m. In general, they are more complex than OWCs, particularly with regards to power takeoff (PTO) system by which mechanical energy is converted into electrical energy. An example of oscillating body power converters are Pelamis Tidal Turbine (Figure 10.7, top left), and wave turbine by Ocean Power Technologies (Figure 10.7, top right).

Overtopping converters (or terminators) consist of a floating or bottom fixed water reservoir structure, and also usually reflecting arms, which ensure that as waves arrive, they spill over the top of a ramp structure and are restrained in the reservoir of the device. The potential energy, due to the height of collected water above the sea surface, is transformed into electricity using conventional low-head hydro turbines (similar to those used in mini-hydro plants). The main advantage of this system is the simple concept—it stores water and when there is enough, lets it pass through a turbine. Key downsides include the low head (in the order of 1–2 m) and the vast dimensions of a full-scale overtopping device. Some representative devices—wave turbines by WaveCat (Spain) and by WaveDragon (Denmark)—are shown in Figure 10.7, bottom left and right.

10.1.3   Wave and Tidal Turbines Challenges

Tidal turbines can deliver a lot of power predictably twice a day. However, to get this energy an unmanned submarine should be effectively built that will work for 5 years under high pressure in a corrosive environment. For this period, the turbine should not need any maintenance, and there should be no leaks despite having a rotating shaft and an electrical connection.

The role of the seals between the moving shaft and the stationary generator is at the danger to be broken as these are mechanically stressed under pressure of the seawater. The new solution of flooding the electrical generator and the air gap, respectively, may help overcome seal failure by reducing mechanical stress on the seals. The idea of running an electrical machine with a fully flooded air gap is used in the oil and gas industry already. With this solution, the strains on the seals

FIGURE 10.7 Oscillating body power converters: Pelamis turbine (top left), Ocean Power Technologies (top right); Overtopping wave power converters: WaveCat (bottom left) and Wave Dragon (bottom right). (Adapted from Wave Power Technologies Brief 4, International Renewable Energy Agency (IRENA), June 2014, Online. Available: http://www.irena.org/documentdownloads/publications/wave-energy_v4_web.pdf.)

can be reduced as they would have to seal nonmoving parts against the seawater only. The lifetime of seals could be prolonged with this measure. On the other hand, this will also result in changes to the electrical behavior of the windings and the electrical machine, respectively [5].

The turbine needs strong underwater foundations that have to be built cost-effectively in a fast tidal environment. Its support structure should have to allow for maintenance either by raising the entire generator assembly free of the water or through quick release so that it can be lifted off between tides.

Other conditions may reduce durability and efficiency of wave and tidal turbines. Strong ocean storms and saltwater corrosion can damage the devices, which could increase the cost of construction to increase durability and/or cause frequent breakdowns. Changes in tidal movement could substantially reduce efficiency of tidal energy generators. Sea life could be harmed by the blades in the open and venturi turbines. The floor mounting of tidal generators could also disrupt the habitats of different sea life and plants.

The high cost of the different tidal generating stations and cost of installing power lines underwater could lengthen the payback period and be cost prohibitive based on the characteristics and size of each project. There are relatively few commercial installations as compared to other technologies, such as wind and solar farms. Additional difficulties implementing tidal energy generating devices could also arise.

10.1.4 OTHER METHODS OF EXTRACTING OCEAN ENERGY

10.1.4.1 Salinity Gradient

The locations where river water mixes with seawater are called estuaries. They are found along coastlines throughout the world, and it was understood that they have the significant potential to produce clean energy from these mixing water streams. The salinity gradient between the two streams of water contains a large amount of osmotic power, which can be thought of as the available energy (or chemical potential) from the differences in salt concentration between the fresh water and seawater.

The enormous amounts of energy released where freshwater and seawater meet can be utilized for the generation of power through osmosis, which is defined as the transport of water through a semipermeable membrane. The research on extracting power from salinity gradient is in early phase. The devices are installed in several locations around the globe, but most are experimental.

The idea of obtaining energy from osmosis, or *salinity gradient power*, has been studied for decades, but in early 2009 two teams, one in Netherland and another in Norway, were racing to be the first to build a working prototype power plant making salinity gradient power a feasible method for renewable energy generation. Both teams have been working on the development and implementation of a membrane-based osmotic process, but their approaches for generating the electricity are very different [13].

Westus (The Centre for Sustainable Water Technology), located in the Netherlands, is using the *Reverse Electrodialysis (RED)* method to produce electricity. They claim that they will utilize fresh water from the Rhine river and saltwater from the North Sea to construct a type of battery by employing two membranes permeable to ions, but not to water. Utilizing the saltwater, one membrane will allow the passage of positively charged sodium ions into a stream of fresh water and the other membrane will allow the passage of negatively charged chloride ions into another channel of freshwater. The separated charged particles with electrodes placed in both streams make up the chemical battery, which directly produces electricity.

Statkraft, a leading renewable energy group located in Norway, is focusing on *pressure retarded osmosis (PRO)* as their method to extract electricity from salinity gradients. This method utilizes a membrane, permeable to water, to draw fresh water into the concentrated saltwater, thus increasing the pressure in the saltwater chamber. The resulting pressure can then be used to drive a turbine to produce electricity. In November of 2009, Statkraft opened the world's first prototype osmotic power plant in Tofte, Norway [13].

There are two other popular methods to extract energy from salinity gradient. One is *solar pond*. The second technique is *Doriano Brogioli's capacitive method* which is relatively new and has so far only been tested on lab scale.

Harvesting free energy from estuaries and salinity gradients has both pros and cons. On one side, the technology is considered "as green as it gets," with the only waste product being brackish water, which flows into the sea mixing with the seawater. It is a constantly flowing source of renewable energy, unlike the intermittent energy provided by sources such as solar or wind power. It can also easily be combined with existing power plants and industries and can be built underground, thus reducing costs and visual pollution.

On the other side, the membrane technology still has a long way to go. The membranes are prone to biofouling from algae and silt (see Section 10.2), which reduce the membrane's lifetime and efficiency. Salinity gradient power is mainly suitable only for places where there is an abundant supply of freshwater meeting saltwater, which clearly favors countries with a large coastline [13].

Also, the environmental impact and environmental policy should be considered for future plants of this type. There are many species of aquatic life that are adapted to survive in waters with a specific range of salinity concentrations, and these power plants could affect the salinity of an area of water. One must consider the environmental policy and impact of structures that intake such large volumes of river water and seawater. These power plants must conform to strict construction permits and environmental regulations.

10.1.4.2 Ocean Thermal Energy Conversion (OTEC)

OTEC technologies make use of a heat engine which uses the temperature difference between cold and hot water. Due to heat from sun, the water is warmer on the top and gets cooler as the depth increases. When the temperature difference between hot and cold water is 20°C, the conditions for OTEC are most promising. These conditions are found near the equator. The efficiency of heat engine increases with increase in difference in temperature. OTEC is still an emerging technology. Sagar-Shakthi is a closed cycle OTEC plant with a capacity of 1 MW in India [14].

10.2 BIOFOULING AND CORROSION OF TIDAL AND WAVE POWER EQUIPMENT

Tidal and wave generation systems and supporting structures installed mostly in saltwater seas. Therefore, this equipment is exposed to a very corrosive environment and it becomes vulnerable to natural processes. For tidal power spanning from the lowest tidal mark in a given year to as many as 3 m (10 ft) above the year's highest tidal mark, the splash zone of a structure suffers the most severe corrosion.

Additionally, location of tidal power equipment leads to device breakdown. Strong ocean storms and saltwater corrosion can damage the devices, which could increase the cost of construction to increase durability and/or cause frequent breakdowns. High mechanical forces paired with comparable slow movements in a highly corrosive environment (contact with seawater) are leading to a reduced time in service [4].

The majority of generating devices are at least partially immersed in seawater. Some devices are completely submerged during operation. This mode of operation implicates several issues which are reducing the lifetime and/or the service intervals of such machines significantly. These issues include corrosion, biofouling, degradation of materials, and others.

10.2.1 BIOFOULING

Biofouling is the attachment of an organism or organisms to a surface in contact with water for a period of time. There are several organisms that cause biofouling, many different types of surfaces affected by it. The complex process often begins with the production of a *biofilm* which is a film made of bacteria. The growth of a biofilm can progress to a point where it provides a foundation for the growth of seaweed, barnacles, and other organisms. In other words, microorganisms such as bacteria, diatoms, and algae form the primary slime film to which the macroorganisms such as mollusks, sea squirts, sponges, sea anemones, bryozoans, tube worms, polychaetes, and barnacles attach [15,16]. An example of biofouling is shown in Figure 10.8.

Actually regardless of the location and the season, every immersed surface will be covered in a matter of seconds by a layer of adsorbed organic compounds (e.g., polysaccharides, lipids, and proteins). Less than 24 hours after the formation of this "conditioning layer," the biofouling process starts. Primary colonizers mainly consist of bacteria, yeasts, and diatoms which establish themselves within protective biofilm. Secondary colonizers comprise the spores of macroalgae, fungi, and protozoa which, according to the literature, settle roughly about 1 week after immersion when the environmental conditions are favorable. Invertebrate larvae are often regarded as the last stage of the marine biofouling process, their arrival to the surface take place on average after 2–3 weeks of immersion during the spawning season.

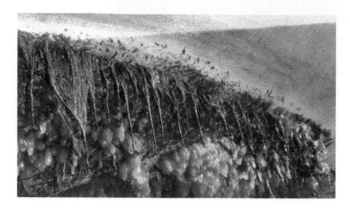

FIGURE 10.8 Example of biofouling in marine renewable device. (Courtesy of Dr. Christopher Harling.)

Marine biofouling can significantly increase the weight of floating structures and increase the drag resistance of moving parts such as tidal turbine blades, which results in the loss of efficiency of a tidal turbine. Based on the estimates shown in Ref. [17], the cost associated with the loss in efficiency in the turbine are in the order of several tens of millions of euros for large tidal farms.

To find the ways to protect tidal power equipment from deteriorating effects of biofouling, several studies performed an extensive research of various materials used in underwater equipment exposed to seawater and organisms that live there. Some studies in Ref. [5] have been done to investigate the influence of fouling and sea organisms on tidal equipment.

10.2.1.1 Observation of Biofouling and Corrosion Based on Testing

It was determined that the quality of surfaces of the equipment biofouling progress differently. For example, smooth surfaces are not prone to fouling on the time scales investigated in Ref. [18] (at least up to 10 months). However, rough surfaces, edges, and crevices are prone to fouling. As was found, the barnacles and starfish colonies are able to establish themselves on the edges of composite and metallic coupons.

It was also determined that biological fouling is cumulative, which means that once a surface is roughened by some degree of fouling or corrosion, it is more susceptible to farther fouling. If edges and crevices are fouled, fouling of adjacent smooth surfaces is observed. This emphasizes the need to minimize crevice spaces on tidal energy devices to minimize the risk of cumulative fouling.

Biological fouling is seasonal. In particular, the crevices between all coupons and the substrate plates were heavily fouled with sediment and krill after the April–August and August–November, 2009 deployments [18], coinciding with annual period maximum of biological productivity. These observations are likely to change for deployments in different climates, geographic locales, or depth.

10.2.1.2 Performance of Different Materials in Seawater

In Ref. [18], coupons of materials which could be used in the rotor, drive train, or foundation of tidal energy devices were deployed *in situ* on the seabed at a prospective tidal energy site from April 2009 to February 2010 to screen for biofouling and corrosion. The result of this study are supported by previous research made earlier on corrosion and fouling of different metals (such as steel, copper, copper-based alloys, and aluminum) used in constructions of underwater marine devices [19].

The materials tested in Ref. [18] are various metals and polymeric materials that may be used in a tidal energy device:

- Carbon fiber and glass fiber composite: rotor, hub, or duct/shroud
- Aluminum (6061): rotor
- Stainless steel (314 and 316): hub or shroud

- Steel (1018 and 539): support structure
- High density polyethylene (HDPE): bearings
- Fiber reinforced phenolic resin (Feroform T14 from Tenmat): bearings
- Low friction liner on stainless backing (Feroglide 700 from Tenmat): bearings

Coupons made of these materials which might be used in the rotor, drive train, or foundation of tidal energy devices were deployed *in situ* on the seabed in Admiralty Inlet, Puget Sound, Washington. This location has been selected by Snohomish Public Utility District for the deployment on an OpenHydro tidal turbine. Coupons were deployed for 3 months minimum and for up to 10 months. Deployment depths vary from 55 to 75 m and tidal currents close to the seabed approach 3 m/s on strong spring tides. The purpose of this testing was to screen materials for biofouling and corrosion and develop an indication of how tidal energy devices deployed at this location might be affected. Table 10.2 shows the results of testing these materials in 3 and 10 months of exposure.

TABLE 10.2
Performance of Materials Used in Tidal Equipment in Sea Water

No	Material	Changes after 3 Months of Exposure	Changes after 4 Months of Exposure	Changes after 10 Months of Exposure
1	Glass fiber composite	Minimal fouling on surface	Minimal fouling on surface, one barnacle on edge	Minimal fouling on surface, surface color has darkened
2	Carbon fiber composite	Minimal fouling on surface	Minimal fouling on surface	Minimal fouling on surface
3	Aluminum (6061)	>90% surface and edge corrosion	>90% surface and edge corrosion	>95% surface and edge corrosion, coupon edges embrittled
4	Stainless steel (314)	Minor fouling on surface with origin in edge/crevice fouling	Minimal fouling on surface	Minimal fouling on surface
5	Stainless steel (316)	Minimal fouling on surface, one barnacle on edge	Minimal fouling on surface, one barnacle on edge	Superficial corrosion on one edge, concentrated at fiberglass/steel interface
6	Common steel (1018), no protection	Heavy surface oxidation (100% coverage), bioaccumulation within oxidation blisters		
7	Common steel (1018), Zn anode protection	100% surface oxidation, but not as pronounced as unprotected surface		
8	Structural steel (539), no protection	Heavy surface oxidation (85% coverage)		
9	Structural steel (539), Zn anode protection	Corrosion on 15% of surface and edges		
10	Bearing materials HDPE	Minimal fouling on surface and edges		
11	Bearing materials Feroform T14	No visible fouling on surface or edges		
12	Bearing materials Feroglide 700	No visible fouling on surface or edges		

Source: Adapted from Polagye B. and Thomson J. *Screening for Biofouling and Corrosion of Tidal Energy Device Materials: In-situ results for Admiralty Inlet*, Puget Sound, Washington, Northwest National Marine Renewable Energy Center Technical Memorandum University of Washington, Seattle, WA, United States, April 2010, Online. Available: http://depts.washington.edu/nnmrec/docs/20100408_PolagyeB_report_BiofoulingCorrosion.pdf.

Coatings, including high copper antifouling, low-copper antifouling, and inert foul release are also evaluated for their ability to control biological fouling. For smooth surfaces, there is limited biological fouling at this particular site, which is below the photic zone. Stainless steel shows excellent corrosion resistance, while common and structural steels experience major surface oxidation after three months of exposure to the marine environment—even with sacrificial anodes.

The testing in Ref. [19] allowed to define the behavior of different materials in environment leading to corrosion and degradation in tidal energy devices. According to results of the study, it was found that stainless steel, glass fiber composite, and carbon fiber composite did not visually degrade over the test duration (up to 10 months).

Structural steels experience major surface oxidation after 3 months of exposure to the marine environment—even with sacrificial anodes. It was recommended to provide more quantitative work to evaluate corrosion rates and the potential strength degradation of glass and carbon fiber composites over long-term exposure to the marine environment.

The study in Ref. [19] lasted much longer, from 3 up to 60 months. Materials that have been exposed to seawater to determine biofouling and corrosion effects have been divided into three categories: the corrodible metals (carbon steel), the passive metals (aluminum 5086-0), and the toxic film formers (copper, 90–10 Cu–Ni, and 70–30 Cu–Ni). An example of carbon steel corrosion in seawater is shown in Figure 10.9 for different periods of exposure.

FIGURE 10.9 Biofouling and corrosion of carbon steel in seawater: Top (from left to right)—3 month, 9 months, 18 months; Bottom (from left to right)—36 months, 48 months, 60 months. (From Efird K. D. *Materials Performance*, 15(4), April 1976, Online. Available: www.copper.org/applications/marine/cuni/pdf/EfirdFouling.pdf. With permission.)

FIGURE 10.10 Galvanic corrosion in steel structure of tidal turbine. (From Calcagno G. and Moroso, A. The Kobold marine turbine: From the testing model to the full scale prototype. *Tidal* Energy Summit, November 28–29, 2007, London, *Tidal Today*, 2007, Online. Available: http://www.tidaltoday.com/tidal07/presentations/GuidoCalcagnoMoroso.pdf. With permission.)

The passive metals are characterized by the formation of a tightly adherent passivating film on exposure to seawater. These materials include the stainless steel, many Ni-based alloys, and some aluminum alloys (5000 series). The alloys in the third category form film on the surfaces that are toxic to living sea organisms, which prevents biofilm attachment of the fouling organisms, and they are not prone to biofouling.

10.2.2 MARINE CORROSION EFFECTS ON TIDAL POWER EQUIPMENT

Marine corrosion depends on numerous factors such as temperature, galvanic interactions, alloy surface films, biofouling, water chemistry, alloy composition, microbiological organisms, geometry and surface roughness, and other conditions. To design a robust support structure for tidal current turbine, it is important to understand how these factors affect marine corrosion.

The presence of carbon, having a very high electrolytic potential (similar to gold) generates strong galvanic currents in the steel structure of the turbine and thus can cause corrosion problems if not correctly protected [7] (Figure 10.10).

Generally, corrosion in seawater accelerates with increase in temperature, concentration of oxygen, and marine biological activity. Corrosion of metals in seawater is also effected by the turbulent or laminar flow taking off the protective film or increasing migration of deleterious species or by enhancing diffusion. However, increased fluid flow may also help decrease corrosion by removing the aggressive ions that begins to accumulate on metal surface. Generally, cavitation and erosion–corrosion are forms of flow influenced corrosion [20–22].

10.3 TECHNIQUES TO MITIGATE BIOFOULING AND CORROSION OF TIDAL AND WAVE EQUIPMENT

One of the ways to protect the metal surfaces from biofouling is to apply a protective and sealing coating. The sealing elements in tidal and wave equipment have to deal with harsh environment. A protective coating needs to be applied also onto the windings to prevent early outages caused by corrosion. Suitable materials are available, but the influence of such coatings on other properties, such as electrical, magnetic, and thermal properties need to be investigated. The use of marine coatings to reduce corrosion should be further investigated to determine what coatings may provide long-lasting corrosion and biofouling protection for tidal power equipment. More quantitative work is needed to evaluate corrosion rates and the potential strength degradation of glass and carbon fiber composites over long-term exposure to the marine environment.

As was stated in Ref. [20], practically all surfaces are colonized with biofilm sooner or later both on metals and composites. Polymers contain additives, pigments, stabilizers to improve physical and chemical properties, but they may leach out and become nutrients for marine inhabitants. Often biofouling on polymers was reported to be as much as four times higher than on stainless steel. The extent of marine grows depends on number of different factors such as geographic location, season of the year, water chemistry, temperature, substratum type, sunlight, distance from the shore line, and condition of turbulence. All these conditions should be considered when choosing a proper coating onto tidal equipment part to protect them from biofouling and corrosion.

10.3.1 Biofouling Control Products

In general, biofouling control products, originally designed for ship hulls, can be classified into two broad categories [17]: (a) chemically active antifouling (AF) paints and (b) nonstick, fouling release coatings.

10.3.1.1 AF Paints

This category includes technologies based on seawater-reactive binders in which the AF activity is achieved by the controlled-release of active ingredients (i.e., biocides) aided by a controlled surface erosion. On the second category, physical properties of the coating, such as a tailor-made surface energy and elasticity, contribute to a low fouling settlement and its subsequent easy release while sailing.

For biocide-based paints, the specific type of AF paint is determined from the expected trading pattern of the vessel. Low speed and low activity requires a faster polishing paint type to provide the correct amount of leached biocide and high speed and high activity on the other hand requires a low polishing type. The temperature and fouling intensity also have to be taken into consideration before the AF paint is specified. Depending on the type and trading pattern, the total thickness of the AF system typically is in the range from 200 to 400 μm and typically applied in 2–3 turns (for 5 years lifetime).

The service life of these coatings is obviously defined by the coating thickness (i.e., biocide amount) and by the leaching rate of these biocides (also related to the erosion rate of the paint). In completely static structures, such as, for example, buoys, the paint erosion is lower so fouling is likely to happen before the entire biocide load has been released. Once fouled, these structures need to be cleaned by means of harsh mechanical methods such as high-pressure water hosing.

10.3.1.2 Fouling Release Paints

These paints do not contain biocides and their function is that this material is nonstick, which prevents bio-organisms to attach to the surfaces for a relatively long time. A drawback of these biocide-free technologies is that any mechanical damage to the surface renders the coating largely ineffective.

However, on static structures, which are tidal power installations, fouling release coatings are commonly used because of its ease of cleaning. In fact, the coating often is "self-cleans" when the accumulated fouling exceeds a threshold weight after long exposure times. Hence, for long-term immersion of static structures, fouling release coatings will remain cleaner on average than conventional AF solutions. An added interesting feature of these coatings, on top of their remarkable green profile, is that these coatings are significantly smoother than the biocide-based counterparts, which favor measurable lower friction to seawater in the nonfouled condition.

10.3.2 Systematic Approach to Corrosion Protection of Tidal Power Plant

When a large tidal power plant is designed and built, the systematic approach to protecting different components of power generating equipment from corrosion is needed. An example of such approach is presented in Ref. [23] for one of the world's largest tidal power plant in South Korea, consisting of 10×26 MV bulb turbine units, with VA Tech Hydro as supplier of technology and EM equipment.

The means to protect tidal equipment include the system for *cathodic protection*, which makes the potential of the protected components more negative to avoid the dissolution (corrosion) of metal (see Section 12.3.3). The necessary protective current depends on oxygen and salt content of seawater, alkalization of protected surfaces, temperature, and flow velocity. Cathodic protection is very effective in zones of structures that are permanently immersed in water.

In transition and splash zones where the metal is not continuously in contact with seawater, cathodic protection is largely ineffective. Special treatment should be applied to areas which cannot be protected by cathodic protection system and contacted by seawater. These areas are seawater resistant stainless steel material, overlay welding with seawater resistant electrodes, tight welding seam of flange connections, and sacrificial anodes.

Seawater resistant sealing materials are usually chosen to protect these areas from corrosion. Generator interior parts are coated by special coating to avoid any corrosion from salt contaminated air. Steel components are protected with *polyurethane or epoxy resin-based coating* systems that have a life of roughly 15 years. Thing that these systems have in common is that they have to be repaired and partially renewed regularly to achieve their designed lifetime.

Another method of corrosion protection was first tried out successfully in 1949 in the Gulf of Mexico. It involves *sheathing the steel supports with alloys containing nickel and copper*. In the first installations, alloys containing high levels of nickel were used. One of the first large projects in which CuNi 90/10 was used as corrosion protection in tidal and splash zones was in 1984 on the columns of the platforms in the Morecambe Bay Gas Field, a large gas deposit in the Irish Sea.

Regular inspections have found no indication of corrosion on the steel or on the 90/10 copper–nickel cladding. No repairs in the zone protected by the 90/10 copper–nickel were necessary. Mechanical damage to the corrosion protection, such as may occur in service with conventional coatings, were prevented by the robust nature of the copper–nickel plates.

Thanks to the absence of the need for a corrosion coating on the steel columns in the transition and splash zones, and the lack of repairs or maintenance of any kind required in this area, the cladding of offshore load-bearing structures with 90/10 copper–nickel plate has been demonstrated to be a more durable and economic alternative than conventional protection methods [24].

Proper *material selection* plays extremely important role in corrosion protection. For components like runner blades and runner hub, special steel with higher grade of chromium and molybdenum have been chosen. Components usually made of carbon steel like runner cone, gate barrel cones, wicket gates, water passage shield are made of stainless steel, and coolers are made of plated stainless steel.

Coatings and paints are also important as well. For all surfaces made of carbon steel components which are in physical contact with seawater, painting material should be resistant to seawater and reliable for cathodic protection system.

10.3.3 Coatings for Marine Corrosion Control

10.3.3.1 Coatings

The most common method used for the protection of materials in offshore environments is the use of various types of coatings and, for the immersion zone, coatings combined with cathodic protection. If composite materials are used, corrosion is no longer an issue.

However, water molecules can still diffuse into the network of composites to affect the mechanical properties. When moisture diffuses into composites, it can degrade the fiber-matrix interfacial bonding, lower the glass transition temperature, swell, plasticize, hydrolyze, and sometimes microcrack the matrix. The most important standards that apply within the field of offshore coatings are the following:

- NORSOK M501-revision 5
- ISO 12944-1998/2007
- ISO 20340-2009

10.3.3.2   Rules of the Coating Selection

For proper performance of a coating system and thus durable protection under such extreme conditions, it is very important to follow the rules of choosing the correct system.

These rules include [17]

- Determine type and condition of the substrate
- Evaluate environment and possible additional stresses
- Prepare the surface
- Select proper coating systems (generic types, thickness, etc.)
- Apply the coating correctly
- Provide quality control

Substrate: Steel is an important material in offshore structures. There are different grades or qualities of steel, and improper storing (or selection of shop primer) could also modify its properties, for example, incipient pit corrosion. Too many laminations in the steel and/or too much corrosion of steel (grade D according to ISO 8501-1) will make the surface preparation process more difficult and not even the best coating or the highest quality of workmanship can make up for this in the later coating process.

Exposure Environment: The offshore environment is one of the most corrosive naturally occurring environments that can be found on earth. This factor must be considered by avoiding impact and abrasion in the splash zone or in working areas which could eventually accelerate corrosion rates. Other special stresses encountered on offshore structures may be of thermal or chemical nature, for example, from equipment working at high temperatures or areas subject to spillage of chemicals.

Surface Preparation: The surface preparation is the single most important parameter in relation to the performance of any coating system. It is the degree of cleaning (removal of rust, mill scale, oil/grease, soluble salts, etc.) and the roughness (anchor pattern) as well as preparation (rounding and grinding) of sharp edges, welding seams, and weld spatter and other imperfections in the steel work, that are critical in this phase. Paint adheres better to a clean and rough surface and will therefore also last longer.

Coatings: The coatings selected for the job should be of a good quality (based on quality raw materials), and produced according to strict guidelines, ensuring a uniform high quality in each delivery. To document the quality of the coatings and the production procedures, the coating manufacturers should normally be able to present references, third party test results as well as ISO 9000 certification of production facilities. In Europe, anticorrosion coating should be certified according to NORSOK standard M-501, Rev.5, June 2004. The NORSOK standards are a series of standards relevant to offshore installations developed by Norwegian petroleum industry. NORSOK M-501 specifically deals with anticorrosive coating systems and the processes related to their application [25].

Application: The application is also crucial for the process—key parameters here are application equipment (type and condition), microclimate during the application and curing of the coatings and, most importantly, workmanship. It is important that the correct film thickness is applied (within the normal tolerances of a quality paint application). Too thin film will result in premature corrosion either because of pinholes in the film or just because of insufficient thickness, but too thick film either also can result in adverse effects such as solvent retention, reduced adhesion, cracking, and so on.

Due to the variations in film thickness during the application, which cannot be avoided as long as the paint is applied manually, more coats will generally give better protection than, for example, one coat in the same total film thickness. The variations in the application of each single coat are levelled out with the application of the additional coats. Stripe coating on welding seams, edges, corners, and areas that are difficult to reach by the airless spray is mandatory for a later high durability of the coating system.

Quality control: The final parameter and again one of the most important ones is quality control throughout the process. Special focus needs to be put on the surface preparation and coating application processes, where several check points need to be confirmed to ensure the proper final result. Coating inspection is an expertise in itself, and a certified inspector (NACE or FROSIO) should be preferred as the foundation for professional supervision.

10.3.3.3 Coating Systems for Various Environment

The coating systems should be selected with due consideration to the environment as well as the special stresses [17].

For the *atmospheric* zone, this will typically mean a zinc-rich primer followed by epoxy intermediate coats and a UV durable topcoat (e.g., epoxy siloxane), minimum 320 µm *dry film thickness (DFT)* in no less than three coats.

The *splash* zone will often be protected by epoxy or polyester coatings—in a thickness that takes into account the special stresses—normally more than 600 µm DFT in total. For optimum protection against impact damages in such areas, zinc-rich primers are normally avoided while fiber reinforced coatings are recommended as a means of increasing the impact resistance.

Finally, *immersed* areas will be coated with epoxy barrier coatings in a film thickness of no less than 450 µm DFT in minimum two coats.

It is important that the epoxy coating system is compatible with the used cathodic protection system. Anticorrosion coating Ecospeed [25] widely used in Europe is a barrier coating for fouling. It limits mechanical damage and the effects of other chemical and gaseous solutions that may occur in seawater. These properties are provided by a combination of high layer thickness, vinyl-ester properties, and a very high concentration of glass platelets. This coating helps avoiding the effects of corrosion, erosion, and galvanic action as opposed to most other coatings which are water resistant to some degree, however, not water-tight.

10.3.4 Corrosion Protection of Bearings in Tidal and Wave Power Systems

Very important issue is to protect *bearings* that are used in tidal and wave equipment from corrosion [26]. The ambient conditions in water and subsea environments place special requirements on the bearings. In these applications, the water itself acts as the lubricating medium. The rolling bearing rings can be made from special, corrosion-resistant steel, with ceramic rolling elements (balls).

Bearing cages that guide these rolling elements can also be manufactured from special, water-resistant plastics. Nonhermetic, lightweight seals can be fitted to the bearings, which enable water to enter the bearing, but which prevent small particles from reaching the rolling contacts.

Tidal power generation schemes are protected by various sealing, for example, James Walker (JW) products [27]. Such sealing has been designed into, and specified for, a wide range of applications, including horizontal and vertical axis turbines, oscillating and pressure differential turbines and venturi turbines.

The prototype SeaGen turbine with JW sealing was installed in Strangford Lough in May 2008. It is the first tidal current or wave energy system in the world to have exceeded 1000 hours of operation delivering more than 800 MWh into the National Grid. Two 1100 mm diameter Walkersele® lip seals fitted back to back protect the main bearings from seawater ingress and prevent pollution of the environment by stopping lubricant seepage into the sea. These have already worked maintenance-free for over 3 years.

For example, the design of plain bearings by Shaeffler [26] is optimized for slow swivel movements and high forces that are commonly found in wave and tidal energy applications. The bearing is designed specifically for water lubrication and so does not require seals of the type normally found in conventional rolling and plain bearings. As the bearings do not require oil or grease when used in water, there is no risk of lubricant leakage that may contaminate the environment.

The bearings' material comprises two layers that are wound onto each other. The internal sliding layer, which is embedded in a resin matrix with fillers and solid lubricants, is made of continuous synthetic and PTFE fibers. The external layer also comprises continuous glass fibers (glass filament) in epoxy resin. A specific winding angle stabilizes the layers, significantly increasing the strength of the bush. This combination of materials enables plain bearings to resist corrosive media and allows a constantly low friction value with low wear and zero maintenance.

The bearings have a high-radial load carrying capacity, can be used for axial movements, and are insensitive to shocks and edge pressures. Although the bearings are not sealed, if required, they can be supplied with internal seals for additional protection against abrasive contaminants.

10.4   COMPOSITE MATERIALS USE FOR TIDAL AND WAVE EQUIPMENT

Use of composite materials in marine application is not studied well enough yet. Three-year study specifically designed to evaluate the coupling between seawater diffusion and fatigue loading in composites is presented in Refs. [28,29]. Results of this study allow to conclude that the use of reinforcing fibers and matrix resins in marine applications such as tidal and wave equipment should be considered. Manufacturing the important parts of tidal and wave tidal turbine blades may be critical for life extension of this equipment.

The standard *E-glass fibers* are known to be sensitive to stress corrosion in water under prolonged loading, but improved fibers are now available to limit this phenomenon. First, results showing how this improvement transfers to seawater aging and fatigue were presented in Ref. [28]. Two years later, damage mechanisms and fatigue lifetimes in seawater have been characterized in detail, higher performance glass fibers have been tested, and conclusions on material selection are presented in Ref. [29]. Considerations for the use of these materials in a tidal turbine blade and careful material selection is primal in the design of ocean energy devices if costly failures are to be avoided.

Another reason for the standard E-glass fibers being sensitive to water is because of the presence of boron in their composition. Long-term exposure to water in combination with tensile stress results in a stress corrosion phenomenon and premature failure due to crack development at the fiber surface. For this reason, other types of glass fibers such as the Advantex® (a registered trademark of Owens Corning used under license by 3B The Fibreglass Company) and HiPer-tex® (a trademark of 3B The Fibreglass Company) glass fiber grades are to be preferred for ocean energy applications, as they are boron free.

In order to guarantee the long-term durability of composites used for ocean energy structures four conditions must be satisfied. First, the reinforcements (fibers and sizings) must be chosen correctly. Second, an appropriate matrix resin (polyesters, vinylesters, or epoxies) must be selected. Third, the manufacturing process and quality must be suited to the application. Finally, the structure must be designed to resist the loading conditions with an appropriate safety coefficient. These conditions are clearly interdependent, but the designer can use a large available database to help in developing a reliable composite structure [29,30].

10.5 TIDAL, WAVE, AND OTHER SOURCES OF MARINE POWER GLOSSARY

The glossary is adapted from the following online sources:

1. http://tidalpower.co.uk/glossary
2. Wave Energy Prize Glossary of Terms. U.S. Department of Energy
3. Ocean Energy Glossary, IEA Wave Energy Centre, 14 pp., 2007, Online. Available: https://www.ocean-energy-systems.org/library/oes-technical-reports/guidelines/document/ocean-energy-glossary-2007-/

A–B

Absorbed Power: The hydrokinetic to mechanical power conversion. It is product of the dynamic (forces, pressures, torques, etc.) and kinematic (velocities, flows, rotational velocities, etc.) parameters for a hydro-kinetically excited device.

Acoustic pollution: Adverse effects that result from noise generated by a turbine or turbine blades.

Adaptive system control: Control of overall system state typically conducted at longer time scales (not wave by wave), excluding controls of power converting forces (e.g., configuration, orientation, ballasting).

Aquatic: Anything that takes place in water.

Axial flow water turbine: The axial flow turbines are used for low head and relatively high–low rates in hydroelectric plants. Consequently, they are suitable for tidal energy barrages or wave energy converters using overtopping. There are many types of axial flow turbines as tubular, rim, bulb, and so on. The type depends to the arrangement of the electrical generator. The axial flow water turbines could be equipped with adjustable runner blades.

Bandwidth: A bandwidth describes the range of wave frequencies over which a wave energy device responds.

Bathymetry: The measurement of water depth and the shape of seabed—often as shown on a map of the sea or hydrographical chart.

Barrage: A dam used to increase the depth of water or to divert water into a channel for navigation, irrigation, or flood control.

Benthic: Of or relating to the bottom of a body of water. The seafloor is a unique benthic habitat that must be protected during tidal power scheme installation.

Blades: The board, flat (or slightly rounded) arms of some turbines that capture kinetic energy from a fluid and convert it to mechanical energy. The force of moving water turns the blades of a tidal turbine (some turbine designs have vanes instead of blades). Tidal turbine blades can be nearly 100 ft across and turn as slowly as 20–30 rpm. Initial studies are indicating that this relatively slow speed helps underwater marine life maneuver the blades safely.

Bulb turbine: It is a type of axial flow turbine. A type of turbine for use in a tidal barrage. The bulb turbine is derived from Kaplan turbines with the generator contained in a water proofed bulb submerged in the flow. The La Rance tidal plant near St. Malo on the Brittany coast in France uses bulb turbines.

Buoys: An anchored floating device. Traditionally these have served as navigation marks or for mooring but now can be incorporated to wave energy devices. They are typically small compared to the incoming wavelengths, thus are a common form of point absorber.

C–D

Capture width: The power absorbed from the waves by the device in kW divided by the incident wave energy flux per meter crest width in kW/m.

Current: A continuous movement of ocean water in a specific direction. Currents have been likened to rivers within the larger ocean itself. Famous currents include the Gulf Stream (Eastern North America and Western Europe), the Antarctic Circumpolar, the California, The South Equatorial, the Monsoon, and the North Equatorial.

Controllability with fast wave by wave control: Deterministic control of WEC in millisecond time scale for adaptation to instantaneous and predicted observable signals.

Controllability and adaptability with slow sea state by sea state control: Stochastic control of WEC hour time scale for adaptation to sea state.

Directional wave spectrum: A two-dimensional spectrum that shows how energy is distributed between various directions of incidence, in addition to how it is distributed among various frequencies.

Displacer: The part of a wave energy device that moves in response to the waves. Power is usually taken off from the relative motions of the reactor and displacer.

Drag: The retarding force exerted on a body moving relative to a fluid. Drag is usually an energy loss process. It can arise in water movements as friction on wetted surfaces or as vortex shedding from fluid flowing past solid object corners.

Dredging: Excavation underwater to gather up bottom sediment. Dredging is often necessary during building and maintenance of barrages.

Duct: With particular application to tidal stream turbines; a duct is a cowling placed around a turbine to enhance the flow through the rotor. The term duct can also apply to the part of oscillating water columns where the air turbine is placed.

Dynamic tidal power: A type of tidal power in which a long (30 km or more) dam is built perpendicular to the land and jutting out into the ocean. It interferes with tidal flows to create head and thus generate electricity.

E–F–G

Ecosystem: A community of living organisms along with their related physical and chemical environment.

Energy period (Te): Real sea waves can be described as a series of superimposed waves of different periods and amplitudes. The energy period is the period of a monochromatic (single frequency) wave containing the same energy as the real sea state.

Entrainment: The trapping of smaller aquatic life in filters, sluices, and so on.

Estuary: Any location where fresh river water meets salty ocean water. Estuaries are critical habitat for spawning and early life cycle stages of many ocean organisms as well as birds.

Fast tuning: Fast tuning requires changing characteristics of a device to adjust (or ideally to maximize) the energy capture. Fast tuning means adjustments for each wave or loosely over a period of around 1 second for real-sea waves. Also known as wave-by-wave tuning.

Flow: The movement of a fluid in a stream. Tidal turbines are basically similar to wind turbines that can be located nearly anywhere there is strong tidal flow. Tidal turbines are physically much smaller than wind turbines; however, they can extract much more energy for their size by virtue of the higher density of water.

Force: The force of moving water, such as tidal currents, turns the blades of a tidal turbine to generate electricity.

Force flow: The way forces and load penetrate the system.

Francis turbine: A reaction turbine in which water moves through a spiral trajectory to push curved blades connected to the rim or a wheel.

Generator: A device that transforms mechanical energy into electrical energy. Tidal generators are integrated into the tidal turbine structure just as in a wind turbine.

Gigawatt: One billion watts.

Gorlov turbine: A turbine that makes use of a helical blade or foil. It is not directional and can be mounted horizontally or vertically.

H–I–K–L

Head: The highest part of a body of water where water pressure is measured from.

Horizontal axis turbine: A turbine in which the axis is parallel to the ground. A tidal stream turbine mounted such that it rotates about a horizontal axis, typically running parallel with the flow direction.

Hydrofoil: A wing-shaped structure designed to alter the flow of water and react to it.

Hydrokinetic: The energy of moving fluid.

Impulse turbine: A turbine in which rotation is derived by jets of fluid impinging against the blades. There is no pressure difference for the fluid between inflow and outflow of an impulse turbine, but there is a velocity difference.

Kaplan turbine: A reaction turbine with adjustable blades.

Lagoon: A shallow pond that can connect with or communicate with a larger body of water. A tidal lagoon is constructed of a 360 degree barrage in a tidal basin or estuary.

Lunar: Having to do with the moon.

Lunar tide: The part of the tide that results from the gravity of the moon.

M–N–O

Marine: Anything having to do with the sea or seashore.

Mean wave power: Mean power is the average power in a real (polychromatic) sea. It is usually measured in kilowatts or megawatts.

Monochromatic wave: Wave with the same length and period.

Neap tide: A tide that occurs when the difference between high tide and low tide is the least. In other words, the lowest high tide in a month that occurs when the moon is 1/4 or 3/4 full. Neap tides occur when gravitational forces from the sun and moon are at right angles (perpendicular) to one another.

Nearshore: (1) In beach terminology an indefinite zone extending seaward from the shoreline well beyond the breaker or surf zone. (2) The zone which extends from the swash zone to the position marking the start of the offshore zone, typically at water depths of the order of 20 m.

Ocean energy: Ocean energy covers a series of emerging technologies that use the power of waves, ocean currents, tides, ocean thermal energy gradient, and salinity gradient to generate energy.

Ocean thermal energy conversion (OTEC): Ocean thermal energy conversion (OTEC) is an energy technology that converts solar radiation to electric power. OTEC systems use the ocean's natural thermal gradient—the fact that the ocean's layers of water have different temperatures—to drive a power-producing cycle. As long as the temperature between the warm surface water and the cold deep water differs by about 20°C (36°F), an OTEC system can produce a significant amount of power.

Ocean thermal power: Power derived by exploiting differences in water temperature between the surface of the ocean and deeper parts of the ocean.

Osmotic power: Power derived from water flowing between areas of low solute concentration and high solute concentration.

P–R

Pelton wheel: An impulse turbine in which cup-shaped buckets attached to the rim of a wheel are pushed by gets of water moving at high velocity.

Predictability: The electricity that tidal turbines can generate is more predictable than the electricity that wind turbines can generate because tides are much more predictable than wind.

In fact the tides can be accurately predicted for decades into the future. This predictability makes it much easier to integrate tidal energy into the electrical grid.

Pressure retarded osmosis (PRO): It is a salinity gradient energy conversion technique that uses the osmotic pressure difference between seawater and fresh water to pressurize the saline stream, thereby converting the osmotic pressure of seawater into a hydrostatic pressure. Semipermeable membranes are in this process. Other technique is the Reverse Electrodialysis (RED).

Rance river: A river and estuary system in France that is the site of the first commercial-scale tidal barrage power system.

Reaction turbine: A turbine with curved blades in which rotation results from the pressure of a fluid. There is a pressure drop in the fluid between inflow and outflow.

Reverse electrodialysis (RED): Salinity gradient energy conversion technique in which ion selective membranes are used in alternate chambers with freshwater and seawater, where salt ions migrate by natural diffusion through the membranes and create a low voltage direct current. Other technique is the pressure retarded osmosis (PRO).

Rim turbine: It is a type of axial flow turbine. In rim turbines, the generator is mounted on the barrage at right angles to turbine blades (this turbine is used in Annapolis Royal in Nova Scotia).

S

Salinity: The level of salt dissolved in a given quantity of water.

Salinity gradient: Energy can be extracted from the sea where large changes or salinity gradients exist. A semipermeable membrane is placed between the two bodies of water. Slowly the less salty water moves into the salty water by osmosis.

Salinity gradient energy: Energy that can be captured by exploiting the pressure difference at the boundary between freshwater and saltwater.

Saltwater intrusion: The movement of salty ocean water into fresh lake or groundwater.

Scheme: In tidal energy, a scheme is a reference to the specific type of technology used to harness tidal energy. For instance, a tidal barrage scheme.

Seabed: The floor of the sea or the ocean. Tidal turbines can be attached directly to the seabed so they are out of sight on the surface of the water.

Sediment damming: The trapping of sediment, by a dam or barrage that would normally have flowed into the open ocean.

Semipermeable membrane: Also termed a selectively permeable membrane, it is a membrane which retains the salt ions but allows water through. It is used to extract the power from salinity gradient with the pressure retarded osmosis (PRO) process.

Silt: Sand and clay that is captured by moving water and eventually deposited by it.

Sluice: An artificial passage for water that usually has a gate to obstruct flow as desired.

Solar tide: The part of the tide that results from the gravity of the sun.

Spring tide: The tide that occurs when the difference between high and low tides is greatest. In other words, the highest high tide in a month. It occurs when the moon is new or full. Spring tides result when the gravitational forces of the sun and moon are parallel to one another.

Shallow water: (1) Commonly, water of such a depth that surface waves are noticeably affected by bottom topography. It is customary to consider water of depths less than one-half the surface wavelength as shallow water. (2) More strictly, in hydrodynamics with regard to progressive gravity waves, water in which the depth is less than 1/25 the wavelength.

Shoreline: The line along which a large body of water meets the land.

Shoaling: The influence of the seabed on wave behavior. Manifested as a reduction in wave speed, a shortening in wavelength and an increase in wave height.

Significant wave height: The average height of the one-third highest waves of a given wave group or sample. It is usually approximately equal to four times the square root of the zero order moment of wave energy spectrum (see spectral moment).

T

Terawatt: One trillion watts.

Temperature gradient: In the oceans, there can often be found a temperature difference between water near the surface and that deeper down. Where this temperature difference occurs over a relatively short distance it can be used to capture energy using a Rankine cycle.

Tidal barrage: Tidal barrage works in a similar way to that of a hydroelectric scheme, except that the dam is much bigger and spans a river estuary. A hard barrier is placed at a strategic point in an estuary with a high tidal range, thus creating an impoundment upstream of the barrage in conjunction with the banks of the estuary.

Tidal current: The rise and fall of the tides create horizontal movements of water. Usually these are of fairly low velocity, but local topography can greatly magnify them, for example, in the straits between islands.

Tidal energy: The most notable ways to extract electrical energy from them tides are (a) tidal barrages and (b) tidal stream turbines.

Tidal kite: A buoyant hydrofoil is tethered to the ocean floor by a cable and can "fly" or move about in the water as currents and waves push it.

Tidal lagoon: Offshore tidal impoundment, or "tidal lagoon" is a completely artificial impoundment that would be constructed in shallow water areas with a high tidal range. See also tidal barrage.

Tidal mill: A system that converts energy from the tides into mechanical energy to do work.

Tidal phase: The phases of the tides occur as the Earth rotates and the oceans are acted upon by solar (sun) and lunar (moon) gravitational forces. A reference to whether water is at its highest point during a 24 hour cycle (high tide) or at its lowest point (low tide).

Tidal power technologies: It includes tidal range technologies and tidal stream technologies.

Tidal range: The vertical difference between high tide and low tide.

Tidal range resource: The tidal range resource refers to the "gravitational potential energy" that is created as a result of impounding a large volume of water on the high tide. This water is then passed through low-head turbines once a height difference is created on either side of the impoundment, generating electricity. There are two principal concepts for the design and placement of a tidal impoundment: tidal barrage and tidal lagoon.

Tidal stream: The tides are generated by the rotation of the earth within the gravitational fields of the moon and sun. The relative motions of these bodies cause the surface of the oceans to be raised and lowered periodically, producing the bulk movement of water. Where these moving bodies of water meet land masses, channels, or other underwater features they can be enhanced forming a tidal stream.

Tidal stream generator: A tidal power scheme that uses the kinetic energy of water moving with tides to produce energy.

Tidal stream technologies: Tidal stream technologies work by extracting some of the kinetic energy from fast-flowing tidal currents and converting the kinetic energy to mechanical energy before being further converted to typically electricity. To do this they cannot completely block the path of the tidal currents, as otherwise there would be no energy to extract. Instead, they are designed to extract the maximum possible.

Tide: The regular rise and fall in the surface level of the Earth's oceans, seas, and bays caused by the gravitational attraction of the Moon and to a lesser extent of the Sun as well as other astronomical bodies acting upon the rotating Earth. Tides vary greatly by region and are influenced by sea-floor topography, storms, and water currents. There are actually over 100 constituents that govern the tides; the exact same tidal cycle repeats itself only about every

18.5 years. Although the accompanying horizontal movement of the water resulting from the same cause is also sometimes called the tide, it is preferable to designate the latter as tidal current, reserving the name tide for the range of vertical movement.

Total surface area: Total surface area (m^2) at full scale is identified as all structural surface area that is subject to loading and/or is inherent to the production of power. For this prize, only surface areas that define the profile of the device are considered (i.e., it is not the surface area of all components that are needed to physically construct a device, like the underlying girders and stiffeners). Included in this area are structural surface areas below and above the water line when the system is installed with the mooring attached and in still water. Included is the station keeping mechanism. Not included are anchor lines.

Turbidity: The measure of the clarity of water as a result of debris and silt.

Turbine: A device with blades (or vanes), which are turned by a force such as wind, water, or high-pressure steam. The mechanical energy of the spinning turbine is converted into electricity by a generator.

Turbine death: The death of wildlife as a result of direct contact with the blades of a turbine.

U–V

Useful power: The useful power which is delivered by a wave-energy converter. The difference between absorbed-wave power and power that is lost due to dissipative effects, such as friction and viscosity, and so on.

Variable pitch turbine: Wells turbine use symmetrical profile blades with their chords in the plane of rotation. The possibility of the blade being able to change pitch so as to prevent the angle of incidence exceeding some maximum angle has been demonstrated numerically to be more productive than a fixed pitch turbine.

Venturi tube: A simple, hollow structure that is widest at both ends and narrowed in the middle. It accelerates a fluid flowing through it.

Vertical axis turbine: A turbine in which the axis is perpendicular to the ground.

Vertical axis tidal stream turbine: A tidal stream turbine mounted such that it rotates about a vertical axis perpendicular to the flow of water.

Viscous drag: Drag caused by interaction with viscous fluid such as water.

W

Wave energy: Kinetic energy that results from the oscillation of water.

Wave energy converter: A technical device or system designed to convert wave energy to electrical energy or another kind of useful energy.

Wave energy spectrum: A mathematical or graphical description of how a wave state of irregular waves is distributed among the various frequencies.

Water depth: Distance between the seabed and the still water level.

Wave frequency: The inverse of wave period.

Wave farm infrastructure: Non WEC device—for example, interconnectors of device umbilicals, cables to shore, grid connection, and anchoring system.

Wave height: The vertical distance between a wave crest and the previous wave trough.

Wave load: The forces which waves exert on floating, submerged or bottom-standing structures.

Wave period: The time for a wave crest to traverse a distance equal to one wavelength. The time for two successive wave crests to pass a fixed point.

Wave power: Mechanical power from waves, normally expressed in kilowatts per meter of wave crest length.

Wave power device: See wave energy converter.

Wave power plant: Power plant run by wave energy.

Wave-powered generator: Electrical generator run by wave energy.

Wave-rider buoy: A device used to measure wave properties. The buoy rides the waves and estimates the wave positions and directions based on measurements of its own accelerations in different directions.

WEC: Wave Energy Converter.

Wave front: An envisaged plane which is perpendicular to the direction of wave propagation, and which moves with the propagation speed of the wave.

Wells turbine: Air turbine using symmetrical profile blades with their chords in the plane of rotation. This turbine is self-rectifying, that is, its sense of rotation is the same for both of the two opposite air-flow directions. It is usual to equip the OWC wave energy device with such a turbine.

REFERENCES

1. Mehmood N. Tidal current technologies: Green and renewable, *Proceedings of the 4th IEEE International Conference on Computer Science and Information Technology (ICCSIT)*, Vol. 08, Chengdu, Sichuan, China, June 2011, Online. Available: https://www.academia.edu/1943989/Tidal_Current_Technologies_Green_and_Renewable.
2. Grzelak D., Brose N., and Schünemann L. *Ocean Power: Tidal and Wave Energy*, Online. Available: eeeic.eu/proc/papers/124.pdf.
3. Goodall C. Tidal energy – the UK's best kept secret, *The Guardian*, May 2011, Online. Available: http://www.theguardian.com/environment/2011/may/18/tidal-energy-uk-best-secret.
4. Schmitt G. Global needs for knowledge dissemination, research, and development in materials deterioration and corrosion control, *The World Corrosion Organization*, May 2009, 44pp., Online. Available: http://www.corrosion.org/wco_media/whitepaper.pdf.
5. Judendorfer T., Fletcher J., Hassanain N., Mueller M., and Muhr M. Challenges to machine windings used in electrical generators in wave and tidal power plants, *Proceedings of the IEEE Conference on Electrical Insulation and Dielectric Phenomena (CEIDP)*, October 18–21, 2009, Online. Available: https://online.tugraz.at/tug_online/voe_main2.getvolltext?pCurrPk=46153.
6. Pérez Sandra Royo. Thermoelectric energy harvesting for wireless self-powered condition monitoring nodes, *School of Engineering Department of Power and Propulsion*, Cranfield University, May 2013, Online. Available: https://dspace.lib.cranfield.ac.uk/bitstream/1826/8049/1/Sandra_Royo_Perez_Thesis_2012.pdf.
7. Calcagno G. and Moroso, A. The Kobold marine turbine: From the testing model to the full scale prototype. *Tidal* Energy Summit, November 28–29, 2007, London, *Tidal Today*, 2007, Online. Available: http://www.tidaltoday.com/tidal07/presentations/GuidoCalcagnoMoroso.pdf.
8. *Lunar Energy Tidal Power.* Lunar energy Ltd. 2010, Online. Available: http://www.lunarenergy.co.uk/ http://www.reuk.co.uk/Lunar-Energy-Tidal-Power.htm.
9. Kai J. Atlantisstrom seeks finance take innovative design to next level, Atlantisstrom GmbH & CoKG, *Tidal Today*, 2015, Online. Available: www.atlantisstrom.de/files/Atlantisstrom_paper_Draft3.pdf.
10. DeltaStream Tidal Energy Solution, Tidal Energy Limited, Cardiff, Online. Available: http://www.tidalenergyltd.com/cms/wp-content/uploads/downloads/2012/10/DeltaStream_White_Paper_Aug12.pdf
11. The Wave of the Future? *Nickel*, 23(1), 4, December 2007, Online. Available: http://www.nickelinstitute.org/~/media/Files/Magazine/Volume23/Vol23-02Dec2007.ashx#Page=3.
12. *AWS-III Multi-Cell Wave Power Generator*, Online. Available: http://awsocean.com/technology/aws-iii-multi-cell-wave-power-generator/.
13. Kennedy M. Salinity power as renewable energy, *New Atlas Online Magazine*, March 2009, Online. Available: http://newatlas.com/salinity-power-renewable-energy-osmosis/11206/.
14. Mestry D. The Blue Revolution – ocean of energy in "Ocean" (A review), DSpace at VPM (Thane), V.P.M's Polytechnic, 5pp., October 2013, Online. Available: http://dspace.vpmthane.org:8080/jspui/bitstream/123456789/3184/1/The%20Blue%20Revolution%20%E2%80%93%20ocean%20of%20energy%20in%20%E2%80%98Ocean%E2%80%99%20(A%20review).pdf.
15. Langhamer O., Wilhelmsson D., and Engström J. Artificial reef effect and fouling impacts on offshore wave power foundations – A pilot study, *Estuarine, Coastal and Shelf Science*, 82(3), 426–432, April 30, 2009.
16. Stanczak M. Biofouling: It's not just barnacles anymore, *ProQuest*, March 2004, Online. Available: http://www.csa.com/discoveryguides/biofoul/overview.php.

17. Meseguer Y. D., Nyborg R. S., Weinell C., and Thorslund P. L. Marine fouling and corrosion protection for off-shore ocean energy setups, *Proceedings of the 3rd International Conference on Ocean Energy (ECOE)*, Bilbao, Spain, October 6, 2010, Online. Available: http://www.icoe-conference.com/publication/marine_fouling_and_corrosion_protection_for_off_shore_ocean_energy_setups/.

18. Polagye B. and Thomson J. *Screening for Biofouling and Corrosion of Tidal Energy Device Materials: In-situ results for Admiralty Inlet*, Puget Sound, Washington, Northwest National Marine Renewable Energy Center Technical Memorandum University of Washington, Seattle, WA, United States, April 2010, Online. Available: http://depts.washington.edu/nnmrec/docs/20100408_PolagyeB_report_BiofoulingCorrosion.pdf.

19. Efird K. D. The interrelation of corrosion and fouling of metals in seawater, *Materials Performance*, 15(4), April 1976, Online. Available: www.copper.org/applications/marine/cuni/pdf/EfirdFouling.pdf.

20. Turnock S. R., Nicholls-Lee R., Wood R. J. K., and Wharton J. A. *Tidal Turbines that Survive*, Marine Corrosion Forum, Birmingham, UK, July 2009. 39pp. Online. Available: eprints.soton.ac.uk/66727/1/Gold_platedv4.pdf.

21. Mehmood N., Qihu S., Xiaohang W., and Liang Z. Tidal current turbines, *Proceedings of the 3rd International Conference on Mechanical and Electrical Technology (ICMET)*, 3 Volumes; China, August 2011, Online. Available: https://www.academia.edu/1944011/TIDAL_CURRENT_TURBINES.

22. Mehmood N., Liang Z., and Khan J. Diffuser augmented horizontal axis tidal current turbines, *Research Journal of Applied Sciences, Engineering and Technology*, 4(18), 3522–3532, 2012, Online. Available: maxwellsci.com/print/rjaset/v4-3522-3532.pdf.

23. Schneeberger M. *Sihwa Tidal Turbines and Generators for the World's Largest Tidal Power Plant*, Tidal Energy Summit, London, 27pp., November 2007, Online. Available: www.tidaltoday.com/tidal07/presentations/markusschneeberger.pdf.

24. Grolm M. Copper-nickel sheets against corrosion attacks, *Material Views*, November 2013, Online. Available: http://www.materialsviews.com/copper-nickel-sheets-against-corrosion-attacks.

25. Muñoz A. Once in a lifetime coatings-anticorrosion and underwater maintenance solutions, *Proceedings of the 3rd International Conference on Ocean Energy*, Bilbao, Spain, 5pp., October 6, 2010.

26. *Corrosion-Resistant and Media Lubricated Bearings for Water Power Applications*, Schaeffler (UK) Ltd, Sutton Coldfield, Press Release 000-004-654 GB-EN, November 11, 2013, Online. Available: http://www.machinebuilding.net/ta/t0399.htm.

27. Tidal & Wave Energy, James Walker & Co Ltd., *Renewable Energy Support Team*, issue 1, Cheshire CW1 6AY, UK, 12pp., 2011, Online. Available: www.jameswalker.biz/en/pdf_docs/150-tidal-wave-energy-issue-1.

28. Boisseau A. Long term durability of composites for ocean energy conversion systems, PhD thesis, January 2011, Online. Available: archimer.ifremer.fr/doc/00031/14247/11529.pdf.

29. Davies P., Boisseau A., Choqueuse D. et al. Marine composites for ocean energy applications: Ensuring long-term durability, *Proceedings of the 3rd International Conference on Ocean Energy (ICOE)*, 6pp., Bilbao, Spain, October 6, 2010, Online. Available: http://www.icoe-conference.com/library/conference/icoe_2008/.

30. Boisseau A., Davies P., Choqueuse D. et al. Seawater ageing of composites for ocean energy conversion systems, *Applied Composite Materials*, 19(3), 459–473, June 2012, Online. Available: http://iccm-central.org/Proceedings/ICCM17proceedings/Themes/Behaviour/AGING,%20MOIST%20&%20VISCOE%20PROP/F1.2%20Boisseau.pdf.

31. Bearings for "underwater windmills" keep gearboxes operating, *Design World On-Line*, August 15, 2012, Online. Available: http://www.designworldonline.com/bearings-for-underwater-windmills-keep-gearboxes-operating/#_.

32. *Wave Power Technologies Brief 4, International Renewable Energy Agency (IRENA)*, June 2014, Online. Available: http://www.irena.org/documentdownloads/publications/wave-energy_v4_web.pdf.

11 Wind Energy Equipment

11.1 RENEWABLE WIND ENERGY

Like other renewable energy sources, wind energy has many advantages. It reduces greenhouse gas emissions by using turbines, which produce energy and electricity when moved by the wind, and can reduce electricity costs. Today, wind power is considered one of the fastest growing sources of new electricity, globally. There are over 150,000 wind turbines (WTs) operating around the world in over 90 countries.

Wind power has become one of the fastest growing renewable energy sources around the world. According to Global Wind Energy Council [1], in 2014 global wind power capacity expanded to almost 370,000 MW. Yearly wind energy production is growing rapidly and has reached around 4% of worldwide electricity usage.

According to the North American Electric Reliability Corporation (NERC), approximately 260 GW of new renewable nameplate capacity was projected in the United States during 2009–2018. Roughly 96% of this total is estimated to be wind energy. In fact, NERC predicts that wind power alone will account for 18% of the U.S. total resource mix by 2018. The United States ranks second in global wind power, with more than 13 GW of new wind power installed in 2012 to pass the 60 GW milestone for installed wind power capacity. The United States wind base experienced 28% growth in 2012 and wind power was the source of more than 40% of all new U.S. power capacity in 2012. The U.S. Department of Energy has indicated that wind could supply 20% of the U.S. electricity by 2030—requiring some 300 GW of new wind generating capacity.

Abundant, renewable, clean, and cost-effective, wind energy has many advantages. It is actually one of the lowest-priced renewable energy technologies available today, costing between 4 and 6 cents per kilowatt-hour, depending upon the wind resource or project [2].

For 30 years since 1980, WT size increased 200 times, with the largest WT by 2010 producing 20 MW of electrical energy. The progress in developing larger and larger WTs is very fast [3]. Worldwide increase of energy demand and supportive environmental and renewable energy policies have been key factors resulting in the wind power installed capacity growing each year during last two decades for about 30% annually. International organizations are expecting similar figures of growth until the year 2030.

There are some disadvantages to wind energy as well. Like all renewable technologies, land use for implementing new facilities needs to be considered. Many of the power plants that must be constructed for these resources to be utilized like wind energy, are location-specific and the capital costs of construction are extremely high, and sometimes polluting. Also, similar to solar energy and hydropower energy, wind energy is intermittent.

Wind, sun, and rain are not always consistent and are often hard to predict, which cannot always make for a reliable energy source unless the energy can be stored (Energy Informative 2012). Wind energy is one of the lowest-priced renewable energy technologies available today, costing between 4 and 6 cents per kilowatt-hour, depending upon the wind resource and project financing of the particular project.

Wind energy parks are presently installed at many places in the world. In Germany [4], for example, about 20,000 windmills produce about 3% of the German electric energy. Registered turbines in December 2012: ~5000 in Denmark, ~24,000 in Germany, and ~1200 in Sweden. The installation of another 20,000 windmills is planned in the North and Baltic Seas in the next 10 years [5].

The new application of WT is successfully developing on a large scale in Australia [2]. By using WTs to convert saltwater to freshwater instead of just for electrical use, Australia has been able to smooth the peaks and troughs that are characteristic of the wind. Instead of storing excess energy in batteries, plant operators store it in freshwater. When the turbines are not spinning, Australia can draw on this water to provide a consistent supply to homes and businesses. A wind farm consisting of 48 turbines in Perth routinely converts saltwater into freshwater, generating as much as 40 million gallons of drinking water each day. In Sydney, another 63 turbines power the desalination process, accounting for 15% of the city's water supply. This model becomes increasingly significant as the world faces greater water shortages. Climate change, combined with increasing populations, is putting mounting pressure on existing water supplies worldwide. According to the United Nations, 85% of the world population lives in the driest half of the planet. Not surprisingly, many countries are looking for new sources of freshwater. One of the most popular alternatives is to derive freshwater from the ocean. And now, perhaps one of the most renewable ways of generating clean water is via the wind.

11.2 WIND POWER EQUIPMENT

WTs produce a significant amount of energy for a relatively small cost and with very little disruption to the environment. Right now, the industry is virtually exploding in growth, enjoying high public acceptance (with some complaints) and full government support. A typical wind farm consists of anywhere from ten to hundreds of turbines installed in arrays perpendicular to the prevailing wind direction [6]. General structure of WT is shown in Figure 11.1.

The turbines sit on towers up to 300 ft above the ground to take advantage of less turbulent but faster wind with rotor diameters often reaching more than 250 ft (~75 m).

The turbine tower is anchored in a platform of over 1000 tons of concrete and steel rebar that is 30–50 ft (9–15 m) across and 6–30 ft (2–9 m) deep. The tower alone weighs about 140,000 pounds

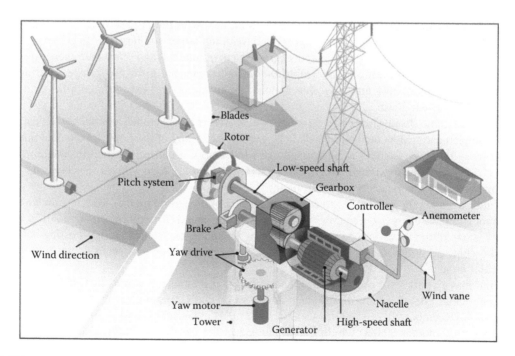

FIGURE 11.1 General construction of wind tower. (Adapted from *The Inside of Wind Turbine*, Online. Available: http://energy.gov/eere/wind/inside-wind-turbine-0.)

tons, and the nacelle weighs about 110,000 pounds. The nacelle sits atop the tower and contains the gearbox, shafts, and generator of a WT. Some nacelles are large enough for a helicopter to land on. Just in case the tower does not have enough weight bearing on it already, some nacelles are equipped with a helicopter landing pad.

WTs consist of a rotor with wing-shaped blades that are attached to a hub. The hub is attached to the nacelle, which houses the gearbox, connecting shafts, support bearings and a generator. Modern WTs are designed to work most efficiently at wind speeds of about 11 m/s (25 mph). Because the wind is often stronger than this, a WT must adapt itself to the prevailing wind speed to operate most efficiently, by automatically adjusting the blade speed or blade angle with a computer-controlled yaw system that turns the turbine and aligns the blades for maximum efficiency [7].

11.3 SUBASSEMBLIES AND COMPONENTS OF WIND TURBINES (WT)

WT is a very complicated machine made of the assemblies/components each of them being very complicated. Four types of systems can be identified in WT: electrical, structural, mechanical, and auxiliary systems.

Electrical systems include the main power converter and control units, but not the electrical generator. The structural systems include the rotor and tower structures; the mechanical systems include the gearbox, the mechanical brake, and the drive train; the auxiliary systems include the yaw system, the hydraulic system, and the various sensors [8]. The WT structural system (Figure 11.2) is designated in a standardized way. In Ref. [9], the components and parts of WT are defined slightly differently (Table 11.1).

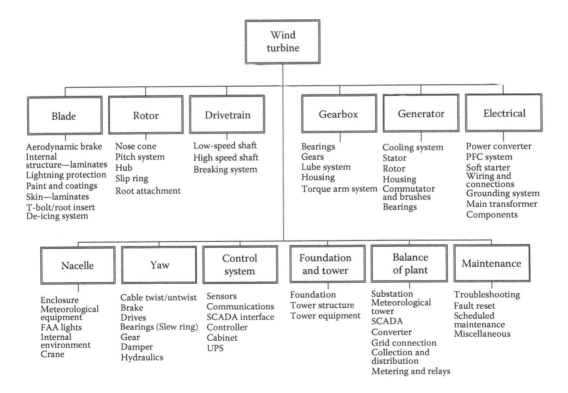

FIGURE 11.2 Subassemblies and components of wind turbine. (Adapted from Hill R. R. et al. *Wind Turbine Reliability Database Update*, Sandia National Laboratories, Sandia Report SAND2009-1171, March 2009, Online. Available: http://windpower.sandia.gov/other/091171.pdf.)

TABLE 11.1

Parts Categories by Wind Turbine System

System	Component Category
Rotor	Blades, pitch bearings, pitch mechanism actuators
Drivetrain	Main bearing, seals, couplings
Generator and cooling	Generator, power converter, cooling system
Brakes and hydraulics	Hydraulics, calipers, shoes
Yaw system	Calipers, wear pads
Control system	CPU, interface modules, sensors
Electrical and grid	Contactors, circuit breakers, relays, capacitors
Miscellaneous	Hardware, other small mechanical, hydraulic, and electrical parts not identified specifically

Source: Adapted from Martin-Tretton M. et al. *Data Collection for Current U.S. Wind Energy Projects: Component Costs, Financing, Operations, and Maintenance,* January 2011 – September 2011, Subcontract Report NREL/SR-5000-52707, January 2012, Online. Available: http://www.nrel.gov/docs/fy12osti/52707.pdf.

The components included in the generic configuration represent the current state of the art for modern turbines currently being supplied. The assumptions for each component category are described in Ref. [9]. General components of WT with horizontal axis are described in more details below.

Rotor: The rotor is three-bladed and each blade has independent pitch. The pitch bearings are the rolling-element type, and are periodically lubricated with grease. The pitch mechanism may be one of two types. *Hydraulic pitch system* includes a pitch cylinder, proportional valve, crank arm mechanism, accumulator, and displacement transducer for each blade. *Electric pitch system* includes a motor with a position encoder, gear reducer, electronic drive, and backup battery bank for each blade. The pump and position controller are common for all three mechanisms (i.e., one per blade).

Drive train: The drive train consists of a main shaft supported by two main bearings, coupled to the gearbox using a hydraulic shrink coupling. A composite tube, with flexure connections, is used to couple the gearbox to the generator.

Gearbox and lubrication: Gears connect the low-speed shaft to the high-speed shaft and increase the rotational speed of the shaft to the speed required by the generator. The gearbox is heavy and power losses from friction are inherent in any gearing system. The gearbox is a combination planetary/helical unit, with an integral lubrication system and fluid cooler system. The gearbox is suspended from the bedplate with elastomeric bushings.

Generator and cooling: A device that produces electricity from mechanical energy, such as from a rotating turbine shaft. The generator is a single-speed, induction type. The variable-speed machine includes a wound rotor and slip rings. Cooling is provided by an integral forced-air system.

Brakes and hydraulics: The brake is a caliper-type system located on the high-speed shaft of the gearbox. A dedicated hydraulic system provides pressure for the calipers. The brake is used only for parking, since the primary rotor brake is the blade pitch system.

Yaw system: The yaw bearing is a sliding-bearing type, with spring-applied calipers for stabilization. The surfaces are periodically lubricated with grease. The yaw drives are electric motor driven in a multiple-reduction gearbox. The number and size of the yaw drives increases with turbine size.

Control system: The control and monitoring system consists of a main controller in the turbine base, a remote controller in the nacelle, and another in the hub. The base unit contains a user interface and display. Control sensors include wind measurement instruments, rotor speed control, and power/grid monitoring transducers. Sensors specific to monitoring component condition are included in that category. The wind speed is measured by anemometer, which transmits wind speed data to the controller. The controller starts up the turbine generator at wind speed of about 3.6–7.2 m/s (8–16 mph) and shuts off the generator at about 29 m/s (65 mph).

Electrical and grid: The turbine switchgear consists of a main breaker/disconnect, a line contactor for the generator power, and smaller contactors and circuit breakers for ancillary systems and power factor correction capacitors. A soft starter is included for connecting to the grid for constant-speed machines. Wind farm is connected to the electricity transmission grid via transformers and substations as any traditional energy generation systems.

Miscellaneous: This category includes a value for miscellaneous parts not specifically identified elsewhere, such as hardware, small hydraulics, and small electrics.

11.4 DESIGN TRENDS OF WT EQUIPMENT

Over the past two decades, WT manufacturers, gear designers, bearing manufacturers, consultants, and lubrication engineers have been working together to improve load prediction, design, fabrication, and operation of WTs. However, despite this, most systems still require significant repair or overhaul well before their intended life is reached.

For design of gearboxes, this collaboration resulted in an internationally recognized WT gearbox design standard (International Organization for Standardization 2005) [10]. For further improvements in design process, the Gearbox Reliability Collaborative (GRC) project was established at National Renewable Energy Laboratory (NREL). In this project, turbine manufacturers, gearbox designers, bearing experts, universities, consultants, national laboratories, and others will jointly investigate issues related to WT gearbox reliability and to share results and findings.

Manufacturers continue to modify and redesign existing turbines and it is difficult to validate the modifications fast enough to prevent multiple units with unsatisfactory "solutions" being deployed. Under current procedures, many years may be needed to develop confidence in a solution. By that time, the industry may have moved to larger turbines or different drivetrain arrangements.

11.5 DEVELOPMENT OF MEDIUM VOLTAGE EQUIPMENT FOR WIND FARM APPLICATIONS

The expected growth of wind power along with the development of bigger WTs have allowed to build wind power plants in such places as in the sea, where requirements on MV equipment are very high. Reliability and safety features are vital in order to guarantee the maximum continuity of supply during operation and personnel's safety while inside the WT tower. As a consequence of this new reality and particular needs, MV switchgear should be specifically designed for this application [11], to operate under much harder nominal voltage, salinity corrosion, and humidity and temperature conditions than those established by IEC Standard 62271-1 [12].

International IEC 62271-1 Standard "High Voltage Switchgear. Part 1: Common specifications" establishes the normal temperature range for operation of this equipment from −5°C to 40°C (23–104°F). Development of wind energy on a worldwide scale and the construction of wind farms in very aggressive climate locations (offshore installations, high altitudes, very low temperatures, etc.) has made necessary the reengineering of MV equipment, which is responsible of operating and protecting the network under these conditions.

Meanwhile in some wind farms located in geographical places like the United States, China, Canada, or Mongolia, the temperature can be as low as −30°C (−22°F) during operation and −40°C (−40°F) while equipment is being stored. Under these conditions, the operation of the driving mechanisms in SF_6-insulated switchgear must be guaranteed and leakage levels from the MV gas tank do not increase.

Another problem arises for electrical equipment due to high corrosive environment where WT is installed. IEC 62271-1 International Standard for indoor switchgear states that "air must not be significantly contaminated with dust, smoke, corrosive/explosive gas, steam or salt." The environmental conditions in any offshore wind farm obviously cannot guarantee the above statement due to the high levels of salinity in the atmosphere. It is necessary to apply special surface treatments for the galvanized steel elements and most important to extend the driving mechanisms life cycle against corrosion to guarantee that it will accomplish the number of operations it was designed for.

The case study conducted in Spain [11] allowed to implement special solutions for the MV switchgear used in environmental conditions of wind farms. A new family of fully insulated MV switchgear was offered based on the use of new surface treatments, materials, new designs to accomplish IAC internal arc fault resistance installation levels inside the WT within different tower and new insulating levels of 40.5 kV. According to the certification through a whole set of type tests, these SF_6-insulated MV switchgear satisfy the international standard IEC 62271-200 [13] and special requirements needed for best performance in wind farms which are operating under very harsh climatic conditions.

11.6   WT STANDARDS, CERTIFICATION, AND CLASSES

11.6.1   Standards and Certifications

While wind power is still developing, there are already a number of formal standards, and guidelines around the world, which WTs must meet [14,15]. Multiple standards for every aspect of WT are developed by the IEC, International Standards Organization (ISO), or ASTM. Depending on the wind energy system technology and component, it may also need to meet American Gear Manufacturers Association (AGMA) standards, as well as the standards developed and approved in different countries, such as Canada, Denmark, Norway, Germany, and the European Union (EU).

11.6.1.1   IEC WT Standards

The IEC 61400 standards family is an internationally recognized standard for WT design and performance. The IEC 61400 standards have been adopted by the ISO and are recognized as formal EU standards for WTs. Aside from the ISO/IEC 61400 standards, there are no other WT specific standards by the ISO. The 61400 is a set of design requirements made to ensure that WTs are appropriately engineered against damage from hazards within the planned lifetime.

The standard concerns most aspects of the turbine life from site conditions before construction, to turbine components being tested. Some of these standards provide technical conditions verifiable by an independent, third party, and as such are necessary in order to make business agreements so WTs can be financed and erected. IEC started standardizing international certification on the subject in 1995, and the first standard appeared in 2001. The common set of standards sometimes replace the various national standards, forming a basis for global certification.

IEC WT Standards:

- IEC 61400-1 lists the design requirements for WTs. It is also possible to use the IEC 61400-1 standard for turbines of less than 200 m² swept area. Created and shared 100-year loads database for 5 MW onshore turbine and 64-year loads database for shallow water offshore turbine. The standard developed new, simpler extreme load extrapolation procedure that reduces ambiguity from extrapolating a short-term distribution to an extreme 50-year value.
- IEC 61400-2 gives the IEC's design requirements for smaller WTs. Small WTs are defined as being of up to 200 m² swept area and a somewhat simplified IEC 61400-2 standard addresses these.
- IEC 61400-3 gives the design requirements for WTs placed offshore.
- IEC 61400-3-2 standard is for design requirements for floating offshore WTs.
- ISO/IEC 61400-4 gives design requirements for WT gearboxes.

- IEC 61400-5 is standard for WT rotor blades.
- IEC 61400-11 describes the methods used to determine the noise level produced by WTs. WTs can create a loud, recurrent noise that disturbs the sleep of nearby residents and has been blamed for triggering recurring headaches in those nearby.
- IEC 61400-12 gives the power-performance requirements for WTs used to produce electricity.
- IEC 61400-13 outlines the procedure for measuring the mechanical loads of WTs such as the stresses the turbine places on the supports.
- IEC 61400-16 is the specification for power transformers used with WTs.
- IEC 61400-21 is the standard for assessing the power quality of WTs connected to the power grid.
- IEC 61400-22 is the standard for conformity testing and certification of WTs.
- IEC 61400-23 is the standard for full-scale structural blade testing.
- IEC 61400-24 states the lightning protection requirements for WTs.
- IEC 61400-25 describes the communication profile used for monitoring and controlling WTs. Subsections of 61400-25 like 61400-25-1 give the logical nodes and data classes used in the communication profile.
- IEC 61400-26 (Part 1) outlines how the time-based availability of WTs is found.
- IEC 61400-26 (Part 2) outlines how the production-based availability of WTs is found.
- IEC 61400-26 (Part 3) outlines how the plantwide-based availability of WTs is found (in development).
- IEC 61400-27 is the standard for electrical simulation models for wind power generation.

IEC 60050-415 standard lists the terms and definitions used in all WT standards.

11.6.1.2   Underwriter's Laboratory (UL) Standards for Wind Power Systems

UL does not have any standards for the construction or operation of wind power turbines. The following standards are applied to small WTs and converters and interconnection systems equipment for WTs:

- UL 6141: Outline of Investigation for Wind Turbine Converters and Interconnection Systems Equipment
- UL 6142: Standard for Safety for Small Wind Turbine Systems (2012). Co-published with AWEA, AWEA 6142

11.6.1.3   ISO Standards for Wind Power Systems

The U.S. NREL participates in IEC standards development work, and tests equipment according to these standards. For U.S. offshore turbines, however, more standards are needed, and the most important are ISO Standards:

- ISO 19900, General requirements for offshore structures
- ISO 19902, Fixed steel offshore structures
- ISO 19903, Fixed concrete offshore structures
- ISO 19904-1, Floating offshore structures—mono-hulls, semisubmersibles and spars
- ISO 19904-2, Floating offshore structures—tension-leg platforms
- ISO 81400-4 was the standard for the gearboxes used in WTs with power ratings over 500 kW. This standard was revised by IEC standard 61400-4.
- API RP 2A-WSD, Recommended practice for planning, designing and constructing fixed offshore steel platforms—working stress design.

11.6.1.4   ASTM Standards for Wind Power

ASTM International has not issued standards specific to WTs. However, a number of existing standards relate to WTs:

- ASTM E1780-12 describes the method of measuring outdoors sound generated by a nearby fixed source like a WT. ASTM E1503-12 outlines the test procedure using a digital statistical sound analysis system.
- ASTM E1779 gives ASTM's recommendations for preparing a measurement plan before measuring outdoor sound levels.
- ASTM E1014-12 gives the recommended steps for recording the outdoor, A-weighted sound level.
- ASTM E1240-88 was the ASTM's method of measuring the performance of WTs. This standard was issued in 1996.
- The ASTM E1240-88 standard was withdrawn in 2001 and was not replaced by another ASTM standard.
- ASTM E1780-12—Standard Guide for Measuring Outdoor Sound Received from a Nearby Fixed Source.

11.6.1.5   AGMA Standards for WTs

An AGMA developed the following Standards:

- AGMA 10FTM13 is a standard for designing gearboxes for WTs in order to minimize the noise produced. AGMA 10FTM13 was written so that gearbox designs would meet the noise limits specified in IEC 61400-11.
- AGMA 10FTM02 is a standard for improving the heat-treating flexibility of WT gears. This standard allows the gears to be carburized or quenched.

11.6.1.6   American Wind Energy Association (AWEA) Standards

AWEA has worked to develop wind industry standards in partnership with the U.S. Department of Energy and other organizations that have a stake in the development of wind energy technology:

- AWEA 9.1-2009 is Small Wind Turbine (SWT) Performance and Safety Standard, which can be used to verify the performance, safety, and durability of small WTs with swept areas of 200 m² or less (approximately 65 kW of power capacity and under).
- ASCE/AWEA RP2011 "Recommended Practice for Compliance of Large Land-based Wind Turbine Support Structures" is the document developed by AWEA together with American Society of Civil Engineers (ASCE), which details prudent recommendations for designs and processes for use as a guide in the design and approval process in order to achieve engineering integrity of WTs in the United States.

11.6.1.7   Other International Standards and Guidelines

Canada:

- Canadian Standard Association (CSA): CAN/CSA-C61400-1-08 Wind Turbines—Part 1: Design Requirements (Adopted IEC 61400-1:2007, third edition, 2005-08, with Canadian deviations)

Germany:

- Germanische Lloyd—GL Rules and Regulations, II Materials and Welding. Part 1 Metallic Materials

- Germanische Lloyd—GL Rules for Classification and Construction, III Offshore Technology Guideline for Certification of Offshore Wind Turbines
- Germanische Lloyd (GL) Guideline for the Certification of Wind Turbines
- TÜV NORD Standard for Certification of Wind Turbines Reference No P20-001, Rev. 0, March 2009

Denmark:

- The Danish Energy Agency's Approval Scheme for Wind Turbine. Recommendations for Technical Approval of Offshore Wind Turbines

Norway:

- DNV-OS-J101 (Det Norske Veritas). Design of Offshore Wind Turbine Structure

Europe:

- EU-Project RECOFF. Contract No. ENK-CT-2000-00322. Recommendations for Design of Offshore Wind Turbines

11.6.2 WT DESIGN CLASSES AND PACKAGES

Design classes of WTs consider multiple environmental conditions, such as long-term wind speed, extreme gust wind speed, turbulence intensity, extreme wind direction change, extreme wind shear, and temperature range. EIC 61400-1 suggests consideration of the following environmental conditions in WT designs: temperature, humidity, air density, solar radiation, rain, hail, snow, and ice. Other conditions are chemically active substances, mechanically active particles, salinity, lightning, and earthquakes. WT classes are presented in Ref. [14].

WTs are designed for specific conditions. During the construction and design phase, assumptions are made about the wind climate that the WTs will be exposed to. Turbine wind class is just one of the factors which need to consider during the complex process of planning a wind power plant. Wind classes determine which turbine is suitable for the normal wind conditions of a particular site.

Turbine classes are determined by three parameters—the average wind speed, extreme 50-year gust, and turbulence (Table 11.2). Wind speed is measured in meters per second (m/s) or in miles per hour (mph).

Turbulence intensity quantifies how much the wind varies typically within 10 min. Because the fatigue loads of a number of major components in a WT are mainly caused by turbulence, the

TABLE 11.2
Wind Turbine Generator (WTG) Wind Classes

Wind Class at Higher Turbulence—18%	Wind Class at Lower Turbulence—16%	Wind Speed[a] (m/s)	Extreme (m/s)	50-year Gust (Mph)
IA high wind		10	70	156
	IB high wind	10	70	156
IIA medium wind		8.5	59.5	133
	IIB medium wind	8.5	59.5	133
IIIA low wind		7.5	52.5	117
	IIIB low wind	7.5	52.5	117
IV	IV	6.0	42.0	94

Source: Adapted from IEC Standard 61400-1, edition 2.

[a] Annual average wind speed at hub height

knowledge of how turbulent a site is of crucial importance. Normally, the wind speed increases with increasing height.

On flat terrain, the wind speed increases logarithmically with height. In complex terrain, the wind profile is not a simple increase and additionally a separation of the flow might occur, leading to heavily increased turbulence. The extreme wind speeds are based on the 3 s average wind speed. Turbulence is measured at 15 m/s wind speed. This is the definition given in IEC Standard 61400-1 edition 2.

For the U.S. waters, several hurricanes have already exceeded wind class IA with speeds above 156 mph (70 m/s), and efforts are being made to provide suitable standards.

International standards for turbine designs (IEC and GL) specify minimum normal ambient temperature of −10°C (14°F) operational and −20°C (−4°F) standstill. However, these specifications are not sufficient for cold climate of northern countries, such as Canada, where operational ambient temperatures are extended to −30°C (−22°F) operational and to −40°C (−40°F) standstill.

There are special cold climate design packages [14], which include special alloyed materials for hub and machine frame, tower steel, main and planetary shafts. Special sealing/insulation for nacelle should be used in cold climates, as well as low-temperature lubricants. Cold weather packages also include heating for nacelle space, yaw drive and pitch motors, gearbox, controller and control cabinets, and other components. Temperatures to which some components must be heated are specified. For example, nacelle temperature should not be lower than −30°C (−22°F), hydraulic oil should not be colder than 10°C (50°F), and gearbox oil not colder than 30°C (86°F).

11.7   NEW TRENDS IN WT TECHNOLOGY

11.7.1   Vertical-Axis Wind Turbines (VAWTs)

The VAWTs tend to come in two main configurations: the *Savonius* and the *Darrieus* turbines [16]. In the Savonius turbine, power is generated using momentum transfer (a drag device) (less than half the Betz limit). Betz's law (or limit), developed in 1919 by the German physicist Albert Betz, states that no turbine can capture more than 59.3% of the kinetic energy in wind. The Savonius is characterized by its high torque, low speed, and low efficiency.

In the Darrieus turbines, power is generated using aerodynamic forces (the lift force on an airfoil). The Darrieus rotor is characterized by its high speed and high efficiency (approaching the Betz limit). The full-Darrieus-rotor VAWT is a high-efficiency design (Figure 11.3).

VAWTs of relatively small size have a direct connection on the generator which simplifies the structure, minimizing the number of components and the weight and favoring the sensitivity even to the light wind. VAWT produce less noise and have minimal environmental impact. They require minimal maintenance and produce electricity continuously.

What would be the "best" configuration for offshore WTs still remains an open question. HAWT designs have a distinct advantage that their designs have continued to advance since development of VAWT technology was halted in mid-1990s. Large multimegawatt land-based turbines have been built and tested, and these turbines are economically viable in today's market.

As for VAWTs, there was the lack of development over the past 15 years; however, they do have significant advantages over HAWTs in offshore applications due to the location of the heavy drivetrain components at or below grade (sea level) in the support platform. Their primary disadvantage remains the longer blade length required by the full-Darrieus VAWT configurations and the lack of a proven, reliable aerodynamic braking system.

11.7.2   Gearless Technology

For decades, WT design has been based on a conventional gearbox architecture. Many turbine manufacturers still use the old gearbox designs and many of today's wind farms are full of gear-driven WTs. However, the need to move away from gearbox-laden designs is well known throughout

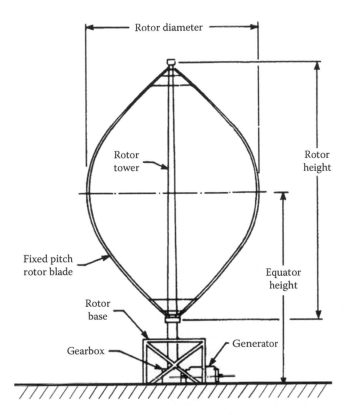

FIGURE 11.3 Vertical-axis wind turbine (VAWT): typical full-Darrieus turbine layout. (Adapted from Sutherland H. J. et al. *Retrospective of VAWT Technology, Sandia Report, SAND2012-0304, January 2012, Prepared by Sandia National Laboratories*, Albuquerque, New Mexico and Livermore, California, Online. Available: http://energy.sandia.gov/wp-content/gallery/uploads/SAND2012-0304.pdf.)

the industry. High failure rates have had far reaching effects and caused concern for many years. Extending back to the late 1990s, the U.S. Department of Energy commissioned a subcontract project (YCX-1-30209-02) to identify, design, and test a megawatt scale drivetrain with the lowest overall life-cycle cost.

The company Northern Power Systems played a major role in the project developing gearless WT design [17]. The NREL commented in Ref. [18] that WT gearboxes have yet to achieve their design life goals of 20 years, with most systems requiring significant repair or overhaul well before the intended life is reached. Since gearboxes are one of the most expensive components of the WT system, the higher than expected failure rates are adding to the cost of wind energy.

Gearless technology is used in the design of the NPS 100 & NPS 2.3 WTs by Northern Power Systems. By eliminating the gearbox architecture and the need for the many subsystems that supports its functions, Northern introduced a vastly simplified and ultrareliable gearless technology that results in better energy capture, quieter operation, and significantly lower operation and maintenance costs, which is an ideal combination for the community wind application.

Northern Power WT were originally designed according to the IEC 61400-1 standard for a 30-year life at an IEC WTGS Class I site (i.e., 10 m/s wind speed for average hub height).

An example of practical use of gearless WT is given in Ref. [19]. Brazil's first commercial wind farm was installed on the sandy seaside pier in Fortaleza, the windy capital of Ceara state in the north east.

The Mucuripe project, which started operating in 1996 with four 300 kW geared Tacke turbines, was a pioneering cooperation contract with the German government, but the climate and conditions were radically different from anything the Germans were used to. Temperatures hit 40°C in summer and fine grains of sands are whipped up from the beaches by constant wind.

In 2002, just 6 years after it was constructed, Mucuripe was totally refurbished by Wobben Windpower, the Brazilian subsidiary of German turbine manufacturer Enercon, doubling the installed capacity to 2.4 MW. The old turbines were dismantled and handed over to a local university for research. These turbines have been replaced with direct-drive E44-600 kW units, which, because they had no gearboxes, are more appropriate to sandy conditions, which lead to the failures of gearboxes. Maintenance of the turbine is easier because of its gearless technology and aluminum nacelle, but the care still needs to be taken to cool the equipment, keep salinity out, and repair abrasion from sand, doing preventive maintenance to the concrete or steel towers.

11.7.3 Offshore Wind and Wave Turbines Together

Offshore wind and wave power can be complimentary. Economies of renewable power production and energy storage are possible when using extended, stable floating platform to share. The Pneumatically Stabilized Platform (PSP) is developed with the U.S. DARPA support. In model scale, it was proved that the PSP may achieve its at-sea motion stability and structural loads mitigation by decoupling the "hull" from ocean wave pressures through the partial use of mobile air buoyancy.

In addition to supporting an array of WTs, the PSP deploys along its seaward edge the "Rho-Cee" wave energy converter (WEC), named the "Impedance-Matched-Terminator"; comprising a nested set of tuned oscillating water column (OWC) absorbers, resonant across a selected frequency band. By means of impedance matching, highly efficient wave energy absorption was achieved [20].

The PSP and Rho-Cee WEC are constructed, modularly, in prestressed reinforced concrete, which is found degradation free in long-term exposure to seawater—and only concrete touches seawater in the platform, WEC, or WT systems. With integral foundations, WTs deployed upon such a floating platform can be positioned offshore in the greater water depths favorable to the WECs; avoiding bottom losses.

Both wind and wave systems are uniquely capable of storing substantial amounts of potential energy in their common supporting PSP structure. This potential energy is embodied in compressed air residing in closed volumes and buoyancy cylinders of the PSP. These volumes and cylinders are charged and tapped by reversible motor-driven Roots blowers that are already part of the platform system. That energy can be tapped during intervals of low wave or wind activity to better match varying demands of the load infrastructure, thereby avoiding the principle objection to renewable energy sources.

11.7.4 Wind Energy Glossary

The glossary is adapted from the following sources: Wind Energy Glossary, Technical Terms and Concepts, West Michigan Wind Assessment, MICHU-10-727, 2010 [21]; Wind Energy Glossary [22], Wind Energy Foundation, and Wind Energy Glossary, Own Energy [23].

11.7.4.1 Wind Turbine Components

A

Airfoil: The cross section profile of the leeward side of a wind generator blade, designed to provide low drag and good lift. Also found on an airplane wing.

Anemometer: Measures the wind speed and transmits wind speed data to the controller.

Angle of attack: The angle of relative air flow to a wind turbine's blade.

Armature: The moving part of an alternator, generator, or motor. In many alternator designs, it carries the magnets and is attached to the blades and hub. Also called a rotor.

Availability factor: The percentage of time that a wind turbine is able to operate and is not out of commission due to maintenance or repairs.

Average capacity (see capacity factor): A measure of a wind turbine's productivity, calculated by the amount of power that a wind turbine produces over a set time period, divided by the amount of power that would have been produced if the turbine had operated at full capacity during that same time interval.

Average wind speed (velocity): The mean wind speed over a specified period of time.

B

Balancing: Adjusting wind turbine blades' weight and weight distribution through two axes so that all blades are the same. Unbalanced blades create damaging vibration.

Beaufort scale: A scale used to classify wind speeds, devised in 1805 by Admiral Francis Beaufort of the British Navy.

Battery: For most wind energy projects, battery systems are cost prohibitive and not considered commercially viable to include as part of a commercial or utility-scale wind farm development project for alternative energy. Also called galvanic battery or voltaic battery.

Betz coefficient—59.3%: This is the theoretical maximum efficiency at which a wind generator can operate, by slowing the wind down. If the wind generator slows the wind down too much, air piles up in front of the blades and is not used for extracting energy.

Blade: The aerodynamic surface that catches the wind. Most commercial turbines have three blades.

Blade passing frequency: The frequency at which the blades of a wind turbine pass the tower.

Blade element-momentum theory (BEM): An aerodynamic theory linking the drag and lift forces experienced by each section of a wind turbine blade to the change in momentum of air passing through the rotor disc.

Braking system: A device to slow a wind turbine's shaft speed down to safe levels electrically or mechanically.

C

Capacity factor: The average power output of a wind development divided by its maximum power capability, its rated capacity. Capacity factor depends on the quality of the wind at the turbine. Higher capacity factors imply more energy generation. On land, capacity factors range from 0.25 (reasonable) to over 0.40 (excellent). Offshore, capacity factors can exceed 0.50.

Chord: The width of a wind turbine blade at a given location along the length.

Commercial-scale wind: Wind energy projects greater than 100 kW in which the electricity is sold rather than used on-site. This category includes large arrays of 100 or more turbines owned by large corporations and a single locally owned wind turbine greater than 100 kW in size.

Community wind: Locally owned, commercial-scale wind projects that optimize local benefits. Locally owned means that one or more members of the local community has a significant direct financial stake in the project other than through land lease payments, tax revenue, or payments in lieu of taxes.

Controller: The controller starts up the turbine generator at wind speed of about 8–16 mph and shuts off the generator at about 65 mph.

Converter: A piece of equipment found within wind turbine generators that converts a direct current (dc) voltage to another dc voltage.

Cut-in speed: The wind speed at which the turbine blades begin to rotate and produce electricity, typically around 10 mph.

Curtailment: The forced shut down of some or all of the wind turbine generators within a wind farm to mitigate issues associated with turbine loading export to the grid or certain planning conditions. Curtailment is controlled by the regional transmission operator.

Cut-out speed: The wind speed at which the turbine automatically stops the blades from turning and rotates out of the wind to avoid damage to the turbine, usually around 55–65 mph.

D–F

Direct drive: A new generation of wind turbines that has recently emerged where the rotor is connected directly on a single shaft to a special high-torque, low speed generator without the use of a gearbox.

Distributed generation: A small-scale power generation technology that provides electric power at a site closer to customers than the central station generation. The term is commonly used to indicate nonutility sources of electricity, including facilities for self-generation.

Downwind turbine: Refers to a horizontal-axis wind turbine in which the hub and blades point away from the wind direction; the opposite of an upwind turbine.

Dump load: A device to which wind generator power flows when the system batteries are too full to accept more power, usually an electric heating element. This diversion is performed by a shunt regulator, and allows a load to be kept on the alternator or generator.

Dynamically scaled wind energy research: The LEC facility is comprised of a water-towing tank, having dimensions of 40 m × 1 m × 1 m and a carriage that tows a model wind turbine through the stationary water.

Freewheeling: A wind generator that is NOT connected to a load is freewheeling, and in danger of self-destruction from over speeding.

Full-scale wind turbine research: The LEC research approach uses an instrumented UAV, incorporating the use of fast response aerodynamic probe technology provides high spatial resolution (~5 m) measurements of wind (speed, direction, and turbulence).

Furling: The act of a wind generator yawing out of the wind, either horizontally or vertically, to protect itself from high wind speeds.

Furling tail: A wind generator protection mechanism where the rotor shaft axis is offset horizontally from the yaw axis, and the tail boom is both offset horizontally and hinged diagonally, thus allowing the tail to fold up and in during high winds. This causes the blades to turn out of the wind, protecting the machine.

G–H

Gearing: Using a mechanical system of gears or belts and pulleys to increase or decrease shaft speed. Power losses from friction are inherent in any gearing system.

Gearbox: A wind turbine generator's protective casing for the system of gears. Gears connect the low-speed shaft to the high-speed shaft and increase the rotational speed of the shaft to the speed required by the generator. The gearbox is heavy and power losses from friction are inherent in any gearing system.

Generator: A device that produces electricity from mechanical energy, such as from a rotating turbine shaft.

Gigawatt: A unit of power equal to 1 billion W, 1 million kW, or 1000 MW.

Governor: A device that regulates the speed of a rotating shaft, either electrically or mechanically.

Guy anchor: A concrete or metal base that secures wind tower guy wires to the Earth.

Guy wire: A strong metal cable or wire that attaches some towers (typically those of small residential wind turbines) to the ground.

High-speed shaft: The shaft within a wind turbine generator that is driven by the rotation of the turbine blades when propelled by wind to convert into wind energy.

Horizontal-axis wind turbine: A wind turbine design in which the shaft (axis of rotation) is parallel to the ground and the blades are perpendicular to the ground.

Hub: The central part of the wind turbine, which supports the turbine blades on the outside and connects to the low-speed rotor shaft inside the nacelle.

Hub height: Measuring from the ground, the tower height of the hub, or central part of a horizontal-axis wind turbine.

Hybrid system: The combination of multiple energy-producing technologies such as Photovoltaic solar electric systems combined with small wind turbine systems.

I–K–L

Independent power producer: An electricity generator that sells electricity but is not owned by a utility.

Installed capacity: The total capacity of electrical generation devices in a system. Learn more about the installed capacity of wind energy in the United States here.

Inverter: A device that converts direct current electricity to alternating current, either for stand-alone systems or to supply power to an electric utility system.

Kilowatt: A standard unit of electrical power, equal to 1000 W.

Kilowatt-hour: A unit or measure of electricity supply or consumption of 1000 W for a period of 1 h.

Leading edge: The surface part of a wind turbine blade on a wind farm that first comes into contact with the wind.

Lift: The force exerted by moving air on asymmetrically shaped wind generator blades perpendicular to the direction of relative movement. Ideally, wind generator blades should produce high lift and low drag.

Low-speed shaft: Low-speed shaft connects the rotor to the gearbox in wind turbines with gearboxes.

M–N

Mean power output (of a wind turbine): The average power output of a wind farm at the mean wind speed of the wind farm.

Mean wind speed: The average wind speed over a time period at a specific height. The mean wind speed is used to determine the average amount of electricity produced over a time period for alternative energy potential.

Median wind speed: The wind speed with 50% probability of occurring.

Megawatt: A standard measure of electric power plant generating capacity equal to 1000 kW or 1 million W.

Met tower: A tower with a group of instruments (including anemometers and wind vanes) attached that collectively measure various meteorological parameters such as wind speed, wind direction, and temperature at various heights above the ground. The term *met* is short for meteorological.

Nacelle: The nacelle sits atop the tower and contains the gearbox, shafts, and generator of a wind turbine. Some nacelles are large enough for a helicopter to land on.

P–R–S

Park effect: The effect whereby wind turbines positioned together in large wind parks each produce less energy than they would if in the same position on their own, due to the wind shadows of the other wind turbines in the park.

Peak wind speed: The maximum instantaneous wind speed (or velocity) that occurs within a specific time period.

Pitch: The angle between the edge of the blade and the plane of the blade's rotation. Blades are turned, or pitched, out of the wind to control the rotor speed.

Power curve: A graphic displaying the instantaneous power output of a specific turbine design at various wind speeds; used with wind resource data to determine the potential for electricity generation at a project site.

Prevailing wind direction: The direction from which the wind predominantly blows as a result of the seasons, high and low pressure zones, the tilt of the Earth on its axis, and the rotation of the Earth.

Rated power: The maximum power output that can be generated by a wind turbine. This is dictated by the generator size and loads that the wind turbine can bear. Choice of rated power for a site is a balance dictated by the amount of energy available in the wind at different wind speeds and the cost of increasingly large and powerful WTs.

Rated wind speed: The wind speed at which the turbine is producing its nameplate-rated power production. For most small wind turbines, this is around 30–35 mph (13.4–15.5 m/s).

Root: The area of a blade nearest to the hub. Generally, the thickest and widest part of the blade.

Rotor: The visible spinning parts of a wind turbine, including the turbine blades and the hub.

Rotor hub: The center of a turbine rotor, which holds the blades in place and attaches to the shaft. The rotor refers to both the turbine blades and the hub.

Shaft: The rotating part in the center of a wind turbine or motor that transfers power. A high-speed shaft drives the generator. A low-speed shaft is turned by a rotor at about 30–60 rpm.

Start-up speed: The wind speed at which a wind turbine rotor starts to rotate. The turbine does not necessarily produce any power until the wind reaches cut-in speed.

Swept area: The area swept out by the wind turbine blades as the rotor rotates—the area of the rotor disc—is also known as the rotor area.

T–U–V

Tail boom: A strut that holds the tail (vane) to the wind generator frame.

Tip: The end of a wind generator blade farthest from the hub.

Tip speed ratio: Ratio of blade tip speed to speed of incoming wind speed.

Thrust bearing: A bearing that is designed to handle axial forces along the centerline of the shaft; in a wind generator, the axial force is the force of the wind pushing back against the blades.

Tower: The base structure that supports and elevates a wind turbine rotor and nacelle.

Turbine: A device for converting the flow of a fluid (air, steam, water, or hot gases) into mechanical motion that can be utilized to produce electricity.

Turbine lifetime: This is the expected total lifetime of the turbine (normally 20 years).

Turbulence: A swirling motion of the atmosphere that interrupts the flow of wind. Wind turbulence has a direct impact on the siting of a wind farm and the layout of turbines across the project footprint.

Twist: In a wind generator blade, the difference in pitch between the blade root and the blade tip. Generally, the twist allows more pitch at the blade root for easier startup and less pitch at the tip for better high-speed performance.

Utility-scale wind: Wind energy projects greater than 100 kW in capacity in which the electricity is sold rather than used on-site. This category includes large arrays of turbines owned by corporations and a single locally owned wind turbine greater than 100 kW in size.

Vane: A large, flat piece of material used to align a wind turbine rotor correctly into the wind. Usually mounted vertically on the tail boom. Sometimes called a tail.

Variable pitch turbine: A type of wind turbine rotor where the attack angle of the blades can be adjusted either automatically or manually.

Variable-speed wind turbines: Wind turbine generators in which the rotor speed increases and decreases with changing wind speeds. Complex power control systems are required on variable-speed turbines to insure that their power maintains a constant frequency compatible with the grid.

Vertical-axis wind turbine: A wind generator design in which the rotating shaft (axis of rotation) is perpendicular to the ground and the cups or blades rotate parallel to the ground.

W–Y

Wind Farm: Wind farm is used in reference to the land, wind turbine generators, electrical equipment, and transmission lines for the purpose of generating wind energy and alternative energy.

Wind monitoring system: An instrument or group of instruments (including anemometers and wind vanes) that collectively measure various meteorological parameters, such as wind speed, wind direction, and temperature at various heights above the ground.

Wind power class: A system designed to rate the quality of the wind resource in an area, based on the average annual wind speed. The scale ranges from 1 to 7 with 1 being the poorest wind energy resources and 7 representing exceptional wind energy resources. Class 3 is typically the minimum required for utility-scale wind development.

Wind power density: A way to define the amount of wind power contained in a given area for use by a wind turbine, measured in watts per square meter.

Wind resource: The wind energy available for use based on historical wind data, topographic features, and other parameters.

Wind resource assessment: The process of characterizing the wind resource and its energy potential for a site of geographical area. Wind resource maps for the United States are available.

Wind rose: A circular plot used to portray certain characteristics about wind speed and direction observed at a monitoring location.

Wind shear: A term and calculation used to describe how wind speed increases with height above the surface of the Earth. The degree of wind shear is a factor of the complexity of the terrain as well as the actual heights measured. Wind shear increases as friction between the wind and the ground becomes greater. Wind shear is not a measure of the wind speed at a site.

Wind speed: The rate at which air particles move through the atmosphere, commonly measured with an anemometer.

Wind speed frequency curve: A curve that indicates the number of hours per year that specific wind speeds occur.

Wind speed profile: A profile of how the wind speed changes at different heights above the surface of the ground or water.

Windmill: A device that uses wind power to mill grain into flour. Informally used as a synonym for wind generator or wind turbine, and to describe machines that pump water with wind power.

Wind turbine: A machine that captures the force of the wind. Also called a wind generator when used to produce electricity. Most commercial wind generators are horizontal axis wind turbines. If wind energy is used directly by machinery, such as for pumping water, cutting lumber, or grinding stones, the machine is called a windmill.

Wind turbine-rated capacity: The amount of wind energy a wind turbine can produce at its rated wind speed.

Wind velocity: The wind speed and direction in an undisturbed flow.

Windmill: A wind energy conversion system that is used to grind grain. However, the word windmill is commonly used to refer to all types of wind energy conversion systems.

Wind power profile: The change in the power available in the wind due to changes in the wind speed or velocity.

Wind vane: Measures wind direction and communicates with the yaw drive to orient the turbine properly with respect to the wind.

Yaw: To rotate around a vertical axis, such a turbine tower. The yaw drive is used to keep a turbine rotor facing into the wind as the wind direction changes.

Yaw drive: Upwind turbines on a wind farm that face into the wind; the yaw drive is used to keep the rotor facing into the wind as the wind direction changes. Downwind turbines do not require a yaw drive, the wind blows, the rotor downwind.

Yaw motor: Powers the yaw drive.

11.7.4.2 Wind Energy Challenges, Issues, and Solutions

A–B

Aggregation: Bundling several wind energy projects together so that they are treated as one larger project (e.g., when purchasing turbines, interconnecting, or maintaining a project) in order to spread out costs over more turbines or projects. This can have the effect of improving project economics.

Alternative energy: Energy that is produced from alternative energy sources such as solar, wind, or nuclear energy that serve as alternative energy forms that produce traditional fossil-fuel sources such as coal, oil, and natural gas.

American Wind Energy Association (AWEA): Formed in 1974, American Wind Energy Association (AWEA) is a Washington, D.C.-based national trade association representing wind power project developers, equipment suppliers, service providers, parts manufacturers, utilities, researchers, and others involved in the wind energy industry.

Bird mortality: Mortality from bird collisions with the turbine blades, towers, power lines, or with other related structures, and electrocution on power lines.

E–F

Electric Reliability Council of Texas (ERCOT): Formed in 1970; one of eight Independent System Operators in North America and the successor to the Texas Interconnected System (TIS). TIS originally formed in 1941 when several power companies banded together to provide their excess generation capacity to serve industrial loads on the Gulf Coast supporting the U.S. World War II effort. ERCOT is one of nine regional electric reliability councils under North American Electric Reliability Corporation (NERC) authority. NERC and the regional reliability councils were formed following the Northeast Blackout of 1965.

Energy payback: The time period it takes for a wind turbine to generate as much energy as is required to produce the turbine, install it, maintain it throughout its lifetime and, finally, scrap it.

Feed-In Tariff (FIT): Feed-In Tariffs, also known as renewable energy payments, are often included in national, state, or local policy to encourage the adoption of renewable energy sources. FITs typically include three key provisions: (1) guaranteed grid access, (2) long-term contracts for the electricity produced, and (3) purchase prices that are methodologically based on the cost of renewable energy generation and tend toward grid parity.

Full-scale Wind Turbine Research: The LEC research approach uses an instrumented UAV, incorporating the use of fast response aerodynamic probe technology provides high spatial resolution (~5 m) measurements of wind (speed, direction, and turbulence).

G

Geographic information system (GIS): A software system which stores and processes data on a geographical or spatial basis. Levelized costs: the present-day average cost per kilowatt-hour produced by the turbine over its entire lifetime, including all costs—(re)investments, operation, and maintenance. Levelized costs are calculated using the discount rate and the turbine lifetime. GIS can be used to evaluate potential sites for win farms and consider a variety of geographic data simultaneously, such as wind conditions, population, and bird migration routes.

Green pricing: A practice utilized by some power providers in which electricity produced from clean, renewable resources is sold at a higher cost than electricity produced from conventional fuels to buyers willing to pay a premium for clean power.

Grid: An electricity transmission and distribution system.

Grid-connected system: A residential electrical system, such as solar panels or wind turbines, which is connected to the electric utility system. The utility system serves as a backup source of electricity if the residential system is not producing power.

Green credit: One way to purchase renewable electric generation. Green credits divide alternative energy generation into two separate products: the commodity energy and the renewable attributes associated with the generation of the commodity energy. The green credit represents the renewable attributes of a single megawatt of renewable energy. Also known as green tags, renewable energy credits, or renewable energy certificates.

Green power: A popular term for alternative energy produced from renewable energy resources such as wind energy.

Green pricing: A practice utilized by some power providers in which electricity produced from clean, renewable resources is sold at a higher cost than electricity produced from conventional fuels to buyers willing to pay a premium for clean power.

H–I

Hybrid system: The combination of multiple energy-producing technologies such as photovoltaic solar electric systems combined with small wind turbine systems.

Installed/Rated/Nameplate capacity: A wind farm's capacity is the most common term used to describe the maximum possible electricity output from the wind farm. Generally, wind farms operate at capacity approximately 30%–40% of the time, depending on the average wind speed and profile of the wind at a specific site.

Interconnection: The process of linking a wind farm to the electric grid for the purpose of distributing electricity to a purchaser of the power produced. Interconnection rules vary by region and require permission from the local utility and regional transmission operator.

Investment tax credit: A tax credit granted for specific investment types, such as investment in wind projects.

Investor-owned utility (IOU): A power provider owned by stockholders or other investors rather than government agencies or cooperatives.

L–N

Levelized costs: the present-day average cost per kWh produced by the turbine over its entire lifetime, including all costs—(re)investments, operation, and maintenance. Levelized costs are calculated using the discount rate and the turbine lifetime.

Life-cycle assessment: An evaluation of the environmental impacts of a given product or service, such as a wind farm, throughout its life cycle, including manufacturing the parts, installation, operation, and disposal.

National Renewable Energy Laboratory (NREL): A U.S. Department of Energy research facility funded to research renewable and alternative energy technologies including solar, biomass, hydro, geothermal, and wind energy.

Negative power prices: Wind energy generators face very small costs of shutting down and starting backup, but they do face another cost when shutting down: loss of the Production Tax Credit and state Renewable Energy Credit revenue which depend upon generator output. It is economically rational for wind power producers to operate as long as the subsidy exceeds their operating costs plus the negative price they have to pay the market. Even if the market value of the power is zero or negative, the subsidies encourage wind energy power producers to keep churning the megawatts out.

Net metering: A term used to describe grid-connected alternative energy generation in which the local electrical source, including wind turbines or solar panels, is connected to the electrical meter so that excess generated electricity passes to the grid and causes the meter to run backward.

O–P

Offshore wind developments: Wind projects installed in shallow waters off the coast. Turbine construction has to be modified to accommodate the depth of the water. A number of countries have built offshore wind developments, but currently there are no offshore wind farms in the United States.

Onshore Wind Developments: Wind farms installed on land. Land-based wind farms are significantly cheaper to build than offshore facilities, but the wind is generally stronger and more offshore.

Participatory planning: A planning process open to high levels of public engagement. The success of a wind farm development is influenced by the nature of the planning and development process and public support tends to increase when the process is open and participatory. Thus, collaborative approaches to decision-making in wind power implementation can be more effective than top-down, imposed decision-making.

R

Rated power: The maximum power output that can be generated by a wind turbine. This is dictated by the generator size and loads that the wind turbine can bear. Choice of rated power for a site is a balance dictated by the amount of energy available in the wind at different wind speeds and the cost of increasingly large and powerful WTs.

Renewable energy: Energy which comes from renewable resources such as sunlight, wind, rain, tides, and geothermal heat, which are naturally replenished. Fossil fuels, such as coal and oil, are considered nonrenewable resources because they are consumed much faster than nature can create them.

Renewable energy certificates: Also known as green tags, certificates representing the environmental attributes of power produced from renewable resources. By separating the environmental attributes from the power, clean power generators are able to sell the electricity they produce to power providers at a competitive market value. The additional revenue generated by the sale of the certificates can be applied to the above-market costs associated with producing power made from renewable energy sources.

Renewable energy credits (RECs): An REC represents the property rights to the environmental, social, and other nonpower qualities of renewable electricity generation. An REC can be sold separately from the electricity associated with a renewable energy generation source.

Renewable Energy Standard: The Renewable Electricity Standard (RES), also known as a Renewable Portfolio Standard (RPS), uses the free market to ensure that an increasing percentage of alternative energy and electricity are produced from renewable sources, like wind energy. The RES provides a predictable, competitive market, within which renewable energy generators compete with each other to lower prices. RES policies currently exist in 28 U.S. states, but not at the national level.

S–U

Shadow flicker: Shadow flicker is the term used to describe what happens when rotating turbine blades come between the viewer and the sun, causing a moving shadow. Shadow flicker is almost never a problem for residences near new wind farms, and in the few cases where it could be, it is easily avoided. For some who have homes close to wind turbines, shadow flicker can occur under certain circumstances and can be disruptive when trying to read or watch television. However, the effect can be precisely calculated to determine whether a flickering shadow will fall on a given location near a wind farm, and how many hours in a year it will do so. Potential problems can be easily identified using these methods, and solutions range from providing an appropriate setback from the turbines to planting trees to disrupt the effect. Normally, shadow flicker should not be a problem in the United States because at U.S. latitudes (except Alaska) the sun's angle is not very low in the sky. If any effect is experienced, it is generally short-lived, as in a few hours over a year's time.

UAV: The UAV is an acronym for unmanned aerial vehicle, an aircraft with no pilot on board. UAVs can be flown remotely by a pilot at a ground control station, or can fly autonomously on preprogrammed flight plans to map technical data over a fixed location. The UAV is often referred to as a drone; however, the term drone actually refers to an unmanned air vehicle that is not generally intended to be reused, such as a vehicle used for target practice or a vehicle that is intended to be crashed.

W

Wind lease: An agreement signed by a landowner that grants a developer the right to use their land for wind development, and in return, provides compensation to the landowner. Typically, the developer owns any turbines that are put up and does all of the work of developing the project. Wind leases are binding legal documents that typically cover 30–60 years or more. These agreements can allow turbines to be constructed on privately owned, actively farmed land.

Wind powering America (WPA): A U.S. Department of Energy initiative designed to promote the use of wind energy across the country, with the goal of quadrupling U.S. wind energy capacity by 2010.

Wind resource assessment: The process of characterizing the wind resource and its wind energy potential for a specific site or geographical area.

Wind power class: A way of quantifying on a scale, the strength of the wind at a project site. The Department of Energy defines the wind class at a site on a scale from 1 to 7 (1 being low and 7 being high) based on average wind speed and power potential to offer guidance about where wind projects might be feasible. Class 3 is typically the minimum required for utility-scale wind development.

Wind turbine Noise: An operating modern wind farm at a distance of 750–1000 ft is no noisier than a kitchen refrigerator or a moderately quiet room. The sound turbines produce is similar to a light whooshing or swishing sound, and is quieter than other types of modern-day equipment. Even in rural or low-density areas, where there is little additional sound to mask that of the wind turbines, the sound of the blowing wind is often louder. Exceptions to quiet operating turbines can occur in two instances: with older turbines from the 1980s and with contemporary turbines in some types of hilly terrain. Modern wind turbines have been designed to drastically reduce the noise of mechanical components so the most audible noise is the sound of the wind interacting with the rotor blade. However, in some hilly terrain where residences are located in sheltered dips or hollows downwind from turbines, turbine sounds may carry further and be more audible. This effect can generally be anticipated and avoided in the development process through adequate setbacks from homes.

REFERENCES

1. Global wind statistics 2014, *Global Wind Energy Council (GWEC)*, February 2015, Online. Available: http://www.gwec.net/wp-content/uploads/2015/02/GWEC_GlobalWindStats2014_FINAL_10.2.2015. pdf.
2. Matsumoto I. and Chrissie L. Catching Waves: A breakthrough in wind power generation, *Clean Energy*, January/February 2014, Online. Available: http://www.nacleanenergy.com/articles/17313/catching-waves-a-breakthrough-in-wind-power-generation.
3. Wood Robert J. K. Tribology and corrosion aspects of wind turbines, *Wind Energy – Challenges for Materials, Mechanics and Surface Science*, IoP, London, 28th October 2010, Online. Available: https://www.iop.org/events/scientific/conferences/y/10/wind/file_45313.pdf.
4. Schmitt G. *Global Needs for Knowledge Dissemination, Research, and Development in Materials Deterioration and Corrosion Control*, The World Corrosion Organization, May 2009, 44 pp. Online. Available: http://www.corrosion.org/images_index/whitepaper.pdf.
5. *Windstats Reports*, Vol. 16–Vol. 25, Q1 2013.
6. Hogg P. Durability of wind turbine materials in offshore environments, *Forth Supergen-Wind Seminar*, June 2010, Online. Available: https://community.dur.ac.uk/supergen.wind/docs/presentations/4th_Seminar_Presentations/PaulHogg.pdf.

7. Van Rensselar J. The elephant in the wind turbine, *Tribology & Lubrication Technology*, 66(6), 38–48, June 2010, Online. Available: http://www.onlinedigitalpublishing.com/publication/index.php?i=37528 &m=&l=&p=3&pre=&ver=html5#{\"page\":2,\"issue_id\":37528}.

8. Avia F. *Summary of IEA RD&D Wind – 65th Topical Expert Meeting*, Online. Available: http://www. ieawind.org/task_11/TopicalExpert/Summary_65.pdf.

9. Martin-Tretton M., Reha M., Drunsic M., and Keim M. *Data Collection for Current U.S. Wind Energy Projects: Component Costs, Financing, Operations, and Maintenance*, January 2011 – September 2011, Subcontract Report NREL/SR-5000-52707, January 2012, Online. Available: http://www.nrel.gov/docs/ fy12osti/52707.pdf.

10. Appleyard D. Wind turbine gearbox study raises reliability hopes, *Renewable Energy World Magazine*, November 2011, Online. Available: http://www.renewableenergyworld.com/rea/news/article/2011/11/ wind-turbine-gearbox-study-raises-reliability-hopes.

11. Novalbos José María Torres, Bartolomé Iñaki Blanco. Evolution and development of medium voltage equipment for special wind farm application, *21st International Conference on Electricity Distribution*, Paper 0077, CIGRE, Frankfurt, Germany, June 2011, p. 4, Online. Available: http://www.cired.net/ publications/cired2011/part1/papers/CIRED2011_0077_final.pdf.

12. *International IEC 62271-1 Standard High Voltage Switchgear. Part 1: Common Specifications*, Online. Available: http://www.highvoltage.org.tw/news_file/08.pdf.

13. *IEC 62271-200 International Standard for Indoor Switchgear*, Online. Available: https://law.resource. org/pub/in/bis/S05/is.iec.62271.200.2003.pdf.

14. Garrad H. Wind energy development in harsh environment. *Subtopic 3: Design and Installation Challenges*, 2010, Online. Available: http://www.gl-garradhassan.com/assets/downloads/Design_and_ Installation_Challenges_in_Harsh_Environments.pdf.

15. Bjørgum A. and Øystein K. O. Corrosion protection of offshore wind turbines, *Wind Power R&D seminar – deep sea offshore wind*, Trondheim, January 2010, Online. Available: http://www.sintef.no/ project/Nowitech/TR%20A6920.pdf.

16. Sutherland H. J., Berg D. E., Ashwill and Thomas D. A. *Retrospective of VAWT Technology, Sandia Report, SAND2012-0304, January 2012, Prepared by Sandia National Laboratories,* Albuquerque, New Mexico and Livermore, California, Online. Available: http://energy.sandia.gov/wp-content/gallery/ uploads/SAND2012-0304.pdf.

17. The gearbox problem, *Northern Power Systems*, July 2009, Online. Available: http://getmore. northernpower.com/downloads/the-gearbox-problem.pdf.

18. Butterfield S. Musial W., Jonkman, J., and Sclavounos P. Engineering challenges for floating offshore wind turbines, Presented at the *2005 Copenhagen Offshore Wind Conference*, Copenhagen, Denmark, October 2005, Online. Available: http://www.nrel.gov/docs/fy07osti/38776.pdf.

19. Spatuzza A. Avoiding excess erosion in a sandy environment, *Wind Power Monthly*, February 2014, Online. Available:http://www.windpowermonthly.com/article/1281868/avoiding-excess-erosion-sandy-environment.

20. Brown N. A. and Martin F. Offshore wave and wind together – Afloat, *Proceedings of the 4th International Conference on Ocean Energy*, Dublin, Ireland, October 2012.

21. Nordman E. Wind energy glossary, technical terms and concepts, *West Michigan Wind Assessment*, MICHU-10-727, 2010, Online. Available: http://scholarworks.gvsu.edu/cgi/viewcontent.cgi?article=10 05&context=bioreports.

22. Wind energy glossary, *Wind Energy Foundation*, Online. Available: http://windenergyfoundation.org/ about-wind-energy/glossary/.

23. Wind energy glossary, *Own Energy*, Online. Available: http://www.ownenergy.net/knowledge-center/ wind-energy.

24. Hill R. R., Peters V. A., Stinebaugh J. A., and Veers P. S. *Wind Turbine Reliability Database Update*, Sandia National Laboratories, Sandia Report SAND2009-1171, March 2009, Online. Available: http:// windpower.sandia.gov/other/091171.pdf.

12 Wind Energy Equipment Corrosion

12.1 ISSUES OF WIND POWER EQUIPMENT CORROSION

Wind energy parks are presently installed at many places in the world. For example, about 20,000 windmills produce about 3% of the German electric energy [1]. The installation of another 20,000 windmills is planned in the North and Baltic Seas in the next 10 years.

Gigantic corrosion problems will arise due to the hostile environment in which offshore windmills have to work. Corrosive seawater attacks fasteners, mast platforms, and anchoring, and there is mechanical impact from wind and waves. Therefore, this equipment is exposed to a very corrosive environment and they become vulnerable to natural processes. Wind turbines are installed onshore and offshore.

Wind power equipment is suffering from corrosion, and types of corrosion affecting wind towers differ for onshore and offshore equipment significantly. The main differences between onshore and offshore locations are the environmental corrosive factors wind equipment is exposed to. Onshore, the wind generating equipment is installed in the windiest places such as tops of mountain ranges, the plains, and so on. Energy generation systems and supporting structures are also installed in other areas known for extreme conditions such as deserts. For onshore equipment, it is generally cyclic dew/condensation with or without minor salinity and moderate exposure to sunlight, resulting in moderate corrosion at weak points and damaged areas of the coating system.

Offshore wind farms are in an even more corrosive environment because of the salt water. For offshore equipment, it is long-term exposure to humidity with high salinity, intensive influence of UV light, wave action and the presence of a splash zone (SZ) area, and high corrosive stress, which all together give rise to dramatic, very fast corrosion. The onshore corrosivity category is evaluated as being about C3, according to ISO Standard 12944. In Europe, it is the Standard DIN EN ISO 12944: Corrosion protection of steel structures by protective paint systems.

Corrosivity category C3 corresponds to urban and industrial atmospheres, coastal areas with low salinity. It is typical location for an *onshore* wind turbine. Corrosivity category C5-Marine corresponds to coastal and offshore areas with high salinity, typical environment for *offshore* wind turbine. This difference in exposure severity is reflected in the mass and thickness loss per unit surface of low-carbon steel and zinc in the first year of their unprotected exposure to corrosive environment in onshore and offshore installations [2] (Table 12.1).

Comprehensive system monitoring, which still has to be developed, will be mandatory to prevent failure caused by corrosion. Otherwise, no company will take the risk of insuring these systems. Failure of corrosion protection systems on offshore windmills will eliminate profit over the expected lifetime of the structure. This is because of the difficulty in scaffolding for repair work. Various systems for condition monitoring of wind power equipment will be presented and discussed in Chapter 18.

12.2 CORROSION OF OFFSHORE WIND TURBINES

Offshore equipment being installed in seawater is exposed to a strongly corrosive environment, including flowing seawater, sea spray, temperature variations, biofouling, and deaerated seawater on inner side. Practically, all parts of offshore wind turbines showed the signs of corrosion: control units, cooling and ventilation systems, main shaft bearings, gearboxes, brakes, tower, stairs, doors,

TABLE 12.1

Effect of Corrosion on Low-Carbon Steel and Zinc in the First Year of Their Unprotected Exposure for Offshore and Onshore Installations

Corrosivity Category	Locations	Thickness Loss (μm)	Mass Loss (g/m^2)
C3 (onshore)	Urban and industrial atmospheres, coastal areas with low salinity	25–50	200–400
C5 (offshore)	Marine, coastal, and offshore areas with high salinity	80–200	650–1500

Source: Adapted from Standard DIN EN ISO 12944.

internal monopole, boat landing, J-tubes, and so on [3,4]. The corrosive stress includes features such as seawater exposure, wet-dry cycles, temperature variations, construction details (joints, bolts, welds), and construction materials (material combinations).

Corrosive stress depends largely on the location of a structure. An offshore wind energy tower, as a sea-based construction, has significant exposure in several zones, including (a) underwater zone (UZ), the area permanently exposed to water; (b) intermediate zone (IZ), the area where the water level changes due to natural or artificial effects, and the combined impact of water and atmosphere increase corrosion; and (c) SZ, the area wetted by wave and spray action, which can cause exceptionally high corrosion stresses, especially with seawater [5–7].

The corrosion process will be also affected by type and mass of dissolved salt, mass of dissolved oxygen, temperature of the seawater, and movement of the seawater. In atmospheric zone (AZ), which is an area above the SZ which includes the turbine tower, steel corrosion rate is about 0.1 mm per year.

The corrosion rate of steel in saline environment can be greater than 2.5 mm per year. It is already known that corrosion rates of steel are highest in the SZ. In AZ, which is an area above the SZ which includes the turbine tower, steel corrosion rate is about 0.1 mm per year.

Table 12.2 summarizes results reported in Refs. [7,8]. The values in Table 12.2 are based on the corrosion of unprotected steel, though the majority of steel parts in modern wind towers are built with protective coating systems over steel, and these systems are intensely studied [8].

Corrosion types include *uniform corrosion, local corrosion, microbiologically influenced corrosion* (MIC), pitting, and others. MIC is initiated and accelerated due to the interaction between construction materials and microbial activity. There are many examples of offshore wind tower corrosion [9–12], some are shown in Figures 12.1 through 12.3.

TABLE 12.2

Corrosion Rates of Steel in Offshore Service in Different Marine Zones

Marine Zones	Steel Corrosion Rates (mm/year)
Atmospheric zone	0.1
Sea mud zone (buried in soil)	0.1
Immersion or underwater zone (UZ)	0.2
Tidal or intermediate zone (IZ)	0.25
Splash zone (SZ)	0.4

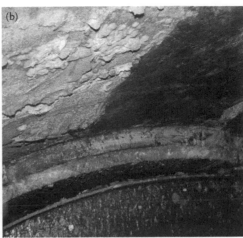

FIGURE 12.1 Corrosion: on offshore wind tower (a) (From http://www.abfad.co.uk/rope-access/renewables/. With permission.); of offshore wind farm monopile foundation (b). (From Låte L. *FORCE Technology: Wind Power Services*, Norway AS, 27 pp., Online. Available: http://windcluster. no/files/2014/12/Leiv-Late-WCN-Forjukstreff-2014-Force-technology.pdf. With permission.)

Among materials a serious problem is found for coatings, which degrade and do not perform properly. Most severe corrosion is found in SZ; however, underwater construction suffers from the presence of seaweed, mussels, and hydroids. This part of the tower experiences *accelerated low-water corrosion* (ALWC), local corrosion under clusters of mussels and under soil [2]. ALWC occurs on steel surfaces below the tidal low water mark to seabed influenced by growth of bacterial colonies (MIC). Example of MIC is shown in Figure 12.4 [4].

Unprotected steel structures in the fully immersed area, and those situated above water in the AZ, corrode at four times the rate of similar structures located inland due to the prolonged wet conditions and high chloride concentrations. Rates are higher in the SZ: up to 10 times the rate for inland structures, with structures also subject to water erosion, winter ice, impact, and abrasion.

FIGURE 12.2 Corrosion at the splash area of a tower base. (Adapted from Sheppard R. E., Puskar F., and Sheppard R. E. *Inspection Guidance for Offshore Wind Turbine Facilities*, Final Report, Energo Engineering, Inc., January 30, 2009, Online. Available: http://www.bsee.gov/Technology-and-Research/Technology-Assessment-Programs/Reports/600-699/627AA/.)

FIGURE 12.3 Corrosion of wind turbine after 2 years in service. (Adapted from Rasmussen S. N. Corrosion protection of offshore wind turbines, *Proceedings Copenhagen Offshore Wind International Conference and Exhibition,* Copenhagen, Denmark, October 26–28 2005, Online. Available: http://wind.nrel.gov/ public/SeaCon/Proceedings/Copenhagen.Offshore.Wind.2005/documents/papers/Critical_design_aspects/ S.Nyborg_RasmussenCorrosion_protectionofoffshorewind_pp.pdf.)

FIGURE 12.4 Welded steel exposed for 1 year in natural seawater in a deaerated environment inside the monopile. (Courtesy of Harald Van der Mijle Meijer, Maritime and Offshore, TNO Norway.)

12.3 CORROSION PROTECTION MEANS FOR WIND POWER EQUIPMENT

There are many components in proper corrosion protection of wind power equipment from corrosion, and they are constructive corrosion protection, coating systems for corrosion protection, and cathodic corrosion protection (CCP).

12.3.1 CONSTRUCTIVE CORROSION PROTECTION

The way to take care of the areas that are at risk of possible corrosion is the use of the design [9]. This constructive corrosion protection should avoid any mistakes by choosing the material and its structural and geometrical design detail. The main constructive corrosion protection is to prevent the direct contact between two different precious metals (mixed construction) by using insulation. With the insulation, there will be no electrochemical reaction between both different metals. Mixed construction should, if possible, be avoided.

Moisture facilitates the formation and propagation of corrosion, therefore, the points at which moisture tends to concentrate such as gaps and sumps, must be avoided as far as possible. Providing sufficient venting possibilities helps to avoid accumulations of condensed water in steel structural elements.

Blowholes and pores caused by the welding should be removed from the surface of the steel structure, which provides good coating condition for the corrosion protection system. Burrs and sharp edges should be rounded off, in order to facilitate the coating work and to increase the durability of the coating. The minimum radius should be 2 mm.

12.3.2 PROTECTIVE COATING SYSTEMS USED BY OFFSHORE WIND ENERGY INDUSTRY

Using of coating system for protection of wind power equipment is the "passive corrosion protection." This is the application of coating systems which have the properties of a barrier and shield the steel from oxidation. It is mostly and essentially used in the atmospheric and the SZ.

Corrosion protection of offshore structures in the United States and Europe is regulated mainly by the following standards:

- DIN EN ISO 12944: Corrosion protection of steel structures by protective paint systems
- ISO 20340: Paints and varnishes—Performance requirements for protective paint systems for offshore and related structures
- NORSOK M 501: Surface preparation and protective coating

Some companies, especially those in the gas and oil industry (Shell, BP, Total, Texaco, etc.), also have their own standards. In addition, further recommendations are mentioned in the company standards for painting hot-dip galvanized and metalized surfaces. According to Ref. [13], since 1991 more than 20 offshore wind projects have been installed and successfully painted in Europe for the following 15 years. The paint systems are most commonly used to protect the various areas of these structures.

A selection of the recommendations of the above mentioned standards for painting carbon steel are summarized in Table 12.3. This table presents the recommendations for using a specific paint system for offshore wind turbines for different environmental zones: *AZ*, and *Immersion and SZs* . In this table, "*zinc rich*" means that the primer contains minimum 80% by mass of zinc dust in the nonvolatile part of the paint as defined in ISO 12944-5:2007. The zinc dust pigment should conform to ISO Standard 3549. Environment Im^2 is a corrosivity category for immersion in brackish or seawater. ISO 12944 defines durability of the paint system; *high (H)* means that desired time period until first maintenance is more than 15 years. Other acronyms used in the table are EP for *Epoxy*

TABLE 12.3

Corrosion Protection for Wind Turbine Parts Exposed to Different Environments: In Marine Atmosphere and in Underwater and Splash Zones

Standard	Atmosphere: Prime Coat	C5-Marine, # of Layers	High DFT (µm)	Under Water Prime Coat	& Splash Zone: # of Layers	Im², High DFT (µm)
DIN EN	EP, PUR	3–5	320	EP (zinc rich)	3–5	540
ISO	EP, PUR	2	500	EP, PUR	2	600
12944	EP, PUR (zinc rich)	4–5	320	EP	4–5	320
ISO	EP, PUR (zinc rich)	Min. 3	>280	EP, PUR (zinc rich)	Min. 3	>450
20340	EP	Min. 3	>350	EP, PUR	Min. 3	>450
				EP	Min. 2	>600
NORSOK	EP, PUR (zinc rich)	Min. 3	>280	EP	Min. 2	Min 350[a]
M 501	EP	Min. 2	>1000			

Source: Adapted from Mülberg, K. *J. Prot. Coatings and Linings*, 21(4), 30–35, 2004.

[a] The coating system shall always be used in combination with cathodic corrosion protection (CCP).

Paint, PUR for *Polyurethane Resin*, DFT for *Dry Film Thickness*. More detail presentation of the recommendations in Table 12.3 is presented in the following Sections.

12.3.2.1 Exterior Atmospheric Corrosion Protection

The majority of steel towers for wind turbines located offshore are metallized and painted on their outer surfaces. Inside the towers, only pure paint systems are typically used.

For the tower lower part, some specifications call for metallization plus paint. Metallization is the thermal spray process of an alloy of 85% zinc with 15% aluminum, for example, Zn/Al, 85/15, with the thickness 60–100 µm. It is followed by *EP* two coats with thickness 100–120 µm (including flash coat). Final step is an application of PUR paint 50–80 µm

However, due to the demands for less time spent on painting, cost reduction, and good experience over years with paint systems, wind turbine structures may be metallized less often. More and more, high-quality paint systems without metalizing at 320 µm DFT for external protection are recommended to be used (according to DIN EN ISO 12944, C5—Marine).

An example of such a paint system is an epoxy zinc dust primer applied at 60 µm DFT; an epoxy midcoat at 200 µm; and a polyurethane topcoat at 60 µm. Inside of steel towers, paint systems at 240 µm DFT (C4) are recommended to be used. For example, epoxy zinc dust primer at 60 µm; EP at 180 µm; and another coating system used: epoxy zinc dust primer at 60 µm; epoxy midcoat at 120 µm; and a polyurethane topcoat at 60 µm [2].

12.3.2.2 Immersion and Splash Zone Corrosion Protection

For many years, the foundation structures of wind turbines have been mainly monopole construction, although jackets, tripods, and triples have also been used for construction.

The following coating systems have most commonly been used on European wind farms for immersion and SZ [2,14]:

> *External (immersion and SZ)*: specialized epoxy coating, two to three coats each at 200–250 µm; polyurethane topcoat at 50–70 µm; specialized epoxy coating, two coats at 500 µm; and polyurethane topcoat 50–70 µm.
> *Permanently immersed areas*: epoxy coating, two coats at 200–250 µm.

Foundation structures insides (not air tight closed): epoxy coating, two coats, each at 200–250 μm are normally used.

The epoxy and specialized epoxy coatings used for wind turbines, depending on the design of the relevant structure, must be compatible with impressed current cathodic protection (ICCP) (see Section 12.3.3).

Comparing the recommendations in Table 12.3 for different environments, one can see that for corrosion protection of offshore turbines, generally the same coating systems and coating materials can be used as for the protection of onshore wind turbines. They differ mostly in film thickness and in number of layers. This statement is valid if only top-quality products on the basis of epoxy and PURs have been used for the protection of wind turbines from corrosion [14].

12.3.2.3 Qualification of Paint System for Offshore Corrosion Protection

Paint systems to be used for offshore corrosion protection should be qualified by external testing according to the following standards [14]:

- EN ISO 12944, Part 6, Corrosivity categories C5-Marine and Im^2 (durability high, >15 years)
- ISO 20340, C5-Marine, Im^2, tidal and splash zone
- NORSOK M 501, coating systems 1 and 7

The tests include the exposure of painted test panels to over 4200 hours of different stresses such as UV light, condensation, salt spray, and freezing. Further tests include immersion in seawater (ISO 2812-2) and cathodic disbanding (ISO 15711:2003, method A).

Though, even a qualified coating system would not guarantee a successful corrosion protection. To make a correct decision for successful corrosion protection, there are other important factors to consider.

12.3.2.4 Factors to Consider for Successful Offshore Corrosion Protection

Before the coating application begins, the steel builder has to provide a structure that is suited to be painted successfully. It is well known that corrosion always starts at edges and weld seams [2]. Therefore, it is necessary to evaluate a design of the structure, of edges, and weld seams. Coatings can protect only accessible surfaces; therefore, it is important to analyze if offshore structure provides a certain access and surface conditions for successful corrosion protection. It is important to examine condition of the steel before surface preparation and exposure of the paint system immediately after completing application.

Offshore structures present some of the toughest corrosion protection challenges for coating systems; therefore, the best possible equipment and set up should always be used for the painting job. Automatic blast facilities should be used to minimize variations in surface cleanliness and surface roughness. Painting booths with climate control and temperature regulation are recommended [2], two-component spray equipment should be used to minimize mixing errors, and qualified people should be hired for the job. The workmanship of the applicator is very important for the long service life of a coating system.

Additional information can be taken from the rules, codes, and the standards of institutions such as NACE [15], STG [16], GL [17,18], NORSOK [19], and ISO [20]. These standards and codes may be used for the design as well.

12.3.3 CATHODIC CORROSION PROTECTION

Although a good quality coating system provides the primary defense against corrosion, there will be defects that occur during installation and general coating degradation over time. Corrosion of steel in salt water exists when two dissimilar metals (such as steel and aluminum) are immersed in electrolyte, which is water with any type of salt or salts dissolved in it. With the presence of electrolyte, there is a conducting metal path between the dissimilar metals.

If these conditions exist, at the more active sites at iron surface a chemical reaction takes place, which produces two iron ions plus four free electrons. The free electrons travel through the metal path to the less active sites. Oxygen gas is converted to oxygen ion by combining with the four free electrons, which in turn combine with water to form hydroxyl ions. Recombination of these ions at the active surface (iron combining with oxygen and water) produce the iron-corrosion product ferrous hydroxide Fe $(OH)_2$.

This reaction is more commonly described [21] as "current flow through the water from the anode (more active site) to the cathode (less active site)."

Corrosion mitigation can be achieved by utilizing a combination of a good-quality coating and a cathodic protection system designed to protect the submerged and buried steelwork for the lifetime of the installation, which is typically up to 25 years and beyond.

There are two types of cathodic protection systems available: impressed current cathodic protection (ICCP) systems and galvanic (sacrificial) cathodic protection systems. CCP for the submerged and the buried zone is cathodic protection called "active corrosion protection" [13].

Cathodic protection prevents corrosion by converting all of the anodic (active) sites on the metal surface to cathodic (passive) sites by supplying electrical current (or free electrons) from an alternate source. Usually, this takes the form of *galvanic anodes*, which are more active than steel. This practice is also referred to as a sacrificial system, since the galvanic anodes sacrifice themselves to protect the structural steel or pipeline from corrosion.

In the case of *aluminum anodes*, the reaction at the aluminum surface is producing 4 aluminum ions plus 12 free electrons. At the steel surface, oxygen gas is converted to oxygen ions, which combine with water to form hydroxyl ions OH^-. As long as the current (free electrons) arrives at the cathode (steel) faster than oxygen is arriving, no corrosion will occur.

12.3.3.1 Impressed Current Cathodic Protection (ICCP)

There are many different system developed to provide CCP for submerged steel structures. One of them called *ICCP* system consists of a transformer rectifier unit, junction boxes, cables, a number of anodes, and reference electrodes.

ICCP units are designed to suit the particular application. Housed in a stainless steel enclosure, the unit can be fitted in the turbine switch room; various input voltage options including single or three phase supply can be accommodated. The unit can be supplied with remote monitoring and controlled via the Internet or interfaced with an existing SCADA system [22].

The result will be an electrochemical protection. The first way to reach this protection is to impress external voltage or the second way is to use a galvanic (or sacrificial) anode. Sacrificial anodes for offshore applications are predominantly made of either aluminum or zinc. The existence of seawater is substantial because of its function as electrolyte. The value of the potential against the reference electrode (Cu/CuSO$_4$) should be between −850 and −1100 mV. To be able to ensure sufficient protection, the structure has to be polarized enough.

It is essential to apply cathodic protection ("active corrosion protection") in the UZ and in the buried zone. A coating system ("passive corrosion protection") is efficient in reducing the current demand for cathodic protection. The cathodic protection system must be compatible with the coating system that is used, that is, their application must not lead to an impairment of functionality and quality of the coating.

12.3.3.2 Galvanic Anode Cathodic Protection System

A galvanic, or sacrificial, cathodic protection system utilizes the galvanic effect of coupling metals of different types with differing electrical potentials. The more electronegative metal will be anodic with respect to the more electropositive metal.

Typically alloys of aluminum are used to protect steel structures in a marine environment. Aluminum Anode Alloy 778S specifically meets the rigorous demands of the offshore industry [22]. The anode alloy was introduced following extensive trials and independent verification of the alloy's long-term electrochemical performance, making it very suitable for wind farm applications. The 778S alloy anodes are suitable for use in a wide range of environments, water depths, and temperatures.

Used in conjunction with a good quality coating system, galvanic anode cathodic protection systems offer a cost effective maintenance-free solution to corrosion mitigation of offshore wind farm structures.

12.3.3.3   Use of Aluminum in Construction of Offshore Wind Turbines

Aluminum is a viable solution for using in construction of offshore wind turbines together with steel [23]. The parts which could be made of Al (such as stiffeners in nacelle), would weigh 40%–50% less than steel stiffeners. Al could be used in other parts, such as electrical cabinets, floor sheets, steps and ladders, transformers and cabling.

There are many advantages in using Al, since it is incombustible and does not require additional lightning protection. Since Al has thermal conductivity 3–4 times less than that of steel and 100 times less than plastic, it provides better heat dissipation through the outer skin of nacelle and, therefore, allows smaller cooler sizes, it is also extensively used in heat exchangers [23].

Due to natural oxide layer, an aluminum surface does not corrode and does not need to be painted for surface protection. All Al alloys are compatible with seawater proof stainless steel (with Mo), and Al alloys are couples with unalloyed steel or stainless steel without Mo, the cathode (steel) should be coated to avoid galvanic corrosion. To minimize galvanic corrosion rate, it is necessary to have the cathodic (steel) surface Mach smaller compared with Al surface in coupling.

12.4   PROTECTION OF WIND TURBINE ROTOR BLADES FROM ENVIRONMENTAL AND OPERATIONAL IMPACT

The rotor blades are fundamental, essential components of a wind turbine and they are optimized for maximum efficiency. The blades of typical modern turbines are between 40 and 90 m long. The largest and most modern blades are made from bonded glass and carbon fiber mats into which epoxy resin is injected in vacuum.

Composite materials have become the wind industry standard. The blades are built according to the sandwich construction principle and are stabilized with reinforcing spars and bars on the inside. This high-tech construction technique also provides for exceptional stability and flexibility.

Rotor blades' surfaces are exposed to different harsh environment and could be damaged by exposure to sand and rain, and the coatings should provide a reliable and long-lasting protection of rotor blades from sand and rain erosion [24]. Example of heavily eroded turbine blade is shown in Figure 12.5.

The finish of rotor blades consists of multilayer polyurethane-based coats, with different erosion and UV resistance depending on requirements [25]. The coatings should provide an excellent adhesion property to minimize the risk of stress cracking. In addition, they should provide excellent flexibility to prevent them from flaking off despite rotor tip vibrations that cause them to bend by several meters. The coatings should provide lasting protection for operating times of up to 20 years.

12.5   CODES AND STANDARDS FOR WIND POWER EQUIPMENT CORROSION PROTECTION, COATINGS AND PAINTING

The following codes and standards are used worldwide to control and protect wind power equipment from damaging effects of corrosion [26]:

FIGURE 12.5 Examples of erosion of leading edge turbine blades (a and b). (From Kanaby G. *Blade Repair Challenges, Wind Energy Operations & Maintenance Summit, Wind Energy Update*, April 25–27, 2012, Dallas, TX Online. Available: http://www.mfgwind.com/sites/default/files/images/pdfs/WES_blade-repair-challenges.pdf. With permission.)

- NACE RP 01-94. Recommended Practices Cathodic Design.
- DNV RP-B401. Cathodic Protection Design.
- NEN-EN 12495 CP for fixed steel offshore structures.
- Shell Expro ES/115, Corrosion Protection of Fixed Steel Structures: Offshore Installations.
- DNV-OS-C40.1. Fabrication and testing of offshore structures.
- NORSOK M501: Surface preparation and protective coating.
- ISO 8501-1. Preparation of steel substrates before application of paint and related products, visual assessment of surface cleanliness.
- ISO 8502-6. Preparation of steel substrates before application of paints and related products—Tests for the assessment of surface cleanliness—Part 6: Extraction of soluble contaminants for analysis—The Bresle method.
- ISO-8502-9. Preparation of steel substrates before application of paints and related products—Tests for the assessment of surface cleanliness—Part 9: Field method for the conductometric determination of water-soluble salts.
- ISO 11127-6. Preparation of steel substrates before application of paints and related products—Test methods for non-metallic blast-cleaning abrasives—Part 6: Determination of water-soluble contaminants by conductivity measurement.
- SSPC-SP10. Near-White Blast Cleaning—Surface Preparation Standards of Steel Structures Paining Counsel—complies with NACE No. 2.
- ISO 8503-1. Preparation of steel substrates before application of paint and related products, surface roughness characteristics of blast cleaned steel substrates.
- BN 11b Rugotest No. 3. Comparison standard for blasted surfaces. The method complies with ASTM D 4417/A.
- ISO 2813. Paints and varnishes—Determination of specular gloss of non-metallic paint films at 20°, 60°, and 85°.
- ASTDM D523 Standard Test Method for Specular Gloss.
- ISO 4624. Paints and varnishes—Pull-off test for adhesion.
- ISO 12944. Corrosion protection of steel structures by protective paint.
- ISO 2063. Thermal spraying–Metallic and other inorganic coatings—Zinc, aluminum and their alloys.

- ISO 4828. Paint and varnishes—Evaluation of degradation of coatings—Design quantity and size of defects and of intensity of uniform changes in appearance.
- ISO 20340. Paints and varnishes—Performance requirements for protective paint systems for offshore and related structures.
- NACE TM 0204, 0304 and 0404 (first editions in 2004).

REFERENCES

1. Schmitt G. *Global Needs for Knowledge Dissemination, Research, and Development in Materials Deterioration and Corrosion Control*, The World Corrosion Organization, 44 pp., May 2009, Online. Available: www.corrosion.org/images_index/whitepaper.pdf.
2. Mülberg K. Corrosion protection of offshore wind turbines, Hempel protective, *J. Prot. Coatings and Linings* 21(4), 30–35, 2004, Online. Available: http://www.hempel.co.uk/en-gb/protective/~/media/300 CF7728E9C41CF804F56FF8A0456D4.pdf.
3. Hogg P. Durability of wind turbine materials in offshore environments, *Forth Supergen-Wind Seminar*, June 2010, Online. Available: http://www.supergen-wind.org.uk/docs/presentations/4th_Seminar_Presentations/PaulHogg.pdf.
4. Van der Mijle Meijer H. Corrosion in offshore wind energy "a major issue," *Essential Innovations*, Den Helder, The Netherland, February 12–13, 2009, Online. Available: http://www.we-at-sea.org/wp-content/uploads/2009/02/3-Harald-vd-Mijle-Meijer.pdf.
5. ISO 20340. Performance requirements for protective paint systems for offshore and related structures, *International Organization for Standardization*, Genève, Switzerland, 2005.
6. ISO 12944-2. Paints and varnishes—Corrosion protection of steel structures by protective paint systems—Part 2: Classification of environments, *International Organization for Standardization*, Genève, Switzerland, 1989.
7. Ault J. P. The use of coatings for corrosion control on offshore oil structures, *Journal of Protective Coatings and Linings* 23(4), 42–47, 2006, Online. Available: http://www.elzly.com/docs/The_Use_of_Coatings_for_Corrosion_Control_on_Offshore_Oil_Structures.pdf.
8. Momber A. W., Plagemann P., and Schneider M. Investigating corrosion protection of offshore wind towers part 1: Background and test program, *Journal of Protective Coatings and Linings* April 2008, Online. Available: www.vliz.be/imisdocs/publications/237912.pdf.
9. High-durability systems for the protection of offshore wind farms, PPG Protective and Marine Coatings, *Wind Energy Network*, Online. Available: http://www.windenergynetwork.co.uk/enhanced-entries/ppg-protective-marine-coatings/.
10. Låte L. *FORCE Technology: Wind Power Services*, Norway AS, 27 pp., Online. Available: http://windcluster.no/files/2014/12/Leiv-Late-WCN-Forjukstreff-2014-Force-technology.pdf.
11. Sheppard R. E., Puskar F., and Sheppard R. E. *Inspection Guidance for Offshore Wind Turbine Facilities*, Final Report, Energo Engineering, Inc., January 30, 2009, Online. Available: http://www.bsee.gov/Technology-and-Research/Technology-Assessment-Programs/Reports/600-699/627AA/.
12. Rasmussen S. N. Corrosion protection of offshore wind turbines, *Proceedings Copenhagen Offshore Wind International Conference and Exhibition*, Copenhagen, Denmark, October 26–28, 2005, Online. Available: http://wind.nrel.gov/public/SeaCon/Proceedings/Copenhagen.Offshore.Wind.2005/documents/papers/Critical_design_aspects/S.Nyborg_RasmussenCorrosion_protectionofoffshorewind_pp.pdf.
13. Lossin M. *Corrosion Protection for Offshore Wind Turbines*, 5 pp., Online. Available: http://www.risoe.dk/vea/recoff/documents/sec_5/recoffdoc034.pdf.
14. Mülberg K. *Corrosion Protection for Windmills Onshore and Offshore*, 12 pp., Online. Available: http://www.galvinfo.com/Thermal_Spraying/CaseStudy/windmills-1.pdf.
15. NACE Standard RP0176-2003, Corrosion Control of Steel Fixed Offshore Structures Associated with Petroleum Production, 2003.
16. STG 2215:1998 Richtlinien Fuer Korrosionsschutz Fuer Schiffe Und Seebauwerke (STG Directive no. 2215 "Corrosion Protection for Ships and Marine"), 1998.
17. Germanischer Lloyd Rules and Regulations IV– Non-Marine Technology Part 2. Regulation for the Certification of Offshore Wind Energy Conversion Systems, Part 10, 1999.
18. Germanischer Lloyd VI. Additional Rules and Guidelines, Part 10. Corrosion Protection, Section 2. Guidelines for Corrosion Protection and Coating Systems, 2010, Online. Available: http://www.gl-group.com/infoServices/rules/pdfs/gl_vi-10-2_e.pdf.

19. NORSOK Standard M-501, *Surface Preparation and Protective Coating*, Rev. 4, 1999.
20. ISO/CD 19902, *Draft, Petroleum and Natural Gas Industries – Fixed Steel Offshore Structures*, 2001.
21. Baxter R. and Britton J. *Offshore Cathodic Protection 101: What It Is and How It Works*, Online. Available: http://www.cathodicprotection101.com/.
22. *Cathodic Protection for Offshore Wind Farms*, Corrpro Companies Europe Limited, Online. Available: http://corrpro.com/Resources/~/media/Corporate/Files/Corrpro%20Literature/5-29-Lit/Corrpro_Wind%20Farms.ashx.
23. Jupp S. Aluminum is a viable solution for offshore wind turbines, *Deep Sea Offshore Wind R&D Seminar*, January 2011, Online. Available: https://www.sintef.no/globalassets/project/nowitech/wind-presentations-2011/a1/jupp-simon_hydro.pdf.
24. Kanaby G. *Blade Repair Challenges, Wind Energy Operations & Maintenance Summit, Wind Energy Update*, April 25–27, 2012, Dallas, TX, Online. Available: http://www.mfgwind.com/sites/default/files/images/pdfs/WES_blade-repair-challenges.pdf.
25. *RELEST® Wind Coating Systems for Windenergy*, BASF, RELIUS COATINGS GmbH & Co. KG Donnerschweer Str. 372, 26123 Oldenburg/Germany, Online. Available: http://www.windenergy.basf.com/group/corporate/wind-energy/en/function/conversions:/publishdownload/content/microsites/wind-energy/downloads/BASF_Coatings_Wind_Brochure.pdf.
26. de Jong M. P. Adaptations to a marine climate, salt and water OWEZ_R_111_20101020. *Results of Corrosion Inspections Offshore Wind Farm Egmond aan Zee*, 2007–2009, KEMA Nederland B. V., 50863231-TOS/NRI 10-2242, Arnhem, The Netherlands, October 20, 2010, Online. Available: http://www.noordzewind.nl/wp-content/uploads/2012/02/OWEZ_R_111_20101020_corrosion_2009_R01.pdf.

13 Wind Turbine Gearboxes and Bearings

13.1 GEARBOX PROBLEMS IN WIND POWER TURBINES

Gearboxes in WTs are located in nacelle between rotor and generator (main drive). It is a heavy part of gondola (nacelle); the mass is up to 60,000 kg or 60 tons. Bearings are used in various places of the nacelle: rotor shaft, gearbox (step-up gear), generator, yaw gearbox (reduction), yaw slewing table, blade pitch revolving seat, and hydraulic pump. Schematic locations of gearboxes and bearings in nacelle are shown in Figure 13.1.

From the very first days of the WT industry, high gearbox failure rates have been observed. While damage and failure rates of WT gearboxes may vary according to the source of published data, a benchmark has been reported by the EU-funded Reliawind study, which is around 6% per year. That study also quoted publicly available data on the failure rate and downtime from large reliability surveys performed by Landwirtschaftskammer in Schleswig-Holstein, Germany, and the Wissenschaftliches Mess- und Evaluierungsprogramm (WMEP) of the Fraunhofer Institute in Kassel, Germany, and noted failure rates of around 10%. While these figures may be considered low when compared with other causes of breakdowns in the field, such as electrical systems, the consequence of that failure can be heavy. Damage and failures in mechanical drive trains can result in high costs due to long downtimes [1].

Since gearboxes are extremely expensive, their high failure rates are resulting in a higher cost for wind energy, exasperated by price escalation due to the uncertainty of gearbox life expectancy.

Multimegawatt WT gearboxes operate under demanding environmental conditions, including considerable variation in temperature, wind speed, and air quality. As WTs increase in size, their operating conditions become ever more extreme. Wide variations in speed and load, changing wind direction, climatic extremes, and limited space inside nacelles conspire to reduce WT reliability and increase maintenance costs. The primary problem of WTs is the gearbox, followed by blades and generators. These gearboxes experience several types of repairable damage, including *micropitting* or *gray staining, abrasive wear, foreign object debris* (FOD) *damage, surface corrosion, and fretting corrosion* [2].

Gears fail for several reasons. WT gears operate under extreme environmental conditions including highly variable temperature, wind speeds, and air quality. These conditions cause variable high loading and torque. During periods of low or no wind, the loading on slowly moving or stationary gears is intensified. Moreover, moisture can contaminate the lubricant and condense on the gear surfaces forming sludge, corrosion, and micropitting. Finally, dust and other foreign debris in the air can contaminate the lubricant during maintenance, leading to abrasive wear. Examples of different kind of damage are described in Refs. [2–4] and shown in Figures 13.2 through 13.4.

Operation and maintenance (O&M) costs of WTs often start escalating in the fourth year of operation, with gearboxes being the single largest issue. While WTs have design lives of 20–30 years, gearbox warranties are often as short as 2 years. The most problematic of all gearbox issues is bearing failures, a challenge expected to increase as turbines get larger. A recent NREL study concluded that the majority of WT gearbox failures start in the bearings [5].

Gearbox reliability problems are being addressed with advanced designs, better materials, and with another important strategy: oil contamination control. Next-generation contamination control technologies, already proven in the industrial plants and mobile equipment, can significantly reduce particle and water contamination problems.

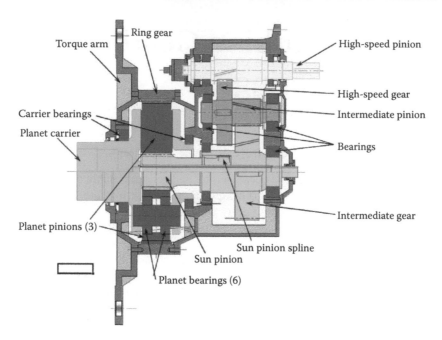

FIGURE 13.1 Schematic locations of gearboxes and bearings in nacelle. (Adapted from Milburn A. Wind turbine gearbox wear and failure modes and detection methods, *NREL—Wind Turbine Condition Monitoring Workshop*, September 2011.)

13.2 PARTICLE CONTAMINATION OF GEARBOX OIL

13.2.1 INTERNAL CONTAMINATION

Contaminants in gearbox oil may be generated internally. These particles are usually wear debris from gears, bearings, splines, or other components resulting from micropitting, macropitting, adhesion, abrasion, or fretting corrosion wear modes.

FIGURE 13.2 Bearing race macropitting. (Adapted from Milburn A. Wind turbine gearbox wear and failure modes and detection methods, *NREL—Wind Turbine Condition Monitoring Workshop*, Broomfield, CO, September 2011.)

FIGURE 13.3 Bearing roller scuffing. (Adapted from Milburn A. Wind turbine gearbox wear and failure modes and detection methods, *NREL—Wind Turbine Condition Monitoring Workshop*, September 19–21, 2011.)

FIGURE 13.4 Gear tooth micropitting. (Adapted from Milburn A. Wind turbine gearbox wear and failure modes and detection methods, *NREL—Wind Turbine Condition Monitoring Workshop*, September 2011.)

Micropitting: Although the term micropitting accurately describes the appearance and mechanism of the problem, it is sometimes also referred to as fatigue scoring, flecking, frosting, glazing, gray staining, microspalling, peeling, and superficial spalling. Micropitting is widespread in WT gearboxes and is detrimental because it reduces gear accuracy, may cause gears to be noisy, and may escalate into other failure modes such as macropitting, scuffing, or bending fatigue. Micropitting contaminates the lubricant with large amounts of iron particles. Particle contamination of gearboxes includes cutting away material from component surfaces, which results in loss of clearance and high friction in gear. Chemically reactive surfaces of fresh metallic wear debris lead to oil thickening, acidity, and formation of fouling gums. All together result in accelerated oil oxidation and abrasive wear. It is likely that the large amount of steel wear debris in gearboxes is the major problem in WTs.

Fretting corrosion: When the bearing shaft interface (inner ring bore—shaft seat) or the bearing housing interface (outer ring outside surface—housing seat) are subjected to micromovements under varying loading conditions, the native oxide on the steel surfaces can be removed. Furthermore, surface asperities can corrode and are torn off. These particles become trapped in the contact, and if oxidizing agents such as moisture are also present, further corrosion happens at the surface. Under load, the trapped air and moisture corrode the surface further and can lead sometimes to further particle generation. These particles will act as grinding paste resulting in further loss in interference and increased ring creep, or in the worst case (if corrosion particles remain trapped at the seat) to ring through cracking by increased and too high local stress.

Fretting corrosion also can occur when a WT is rotating. It occurs on components such as splines or blade pitch bearings that are subjected to small-amplitude, vibratory motion. Splines are especially vulnerable to fretting corrosion because, unlike gears, whose rotation entrains lubricant between mating gear teeth, splines have small sliding motion with essentially no motion to entrain lubricant between mating spline teeth. Blade pitch bearings are susceptible to fretting wear if the blades remain at one pitch angle for too long, and the lubricant is not replenished by movement of the rolling elements.

13.2.2 Contamination Ingested from Environment

WTs in desert environments are exposed to airborne dust during the hot season and moisture during the rainy season. Solid particle contaminants vary in hardness, friability, and ductility, depending on the composition of the particle [6].

Common contaminants like rust and black iron oxides have a Mohs hardness rating of 5–6 (on a scale of 1–10, with 10 being the hardest). Environmental dust, mostly silica, has a hardness rating of 2–8. Quarry dust has a rating of 5–9. From the manufacturing process, tool steel has a hardness rating of 6–7. Silicon carbide and aluminum oxide have a hardness rating of 9. Diamond tops the scale with a hardness rating of 10.

Contamination can enter gearboxes during manufacturing, be internally generated, ingested through breathers and seals, and inadvertently added during maintenance. All of these sources must be addressed to limit the impact that contamination can have on components.

To keep gear oil clean, contaminants should be reduced to negligible amounts. Many gearbox manufacturers and bearing experts believe that to eliminate the negative effects of particle contamination oil, cleanliness levels should be maintained at ISO 14/12/10 max, which could bring gearboxes closer to a more than 20-year design life [4]. ISO 14/12/10 is a *code based on ISO 4406*, shorthand notation for particle counts, where the first number is a code for the number of particles greater than 4 µm/mL of fluid, not the actual number; the second number is the count of particles >6 µm; and the last number the count of particles >14 µm.

13.3 WATER CONTAMINATION OF GEARBOXES

Small amounts of water are usually found in lubricating oils. Offshore turbines are constantly exposed to moisture, which may penetrate into gearboxes. The maximum amount of water that can dissolve in any specific oil is called 100% saturation, depends on base stock and additives. For many gear oils, this value is in the range of 400–600 ppm.

Small amounts of either free and/or dissolved water contamination lead to corrosion and to accelerated abrasive and fatigue wear. Water also reacts with the liquid component of gearboxes, the lubricant. It quickens oil oxidation, generating overly viscous lubes, acids, and premature oil replacement. Contaminant water decomposes ester base fluids and additives and free water can break emulsions of finely divided antiwear particle additives, leading to massive additive dumping accompanied by simultaneous fouling of sensors, flow controllers, and filters. Free water results in corrosion, additive dropout, microbial growth, and oil oxidation and surface-initiated fatigue [5].

Corrosion results in pitting, leakage, weakening and breaking of parts, and the release of abrasive particles into the oil such as iron oxide, better known as rust. The damage by moisture or standstill corrosion on a raceway is shown in Figure 13.5 [1].

Additive dropout is a result of certain additives having strong affinity for water. These particles are congregating in and around water droplets. Water also can break colloidal suspensions of finely divided powders sometimes used as antiwear additives, resulting in dumping of massive amounts of

FIGURE 13.5 Damage caused by moisture or standstill corrosion on a bearing raceway. (From Stadler K. and Dvorak P. How black oxide coated bearings can make an impact on cutting wind turbine O&M costs, *Wind Power Engineering*, September 2014. With permission.)

material. Not only are these additives inactivated by water but additive dropout can also completely foul components, taking them out of action.

Microbial growth develops when common strains of bacteria and molds multiply and thrive in oil, providing free water is present and temperatures of oil ranging from 15°C to 52°C (from 60°F to 125°F). Results include accumulation of acids promoting corrosion since free water is present and the formation of biological slimes that foul flow passages and moving parts. Microbial growth in oils is also associated with fetid odors, asthma, and skin allergies.

Oil oxidation is also associated with dissolved water. In the study of Ref. [7], it was found that water accelerates oil oxidation as rapidly as copper. When water is present along with either copper or iron, rates of oil oxidation increased up to 120 times.

Foaming: A potentially damaging problem is foaming, where the high speeds and loads that are consistent with WTs can cause severe churning, pushing air into the oil. That means, the lubricant does not pump or circulate which reduces its effectiveness.

Surface-initiated fatigue is another aging process that shortens service life of gearboxes [5]. During this process, cracks start at the surface and spread underneath the steel. Gear oil carries dissolved water to the tip of these cracks, where the metal is highly reactive. Water disassociates into its component chemical elements, hydrogen and oxygen gas. The oxygen reacts with either the lubricant or the metal surfaces, but it is the hydrogen that plays the major role. Hydrogen's small size allows it to diffuse through grain boundaries and into the metal, where it weakens the steel by *hydrogen embrittlement*. This accelerates the propagation of cracks through the bearing steel and shortens the time to spall formation and component failure.

In summary, gearbox failure modes from water contamination include hydrogen-induced fractures, corrosion, oxidation, lubricant additive depletion, oil flow restrictions, aeration/foaming, impaired lubricant film strength, microbial contamination, and water washing. To minimize water contamination when manufacturing and shipping the WT, the manufacturing environment must be dry and clean, with components sealed to the atmosphere during shipping [8]. However, WTs can still be affected by water contamination during service [9].

A list of all the detrimental effects to gearbox lubricant from water contamination can be found in the ANSI/AGMA/AWEA Standard 6006-A03 [10], which was developed for the specification of WT gearbox design. This standard contains maintenance specifications based upon an expected gearbox failure lifetime of 20 years. However, a number of investigations have demonstrated that repair and/or replacement of gearboxes is required over significantly smaller time intervals than that which is expected. The additional cost of providing unexpected maintenance and/or repair to WT systems significantly affects the cost of wind energy as a whole [11].

13.4 TECHNIQUES TO MINIMIZE GEARBOXES CONTAMINATION

Depending on the source of contamination, different techniques are recommended to minimize gearboxes contamination and to avoid a premature failure of WTs. Micropitting and smearing are life-limiting wear modes of critical bearing positions in WT gearboxes. These wear modes are caused by roller/raceway sliding in thin-film lubricant conditions.

Whereas bearings experiencing low load conditions at high speeds and rapid accelerations are more at risk of smearing, bearings operating for extended periods of time at low loads, slower speeds, and in boundary layer lubrication are more at risk of micropitting. Micropitting on the bearing has led to peeling which is the light gray band in the raceway [12], shown in Figure 13.6. The peeling has compromised the raceway profile creating areas of high contact stress at the edges of the roller/raceway contact. The bearing failed from fatigue spalls that initiated in these areas of concentrated stresses.

13.4.1 INTERNALLY GENERATED CONTAMINATION

To minimize internally generated wear debris, it is important to use accurate and smooth surfaces, surface-hardened gears and splines, and high-viscosity lubricants. External and internal spline teeth should be nitrided and force-lubricated to prevent fretting corrosion. Annulus gears should be carburized or nitrided rather than through hardened because through-hardened gears are relatively soft and prone to generating wear debris [6].

Important design considerations are material selection and heat treatment choices that determine material hardness. Filters cannot remove particles once they become embedded in softer gearbox components. Hard particles embedded in through-hardened annulus gears can cause polishing (fine-scale abrasive wear) on mating planet gears. This degrades gear accuracy and increases the amount of wear debris. Nonmetallic bearing cages should not be used because they are susceptible to particle embedment that can result in severe abrasive wear on rollers.

FIGURE 13.6 Micropitting on the bearing. (Adapted from Gary L. Doll. Tribological challenges in wind turbine technology, *Wind Turbine Tribology Seminar NREL-Argonne-DoE*, Broomfield, CO, November 2011.)

WTs should not be parked for extended periods. Otherwise, fretting corrosion may occur on gear teeth, splines, and bearings. In addition, wet clutches and brakes should have separate lubrication systems to avoid contaminating the gearbox with their wear debris.

13.4.2 MINIMIZING INGRESSED CONTAMINATION

Breathers are used to vent internal pressure when air enters through seals or when air within the gearbox expands and contracts during normal heating and cooling. The breather should have a filter and desiccant to prevent ingress of dust and water. It should be located in a clean, nonpressurized area away from contaminants such as brake dust and water. In especially harsh environments, the gearbox should be completely sealed and have an expansion chamber with a flexible diaphragm to accommodate pressure variation [6].

Most WT gearboxes have labyrinth seals that provide long life and adequately seal in oil, but may allow contaminant ingression. Therefore, V-rings should be used as external seals. They are effective, but should have metal shields to protect them from damage. See also Section 13.8 for more information on sealing technology.

Most large WTs have an inline filter located in the cooling system. However, these filters must out of necessity have a larger pore size than the oil film thickness, typically 10 micron or larger. Because the oil flow rates required by the cooler are high, a finer filter is not an option as this would make the inline filter too large for the nacelle. As a consequence, these filters have a low dirt holding capacity and in some cases require frequent changes.

The solution presented in Ref. [13] is to supplement the inline filter with an offline filter. Offline filters are installed independent of the gearbox. Here a finer filter can be used, typically around 3 micron, because the oil flow requirements are less than 1 gallon per minute. Offline filters are depth-type filters, meaning that they have a larger surface area than inline filters. Therefore, they have a higher dirt holding capacity, providing a longer service life.

Furthermore, the offline filter can run continuously, even during shutdown. Cellulose-based offline filters have the added capability of removing moisture via absorption. Portable filters are not the ideal solution to maintain the oil cleanliness level. While a filter cart will clean the oil for a limited time, experience shows that when the filter cart is disconnected, the particle count quickly rises, effect known as the "sawtooth curve." A well-designed contamination control system incorporating inline and offline filters will reach oil cleanliness targets and provide operational economy. Offline and inline filters are common in the wind industry and provide the optimum level of contamination control.

13.4.3 MINIMIZING CONTAMINATION ADDED DURING MAINTENANCE

Maintenance in WTs is difficult and risky. Therefore, the gearbox, lubrication system, work platforms, and nacelle housing should be designed to ensure maintenance tasks are readily accomplished in a safe manner. Breathers, filters, drain ports, fill ports, sample ports, dip sticks, magnetic plugs, and inspection ports should be designed and located for easy maintenance and minimization of contamination.

All maintenance that involves opening the gearbox or lubrication system should be performed with good housekeeping procedures. Oil should be added from a filter cart connected to the gearbox with quick connect couplings to minimize contaminants in the new oil and to minimize contaminant ingression during the transfer.

13.5 FAILURE MODES OF WT BEARINGS

13.5.1 FAILURE STATISTICS

A survey of field data from the U.S. wind industry generally shows a distribution of failure causes by component in the WT generators (WTGs). The comparison of failure causes for different size of generators based on survey of 800 failed WTGs is presented in Figure 13.7 based on the data from Ref. [14].

This survey shows that failure of the bearings in medium size generators (1 MW < P = <2 MW) prevails on any other parts/systems failure cause (70%). Bearing failure caused 58% of failures of all large generators (P > 2 MW) and much less percent (21%) of all small generators (P < 1 MW) fail because of the bearings failure.

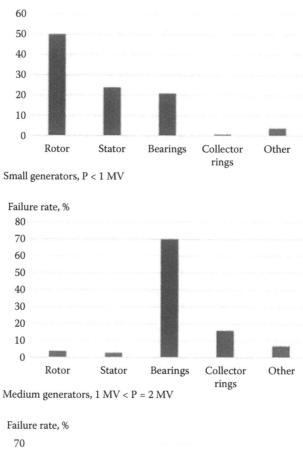

FIGURE 13.7 Failure statistics based on survey of 800 failed wind turbine generators. (Adapted from Alewine K. and Chen W. Wind turbine generator failure modes analysis and occurrence, *NREL Wind Power Conference and Exhibition*, Dallas, TX, May 2010.)

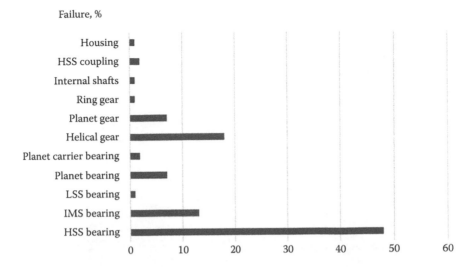

FIGURE 13.8 Failure statistics graph based on gearbox failure database. (Adapted from Wind Energy Technologies, Reliability and A2e, *Wind Power Program Peer Review*, US Department of Energy, March 2014.)

Based on released gearbox failure statistics on 289 gearbox failure incidents, it was found that top failure mode is bearing axial cracks in high-speed shaft (HSS) or intermediate-speed shaft (IMS) (Figure 13.8), confirming the importance of instrumenting HSS stage in the GRC [15]. According to this statistics, 70% of all failures are in bearings and 26% in gears.

The rolling element bearings play a crucial role for the proper functioning of gearbox used in WTG. The reliability of bearing in special environments like corrosive, high temperature and power, and high speed and high vacuum zone is very important. Many problems are arising in WTG gearbox operation even after conducting scheduled maintenance at stipulated intervals. Oil analysis, drive train alignment, and tower torque are linked to bearing faults.

In wind industries, bearing failure cannot be tolerated because it leads to catastrophic losses in power production due to down time, cost of repairing, replacement of parts, and so on. Study reveals that defects in bearings and bearings failure are the main reasons for gearbox failure in WTG. The persistent running of the gearbox with defective bearing [16] causes damage on intermediate pinion shown in Figure 13.9 due to tooth loading. Damaged outer race of bearing is shown in Figure 13.10.

The summary of bearing and gear failure modes is given in Ref. [17]. Current failure modes that are prevalent in WT gearboxes include

- Bending fatigue originating from nonmetallic inclusions
- Micropitting due to rough surfaces or lubricants with inadequate micropitting resistance
- Subcase fatigue due to grind temper or inadequate case depth
- Adhesion or abrasion due to contaminated lubricants
- Fretting corrosion during parking
- Case/core separation due to excessive case depth at tips of teeth
- Axial cracks in bearing inner rings

13.5.2 Typical Failure Modes of Gears and Bearings in WTs

Detailed studies, including visual inspection, SEM, and energy-dispersive spectrum (EDS) analysis is recommended on the damaged bearing surface to determine the root cause(s) for failure.

FIGURE 13.9 Damaged intermediate pinion. (Adapted from Sankar S., Nataraj M., and Prabhu Raja V. *Journal of Information Systems and Communication*, 3(1), 302–309, 2012.)

Any abnormalities in WTG may cause bearing failures; therefore, the peak temperature parameter in the gearbox at various places and power of the WT monitoring is recommended.

The recently revised standard for WT gearbox designs, ISO 61400–4, requires sizing gearbox bearings per ISO 281 and ISO 76, which includes calculations for two failure modes: (1) subsurface initiated, rolling contact fatigue (RCF) and (2) yielding under maximum stress. ISO 15243 is a useful standard to classify the bearing damage. Table 13.1 provides a few examples of bearing failures in the wind industry. However, that damage classification is not a root cause of failure. One or more initiating factors are often behind the observed damage. According to Ref. [18], associated costs of bearing damage repair run from $10K to $300K.

FIGURE 13.10 Damaged outer race of bearing. (Adapted from Sankar S., Nataraj M., and Prabhu Raja V. *Journal of Information Systems and Communication*, 3(1), 302–309, 2012.)

TABLE 13.1
Common Bearing Failure Modes in Wind Turbine Drive

Component	Type of Damage	Term in Use
Spherical rollers in main rotor shaft bearings	Micropitting	Gray staining
Through hardened cylindrical roller bearing on gearbox parallel stage shafts	White etching cracking (WEC)	Axial cracks
Generator bearings	Electrical erosion	Fluting
Tapered roller bearings on gearbox parallel stage shafts and planet bearings	Fatigue	Spalling

Source: Adapted from Sankar S., Nataraj M., and Prabhu Raja V. *Journal of Information Systems and Communication*, 3(1), 302–309, 2012.

13.5.3 MICROPITTING AND SPALLING

The detailed study in Ref. [18] revealed, for example, that an intermediate nondrive end bearing failed in the gearbox due to excessive material removal on the rollers and bearing races due to contact wear (*scoring*) followed by contact fatigue (*spalling*). The deterioration of the rollers generally starts with micropitting, also called "gray staining" or "frosting." This consists of microscopic cracks only a few microns (~2.5 micron or 0.0001″) deep. Individually these cracks are too small to be visible. As they accumulate, they appear as gray stains on the roller surface.

Eventually, the bearing roller starts to shed its cracked and weakened surface losing a small bit of its precision tolerance. Furthermore, this contaminates the oil with microscopic super hard steel particles most of which are too small to be filtered out. The gray staining begins when the oil film that separates the rollers from the races breaks down. An example of bearing rollers with severe fatigue spalling damage due to overloading [19] is shown in Figure 13.11.

FIGURE 13.11 Overloading resulted in severe fatigue spalling on the tapered roller bearing. (Adapted from Evans R. D. Classic Bearing Damage Modes, *Wind Turbine Tribology Seminar NREL-Argonne-DoE*, Broomfield, CO, USA, November 2011.)

FIGURE 13.12 Sliding damage and smearing. (From Stadler K. and Dvorak P. How black oxide coated bearings can make an impact on cutting wind turbine O&M costs, *Wind Power Engineering*, September 8, 2014. With permission.)

The types of damage to bearing components, such as the rings and rollers, generally fall into categories such as cracks, sliding (smearing) shown in Figure 13.12, and surface distress, as well as environmental effects, such as moisture and chemical attack [1].

13.5.4 Smearing, Surface Distress, and Microspalling

In lightly loaded roller bearings, pure sliding between rolling elements and the inner ring can occur when there is a large mismatch between the inner ring and roller set rotational speed. For demanding applications such as wind gearbox, HSSs, idling conditions, and changing of load zones can sometimes lead to high sliding risk.

In radially loaded roller bearings, the most critical zone where sliding can occur is the entrance of the rollers into the load zone. While rotating, the rollers slow down in the unload zone of the bearing because of friction and subsequently have to be suddenly accelerated as they re-enter the load zone. Resultant conditions can cause *smearing*. The microstructure of rollers and raceways is altered, and this results in local stresses that ultimately cause spalling and bearing failure [1].

Surface distress and microspalling: Many machine elements having rolling and sliding contacts (e.g., rolling bearings, gears, and cam followers) can sometimes suffer from various types of damage. Among these are mild abrasive wear and microspalling. Surface distress or microspalling occurs because of an insufficient oil film to separate the moving contacts. It is a form of localized surface damage that occurs on both gear teeth and in bearings and is a common phenomenon found in WT gearboxes. Gear teeth are usually more affected than bearings. Nevertheless, if it happens to bearings, it can be particularly detrimental to the bearing function. It alters the geometry of rollers and raceways, increasing internal clearance and resulting in local stresses that ultimately cause spalling and bearing failure. Contamination by water in WT gearboxes could also be a contributing factor [1].

The oil analysis, EDS, temperature, and power analysis confirmed that contamination, presence of bauxite in the lubricant, and overloading due to continuous peak power generation in high-wind season were the reasons for bearing failure. Further, the SEM study inferred that the mode of failure was fatigue fracture due to high cyclic fatigue phenomenon.

Despite these failures having been observed for two decades in various industries, the detailed reasons and mechanisms for their formation are not fully understood. It was found that subsurface flaking is originated under clean lubrication condition, while surface flaking is originated under contaminated lubrication condition.

13.5.5 White Structure Flaking (WSF) and Axial Cracking

A lot of premature wind gearbox bearing damage results in a failure mode that does not follow classical RCF mechanisms. One of the prevalent failure modes in gearbox bearing raceways is *WSF* by the formation of "butterflies" and white etching cracks (WECs) with associated microstructural change called white etching areas.

Butterflies are cracks with microstructural changes induced around stress raisers (typically inclusions) and WECs are metallographically similar but are distributed in large irregular-shaped branching networks [20]. These classical mechanisms are subsurface-initiated fatigue as well as surface-initiated fatigue and can be predicted by standard bearing life calculations, whereas premature WEC failures experienced in WT bearings are not covered.

WEC bearing failure also called *axial cracking* [15] refers to the appearance of the altered steel microstructure when polishing and etching a microsection [21]. Images of WEC are shown in Figures 13.13 and 13.14.

Axial cracks in bearing raceways have become a major cause of premature gearbox failures in the latest generation of WTs. However, it is rare to find axial crack failures in gearbox bearings in other industries. Why damage is so common in WTs has been a mystery and the subject of intense research in Ref. [22]. Failures can be found at several bearing locations, namely, the planet bearings, intermediate shaft, and HSS bearings [1]. For reasons that remain unclear, axial cracking failures usually arise in bearings that support gears of the intermediate and high-speed stages.

Cylindrical roller bearings are more vulnerable than other types such as tapered roller bearings. The occurrence of premature failures due to WEC is widely discussed within the wind industry and is being independently investigated by WT manufacturers, gearbox manufacturers, and bearing suppliers as well as universities and independent institutes.

Axial cracking failure mode of a bearing mainly occurs on the inner ring typically mounted to a shaft with an interference fit. The ring is heated during assembly onto the shaft. When the ring cools, it shrinks holding the ring in place. But the shrinking also creates tensile stress in the ring that

FIGURE 13.13   White etching crack (WEC) bearing failure. (Adapted from Errichello R. Wind Turbine Gearbox Failures, *Wind Turbine Tribology Seminar NREL-Argonne-DoE*, Broomfield, CO, November 2011.)

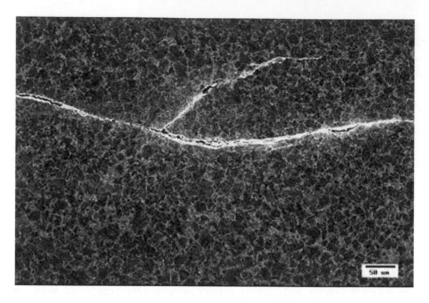

FIGURE 13.14 Damaged bearing chemically etched to highlight damage—white etching crack (WEC). (Adapted from Evans R. D. Classic Bearing Damage Modes, *Wind Turbine Tribology Seminar NREL-Argonne-DoE*, Broomfield, CO, USA, November 2011.)

increases the possibilities for axial cracking. An axial crack failure of the inner ring can be caused by excessive hoop stresses during installation, but this does not appear to be the general case in the wind industry.

Microscopic, white-colored areas at the edge of the cracks when the bearings are sectioned and etched are key indicators that this is not merely standard hoop stress failures. These cracks (WEC) are commonly found in through-hardened bearing races. When WEC damage causes flaking or spalling; it is often termed white structure flaking (WSF). This generally occurs in carburized raceways. Either way, the damage starts with the creation of *white etch areas* or WEAs [22].

Current hypotheses are focused on either hydrogen-enhanced WEC developments or purely load/stress-related WEC developments, preferably at inclusions as well as at the surface, or combinations of these [21]. Many studies suggest that WSF is caused by hydrogen embrittlement, based on the fact that more hydrogen was detected in the bearing steel with WSF [23].

Hydrogen embrittlement occurs when steel absorbs hydrogen atoms and becomes less ductile. Potential sources of hydrogen that makes steel bearing parts brittle and, therefore, more prone to axial cracking, include gearbox oil, moisture in the oil, and common oil additives. Electrostatic discharge off the turbine's electrical or lubrication systems may also provide the energy to disassociate water and oil in the gearbox of their hydrogen, causing hydrogen embrittlement of the bearing material.

In Ref. [20], white etching area formation mechanisms are linked to grain refinement and effects of carbon/carbide in a range of bearing steels of widely differing carbon content. This review also highlights the severe transient, cyclic loading, and tribochemical operating conditions of gearbox bearings and explains how these may act as drivers to produce WSF.

Other theories consider *allowable loads*. In bearings loaded to their minimum or higher, rollers do not slide but roll as intended. Underloaded rollers that slide may risk bearing surface damage, including axial cracking. It is possible that *adiabatic shear*, which happens to steel during impact, causes axial cracking if the steel's microstructure is plastically deformed. The deformed sections under normal cyclic loading sometimes form fatigue cracks.

Axial cracking in bearings occur at stresses well below those causing RCF. This suggests a possible link to *corrosion fatigue*, which reduces the allowable stress on mechanical components. The harsh outdoor environments that WTs endure put their parts at risk of corrosion.

13.5.6 ROLE OF THE LUBRICANTS IN BEARING FAILURES

It is reported that lubricant is decomposed by a chemical reaction with a fresh metal surface, which is formed by local metal-to-metal contact and thereby generates hydrogen [24,25]. Bearing life of WSF is depending on additives included in the lubricant. Some additives decrease bearing life and other additives increase it. The effect of extended life is more likely due to the oxidation film formed by a tribochemical reaction. Oxidation film can prevent a fresh metal surface from being exposed to lubricants and keeping the raceway surface chemically stable as some kinds of additives enhance to form the oxidation film on the fresh metal surface [26].

Cylindrical roller bearings do not always have a separating lubricant film built up between the contacting rollers due to opposing surface speed. Thin film or even mixed lubrication under high roller-to-roller contact pressures leads to metallic contact between neighboring rollers, and this then increases friction, which consequently can lead to smearing and surface destruction.

Important information on failure cause of gearboxes and bearings may be obtained by performing wear debris analysis and online oil analysis [27]. In the gearbox, the gears produce small particles during the wear. The friction and the tear between the gears make the particles come loose. These particles are called *wear debris*, and as they come loose, they mix up with the lubricating oil. By examining the wear debris, one can find out about the current status of the gearbox. Wear debris analysis is undertaken in different ways, but generally the particles are categorized in terms of their quantity or concentration, size, morphology, and composition. The associated wear characteristics are the severity, rate, mode, and source of the wear.

The oil analysis can be utilized as a part of a proactive maintenance strategy, but wear debris analysis can only be used to monitor active primary wear. Examples of typical wear debris produced by machines with rolling bearings and gear teeth, which undergo a nonconformal, rolling–sliding type of contact situation, are ferrous particles of varying shapes and sizes of between 10 and 1000 μm. More about failure causes and failure statistics will be described in Chapter 15.

13.6 EFFECT OF OPERATING CONDITIONS ON WT GEARBOXES

WT gearboxes are subjected to a wide variety of operating conditions (load, speed, lubrication, and combinations of these conditions) that may push the bearings beyond their limits. The wind energy segment faces some of the toughest challenges for extending bearing life and reducing the occurrence of premature failures while at the same time reducing the overall cost of energy [28].

Among severe operating conditions leading to premature failures in WT applications are periods of heavy and dynamic loads/torques. These periods are leading to vibrations and rapid load changes, additional radial and axial forces by the rotor, and axial motion of the main shaft. They are also leading to dynamical loading, higher stresses of gearbox components. Occasional connecting and disconnecting of the generator from the power grid are leading to torque reversals, bouncing effects, and so on.

Lubrication compromise between needs of gears and bearings as well as between low- and high-speed stages, insufficient oil drains, and refill intervals also may cause premature failure of gearboxes [29]. Negative effect on gearbox condition is produced by harsh environmental and operating conditions. These conditions are eventual large temperature changes and consequently larger temperature differences between the bearing inner ring and housing than expected when starting up. These conditions also include dust, cold climate, and moisture, particularly for offshore WTs [30].

As pointed in Refs. [31,32], a WT gearbox is susceptible to *fretting wear* and another failure mechanism called *false brinelling*, both mechanisms are in effect in the presence of light and strong winds. False brinelling is damage caused by fretting, with or without corrosion, which causes imprints that look similar to brinelling, but are caused by a different mechanism. Shown in Figure 13.15 is the high-speed pinion from a modern WT. The four faint lines in Figure 13.15 (top picture) are false brinelling that occurred when the WT was parked for a short time under light winds. The three prominent red lines are fretting corrosion that occurred when the WT was parked for an extended period

FIGURE 13.15 False brineling and fretting corrosion: *Top*—on teeth of a wind turbine (WT) pinio from a modern WT. (Adapted from Errichello R. Wind Turbine Gearbox Failures, *Wind Turbine Tribology Seminar NREL-Argonne-DoE, Broomfield*, CO, November 2011.). *Bottom*—in the outer raceway of a rolling element bearing from WT. (Adapted from Sheng S. Improving component reliability through performance and condition monitoring Data Analysis, *NREL Wind Farm Data Management & Analysis North America*, Houston, TX, March 2015.)

under heavy winds. The outer raceway of a rolling element bearing from a WT shown in Figure 13.15 (bottom picture) suffered from false brinelling and fretting corrosion when the WT was parked.

An example of fretting corrosion in the bore of an inner bearing ring [1] is shown in Figure 13.16. With the high-speed pinion stopped by the brake, and the rotor buffeted by the wind, the mating gear rocks back and forth through small amplitude motion.

FIGURE 13.16 Fretting corrosion in the bore of an inner ring. (From Stadler K. and Dvorak P. How black oxide coated bearings can make an impact on cutting wind turbine O&M costs, *Wind Power Engineering*, September 2014. With permission.)

13.7   BEARING TREATMENT TECHNIQUES TO EXTEND BEARING AND GEARBOX LIFE

Surface engineering is the practice of altering the chemical and/or topographical device. One surface engineering technique being promoted to mitigate micropitting and smearing-type wear on gearbox cylindrical roller bearings is the chemical conversion of the surfaces of the rollers and raceways from steel to magnetite (Fe_3O_4), commonly referred to as black oxide.

Since black oxide provides no protection against smearing, and it does not provide a permanent barrier to the mechanism of micropitting, a more technologically advanced engineered surface (ES) is required to ensure that gearbox bearings attain their desired fatigue life, such as a combination of two ES technologies.

13.7.1   THE ENHANCED BLACK OXIDATION PROCESS

Black oxide is surface treatment that has been employed on bearings for many years to inhibit corrosion and in some cases to facilitate a break-in of rollers and raceways that do not have adequate surface finishes or geometries. Black oxide surface treatments are considered to be sacrificial since they wear away rapidly once the bearing is in operation. For this reason, black oxide has not usually been associated with improving the durability of bearings.

However, there are at least two reasons to believe that black oxide surface treatments might mitigate micropitting wear. While the black oxide is on the raceways micropitting cannot occur, and after the black oxide wears away, the raceway surfaces are usually smoother than the as-ground raceways.

The black oxide coating used by SKF [1] on the bearing functional surfaces provides a significant degree of protection. This layer adds beneficial properties to the bearing operation, such as an improved running-in phase, and results in equally improved surface properties after running-in, better performance under poor lubrication regimes (low conditions), and better lubricant adhesion, as well as enhanced smearing resistance. The risk of fretting, microspalling, and crack formation can be reduced. Black oxide-coated bearings from SKF in WT gearboxes are shown in Figure 13.17.

Furthermore, the black oxide layer offers an elementary corrosion resistance as well as an enhanced chemical resistance when compared with untreated surfaces. The moderate corrosion resistance of black oxide is sufficient to suppress standstill corrosion and fretting corrosion, and the chemical resistance reduces detrimental effects from aggressive oil ingredients. It improves frictional behavior and reduces wear, particularly under mixed lubrication conditions. Recent R&D results indicate that black oxide acts as a barrier to hydrogen permeation into the steel.

To give a comparison of the potential improvement in failure rates, a wind gearbox manufacturer has reported that in a sample of 1000 standard cylindrical roller bearings in a gearbox application, a failure rate ranges from 40% to 70% after 2 years. Subsequently, in a sample of 1150 black oxidized cylindrical roller bearings for a similar application, the failure rate was 0.1% over the same period.

Compared with untreated bearings, black oxide-coated bearings from SKF in WT gearboxes can offer the following benefits [1]:

* Improved running-in behavior
* Better corrosion resistance
* Improved resistance against smearing damage
* Better performance in low lubrication conditions
* Increased oil and grease adhesion
* Reduction of chemical attack from aggressive oil additives on the bearing steel
* Reduced hydrogen permeation in the bearing steel
* Decreased risk of fretting corrosion in the fits

FIGURE 13.17 Black oxidized wind turbine gearbox bearings: the cylindrical roller bearings come without outer ring for use in planetary wheels (top); separable high-capacity cylindrical roller bearings bearing used on high-speed intermediate and high-speed output shafts (bottom). (From Stadler K. and Dvorak P. How black oxide coated bearings can make an impact on cutting wind turbine O&M costs, *Wind Power Engineering*, September 2014. With permission.)

The treatment method as well as the size and weight range is proprietary to SKF. Black oxide is a surface treatment formed by a chemical reaction at the surface layer of the bearing steel and is produced when parts are immersed in an alkaline aqueous salt solution operating at a temperature of approximately 150°C (302°F). The reaction between the iron of the ferrous alloy and the reagents produces an oxide layer on the outer surface of bearing components consisting of a well-defined blend of FeO and Fe_2O_3, resulting in Fe_3O_4.

The result is a dark black surface layer of approximately 1–2 μm in thickness. The company recommends that both inner and outer rings as well as the rolling elements are coated for the best performance. The black oxide can be applied to all bearing types used in key WT systems. Cylindrical and tapered roller bearings in particular have been successfully treated and put into operation in recent years.

13.7.2 A Combination of Two Engineered Surfaces (ES) Technologies

A combination of two ES technologies, isotropic superfinishing and a durable, wear-resistant coating applied to rollers, provides very smooth surfaces on rollers and permanent barriers to micropitting and smearing wear. The superfinishing process produces surfaces on the rollers with very low roughness and asperity slopes. The benefits of this isotropic surface include reduced asperity contact and shear stresses. The wear-resistant coating has been specially designed for rolling elements in bearings.

This coating consists of nanocrystalline tungsten carbide (WC), which precipitates homogeneously dispersed in an amorphous hydrocarbon matrix [12,33]. Although chemically similar to other diamond-like carbon coatings, this coating has a microstructure that has been specifically engineered for maximum durability on rolling elements. The combination of these two technologies produces smooth, hard, and extremely durable surfaces on rollers that provide exceptionally good wear resistance and additional benefits to rolling element bearings.

In addition to eliminating micropitting and smearing wear, ES treatments on bearing rollers provide additional desirable benefits. Gearbox roller bearings can suffer intermittent loss of lubrication due to malfunctions or maintenance issues regarding the WT lubricant delivery systems. During intervals when the bearings operate in a lubrication-starved environment, the opportunity for adhesive wear between the rollers and the raceways is greatly increased. Results of full-scale bearing tests indicate that bearings with ES-treated rollers can survive about 10 times longer than untreated bearings in lubricant-starved situations.

More technologically advanced surface treatments applied to the rolling elements of bearings eliminate micropitting and greatly reduce the risk of smearing. Additionally, bearings with advanced ES have more than a 3.5 times greater fatigue life than bearings with untreated rollers, have about 15% less frictional torque, and are twice as resistant to damage from gearbox debris than bearings with untreated rollers [12].

13.8 BEARING SEALS

The wind power industry continues to seek new ways to increase the lifespan and reliability of WTs. WTs not only require extreme protection against weather and mechanical forces but also solutions that are long lived and robust to provide longer maintenance periods for an offshore and onshore system. As many engineers know, machines are only as good as their parts, and it is often the case that the smallest of components have the potential to have the largest influence on the lifespan and reliability of the entire system [34].

Sealing—the act of preventing something from escaping or entering—is just one example of how a relatively simple concept can play a critical role in the successful functioning of the overall system. From a basic O-ring seal to more complex multipart sealing systems, the simple act of "keeping something in or out" is a vital aspect in the proper function of WTs.

The effort to develop and implement sealing solutions that maximize service life, while also minimizing costly downtime, can be influenced by a number of outside factors, including environmental conditions (extreme temperatures, ozone containments, UV rays, and salty air) and surface characteristics, as well as external media and lubrication requirements.

Whether onshore or offshore, WTs need to withstand large-scale temperature changes. As a result, sealing components need to be made with materials that are able to withstand these extreme fluctuations, especially extreme cold, without comprising shape, structure, or sealing ability. This makes the choice of material a critical component in the development of a proper sealing system.

In addition to extreme temperatures, these machines are also subject to a variety of harsh conditions on a daily basis. UV rays that affect a component's rubber matrix and material characteristics and ozone containments that cause material cracks and breakdown, as well as geographic-specific conditions such as salty air are all at odds against the proper sealing of a WT. Further, the system's surface conditions can also impact performance and seal behavior. The accurate evaluation of surface characteristics can provide engineers the needed information to better select or develop the best solution for the given application [35].

From the other hand, internal media such as the oil and grease used for lubrication pose a particular challenge to sealing systems. Improper lubrication can cause increased friction and wear, which can result in decreased seal life as well as increased maintenance requirements.

13.8.1 Seals and Sealing Systems Challenges

13.8.1.1 Seal Design

A seal's design is evolving to meet the sealing requirements of the wind power industry. Multiple companies around the world design and manufacture various types of bearing seals: Garlock Klozure [34], Simrit [36], Shaeffler [37], Parker [38], Telleborgen [39], Timken [40], and so on.

As one example, *Simrit* has developed customer-specific profile seals targeted for large bearing movement that, due to a more subtle sealing lip contour, offers increased flexibility. This flexibility is particularly important as it helps ensure the proper performance during large changes in sealing gap caused by extreme operating conditions, for example, windstorms or strong gusts of wind. The design incorporated an enlarged sealing bead for the rotating bearing rings, which reduces the profile movement considerably. This next-generation profile provides an enhanced groove fit to also increase the profile's pressure resistance.

Labyrinth seals work as a contactless and frictionless solution that provide extreme service life (20 years in some cases) and require little if any service/maintenance. For decades, metal labyrinths have been an established solution for a variety of industries. For the wind industry, metal labyrinths have shown to be difficult to assemble, especially at the bigger shaft diameter used in turbines.

One example of supplier innovation is the recently developed PTFE labyrinth seal (or rotary seals), which was specifically designed for large gearbox applications. Simrit is manufacturing this type of seal [36]. The Radiamatic RCD seal—which stands for *reject, collect, and drain*, the three key principles of function in what are known as liquid-collecting labyrinth seals—is a highly efficient solution that offers the prospect of a service life spanning several decades. The seal consists of two parts and features a rotating inner ring and a stationary outer ring. The inner ring is open (optionally) and kept in place on the shaft by springs that adjust to axial movements. The outer ring sits securely in the housing bore. Designed specifically to meet customers' application requirements, the seal is adaptable to various designs and easy to assemble.

One example of an evolved solution that addresses the extreme environmental challenges associated with the wind industry is the Simmerring Enviromatic seal, which offers excellent protection against both dust and salty air. This innovative solution was created using a small line contact of seal lip, contrary to the standard use of V-rings, to guarantee protection of the main seal. Uniquely, the seal operates with the separate functions in individualized, optimized parts.

Working to strengthen the sealing lip of previous V-ring designs, Simrit developed its Enviromatic seal as a premium solution. By incorporating the improved hydrogenated nitrile butadiene rubber (HNBR) material Ventoguard™ 467 and optimizing the design using finite element analysis, the new solution offers improved stiffness, contour, and function that results in more reliable protection and increased wear resistance.

13.8.1.2 Sealing Materials

Many factors can influence sealing ability of a material and the service life of the overall system, including mechanical loads, temperatures, greases and other media, ozone, and salt water. As a result, materials for wind power sealing need to be durable, compatible with oil and grease lubricants, resistant to UV rays, the ozone, salty water, and more.

New sealing material and design innovations are driven by the consistently evolving needs of the industry. As a result, there is an array of pitch seal options available to fit the specific geometries of various pitch bearings applications. A crucial element in addressing pitch and yaw bearing sealing challenges is improved materials solutions. Even more than design enhancements, material innovations are key to properly addressing wind power sealing challenges, as the quality of elastomeric sealing materials is decisive for the functional reliability of the WTG.

An example of industry-specific material solution is the newly enhanced line of nitrile butadiene rubber (NBR) Ventoguard materials [36] that have been specifically designed to offer superior low compression set (which is a direct measure for its durability as it determines the elastomeric lasting), grease and ozone resistance, and low-temperature capability.

Ventoguard 453 and Ventoguard 454 are the next generation of Simrit's line of established NBR materials designed for use in slewing bearing profile seals in WTs. Each material was developed to address a specific wind power-related challenge with the ultimate goal of helping prolong the maintenance cycle of the sealing application. Ventoguard 453 is a premium material that offers excellent cold flexibility, very good compression set, and demonstrates superior aging in several greases, including Klüber, Fuchs, and Mobil. Simrit's Ventoguard 454, also a new premium material, provides superior compression set and is optimized for use in Shell Rhodina BBZ greases.

REFERENCES

1. Stadler K. and Dvorak P. How black oxide coated bearings can make an impact on cutting wind turbine O&M costs, *Wind Power Engineering*, September 2014, Online. Available: http://evolution.skf.com/us/how-black-oxide-coated-bearings-can-make-an-impact-on-cutting-om-costs-for-wind-turbines/.
2. Michaud M., Sroka Gary J., and Benson Ronald E. Refurbishing wind turbine gears, *Gear Solutions*, 28–39, June 2011, Online. Available: http://www.gearsolutions.com/article/detail/6095/refurbishing-wind-turbine-gears.
3. Errichello R. How to analyze gear failures, *Power Transmission Design*, 35–40, March 1994, Online. Available: http://www.researchgate.net/publication/225602159_How_to_analyze_gear_failures.
4. Franke J.-B. and Grzybowski, R. Lifetime prediction of gear teeth regarding to micropitting in consideration of WEC operating states. *The International Technical Wind Energy Conference*, or *7th German Wind Energy Conference (DEWEK)*, Wilhelmshaven, Germany, October 2004.
5. Needelman W. M., Barris M. A., and LaVallee G. L. Contamination control for wind turbine gearboxes, *Power Engineering*, November 2009, Online. Available: http://citeseerx.ist.psu.edu/viewdoc/download?doi=10.1.1.470.7228&rep=rep1&type=pdf.
6. Muller J. and Errichello R. Oil cleanliness in wind turbine gearboxes, *Machinery Lubrication*, July 2002, Online. Available: http://www.machinerylubrication.com/Read/369/wind-turbine-gearboxes-oil.
7. Cantley R. E. The effect of water in lubricating oil on bearing fatigue life, *ASLE Transactions*, 20(3), 244–248, 1977.
8. Wood Robert J. K., Bahaj AbuBakr S., Stephen R. Turnock et al. Tribological design constraines of marine renewable energy systems, *Philosophical Transactions of the Royal Society. A 28*, 368(1929), 4807–4827, October 2010, Online. Available: http://rsta.royalsocietypublishing.org/content/roypta/368/1929/4807.full.pdf.
9. Kotzalas M. N., Needelman W. M., Lucas D. R., and LaVallee G. L. Improving wind turbine gearbox life by minimizing oil contamination and using debris resistant bearings. *Proc. AWEA Wind Power Conference*, Houston, 2008.
10. American Gear Manufacturer's Association, *Standard for Design and Specification of Gearboxes for Wind Turbines*. ANSI/AGMA/AWEA 6006-A03, 2004.
11. Terrell Elon J., Needelman William M., and Kyle Jonathan P. Tribological challenges to the system, *Wind System*, March 2010, Online. Available: http://www.windsystemsmag.com/view/article.php?articleID=71.
12. Doll G. L., Kotzalas M. N., and Kang Y. S. Life-limiting wear of wind turbine gearbox bearings: Origins and solutions. *Proceedings of European Wind Energy Conference & Exhibition 2010 (EWEC 2010)*, vol. 4, p. 2559, Warsaw, Poland, April 2010. Presentation available online: http://studylib.net/doc/5305452/life-limiting-wear-of-wind-turbine-gearbox-bearings-orig.
13. Barrett Michael P. and Stover J. Understanding oil analysis: How it can improve reliability of wind turbine gearboxes, *Gear Technology*, 102–109, November/December 2013, Online. Available: http://www.testoil.com/pdf/Gear%20Technology%20Nov-Dec%202013.pdf.
14. Alewine K. and Chen W. Wind turbine generator failure modes analysis and occurrence. *Wind Power 2010*, Dallas, May 2010. Online. Available: http://www.nrel.gov/wind/pdfs/day1_sessioniv_04_shermco_alewine.pdf.
15. Wind energy technologies, reliability and A2e, *Wind Power Program Peer Review*, US Department of Energy, March 24–27, 2014, Online. Available: http://energy.gov/sites/prod/files/2014/06/f16/eere_wpp_2014_peer_review_ReliabilityandA2e.pdf.

16. Sankar S., Nataraj M., and Prabhu Raja V. Failure analysis of bearing in wind turbine generator gearbox, *Journal of Information Systems and Communication*, 3(1), 302–309, 2012, Online. Available: http://www.bioinfo.in/contents.php?id=45.

17. Errichello R. Bearing and gear failure modes seen in wind turbines, *Wind Turbine Tribology Seminar*, A Recap, Broomfield, CO, November 2011, Online. Available: http://www.nrel.gov/wind/pdfs/day2_sessioniii_2_geartech_errichello.pdf.

18. Hornemann M., Crowther A., and Dvorak P. Establishing failure modes for bearings in wind turbines, *Wind Power Engineering*, August 2013, Online. Available: http://www.windpowerengineering.com/design/mechanical/bearings/establishing-failure-modes-for-bearings-in-wind-turbines/.

19. Evans R. D. Classic bearing damage modes, *Wind Turbine Tribology Seminar NREL-Argonne-DoE*, Broomfield, CO, USA, November 2011, Online. Available: http://www.nrel.gov/wind/pdfs/day2_sessioniii_4_timken_evans.pdf.

20. Evans M-H. White structure flaking (WSF) in wind turbine gearbox bearings: Effects of "butterflies" and white etching cracks (WECs), *Materials Science and Technology*, 28(1), 3–22, January 2012, Online. Available: http://www.maneyonline.com/doi/pdfplus/10.1179/026708311X13135950699254.

21. Budny R. Fixing wind-turbine gearbox problems, *Machine Design*, June 2014, Online. Available: http://machinedesign.com/mechanical-drives/fixing-wind-turbine-gearbox-problems.

22. Herr D. and Heidenreich D. Understanding the root causes of axial cracking in wind turbine gearbox bearings, *Wind Power Engineering*, April 2014, Online. Available: http://www.windpowerengineering.com/design/mechanical/understanding-root-causes-axial-cracking-wind-turbine-gearbox-bearing.

23. Uyama H. The Mechanism of white structure flaking in rolling bearings, *Wind Turbine Tribology Seminar*, Bloomfield, CO, USA, November 2011, Online. Available: http://www.nrel.gov/wind/pdfs/day2_sessioniv_1_nsk_uyama.pdf.

24. Kohara M., Kawamura T., and Egami M. Study on mechanism of hydrogen generation from lubricants, *Tribology Transactions*, 49, 53–60, 2006.

25. Lu R., Nanao H., Kobayashi K. et al. Effect of lubricant additives on tribochemical decomposition of hydrocarbon oil on nascent steel surfaces, *Journal of the Japan Petroleum Institute*, 53(1), 55–60, 2010.

26. Uyama H. and Yamada H. White structure flaking in rolling bearings for wind turbine gearboxes, September 2013, *American Gear Manufacturer Association (AGMA)*, Paper 13FTM15, Online. Available: http://www.nrel.gov/wind/pdfs/day2_sessioniv_1_nsk_uyama.pdf.

27. Ribrant J. Reliability performance and maintenance – A survey of failures in wind power systems, Master Thesis written at KTH School of Electrical Engineering, 2005/2006, Finland, Online. Available: http://faculty.mu.edu.sa/public/uploads/1337955836.6401XR-EE-EEK_2006_009.pdf.

28. Stadler K. and Stubenrauch A. Premature bearing failures in wind gearboxes and white etching cracks (WEC), *Evolution Business and Technology Magazine from SKF*, March 2013, Online. Available: http://evolution.skf.com/us/premature-bearing-failures-in-wind-gearboxes-and-white-etching-cracks-wec/.

29. Kamchev B. Wind energy encounters turbulence, *Lubes'n'Greases Magazine*, 2011.

30. Heemskerk R. Challenges on rolling bearings in wind turbines, *VDI Gleit- und Wälzlagerungen* 2011.

31. Godfrey D. Fretting corrosion or false brinelling? *Tribology & Lubrication Technology*, 59(12), 28–30, 2003.

32. Errichello R. Another perspective: False brinelling and fretting corrosion, *Tribology & Lubrication Technology*, 60(4), 34–36, 2004, Online. Available: http://gearboxfailure.com/wp-content/uploads/2014/04/55-False_Brinelling_Fretting.pdf.

33. Evans R. D., Cooke E. P., Ribaudo C. R., and Doll G. L. Nanocomposite tribological coatings for rolling element bearings, *Surface Engineering 2002—Synthesis, Characterization and Applications*, Materials Research Society Press, Pittsburgh, PA, 750, 407–417, 2003.

34. Roberts David C. Improved sealing technology extends equipment life, *Power-Gen International Conference*, New Orleans, December 2007, Online. Available: http://www.powertransmission.com/articles/1208/Improved_Sealing_Technology_Extends_Equipment_Life.

35. Moskob F. Sealing solutions for wind, *Wind Systems Magazine*, pp. 26–31, October 2011, Online. Available: http://www.windsystemsmag.com/media/pdfs/Articles/2011_Oct/1011_Simrit.pdf.

36. Industry solutions for wind turbines, *Freudenberg Sealing Technology*, Simrit, Online. Available: http://www.o-pak.com.tr/pdf/DLFE-226.pdf.

37. Optimized bearing seals for wind turbines, *Shaeffer Technologies*, Online. Available: http://www.schaeffler.com/remotemedien/media/_shared_media/08_media_library/01_publications/schaeffler_2/reprint/downloads_16/ssd_29_de_en.pdf.

38. Rotary seals for wind energy, *Parker Hannifin Corporation*, Online. Available: http://www.parker. com/literature/TechSeal%20Division/Literature%20PDF%20Files/Parker_TechSeal_Pitch_and_Yaw_ Bearing_Seals_TSD%205437.pdf.
39. Innovative sealing in wind power, *Trelleborg Sealing Solutions*, Online. Available: http://www.tss. trelleborg.com/remotemedia/media/globalformastercontent/downloadsautomaticlycreatedbyscript/ catalogsbrochures/wind-power_en.pdf.
40. Timken® wind energy seals, *The Timken Company*, Online. Available: http://www.timken.com/en-us/ products/seals/Documents/Seals_0.pdf.

14 Wind Turbine Lubrication

14.1 LUBRICATION CHALLENGES FOR WIND TURBINES

Challenges for wind turbine lubrication are multifold. First, the wind farms are usually located in remote areas; secondly, the turbines are very tall; and thirdly, the ambient conditions in which they operate and the loads and vibrations to which components are exposed, are relatively harsh. All these conditions represent serious reliability challenges for installed wind turbine fleet. Newer multimegawatt wind turbine with larger blade diameters and tower heights, increased pitch control, and greater focus on offshore siting impose new challenges to bearing and lubricant manufacturers for wind turbines [1].

Multiple factors should be considered and analyzed to properly select the greases for wind turbine bearing lubrication. Among these factors are an understanding of the design and typical failure modes of the bearings, lowest and highest ambient operating temperatures, servicing frequency, and method of lubricant application. Among other considerations is an issue of compatibility of greases applied at the factories with the greases applied during maintenance.

Tribological challenges for wind turbines include complex failure modes involving bearings, gears, and lubricants inside wind turbine gearboxes [2]. Added are variable and harsh operating conditions: (a) changing wind speeds and directions; (b) intermittent operation with many starts/stops; (c) high transient loads from wind gusts, grid engagement, and braking; (d) high torque and low-speed input.

Remote locations of wind turbines create maintenance difficulty. Harsh environmental conditions include contamination from dust and wear debris, wide temperature range, high humidity and water ingress, and offshore environments that cause corrosion.

The lubricants have to meet the necessary requirements from different subsystems for main bearings, generator bearings, pitch bearings, yaw bearings, and yaw gears in diverse operating conditions. Gear oil has to meet higher requirements than for typical industrial applications. When choosing a lubricant suitable for a particular wind power plant, the operator has to consider several important factors. Bearing greases, for example, should be easy to pump and allow precise metering in centralized lubricating systems, thereby attaining a good grease distribution.

Good wear protection even under vibration increases the lifetime of the bearings during periods of idleness, the "false brinelling" is always a major cause of concern (see Section 13.6). Also, when the power station runs at low speed, wear is provoked due to the lack of a sufficient hydrodynamic lubricant film. A good lubricant must contain suitable additives to counteract these effects. Finally, it has to be ensured that the lubricant is compatible with the elastomers involved and covers the wide service temperature range of −40°C to ~150°C (−40°F to 300°F). With the operating temperature being approximately 90°C (190°F), a service temperature range up to 150°C (300°F) results in extended relubrication intervals [3].

There are currently two approaches to the development of greases for main bearing, blade pitching, yaw bearings, and generator bearings: developing a multipurpose grease suitable for the lubrication of all components, or developing a number of greases designed for each given component. Each approach has its pros and cons and the balance between ultimate bearing reliability and ease of use and ordering for service technicians has to be considered carefully [1].

14.2 TYPE OF LUBRICANTS USED IN WIND TURBINES

In wind turbine, there are multiple points that require lubrication and depending on location and function of the parts there are different types of lubricants that should be used at every specific location. Type of lubricants to use in wind turbine depends on application, specifically on the part the lubricant is applied. They should be chosen together with coatings if any are applied.

14.2.1 THREADED CONNECTIONS

Strong threaded connections are found practically everywhere in wind turbines, in nacelle frame joint, in front and rear main journal bearings, in pitch bearing fasten bolts, in gearbox, drives, and cooler units. Yaw bearing fasten bolts, tower segment flange bolts, foundation anchor bolts all have threaded connections.

Safe, reliable wind turbines depend on strong threaded connections. However, frequent problems with threaded connections often result in failures, such as tightening difficulties and thread damage, inconsistent or loose clamping force, broken bolts and base parts, such as flanges or plates, as well as difficult disassembly and destroyed threads.

Good lubricants for bolts and fasteners can help solve problems and increase reliability by reducing the effects of root causes, including: seizure and abrasive wear, uneven coefficient of friction, fretting corrosion, embrittlement failures of substrates due to use of low-melting-point metals, such as lead, tin and copper, and stress corrosion cracking [4].

Threaded connections on pitch and yaw bearings, tower, and other parts should be treated with antifriction coatings and lubricants. Requirements for such materials include: providing consistent tightening torque and friction coefficients, high load-carrying capacity, anti-seizure and corrosion protection, and wide temperature range. Lubrication products for this particular application enable precise conversion of tightening torque into force during assembly and nondestructive dismantling of connections even after long use under high temperature, as well as equipment corrosion protection [5]. Thread pastes and antifriction coatings (AFCs) are materials for application to threaded connections.

14.2.2 BEARINGS

Bearings in wind turbines are everywhere, such as blade bearings, pitch bearings, yaw bearings, main shaft bearings, generators bearings, and so on. Bearings are in high-speed/low-speed gearboxes [6]. Lubrication points for wind turbine grease lubrication of different bearings [7] are shown in Figure 14.1. The essential bearings of a wind power plant operate under very different operating conditions and therefore pose very different requirements regarding lubrication [3]. The main bearing rotates slowly but is subject to high loads and vibration. The generator bearing, by contrast, needs to cope with high speeds and high temperatures. Pitch and yaw bearings are subject to high loads as well, but they also perform oscillating motion under strong vibration.

Due to these varying requirements, wind power plant operators have often had to resort to a variety of greases up to this point. In addition, most wind parks use turbines from more than a single manufacturer, so different lubricant recommendations have to be taken into consideration. And most manufacturers offer various turbine models, which are often used in parallel. For the operator this means that he has to spend a lot of resources on logistics, storage, and grease disposal, plus an increased risk of lubricants being mixed up. Most turbines are still lubricated manually, so service technicians have to carry a variety of lubricants. Another problem is that not all lubricants are available worldwide, at least not in the same quality [3].

The lubricants in bearings application should provide high load-carrying capacity, low friction, and protection against wear, fretting corrosion, groove formation (Brinell effect), and moisture. Lubricants should perform properly at low temperatures up to −40°C in cold climates and protect from offshore corrosion. The wind turbine bearing grease lubrication performance challenges [7] are summarized in Table 14.1.

Special bearing greases provide long-term lubrication of the bearing, wear protection in mixed friction due to solid lubricants and EP additives, as well as corrosion protection. Important issue for

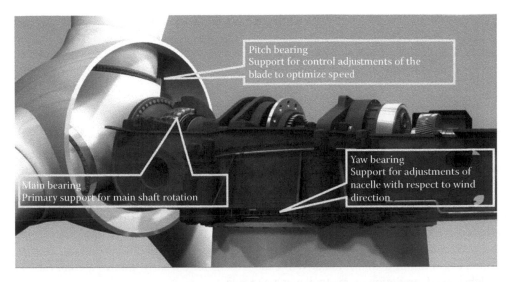

FIGURE 14.1 Lubrication points of different bearings in wind turbine. (Adapted from Braun H. R. Wind turbine grease lubrication, *Wind Turbine Tribology Seminar NREL-Argonne-DoE*, Broomfield, Colorado, November 2011, 45pp.)

bearings lubrication is high loads occurring due to the wind gusts and inner vibrations in the turbine. The intermediate bodies of the bearings like balls or rollers are pressed against the outer and inner rings under vibrations so that the grease is squeezed out of the highly loaded contact zones. Wear marks in form of the contact area of the intermediate bodies against the rings are generated, which damages the bearings severely [8,9].

14.2.3 BRAKES, SHRINK DISCS, AND SERVICE PARTS

For brakes and shrink discs, various pastes and AFCs provide corrosion protection, friction reduction, noise reduction caused by stick slip, and wear protection. Clean and dry AFCs provide excellent

TABLE 14.1
Wind Turbine Grease Lubrication Performance Challenges Summary

Condition	Challenges	Bearing Pitch/Yaw	Location Main	Generator
Wide temperature range	Lubricant viscosity	−30°C to 50°C	−30°C to 70°C	−30°C to 100°C
Load capacity	Composition	X	X	
Shock loads	Working T range	X	X	
Slow speed	Oscillation	X	X	
Moderate/high speed				X
Vibration	Fretting wear	X	X	
Salt water, sand	Corrosion, abrasion	X		
Humidity/condensation	Corrosion		X	X
Long life	Wear		X	X

Source: Adapted from Braun H. R. Wind turbine grease lubrication, *NREL Wind Turbine Tribology Seminar*, Broomfield, Colorado, November 15–17, 45pp. 2011.

combination of corrosion protection and lubrication, while pastes suppress stick slip, prevent seizure and scoring, reduce fretting corrosion.

For parts difficult to reach and for added service safety, dry-film lubrication is recommended such as AFCs. These coatings are easy in application, they dry in air, providing strong adhesion and good combination of corrosion protection and lubrication.

14.3 ROLE OF ENVIRONMENT IN LUBRICANT SELECTION

There are several environment factors, that may adversely affect the performance of various lubricants used in wind turbines whether it is main gearbox oil, hydraulic fluid, or grease. The wind turbines operate almost 24/7 in a wide variety of environments, such as at subzero temperatures, in scorching desert heat and high humidity with saltwater seeping into the tower, and with sand buffeting the blades and mechanisms. Therefore, the choice of the lubricants is an extremely important issue in providing a long servicing life of the equipment.

The gearboxes, which bear the greatest environmental assault, are designed to last 20 years but they often begin to fail within 7 years. While the reasons for the early failures are not fully understood, lubricant degradation seems to play a major role.

14.3.1 TEMPERATURE EFFECT ON THE LUBRICANTS

One of the most critical requirements is the ability of the lubricant to function over a wide operating temperature range, from extreme climate (Arctic) conditions to high ambient temperature. Temperatures from $<-40°C$ to $>50°C$ ($<-40°F$ to $>122°F$) are not uncommon, often with wide seasonal and daily swings.

In modern wind turbines, centralized greasing systems of progressive or single-line type have narrow-diameter feeder lines. It is important for the lubricants to remain fluid and have excellent low-temperature pumpability even at lower ambient temperatures and improved wear protection and oxidation stability at higher operating temperatures.

14.3.2 LUBRICATION IN WET AND CORROSIVE ENVIRONMENT

Wind farms are commonly located in coastal areas onshore and offshore surrounded by sea water, which results in an increased potential for salt water ingress. As a result, greases for wind turbine lubrication must be able to demonstrate excellent rust and corrosion resistance even in a saltwater corrosion environment.

The SKF EMCOR test (ASTM D6138) "Corrosion-Preventive Properties of Lubricating Greases Under Dynamic Wet Conditions" is an industry standard test used to determine the anticorrosion properties of greases when exposed to waters of varying quality. This test is conducted in contact with double-row self-aligning ball bearings. At the end of the test, the bearing raceways are examined and degree of corrosion is rated against a defined rating scale of 1–5, with 0 being no corrosion and 5 representing heavy corrosion with corroded areas covering more than 10% of the running track surface. Advanced corrosion-preventive greases would provide a rating of 1 or less than 1 with synthetic seawater [2].

14.4 REQUIREMENTS FOR GEARBOXES OIL USED IN WIND TURBINE APPLICATION

Wind turbines are highly engineered, sophisticated pieces of machinery operating in harsh environments. For example, to maximize life of the main gearbox in wind turbine, one of the most important components of wind turbine, the gear oil should offer the following qualities:

- Superb long-term gear and bearing wear protection
- Excellent oxidation stability to extend service life
- Outstanding rust and corrosion protection
- Exceptional filterability and keep clean performance

The correctly selected oil can potentially extend the life of the gearbox, reduce downtime and lower maintenance costs to enhance competitiveness, and help improve the productivity of the wind turbine.

However, lubricating the gearbox of a wind turbine poses many challenges, which have a direct impact on the required properties of the lubricant. Weight restrictions such as compact design and high load handling capabilities require excellent protection against micropitting and scuffing. Extended oil drain interval demands make lubricant property retention important as well as long-term protection against aging. Fine filtration systems with mesh size of 10 μm or smaller make oil filterability under dry and wet conditions critical.

14.4.1 Role of the Gearbox Oil Cleanliness in Gearbox Life

Cleanliness of gearbox oil plays a crucial role in life of the gearbox. For example, British researchers [10] showed that rolling element bearing life can be increased up to seven times by changing from a 40-mm filter to a 3-mm filter thus providing better cleanliness of the oil. Their results also show that a gearbox must be clean after assembly, otherwise a fine filter will not be effective. It happens that even before a filter can remove built-in contamination, gears and bearings can suffer permanent damage in as little as 30 min during run-in. There are many sources of contamination of the oil [11].

14.4.1.1 Types and Sources of Oil Contamination in Wind Turbine Gearboxes

Contamination can enter gearboxes during manufacturing, be internally generated, ingested through breathers and seals, and inadvertently added during maintenance. All of these sources must be addressed to limit the impact that contamination can have on components.

Location of the wind turbines often defines the type of oil contamination. In desert environments, these are exposed to airborne dust during the hot season and moisture during the rainy season. Offshore turbines are constantly exposed to moisture.

The size, hardness, and friability or ductility of the particle influence the amount of damage of gears and bearings that the particles can cause. Solid contaminants vary in hardness, friability, and ductility, depending on the composition of the particle.

Environmental dust, mostly silica, has a hardness rating of two to eight. Quarry dust has a rating of five to nine. Common contaminants like rust and black iron oxides have a Mohs hardness rating of five to six (on a scale of 1–10, with 10 being the hardest). From the manufacturing process, tool steel has a hardness rating of six to seven. Silicon carbide and aluminum oxide have a hardness rating of nine.

14.4.1.2 Built-In Contamination

Contamination can enter gearboxes during manufacturing. Filters do not immediately remove built-in manufacturing debris. Consequently, permanent debris dents and other damage may occur during run-in, unless gearboxes are assembled in a clean room using clean assembly procedures and then filled with clean lubricant.

There are many sources of contamination to eliminate long before the gearbox is placed into service. For instance, the interior of gear housings should be painted with white epoxy sealer to provide a hard smooth surface that is easy to clean, seals porosity, and seals in debris like casting sand. All components for assembly should be properly stored in a dry area prior to assembly. All gears and bearings should be covered, and bearings should be stored on their sides.

All components should be cleaned prior to assembly. Initial cleaning should be done in an area separate from the clean room, followed by final cleaning in the clean room just prior to assembly. All components should be carefully inspected to ensure they are clean and rust-free before assembly. Pay special attention to bolt holes, oil passages, and other cavities that may contain dirt.

Gearboxes should be assembled in a clean room separate from any manufacturing processes such as machining, grinding, welding, or deburring. Windows and doors should be adequately sealed to prevent contamination ingression and the ventilation system should be filtered so it provides clean, draft-free air. The floor should be painted and sealed so it is easily cleaned and the overhead structure should be painted and dust-free. No tow motors should be allowed in the clean room because they invariably introduce contaminants.

14.4.1.3 Internally Generated Contamination

Contamination particles that are generated inside gearboxes are usually wear debris from gears, bearings, splines, or other components. The contamination results from various wear modes, such as micropitting, macropitting, adhesion, abrasion, or fretting corrosion (see Chapter 13).

Fretting corrosion may occur on gear teeth, splines, and bearings in wind turbines which are parked for extended periods. Wet clutches and brakes also can produce wear debris. To avoid gear oil contamination, wet clutches and brakes should have separate lubrication systems.

To minimize internally generated wear debris, it is recommended to use high-viscosity lubricants. The use of the following features in gearbox can also minimize generation of contamination particles: accurate and smooth surfaces, surface-hardened gears and splines. To prevent fretting corrosion, external and internal spline teeth should be nitrided and force-lubricated. Annulus gears should be carburized or nitrided rather than through-hardened because through-hardened gears are relatively soft and prone to generating wear debris [11].

14.4.1.4 Required Oil Cleanliness for Wind Turbine Gearboxes

Lubrication systems should be properly designed and carefully maintained to ensure gears receive an adequate amount of cool, clean, and dry lubricant. Modern filters are compact and provide fine filtration and long life without creating large pressure drops. Off-line filters provide fine filtration during operation and during turbine shutdown. Once the oil is clean, it should stay clean provided the gearbox and lubrication system were properly designed and seals, breathers, and maintenance are adequate.

An industrial standard AGMA/AWEA 6006-A03 [12] is written by the American Gear Manufacturers Association in cooperation with the American Wind Energy Association. It provides guidelines for specifying, selecting, designing, manufacturing, procuring, operating, and maintaining gearboxes for use in wind turbines.

AGMA/AWEA 60062 discusses many elements of the lubrication system that determine oil cleanliness. This Standard sets up the oil cleanliness levels in gearbox (Table 14.2), which comply with international ISO 4406:1999 Fluid Cleanliness Standard [13].

To improve oil cleanliness in wind turbine gearboxes, it is recommended to use off-line filter system, which is simple and easy to install [14]. The benefits of the off-line filter include improved lifetime of the gear oil and gearbox bearings, reduced wear and tear on gearbox bearings, and reduced risk of bearing damage due to poor oil cleanliness.

14.4.2 Role of Water Contamination of Oil

Many experiments have shown water in oil promotes both micro- and macropitting through loss of oil film. Water interferes with the pressure–viscosity coefficient of the oil, diminishing its ability to momentarily solidify in the contact area.

It was found that the primary cause of reduction of bearing life was occasional passage of microscopic water droplets under high pressure through the lubricating zone, resulting in local lubricating

TABLE 14.2
Required Oil Cleanliness for Wind Turbine Gearboxes

Source of Oil Sample	Required Cleanliness[a]
Oil added to gearbox	16/14/11
Oil from gearbox after factory testing	17/15/12
Oil from gearbox after 24 hour in service	17/15/12
Oil from gearbox in service	18/16/13

Source: Adapted from ANSI/AGMA/AWEA 6006-A03 Standard for design and specification of gearboxes for wind turbines; Section 6—Lubrication, American Gear Manufacturer's Association, 2004 and ISO Standard 4406:99 Hydraulic fluid power—Fluids—Method for coding the level of contamination by solid particles, International Organization for Standardization, 12/01/1999.

[a] Three numbers a/b/c correspond to the number of particles larger than 4 micron (a), 6 micron (b), and 14 micron (c), respectively, in particle content of 1 milliliter of oil.

film breakdown. The number and the size of the microscopic droplets increase with the amount of water in lubricating oil, thus increasing the probability of water passing through the lubricating zone.

Excess water in wind turbine gear oil is associated with many negative effects. Some of these listed in the 2003 ANSI/AGMA/AWEA 6006-A03 wind turbine documents are

- Accelerated additive depletion
- Accelerated oxidation
- Interference with oil film formation
- Contributes to foaming
- May plug filters
- May cause corrosion etch pits and initiate fatigue cracks
- May lead to hydrogen embrittlement promoting fatigue cracks

Another role of water presence in oil, particularly with water concentrations above 300 ppm is the tendency of oils to form residues at high temperatures, in the form of sludge and varnish. This not only accelerates the aging of the base oils but also causes additives to precipitate out or reduces their effectiveness. Thickened oil/water emulsions also suspend abrasive particles in lubricants and cause surface damage by indenting and scratching the metal surface, causing stress concentrations, and disrupting the lubricant film.

Therefore, it is important to avoid using oil which is susceptible to water contamination. Water presence in oil shortens the life of bearings significantly. The curve showing the effect of water concentration in oil on bearing life is shown in Figure 14.2. It was found that ester-based lubricants and mineral oils with EP or antiwear additives are especially prone to absorbing water [15].

14.4.3 BASIC REQUIREMENTS TO QUALITIES OF GEARBOX OIL

Additional qualities are required for offshore applications, such as outstanding rust and corrosion protection from salt water. Since wind turbines operate in a wide range of environments: extreme temperatures and intermittent operations, the lubricants should provide flowability at low temperatures and antiwear capability at high temperature.

The minimum requirements for a good wind turbine gear oil [2,16] include long oil life (minimum 3–5 years), constant wear performance as the oil ages, thermal and oxidation stability, and

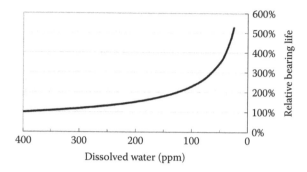

FIGURE 14.2   Effect of water concentration on bearing life. (Adapted from Majumdar S. et al. A new membrane system for efficient removal of water from gear oil – road to commercialization, *NREL Wind Turbine Condition Monitoring Workshop*, Broomfield, CO, September 2011.)

resistance to sludge formation. This oil should perform in a wide temperature range, providing cold startup and high operating temperatures.

Good oil should display stability with water and condensation thus provide rust and corrosion protection and oil should not affect the filter service interval. Cleanliness of new oil should be no less than 16/14/11 and 18/16/13 of used oils (see Table 14.2). Such oils should have extreme pressure additives with a high load-carrying capacity and additives that prevent micropitting. Wind turbine oil should be compatible with elastomers and paints, passing static, dynamic, and long-duration tests (at least 1000 h), foam tests, micropitting tests, and bearing antifriction tests. The oils should be tested on formation of residues under the influence of water and temperature.

14.4.4   Types of Wind Turbine Gearbox Oils Meeting Requirements

Mineral oils cannot meet the basic demands, and operators of wind power plants are turning to synthetic oils for the range of benefits they offer, including improved thermal resistance, better viscosity characteristics, product longevity, and longer machine component life. Different base oils (polyalphaolefin (PAO), polyglycol, or rapidly biodegradable ester) are used to formulate these gear oils.

Today, however, many of these traditional synthetic oils also cannot meet the new requirements created by changes in the wind power industry. As a result, operators are turning with increasing frequency to new, higher performance synthetic oils. The lubricant industry is responding with specialty products that meet and even exceed the standards set before them. Lubricants' manufacturers are looking for new lubricants that offer high thermal resistance and resistance to oxidation, lower change in viscosity at rising or falling temperature, lower friction coefficients, high wear protection for bearings and gears, good load-carrying capacity in bearings and gears, and low residue formation. These lubricants should offer extended service life and economical operation. Developing these lubricants requires knowledge of additives chemistry, and also which combination of additives and base oils to use. The more pure the molecular structure of the base oil, the better the lubricant [17].

Like the traditional synthetic oils, new, high-performance synthetic oils are subject to the tests of original equipment manufacturers (OEMs) and must meet a number of universal standards. For example, industrial gear oils are classified in accordance with German Standard DIN 51 517 [18]. Part 3 of this standard defines the requirements for gear oils that are exposed to high loads. In addition to the usual tests on viscosity, pour point, foaming characteristics, steel and copper corrosion, the scuffing load characteristics of the oils are determined in the FZG (Gear Research Center) scuffing load test. DIN 51 517, Part 3, stipulates a scuffing load stage greater than 12 for gear oils.

According to these elevated requirements, the gear oil testing includes multiple tests, which the oil must pass prior to choosing a particular oil to use in gears [16,19]. The list of some of these tests is presented in Table 14.3.

Because gear oils should also be suitable for lubricating the rolling bearings in the gearbox, the revised standard DIN 51 517, Part 3, also contains the FE 8 rolling bearing test rig developed by the rolling bearing manufacturer FAG. The FAG FE 8 test rig can be used to assess the antiwear properties of oil and its effect on the rolling bearing service life. In this test, the wear of the rolling elements should not exceed 30 mg.

The assessment of gear oil performance for wind turbines also includes tests that measure scuffing load resistance and micropitting resistance. A test developed by FZG, measures antiwear properties of the lubricant at low gear speeds, as the planetary gear stage is run at the lowest speed. In this test, better performing lubricants fall within the low wear category.

TABLE 14.3
Tests Qualifying the Oil for Use in Wind Turbine Application

No	Property	Test Name	Standards	Test Condition	Test Passed
1	Micropitting protection	FZG micropitting Test	FVA Proc No. 54	@ 60°C @ 90°C	
2	Lubricant life	Test method for life of lubricating greases in ball bearings at elevated temperatures	ASTM D3336 −05e1	Number of hours until failure of grease at temperature up to 371°C @ 10000 rpm	Product life estimated as number of hours before 10% of grease failed or hours to 50% failure
3	Oxidation control, stability, viscosity index	Oxidation test, L-60-1 thermal oxidation test, viscosity index	ASTM D2893, ASTM D5704, ASTM D2270	Hours required for % viscosity and TAN increase	40–120 hours
4	Low temperature flow	Low T viscosity Pour point	ASTM D6821 ASTM D2983 ASTM D97	Viscosity measured at temperature from −40°C to −26°C	
5	Filterability	Filterability of lubricating oils	ISO 13357	Oil with 1% water preheat to 70°C (158°F) pass through a 3 micron filter	Pass determined by pressure differential, oil unacceptable if oil blocks filter
6	Water tolerance	Demulsibility of industrial gear oils	ASTM D1401 ASTM D2711	Water added to oil at 130°F, stir 5 min at 1500 rpm, time to water/ oil separation in min	15 min
		Water tolerance	ASTM D6422	Residue after contact with 1% water	No residue in 24 hours @ 71°C (160°F)
7	Foam control	Foam tendency test	ASTM D-892	Measure volumetric % of foam formed in testing cell filled with oil and turned gears for a certain time at 25°C	Pass if < Vol. 15% of foam
8	Air release	Air release test	ASTM D3427		<20 min
9	Rust and corrosion protection	Moisture corrosion resistance of automotive gear lubricants	ASTM D7038 −15	Signs of rusting in roller bearings rotating for 110 hours in oil in the presence of water (distilled, salt, or synthetic sea water)	On rating scale from 0 (no rust) to 5 (heavy rusting), 5 is not acceptable

Gear efficiency is determined to a large extent by the friction characteristics of the lubricating oil. The friction coefficients of different base oils can be seen in the result of the FZG test rig. Today's new gear oils can reduce temperatures by as much as 68°F (20°C) and power losses by as much as 18% when compared with standard gear oils. Lowering the friction component in a wind station improves efficiency, increases power output, and generates additional income.

14.5   LUBRICATION-RELATED FAILURES OF BEARINGS AND GEARBOXES

Lubricants are supposed to reduce the susceptibility to failures, but if they become highly contaminated (e.g., if they accumulate dust, dirt, water, salt, etc.), they will progressively lose their effectiveness. Even if a bearing or gearbox is designed and manufactured for a specific application, premature damage and wear may occur because of contamination, misuse, starved lubrication, or all three.

Many of the lubrication issues that face the wind industry are very similar to those that exist in general electrical industry (see Chapter 8). Many lubrication pitfalls are noted while analyzing used wind generator oil and grease samples. Some of the factors that may play a significant role in promoting bearing failures are summarized in [9].

Lubrication quantity: An insufficient amount of grease and oil will lead to excessive wear and overheating, eventually leading to catastrophic failure. Excessive amounts of grease allow the rollers to slide in the journal, causing heating and scoring. Conversely, large amounts of grease can migrate into the generator windings creating potential contamination and electrical insulation issues.

Lubrication storage: Humidity, moisture, sand, and other contaminants are typically found in many samples. A drum of oil stored in the sun and allowed to cool during the night can collect as much as a gallon of water within a few weeks of being stored in such conditions. These same conditions will occur within lubricated systems like hydraulic reservoirs and gearcases. In these conditions, water-absorbing filtration systems should be considered.

Lubricant corrosion: Red/brown areas on balls, raceways, cages, or bands of ball bearings are symptoms of corrosion. This condition results from exposing bearings to corrosive fluids or a corrosive atmosphere. In extreme cases, corrosion can initiate early fatigue failures.

Lubrication sampling/trending: Contamination of lubricants by dirt, wear metals, water, or other particulate matter cause over 70% of lubrication related equipment failures.

Lubricant incompatibility: Incompatibility can be caused by something as simple as mixing two different greases or oils to the same lubricated component. Incompatible greases may be indicated when the oil in the grease begins to leak past the bearing seal within minutes of regreasing a bearing. Incompatible gear oils may cause oil seals to soften, shrink, or harden, resulting in a leak. Excessive foaming in a hydraulic reservoir may be the result of mixing two incompatible hydraulic oils.

Lubricant color: As oils age in service, it is normal for them to thicken and become darker in color. If an industrial oil becomes thicker and turns brown or almost black in color in an abnormally short period of time, the cause is almost always temperature related. Excessive operating temperatures will cause the oil to oxidize prematurely causing the viscosity to increase and the color to darken. Greases can also oxidize if a bearing is continually over greased.

Lubricant failure: Discolored (blue/brown) ball tracks and balls are symptoms of lubricant failure. Excessive wear of balls, rings, and cages will follow, resulting in overheating and subsequent catastrophic failure. Ball bearings depend on the continuous presence of a very thin film of lubricant between balls and races and between the cage, bearing rings, and balls. Failures are typically caused by restricted lubricant flow or excessive temperatures that degrade the lubricant's properties.

Lubricant contamination: Many of the gearbox and generator manufacturers utilize epoxies and specialty glues to attach bearing covers, inspection ports, and filler caps. Many times traces of large amounts of these substances are found in grease and oil samples, or worse implicated in bearing failures of gearboxes and generators.

See also Section 14.4.1 for the sources of oil contamination. The largest majority of sample data have shown that high ferrous and ferrous particles are usually present. Large amounts of sand and epoxies contribute to excessive wear, overloading, and overheating that eventually leads to catastrophic bearing failure.

14.6 LUBRICANTS FOR WIND TURBINE MECHANICAL PARTS RECOMMENDED BY LUBRICANT MANUFACTURERS

Based on general recommendations and challenges for lubrication of specific components in wind turbines, lubricant manufacturers worldwide develop multiple products. These companies define the properties of their products based on requirements for the greases and pastes to provide long life of the lubricants in various climates and environments where the wind turbines are installed.

14.6.1 FUCHS PETROLUB AG

The two sister companies Fuchs Europe Schmier-Stoffe and Fuchs Lubritech have joined forces in the wind power sector under the name of Fuchs Windpower Division. The two companies are wholly owned subsidiaries of Fuchs Petrolub AG, an independent lubricant manufacturer. Fuchs Lubritech manufactures greases and grease pastes and another subdivision Fuchs specializes in producing gear and hydraulic oils, both companies are parts of Fuchs Windpower Division. Table 14.4 presents multiple lubrication products manufactured by Fuchs. More information about each product from this table could be found in Ref. [20]. The location of lubrication points for Fuchs lubricants' applications in wind turbines are given in Ref. [21].

Various oils by Fuchs demonstrate excellent properties. *RENOLIN MR 90* is a special rig oil providing corrosion protection; exhibiting excellent cleaning and flushing properties, and very good wear protection. *RENOLIN CLP VCI* is an EP gear oil with excellent corrosion protection properties, it contains specially developed VCI-components (VCI = volatile corrosion inhibitors) for safe storage and transport of machines and components. Machine elements and gears are protected against corrosion also in the vapor phase without direct contact of the oil with the metal surface.

RENOLIN HighGear synthetic gear oil is an industrial gear oil based on PAO and newest additive technology and plastic deformation (PD) technology; this oil is recommended for predamaged gear sets and critical applications.

RENOLIN Unisyn CPL range from Fuchs is one of the most widely used wind turbine gear lubricants. These are industrial gear lubricants based on PAO, developed in conjunction with Flender and approved by several leading OEMs such as FAG and Hansen. These lubricants are specially developed for use in environments where extreme temperatures occur, even with short-term peak temperatures up to 150°C.

Fuchs-manufactured greases developed for use in bearings. For example, special lubricant for pitch and yaw bearings (tooth system and bearing) is *Gleitmo 585 K*. Gleitmo 585 K is a fully synthetic special lubricant containing a synergistic combination of white solid lubricants. This combination offers excellent protection against wear, especially under critical operation conditions like vibrations and oscillation movements under high load which are typical for pitch and yaw bearings on wind turbines. Gleitmo 585 K is well known in the wind power industry and used as OEM first fill and service lubricant since many years with best results. It is also recommended for use for the tooth system of the yaw and pitch bearings.

TABLE 14.4

Fuchs Lubricants for Various Wind Turbine Applications

Lubricant	Application
STABYL EOS E2	Pitch adjustment bearing, tooth system, rotor bearing, clutch, yaw bearing
STABYL LT 50	Pitch adjustment bearing, tooth system, yaw bearing
Gleitmo 585 K	Pitch adjustment bearing, tooth system, yaw bearing, clutch
Gleitmo WSP 5040	Fasteners, assembly aids
CEPLATTYN BL White	Tooth system
CEPLATTYN BL	
CEPLATTYN 300	Chain hoist
ECO HYD S-Range	Hydraulic system
PLANTOHYD-Range	
RENOLYN UNISYN CPL 220	Yaw system reduction gear
RENOLYN PG 220	Yaw system reduction gear
RENOLYN CLP 220	
RENOLYN UNISYN OL 46	Hydraulic system
RENOLYN HVLP-Range	
Cleaners, Rust Removers and Preventives	**Application**
Rivolta S.L.X. Top	Slip ring cleaner
FERROFORM LOCC	Rapid rust removers
FERROFORM ECO LOCC	
ANTORIT CPX	Waxy rust preventive
DECORDYN HF 91 DECORDYN 350	Waxy rust preventive

Another product *STABYL EOS E 2* is a high-performance grease for wind power applications (tooth system and bearing). STABYL EOS E 2 is based on a fully synthetic ester with lithium soap as thickener. It fulfils the highest technical requirements for modern lubricants used in wind turbines. STABYL EOS E 2 was developed in a perennial research project in cooperation with leading bearing manufacturers and is successfully in use on wind turbines as general purpose lubricant. It is also recommended for the tooth system of the yaw and pitch bearings. Both greases work in a wide temperature range from −45°C up to 130°C, with NLGI consistency grade 2, suitable for all types of climate conditions.

14.6.2 Dow Corning, Molykote

Wind turbines require different types of special lubricants for the hydraulic circuits and brakes, greases for the slewing rings and bearings. Among the most critical components of the nacelle are the blade bearings which have to operate under extreme stresses and operating conditions. Together with design changes and component development, improved lubricants and adapted maintenance intervals with automatic lubricant dispensing systems have contributed to the growth of the higher power of the latest wind energy technology.

Among many lubrication products, Dow Corning Corporation manufactures pastes for threaded connections: Molykote®1000 and Molykote® G-Rapid Plus pastes. Typical properties of these pastes [8] are presented in Table 14.5.

Specialty lithium or calcium soap greases with mineral, PAO, or ester base oils are in use for wind turbines. Synthetic greases are used when compatibility to the plastic materials is involved. In

TABLE 14.5
Typical Properties of Molycote Pastes[a] for Threaded Connections

Property, Unit	Molycote	Pastes	Test	Standard
Paste	1000	G-Rapid Plus		
Color	Brown	Black	Saybolt color code, colorimeter	DIN 51 411 ISO 2049
Consistency, mm/10	280–310	255–275	Unworked penetration	DIN ISO 2137
Density, g/mL	1.26	1.40	Density at 20°C (68°F)	ISO 2811 DIN 51 757
Working temperature, °C (°F)	−30 to 650 (−22 to 1202)	−35 to 450 (−31 to 842)	Temperature resistance of solid lubricants	
Extreme pressure properties, N	4800	5300	Four-ball tester welding load	
Adhesive properties of lubricant, N	20,000	>20,000	Almen-Weiland machine, OK Load	DIN 51 350, Pt. 4
Coefficient of friction in bolted connections	0.08–0.13	0.06–0.10	Screw test-μ head	

[a] Both pastes are solid lubricants in mineral oil.

addition, solid lubricants and anticorrosion additives can ensure high wear reduction and less corrosion leading to the long service intervals needed.

With a trend in wind farms moving offshore so as to capture maximized wind speeds and therefore run more efficiently, maintenance will increasingly become more difficult. This emphasizes the importance of proper lubrication at the design stage. To date, very poor maintenance of wind turbine lubrication has been seen in the field. But this could be improved by using the most suitable lubricant for every particular environment and operating conditions. For example, if the life of the bearings which operate under high load in a wet, corrosive environment in wind turbine is threatened to shorten, the equipment life can be extended by applying Molykote G-0102 heavy duty bearing grease, which is well suited in the presence of water.

14.6.3 ExxonMobil, Mobil

Mobil (ExxonMobil) manufactures multiple products for lubrication on different parts of wind turbines [7,22–24]. One of the most recommended products by Mobil is synthetic gear oil Mobilgear SHC XMP series for use in main gearbox. This oil provides all-around protection against gear and bearing wear, such as micropitting and scuffing wear. It provides excellent foam and air release, outstanding stability in the presence of water contamination and corrosion protection.

Another available oil is synthetic hydraulic oil Mobil SHC, which provides maximum antiwear protection while being shear stable, working in a wide temperature range, and resistant to deposit formation. Mobil SHC 600 Series synthetic gear and bearing oils are recommended for use in ancillary wind turbine gearboxes. These oils demonstrate excellent high- and low-temperature capability and provide superior wear protection; they are highly resistant to oxidation and slugging and have low traction coefficient.

Mobil also produces a number of synthetic greases for use in open gears, such as Mobil SHC Grease 460WT for lubrication of main, pitch, and yaw bearings and Mobil SHC 100 for generator

TABLE 14.6
Lubrication Points for Mobil Lubricants

Lubrication Points	Mobil Products
Generator bearings	Mobilith SHC 100
Gear box	Mobilgear SHC XMP 320
Main shaft bearing	Mobil SHC Grease 460WT
Pitch gear	Mobilgear SHC XMP Series
	Mobil SHC 600 Series
Pitch bearing	Mobil SHC Grease 460WT
Yaw gear	Mobilgear SHC XMP Series
	Mobil SHC 600 Series
Yaw bearing	Mobil SHC Grease 460WT
Hydraulic syatem	Mobil SHC 524
Open gear	Mobilac 375 Non Synthetic
Open gear	Mobilgear OGL 007

bearings. For open gears, Mobil recommends nonsynthetic lubricant Mobiltac 375 NC, which is nonleaded, diluent-type lubricant for heavily loaded open gears. It provides excellent protection of gear teeth and other machine elements under boundary lubrication conditions. Another product for the same application is Mobilgear OGL 007, which is manufactured based on advanced technology, contains extreme pressure additives and graphite for heavy loads, it is lead and chlorine free and adhesive in nature, and use of this grease reduces environmental impact. Parts where Mobil lubrication products are applied [24] are listed in Table 14.6.

14.6.4 SHELL LUBRICANTS

Shell manufactures several lubrication products for wind turbines and distributes them through European blade bearing suppliers such as IMO, Liebherr, Rollix, and Rothe Erde. Shell lubricants are used by wind turbine manufacturers including Vestas, Acciona, Gamesa, Dong Fang New Energy Equipment, Sinovel Wind Group, and Siemens Wind Power. Recently, Shell introduced wind turbine portfolio to North America [25].

Lubrication products from Shell recommended for use in wind turbines include greases, gearbox oils, and hydraulic fluids.

Shell Rhodina BBZ blade-bearing grease is designed to provide protection to bearings against fretting corrosion, moisture contamination, and false brinelling at temperatures as low as −55°C. This grease is lubricating the blade bearings of many wind turbines globally. *Shell Omala S4 GX 320* synthetic gearbox oil provides excellent protection against common failure modes, including micropitting and bearing wear. Offering excellent low-temperature fluidity and long oil life, Shell Omala S4 GX 320 provides benefits for difficult-to-maintain wind turbine gearboxes. Other Shell lubrication products are Shell Gadus S5 V100 2—high-speed, low-temperature generator bearing grease, and Shell Tellus S4 VX hydraulic oil.

Shell Tellus Arctic 32 is used as the hydraulic fluid for extreme-climate wind turbines, used by wind turbine OEMs including GE Wind, Voith Wind, Vestas, Dongfang Wind Turbines, Sinovel, RePower, Nordex, and DHI. The product has demonstrated its performance in the harsh winters of Mongolia, Scandinavia, and the Americas at temperatures as low as −40°C (−40°F).

In addition, Shell Lubricants also offers Shell Tivela S 150 & 320 synthetic gear oil for yaw and pitch drives, Shell Albida EMS 2 electric motor bearing synthetic grease, Shell Stamina HDS main bearing grease, and Shell Malleus GL & OGH premium quality open-gear grease.

14.6.5 Total Lubricants

Total Lubricants manufactures many gear oils, hydraulic and transformer fluids, and greases and coolants specifically designed for wind turbine applications [26]. These lubricants are presented in Tables 14.7 and 14.8.

14.6.6 Castrol

Castrol [27,28] offers a full range gearbox oils and greases for lubricating bearings and gear in multiple locations in wind turbines as shown in Figure 14.3. Castrol lubricants are listed in Tables 14.9 and 14.10.

14.6.7 Klüber Lubrication

Companies like Midland and Texas-based Global Wind Power Services, providing service to the wind power industry, recommend Klüber lubricants to their customers. Global uses as many as 10 different Klüber Lubrication products in the wind turbines it services. Klüber synthetic oils perform well, reducing heat in gearboxes and prolonging the use of the oils.

According to the customers [17], these oils keep gears running smoothly and extend the interval between oil changes. The Klüber lubricants line include *Klübersynth GEM 4 N*—polyalphaolefin, *Klübersynth GH 6*—polyglycol, and *Klübersynth GEM 2*—rapidly biodegradable ester. Each of the products complies with or exceeds performance parameters stipulated in the standards currently in place.

TABLE 14.7
Gearbox Oils from Total Lubricants

Carter Gearbox Oil	Type	Corrosion Protection in Sea Water	Interval between Oil Change	Protection against Teeth Micropitting	Special Properties
XER 320	Mineral	Yes	Up to 3 years	Yes	
SH 320	Synthetic PAO		Up to 5 years	Yes	Excellent for very low temperatures, miscible with mineral oil
SY WM 320	Synthetic PAG			Yes	Oxidation stability
BIO 320	Synthetic	Yes			Biodegradable >75% after 28 days

TABLE 14.8
Bearing Greases from Total Lubricants

Grease	Thickener, Complex	Working T Range, °C (°F)	Application	Specifications
Multis Complex SHD 100	Lithium	−50 to 160 (−58 to 320)	Generator bearings High rotation speeds	ISO 6743-9: L-XEEHB 00 DIN 51502: KP00P-50
Multis Complex SHD 220	Lithium	−50 to 160 (−58 to 320)	Generator bearings Moderate rotation speeds	ISO 6743-9: L-XEEHB 2 DIN 51502: KP2P-50
Multis Complex SHD 460	Lithium	−40 to 160 (−40 to 320)	Main shaft, pitch and yaw gear, slow to moderate rotation speeds	ISO 6743-9: L-XDEHB 1/2 DIN 51502: KP1/2P-40
COPAL OGL 0	Aluminum		Solid lubricant: reduces friction, protects parts from wear	ISO 6743-9: L-XBDHB 0 DIN 51502: OGPF0N-20

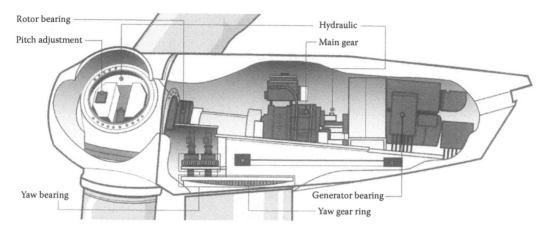

FIGURE 14.3 Lubrication points in wind turbine recommended by Castrol. (Adapted from Tschauder K. and Leather J. Fundamentals of lubrication gear oil formulation, *Wind Turbine Tribology Seminar NREL-Argonne-DoE*, Broomfield, CO, November 2011.)

TABLE 14.9
Castrol Oils for Wind Turbines

Castrol Oils	Base Oil	Antiwear Additives	ISO Viscosity Grade (mm²/s)	Working T Range °C (°F)	Application
Optigear Synthetic X	PAO	2nd Generation MicroFlux Trans, MFT 2	100–680	−30 to 95 (−22 to 203)	Cylindrical, bevel gears and planetary gears, oil-lubricated rolling bearings
Aptigear Synthetic A	PAO	MicroFlux Trans, MFT	100–320	−30 to 95 (−22 to 203)	Sliding and rolling bearings, industrial gears at high temperatures and under high mechanical loads
Tribol 1710	Mineral/PAO	Tribol Grease and Oil Additive, TGOA	220–460	−30 to 95 (−22 to 203)	Gear oil with special high-performance additives for wind-power gears
Tribol Bio Top 1418	Ester	Extreme Pressure, Anti-Wear, EP/AW	150–460	−25 to 90 (13 to 194)	Gears, rolling and sliding bearings, as well as in revolving equipment, biodegradable
Tribol 1100	Mineral	TGOA	68–1500	−20 to 90 (−4 to 194)	Modern cylindrical and worm gear pairs, rolling and sliding bearings, revolving equipment
Optigear BM	Mineral	MicroFlux Trans, MFT	68–1500	−10 to 90 (14 to 194	Spur and bevel gear units, even under severe operating conditions, worm gear units, rolling and sliding bearings, gear couplings, circulating systems
BP Bartran HV 32	Mineral	Zinc-free	32	−20 to 80 (−4 to 176)	Severely stressed hydraulic systems requiring a high level of antiwear performance or ultrafine filtration

TABLE 14.10
Castrol Greases for Wind Turbines

Castrol Greases	Base Oil	Thickener, Additive	NLGI Grade	ISO Viscosity Grade, mm²/s	Working T Range, °C (°F)	Application
LONGTIME PD 2	Mineral	Lithium-12-Hydroxystearate, MTF	2	100	−35 to 140 (−31 to 284)	High-speed rolling and sliding bearings subjected to high mechanical load
TRIBOL 4020	Mineral	Li Complex	1 2	220 460	−30 to 150 (−22 to 302)	Sliding and rolling bearings, medium and high mechanical load conditions, water resistant
TRIBOL 3020/1000	Mineral	Li	000, 00, 0, 1, 2	1000	−40 to 120 (−40 to 248)	Rolling and sliding bearings at low speeds, gears that are not oil-tight, general grease lubrication
OPTIPIT	Mineral	Lithium-12-Hydroxystearate, MTF	2–3	1350	−10 to 140 (14 to 284)	Rolling and sliding bearings with MFT, large units with low peripheral speeds in damp, dusty environments, as well as for open gears
MOLUB-ALLOY 936 SF HEAVY	Mineral	Mixed Soap, MoS2	0	1660	−15 to 100 (5 to 212)	Parts with high mechanical loads and at low speeds, water-resistant, solvent-free
ORTITEMP TT1	Synthetic	Organic/inorganic	1	15/22	−60 to 120 (−76 to 248)	Long-term lubrication at high speeds, low T or where large T differences exist
MOLUB-ALLOY 3036/680-1	Mineral	Li	1	680	−20 to 120 (−4 to 248)	Rolling and sliding bearings at low speeds and high loads or shock loads in unfavorable ambient conditions

14.7 AUTOMATIC LUBRICATION FOR WIND TURBINES

Like any mechanical system, a wind energy system needs proper lubrication to function optimally. Vibration, high mechanical loads, contamination, and moisture are all threats to bearing and gear service life. However, wind turbines can be challenging and expensive to service. They can reach more than 100 meters off the ground and are often in remote and difficult-to-access locations. The solution is an automatic lubrication system.

Compared with manual lubrication, automatic lubrication systems provide lubricant supply more reliably and precisely to moving components in the nacelle. By delivering the smallest effective amount of lubricant reliably to all friction points while the machine is running, automatic lubrication systems reduce friction inside bearings and help prevent contamination. The result is optimized bearing service life over the long term, more turbine uptime, and reduced manpower costs—all combining to help make wind farms more profitable.

Automatic lubrication systems can provide a quick return on investment by increasing turbine system availability, extending maintenance intervals, and preventing failures of major components [28]. Additional savings can be achieved through proper lubricant handling and consumption. SKF and Lincoln automatic lubrication systems are unmatched in their ability to deliver these benefits and many more.

REFERENCES

1. Guerzoni F. Challenging lubrication application, *World Wind Technology*, May 2013, Online. Available: http://www.windpower-international.com/features/featureworld-wind-technology-shell-lubricants-dr-felix-guerzoni-bearing-grease/.
2. Van Rensselar J. Extending wind turbine gearbox life with lubricants, *Tribology and Lubrication Technology*, 69(5), 40–49, May 2013, Online. Available: http://callcenterinfo.tmcnet.com/news/2013/05/25/7161084.htm.
3. Holm A. Specialty lubricants for optimum operation, *Wind Systems Magazine*, September-October issue, 42–47, 2009, Online. Available: http://windsystemsmag.com/article/detail/23/specialty-lubricants-for-optimum-operation.
4. Specialty lubricants for the wind energy industry, Dow Corning Corporation, 2010, Form No. 80–3452A-01, Online. Available: http://www.dowcorning.com/content/publishedlit/80-3452.pdf.
5. Vanecek R. *Lubrication Technology and Increasing the Reliability of Wind Turbine Components*, Dow Corning Corporation, Form No. 80-3682-01, 2011, Online. Available: http://www.dowcorning.com/content/publishedlit/80-3682.pdf.
6. Lugt P. M. *Innovative Bearing Technology for Wind Turbines*, SKF Engineering & Research Center, November 7, 2012, Online. Available: http://www.ltu.se/cms_fs/1.99723!/file/Lugt-Tribodays%202012%20Wind%20Turbines.pdf.
7. Braun H. R. Wind turbine grease lubrication, *NREL Wind Turbine Tribology Seminar*, Broomfield, Colorado, November 15–17, 45pp., 2011, Online. Available: http://www.nrel.gov/wind/pdfs/day1_sessionii_3_exxonmobil_braun.pdf.
8. Key wind turbine lubrication points. *Lubricants Improve Efficiency and Longevity of Wind Turbines*, Dow Corning, July 2013, Online. Available: http://www.azom.com/article.aspx?ArticleID=4891.
9. Moore Mike. Lubricating wind components, *Wind Systems Magazine*, pp. 52–57, September 2011, Online. Available: http://windsystemsmag.com/media/pdfs/Articles/2011_Sept/0911_Shermco.pdf.
10. Sayles R. and Macpherson P. Influence of wear debris on rolling contact fatigue. *Rolling Contact Fatigue of Bearing Steels*. ASTM STP 771, pp. 255–274, 1982.
11. Muller J. and Errichello R. Oil cleanliness in wind turbine gearboxes, *Machinery Lubrication*, pp. 34–40, July/Aug 2002, Online. Available: http://www.machinerylubrication.com/Read/369/wind-turbine-gearboxes-oil.
12. ANSI/AGMA/AWEA 6006-A03 *Standard for Design and Specification of Gearboxes for Wind Turbines; Section 6—Lubrication*, American Gear Manufacturer's Association, 2004.
13. ISO Standard 4406:99 *Hydraulic Fluid Power—Fluids—Method for Coding the Level of Contamination by Solid Particles*, International Organization for Standardization, 12/01/1999.
14. Barrett M. P. and Stover J. Understanding oil analysis: How it can improve reliability of wind turbine gearboxes, *Gear Technology*, 102–109, November/December 2013, Online. Available: http://www.geartechnology.com/articles/1113/Understanding_Oil_Analysis:_How_it_Can_Improve_Reliability_of_Wind_Turbine_Gearboxes.
15. Cummins J. Wind turbine lubrication, Hydrotex Lubrication University, *STLE Houston Chapter Meeting*, 66pp., September 11, 2013, Online. Available: http://www.stlehouston.com/2HoustonSTLE/2013-2014/Program/Wind%20Turbine%20Lubrication%20STLE%20Sept%20Meeting%202013%20Rev.pdf.
16. Leather J. Fundamentals of Lubrication Gear Oil Formulation, US Department of Energy, National Renewable Energy Laboratory (NREL), *Wind Turbine Tribology Seminar*, A Recap, Broomfield, Colorado, USA, November 15–17, pp. 8–9, 2011, Online. Available: http://www.nrel.gov/docs/fy12osti/53754.pdf.
17. Hermann S. Wind turbines power up with oil, *Lubrication & Fluid Power*, September-October Issue 2006, Online, Available: http://www.klubersolutions.com/pdfs/_misc/wind_turbines_power_up_with_oil.pdf.
18. Standard DIN 51 517 "Lubricants – Lubricating Oils"—Part 1: Lubricating oils C, Minimum Requirements, PART 2: Lubricating oils CL, Minimum Requirements; PART 3: Lubricating oils CLP, Minimum Requirements, Publsher Deutsches Institut fur Normung E.V. (DIN), Revision/Edition: 14, Germany, Date: 02/00/14.
19. Daschner E. *Wind Turbine Lubrication Challenges and Increased Reliability*, MDA Forum, Hannover Messe, 2011, 15pp., Online. Available: http://www.vdma.org/documents/105806/139201/MDA_Forum_Vortrag_%20Daschner_ExxonMobil.pdf/a2146a9b-c880-4d08-a9ed-0ece26c2ca33.

20. *Special Lubricants for Wind Power Plants*, Fuchs Windpower Division, Online, Available: http://fuchs. pt/upload/mercados/21901_Windpower_Industry_Brochure.pdf.

21. Brazen D. Lubrication of wind turbines, *Kansas Renewable Energy Conference*, 17pp., October 2009, Online. Available: www.kcc.ks.gov/energy/kwrec_09/presentations/A2_Brazen.pdf.

22. *Mobilgear SHC XMP 320: Taking Wind Turbines to New Heights*, ExxonMobil Lubricants and Specialties, 2007, Online. Available: http://www.windenergynetwork.co.uk/enhanced-entries/ exxon-mobil/.

23. Taking on the challenge of grease lubrication in wind turbines, *Power & Energy Solutions (PES) Magazine, Europe*, pp. 109–110, February 2014, Online. Available: www.pes.eu.com/assets/articles/ exxonmobil.pdf.

24. *Mobil Lubricants and Greases for Wind Turbines*, Mobil Industrial, Online. Available: http://www. mobilindustrial.com/IND/english/files/br_wind.pdf.

25. *Shell Introduces Wind Turbine Lubricant Portfolio to North America*, 2011, On-Line. Available: http:// www.machinerylubrication.com/Read/25079/Shell-wind-turbine-lubricant-portfolio.

26. *Wind Energy Powered by Total Lubricants*, Online. Available: http://www.total-lubrifiants.ca/Pages/ content/NT00006FD6.pdf.

27. Tschauder K. and Leather J. Fundamentals of Lubrication Gear Oil Formulation, *NREL Tribology Seminar*, Broomfield, CO, November 2011, Online. Available: http://www.nrel.gov/wind/pdfs/day1_ sessionii_1_castrol_leather.pdf.

28. *Focus on Wind Turbines*, 8pp., Castrol, Online. Available: http://www.trademarkoil.com/PDF/ Castrol%20wind_turbines_brochure_EN.pdf.

29. *Optimizing Wind Farm Performance*, SKF-Lincoln, PUB LS/S2 14161 EN. January 2014, 11pp., Online. Available: http://www.skf.com/binary/56-141953/14161-EN_LBU-Wind-Support-Brochure_ spreads.pdf.

15 Wind Turbines Failures

15.1 CATEGORIES OF WT FAILURES

Two types of failure rates are suggested in Ref. [1] to apply to selected components, or categories of components (see Chapter 11). The first type of failure event is random, and is represented by a constant failure rate. The model assumes by default that 5% of the blades and gearboxes, and 10% of the generators, will fail over the 20-year life of the project because of uncontrollable circumstances such as lightning strikes, manufacturer defects, operational errors, or servicing omissions and errors.

The second type of failure event is wear out or deterioration, and is a two-parameter Weibull distribution. This distribution is commonly used in reliability studies as it allows for variation of the scale as well as the shape of the failure distribution. Weibull distributions are intended to describe failure rates for a given population of like components. Generally, the most reliable data are obtained from exercising the components in actual or simulated conditions that are consistent over time. In an actual application, however, the parts that fail are replaced, so that the population eventually becomes a combination of components with varying periods of operation. At some point in time past the characteristic life, the instantaneous failure rate will oscillate about, and finally approach, a constant value.

15.2 CAUSES OF FAILURES OF WTs

Every of the WT subassemblies and components may fail in course of operation. It was found that 17% of all failures happen in gearbox (12%) and generator (5%), yaw system fails in 8% of cases, and hydraulic system is responsible for 5% of all failures [1]. The overall account for failures of different parts of WT are graphically presented in Figure 15.1.

The "*failure rate*" is usually expressed by the number of failures per unit of time, which also could be used as a measure of *reliability*. The estimated failure rate of the electrical system is more than those of the structural and mechanical systems combined.

There are a couple of differences between failures and their prevention between onshore wind farms and offshore wind farms. The first major difference is one of the main arguments in favor of going offshore: wind flows tend to be stronger, more predictable, and more consistent offshore. Thus, it is possible to harness more energy with larger turbines and to better plan the operation of the remaining other sources in the power system with which the wind farm is integrated.

Since offshore winds are more consistent and have fewer gusts than onshore winds, the stresses on the structural and mechanical parts tend to be lower resulting in lower failure rates. It should, however, be noted that structural and mechanical failures result in much higher downtime and lost energy production than electrical failures, which are responsible for more failures than mechanical or structural systems, but are also comparatively easier to repair. Almost every turbine will require some sort of repair every year. The results from several reviews [2–4] are summarized in Ref. [5] and shown in Figure 15.2, which represents the probability of the location of the failure, given that a failure has occurred.

The second major difference is also one of the main technical challenges to ensure profitable operation of offshore farms: the operating environment of offshore farms is tougher than that of onshore farms, and the accessibility of offshore farms for maintenance is considerably more expensive than that of onshore farms. It is expensive to send crews on a helicopter or a boat to repair failures and perform scheduled maintenances, and these costs combined with the resulting lost

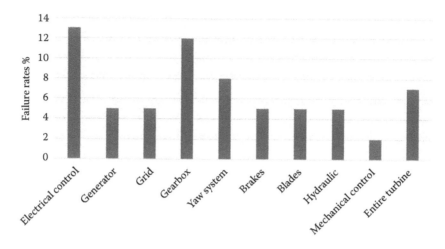

FIGURE 15.1 Causes of failures of wind turbines. (Modified from Poore R. *Development of an Operations and Maintenance Cost Model to Identify Cost of Energy Savings for Low Wind Speed Turbines*: July 2, 2004–June 30, 2008. NREL/SR-500-40581. Work performed by Global Energy Concepts, LLC, Seattle, Washington, Golden, CO: National Renewable Energy Laboratory, January 2008, Online. Available: http://www.nrel.gov/docs/fy08osti/40581.pdf.)

energy production reduce the profitability of the farm. Therefore, reducing failure rates and improving serviceability are greater concerns for offshore wind farms than they are for onshore farms [2].

There are other interesting conclusions from multiple research projects focused on collecting historical data on offshore and onshore wind farms [6,7]. For example, larger WT have been found to have a lower reliability (higher failure rate) than smaller turbines but downtime also decreases with size—so annual downtime is about the same. It was estimated that 75% of failures cause only

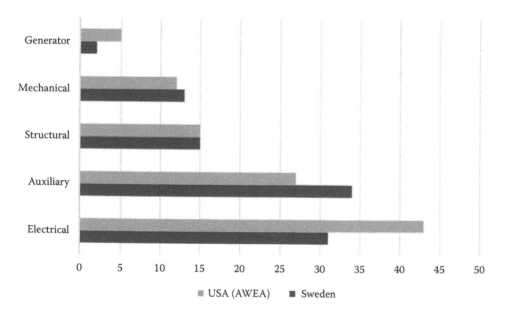

FIGURE 15.2 Distribution of failure type in wind turbine components in the United States and Sweden. (Adapted from Ragheb A. M. and Ragheb M. *Wind Turbine Gearbox Technologies*, Chapter 8, InTech, pp. 189–206, 2011, Online. Available: http://cdn.intechopen.com/pdfs-wm/16248.pdf.)

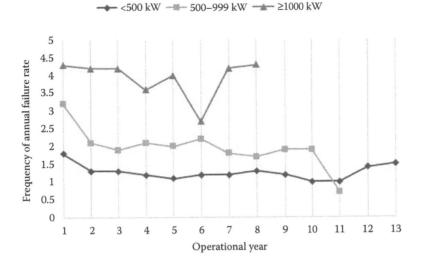

FIGURE 15.3 Frequency of failure rate of wind turbines with tncreasing operational age. (Adapted from Hahn B., Durstewitz M., and Rohrig K. *Reliability of Wind Turbines*, Online. Available: http://renknownet2. iwes.fraunhofer.de/pages/wind_energy/data/2006-02-09Reliability.pdf.)

5% of downtime and the other 25% cause 95% of downtime. Another observation is that the failure frequency of direct drive generators is generally higher than that of geared turbines (possibly due to the larger number of coils), but permanent magnet generators are likely to improve this. It was found that WT converters are unreliable compared to converters in other industries and this is a challenge the industry needs to face. Also, failure rates increase with higher wind speeds due to cycling, and the variability of wind, particularly so for electrical components.

The age of WT is also an important factor in the failure rate [8]. The failure rates of WTs depending on their operational age are presented in Figure 15.3, which shows that failure rates decrease with operational age for practically all types of WTs. This graph is based on over 60,000 reports on maintenance and repair, which have been submitted to Institut für Solare Energieversorgungstechnik (ISET) in Germany [9]. It also shows that for the WTs under 500 kW it can be expected that the failure rate due to "wear-out failures" does not increase before the 15th year of operation.

15.3 DATABASES OF COMMON WT FAILURES

For several decades of building and operating WTs all over the world, information on multiple failures was collected and published. A very important progress in building a reliable database of common WT failures was done by Fraunhofer Institute for Wind Energy and Energy System Technology (IWES) [10]. Information from several European countries was summarized in this study. To identify, quantify, and understand WT critical failures and their mechanism through quantitative studies of detailed wind farm data, the Reliawind project was created. It is a European Union 7th Framework Integrated Project with overall budget of 7.7 Mil euro involving 10 industrial and academic partners [11].

15.3.1 FAILURES IN WTs IN FINLAND

Failure statistics of WTs started in 1996 in Finland was analyzed in Ref. [12]. According to this analysis, gear failures make up the greatest part of the downtime caused by technical failures where the gear has been taken down from the nacelle and transported for repair. Manufacturing defects in the gear have also contributed to downtime, especially during the first 4 years of operation. The most common cause of downtime in connection with generators is the changing of slip rings, due to wear and tear. Rotor blade pitch mechanism faults are often connected to failures in the hydraulic system. Other sources of downtime in the pitching mechanism are failures in the bearings of the motors and sensor equipment.

Typical failures in the electrical system are defect components or fuses and burned cables. There are two different kinds of brakes in a WT: the mechanical brake and the aerodynamic brake (tip brake). These two are treated separately because the failures and causes of failures are different from each other. Yaw motor failure is one of the most common reasons for downtime and also the reason for the largest value of downtime, due to long waiting period for a new motor. In cold climate of Finland some icing problems are also connected to the yaw system.

15.3.2 Failures in WTs in Germany

Based on historical data collected on wind farms in Germany [8], the total number of failures of main components are shown in Figure 15.4.

The duration of downtimes, caused by malfunctions, are dependent on necessary repair work, on the availability of replacement parts, and on the personnel capacity of service teams. In the past, repairs to generator [13], drivetrain, hub, gearbox, and blades have often caused standstill periods of several weeks. The average failure rate and the average downtime per component are presented in Figure 15.5.

These data of *Failure* per turbine per year and *Downtime* are coming from two large surveys of onshore European WTs for the period 1993–2006. The scientific measurement and evaluation program WMEP database was compiled from 1989 to 2006 and contains failure statistics from 1500 WTs. Failure statistics published by the agricultural chamber of the German federal state of Schleswig-Holstein (LWK) cover 1993–2006 and contain data from more than 650 WTs.

As follows from the diagram in Figure 15.5, 75% of faults cause 5% of downtime, 25% of faults cause 95% of downtime. It gets clear, that the downtimes declined in the past 5–10 years. So, the

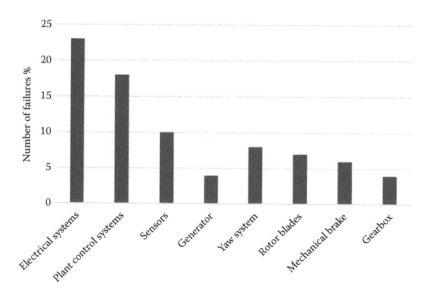

FIGURE 15.4 Total number of failures of main components in wind turbines. (Adapted from Hahn B., Durstewitz M., and Rohrig K. *Reliability of Wind Turbines*, Online. Available: http://renknownet2.iwes. fraunhofer.de/pages/wind_energy/data/2006-02-09Reliability.pdf.)

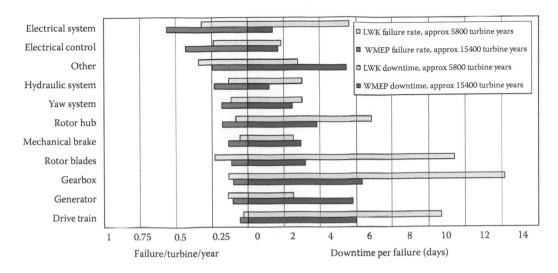

FIGURE 15.5 Failure frequency and downtimes of wind turbine components. (From Peter Tavner. A survey of wind turbine condition monitoring experience in Europe, *2011 Wind Turbine Condition Monitoring Workshop* September 19–21, 2011, Online, Available: http://www.nrel.gov/wind/pdfs/day1_sessioniii_01_nrelcmworksop_pjt.pdf.)

high number of failures of some components is now balanced out to a certain extent by short standstill periods. But still, damages of generators, gearboxes, and drivetrains are of high relevance due to long downtimes of about 1 week as an average.

15.3.3 ANALYSIS OF BREAKDOWN RISK AND FAILURES IN WTs IN INDIA

India is the fourth largest wind power-generating country in the world after Germany, Spain, and the United States. According to a study in Ref. [9] based on the data from international and Indian windmills, the most often failure happens due to mechanical breakdown (about 56%), while the lightning and fire cause about 22% of claims.

Result of heavy surge following the lightning may lead to shorting of phase winding and bursting of insulation in a transformer, shown in Figure 15.6 [14]. Electrical breakdowns cause about 20% of all failures, which includes damaged generator windings, short circuit, and over-voltage damage to controllers and electronic components as well as damage to transformers and wiring.

Mechanical components which are the most vulnerable parts of WT are rotor blades, rotor, generator and transmission bearings, rotor shaft, and gearboxes. The latter is the most frequently occurring type of mechanical damage. Damage to gearbox arise due to wear (pitting), backlash, and tooth breakage. Gearbox breakdown is also often caused by foreign particle contamination in gear oil resulting from chipping. Highly loaded conditions can also result in chipping or micropitting. Frequent stoppage and starting of windmill may also result in gear wheels and pinions being shifted from their original position (Figure 15.7). Frequent stoppages of WT may also leave scoring marks on the gear wheel, shown in Figure 15.8.

15.4 EFFECT OF WEATHER CONDITIONS ON WT FAILURES

Weather conditions can affect the reliability of WTs substantially. Harsh weather events such as storms and gusts will generate nonstationary loading (or stress) and reduce the fatigue life of key components. There is a strong relationship between the number of failures and wind speeds [15]. Analysis of field data from Denmark and Germany confirms a clear 12-month periodicity of the number of failures, which coincides with annual weather seasonality.

FIGURE 15.6 Failure in transformer resulting from lightning and following heavy surge. (Modified from Ramesh Babu J. and Jithesh S. V. *Breakdown Risks in Wind Energy Turbines*, Cholamandalam MS, Risk Services, 20pp., Online. Available: http://www.cholarisk.com/uploads/Breakdown%20Risks%20in%20 Windmills%20-%20Final.pdf.)

FIGURE 15.7 Intermediate pinion teeth broken as a result of a frequent stoppage and starting of the windmill. (Modified from Ramesh Babu J. and Jithesh S. V. *Breakdown Risks in Wind Energy Turbines*, Cholamandalam MS, Risk Services, 20pp., Online. Available: http://www.cholarisk.com/uploads/Breakdown%20Risks%20 in%20Windmills%20-%20Final.pdf.)

FIGURE 15.8 Scoring marks on the gear wheel resulting from frequent stoppages of the windmill. (From Ramesh Babu J. and Jithesh S. V. *Breakdown Risks in Wind Energy Turbines*, Cholamandalam MS, Risk Services, 20pp, Online. Available: http://www.cholarisk.com/uploads/Breakdown%20Risks%20in%20 Windmills%20-%20Final.pdf.)

Stochastic weather conditions affect the reliability of WTs in several ways [16]. Maintenance activities can be constrained by the stochastic weather conditions. Harsh weather conditions may reduce the feasibility of maintenance. This is in contrast with traditional fuel-based power plants which operate under relatively stationary operating conditions.

To maximize potential power generation, wind facilities are built at locations with high wind speeds. However, climbing a turbine during wind speeds of more than 20 m/s is not allowed; when speeds are higher than 30 m/s, the site becomes inaccessible [17]. Moreover, some repairing work takes days (and even weeks) to complete due to the physical difficulties of the job. The relatively long duration of a repair session increases the likelihood of disruption by adverse weather.

A total of 11 out of 15 cases of turbine failures in the United States were caused by lightning strikes or another ignition source causing blazes that begin hundreds of feet in the air. Since WTs are sited on towers or pylons, often on hilltops, with their structural towers, they constitute a ready pass for static electricity to the ground. Lightning protection of WTs follows the international standard IEC 1024-1 and Danish Standard DS 453 [18].

Lightning strikes are often causing WT failures and may result in significant damage to environment. Shown in Figure 15.9 is an accident in Ardrossan, Ayrshire, the United Kingdom, in December 2011 when a WT caught fire during a heavy storm in North Ayrshire despite being nonoperational. The WT was completely burnt out and burning debris were scattered across long distances due to the strong wind [19].

The cause of the fire was said to have been a lightning strike to the turbine. The turbine was completely destroyed. Secondary ignition of nearby vegetation and property was avoided due to timely fire service intervention. The wind farm lost about 1210 MWh of energy in the weeks after the fire due to downtime.

15.5 WT FIRES

WTs often catch fire and burn much more frequently than is reported, a study from the United Kingdom and Sweden maintains. Researchers at Imperial College London, Edinburgh University and at SP Technical Research Institute of Sweden report in Refs. [18,19] that while an average of about 12 turbine fires are reported annually, more than 117 fires actually occur worldwide. Judging from the accident statistics and analyzing the yearly evolution of the number of WT-related accidents, including fire incidents, the fire problem for WTs becomes apparent. Fire accounts for a substantial fraction of the accidents of any year, between 10% and 30% (Figure 15.10). Because the

FIGURE 15.9 Wind turbine burning in Scotland after lightning strike. (From *Wind Turbine Bursts into Flames as Hurricane-Force Winds Hit Scotland*, December 8, 2011, Online. Available: http://www.epaw.org/multimedia.php?article=all.)

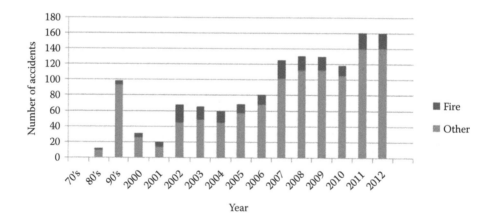

FIGURE 15.10 Accident statistics worldwide, including fire, since 1980s as collected by the UK Lobby Group [20]. (From Uadiale S., Urban É., Carvel R., Lange D., and Rein G. Overview of problems and solutions in fire protection engineering of wind turbines, *Proc. 11th Symposium on Fire Safety Science*, vol. 11, pp. 983–995, University of Canterbury, New Zealand, February 10–14, 2014. With permission.)

absolute number of fire accidents tend to increase with the number of installed turbines, the expected growth in the installation of WTs also brings the expectation of an increase in the number of turbine fires. However, the ratio of fire accidents per turbine installed has decreased significantly since 2002.

In comparison to other energy industries, the WT industry is relatively fire free. For example, in the oil and gas industry, thousands of fires are reported each year. The difference is, according to the report [18], that the economic impact of a WT fire can be significant.

The extreme difficulty is in fighting a turbine fire, owing to their height and their often remote locations. As an example, the report gives the case of a turbine on an Australian wind farm containing 112 turbines. When one of them caught fire during a heat wave, the entire farm was shut down, cutting power to 63,000 homes. In addition, burning debris from the turbine ignited ground fires that destroyed 80,000 hectares of a national park. The cause of the fire was found to be electrical failure in the nacelle. Lightning strikes are the most common cause of WT fires.

The nacelle is typically the main "fuel load" within a turbine. A large amount of highly flammable materials (hydraulic oil and lubricants, composite materials, insulation, and polymers) are confined within the nacelle of the WT. A single 1.5 MW WT nacelle can contain up to 900 L of lubricating oil including cooling and cleaning fluids, the report [18] says. The transformer, located at the base of the tower, can easily contain an additional 2200 L of transformer oil. All these flammable materials are packed in close proximity to potential ignition sources such as overheated mechanical components (hot surfaces) and electrical connections that could fail.

The fuel load in a WT nacelle is frequently high, but depends on the type of turbine, and is made of four main components: polymers, insulation, cables, and oils. The two main existing types of turbines are the VAWT and the Horizontal Axis Wind Turbine (HAWT) (see Chapter 11). Most turbines currently installed are HAWT for commercial power generation. The HAWT is further divided into two types; the direct drive HAWT and the gear-driven HAWT. A gear-driven HAWT typically houses an even greater fuel load because of the large amount of oil and lubricants stored. Understanding and quantifying the source of the high fuel loads typically found inside a WT is an essential step, since the fuel control and the design alteration are needed for improved fire protection.

Electrical/electronic equipment failure or malfunction is a common cause of fire in turbines, often caused by overheating or overloading. Electrical installations and thick long electrical cables spanning from the nacelle to the base of the tower also constitute a great amount of polymers. Short circuits, arcs, and inadequate electrical protection are common causes of fire in WTs. Besides electrical malfunction, overheating surfaces such as bearings, gearboxes, and mechanical brakes can pose a high fire risk in WTs if flammable materials come in contact with them.

The Groß Eilstorf wind farm project in Lower Saxony, Germany began operation in 2011/2012 and has a 51 MW capacity with a total of 17 V-112, 3 MW WTs from the Danish WT manufacturing company Vestas. On March 30, 2012 one of the newly installed WTs caught fire (see Figure 15.11).

The machine house of the turbine and one of the blades burned down. The time to repair the turbine was estimated to be at least a year with a potential loss of €300,000–400,000. The cause of the fire was identified to be a loose connection in the electrical system in the harmonic filter cabinet which caused an arc flash [22].

Typically, WT nacelles are made from fiberglass reinforced plastic (FRP). This is the main fuel element during flame spread over the nacelle. FRP is flammable and difficult to extinguish due to the epoxy resins used in fusing the fiberglass [23]. Materials used for internal acoustic insulation of the nacelle, which are often made from foam, are also a major part of the load. These insulation materials are highly flammable as well and can easily soak up spilled oil, thus giving them an even greater fire potential.

Hot surfaces such as overheated bearings, gearboxes, and mechanical brakes can also pose a high fire risk in WTs if flammable materials come in contact with them. A WT at the Tir Mostyn & Foel Goch wind farm in Nantglyn, Denbighshire, Wales, the United Kingdom caught fire and suffered a total loss due to an overheated gearbox [19]. Maintenance and repair activities involving "hot work" such as flame cutting, welding, soldering, and abrasive cutting can result in a high fire risk especially in a gear-driven HAWT. Flammable materials at distances up to 10 m or more could easily be ignited by sparks from grinding, welding, and cutting activities, and most fires have been reported to occur several hours after completion of hot work inside the nacelle.

Among the recommendations are passive fire protection measures including comprehensive lightning protection systems, use of noncombustible hydraulic and lubricant oil. Fire protection measures may also include installation of a radiant barrier to protect combustible solids in the nacelle, and avoiding the use of combustible insulating materials in the turbine nacelle where possible.

FIGURE 15.11 Wind turbine on fire at the Groß Eilstorf wind farm. (From Wind turbine on fire at the Groß Eilstorf wind farm, *European Platform against Windfarms*, Online. Available: http://epaw.org/photos/ WKA_ausgebrannt_5.jpg.)

15.6 WTs CHALLENGES IN HOT CLIMATE AND DESERTS

An exposure of WT to hot climate and to desert environment leads to many problems that are not present in other climatic zones. Among them are, for example, accelerated corrosion due continuous elevated humidity in tropical climate, erosion and abrasion due to the extensive presence of sand in deserts, premature lubricant failure, and others [24].

Rotor blades surfaces exposed to sand in desert and rain in tropical zones are damaged by sand and rain erosion (see Section 12.4). The adverse effects of sand storms on the nacelle cooling system are observed frequently in desert areas. Specific attention should be given to all metal parts, such as bolts, that will suffer from corrosion in harsh environment and generate extra maintenance costs.

A very serious consideration should be given to properly installed and functioning cooling systems. In mild climates, the following components are typically cooled: converter/inverter, control system electronics, transformer and generator. In hot climates, it may not be enough, it may require additional cooling systems.

15.6.1 MATERIALS TEMPERATURE LIMITS

Special attention should be focused on the choice of the construction and insulating materials that will not deteriorate in hot climate. The temperature limits for polymeric and composite material used in various parts of the WTs [24] are the temperatures at which the material may convert into different phase which does not perform as it is required by standards. One of the most important properties of any epoxy is the temperature at which the polymer transitions from a hard, glassy material to a soft,

TABLE 15.1
Temperature Limits for Polymeric Materials

Polymer	Material Name	Temperature Limit, °C (°F)
PA-6 (dry)	Polyamide 6 (nylon-6)	78 (172)
PVC	Polyvinylidene chloride	80 (176)
ABS	Acrylonitrile butadiene styrene (high heat)	110 (230)
Epoxy resin	Polyepoxide	Based on cure schedule: 50–260 (122–500)
Polyester resin	Composite material, fiberglass	Based on composition: 80–250 (176–482)

rubbery material (Glass Transition Temperature, Tg). The maximum temperatures above which a specific polymeric or composite material should not be used are shown in Table 15.1.

Other WT components and systems also have working temperature limit, for example, generators should not work at the temperatures above 180°C, encapsulated electrical components maximum working temperature is 40°C.

15.6.2 LUBRICATION

The damaging consequences are unavoidable if the lubricants used for lubrication of different parts of WT are exposed to high temperatures for extended periods of time. Gearbox of turbine is always lubricated, but high oil temperatures reduce lifetime of oil. It is a known fact that if maximum rated oil temperature constantly exceeded by 10°C, oil lifetime is reduced by a half. Therefore, mineral lubricating products can be used only up to 80°C of oil temperature. For hot climates, oil cooler may be needed for lubrication system. Addition risk arises in deserts because of sandy environment. If sand penetrates into cooling system, it can be blocked.

15.7 COMPARATIVE STUDY OF THE OFFSHORE AND ONSHORE WT FAILURES

The failure rates for the subassemblies of the onshore and offshore WT systems have been compared in Refs. [25–27]. In Table 15.2, the failure rates are compared for different subassemblies in onshore and offshore WTs.

Based on the data from studies [25–27], the comparison of failure rates of different subassemblies shows that for some systems, such as rotating and electrical systems, the failure rates are practically always higher for offshore WTs than for onshore ones, though some differences are not significant or reverse (for main shaft failures). In Figures 15.12 through 15.14, the failure

TABLE 15.2
Failure Rates for the Subassemblies of the Onshore and Offshore Wind Turbine Systems

Subassembly	Onshore	Offshore	Subassembly	Onshore	Offshore
Brake system	0.012	0.013	Rotor bearings	0.009	0.01
Cables	0.009	0.008	Rotor blades	0.124	0.174
Gearbox	0.134	0.179	Rotor hub	0.121	0.139
Generator	0.11	0.15	Screws	0.005	0.005
Main frame	0.012	0.011	Tower	0.151	0.144
Main shaft	0.051	0.043	Transformer	0.121	0.14
Nacelle housing	0.012	0.012	Yaw system	0.011	0.013
Pitch system	0.012	0.013	Others	0.313	0.258
Power converter	0.013	0.068			

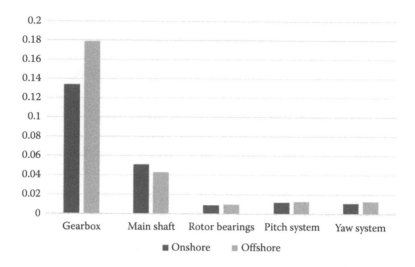

FIGURE 15.12   Failure rates for the *rotating subassemblies* of the offshore and onshore wind turbines.

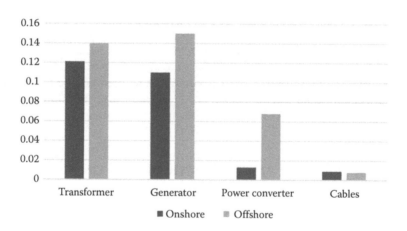

FIGURE 15.13   Failure rates for the *electrical subassemblies* of the offshore and onshore wind turbines.

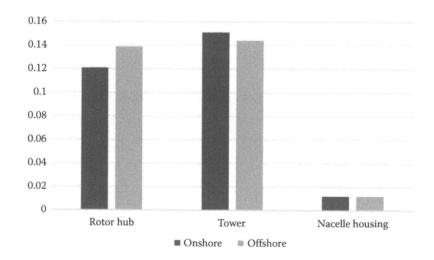

FIGURE 15.14   Failure rates for the *structural subassemblies* of the offshore and onshore wind turbines.

rates are presented for different types of subassemblies. For example, the gearbox and generator failure rate are about 25% higher for offshore turbines, but the largest difference is found for power converter with five times higher failure rate for offshore turbines than for onshore ones (Figure 15.13).

REFERENCES

1. Poore R. *Development of an Operations and Maintenance Cost Model to Identify Cost of Energy Savings for Low Wind Speed Turbines*: July 2, 2004–June 30, 2008. NREL/SR-500-40581. Work performed by Global Energy Concepts, LLC, Seattle, Washington, Golden, CO: National Renewable Energy Laboratory, January 2008, Online. Available: http://www.nrel.gov/docs/fy08osti/40581.pdf.
2. Watson J. and Infield D. Condition monitoring of offshore wind farms – CONMOW, EU contract ENK5-CT-2002-00659, 2005.
3. Ribrant J. Reliability performance and maintenance – A survey of failures in wind power systems, Master thesis, 2006, Online. Available: http://faculty.mu.edu.sa/public/uploads/1337955836.6401XR-EE-EEK_2006_009.pdf.
4. Ribrant J. and Bertling L. Survey of failures in wind power systems with focus on Swedish wind power plants during 1997–2005, *Proc. IEEE Power Engineering Society General Meeting*, Tampa, FL, pp. 1–8, June 2007.
5. Ragheb A. M. and Ragheb M. *Wind Turbine Gearbox Technologies*, Chapter 8, InTech, pp. 189–206, 2011, Online. Available: http://cdn.intechopen.com/pdfs-wm/16248.pdf.
6. Bala S., Pan J., Das D. et al. *Lowering Failure Rates and Improving Serviceability in Offshore Wind Conversion-Collection Systems*, Online. Available: http://www05.abb.com/global/scot/scot232.nsf/verity display/8d1772a30d4e8ea5c1257a8b00583093/$file/Lowering-Failure-Rates-and-Improving-Service.pdf.
7. Buckley S. *Forecasting Wind Farm Component Failures and Availability Post-Warranty*, WindTech International, April 2013, Online. Available: http://www.windtech-international.com/articles/forecasting-wind-farm-component-failures-and-availability-post-warranty.
8. Hahn B., Durstewitz M., and Rohrig K. *Reliability of Wind Turbines*, Online. Available: http://renknownet2.iwes.fraunhofer.de/pages/wind_energy/data/2006-02-09Reliability.pdf.
9. Ensslin C., Durstewitz M., Hahn B. et al. *German Wind Energy Report 2005*, ISET, Kassel, Germany, 2005.
10. Lyding P., Faulstich S., and Kuhn P. *Establishing a Common Database for Turbine Failures*, Fraunhofer Institure for Wind Energy and Energy System Technology (IWES), Online. Available: http://energy.sandia.gov/wp/wp-content/gallery/uploads/2-A-4-Lyding.pdf.
11. Wilkonson M., Harman K., and Hendriks B. *Measuring Wind Turbine Reliability – Results of Reliawind Project*, Online. Available: http://www.gl-garradhassan.com/assets/downloads/Measuring_Wind_Turbine_Reliability_-_Results_of_the_Reliawind_Project.pdf.
12. Stenberg A. and Holttinen H. *Analyzing Failure Statistics of Wind Turbines in Finland*, Online. Available: http://proceedings.ewea.org/ewec2010/allfiles2/442_EWEC2010presentation.pdf.
13. Durstewitz M. and Wengler R. *Analyses of Generator Failure of Wind Turbines in Germanys "250 MW Wind" Programme, Study*, ISET, Kassel, 1998.
14. Ramesh Babu J. and Jithesh S. V. *Breakdown Risks in Wind Energy Turbines*, Cholamandalam MS, Risk Services, 20pp., Online. Available: http://www.cholarisk.com/uploads/Breakdown%20Risks%20in%20Windmills%20-%20Final.pdf.
15. Tavner P. J., Edwards C., Brinkman A., and Spinato F. Influence of wind speed on wind turbine reliability, *Wind Engineering*, 30, 55–72, 2006.
16. Eunshin B., Lewis N., Chanan S., and Yu D. Wind energy facility reliability and maintenance, In: *Handbook of Wind Power Systems, Energy Systems*, P. M. Pardalos et al. (eds.), Springer-Verlag, Berlin, Heidelberg, pp. 639–672, 2013, Online. Available: http://ise.tamu.edu/metrology/publications/R3.pdf.
17. McMillan D. and Ault G. W. Condition monitoring benefit for onshore wind turbines: Sensitivity to operational parameters, *IET Renewable Power Generation*, 2, 60–72, 2008.
18. Ragheb M. *Safety of Wind Systems*. 2/20/2014, Online. Available: http://mragheb.com/NPRE%20475%20Wind%20Power%20Systems/Safety%20of%20Wind%20Systems.pdf.
19. Uadiale S., Urban É., Carvel R., Lange D., and Rein G. Overview of problems and solutions in fire protection engineering of wind turbines, *Proc. 11th Symposium on Fire Safety Science*, vol. 11, pp. 983–995, University of Canterbury, New Zealand, February 10–14, 2014.

20. *Caithness Windfarm Information Forum*, Summary of Wind Turbine Accident data to 31 December 2012, 2013, Online. Available: http://www.caithnesswindfarms.co.uk/accidents.pdf.

21. Wind turbine on fire at the Groß Eilstorf wind farm, *European Platform Against Windfarms*, Online. Available: http://epaw.org/photos/WKA_ausgebrannt_5.jpg.

22. Vestas identifies cause for V112 wind turbine fire, Renewable Energy Focus, April 2012, Online. Available: http://www.renewableenergyfocus.com/view/25458/vestas-identifies-cause-for-v112-wind-turbine-fire/.

23. Morchat Richard M. Technical Memorandum 93/214: *The Effects of Zinc Borate Addition on the Flammability Characteristics of Polyester, Vinyl Ester and Epoxy Glass Reinforced Plastic*, October 1993, Online. Available: www.dtic.mil/dtic/tr/fulltext/u2/a273164.pdf.

24. Roscheck F. and Rudolf R. Technical considerations for wind power in hot climates, *INEES Workshop*, September 24, 2012.

25. Dinmohammadi F. Shafiee M. A fuzzy-FMEA risk assessment approach for offshore wind turbines. *Int. J. Progn. Health Management*, 4, 1–10, 2013, Online. Available: http://www.phmsociety.org/sites/phmsociety.org/files/phm_submission/2013/ijphm_13_013.pdf.

26. Dinmohammadi F., Shafiee M. An economical FMEA-based risk assessment approach for wind turbine systems, *Proc. European Safety, Reliability and Risk Management (ESREL) Conference*, Amsterdam, The Netherlands, pp. 2127–2136, September 30–October 2, 2013.

27. Shafiee M. and Dinmohammadi F. An FMEA-Based risk assessment approach for wind turbine systems: A comparative study of onshore and offshore, *Energies*, 7, 619–642, 2014, Online. Available: www.mdpi.com/1996-1073/7/2/619/pdf.

28. Peter Tavner. A survey of wind turbine condition monitoring experience in Europe, *2011 Wind Turbine Condition Monitoring Workshop* September 19–21, 2011, Online, Available: http://www.nrel.gov/wind/pdfs/day1_sessioniii_01_nrelcmworksop_pjt.pdf.

29. *Wind Turbine Bursts into Flames as Hurricane-Force Winds Hit Scotland*, December 8, 2011, Online. Available: http://www.epaw.org/multimedia.php?article=a11.

16 Life Expectancy and Life Cycle Assessment (LCA) of Renewable Electrical Equipment

16.1 WIND TURBINES

Evaluating and predicting life expectancy for new sources and technologies for producing electrical energy is an extremely important issue. Electrical equipment either produced by conventional technologies or using renewable energy sources is continuously aging, a natural process based on aging and deterioration of the materials these machines made of, which is the base for evaluating technical life of the electrical machines.

16.1.1 DESIGN LIFE OF WT

The expected lifetime estimated in the earlier years of WT technology development has been set to 20 years. Certain components in the WT have estimated lifetimes of up to 50 years, including the foundation and transmission cables. However, considering technology improvements and other maintenance and replacement related factors, a 20 year lifetime is reasonable.

In the relevant regulations (DiBT guidelines/DIN EN 61400-1) [1], it is stated that "The design life of a wind turbine must be at least 20 years." In general, WTs for onshore are designed for an operating period of exactly 20 years, because each design year more than 20 years would mean increased manufacturing costs and hence a higher sales price for a WT.

The design life is a theoretical parameter and is also termed the planned service life. The load-bearing capacity of the structure and suitability for use must be demonstrated for this period. At the end of the 20-year period there is generally an interest in continuing to operate the WT beyond this period. After all, disassembly and disposal at the end of the planned service life goes against the philosophy of sustainable usage. Further operation can, however, only be tolerated if safe operation is guaranteed within the extended service life. On the other hand, further operation is only desirable if the resulting revenues exceed the costs for operation and maintenance [2].

Just as with conventional forms of power generation, the energy produced by a wind farm also gradually decreases over its lifetime, perhaps due to falling availability, aerodynamic performance, or conversion efficiency. Understanding these factors is rather complicated since availability of the wind is highly variable.

If load factors (also known as capacity factors) decrease significantly with age, wind farms will produce a lower cumulative lifetime output, increasing the cost of electricity from the plants. If the rate of degradation were too great, it could become worthwhile to prematurely replace the turbines with new models, implying that the economic life of the turbine was shorter than its technical life, further increasing its cost. Therefore, it is significant to estimate life expectancy for WTs/farms (as for any other renewable energy source) to justify the policy implications for the desirability of investing in wind power.

16.1.2 EVALUATION OF LIFE FOR WIND FARMS IN THE UNITED KINGDOM AND DENMARK

In a study "The Performance of Wind Farms in the United Kingdom and Denmark" in Ref. [3], it was shown that the economic life of onshore WTs is between 10 and 15 years, not the 20–25 years projected by the wind industry itself, and used for government projections.

The work has been conducted by one of the U.K.'s leading energy and environmental economists, Professor Gordon Hughes of the University of Edinburgh, and has been anonymously peer reviewed. This groundbreaking study applies rigorous statistical analysis to years of actual wind farm performance data from wind farms in both the United Kingdom and in Denmark.

The results show that after allowing for variations in wind speed and site characteristics, the average load factor of wind farms declines substantially as they get older, probably due to wear and tear. By 10 years of age, the contribution of an average U.K. wind farm to meeting electricity demand has declined by a third. This decline in performance means that it is rarely economic to operate wind farms for more than 12–15 years. After this period, they must be replaced with new machines, a finding that has profound consequences for investors and government alike.

However, another study in Ref. [4] revealed that onshore wind farms in the United Kingdom have aged at about the same rate as other kinds of power station, which challenges the results of Ref. [3]. The fact is that several factors can confound the relationship between age and observed output in a fleet of wind farms, given that a turbine's output is dependent on wind speeds at its site and the efficiency with which it captures the energy in that wind.

For example, if wind speeds have fallen slightly over time, farms would have lower load factors in recent months, when they were at their oldest, giving a false correlation between age and poor performance. If improvements in design increase a turbine's output relative to capacity (its power coefficient), then newer turbines (of the improved design) will have higher load factors than old turbines, so that turbine output appears to decline with age, when really it improves with newer generations. On the other hand, if the best (windiest) sites were occupied first, then old farms could have higher load factors than new ones built on inferior sites, so that turbines would appear to improve with age.

To prove the validity of these concerns in Ref. [4], the rate of aging of a national fleet of WTs was determined using free public data for the actual and theoretical ideal load factors from the U.K.'s 282 wind farms. Actual load factors are recorded monthly for the period from 2002 to 2012, covering 1686 farm-years of operation. Ideal load factors are derived from a high-resolution wind resource assessment made using NASA data to estimate the hourly wind speed at the location and hub height of each wind farm, accounting for the particular models of turbine installed.

The results of study in Ref. [4] showed the evidence of important, but not disastrous, performance degradation over time in a large sample of U.K. wind farms. When variations in the weather and improvement in turbine design are accounted for, the load factors of U.K. wind farms fall by 1.57% (0.41 percentage points per year). This degradation rate appears consistent for different vintages of turbines and for individual wind farms, ranging from those built in the early 1990s–early 2010s.

According to the findings in Ref. [4], the average wind farm has an annual load factor of about 28% when first commissioned, which declines by about 0.4 percentage points per year. After 15 years, the load factor would have fallen to 23%. A total of 40 out of the first 45 wind farms commissioned in the United Kingdom were still operating at this age; four had been repowered. Taking this deterioration into account raises the cost of electricity by around 9% over a 24-year life span, discounting at 10% a year. This study suggested that this aging does not appear to have made developers replace their farms early.

16.1.3 LCA of Wind Farms

LCA is a methodological tool used to quantitatively analyze the life cycle of products/activities. International Standards ISO 14040 "Environmental management—Life cycle assessment—Principles and framework" and ISO 14044 "Environmental management—Life cycle assessment—Requirements and guidelines" provide a generic framework.

LCA purpose is to evaluate the potential environmental impacts caused by an offshore and onshore wind turbine farms (WTF) throughout the whole lifetime and use this knowledge in the

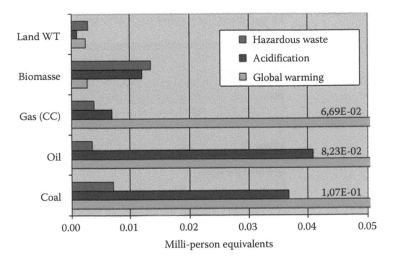

FIGURE 16.1 Comparison of three normalized environmental impact potentials for some production technologies in the Danish electricity and district heating mix per kWh. (Adapted from Properzi S. and Herk-Hansen H. *European Wind Energy Conference and Exhibition*, Copenhagen, Denmark, July 2001, Online. Available: http://tethys.pnnl.gov/sites/default/files/publications/Middelgrund_LCA_2001.pdf.)

planning and improvement of future WTFs. It is important to notice that environmental effect of onshore WTs (or land WT) is the lowest compared with conventional types of producing electrical energy. The environmental effects studied in LCA are hazardous waste, acidification of the environment, and impact on global warming [5], as shown in Figure 16.1.

This figure shows the normalized environmental impact potentials of three selected impact categories for some of the Danish electricity and district heat production technologies. For clarity purposes, only selected environmental impacts are shown here. Global warming and acidification are shown since these are often used and are well-known impacts. The hazardous waste category is shown since WTs have a large contribution to this category. As is often the case, normalized LCA results are shown here in milli-person equivalents (mPE). A milli-person equivalent is 1/1000 of an average European's "environmental footprint" (a person's average contribution to the various impact categories) or allocated emission.

The results for WTs on land in this LCA showed, as expected, that electricity produced by wind had the smallest normalized environmental impacts in most categories. However, WTs on land produced a large amount of hazardous waste per kilowatt-hour electricity amongst the various production technologies. This is attributed to the fact that a WT on land uses a relatively large amount of material (steel) per installed effect and has a relatively low number of operational hours in comparison to the other conventional production technologies.

16.1.3.1 LCA of Offshore and Onshore WT Farms in Denmark

One of the earlier LCA (2001) presented in Ref. [5] applied to the offshore WTF was focused on the advantages and disadvantages in comparison to onshore WTs. The LCA of offshore WT was based on experience from the LCA on Danish electricity and district heating as well as the offshore WTF project at Middelgrunden (Denmark) which is in operation.

It was an important study since land resources in Denmark are limited and experiences with offshore WTs have been good, it was expected that future efforts will be concentrated on offshore WTF. Five areas at sea have been selected which are all suitable for the installation of WTs. Each of these sites are all projected for an electrical effect of 150 MW in their first stage. By the year 2030, a capacity of approximately 4000 MW offshore wind power is expected in Denmark. LCA is a complicated study whose details will not be described in this book, but could be found elsewhere.

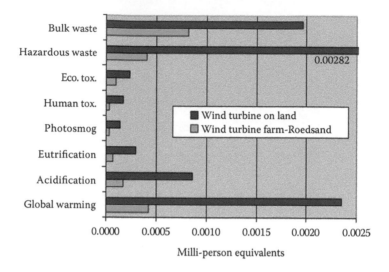

FIGURE 16.2 Comparison of the environmental effect potentials caused by 1 kWh of electricity production delivered to the transmission net from an onshore 600 kW wind turbine and offshore Nysted's 150 MW WTF. (Adapted from Properzi S. and Herk-Hansen H. *European Wind Energy Conference and Exhibition*, Copenhagen, Denmark, July 2001, Online. Available: http://tethys.pnnl.gov/sites/default/files/publications/Middelgrund_LCA_2001.pdf.)

All of the primary processes in the manufacturing, use and decommissioning of the WTF are considered in the LCA. All of the components of the WTF and transmission facilities have been examined and areas of environmental improvement have been identified. The most interesting results of the study in Ref. [5] are shown in Figure 16.2, which presents a comparative LCA for onshore and offshore WTs. It was found that offshore WTF and associated transmission facilities per produced kWh have an improved environmental profile in comparison to a land-based WT.

All studies of LCA show that the major environmental impacts are coming from the manufacturing process of WTs. The largest environmental impact potentials from the offshore WTF at Nysted are *bulk waste, global warming, and hazardous waste*. Bulk waste is mainly generated in connection with coal extraction for electricity generation used primarily in steel production for the tower, foundation, and transmission cable. Land filling of spent WT blades also contributes considerably to this category as well as other processes.

Global warming stems primarily from fossil fuel-based electricity production used in the manufacture of the components in the WTF, especially the blades, tower, and foundation. Hazardous waste is primarily generated in the production of steel for the tower, foundation, and transmission cables, where various metal-rich slag fractions can only be disposed of at special hazardous depots.

16.1.3.2   LCA of 4.5 MW and 250 W WTs in France

The study from France (2009) in Ref. [6] uses LCA, comparing two systems: a 4.5 MW and a 250 W WTs, to evaluate their environmental impact. All stages of life cycle (manufacturing, transports, installation, maintenance, disassembly, and disposal) have been analyzed. The influence of distance and transport or so called sensitivity tests have been performed.

According to the indexes primary energy payback time (PEPBT), CO_2 emissions, and so on, the results show that wind energy is an excellent environmental solution provided that the turbines are high efficiency ones and implemented on sites where the wind resource is good. It is also true if the components transportation should not spend too much energy and recycling during decommissioning should be performed correctly. This study proves that wind energy should become one of the best ways to mitigate climate change and to provide electricity in rural zones which are not connected to the grid.

16.1.3.3   Comparative LCA of 2.0 MW WTs in the United States

The U.S. study (2014) in Ref. [7] examines life cycle environmental impacts of two 2.0 MW WTs. Manufacturing, transport, installation, maintenance, and end of life have been considered for both models and are compared using the ReCiPe methodology for Life Cycle Impact Assessment (LCIA) presented in Ref. [8]. LCA takes into account sourcing of key raw materials (steel, copper, fiberglass, plastics, concrete, and other materials), transport, manufacturing, installation of the turbine, ongoing maintenance through its anticipated two decades of useful life and, finally, the impacts of recycling and disposal at end of life.

In addition, energy payback analysis was conducted based on the cumulative energy demand and the energy produced by the WTs over 20 years. The payback period or *energy payback time* (EPT) is the number of months or years of energy-cost savings it takes to recover an investment's initial cost. To determine the payback, the investor first estimates the WT's total initial cost, annual energy-cost savings, and annual operating costs. Dividing total initial cost by the difference between annual energy-cost savings and annual operating costs gives the payback period.

LCA in Ref. [7] revealed that environmental impacts are concentrated in the manufacturing stage, which accounts for 78% of impacts. The energy payback period for the two turbine models are found to be 5.2 and 6.4 months, respectively. In other words, this analysis shows that the vast majority of predicted environmental impacts would be caused by materials production and manufacturing processes. However, the payback for the associated energy use is within about 6 months. It is likely that even in a worst case scenario, lifetime energy requirements for each turbine will be subsumed by the first year of active use. Thus, for the 19 subsequent years, each turbine will, in effect, power over 500 households without consuming electricity generated using conventional energy sources.

16.2   TIDAL AND WAVE TURBINES

16.2.1   COMPARATIVE LCA STUDY OF WAVE AND TIDAL ENERGY DEVICES IN THE UNITED KINGDOM

LCA was conducted on two types of marine energy devices: wave energy devices and tidal energy devices (see Chapter 10 for details). Wave energy devices extract energy from the movement of waves, which are themselves driven by wind and sea conditions. Waves are unpredictable, making the output from these devices difficult to estimate. A common design of wave energy device incorporates a fixed base mounted on the seabed, with a buoyant flap section moving with wave motion.

Tidal energy devices extract energy from the movement of the tides, and often look like underwater WTs. Tides are governed by gravitational forces and the rotation of the Earth, so are highly predictable. Although there are a number of devices in both categories currently being developed, the two devices chosen for this study were the 1.2 MW SeaGen tidal device (see Figure 10.1 in Chapter 10), and the 315 kW Oyster wave energy device (oscillating body power converter) by Aquamarine Power in Scotland shown in Figure 16.3.

An LCA of the SeaGen device was carried out by Douglas, Harrison, and Chick of the University of Edinburgh in 2007 [9]. The assessment was carried out on the test device installed at Strangford Lough, and included energy and CO_2 assessment of the device manufacture, transport, installation, maintenance, and decommissioning.

In order to allow a comparative analysis of the two devices, a similar assessment was carried out for the Oyster device installed at European Marine Energy Centre (EMEC) in 2008 as a part of the study in Ref. [10].

FIGURE 16.3 Oyster wave device array by aquamarine power. (From Wave Power Technologies Brief 4, International Renewable Energy Agency (IRENA), June 2014.)

TABLE 16.1

Comparative LCA Study of Tidal and Wave Turbines in the United Kingdom

Environmental Effect, Unit	Oyster Wave Device	SeaGen Tidal Device
Net embodied energy, GJ	5366	20,308
Net embodied CO_2, ton	560	1418
Energy intensity, kJ/kWh	236	214
CO_2 intensity, gCO_2/kWh	25	15
Energy payback, months	14	12
CO_2 payback, months	8	8

Source: Adapted from Walker S. and Howell R. Life Cycle comparison of a wave and tidal energy device, *Proc. Institution of Mechanical Engineers, Part M: Journal of Engineering for the Maritime Environment*, 225(4), 325–337, November 2011, Online. Available: http://e-futures.group.shef.ac.uk/publications/pdf/94_15.%20 stuart%20walker%20summary.pdf.

Comparison of LCA studies showed that relative to their size, the Oyster and SeaGen devices exhibited very similar energy and carbon intensity (see Table 16.1). Despite their early stages of development, both devices also demonstrated energy and carbon intensity comparable to that of large WT installations [11].

These results appear to suggest that wave and tidal energy devices do offer a viable future contribution to energy mix in the United Kingdom. However, development must continue in order to ensure these devices are suitable for deployment at a commercial level, and issues which have not been discussed in the study such as survivability must be considered.

16.2.2 Expected Life Span of Tidal and Wave Power Plants

Tidal power plants may have a long expected life span, about 75–100 years. This, however, has yet to be proven since the oldest tidal power plant is the La Rance in France was built in 1966 and still produces electricity as efficiently as in 1966.

A tidal barrage is a dam-like structure used to capture the energy from masses of water moving in and out of a bay or river due to tidal forces. It is predicted to have a life span of 120 years [12], although U.K. engineers estimate 250 years to be more accurate. This compares well with the 60-year life span of a nuclear plant or 15 years for an offshore wind farm, and would mean that for at least 100 years, the barrage would be generating the cheapest electricity in the United Kingdom. However, life span of tidal turbines is limited by the survivability of its components, such as turbine blades and gearboxes. Many of these components do not have a predicted life span longer than 20 years.

A different approach is given in Ref. [13], suggesting that tidal turbine lifetime might be extrapolated from that of current offshore platforms, which typically extend to 30 years or so. Designers of tidal turbines in the United Kingdom and Canada have projected lifetimes of 20–30 years for their designs [14]. Although accurate estimates of operational lifetime of tidal turbines will not be available until more experience is gained with full-scale devices, some cause for optimism lies in the fact that precommercial testing of some near-shore turbines has resulted in turbines operating without failure over a 5-year period, despite the lack of a systematic maintenance program.

From the other hand, this technology might have a strong environmental impact: building a tidal plant affects the marine life of the surroundings. It has an impact on fish, other marine life, and sea birds. An environmental study should always accompany a feasibility and business case associated with a tidal power plant. Additionally, tidal plants may cause disruptions. It is a fact that tidal plants with barrages disrupt the access to open water and the power plant itself disrupts tidal cycles. Also, it was found that tidal power plants affect the salinity of water in the tidal basins.

There are not enough studies and information published about prediction of the life span of wave turbines. Of all different factors that result in different costs for different projects, the life span of wave turbines is the most uncertain one. This is a relatively new technology and the data on how long these wave power inventions realistically can operate are absent [15]. In another study in Ref. [16] of LCA for wave power plants in the United Kingdom, various life spans have been used as a key factor in LCA calculations. For practical reasons, these calculations have been expressed as three different span of lifetime; 40, 60, and 80 years. Using the different lifetime is giving the same results as reducing the material in the same proportion.

The input of materials for specific wave turbine design was the same for the scenarios: present and pessimistic 2025 (40-year life span), 2025 realistic optimistic, 2025 very optimistic and 2050 pessimistic (60-year life span), and 2050 realistic optimistic and 2050 very optimistic (80-year life span). The parts of mooring system, generator, and transformer are replaced onetime within the whole life span, while all other parts are expected to endure the whole life span. However, it is not clear what was used as a base for suggesting such life spans for wave turbines.

16.3 SOLAR PANELS

16.3.1 Life Expectancy for Rooftop Photovoltaic (PV) Panels in Europe

There is widely accepted opinion on life expectancy for PV panels in the world. For example, the U.K. Department of Energy and Climate Change believes that panels will produce energy for at least 25 years. The warranty conditions for PV panels typically guarantee that panels installed on the roofs can still produce at least 80% of their initial rated peak output after 20 (or sometimes 25) years. So, manufacturers expect that their panels last at least 20 years, and that the efficiency decreases by no more than 1% per year.

What makes talking about lifetimes for PV panels difficult is the fact that very few panels have been installed for long enough. In the United Kingdom, more panels have been installed between 2006 and 2008 than in all previous years together. Globally, only a small proportion of all PV panels installed is older than 10 years [17].

From other sources, it shows that PV panels can produce energy effectively even after 20 years in service. The first grid-connected a 10 kW rooftop PV plant in Europe was installed in May 1982 by LEE-TISO testing center for PV components at the University of Applied Sciences of Southern Switzerland. After 20 years in service, the performance of the panels was analyzed in 2002 in Ref. [18]. When the panels were tested in 2002, the average peak output of the panels was 32.9W—11% lower than the nominal value in 1982 and only 3.2% lower than the measured value in 1983. In other words, between 1983 and 2002 the panels' peak output had only degraded by around 0.2% per year since 1983 (0.5% per year against initial nominal rating).

16.3.2 LIFE EXPECTANCY OF SOLAR PV PANELS: STUDY BY NATIONAL RENEWABLE ENERGY LABORATORY (NREL) (U.S.)

Percentage of solar PV systems affected by defective or underperforming panels is very low based on analysis performed by the NREL of U.S. Energy Department. NREL analyzed the data from 50,000 solar energy systems installed between 2009 and 2013. The study in Ref. [19] showed that just 0.1% of all PV systems reported have been affected by damaged or underperforming modules per year, and less than 1% each year had hardware problems. Inverter failures and fuse failures were reported more commonly than panel failure.

Despite hurricanes, hail, shading, vandalism, and hook-up delays, approximately 85% of all systems each year produced 90% or more of the electricity predicted, and the typical system produces more electricity than predicted. An extensive study in Ref. [19] summarized a history of degradation rates using field tests reported in the literature during the last 40 years. Nearly 2000 degradation rates, measured on individual modules or entire systems, have been assembled from the literature and show a mean degradation rate of 0.8%/year and a median value of 0.5%/year. The majority, 78% of all data, reported a degradation rate of <1%/year.

Year to year comparisons suggest that the degradation rate—the gradual loss of energy production—is in the historical range of 0.5%–1% per year. Thin-film degradation rates have improved significantly during the last decade, although they are statistically closer to 1%/year than to the 0.5%/year necessary to meet the 25-year commercial warranties.

Degradation rates define life expectancy of solar panels, which depend on many factors, one of which is *climate effect*. Crystalline silicon modules located in extreme climates showed high degradation rates. For very cold climates, panels subjected to heavy wind and snow loads suffered the most. On the other hand, panels in similar climates that were installed in a facade, eliminating the snow load, had very low rates of degradation.

At the other extreme, panels in desert climates exhibited large decreases in production over time—close to 1% per year—mainly due to high levels of UV exposure. Panels in more moderate climates such as the Northern United States had degradation rates as low as 0.2% per year. Those panels could retain 96% of their production capabilities after 20 years.

Another factor is *period of panel manufacturing* with the border line marked year 2000. It was found in Ref. [19] that the 1% per year rule was somewhat pessimistic for panels made prior to the year 2000, and today's panels, with better technology and improved manufacturing techniques, have even more stamina than their predecessors.

Third factor is the *material* the PV cells are made. PV cells technologies are usually classified into three categories based on basic semiconductor material used in fabrication. First generation of PV cell has been made of crystalline silicon (monocrystalline- Mono C-Si or polycrystalline—Poly C-Si), or EFG ribbon silicon (EFG ribbon-sheet c-SI).

Second generation of PV cell is fabricated from thin films: amorphous silicon (A-Si), copper indium diselenide (CIS), copper indium gallium diselenide (CIGS), or cadmium telluride (CdTe).

The third generation PV cells technologies are based on several novel concepts, such as concentrating PV (CPV), organic PV, and dye-sensitized solar cells (DSSC). Third-generation PV technologies are beginning to be commercialized, but most are still at the pre-commercial stage.

TABLE 16.2
Degradation Rate of Solar Panels of First and Second Generations

Solar Cell Type	Generation	Degradation Rate for Installations Prior 2000, % Per Year	Degradation Rate for Installations After 2000, % Per Year
Mono-C-Si	1	0.47	0.36
Poly-C-Si	1	0.61	0.64
a-Si	2	0.96	0.87
CdTe	2	3.33	0.4
CIGS	2	1.44	0.96

Source: Adapted from Chianese D. et al. Analysis of weathered c-Si PV modules, *Proc. 3rd World Conference on Photovoltaic Energy Conversion*, 3, 2922–2926, Osaka, Japan, May 2003.

For monocrystalline silicon, the most commonly used panel for commercial and residential PV, the degradation rate is less than 0.5% for panels made before 2000, and less than 0.4% for panels made after 2000. That means that a panel manufactured today should produce 92% of its original power after 20 years, quite a bit higher than the 80% estimated by the 1% rule.

Type of PV panel material make a big difference in performance and degradation rate. The rated power output of solar panels typically degrades at about 0.5%/year. However, thin-film solar panels (a-Si, CdTe, and CIGS) degrades faster than panels that are based on mono- and polycrystalline solar panels [19]. Degradation rates solar panels of different generation and/or different cell type are shown in Table 16.2.

16.3.3 EXPECTED LIFE OF SOLAR POWER PLANT

An extensive study of performance and expected life of solar power plant was conducted in India [20]. Solar PV system in power generation consists of many components, which are

- Solar PV modules
- Power conditioning units (inverters)
- Power storage system (battery bank)
- Mounting structures
- Cables/wiring

Each component has its own service life and own degradation rates. All of these should be considered when evaluating expected life of solar PV system. As shown in Ref. [19], estimated life may differ for different types PV cells based on different deterioration rates for the types of the cells, when different materials (crystalline silicon, thin film, or dye-sensitized and organic PV cell) and technology have been used in manufacturing. The estimated life of various components of solar PV system has been estimated in Ref. [21]. The life expectations for different components of solar PV system are presented together with the comments on specific applications and technologies in Table 16.3.

16.3.4 LCA OF SOLAR PANELS

PV modules produce, throughout their lifetime, only a small amount of greenhouse-gas (GHG) emissions, that is, 21–45 g CO_2-eq/kWh and have EPT from 1 to 2.5 years in average U.S. solar conditions. Furthermore, the life-cycle occupational risks of the PV fuel cycle, updated for modern PV plants, are relatively low as found in Ref. [22].

TABLE 16.3

Expected Life of Solar PV System Components

Component of a Solar PV System	Estimated Life (Years)	Comments
PV modules	30	For mature module technologies (e.g., glass–tedlar encapsulation), less for foil-only encapsulation
Inverters	15	For small plants (residential PV)
Inverters	30	For large plants (utility PV) with 10% of part replacement every 10 years
Structure	30	For rooftops and facades
Structure	30–60	For ground mount installations on metal supports
Cabling	30	
Battery	3–5	For lead–acid battery

Source: Adapted from Alsema E. et al. *Methodology Guidelines on Life Cycle Assessment of Photovoltaic Electricity*, Report IEA-PVPS T12-01: 2009.

However, little is known about the risks of accidental events, that is, their frequency and scale that entail human fatalities, injuries, and economic losses in the PV fuel cycle. As discussed in previous studies, perhaps the greatest potential risks of the PV fuel cycle are associated with chemical usage during the stages of materials production and module processing.

The PV manufacturing cycle entails the use of several hazardous substances, although in quantities much smaller than in the process industries. The greatest potential risks of the PV fuel cycle are linked with the accidental release of gaseous materials; however, the risks have not been well quantified in comparison with other electricity-generation technologies. Results of the analysis in Ref. [23] showed that, in terms of statistically expected incidents in the United States, the PV fuel cycle is much safer than conventional sources of energy such as coal, oil, gas, nuclear, and hydro, and by far the safest in terms of potential maximum consequences.

REFERENCES

1. *International Standard IEC 61400-1 Wind Turbines – Part 1: Design Requirements*, Third edition 2005–08, International Electrotechnical Commission, Geneva, Switzerland, 2005, Online. Available: https://webstore.iec.ch/preview/info_iec61400-1%7Bed3.0%7Den.pdf.
2. Wind Energy Report, *Fraunhofer Institute for Wind Energy and Energy System Technology (IWES)*, Germany 2012, Online. Available: http://windmonitor.iwes.fraunhofer.de/opencms/export/sites/wind-monitor/img/SR_2012_Neue_Ansaetze_in_der_Rotorentwicklung_e.pdf.
3. Hughes G. *The Performance of Wind Farms in the United Kingdom and Denmark*, Renewable Energy Foundation, 2012, Online. Available: http://www.ref.org.uk/attachments/article/280/ref.hughes.19.12.12.pdf.
4. Staffell I. and Green R. J. How does wind farm performance decline with age? *Renewable Energy*, 66, 775–786, June 2014, Online. Available: http://www.sciencedirect.com/science/article/pii/S0960148113005727.
5. Properzi S. and Herk-Hansen H. Life cycle assessment of a 150 MW offshore wind turbine farm at Nysted/Roedsand, *European Wind Energy Conference and Exhibition*, Copenhagen, Denmark, July 2001, Online. Available: http://tethys.pnnl.gov/sites/default/files/publications/Middelgrund_LCA_2001.pdf.
6. Tremeac B. and Meunier F. Life cycle analysis of 4.5 MW and 250 W wind turbines, *Renewable and Sustainable Energy Reviews*, 13, 2104–2110, 2009, Online. Available: http://www.ewp.rpi.edu/hartford/~ernesto/S2013/MMEES/Papers/ENERGY/6AlternativeEnergy/Tremeac2009-LCAWindTurbines.pdf.

7. Haapala K. R. and Prempreeda P. Comparative life cycle assessment of 2.0 MW wind turbines, *Int. J. Sustainable Manufacturing*, 3(2), 170–185, 2014, Online. Available: www.ourenergypolicy.org/wp-content/uploads/2014/06/turbines.pdf.

8. Goedkoop M. J., Heijungs R., Huijbregts M. et al. ReCiPe 2008, *A Life Cycle Impact Assessment Method Which Comprises Harmonized Category Indicators at the Midpoint and the Endpoint Level*; First edition Report I: Characterization, January 2009, Online. Available: http://www.lcia-recipe.net.

9. Douglas C. A., Harrison G. P., and Chick J. P. Life cycle assessment of the SeaGen marine current turbine, *Proc. Institution of Mechanical Engineers, Part M: J. Engineering for the Maritime Environment*, 222, 12pp., 2008, Online. Available: http://pim.sagepub.com/content/222/1/1.full.pdf+html.

10. Walker S. and Howell R. Life Cycle comparison of a wave and tidal energy device, *Proc. Institution of Mechanical Engineers, Part M: Journal of Engineering for the Maritime Environment*, 225(4), 325–337, November 2011, Online. Available: http://e-futures.group.shef.ac.uk/publications/pdf/94_15.%20 stuart%20walker%20summary.pdf.

11. Life cycle assessment of offshore and onshore sited wind power plants based on Vestas V90-3 MW turbines – Vestas Wind Systems A/S, Elsam Engineering A/S report No. 200128, October 2004, Online. Available: http://www.apere.org/manager/docnum/doc/doc1252_LCA_V80_2004_uk%5B1%5D. fiche%2042.pdf.

12. Hammond Geoffrey P., Jones Craig I., and Spevack Rachel. The 'Shoots Barrage': An indicative energy technology assessment of a tidal power scheme, *Journal of Sustainable Development of Energy, Water and Environment Systems*, 2(4), 388–407, 2014, Online. Available: http://www.sdewes.org/jsdewes/ pi2014.02.0031.

13. Li Ye. and Florig H. Keith. Modeling the operation and maintenance costs of a large scale tidal current turbine farm, *Proc. Oceans Conference*, Boston, MA, September 2006, Online. Available: http://www. lumina.com/uploads/main_images/Modeling%20the%20Operation%20and%20Maintenance%20 Costs%20of%20a%20Large%20Scale%20Tidal%20Current%20Turbine%20Farm.pdf.

14. *Encana Partners to Enable Pearson College – Encana – Clean Current Tidal Power Demonstration Project At Race Rocks*, BC, Canada, Press Release, February 2005, Online. Available: http://www. racerocks.com/racerock/energy/tidalenergy/pressrelease.pdf.

15. Maehlum Mathias A. Wave Energy Pros and Cons, *Energy Informative*, May 5, 2013, Online. Available: http://energyinformative.org/wave-energy-pros-and-cons/.

16. Sørensen H. C. and Naef S. Report on technical specification of reference technologies (wave and tidal power plant), Project #502687, NEEDS New Energy Externalities Developments for Sustainability, November 2008, Online. Available: http://www.needs-project.org/RS1a/RS1a%20D16.1%20Final%20 report%20on%20Wave%20and%20Tidal.pdf.

17. There is how long do solar electric PV panels last? Centre of Alternative Technology (CAT), UK, Online. Available: http://info.cat.org.uk/questions/pv/life-expectancy-solar-PV-panels.

18. Chianese D., Realini A., Cereghetti N. et al. Analysis of weathered c-Si PV modules, *Proc. 3rd World Conference on Photovoltaic Energy Conversion*, 3, 2922–2926, Osaka, Japan, May 2003.

19. Jordan Dirk C. and Kurtz Sarah R. Photovoltaic degradation rates—An analytical review. *Progress in Photovoltaics: Research and Applications*, Journal Article NREL/JA-5200-51664, 32pp., June 2012, Online. Available: http://www.1sun.info/Claims/Photovoltaic%20Degradation%20Rates%20—%20NREL.pdf.

20. Mathur J. *Methodology for Performance Evaluation of Solar Photovoltaic Power Generation System in India*, US-India Centre for Building Energy Research and Development, June 2013, Online. Available: cberd.org/wp-content/uploads/2015/03/CBERD_4.1TR_004JUN2013.pdf.

21. Alsema E., Fraile D., Frischknecht R. et al. *Methodology Guidelines on Life Cycle Assessment of Photovoltaic Electricity*, Report IEA-PVPS T12-01: 2009.

22. Fthenakis V. M. and Kim H. C. *A Review of Risks in the Solar Electric Life-Cycle*. JRC/TREN Conference on Safety & Security of Energy Infrastructures in a Comparative View (SEIF-CV), Brussels, Belgium, November 2005, Online. Available: http://web.mit.edu/cron/project/urban-sustainability/ Old%20files%20from%20summer%202009/Bjorn/solar/Risks%20in%20the%20Solar%20Electric%20 Life-Cycle%202005.pdf.

23. Fthenakis V. M., Kim H. C., Colli A., and Kirchsteiger C. Evaluation of risks in the life of photovoltaics in a comparative context, *Proc. 21st European Photovoltaic Solar Energy Conference*, Dresden, Germany, September 2006, Online. Available: https://www.bnl.gov/pv/files/pdf/abs_191.pdf.

Section III

Testing, Monitoring, and Diagnostics
of Electrical Equipment

17 Physical Conditions of Electrical Equipment
Testing, Monitoring, and Diagnostics

Equipment reliability might be significantly improved through the effective prediction of equipment degradation. To evaluate the degree to which an electrical machine has degraded, it is necessary that one is able to measure the values of the most important physical parameters characterizing a specific condition. For electrical apparatus, there are multiple parameters that are directly related to the performance of the machine.

In this chapter, the main focus is on the physical conditions of the materials that the components of electrical equipment are made of. The electrical equipment include HV and MV electrical equipment, such as GIS, HV circuit breakers (CBs), power transformers, and MV switchgear and CBs. Not discussed in this chapter are the monitoring and diagnostics of the multiple parameters related to operating conditions of such equipment.

The aging of different materials, both conductive and insulating ones, is a major part of the degradation mechanism of electrical apparatus. For example, conduction problems caused by loose connections or deterioration of contact surfaces result in a local temperature rise (TR), which contributes to a reduction of contact quality. Thermal runaways induced by conduction problems deteriorate insulating material and cause disruptive dielectric discharges, resulting in arcing faults. The consequences of such faults are serious enough to justify the efforts to develop and build condition monitoring systems to protect electric facilities from disaster.

In this chapter, we discuss which physical conditions of the materials are most definitive in providing proper working conditions for different electrical machines and their important parts, and therefore are very important to evaluate. Multiple techniques are developed to measure various physical parameters off-line, online, and remotely. Since some physical conditions may be determined only by off-line testing, a number of testing techniques are discussed.

Various condition monitoring technologies allow condition monitoring of energized equipment continuously (online monitoring). These relatively new techniques give operation and maintenance personnel the means to evaluate the present condition of the equipment, to timely detect abnormal conditions, and to initiate actions for preventing possible forced outages.

17.1 PARAMETERS DEFINING THE PHYSICAL CONDITIONS OF ELECTRICAL EQUIPMENT

The aging of the materials that electric machines are made of is a continuous degradation process that occurs during normal service conditions and depends on the time in service and exposure to various environments. With increasing time of equipment in service, all electrical connections and insulations undergo aging and start losing their designed function. Monitoring and controlling the physical conditions of the components is an essential part of timely evaluation of their performance and making decisions to take appropriate actions for preventing an upcoming failure.

For example, among the many measurable physical conditions of current-carrying parts, the most important are temperature, resistance, and contact load [1]. The list of important parameters can be derived from the well-known failure causes of a specific type of equipment (Chapter 8).

17.1.1 Transformers

The major causes of oil-filled transformer failures related to materials are insulation and bushing deterioration, loss of winding, clamping, overheating, decomposition of winding insulation, oxygen, moisture, and gases in the oil, oil contamination, and partial discharge (PD) activity. This would mean that the important physical parameters for the monitoring and control of transformers are the temperature of electrical contacts, bushings, and windings, the oil temperature and composition, and the presence of PD.

17.1.2 HV Bushings

Bushings are strategically important components for the transmission of electrical energy. They are exposed to high thermal and electrical stresses. Monitoring and controlling basic quantities such as the oil level, pressure, and temperature of bushings, as well as detecting the corona and PD activity, is very important in order to identify dangerous conditions and to prevent thermal runaways.

17.1.3 Circuit Breakers

The choice of parameters to monitor depends on the technology, rating, and application, as well as on the applied insulation media (vacuum, air, SF_6, or oil). Some of the physical conditions are usually measured during scheduled maintenance tests, such as the contact resistance (CR), voltage withstand of insulation, vacuum interrupter conditions, and so on. Many other conditions may be monitored online, when CBs are energized, such as the contact and conductor temperature, presence of PD, audible noise, SF_6 gas level, and so on.

17.1.4 Switchgear

As switchgear distributes electrical current, heat buildup becomes a very important characteristic to monitor. The most significant amount of heat dissipation is on distribution elements such as busbars. As busbars are not made of a single piece of copper, the hot spots usually appear on connectors (used to attach breakers to the installation) or joints, which normally connect busbars and output cables.

Electrical switchgear is dedicated to operating under some maximum load (maximum current flowing through distribution components). The distribution elements, switching elements, and busbar joints may reach a predefined maximum temperature under maximum load to keep the system safe [2]. Unexpected TR at a particular location could indicate corrosion or some other type of defect. If left uncorrected, this defect could result in catastrophic failure, resulting in deactivated loads and potentially hazardous conditions to personnel. Therefore, for switchgear, one of the most important physical conditions to monitor and control is the temperature of the distribution elements.

17.1.5 Power Cables

Several important factors characterize the degree of deterioration of a power cable: the temperature of the cable, presence of PD in polymer-insulated power cables or their accessories, water penetration under the cable shield, leakage in oil-insulated power cables, as well as SF_6 leakage in power cable systems.

Testing, monitoring, and controlling these factors (PD, detection of water in the cable, axial distribution of the cable temperature, etc.) would help to identify and prevent water treeing and cable shield corrosion, and provide recognition, localization, and prevention of premature deterioration of power cable accessories. For transmission overhead lines, it is important to also monitor corona, which might damage the integrity of the cable and insulators. The parameters recommended to be

TABLE 17.1
Parameters to Test and Monitor for Physical Condition Evaluation

Equipment Type	Temperature	PD	Corona	Oil Temperature, Composition	SF$_6$ Level, Composition	Water
Transformer	X	X		X		
Bushing	X	X	X	X		
Switchgear	X					
Circuit breaker	X	X		X	X	
Cables	X	X			X	X
Transmission lines	X	X	X			

tested and monitored in various electrical equipment for the evaluation of physical conditions of electrical equipment are presented in Table 17.1.

17.2 TECHNIQUES FOR TESTING PHYSICAL CONDITIONS OF MV CABLES

Power cables are expected to remain a reliable conductor of electrical power for 30–40 years. It is therefore vital that the correct test procedures are followed and that these cables are correctly installed, diagnosed, and maintained. It is important to analyze and compare different modern techniques of MV cables testing with regard to the application, type of cable insulation and accessories, and type of defect to determine. It is important to determine whether it should be online or off-line testing and what should be the level of testing simplicity and cost.

It is predictable that the evaluation of different techniques by professionals from the companies providing cable testing would not be entirely objective and could often be biased toward a specific method. There are also opposing opinions among users of various techniques of proof testing. Some utilities providing routine testing on regular basis are servicing the customers with increased reliability. However, many utilities continue to experience multiple failures caused by insulation degradation, which is accelerated by excessive proof testing overstressing the cables [3–5].

However, although the data presented in available publications may be affected by the subjective nature of the reviewed material, comparative analysis may still be helpful in determining the test method best suited for the required application.

17.2.1 COMPARISON OF MV CABLE TESTING TECHNIQUES

The most often used techniques for testing MV cables are the high potential (HIPOT) withstand test and PD diagnostics [6].

There are several benefits from using HIPOT to test MV cables. The test requires simple and relatively inexpensive equipment; it uses very simple procedures, and does not require a trained analyst to interpret the results.

On the other hand, HIPOT does not monitor the effect of the test on the cable during the voltage application; it relies on "pass or fail" result, thus exposing the cable to the damaging condition. The test also cannot register certain types of defects. This test also weakens all defects simultaneously, but fails only one at a time.

PD diagnostics are not only the most effective testing methods but also the most sophisticated techniques. Three types of PD diagnostic have been developed: (1) online PD testing, (2) off-line PD testing, and (3) online PD monitoring. Continuous online PD monitoring technology for cables, although promising, requires further development to make it more affordable for industrial users. PD diagnostics are the only cable tests that can locate defects in shielded power cables. PD diagnostic testing has many strengths: it is a nondestructive test and the only testing technology that can detect and locate high-impedance defects such as void, cuts, electrical trees, and tracking. The test can be performed online in limited applications and is effective at locating defects in mixed dielectric systems.

The downside of PD diagnostics is that it is limited to cables with a continuous neutral shield. It requires a trained analyst to interpret measurements. This test cannot detect or locate conduction-type defects, and it becomes complex and time consuming and loses accuracy in branched network applications.

The third category of cable tests is *General Condition Assessment* (GCA), which includes multiple cable diagnostic methods and technologies, all of which determine the overall health of the cable insulation. These tests are quite sophisticated diagnostic tools, which measure the following factors: discharge dissipation factor (DDF)/tangent delta/power factor at 50/60 Hz and at 0.1 Hz—very low frequency (VLF) technique; time and frequency domains (dielectric spectroscopy); return voltage/recovery voltage (depolarization current test); relaxation current (isothermal relaxation current method); measurement of leakage current; and total harmonic distortion [6].

GCA diagnostic tests are considered to be nondestructive, because most tests are conducted at modest voltage levels with a short dwell time. These tests monitor the overall condition of the cable insulation and are effective in detecting and assessing conduction-type defects. Importantly, these tests provide information in the form of three categories—critically aged, moderately aged, and like new.

One disadvantage of GCA diagnostic tests is that they do not provide defect location, which results in complete cable replacement if the "bad" condition is found. The most important downside of these techniques is that the tests are not very accurate, and the results are highly temperature dependent in extruded cables. The tests are blind to high-impedance defects such as cuts, voids, and PD with less than 10,000 pC. The negative economic side is that these tests require pretty costly equipment when compared with HIPOT equipment.

Different testing techniques are compared in Ref. [7] with regard to which test method may be preferable in detecting specific defects in both extruded (cross-linked polyethylene [XLPE] and ethylene propylene rubber [EPR]) and laminated (paper-insulated lead-covered [PILC]) cables, terminations, and joints.

17.2.2 HIGH POTENTIAL WITHSTAND TEST

HIPOT withstand test can be divided by voltage source into three types: (1) direct current (DC), (2) VLF, and (3) power frequency (PF). Each type of HIPOT test has its advantages and weaknesses.

17.2.2.1 DC HIPOT Test

A traditional method to test an MV cable has been to use DC high potential (HP), also known in the industry as DC HIPOT testing. For many years, this test worked well as a factory, field maintenance, and assessment test. However, by the mid-1990s, multiple reports and field data made it evident that DC HIPOT testing could cause damage or prematurely cause the failure of extruded cables, especially field aged XLPE insulated cables.

This observation prompted the Electric Power Research Institute (EPRI) to fund two individual studies relating to the effect of DC HIPOT testing on XLPE and EPR insulated cables. The results were not conclusive on the EPR insulated cable; however, they were definitive on the XLPE insulated cable [8]. It was found that DC HIPOT testing before energizing a new MV cable does not cause any harm to the cable. However, when applied to an XLPE insulated cable that is in service and aged in a wet environment, DC HIPOT testing, using the industry-recommended DC levels and for the recommended length of time (i.e., 25 kV on a 5 kV cable for 15 min), may cause the cable to fail after its return to service. The failure may not have occurred at that point in time if the cable has remained in service and has not been tested with a high DC voltage.

It was also determined that DC HIPOT testing applied at recommended voltage levels cannot detect massive insulation defects in an extruded dielectric insulation cable. Also, applying overvoltage DC to a cable is done to take a snapshot of the cable condition at a given time to determine whether it is a "go" or "no go." It provides no trend data, life expectancy, or condition assessment of the cable.

The latest publications also show that there is nearly unanimous worldwide consensus that DC HIPOT testing is not only damaging to solid dielectric cable insulation but also an ineffective means of determining the insulation quality of a cable.

In North America, the Institute of Electrical and Electronics Engineers (IEEE), EPRI, Insulated Cable Engineers Association (ICEA), and others have denounced DC HIPOT testing for years, with the IEEE having written a new test standard for VLF that is promised to be implemented soon. European standards have been in place for years for VLF cable testing.

High-potential DC testing of XLPE MV cables is admitted to be a good installation check. It is also used for XLPE cables that have been subjected to long periods of exposure to moisture and in which water treeing may exist. Although the test is applicable to the cables in a wet environment, there are some concerns, as confirmed by EPRI studies. Three major industry associations are conservative about DC testing. IEEE, the Association of Edison Illuminating Companies (AEIC), and ICEA all have standards that address DC field testing of MV cables. The three groups are in close agreement about test values for installation tests, but after installation, however, all three reduce the testing stress and add cautions about wet environments.

IEEE-400-2001 Standard, "Guide for Field Testing and Evaluation of the Insulation of Shielded Power Cable Systems," describes the maintenance testing of up to 15 min, but also points out that DC HIPOT testing may be detrimental to XLPE cables in a wet environment in which water treeing may exist. AEIC CS5-94, "Specifications for Cross-Linked Polyethylene Insulated Shielded Power Cables Rated Through 46 kV," describes an after-installation test for the first 5 years, but adds the statement, "After that time, DC testing is not recommended." ICEA S-66-524, "Cross-Linked Polyethylene Insulated Wire & Cable for Transmission and Distribution," says, "If voltage tests are made after installation, they shall be made immediately" and it makes no mention of maintenance testing. A new ICEA S-94-649, "Standard for Concentric Neutral Cables Rated 5 through 46 kV," focused on concentric neutral design specifies maintenance testing limited to 5 min.

17.2.2.2 Very Low-Frequency HIPOT Test

VLF is generally considered to be run at the frequency 0.1 Hz and lower. According to IEEE Standard 400.2, "IEEE Guide for Field Testing of Shielded Power Cable Systems Using Very Low Frequency (VLF)," the main application of VLF HIPOT is in testing solid dielectric cables, installation quality, and accessories, such as splices. The VLF HIPOT test is provided from 25 kV up to 200 kV, and load ratings are available from 0.4 to 50 μF, representing approximately up to 40 miles of 15-kV cable [9].

VLF HIPOT has the advantage of low energy requirements, 1/600 for 60 Hz voltages and 1/500 for 50 Hz voltages, which results in much smaller test sets than those required for normal operating frequency measurements. VLF PD detection can find major defects in cable accessories [8]. VLF has been more effective than DC in causing fault breakdowns on certain 15 kV, paper-insulated, lead-covered systems and on the 5 kV rubber and lead (R&L) cable that is used on service and main circuit taps [10].

It is considered as a good HIPOT test for conductive-type defects and high-impedance defects, which use a portable source with relatively low power requirement. The test lessens the aggravation of defects, because it does not induce as much space charge as DC HIPOT. It works well in aged extruded cables. Although it causes some defects, they grow rapidly, resulting in a shorter test time. However, the VLF HIPOT test has been demonstrated to aggravate defects in aged cable without failing, and therefore is not recommended for aged cables with multiple defects. The test does not replicate service conditions, cannot be directly compared with factory tests, and does not replicate the normal stress distribution in insulation with wet regions.

17.2.2.3   AC Power Frequency HIPOT

There are no widely accepted test standards for AC HIPOT testing of cables, and PF AC HIPOT testing of cable systems is rarely performed. This is primarily due to the availability of other test methods and the high levels of power consumption and the large, heavy units required for AC HIPOT testing. However, this testing technique is good for conductive and high-impedance defects, it does not induce space charge, which minimizes the aggravation of defects in aged extruded cables, and it replicates steady-state service conditions and factory HIPOT tests.

17.2.3   PD Diagnostics

PDs are indications of a degrading insulation capability, which eventually leads to cost-intensive repairs and can result in electrical breakdown of the HV apparatus (see Section 7.2).

PDs are mainly caused by natural circumstances or by accelerated aging of the MV and HV insulation. Manufacturing or installation failures also can lead to harmful PDs. The goal of PD measurements is to detect the PDs at an early stage and to evaluate them in a proper way. Thus, periodic as well as continuous PD measurements have become one of the main means of assessing the quality and condition of HV equipment such as transformers, generators, switchgears, and cable systems. PD diagnostic tests could be performed online and off-line, and they are the only cable tests, which can locate defects in the shielded power cable. There are several methods to provide online PD monitoring, which will be discussed in Section 17.4.1.

Online PD diagnostics can be performed without switching the circuit out of service; the technique can detect and locate some accessory defects and a few cable defects. It does not require an external voltage source. Online PD diagnostic tests are useful where there is no way to take the cable system out of service. They are useful in detecting some defects of cable accessories (joints and terminations), especially in short cables where the PD signals are not greatly attenuated. However, this method has many weaknesses since it detects only 3% or less of the cable insulation defects in an extruded cable. It is not a calibrated test; therefore, the test results are not objective. The online diagnostic test cannot be compared with factory tests or IEEE standards; it requires access to the cable every few hundred feet depending on the cable construction. The major downside is that it requires that manholes be pumped to access the cable and cannot be applied to long, directly buried cables. These days it is only offered as a service and equipment cannot be purchased.

Off-line PD diagnostics requires that the cable be switched out of service for tests to be performed, and the testing equipment is expensive when compared with HIPOT equipment. The equipment for off-line PD diagnostics can be purchased and used by utility personnel and it provides onsite report of the test results. The test can indirectly locate large water trees associated with electrical trees. The test locates all defect sites in one test from one end of the cable. This PD diagnostic technique has been proven to be highly accurate by statistically significant correlation studies (83%–95%). Off-line PD diagnostics detects and locates defects in the cable system during voltage application. It is effective with mixed dielectric cables and can test up to 1–3 miles of cable depending on the cable construction. The off-line PD diagnostics replicate calibrated factory baseline tests and can be readily compared as well, as it replicates steady-state and transient operating conditions.

17.2.4   Choice of MV Cable Diagnostics

In 2002, the IEEE took major steps for revising its "Guide for Field Testing and Evaluation of the Insulation of Shielded Medium Voltage Power Cable" (IEEE 400-2001). This guide lists the various field test methods that are currently available or under development to perform field tests on insulated, shielded power cable systems rated 5 kV through 500 kV, including a summary of the advantages and disadvantages of the methods. The Guide includes six new approved testing methods: direct voltage, PF, PD, VLF (0.1–1.0 Hz), dissipation factor, and oscillating wave. Some of these tests are able to look at the condition of the insulation but cannot locate problems; that is, they are general health tests. Other tests are "partial discharge locator," which can locate PDs that are a sign of incipient faults in the cable. Off-line PD diagnostic testing is considered by IEEE Standard 400 as the most effective test for measuring PDs.

It is important to note that among all these tests, only the PD test can be performed while the cable remains energized and in service, making it the least disruptive test. It recommends that users review technical papers that are included as references and in the bibliography before deciding on whether to perform a test and which test to utilize. In making such decisions, consideration should be given to the performance of the entire cable system, including joints, terminations, and associated equipment. In many cases, this can be the determining factor for deciding on which test to use.

These new test methods have only been developed over the last 5–15 years, which makes them new to the industry and in some cases subject to the interpretation of the test engineer. Even though new tests exist, there is still not an industry standard test for the condition assessment of cable insulation, and none of the tests mentioned above is a clear leader in determining everything the user might want to know about the insulation or remaining life of a cable.

There is no simple recipe as to which technique is preferable for testing MV cables and accessories. To choose a proper testing technique, there is a set of things that need be considered: (1) the type of cable being tested, (2) age of the cable, (3) the type of defect you are looking for, (4) the environment; and (5) the risk tolerance level. Answers to these questions analyzed together with the strengths and weaknesses of each technique will allow one to make a reliable choice.

17.3 TESTING TECHNIQUES TO ASSESS INSULATION CONDITIONS OF HV/MV SWITCHGEAR, CIRCUIT BREAKERS, AND TRANSFORMERS

17.3.1 INSULATION CONDITION: PD TESTING

PD activity has long been accepted as a major cause of failure of HV/MV switchgear. PD is an electrical discharge or spark that bridges a portion of the insulation between two conducting electrodes. The discharge can occur at any location within the insulation system where the electric field strength exceeds the breakdown strength of that portion of the insulating material. Discharge can occur in voids within solid insulation, across the surface of insulating material due to contaminants, moisture or defects, and within gas bubbles in liquid insulation. Discharges known as corona occur around the energized parts in a gas (see Section 7.2).

17.3.1.1 PD Mechanism and Effect on Insulation

PD mechanisms can differ depending on how and where the sparking occurs. In poorly cast resin current transformers (CTs), voltage transformers (VTs), and epoxy spacers, there are many defects such as voids and cavities filled with air. Since air has lower permittivity than insulation material, an enhanced electric field forces the voids to flashover, causing PD. Energy dissipated during repetitive PD carbonizes and weakens the insulation.

Contaminants or moisture on the insulation induces the electrical tracking or surface PD. Continuous tracking will grow into a complete surface flashover. Corona discharge from the sharp edge of an HV conductor is another type of PD. During PDs, ozone and nitrous oxides are produced, which aggressively attack insulation and also facilitate flashover during periods of overvoltage.

When PD activity occurs, it emits electromagnetic energy in the form of radio waves, light, or heat. PDs also produce acoustic waves in audio or ultrasonic frequencies.

PDs may take place on any kind of insulation used in switchgear, including gas, liquid, and solid materials during a temporary overvoltage, during an HV test, or under transient voltage conditions during operation [11,12]. PDs occur in aged, defective, or poor-quality insulation and can propagate and develop until the insulation is unable to withstand the electrical stress and flashover and failure happens.

In HV/MV insulation PD takes place in two forms, surface PD and internal PD. When surface PD is present, tracking occurs across the surface of the insulation. In the presence of airborne contamination and moisture, the insulation erodes. Internal PD occurs within the bulk of insulation materials and is caused by age, poor materials, or poor-quality manufacturing processes. If allowed to continue unchecked, either mechanism will lead to failure of the insulation system under normal working stress, potentially resulting in catastrophic failure of the equipment (see Chapter 8).

PD testing in the field has been performed routinely overseas throughout Europe for many years, and its popularity is growing very quickly in the United States. PD testing has proven to be a very valuable online prediction tool to reduce the failures of MV and HV equipment. Depending on the apparatus that is being evaluated, PD can be detected using electromagnetic, acoustic emission or capacitive sensors to measure the resultant radio frequency signals created by the minute "sparking" that occurs due to PD in the insulation.

According to IEC 60270, "High-Voltage Test Techniques—Partial Discharge Measurements" [13], traditional techniques for the detection of PDs involve taking a plant out of service and energizing using a discharge-free power supply and measuring signals using coupling capacitors and conventional PD detectors. However, very often, owners and operators of electricity distribution networks prefer to use handheld nonintrusive PD detection instruments for the purposes of both condition assessment and enhancing operator safety [14]. Where the cost can be justified, the use of more permanent monitoring options for critical and targeted switchboards is expanding. Many additional monitoring solutions are introduced into the market [15].

17.3.1.2 Ultrasonic Detection of PD

One of the most proper methods for PD detection of surface discharges, which usually have low amplitude, is the method that uses ultrasonic detection. Unlike surface discharges, internal void discharges have consistently high amplitude levels but have much lower discharge rates.

There are two basic types of online solid insulation PD testing equipment. One type is the handheld ultrasonic emission detector, which is used to detect airborne PD or corona in outdoor substations or inside switchgear. This device is useful and has led to the discovery of many electrical insulation problems.

However, there are many disadvantages with this testing. Airborne ultrasonic detectors are successful in detecting surface discharge activity where there is a good acoustic path, and it is an important tool for the detection of surface discharge within air-insulated switchgear. However, this technique is of little use on GIS or on oil-filled transformers.

Since the sensor requires a clear air path to the problem source, it cannot look inside equipment or cables. The test equipment cannot distinguish the difference between corona (discharge into air), which may be relatively harmless, and PD, which is very destructive. In noisy environments such as HV substations or processing lines, this technology is often ineffective due to high background electrical or mechanical noise. Additionally, no true quantitative measurement of PD activity can be made using ultrasonic emission testing only.

17.3.1.3 PD Detection Using Transient Earth Voltages

The other type of online solid insulation testing equipment utilizes sophisticated measurement devices to detect, record, and measure PD. This type of equipment works extremely well to identify and quantify harmful PD activity. This equipment also detects PD within the insulation and does not require a clear line of sight to the defect. One of the most commonly used methods of detection of this type of internal discharge is through electromagnetic techniques and the detection of transient earth voltages (TEV) in the 3–80 MHz frequency range [16].

Suppose that a PD occurs in the phase to earth insulation of an item of HV plant such as a metal-clad switchboard or a cable termination. During this discharge, a small quantity of electrical charge is transferred capacitively from the HV conductor system to the earthed metal cladding. The quantity of the transferred charge is very small and is normally measured in pico coulombs. The transfer occurs typically in a few nanoseconds. When PD occurs, electromagnetic waves propagate away from the discharge site. Owing to the skin effect, the transient voltages on the inside of the metalwork cannot be directly detected outside the switchgear. However, at an opening in the metal cladding such as the joints with the gaskets, the electromagnetic wave can propagate into free space and generate a TEV on the metal surface [12].

There are also relatively low-cost handheld instruments available that can detect both ultrasonic and TEV activity. Using this type of instrumentation as part of routine substation inspection can greatly enhance the condition information collected during this inspection and usually requires minimal additional cost or effort. The PD detection equipment may be applied not only to the most important switchboards' inspection but also to all MV switchgears on the network.

For internal PD problems, it was found that the level of TEV signals is proportional to the condition of insulation for switchgears of the same type and model, measured at the same point. This produced a very powerful comparative technique for noninvasively checking the condition of switches of the same type and manufacturer.

During several decades of PD testing, extensive information on substation PD survey results with over 15,000 entries covering different manufacturers, types of MV switchgear, and associated equipment have been collected [16].

After a large number of air-insulated switchgears with smaller dimensions were introduced into distribution networks, surface PD problems increased. Additional problems occurred when ceramic materials were replaced with polymeric materials for solid insulation in switchgear. Although polymeric materials generally offer good electrical insulation properties, they are more prone to surface aging and degradation processes, and any electrical activity on the surface of such materials eventually causes tracking damage.

17.3.2 DIAGNOSTICS OF OIL CONDITION

To diagnose the condition of oil-filled electrical apparatus, testing the physical conditions of oil provides the most valuable information. This information can be used to detect and identify early faults in apparatus, provide an indication of their severity, and identify long-term aging trends. Both load tap changers (LTCs) and bulk oil circuit breakers (OCBs) can fail mechanically, electrically, and from deterioration because of local overheating.

It was assumed that by-products of the deterioration and overheating could be found in the oil. In recent years, dissolved gas-in-oil analysis and other insulating liquid tests have been used as effective tools to detect overheating problems in LTCs and OCBs. Specific guidelines and algorithms have been developed for evaluating the results from normal and abnormal LTCs and OCBs to help in the transformer condition assessment [17].

17.3.2.1 Dissolved Gases in Oil

For oil-filled transformers, the insulation system consists of oil and cellulose (paper). Under normal operating conditions, transformer insulation deteriorates and generates certain combustible and noncombustible gases. These gases are H_2, CH_4, C_2H_2, C_2H_4, C_2H_6, CO, and CO_2. The main cause of gas formation in the transformer is the heating of the paper and oil insulation due to electrical problems such as corona (low-energy phenomenon) and electric arc (high-energy phenomenon) inside the transformer. The detection, analysis, and identification of these gases can be very helpful in determining the condition of the transformer. Dissolved gas analysis (DGA) is very helpful in the electrical industry for the detection of gases in oil.

Detailed information on the concentration of all gases present in oil and the value of total dissolved combustible gases (TDCG) is given in IEEE Standard C57-1041991, "IEEE Guide for the Interpretation of Gases Generated in Oil-Immersed Transformers" [18]. In this standard, four status conditions are specified to classify risks for an oil-filled transformer. For each status an increasing gas-generating rate indicates a problem in the transformer oil, thus requiring the need to check the gas level in the oil more frequently.

For example, when the measured TDCG level is less than 720 ppm (Condition 1), it represents normal aging of the oil and oil analysis needs to be done in 6 months.

When the TDCG level is found to be higher than 720 ppm but not exceeding 1920 ppm (Condition 2), it indicates decomposition and excess oil aging, which requires oil analysis every 3 months.

Further growth of TDCG up to 4630 ppm indicates excessive decomposition, which requires oil analysis every month (Condition 3).

The highest level of oil decomposition is when TDCG is found to be above 4630 ppm; it indicates excessive decomposition of cellulose insulation and/or oil (Condition 4).

Continued operation could result in failure of the transformer. If TDCG and individual gases are increasing significantly (>30 ppm/day), the fault is active and the transformer should be de-energized when Condition 4 levels are reached.

A sudden increase in key gases and rate of gas production is more important in evaluating a transformer than the amount of gas. One exception is acetylene (C_2H_2). The generation of any amount of this gas above a few ppm indicates high-energy arcing. Trace amounts (a few ppm) can be generated by a very hot thermal fault (500°C). A one-time arc caused by a nearby lightning strike or an HV surge can generate acetylene.

If C_2H_2 is found in the DGA, oil samples should be taken weekly to determine if additional acetylene is being generated. If no additional acetylene is found and the level is below the IEEE Condition 4, the transformer may continue in service. However, if acetylene continues to increase, the transformer has an active high-energy internal arc and should be taken out of service. Operating a transformer with an active high-energy arc is extremely hazardous and may result in catastrophic failure [19].

17.3.2.2 Water, Acids, and Furans in Oil

It is well known that moisture continues to be a major cause of problems in transformers and a limitation to their operation. The presence of *water* in oil is very harmful to both the oil and the solid paper insulation. It can be present in the oil in a dissolved form, as tiny droplets. Since paper is porous it absorbs water, which accelerates the insulation degradation process. Water affects the dielectric breakdown strength of the insulation, the temperature at which water vapor bubbles are formed, and the aging rate of the insulating materials.

In the extreme case, transformers can fail because of excessive water in the insulation. The dielectric breakdown strength of the paper insulation decreases substantially when its water content increases [20]. A more long-term problem is that excessive moisture accelerates the aging of the paper insulation, with the aging rate being directly proportional to the water content. For example, as the water content in the paper doubles, so does the aging rate of the paper.

IEEE Standard C57.106.2002 [21] specifies the limits of water content in the mineral insulating oil, according to various voltage levels. For example, the suggested limits for the continued use of service-aged insulating oil are as follows: in a transformer of voltage class below 69 kV, amount of water should not be higher than 35 ppm; for transformers of voltage class above 69 kV but below 230 kV, moisture content should not be higher than 20 ppm. In the latest edition of this Standard, the amount of water in oil is defined not only according to voltage class but also based on operating temperature (Table 17.2).

Water does not remain at the same concentration in insulations, but rather it is continuously migrating between the solid and liquid insulation. In order to understand the significance of the water-in-oil value, the operating temperature of the transformer at the time of sampling must be known. Most of the water in a transformer system resides in the solid insulation (paper and pressboard) and not in the oil.

As the temperature increases, the water is forced from the paper into the oil. Although the amount of water in the paper will change relatively little, the concentration in the oil may change

TABLE 17.2
Accepted Maximum Water Content in Oil

Temperature (°C)	Up to 69 kV (ppm)	From 69 to 230 kV (ppm)	230 kV and Above (ppm)
		Voltage Class	
50	27	12	10
60	35	20	12
70	55	30	15

Source: IEEE Standard C57.106–2006 *Guide for Acceptance and Maintenance of Insulating Oil Equipment.* © IEEE.

by an order of magnitude or more, depending upon the initial water content of the paper and the temperature increase [20].

Acids are formed in the transformer oil by the oxidation process, and are responsible for sludge and varnish formation. Transformer oils are oxidized under the influence of excessive temperature and oxygen, particularly in the presence of small metal particles. Metal particles act as catalysts in the formation of carboxylic acids, which results in an increase in acid number.

In the worst-case scenario, the oil canals become blocked and the transformer is not cooled well, which further aggravates oil breakdown. The presence of acids in oil causes corrosion, because acids interact with the metallic materials inside the transformer. Acidity of the oil is characterized by acid number or neutralization number, which is the amount of potassium hydroxide (KOH) in milligrams (mg) that it takes to neutralize the acid in 1 g of transformer oil.

The acidity of oil increases with the age of oil in service, and high acidity means that the oil is oxidized and/or contaminated. Furthermore, an increase in the acidity has a damaging effect on the cellulose paper. Oil degradation also produces charged by-products (acids and hydroperoxides), which reduce the insulating properties of the oil.

Furans: Furans are a family of organic compounds which are formed by degradation of paper insulation. Overheating, oxidation, and degradation by high moisture content contribute to the destruction of insulation and form furanic compounds. Changes in furans between DGA tests are more important than individual numbers. The same is true for dissolved gases. Transformers with greater than 250 parts per billion (ppb) should be investigated because paper insulation is being degraded. ASTM D5837-99(2005) "Standard Test Method for Furanic Compounds in Electrical Insulating Liquids by High-Performance Liquid Chromatography (HPLC)" defines technique for determining the amount of furans in transformer oil.

17.3.2.3 Power Factor of Transformer Oil

Power factor indicates the dielectric loss (leakage current) of the oil. This test may be done by the DGA laboratories. A high-power factor indicates deterioration and/or contamination by-products such as water, carbon, or other conducting particles; metal soaps caused by acids attacking transformer metals; and products of oxidation. The DGA labs normally test the power factor at 25°C and 100°C. The Doble Engineering Company, in its *Reference Book on Insulating Liquids and Gases* [22], indicates that the in-service limit for power factor is less than 0.5% at 25°C. If the power factor is greater than 0.5% and less than 1.0%, further investigation will be required. In this case, the oil may require replacement or fuller's earth filtering. If the power factor is greater than 1.0% at 25°C, the oil may cause failure of the transformer; replacement or reclaiming will be required. Above 2%, the oil should be removed from service and reclaimed or replaced because equipment failure has a high probability to occur.

Doble Engineering Company has set up limits for other parameters of in-service oil, such as dielectric breakdown voltage, interfacial tension, water content, and neutralization number (content

TABLE 17.3
Limits for In-Service Oil

Parameter (Max Value)	ASTM Test Method	Voltage Class		
		Up to 69 kV	From 69 to 230 kV	230 kV and Above
Power factor at 25°C	Std D 924	0.5	0.5	0.5
Water content, ppm	Std D 1533	35	25	20
Acidity, mg KOH/g	Std D 974	0.2	0.15	0.15

Source: *Reference Book on Insulating Liquids and Gases (RBILG)-391*, Doble Engineering Company Publishing, Boston, MA, 1993; IEEE Standard C57.106-2006 *Guide for Acceptance and Maintenance of Insulating Oil Equipment*. © IEEE.

of acids in oil) [22]. The Doble limits for some parameters based on the recommendations of IEEE Standard C57.106-2006 are shown in Table 17.3.

17.3.2.4 Techniques of Oil Diagnostics

Transformer oil sampling and analysis has been used for many decades to provide valuable information regarding both the condition of the dielectric and the electrical condition of the transformer itself. Because of the low cost of the analysis and the ease in which samples can be taken, oil sampling has become the most frequent test performed on transformers today. In recent years, progress has been made in the analysis of dielectric fluids contained in LTCs and OCBs. In transformers and CBs, the analysis reveals both the condition of the dielectric fluid and the equipment itself, such as abnormal contact wear. Oil sampling in these devices is being adopted nowadays into more and more maintenance programs.

DGA is used to determine the concentrations of gases dissolved in the oil such as nitrogen, oxygen, carbon monoxide, carbon dioxide, hydrogen, methane, ethane, ethylene, and acetylene, according to ASTM D3612 [23]. The concentrations and relative ratios of these gases can be used to diagnose certain operational problems with the transformer, which may or may not be associated with a change in a physical or chemical property of the insulating oil. For example, high levels of carbon monoxide relative to the other gases may indicate thermal breakdown of cellulose paper, while high hydrogen, in conjunction with methane, may indicate a corona discharge within the transformer.

The detection of water in oil is most often performed in the laboratory by an analytical technique called the Karl Fischer titration described in ASTM Test Method D 1533 or IEC Method 60814. Both methods are very similar and are used to determine the amount of water in the oil sample on a weight-to-weight (mg/kg) basis or what is commonly known as ppm (parts per million). This method does not measure the water content in the paper insulation.

17.3.2.5 Online Monitoring of Transformer Oil Conditions

Equipment to continuously monitor oil condition in service is increasingly being installed on power transformers. The most widely installed systems measure hydrogen content. Systems for measuring other gases and moisture are also available. The hydrogen and composition sensors use semiconductor or fuel cell technology. There are more complex sensors that use infrared technology, gas chromatography, and photo acoustic spectroscopy. These systems can detect several or all dissolved gases in oil.

Several devices are available that provide for continuous online measurement of moisture and dissolved gases in oil [24]. In a simpler solution, the transmitter, which is compatible with any insulating oil, monitors the moisture levels at all ambient and operating conditions. The measurement can be connected to the substation data collection system directly [25]. Other systems can simultaneously monitor moisture, and any number of dissolved gases depending on system complexity [26]. However, the most complex systems are also very expensive; few utilities can afford them on all transformers, so the most sophisticated monitors are commonly found on only the most critical equipment.

There are many types of fairly simple and straightforward analyses that can be done online. However, the collected data can also be used for more complex analysis. This includes analysis of dissolved gas results and variations, even on monitors that detect a weighted sum of gases. Transformer loss of life and overload in real time may be calculated, as well as moisture in oil and moisture in paper at times when transformers are close to equilibrium of temperature and load.

Using online monitoring data allows comparing online condition monitoring results with offline results obtained via traditional methods. It allows to calculate the compensation of oil level for oil temperature, to measure transformer TR above ambient, and to compare the results from transformers of similar design. The focus of this analysis is mainly on the oil/paper insulation systems of the transformer. Online monitoring data, taken together with known theory on equipment behavior, can be used to narrow down information on a transformer's condition to a few possible causes. This is one of the desired outcomes of analysis of data received by online monitoring systems.

17.4 ONLINE MONITORING TECHNIQUES FOR PD OF MV SUBSTATIONS, SWITCHGEAR, AND CABLES

Continuous online monitoring employs special technologies developed for PD detection. For the PD measurement, electrical, high frequency (HF), ultra-high frequency (UHF), acoustical, chemical, and optical techniques are in use. These techniques have been standardized in North America and Europe. For example, the international IEC Standard 60270, "High-Voltage Test Techniques—Partial Discharge Measurements," defines guidelines for conventional PD measurements.

These methods may provide the users with the benefits of early detection of deteriorating insulation components and auxiliaries. Online monitoring will guarantee a reduction of in-service failure rates, that is, reduction of un-scheduled and costly downtimes, help in the planning of preventive or condition-based maintenance.

There are several techniques developed for online continuous monitoring of PD activity of cables, switchgear, CBs, and other MV/HV electrical equipment.

In the simplest case a shunt resistor converts the PD impulse current to a voltage signal. The characteristic of the detector (quadrupole—RLC detection impedance) can be designed for integrating the PD signal to obtain the apparent charge of each discharge. The available measurement systems can detect the apparent charge and also the phase location according to the supply voltage. Furthermore, the number of discharges over a given time interval are recorded [27].

Optical PD detection is based on the detection of the light produced as a result of various ionization, excitation, and recombination processes during the discharge. The amount of emitted light and the wavelength depend on the insulation medium (gaseous, liquid, or solid) and different parameters (temperature, pressure, etc.). The optical spectrum of the emitted light of PDs extends from the ultraviolet to the infrared range [28].

17.4.1 PD Detection in Substations, Switchgear, and Cables

To monitor PD online in *substations* [29], the system uses a multisensor cable loop providing UHF PD detection, which is a sensitive, noncontact method. The single-loop UHF PD location system is applied to an array of *MV switchgear* units. Since all cable lengths are known, the time-difference-of-arrival between acquired signals can be used to locate the defect. The system can locate PD activity and arcing (intermittent or repetitive). The system can be applied to any metal-clad (MC) equipment, not necessarily switchgear, and metal cladding is needed in order to avoid crosstalk between sensors. Another system combines TEV technology and ultrasonic detection (UltraTEV) [30] and is applicable for the continuous monitoring of PD activity in MV substations.

Another continuous online PD monitoring system has been developed for cable joints of *underground cable* circuits, which was based on the optical sensing technique using electro-optic (EO) modulators [31]. This proposed monitoring system does not require any power supply at the site of the cable joints, as the EO modulators are passive. Potentially, new cables could be laid together with optical fibers and new joints could be designed to include optical network-ready PD sensors.

A cost-effective online PD detection and location system for *MV cable* circuits was investigated and the development of a commercial system, named PD-OL (Partial Discharge monitoring, Online with Location), was based on the following concept [32]. If a PD is occurring somewhere in a cable, which is a long homogeneous structure, it induces a voltage/current pulse with a duration in the order of magnitude of 1 ns. The cable is much longer than the wavelengths of a pulse; therefore, the cable is treated as a high-frequency transmission line to predict the behavior. The pulse travels along the cable in both directions. As it propagates through the cable its shape and amplitude will change due to attenuation and dispersion. At the end of the cable the pulse is detected by the measurement unit. The origin of the PD is determined by the difference in time of arrival of the PD pulse at each end of the cable.

The goal of PD location is to determine the physical location of a defect, allowing the utility to replace the defective component. Furthermore, if circuit data are available, the location gives the type of the component. Knowledge of the type of defective component improves the reliability of the interpretation of PD measurements to a large extent. The PD location accuracy must be sufficient to enable one to easily locate the physical location of the defect. A PD detection system locates the defect as a percentage of the total cable length. In order to convert this location to a physical location the exact cable length and cable route need to be known. Due to the frequency-dependent attenuation, it is more convenient to specify the location accuracy relative to the cable length. A relative PD location accuracy is generally valid for a large range of cable lengths. Practical experience shows that a location accuracy of 1% of the cable length is both achievable during field measurements and acceptable for locating the physical defect location [33].

17.4.2 MONITORING PDs WITH FIBER-OPTIC TECHNOLOGY

Acoustic emission monitoring is often used in the diagnosis of electrical and mechanical incipient faults in HV apparatus. A large number of methods are available for the condition monitoring and diagnosis of power transformer insulation, but only a few can take direct measurements inside a transformer. PDs are a major source of insulation failure in power transformers and their detection can be achieved through the associated acoustic emissions. Several studies [34–38] have reported the development of the sensors and systems that are able to monitor the presence of PD in transformers.

For example, the system in Ref. [36] is based on extrinsic and intrinsic fiber Fabry–Perot interferometers for the detection of incipient faults in oil-filled power transformers. The sensors can be placed inside the transformer tank without affecting the insulation integrity, improving fault detection and location. The performances of the sensing heads are characterized and compared with the situations where it operates in air, water, and oil, and promising results are obtained, which will allow the industrial development of practical solutions.

In Ref. [37], detection is based on optic phase changes in an intrinsic Mach–Zehnder fiber-optic interferometric sensor of high sensitivity. The sensor can be placed within transformers and is suitable for the detection of weak acoustic signals associated with the PD. The calibration of the interferometric fiber-optic acoustic sensor at typical frequencies displayed by PDs is presented.

In Ref. [38], a fiber-optic sensor based on a Fabry–Perot interferometry is constructed by a simple micromachining process compatible with microelectromechanical system (MEMS) technology. The sensors are used in a transformer to measure PD acoustic waves. The experimental results show that the sensor not only has an inherent high signal-to-noise capability, but is also able to accurately localize the PD sources inside the transformer.

Fiber-optic sensors are attractive devices for PD detection because of a number of inherent advantages including their small size and high sensitivity. They are electrically nonconductive and immune to electromagnetic interference (EMI). However, until now, not a single system is available on the market for industrial applications, although this technique, if developed further, might present a very precise and safe method of monitoring PD in oil-filled HV electrical equipment.

17.5 TESTING OF HV BUSHING CONDITIONS

The parameters that characterize physical conditions of HV bushings are power factor, radio influence voltage (RIV), and DC insulation resistance. Measuring the moisture content of the oil or compound is also very important since about 90% of all preventable bushing failures are caused by moisture entering the bushing through leaky gaskets or other openings according to operating records [39]. A number of traditional tests have been developed to measure these conditions.

1. *Power-factor test*: The power-factor test is the most effective known field test procedure for the early detection of bushing contamination and deterioration [40]. This test also provides measurement of AC test current, which is directly proportional to bushing capacitance. Bushings may be tested by one or more of four methods depending upon the type of bushing and the power-factor test set available [39].

   These methods are the grounded-specimen test (GST), the hot-guard test, the ungrounded-specimen test (UST), and the hot-collar test. Large variations in temperature have a significant effect on power-factor readings in certain types of bushings. For comparison purposes, readings should be taken at the same temperature, or corrections should be applied before comparing readings taken at different temperatures.

   Complete detailed instructions on the method of testing and test procedure are usually given in the appropriate power-factor test set instruction book. This test has a number of weaknesses. It is affected by temperature and humidity and cannot detect the early stages of PD effects [41].

2. *The RIV test*: The RIV test can provide detection of corona in resin-bonded, solid-core, non-condenser bushings. Methods of measurement of RIV are described in NEMA Publication No. 107 [41]. Liquid-filled bushings generally have a low RIV value. A high RIV value on this type of bushings which cannot be reduced by cleaning the porcelain indicates that the level of filling liquid should be checked.

3. *The DC insulation resistance test*: The DC insulation resistance test generally cannot be relied on for detecting early contamination in bushings. When bushing deterioration can be detected by DC insulation resistance, it is generally at an advanced stage requiring immediate attention. A 2500-V insulation resistance meter may be used for an insulation resistance check, but a high reading should not be completely relied upon as indicating a good bushing. Any bushing testing less than 20,000 $M\Omega$ has questionable insulating value.

4. *Hot-wire test for moisture*: The plastic-type compound used in compound-filled bushings may absorb moisture if there are leaks through the shell or cap. Moisture content of as little as 0.15% in soft compounds can be detected by inserting a red-hot rod into the compound. If moisture is present, a crackling, sputtering, or hissing sound will be heard. If no moisture is present, the compound will melt quietly. Another test is to put some compound on a wire and melt it in the flame of a match. If moisture is present, there will be a sputtering sound and small sparks will be thrown out. In contrast, a dry compound will melt without disturbance.

5. *Testing oil for moisture*: Whenever the presence of moisture in the oil of an oil-filled bushing is suspected or found by a bushing power-factor test, the oil should be drained out and a

sample of it tested by a dielectric or power-factor test. Since the quantity of oil in a bushing is small, the old oil should be discarded and new oil poured in. If moisture is found in the oil, the bushing should be dried out before returning to service.

17.6 THERMAL CONDITIONS OF ELECTRICAL EQUIPMENT AND TEMPERATURE MONITORING

Conduction problems in electrical systems are due to either loose connections or deterioration of contact surfaces. They result in a local TR, and the greater the current, the greater the rise, which contributes to further downgrading of the contact quality. This leads to thermal runaway, and when the temperature becomes excessive, the heat deteriorates the insulating material. Very often, thermal runaway ends up causing disruptive dielectric discharges, which are a disaster for the electrical systems, such as MV CBs, switchgear, and substations.

Overheating is one of the major causes of the failures of transformers and bushings, underground and transmission cables, and other important electrical equipment. In other words, an ability to measure and control the temperature of distribution and transmission equipment contacts and insulation, solid or liquid, is one of the important tools for preventing any equipment's disruptive failure.

Since TR occurs when the current flows through the system, temperature measurement should be made when the equipment is energized, for example, online. It might be periodic or continuous. There are many approaches to measuring temperature, some of which will be described in this chapter.

17.6.1 TEMPERATURE MEASUREMENT USING THERMOGRAPHY

To date, the solution most commonly used to protect against conduction faults consists of periodically carrying out maintenance operations, which include temperature measurement. Typically, at the time of annual servicing, a complete thermal inspection is made of the electrical installation, during which the temperature of the connections is checked. The technique that is usually used for this purpose is infrared thermal analysis. It consists of periodically inspecting the installation by means of an infrared camera, with the aim of detecting thermal faults likely to reveal a conduction fault [42].

In order to have visual access to the conductors, it is necessary to modify electrical installations with the openings/viewing windows. Such openings may be made in the metal plates of MV cubicles, but this downgrades the degree of protection and hence the operating safety. It is possible to make some metal walls "transparent" for infrared (IR) emission by installing inspection windows made of material transparent to infrared rays (quartz), or the so-called IR Windows. Despite such conversions, there are still multiple points to which visual access is impossible.

In addition, the infrared measurement principle is not very accurate since it only allows the emissivity of the radiating bodies to be measured, and not the temperature. To detect a fault by means of infrared thermal analysis, several conditions must be met at the same time: it is necessary to have a substantial fault, a strong current to reveal it, and an expert eye trained to detect differences in emissivity between phases.

The periodic nature of inspection is the main limiting factor with this practice. A fault which appears on the day after the yearly visit will only be detected a year later, but it will continue to degenerate and lead to thermal runaway in the meantime.

17.6.2 CONTINUOUS TEMPERATURE MEASUREMENT

There are several techniques that can be used for continuous temperature monitoring of energized electrical equipment.

17.6.2.1 IR Noncontact Temperature Sensors

The technique for continuously monitoring the temperature using infrared emission of heated surface is to measure the temperature with noncontact IR thermometers. IR sensors are installed in the close vicinity of the target and send signals to a remote PC [43]. An optimal distance between the sensor and the target is determined by the size of the target (diameter, D) and the parameters of the sensor. For each particular type of IR sensor, the ratio field of view (FOV) = X/D is a constant value, where X is the distance between the sensor and the target. The smaller the target area (D) is, the closer the sensor should be located to the target. The solution is relatively inexpensive, but there are several disadvantages. First, the sensor must be directed towards the observed component very precisely and in most cases there is no possibility to install such a system on assembled switchgear. Second, since the components whose temperature is measured are typically enclosed within an insulation boot in the switchgear, the temperature readout is not very precise. This condition contributes to the limits for the number of possible applications and targets of interest. Another downside of using IR noncontact sensors is the need to run the cables from each sensor to receiving units.

17.6.2.2 Electronic Temperature Sensors

Some MV cubicle monitoring systems are based on the use of electronic temperature sensors installed at critical points. This solution offers the advantage of continuous monitoring of the installation and detecting thermal faults very early. The problem is that each sensor must draw its energy from the mains current and transmit digital data via an infrared or optical fiber link. Such complexity leads to a high cost for each measurement point. Another problem with using electronic temperature sensors is that they require the installation of devices whose level of reliability and service life are incompatible with the associated switchgear and the function to be performed. A diagnosis system must be much more reliable than the equipment that it is monitoring, especially if, as is the case here, it is not maintainable without a shutdown of the substation.

17.6.3 FIBER-OPTIC TECHNOLOGY FOR TEMPERATURE MEASUREMENT

Various combinations of fiber-optic technology are in use for temperature measurement and offer solutions for a variety of tasks. In the environment of high field strengths, the usage of optical sensors or fiber-optic technologies offers the possibility to measure important values without any influence of EMI. Also the galvanic separation (isolation) is an advantage [44]. Alternative temperature measuring systems for temperature measurements inside the equipment use thermal sensors based on fiber-optic technologies. This complex method consists of fiber-optic temperature sensors installed in the apparatus. Different types of sensors and functional principle are in use in the electrical industry [45].

17.6.3.1 Optical Fiber Sensing Probe

One possibility is to measure the temperature at one point of the fiber. For example, the temperature measurement principle is based on an optical fiber sensing probe. In such a probe, a small portion of the fiber cladding is removed and replaced by a suitable "reference" liquid whose refractive index versus temperature characteristic is known [46]. Another technique involves an optical fiber with a temperature-sensitive phosphor tip at the end. After excitation with blue-violet light the phosphor fluoresces with red light. The intensity of this light decays exponentially with time. The decay time is measured (the time constant of decay is inversely proportional to the sensor temperature) and correlated to the tip temperature [47].

Various combinations of fiber-optic technology come in many forms and offer solutions for a variety of applications. A fiber-optic sensing device is generally made up of two separate components, namely the amplifier unit and a sensing head assembly. The sensing head assembly comprises a pair of optical fibers, a transmitter and a receiver, terminating in the optical fiber-based head [48,49].

The probe requires no wires or other metal parts. It is electrically nonconducting, unlike thermocouples and resistive temperature detectors (RTD). Therefore, it can be installed in HV environments. Fiber-optic technologies are used for the temperature monitoring of cables and HV transformers.

A fiber-optic wire may be used as a temperature data carrier or as an active sensing element. In an active sensing element, temperature-dependent modification of the optical transmission properties is used in distributed fiber-optic temperature-sensing systems [50]. In other systems, a fiber-optic cable is used as a temperature data carrier, which delivers light to the sensing element installed at the measurement point. After passing through the sensor the signal is modified by the changing temperature in the sensing head. Then another fiber-optic cable carries back the modified signal to the analyzer. The materials in the fiber that communicate with the sensitive element exhibit low thermal conductivity as well as a narrow cross-sectional area. This minimizes heat flow to and from the active sensing element from outside the volume whose temperature is to be measured.

Fiber-optic temperature transducers are compact, immune to EMI and radio-frequency interference (RFI), resistant to corrosive environments, and provide high accuracy and reliability in temperature measurements. The various combinations come in many forms and offer solutions for many applications. Sensitive elements at the fiber tip provide rapid and accurate temperature measurement. Mounting the sensing element at the end of a small optical fiber allows placing the sensor in difficult-to-access locations.

17.6.3.2 Distributed Fiber-Optic Temperature Sensing

Another possibility is to measure the temperature distributions along the length of the fiber-optic cable with the use of thermo-optical effects. The light impulse injected into the fiber is subjected to scattering as it travels and the backscattered light impulse is returned to the detector. Among the returned light pulses, the intensity of Raman scattering is closely related to the temperature of the position of scattering. So the temperature along the fiber can be measured from the intensity of Raman scattering [51].

In the case of the existing power distribution system, applying the fiber-optic modifications to the equipment would require serious modifications with subsequent design verifications and testing. In the case of the draw-out MV CB, for example, it would necessitate designing a special automatic disconnect to comply with the requirements of the ANSI Standards. So the application of such a system, while very reliable and relatively inexpensive for the new equipment, becomes prohibitive for the existing equipment.

17.6.4 Winding Temperature Monitoring of HV Transformers with the Fiber-Optic Technique

The loading capability of power transformers is limited mainly by the temperature of the winding. As part of acceptance tests on new units, the TR test is intended to demonstrate that at full load and rated ambient temperature, the average winding temperature will not exceed the limits set by industry standards.

However, the temperature of the winding is not uniform and the real limiting factor is actually the hottest section of the winding commonly called the winding hot spot (WHS). The temperature of solid insulation is the main factor in transformer aging. When insulation is exposed to high temperature for extended periods of time, the cellulose insulation undergoes a depolymerization process during which the cellulose chains get shorter.

When it happens, the mechanical properties of paper such as tensile strength and elasticity degrade. Eventually, the paper becomes brittle and is not capable of withstanding short-circuit forces and even normal vibrations that are a regular part of the transformer life. Such a condition of the paper indicates the end of life of the solid insulation. Since this condition of the paper insulation is not reversible, it also defines the end of life of the transformer.

Therefore, winding temperature is a prime concern for transformer operators. This variable needs to be known under any and all loading conditions, especially in the unusual condition involving rapid dynamic load changes. Accurate knowledge of the temperature of the WHS is a critical input for the calculation of the insulation aging, assessment of the risk of bubble evolution, and short-term forecasting of the overload capability. It is also critical for efficient control of the cooling banks to ensure that they can be set in motion quickly when needed [52].

The WHS area is located somewhere toward the top of the transformer, and is not accessible for direct measurement with usual methods. For about 40 years, fiber-optic temperature sensors have been available for temperature measurement in HV transformers. The first units were fragile and needed delicate handling during manufacturing and installation. Over the past 20 years, significant development has taken place to improve ruggedness and facilitate connection of the five cables through the tank wall. New fiber-optic probes are protected with a permeable protective PTFE Teflon sheath to withstand manufacturing conditions including long-term immersion in transformer oil.

In one instance of temperature monitoring at Manitoba Hydro, fiber-optic sensors have been installed in more than dozen critical oil-filled HV transformers, each of these transformers contains eight probes to measure hot spots in the windings. Monitoring the winding hot-spot temperature together with measuring the dissolved gas-in-oil and furan-in-oil provides a major support to the operator when the transformer faces overload conditions. Online monitoring of winding temperature can provide a dynamic evaluation of insulation degradation [53,54].

There are several temperature-monitoring systems available on the market that are developed specifically for monitoring WHS in HV transformers.

17.6.5 WIRELESS TEMPERATURE MONITORING

17.6.5.1 Structure, Benefits, and Problems of Wireless Temperature-Monitoring Systems

The most troublesome issue in building a reliable condition monitoring system is wires that are used to get temperature readouts from the sensors. Elimination of wires required for constant communication with the sensor unit provides a great opportunity to build a temperature measurement system that (1) is easily installable, even on the already assembled switchgears, and (2) provides online temperature measurements.

The most important advantage of using wireless technique to monitor thermal condition of the energized equipment is eliminating any cables and wires from the online system. Another important benefit of wireless technology is much lower installation costs than that of any other type of online monitoring equipment [55–59]. Wireless systems should work well in difficult or dangerous-to-reach locations or in moving applications.

An ideal wireless temperature-monitoring system (WTMS) would consist of the specific components of hardware and software. Hardware consists of sensors and receivers or interrogators. To achieve the goal of wireless continuous temperature monitoring of HV and MV energized electrical equipment, there are specific requirements to be met by the sensors. The sensors should be wireless units equipped with unique identification and should be made of miniature and dielectric components. Signal transmissions from multiple sensors should not interfere with each other. The sensing units should be installed at all strategically important points on the equipment, in locations with a very limited space. To make the system work uninterrupted during a long period between maintenance, the sensing units should be equipped either with long-lasting local power source such as batteries or be self-powered by any available technique. For example, it could be the alternating magnetic field of an electrical conductor. Another possibility is that the sensor can be powered from a distance.

Receivers or interrogators should be installed at a significant distance from the sensors in the central location; they should collect the data from all sensors and transfer the data to a PC. They should work independently in series with other receivers. In case a temporary EMI exists, they should easily recover.

Technology progress on wireless systems and devices introduces every year at least a few new wireless solutions so that there are several mature technologies that can be used to implement wireless sensor solutions. Unfortunately, because of the fact that often the wireless sensors are battery-powered, the lifetime of the sensor is limited to at most a few years. For critical applications, the average switchgear lifetime is about 30 years; thus the condition monitoring system should last for the same time and require little maintenance.

However, a major maintenance interval of MV and HV electrical equipment of 5–7 years is not uncommon. There are also solutions based on custom protocol optimized for wireless temperature monitoring [60], but similarly to the standard sensors, the lifetime of the solution is limited by battery capacity.

17.6.5.2 Thermal Diagnostics

Thermal diagnostics has two separate goals. One is an ability to detect an early conduction fault (warning), another would be a thermal runaway detection, which results in forced interruption of electric current (protection). Conduction downgrading phenomena are generally slow and progressive. Early detection of conduction faults is based on an analysis of small changes in temperature. Detection of local overheating right from the start allows scheduling of maintenance operations at a convenient time, when it does not create a significant disturbance of the process.

Timely scheduled maintenance preserves the installation, mainly the contacts and insulating materials, since the thermal phenomena are still weak. A few principles may be used when developing the algorithms for early detection. For example, when monitoring the temperature in electrical switchgear, knowledge of cubicle thermal models is important. Correlation of contact temperature data with the current values and ambient temperature measurements, and temperature measured on all three phases may provide very important information on the thermal processes within the installation.

Since the overheating phenomena are usually rare and slow, the monitoring system should be designed to absolutely avoid false alarms, incorporating all available information on other network fault situations (strong phase unbalance, short circuit) or any temperature measurement fault (noise, incoherence). In the event of a conduction fault, the alarming signal should contain the precise location of thermal fault in order to facilitate and shorten maintenance operations. In the event of a failure, it is necessary to set up a temperature-driven function which, when it observes a measurement greater than a set point value, sets an output signal that controls the tripping of a CB. Such systems should be very reliable to not give false alarms or, even worse, cause nuisance tripping of the CB.

Wireless temperature-monitoring sensors may be installed on electrical equipment at the points that are not accessible with the power on. To provide continuous uninterrupted service, the sensors must have a very high service life and high mean time to failure (MTTF), which is the sensor's reliability calculated over its designed lifetime. Long service life of temperature-monitoring system is defined by a reliable and independent power source for the sensors.

17.6.5.3 Wireless Temperature Sensors: Power Source

In wireless temperature sensing, the search for reliable power supply is very important. A traditional way of supplying energy to a wireless sensor is by using primary (nonrechargeable) batteries, whose main weakness is the limited amount of available energy. The life of the battery depends on ambient temperature. Considering the application of sensors in electrical installation, ambient temperature will be always elevated, thus significantly shortening the battery's useful life. To keep the sensor functioning, the battery should be replaced when depleted.

By definition, wireless autonomous sensors must not depend on an external power supply. Energy harvesting is a means of extending the lifetime of the autonomous sensor beyond the life of a battery [61]. Several dominant energy-harvesting technologies are known. Choosing one or another energy-harvesting technology application depends on the type of equipment where the wireless sensors are used.

One of these technologies uses *photovoltaic* elements [62], which produce electricity from ambient light—either indoors or outdoors. Photovoltaic cells produce electricity from photons by means of a semiconductor p–n junction. This technology is at a relatively advanced stage of development.

Vibration harvesters [63,64] use electromagnetic, piezoelectric, or electrostatic phenomena and may produce electricity from vibrations of the surface the sensor is deployed on. Electromagnetic generators use a resonant magnet and coil arrangement to generate electricity, whereas piezoelectric-based generators use a piezoelectric resonant beam which generates electricity when subjected to strain. Electrostatic generators exploit capacitive effects but, due to machining and practical issues, have not become widespread. Vibration energy harvesters are sensitive to the frequency of vibrations of a surface, and their deployment is generally limited to machinery that vibrates at a constrained range of frequencies and amplitudes.

Thermoelectric elements [65] produce electricity from a temperature gradient. Thermoelectric energy sources exploit the Seebeck effect, in which electricity is generated from a temperature difference across a thermocouple. In general, thermoelectric devices require a large and sustained temperature gradient between two surfaces in order to provide useful power.

For application in the electrical industry, passive/remote power supply was used in temperature wireless sensing: *surface acoustic wave* (SAW) technology [66]. SAW sensors operate with no wire connection or battery; they do not need power supply. They are connected only by a radio frequency link to a transceiver or reader unit. A high-frequency electromagnetic wave is emitted from an RF transceiver and is received by the antenna of the SAW sensor.

The sensors use interdigital transducers (IDTs) that convert the electric field energy to mechanical wave energy and then back to an electric field. The transducers are connected to the antenna to convert the received signal into an acoustic wave, which propagates along the sensor. Depending on the construction of the device, the IDTs can retransmit the signal to the receiver. The received signal is amplified and then converted to a baseband frequency in the RF module and then analyzed by a signal processor. As the operating frequencies are high, SAW sensors are well protected from EMI that often occurs in the vicinity of industrial equipment, such as motors and HV lines.

17.6.5.4 Wireless Temperature-Monitoring Techniques

Traditional methods of measuring temperature have relied on the temperature dependence of resistance (thermistors or RTDs), a variety of different types of thermometers, the temperature dependence of a diode junction (silicon), or the emission of infrared radiation from heated objects (IR sensors).

Passive devices such as thermocouples and RTD require battery-powered transmitters to communicate information. This complicates measuring the temperature of contacts and connections in HV switchboxes and transmission. A standard requirement for these systems is that there be no metallic or fiber-optic cabling from the contact or connection of interest to the supporting structure or frame, as this can cause a dangerous and potentially explosive path to the ground.

Infrared thermometry is sometimes employed, but this requires a direct line of sight to the area of interest, which should be clean for best accuracy and is more often used for spot checking on a periodic basis and therefore not for continuous monitoring. Typically, the infrared measurement systems used for this type of monitoring are cost prohibitive. Battery-powered temperature-transmitting systems have drawbacks related to the typical physical size and the need for inconvenient periodic replacement of the battery. With a few exceptions, batteries are not well suited for high-temperature operation, especially above 150°C.

17.6.5.5 Wireless Temperature Monitoring with SAW Sensors

SAW technology uses the effect of changing the parameters of an acoustic wave, such as velocity and amplitude, when it propagates through or on the surface of the piezoelectric material depending on the physical condition of the material. Any changes to the characteristics of the propagation path affect the velocity and/or amplitude of the acoustic wave.

Changes in velocity can be monitored by measuring the frequency or phase characteristics of the sensor and can then be correlated to the corresponding physical quantity being measured. SAW sensors are very sensitive in measuring mechanical properties such as stress or strain. Special cuts of piezoelectric substrate create SAW devices with a very linear SAW frequency versus temperature dependence. The result is a very high-resolution temperature sensor.

In WTMS based on SAW technology, the number of sensors is limited to no more than a few. In this technology, the measured quantity (temperature) is calculated from the frequency range; thus every sensor must occupy its own frequency range.

This limitation may have significant impact on the applicability of SAW sensors for complex condition monitoring of an electrical system, such as switchgear components. An average switchgear frame contains about 12 hot spots: three each for main bus, for cable terminators, for breaker inputs, and for breaker outputs. Switchgear interiors contain a number of conducting metal plates and other components that reflect and degrade radio signals, which cause difficulties in using this technology. To apply a set of SAW sensors to measure temperature in all 12 locations in each cubicle of switchgear, a special layout of the sensors is required. The solution [67] was found by antenna tuning, so that shielding capabilities of the switchgear cubicle boxes divided the switchgear into sections that contain sensors operating at different frequencies, while frequencies in different sections were duplicated. It allowed installation of 72 sensors on the switchgear to monitor all expected hot spots.

Utilization of miniature SAW sensors enables monitoring of breaker connectors and noninvasive installation inside the switchgear. The small size of sensors allows the system to be extremely flexible in choosing measurement locations. Passive power supply enables indefinite lifetime of the system regardless of the external condition, which is a significant benefit of self-powering devices.

Passive wireless temperature sensors using SAW technology are a promising monitoring and diagnostic solution for critical equipment such as MV and HV switchgears. They provide a very important diagnostic means for the protection of switchgear from catastrophic failure damage. Application of wireless temperature monitoring allows preventing expensive outages and improving the safety of switchgear technicians and maintenance personnel. It may be considered as a valuable replacement for existing traditional temperature-sensing solutions within the switchgear application [68].

17.7 PHYSICAL CONDITIONS OF TRANSMISSION ELECTRICAL EQUIPMENT: ONLINE MONITORING TECHNIQUES

17.7.1 CONDITION MONITORING TECHNOLOGIES IN ELECTRICAL TRANSMISSION

Online condition monitoring is generally achieved through the use of specialized sensors or devices. The technologies used in these sensors have been around for several decades, their application to HV electrical equipment is relatively recent.

Online monitoring devices range from single-function monitors such as dissolved gas in oil and moisture in oil monitors, to more complex devices that monitor a range of parameters from electrical equipment, such as tap changers and CBs. The number of devices available is continually growing, and new ones are regularly developed and marketed [69].

There are online monitoring devices that monitor HV transformer dissolved gas and moisture in oil, and also transformer temperature monitors (including high-accuracy optical fiber temperature sensors). Other devices monitor CT, bushing dielectric dissipation factor, and PD. There are monitors that measure oil level and oil flow for oil-filled transformers. Many monitors are developed for other electrical equipment including CBs, batteries, and backup diesel generators to determine additional information such as loadings and ambient temperature that is also useful information for equipment monitoring.

Some utilities around the world develop integrated online monitoring systems that manage information from a wide range of these devices across an entire HV network. For example, the system described in Ref. [70] is able to handle the following types of information: analog quantities, such as DGA and moisture; digital quantities, such as extended CB alarms; records from protection relays, CBs, tap changers and battery tests; and plain text information, such as flag information from modern protection relays. Even in its early stages, the system has been able to provide a range of information on a variety of equipment, which has served a number of purposes. The main advantages that utilities have found with this system are easy and flexible access to online monitoring data and the ability to use these data for automated analysis as required.

17.7.2 OVERHEAD TRANSMISSION LINES

The safety and benefits of the electric power system highly depend on proper functioning of HV overhead transmission lines (OHTL), which are the arteries of the electric power system.

The temperature of an energized conductor either in overhead transmission or distribution lines, as well as underground cables, is a very important physical property that carries information both on the conditions of a conductor as well as on the load. The major concerns about the condition of OHTL could be divided into two major groups. One is an aging process that leads to the deterioration of all components of transmission lines, including the wires, transmission towers, and insulators. Another concern is the need to increase the power load of existing lines depending on the physical conditions of the lines.

There are several mechanisms that affect the health of transmission lines, and the most serious aging processes are the ones that impact all components of the line. These are extreme temperature changes, wind-induced motions, corrosion, and contamination.

Utilities are running OHTL at operating temperatures exceeding 90°C, especially under emergency conditions. High-temperature operation of OHTL results in degradation of the materials, which affects mechanical strength of the cables and joints. With the deregulation of the power industry, there are increasing pressures on transmission and distribution providers to increase the utilization of their existing assets. This means that increasing power delivery will require raising further the allowable operating temperatures of the conductors in the system. Thus, while 125°C is considered a practical limit to emergency loading, higher temperatures of 150°C, 175°C, or more are used in some cases. The aluminum or copper conductors of overhead lines lose tensile strength and permanently elongate at a high temperature.

Overhead conductors in transmission and power distribution are usually made of aluminum known as all-aluminum conductor or AAC, AAAC, aluminum conductor alloy reinforced (ACAR), and ACSR.

An all-aluminum conductor loses no significant strength at temperatures below 95°C (203°F) but may lose 25% of its tensile strength when subjected to a temperature of 150°C (302°F) for 1000 h. Similarly, the normal sag increase of ACAR, AAAC, or AAC conductors as a result of creep elongation over time may be many times greater after extended operation at high temperature. Corrosion of components is also accelerated by high temperature.

The effect of long exposure to high temperature on mechanical properties of the materials used for transmission cables and joints has been studied in Ref. [71]. The results of the study indicate that operation of a typical transmission conductor (ACSR) at high temperatures will reduce the mechanical integrity of the overhead system. It is clear from the metallurgical and tensile tests that cumulative damage occurs to the aluminum metal in overhead conductors that have been subjected to high operating temperatures. System joints are exposed less to the higher operating temperatures owing to the larger mass and surface area in comparison to the conductor. These factors allow the joint to operate at a lower temperature than the conductor as long as the joint CR remains low. It is possible to extend the full mechanical life of the conductors by minimizing the duration that existing ACSR OHTL are exposed to high operating temperatures.

In Europe, the extra high voltage (EHV) OHTL were mainly built in the early 80s and in the 70s of the last century according to national and international standards. The sag of overhead lines are dimensioned at a specific temperature, which is commonly 60°C for the standard ACSR conductor, because the reversible tension of the conductor depends mainly on the temperature, when there is no additional external load (like ice) on the conductor. The rated current at a conductor temperature of 80°C is defined by specified environmental conditions. These conditions are 35°C ambient temperature, 0.6 m/s wind, and usual solar radiation [72]. It was found that approximately 40–50-year-old conductors have been creeping by several mechanisms [73,74].

To prevent a long duration of transmission line exposure to high temperatures, it is necessary to monitor the temperature of joints and conductors continuously. Having this information will significantly improve the safety of transmission lines and help to prevent unpredicted outages. This information could be used for scheduling a condition-based maintenance and will optimize the power delivery and improve overhead line ratings when conditions allow it. For example, when utilities calculate static rating, they must assume the worst ambient conditions, which is usually the highest summertime temperature. Such an approach results in ratings that are much lower than the true power capacity of the line for most of the year. Therefore, transmission lines often have significant hidden capacity that could be used if information on conductor temperature and sag on a real-time basis (dynamic rating) was available.

The research and development activity in organizations such as EPRI has increased based on a growing demand for the development of a means of monitoring the conditions and dynamic thermal circuit rating of transmission lines. Current EPRI projects are focused on the study of performance and aging characteristics of transmission equipment at increased operating temperature. In other projects, space potential probes for the real-time monitoring of the conductor sag and average core temperature are being developed, as well as SAW sensors and fiber-distributed temperature sensors for transmission line applications, and so on. Other EPRI projects include the monitoring of wind-induced (eolian) vibrations of transmission lines, which cause damage to conductors and towers, and the development of sensors for the detection of defective compression connectors.

The best way to monitor the conditions of transmission lines would be by real-time and permanent monitoring, with sensors that do not require power supply and that may be mounted on hotline without outage. The monitoring should be nondestructive and nonintrusive; it should be secure and controllable. Real-time monitoring of transmission lines allows the detection of imminent failure events. For example, by recording the variation of tension with temperature, the monitoring may remotely detect the breaks in the steel core of ACSR line or structure support failures.

17.7.3 Properties of Transmission Overhead Lines to Monitor, Sensing Elements, and Monitoring Techniques

There are several very important parameters that define transmission line health, such as the *temperature* and *sag* of the lines. Knowledge of these parameters would allow preventing the failures caused by wire damage and increase the power flow. There are other conditions of transmission lines that need to be monitored. Insulators might be severely damaged by *corona*, which might be monitored with corona-discharge and electromagnetic acoustic transducer (EMAT) sensors. UV imaging technology detects UV emitted by corona in damaged insulators. Wind-induced *vibration* of the overhead lines due to strong winds, the so-called eolian vibration, leading to serious damage of the cables, may be measured with vibration recorders. For example, a device called SEFAC is developed for a conductor diameter of 3 cm, but the existing device is too heavy, costly (>$10 k), and not very reliable.

17.7.3.1 Conductor Sag Measurements

There are several known methods to measure or evaluate the sag of the wires, such as measuring wire strain/tension, distance to the target, line temperature, and weather monitoring. *Tension* of the conductor may be measured with the strain gages installed in series with a dead-end insulator string measuring the tension of the line conductor. A transmission line monitoring system (CAT-1) is described in Ref. [75]. CAT-1 stands for "Clearance Assurance of Transmission." CAT-1 measures the various tensions and sags in a transmission line based on actual weather conditions. CAT-1 is using load cells for measuring the strain in the conductor. The price tag on a CAT-1 system is roughly $100,000 per transmission line—more or less depending on the length of the line—and it usually takes one to two days to install the equipment on the lines. Installation required an outage.

The 2003 report "Development of a Real Time Monitoring/Dynamic Rating System for Overhead Lines" [76] described development and testing results for two techniques for sag measurement. In one device, a laser distance meter (LDM) is used. Another one uses machine vision, or Point Source to Tower Measurement (PSTM), based on optical monitoring of the conductor itself or a passive target mounted on a set distance from the vision system. The system called "video sagometer" or just "sagometer" is considered to be the better one. This system was developed by EPRI, EDM International, and Southwest Research Institute (SwRI) [77,78]. EPRI was developing space potential probes for sag measurement [79].

17.7.3.2 Conductor Temperature Measurements

The temperature of the cable is determined by thermal energy that is produced by the current and the heat induced by the solar radiation. At the same time, there are several processes that cool down the cable, such as heat convection due to ambient air movement/wind and through thermal radiation [80]. The total input heat to the conductor in the overhead line is a sum of the heat of input due to line current and the heat input due to solar energy. In stable condition, the total input heat to the conductor must be equal to the total heat output, which is defined as a sum of the heat output due to convection (a function of wind, air temperature, conductor temperature) and the heat output due to radiation (a function of air and conductor temperatures).

The temperature of a conductor may be measured with a device mounted on the conductor. It may be installed on an energized conductor using a hot stick. This device directly measures the line temperature and has a radio transmitter, which transmits the information to a receiver on the ground or on the transmission tower. Air temperature should also be measured to determine the TR on the conductor. Knowing the conductor temperature, air temperature, incident solar radiation, and line current, the rating can be calculated.

17.7.3.2.1 Fiber-Optic Instruments

In 2003, two American companies completed a 2-year field trial of the temperature monitor using a fiber-optic transmission conductor (FOTC) [81]. Communication fibers were stranded into the conventional ACSR conductor providing accurate profiles of the conductor temperature. Optical fibers installed in the conductor communicate with an instrument at one end of the line to report the temperature. FOTC is an application of distributed temperature sensing (DTS) whose first applications were in the aerospace industry. Later, this temperature-monitoring system was applied to *underground cable* systems to identify localized hot spots. DTS determines temperatures by watching the optical fiber change color as the temperature changes. A pulse of light sent down the fiber is partially reflected from every part of the long fiber back to the source. The reflection time determines the location of the elevated temperature. Depending on the desired resolution and based on the number of readings per foot, DTS has a range of up to 30 miles (48 km).

EPRI developed and evaluated another distributed temperature-monitoring system using chirped Bragg grated optic fiber [82]. A similar embedded distributed temperature-monitoring system was developed in Europe. It uses a different version of temperature monitoring of the energized cable—fiber-optic temperature laser radar, also based on fiber-optic technology [83]. A fiber-optic transceiver (FTR) sends a light impulse along the fiber-optic cable, and calculates the temperature and the distance to the point at the distance up to 6 km, with the separation between the points of 2 m. These systems [84], based on fiber-optic DTS, monitor the temperature of 400 and 230 kV underground power cables between the power station and local substations and monitor the whole length of the transmission circuits, including circuits longer than 24 km. The systems are used for reporting and locating temperature excursions, thus enabling the power company to optimize its cable assets and ensure reliable supplies of electricity.

In general, the majority of OHTL temperature-monitoring systems are based on fiber-optic technology or distributed temperature measurement. However, none of the systems have found wide practical application until now. The downsides of such technique are the necessity of fiber-optic cable installation on the power cable and the extremely high cost of the system.

17.7.3.2.2 SAW Temperature Monitoring

SAW sensors (see Section 17.6) could be used for temperature measurement at HV OHTL with a radio link for the connection between the sensor and the measuring unit. The SAW system is using a high-frequency electromagnetic wave at the 2.45 GHz ISM band, which is transmitted to the sensor by the use of a transmitting antenna of the measuring system. By using the transducer, which is connected with the sensor chip, the incoming high-frequency signal is transmitted to acoustic surface wave, which is dispersed along the surface of the crystal and reflected on integrated reflectors. The propagation velocity of the acoustic surface wave depends on temperature. The reflected signals are converted to high transmission frequency by the transducer and sent back to the receiving antenna of the system.

The information of the sensor ID and the temperature information can be determined by the reflected impulses using various algorithms, which compute this information using the time position and the phase relation of the reflected impulses by commonly used signal-processing techniques. The results can be displayed on a personal computer. The sensors are passive and can withstand the common transient and PF stress on OHTL. For usage at overhead lines a suitable clamp was designed so as to be easily installed and to fulfill the mechanical and electrical requirements. The data acquisition unit was designed for installations on EHV overhead line towers. After positive results with this measuring technique, the system was also installed at various OHTLs and substations [85].

17.7.3.3 Combined Monitoring Solutions

There are several systems that are able to monitor multiple parameters of OHTL continuously.

The Power Donut2™ (PD2) is an instrumentation platform whose functions include measuring and monitoring the electrical, thermal, and mechanical parameters of HV overhead conductors in power transmission systems [86]. The device is mounted onto an energized power conductor with a hot stick; its installation can be done with no scheduled outage. The Power-Donut2™ is powered from the electromagnetic field surrounding the energized conductor. It has a long-life lithium battery pack that maintains operation for up to 12 h when current is below nominal, that is, 50 A. PD2 measures conductor load (amperes), conductor temperatures, and conductor sag (inclination angle) in real time, and monitors each parameter against user-defined warning and alarm level set points.

PD2 measures the conductor inclination in real time. The user can enter the conductor specifications including conductor weight, horizontal span for the section of line that PD2 is attached, and the tower height difference. This information allows to compute conductor sag relative to the tower near PD2, and tension at both ends of the conductor. Warning and alarm levels can be monitored for these computed values. The PD2 has on-board flash RAM memory for data logging, thereby providing secure local historical trend data. Historical data are automatically transferred to the communicating device for additional data security. When equipped with license-free radios, PD2 reports warnings and alarms virtually instantaneously. PD2 is supplied with the cell phone communication option, which initiates communication with the central office to report the warning or alarm condition within 60 s of the event.

17.7.3.3.1 LIOS Monitoring System

The system allows simultaneous monitoring of cable temperature and PD activity [87]. It is based on a combined DTS equipment and PD-measuring technique, which provides grid owners a better means of controlling and monitoring their grids. This approach allows most of the failure modes of the apparatus and electrical assets to be diagnosed, thus increasing reliability and decreasing maintenance costs. The system supports different sensors (PD, DGA, DTS, Tan-delta, and Vibrations) and integrates diagnostic algorithms to data coming from these sensors.

REFERENCES

1. Reliability issues in electrical contacts, in *Electrical Contacts: Fundamentals, Applications and Technology*, edited by Braunovic M., Konchits V. V., and Myshkin N. K., Chapter 6, CRC Press, Taylor & Francis Group, Boca Raton, FL, pp. 205–260, 2007.
2. IEEE Standard C37.20.2-1999. *IEEE Standard for Metal-Clad Switchgear*, Section 5.5, IEEE Publishing, New York, February 1999.
3. Doman C. J. and Heyer S. J. AC testing without cable degradation, *T&D World*, 51(7), 36, July 1999.
4. HV Test C.C., *A Handy Guide to the Safe Over-Voltage Pressure Testing and Diagnostic Monitoring of Cable Installations*, Randburg, South Africa, July 2004.
5. Simmons K. L., Pardini A. F., Fifield L. S. et al. *Determining Remaining Useful Life of Aging Cables in Nuclear Power Plants – Interim Study FY13*, US Department of Energy (DOE), Pacific Northwest National Laboratory, PNNL-22812, September 2013, Online. Available: http://www.pnnl.gov/main/publications/external/technical_reports/PNNL-22812.pdf.
6. Lanz B. Power cable diagnostics field application and case studies, Presented at *NETA Conference, Spring 2005, IMCORP*, Online. Available: http://docslide.us/documents/cables-neta-power-cable-diagnostics-field-application-case-studies.html.
7. Wally V. *Putting Hipot Out to Pasture*, EC&M, pp. 20–22, October 2003, Online. Available: http://ecmweb.com/archive/putting-hipot-out-pasture.
8. Sedding H. G., Schwabe R., Levin D., Stein J., and Gupta B. K. The role of AC & DC hipot testing in stator winding ageing, *Proceedings of Electrical Insulation Conference and Electrical Manufacturing & Coil Winding Technology Conference*, Indianapolis, IN, pp. 455–457, September 2003.
9. *Very Low Frequency AC Hipots,* High Voltage, Inc., Online. Available: http://www.hvinc.com/downloads/VLF_faq4.pdf.
10. Orton H. E. Diagnosing the health of your underground cable, *T&D World*, June 2002, Online. Available: http://tdworld.com/mag/power_diagnosing_health_underground/.
11. IEEE Standard 493-1997. *IEEE Recommended Practice for Design of Reliable Industrial and Commercial Power Systems—Gold Book*, IEEE Inc., New York, 1998.
12. Lowsley C. J., Davies N., and Miller D. M. Effective condition assessment of MV switchgear, *Proceedings of the twenty first AMEU Technical Convention*, South Africa, Paper 32, 2006, Online. Available: http://www.maintenanceonline.co.uk/maintenanceonline/content_images/Pages%2046,%2047,%2048,%2049,%2050,%2051.pdf.
13. IEC Standard 60270. *High-Voltage Test Techniques—Partial Discharge Measurements*, Ed 3.0, Publications of International Electrotechnical Commission (IEC), Geneva, Switzerland, 99pp., December 2000.
14. Davies N., Tang J., and Shiel P. Benefits and experiences of non-intrusive partial discharge measurements on MV Switchgear, *CIRED 19th International Conference on Electricity Distribution*, Paper 0475, pp. 1–4, Vienna, Austria, 2007, Online. Available: www.cired.be/CIRED07/pdfs/CIRED2007_0475_paper.pdf.
15. Holmes S., Caruana J., and Goldthorpe S. Low cost monitoring of partial discharge activity in MV substations, *CIRED 20th International Conference on Electricity Distribution*, Prague, Czech Republic, Paper 0804, pp. 1–6, June 2009.
16. Davies N. and Goldthorpe S. *Testing Distribution Switchgear for Partial Discharge in the Laboratory and the Field*, CIRED 2009, Online. Available: https://law.resource.org/pub/in/bis/S05/is.iec.60270.2000.pdf.
17. Lewand L. R. and Paul G. l. Condition assessment of oil circuit breakers and load tap-changers by the use of laboratory testing and diagnostics, *NETA World Journal*, Summer 2004.
18. IEEE Standard C57.104. *IEEE Guide for the Interpretation of Gases Generated in Oil-Immersed Transformers*, IEEE Publishing, New York, 1991.
19. Transformer maintenance, in *Facilities Instructions, Standards, and Techniques (FIST)*, United States Department of the Interior Bureau of Reclamation (USBR), Volume FIST 3-30, 81pp., October 2000, Online. Available: www.usbr.gov/power/data/fist/fist3_30/fist3_30.pdf.
20. Lewand L. R. *Understanding Water in Transformer Systems. The Relationship between Relative Saturation and Parts per Million (PPM)*, Online. Available: http://www.dryoutsystems.com/images/Understanding_Water_in_Tranformer_Systems_-_Lance_Lewand.pdf.

21. IEEE Standard C57.106. *IEEE Guide to Acceptance and Maintenance of Insulation Oil in Equipment*, IEEE Publishing, New York, 2007.

22. *Reference Book on Insulating Liquids and Gases (RBILG)-391*, Doble Engineering Company Publishing, Boston, MA, 1993.

23. ASTM Standard D3612-02(2009)—*Standard Test Method for Analysis of Gases Dissolved in Electrical Insulating Oil by Gas Chromatography*. ASTM Standard D3612-02, 2009.

24. Cargol T. An overview of online oil monitoring technologies, *4th Annual Weidmann-ACTI Technical Conference*, San Antonio, pp. 1–6, 2005.

25. *Online Monitoring of Moisture in Power Transformers*, Online. Available: http://www.vaisala.com/en/industrialmeasurements/applications/power/powertransformers/Pages/default.aspx.

26. *Calisto. Hydrogen-Moisture*, Online. Available: http://www.morganschaffer.com/products.php?id=5.

27. Stone G. C. Partial discharge diagnostics and electrical equipment insulation condition assessment, *IEEE Trans. Dielect. Electr. Insul.*, 12(5), 891–904, October 2005.

28. Muhr M. and Schwarz R. Partial discharge behavior of oil board arrangements by the installation of fiberoptic technology for monitoring, *15th International Symposium on High Voltage Engineering*, University of Ljubljana, Elektroinštitut Milan Vidmar, Ljubljana, Slovenia, p. 6, August 27–31, 2007.

29. Reid A. UHF monitoring of partial discharge in substation monitoring of partial discharge in substation equipment using a novel multi-sensor cable loop, *CIRED 20th International Conference on Electrical Distribution*, Paper 0820, Prague, Czech Republic, pp. 1–4, June 2009, Online. Available: www.cired.be/CIRED09/pdfs/CIRED2009_0820_Paper.pdf.

30. Caruana J., Holmes S., and Goldthorpe S. Low cost continuous monitoring of partial activity in MV substations, *CIRED 20th International Conference on Electrical Distribution*, Prague, June 2009.

31. Tian Y., Lewin P. L., Wilkinson J. S., Sutton S. J., and Swingler S. G. Continuous online monitoring of partial discharges in high voltage cables, *IEEE International Symposium on Electrical Insulation*, Indianapolis, Indiana, USA, pp. 454–457, September 19–22, 2004.

32. Wagenaars P. Integration of online partial discharge monitoring and defect location in medium-voltage cable networks, PhD thesis, Eindhoven University of Technology Library, 180pp., 2010, Online. Available: alexandria.tue.nl/extra2/201010096.pdf.

33. Van der Wielen P. C. J. M., Veen J., Wouters P. A. A. F., and Steennis E. F. Online partial discharge detection of MV cables with defect localisation (PDOL) based on two time synchronized sensors, *18th International Conference on Electricity Distribution (CIRED)*, Turin, Italy, June 2005.

34. Boffi P., Bratovich R., Persia F., Barberis A. M., Martinelli De M. L., Borghetto J., and Perini U. 1550 nm All-fiber interferometer for partial discharge detection in oil-insulated power transformer, *Optical Fiber Sensors (OFS) Conference*, paper TuC5, pp. 1–3, Optical Society of America (OSA) Infobase, Cancun, Mexico, October 2006, Online. Available: http://www.opticsinfobase.org/abstract.cfm?uri=OFS-2006-TuC5.

35. Lazarevich A. K. Partial discharge detection and localization in high voltage transformers using an optical acoustic sensor, MS thesis, Virginia Polytechnic Institute and State University, 70 pp., 2003, Online. Available: https://theses.lib.vt.edu/theses/available/etd-05122003-161802/unrestricted/thesis_fin.pdf.

36. Lima S. E. U., Frazao O., Farias R. G., Araujo F. M., Ferreira L. A., Santos J. L., and Miranda V. Fiber fabry-perot sensors for acoustic detection of partial discharges in transformers, *Microwave and Optoelectronics Conference (IMOC)*, SBMO/IEEEMTT-S International, Belem, Brazil, pp. 307–311, November 2009.

37. Wang X., Li B., Xiao Z., Lee S. H., Roman H., Russo O. L., Chin K. K., and Farmer K. R. An ultra-sensitive optical MEMS sensor for partial discharge detection, *J. Micromech. Microeng.*, 15, 521–527, 2005.

38. Macià S. C., Lamela H., and García S. J. A. Fiber optic interferometric sensor for acoustic detection of partial discharges, *J. Opt. Technol.*, 74(2), 122–126, 2007.

39. Testing and maintenance of high-voltage bushings, in *Facilities Instructions, Standards, and Techniques (FIST)*, US Department of the Interior Bureau of Reclamation (USBR), Volume FIST 3-2, pp. 1–8, November 1991, Online. Available: www.usbr.gov/power/data/fist/fist3_2/vol3-2.pdf.

40. Bleyer J. and Prout P. The value of power factor testing, *T&D World*, November 2005, Online. Available: http://tdworld.com/substations/power_value_power_factor/.

41. *Methods of Measurement of Radio Influence Voltage (RIV) of High-Voltage Apparatus*, National Electrical Manufacturers, NEMA Publication No. 107-1987 (R1993).

42. Lisboa F. The role of thermal imaging in a predictive maintenance program, *Sensors*, pp. 14–17, August 2003, Online. Available: http://archives.sensorsmag.com/articles/0803/14/.

43. Cook R. Smart infrared temperature sensors: Making sense of the new generation, *Sensors*, November 2000, Online. Available: http://archives.sensorsmag.com/articles/1100/48/index.htm.

44. Teunissen J., Peier D., Krijgsman B. M., and Verhoeven R. Prototype-integration of fiber-Bragg-sensors into high-voltage transformer for Online-temperature-monitoring, *13th International Symposium on High Voltage Engineering*, Netherlands, 2003.
45. Bengtsson C. Status and trends in transformer monitoring, *IEEE Trans. Power Deliv.*, 11(3), July 1996.
46. Giovanni B., Pietrosanto A., and Scaglione A. *An Enhanced Fiber Optic Temperature Sensor System for power Transformer Monitoring*, IEE Standard 0-7803-5890-2/00/, 2000.
47. Saravolac M. P. The use of optic fibres for temperature monitoring in power transformer, *IEE Colloquium on Condition Monitoring and Remnant Life Assessment in Power Transformers*, London, UK, pp. 7/1–7/3, March 1994.
48. Alexandre I. Fiber-optic temperature measurement, *Sensors*, 57–58, May 2001, Online. Available: http://archives.sensorsmag.com/articles/0501/57/index.htm.
49. Stokes J. and Palmer G. A fiber-optic temperature Sensor, *Sensors*, August 2002, Online. Available: http://archives.sensorsmag.com/articles/0802/28/.
50. Gockenbach E., Werle P., Wassenberg V., and Borsi H. Monitoring and diagnosis systems for dry type distribution transformers, *7th International Conference on Solid Dielectrics (ICSD)*, Eindhoven, Netherlands, pp. 2.10.1–2.10.8, June 2001.
51. Miyazaki A., Takinami N., Kobayashi S., Nishima H., Nakura Y., Komeda H., Nagano H., and Higashi H. Long-distance 275 kV GIL monitoring system using fiber-optic technology, *IEEE Trans. Power Deliv.*, 18(4), October 2003.
52. Bérubé J. N., Broweleir B. L., and Aubin J. *Optimum Transformer Cooling Control with Fiber Optic Temperature Sensors*, Online. Available http://www.energycentral.com/reference/whitepapers/103284/.
53. Bérubé J. N., Jacques A. J., and McDermid W. *Recent Development in Transformer Winding Temperature Determination*, Online. Available: http://www.neoptix.com/oil-filled-and-dry-type-transformer.asp.
54. Glodjo A., Mueller R., Brown M., and Walsh S. Field experience with multipoint internal temperature measurements of converter transformers, in *Use of Fiber Optic in Transformers*, Publication of Doble Engineering Company, Boston, MA, pp. 1–9, 2004.
55. Manges W. W., Allgood G. O., and Smith S. F. It's time for sensors to go wireless—Part 1. Technological underpinnings, *Sensors*, April 1999, Online. Available: http://archives.sensorsmag.com/articles/0499/0499_10/index.htm.
56. Allgood G. O., Manges W. W., and Smith S. F. It's time for sensors to go wireless—Part II. Take a good technology and make it an economic success, *Sensors*, 70–80, May 1999, Online. Available: http://archives.sensorsmag.com/articles/0599/0599_p70/index.htm.
57. Manges W. W., Allgood G. O., Smith S. F., McIntyre T. J., Moore M. R., and Lightner E. Intelligent wireless sensors for industrial manufacturing, *Sensors*, April 2000, Online. Available: http://archives.sensorsmag.com/articles/0400/44/index.htm.
58. McLean C. and Wolfe D. Intelligent wireless condition-based maintenance, *Sensors*, June, 2002, Online. Available: http://archives.sensorsmag.com/articles/0602/14/.
59. Maxwell D. M. and Williamson R. W. Wireless temperature monitoring in remote systems, *Sensors*, October 2002, Online. Available: http://archives.sensorsmag.com/articles/1002/26/.
60. *Application Notes and Installation Manual, Wireless Temperature Monitoring of Power Distribution Equipment*, Bulletin 0180IB0801R12/09, Schneider Electric, USA, 2008, Online. Available: http://products.schneider-electric.us/support/technical-library/?event=detail&oid=09008926804659a4&cat=0b00892680470585.
61. Weddell, A. S., Merrett, G. V., Harris, N. R., and Al-Hashimi B. M. Energy harvesting and management for wireless autonomous sensors, *Measurement Control*, 41(4), 104–108, 2008.
62. Raghunathan V., Kansal A., Hsu J., Friedman J., and Srivastava M. Design considerations for solar energy harvesting wireless embedded systems, *Proceedings of the Fourth International Symposium on Information Processing in Sensor Networks*, Los Angeles, CA, pp. 457–462, 2005.
63. George S. Development of a vibration-powered wireless temperature sensor and accelerometer for health monitoring, *IEEE Aerospace Conference*, Big Sky, MT, 2006.
64. Torah R. N., Glynne-Jones P., Tudor M. J., and Beeby S. P. Energy aware wireless microsystem powered by vibration energy harvesting, *Power MEMS Conference*, Freiburg, Germany, 2007.
65. Schneider M., Evans J., Wright P., and Ziegler D. Designing a thermoelectrically powered wireless sensor network for monitoring aluminum smelters, *Proceedings of the Institution of Mechanical Engineers—Part E: Journal of Process Mech. Eng.*, 220, 181–190, 2006.
66. Kerem D. *Wireless Temperature Sensing by Acoustic Wave Sensors*, Process Industry News—Editorial Feature Archive, September 2009, Online. Available: http://www.processindustryinformer.com/Editorial-Feature-Archive/Wireless-temperature-sensing-by-acoustic-wave-sensors.

67. Budin M., Karandikar H. M., and Urmson M. G. Switchgear condition monitoring, *CIGRÉ Canada Conference on Power Systems*, Vancouver, October 17–19, 2010.
68. Sabah S., Buehler T., and Buchter F. Temperature monitoring of switchgear utilizing surface acoustic wave wireless sensors, *Power Systems Conference and Exposition (PSCE)*, Phoenix, AZ, p. 1, March 2011.
69. Kingsmill A., Jones S., and McIntyre P. Using online condition monitoring in substations to achieve business benefits, *The Conference of Electrical Power Supply Industry (CEPSI)*, Paper 00011, Japan, Fukuoka, November 2002.
70. Kingsmill A., Jones S., and Zhu J. G. Application of new condition monitoring technologies in the electricity transmission industry, *Proceedings of Sixth International Conference on Electrical Machines and Systems (ICEMS)*, Volume: 2, pp. 1–6, 2003, Online. Available: https://opus.lib.uts.edu.au/bitstream/10453/7132/1/2003001102.pdf.
71. Di Troia G. *Effects of High Temperature Operation on Overhead Transmission Full-Tension Joints and Conductors*, CIGRE Study Committee 22, WG-12, Evreux, France, pp. 1–6, August 2000.
72. *Conductors for Overhead Lines—Round Wire Concentric Lay Stranded Conductors*, DIN Standard EN 50328, September 2003.
73. Muhr M., Pack S., and Jaufer S. Sag calculation of aged overhead lines, *14th International Symposium on High Voltage Engineering*, Tsinghua University, Beijing, China, August 25–29, 2005.
74. A practical method of conductor creep determination, CIGRE Study Committee 22, WG-05, *CIGRE Electra*, No. 24, pp. 105–137, 1972.
75. *CAT-1 Transmission Line Monitoring System*, The Valley Group, Online. Available: http://www.nexans.us/eservice/US-en_US/navigatepub_0_-17373/Learn_about_the_CAT_1_Transmission_Line_Monitoring.html.
76. *Development of a Real Time Monitoring/Dynamic Rating System for Overhead Lines, Consultant Report Prepared by EDM International Inc. for California Energy Commission, Public Interest Energy Research Program*, Online. Available: http://www.energy.ca.gov/pier/reports/500-04-003.html.
77. Chari K. Finding the "hidden" capacity in transmission lines, *Electrical Business*, 38(2), 1–10, February 2002.
78. *The Sagometer, Line Rating System*, EDM International, Online. Available: http://www.energy.ca.gov/reports/2004-04-02_500-04-003.PDF.
79. Olsen R. Space potential probes for real-time monitoring of conductor sag and average core temperature, *36th Annual North American Power Symposium (NAPS)*, Section 1A-2 "Sensors and SCADA," Moscow, Idaho, August 9–10, 2004.
80. Daconti J. R. and Lawry, D. C. The Thermalrate system: A solution for thermal uprating of overhead transmission lines, *Shaw Power Technology*, Newsletter, Issue 95, 4 pp., April 2004, Online. Available: https://w3.usa.siemens.com/datapool/us/SmartGrid/docs/pti/2004April/The%20ThermalRate%20System.pdf.
81. Nandi S., Crane J. P., and Springer P. N. Intelligent conductor system takes its own temperature—Fiber-optic transmission conductor, *T&D World*, September 2003, Online. Available: http://tdworld.com/mag/power_intelligent_conductor_system/.
82. *Fiber-Optic Sensor Technology for Diagnostics of Underground Power Cables*, EPRI Product ID: 3002000868, 142 pp., December 2013, Online. Available: http://www.epri.com/abstracts/Pages/ProductAbstract.aspx?ProductId=000000003002000868.
83. *Fiber Optic Temperature Laser Radar*, Online. Available: http://www.lightoptronics.com.au/ftr-brochure.pdf.
84. *Sensa—Power Circuit Temperature Monitoring System*, Online. Available: http://www.power-technology.com/contractors/condition/sensa/.
85. Bernauer C., Bohme H., Grossmann S., Hinrichsen V., Kornhuber S., Markalous S., Muhr M., Strehl T., and Teminova R. A temperature measurement on overhead transmission lines (OHTL) utilizing surface acoustic wave (SAW) sensors, *CIRED 19th International Conference on Electricity Distribution*, Vienna, May 21–24, 2007.
86. *Power Donut2™ System for Overhead Transmission Line Monitoring*, Online. Available: http://www.usi-power.com/Products%20&%20Services/Donut/donut.htm.
87. *The Global Condition Monitoring Solution for High Voltage Cable Systems*, Online. Available: http://www.lios-tech.com/Menu/EN.SURE/Products/Global+Condition+Monitoring/Global+Condition+Monitoring+Solution+for+High+Voltage+Cable+Systems.
88. IEEE Standard C57.106-2006 *Guide for Acceptance and Maintenance of Insulating Oil Equipment*. © IEEE.

18 Physical Conditions of Renewable Electrical Equipment
Testing and Monitoring

18.1 WIND TURBINES

18.1.1 WIND TURBINE CONDITION MONITORING

Presented in this section, these are few selected examples of wired and wireless condition monitoring systems (CMS) for wind turbines (WTs) developed and tested all over the world in the past decade. There are literally dozens of the systems, which are still in the process of testing in the field and installation on wind farms. The principles of these systems are based on multiple modern techniques, which are very effective in determining damage in all components of WT including its structure. Use of such systems is very important in providing educated base for condition-based maintenance (CBM) and life extension of WTs.

The advantages of continuous monitoring of physical conditions of a WT are manifold. After the capital costs of commissioning wind turbine generators (WTGs), the biggest costs are operations, maintenance, and insurance. WTs often operate in severe, remote environments and require frequent scheduled maintenance. Unexpected failures require unscheduled maintenance which can be costly, not only for maintenance support but also for lost production time. Maintenance costs increase as WTs age, parts fail, and power production performance degrades.

Maintenance costs can be reduced through continuous, automated monitoring of WTs. Monitoring and data analysis enables CBM rather than time-interval-based maintenance. Condition monitoring (CM) detects failures before they reach a catastrophic or secondary-damage stage, extends asset life, and keeps assets working at initial capacity factors. CBM enables better maintenance planning and logistics, and can reduce routine maintenance. Installing CMS allows operators to apply a CBM, which often provides an ability to prevent upcoming failure and therefore reduces costs of both preventive and corrective maintenance strategies [1,2].

Additionally, having CMS also allows the real-time monitoring of assets and optimization of operations and maintenance procedures to increase reliability, safety and maximize cost-effectiveness, which is particularly important for continuous increase of WTs in size and moving the wind farms offshore.

Due to the remote location of wind farms, for both offshore and onshore, environmental conditions, and the vertical height of the nacelle, it is expensive to physically visit WTs for maintenance and repair. Offshore wind farms are remotely located and operate under challenging conditions, which are much worse than the onshore wind farms. In the past and these days, an early failure of their many components has been frequently observed.

As the demand for wind energy continues to grow at exponential rates, reducing *operation and maintenance* (O&M) costs and increasing reliability is now a top priority. Current O&M costs prohibit faster adoption of wind energy and ultimately slow global acceptance of wind energy as an energy supply. It makes the development of a reliable *structural health monitoring* (SHM) and CM strategy particularly necessary [3–5].

With low-cost, online CMS that predict failures and maintenance requirements, it is possible to forecast maintenance activities and lower O&M costs [6].

There are three areas within WT *nacelle* to monitor: electrical system, rotating machine and blade, and pitch [2].

Another important area for continuous monitoring of *structural health* of WT is considered here. As electric utility WTs increase in size, and correspondingly, increase in initial capital investment cost, there is an increasing need to monitor the health of the structure. The size/diameter of WTs grew tremendously over last 30 years from 18 m (60 ft) in 1985 to 120 m (394 ft) in 2007. Built by Siemens in 2012, blade is 75 m (246 ft) long, with total turbine size in action 150 m.

A British company Blade Dynamics announced that it was developing blades of up to 100 m in length (>300 ft)—dwarfing the size of existing technology in the 60-m range. Sitting on top of a tower 170 m high, the structure will be 270 m in all, or 885 ft or one-sixth of a mile. Obviously, for such structures, an early indication of structural or mechanical problems is very important. It allows operators to better plan for maintenance, possibly operate the machine in a derated condition rather than taking the unit off-line, or in the case of an emergency, shut the machine down to avoid further damage [7].

18.1.2 WIND TURBINE DIAGNOSTICS WITH SCADA

Traditionally, CMS for WTs have focused on the detection of failures in the main bearing, generator, and gearbox, some of the highest cost components on a WT. Two widely-used methods are vibration analysis and oil monitoring. These are stand-alone systems that require installation of sensors and hardware [2].

Supervisory Control and Data Acquisition (SCADA)—data-based CMS uses data already being collected at the WT controller and is a cost-effective way to monitor for early warning of failures and performance issues. SCADA systems provide measurements for a WT energy production and confirm that the WT was operational through 5–10 min averaged values transmitted to a central database. SCADA data provide a rich source of continuous time observations, which can be exploited for overall turbine performance monitoring. With appropriate algorithms, performance monitoring can be matured into individual component fault isolation schemes.

For example, SCADA systems can also provide warning of impending malfunctions in the WT drivetrain. According to Refs. [8,9], averaged signals often monitored in modern SCADA systems include

- Active power output and standard deviation over *10 min interval*
- Anemometer-measured wind speed and standard deviation over *10 min interval*
- Power factor
- Reactive power
- Phase currents
- Nacelle temperature (*1 h average*)

Among parameters monitored in modern SCADA systems, there are those which deliver information about specific physical conditions of drivetrain, gearbox, and generator in WTs:

- Gearbox bearing temperature
- Gearbox lubrication oil temperature
- Generator winding temperature

18.1.3 PHYSICAL CONDITIONS AND MAJOR COMPONENTS TO CONTINUOUSLY MONITOR IN WIND TURBINE

Wind turbine CM should serve two major purposes: (1) control current physical conditions of the turbine and structural components and (2) function as preventive measures. Preventive monitoring

TABLE 18.1
Areas of Wind Turbines to Monitor and Condition Monitoring Techniques

Wind Turbine Component/System	Monitoring Techniques
Blades	Vibration, acoustic, and fiber optic strain analysis
Drive train	Acoustic and vibration analysis
Lubrication system	Analysis of oil contaminants and oil quality
Electrical components	Thermographic analysis
Electrical power	Time and frequency domain analysis, power quality

is necessary to extend turbine life cycle, schedule maintenance, and predict fault conditions and failures before they occur. In the first decade of the century, techniques such as vibration and process parameter analysis have been applied to WTs.

From the research work carried out during that period [10], the most promising areas for the application of CM of WTs, particularly offshore ones, have been specified. A proper CMS should comprise combinations of sensors and signal-processing equipment that provide continuous indications of component condition (and hence condition of WT) based on techniques including vibration analysis, acoustics, oil analysis, strain measurement, and thermography. These techniques are used to monitor the status of critical operating major components of WTs such as the blades, gearbox, generator, main bearings, and tower. Monitoring may be (1) *online* which provides instantaneous feedback of condition or (2) *off-line* when data being collected at regular time intervals using measurement systems that are not integrated with the equipment [11].

Table 18.1 shows major components and systems and techniques to use for their CM. In the following sections, more detailed analysis will be presented.

18.1.3.1 Vibration

Vibration monitoring is one of the most important aspects in WT monitoring because it helps determine the condition of rotating equipment. In a WT, this equipment consists of the main bearing, gearbox, and generator.

All rotating equipment vibrates to some degree, but as bearings and gearbox components reach the end of their product life, they begin to vibrate more dramatically and in distinct ways. Sensors placed on the bearing housing or gear case are used to detect characteristic vibration signatures. This signature is unique for each gear mesh or rolling element and depends on the geometry, load, and speed of the components. The data-mining technique then compares the signature during operation with the characteristic signature and flags any anomalies. Ongoing monitoring of equipment allows these signs of wear and damage to be identified well before the damage becomes an expensive problem. Vibration monitoring is applied to the blades, rotors, gearboxes, generators, bearings, and towers [11]. Figure 18.1 shows where it is important to place vibration sensors.

Depending on the applicable frequency range, for vibration measurement different sensors can be used: position sensors (low-frequency range), velocity sensors (middle-frequency range) or accelerometers (high-frequency range), and spectral emitted energy sensors for very high frequencies [12]. These vibration sensors are rigid, mounted on the component of interest and return an analog signal proportional to the instantaneous local motion with an acquisition device having a high sampling rate and high dynamic range.

By monitoring vibrations on the turbine structure at the base and on the nacelle, information concerning structural bending and the aerodynamic effect of the wind is provided. These data may help to determine if any monitored components have problems before they are damaged (e.g., cracked gear tooth, broken bearing, etc.). When used on rotating equipment, the sensor data analysis to display the data harmonics provides insight into the component performance and allows for easier diagnosis [12].

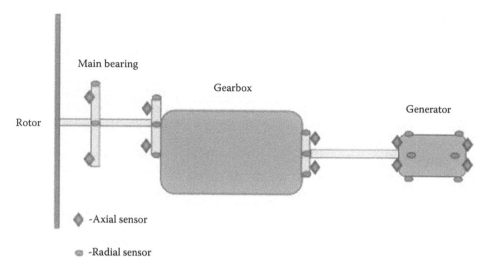

FIGURE 18.1   Position sensors (low range), velocity sensors (middle range), or accelerometers (high range) may be installed on rotor, main bearing, gearbox, and generator to read vibration data in the axial and radial directions.

18.1.3.2   Acoustic Emission and Ultrasonic Testing

Acoustic monitoring has some similarities with vibration monitoring but while vibration sensors are mounted on the component involved to detect movement, acoustic sensors are attached with flexible glue with low attenuation and record sound directly.

Acoustic emission (AE) sensors have been used successfully not only in the monitoring of bearings and gearboxes but also for damage detection in blades of a WT. Its application is also possible to an in-service WT for a real-time rotating blade. Nondestructive testing techniques using acoustic waves to improve the safety of WT blades are discussed in Ref. [13]. The use of AE is gradually growing for both CM of rotating WT components as well as blades, but mostly is used for testing and inspection.

Ultrasonic testing (UT) techniques are used extensively by the wind energy industry for the structural evaluation of WT towers and blades. UT is generally employed for the detection and qualitative assessment of surface and subsurface structural defects. Ultrasonically obtained images make it possible to recognize the geometry of defects and to estimate their approximate dimensions. However, CMS using both AE and UT are not commercially available yet.

18.1.3.3   Oil Quality

There are several physical parameters of oil used in gearbox and bearings in WT, which may be monitored and analyzed to determine the condition of rotating parts of the turbines. These parameters are oil temperature, oil content, and oil debris. By analyzing the composition, content, size, and classification of wear particles in the lubrication oil of WT components, their health conditions can be evaluated (see Chapter 13).

Oil analysis may have two purposes:

- Control of the oil quality (contamination by parts, moisture)
- Control of the physical condition of the components (characterization of parts)

Oil analysis is mostly executed off-line, by taking samples. However, safeguarding the oil quality by application of online sensors is increasing. Besides this, controlling the state of the oil filter (pressure loss over the filter) is mostly applied nowadays for hydraulic and lubrication oil.

Characterization of parts is often only performed in case of abnormalities. In case of excessive filter pollution, oil contamination, or change in component conditions, characterization of parts can give an indication of components with excessive wear. Early detection of changes in oil quality and cleanliness enables the operator to minimize gearbox damage, avoid full failure, and plan the repair, thus, simultaneously reducing the repair cost and business interruption costs. While faults are developing and no vibrations could yet be detected, the oil analysis can provide early warnings [14]. Oil contamination or change in component properties, characterization of the particulates can give an indication of excessive wear.

Oil wear debris monitoring for the WT industry is now a proven, effective CM technique that does provide reliable early detection and quantification of internal damage to gears and bearings of the WT gearbox without the need for expensive hardware or expert data interpretation [10,14]. Oil debris counts of ferrous and nonferrous particles are measured by CMS.

Detailed analysis of measurement results may illustrate the progress of metallic wear debris particle release from benign wear to catastrophic failure. Such analysis summarizes wear debris observations from all the different wear modes that can range from polishing, rubbing, sliding, skidding, abrasion, adhesion, grinding, scoring, pitting, spalling, and so on. According to the chart in Ref. [14], growing number and size of the metal particles in lubricating oil is determined by the degree of deterioration and surface wear of the bearings from benign wear to severe wear leading to advanced failure which may develop into catastrophic failure if no safety measures are taken. Oil/debris analysis is currently one of the important means of CM in wind industry.

However, the use of oil/debris analysis works only for high power rating WTs where oil is used for lubrication and/or cooling of WT bearings and gearboxes. For WTGs whose lubrication of bearings and gearboxes is sealed inside, oil/debris analysis methods are not practical.

Oil contamination. WTs in desert environments are exposed to airborne dust during the hot season and moisture during the rainy season. Solid particle contaminants vary in hardness, friability, and ductility, depending on the composition of the particle (see Section 13.2.2). The size, hardness, and friability or ductility of the particle influence the amount of damage that the particle can cause. The presence of contaminants may be determined during analysis of oil samples. If oil samples do not meet cleanliness requirements during service, there may be one or more failure modes in progress, and seals, breathers, or maintenance procedures need to be improved.

Water contamination of industrial oils plays a major role in bearings and gears deterioration. Since offshore turbines are constantly exposed to moisture [15], high moisture levels can cause components to overheat, corrode, or fatally malfunction.

Oil temperature (and bearing temperatures) as well as *oil pressure level* and *oil filter status* are measured by SCADA system. The gearbox oil temperature rise is assumed proportional to gear temperature rise. Oil temperature rise is an indicator of the failure to happen and it is analyzed to predict gear failure [16].

Figure 18.2 shows the gearbox oil temperature rise against relative power output (%) in three periods: 9 months, 6 months, and 3 months before the known failure. Figure 18.3 shows the binning of average gearbox oil temperature rise for each 50 kW increment of power output in those three corresponding periods. Figures 18.2 and 18.3 clearly show the rise in WT gearbox inefficiency in the 3 months before the failure.

Analysis of in-service gear oil is an important part of an effective CM program. According to Ref. [17], monitoring of used oil condition is recommended by oil sampling every 6 months, during which multiple oil parameters are measured. These parameters and their significance in determining how they affect the oil performance are shown in Table 18.2.

18.1.3.4 Physical Properties and Conditions of the Turbine Blades

The blade on a typical utility-size WT, typically constructed of fiberglass reinforced plastic (FRP), can exceed 40 m in length and weigh over 7 tons, and like the WT, the trend in blades is toward larger, longer, and heavier blades [7]. There are other materials lately used and proven

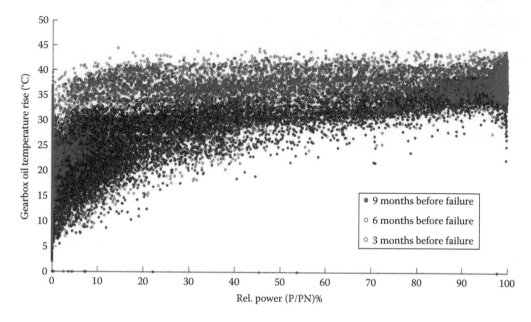

FIGURE 18.2 Gearbox oil temperature rise against relative power output. (From Feng Y., Qiu Y., Crabtree C. J. et al. Use of SCADA and CMS signals for failure detection and diagnosis of a wind turbine gearbox, *Scientific Proceedings of the European Wind Energy Association (EWEA) Conference*, Brussels, Belgium, 2011, Online. Available: https://community.dur.ac.uk/supergen.wind/docs/publications/Feng,%20Qiu,%20 Crabtree,%20Tavner,%20Long_EWEA2011presentation.pdf. With permission.)

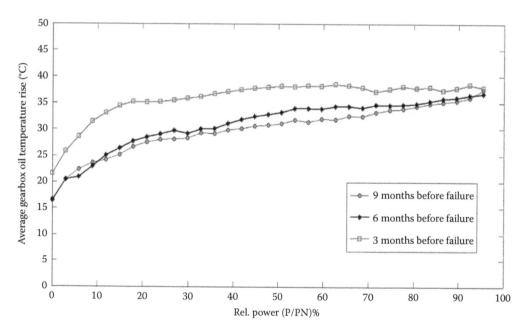

FIGURE 18.3 The trends of gearbox oil temperature rise against relative power output. (From Feng Y., Qiu Y., Crabtree C. J. et al. Use of SCADA and CMS signals for failure detection and diagnosis of a wind turbine gearbox, *Scientific Proceedings of the European Wind Energy Association (EWEA) Conference*, Brussels, Belgium, 2011, Online. Available: https://community.dur.ac.uk/supergen.wind/docs/publications/Feng,%20 Qiu,%20Crabtree,%20Tavner,%20Long_EWEA2011presentation.pdf. With permission.)

TABLE 18.2
Used Oil Monitored Conditions

Parameter, Test	Condition to Monitor
Appearance	Water and/or contamination with Al, Fe, Cr, Cu, Pb, Sn, Si (dust), wear at gears/bearings
Color	Contamination, aging
Total Acid # (TAN), IR Oxidation	Aging—acidic oxidation products
Water	Water contamination (corrosion, emulsion)
PQ index (ferrous particles)	Wear at gears/bearings
Viscosity @40°C	Aging—contamination, shearing
Cleanliness level	Contamination, wear, filtration issue

to perform well in the design of blades. Multiple physical parameters and environmental factors may affect the condition of the blade and register various changes in the blade's structural health, such as strains, loads, cracks, voids, operational dynamics, temperature gradients, lightning, and so on.

Some of these properties if measured and monitored during WT operations may allow predicting and preventing the failure of the component in timely manner. There are multiple techniques and sensors able to measure these properties, already developed or in the process of development. Among them are piezoelectric sensors/actuators, AE sensors, optical strain gages (Fiber Bragg Gratings or FBG), metal-foil strain gages, infrared (IR) thermography, and so on [18]. For example, changes in the condition of the rotor blades caused by cracks and ice formation may be detected via the changes in vibration and can be directly evaluated by the diagnostic electronics. Multiple fatigue tests of large FRP WT blades can be monitored by AE techniques and the monitoring can produce useful information related to the blade condition.

18.1.3.5   Corrosion of the Wind Turbine Towers and Foundation

For either onshore or offshore WTs, corrosion is an important issue that needs to be monitored and controlled (see Chapter 12). The corrosion inside the offshore WT can be thought of as atmospheric corrosion, but with the difference that the surface is not constantly cleaned by rain for chloride contaminations.

The corrosivity of indoor atmospheres is generally considered to be quite mild when ambient humidity and other corrosive components are under control. Nonetheless, some combinations of conditions may actually cause relatively severe corrosion problems. Even in the absence of any other corrosive agent, the constant condensation on a cold metallic surface may cause an environment similar to constant immersion for which a component may not have been chosen or prepared.

The saltwater contains a lot of other ions than sodium chloride, and some of these contaminations are able to immobilize the water at the surface and speed up uniform corrosion at the surfaces [19]. For offshore WT, there multiple relevant types of corrosion in different parts of the system, such as corrosion of main shaft bearing, gearbox, brake, cooling and ventilation system, tower, hub, stairs, and doors.

Significant corrosion is observed on transition piece, boat landing, J-tubes, internal areas of monopole, and in grit connection. There are a number of various sensors that are able to monitor corrosion, such as corrosivity sensors, a remote sensing technique based on *Electrochemical Impedance Spectroscopy* (EIS) for detecting coating degradation, biocorrosion, and biofouling sensors. Applied to offshore WTs for corrosion monitoring, they provide control on the degree and progress of corrosion [20].

TABLE 18.3

Condition Monitoring Techniques for Wind Turbine Components

WT Components	Condition to Monitor	Monitoring Techniques
Gearbox, bearings, generator, blade	Vibration	CMS: position and velocity sensors, accelerometers
Gearbox, bearings, generator	Temperature	SCADA, optical fiber monitoring
Gearbox, bearings	Oil analysis/debris	Sampling oil, oil debris sensors
Gearbox, bearings, blade	Acoustic emission	AE sensors
Blade, shaft, generator	Torque	Torque transducers
Drive train	Friction, shock	Stress wave energy sensors, piezoelectric sensors
Blade, tower, foundation	Torque, strain, bending surface conditions, loads, stresses	Optical fiber monitoring
Bearing, gearbox, brakes, cooling and ventilation systems, tower, hub, stairs, doors	Corrosion	Corrosivity sensors, EIC sensors, biocorrosion and biofouling sensors
Monopile	Corrosion	Corrosion rates and potential, oxygen concentration
Blade, gearbox, bearing, shaft, generator	Current, power monitoring	SCADA, time and frequency domain analysis

18.1.3.6 Summary: Conditions and Components of Wind Turbines to Monitor

The physical conditions and which parts of WT should be monitored are summarized in Table 18.3, which includes also techniques/devices that may be used for monitoring of a specific property. Major disadvantages of these techniques [21] have to be considered when choosing a CMS to use in application. These disadvantages are high cost, high complexity, and the technique may be intrusive.

18.1.4 CONDITION MONITORING SYSTEMS (CMS) FOR WIND TURBINES

There are several prospective projects [1,2] to develop a WT CMS. More than 40 companies around the world have been investigated as suppliers of CMS components and software. The majority of CMS are vibration systems based around the *drivetrain*, which is composed of the gearbox and the generator, the necessary components that a turbine needs to produce electricity [22].

The gearbox is responsible for connecting the low-speed shaft attached to the turbine blades to the high-speed shaft attached to the generator using a series of gears of varying sizes. The gearbox converts the slow rotation of the outer blades (typically 30–60 rpm) to the high-speed rotation (roughly 1000–1800 rpm) that the generator needs to begin producing electricity. Gearbox internal nomenclature and abbreviations are shown in Figure 18.4.

Vibration-based CMS tend to use accelerometers to measure vibrations. These sensors are placed on key components of the drivetrain. The setup used in National Renewable Energy Laboratory's (NREL) customized vibration-based CMS is shown in Figure 18.5. It illustrates most of the typical accelerometer mounting locations for vibration-based CMS seen in the wind industry today. The sensor notations are given in Table 18.4. In fact, a typical commercial CMS uses only a portion of the 12 sensors listed in Table 18.4.

Data acquisition units are normally placed within the nacelle. These units perform signal processing and basic computations including envelope spectra from time series data and spectrum analysis using Fast Fourier Transforms (FFT). These units are programmed with limit values that when exceeded set alarms.

Another approach to monitoring the health of a drivetrain of a WT is high-frequency sound-based *Stress Wave Analysis* (SWAN™) techniques described in Ref. [22]. The system uses AE sensors and

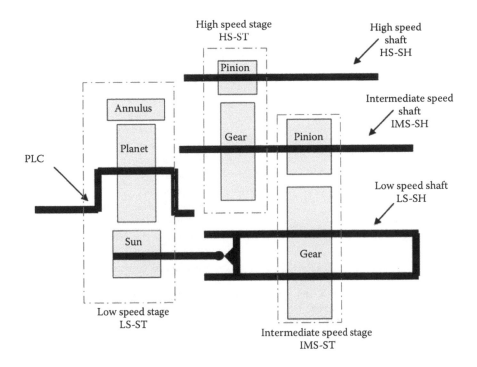

FIGURE 18.4 Gearbox internal nomenclature and abbreviations. (From Sheng S. and Veers P. *NREL Mechanical Failures Prevention Group: Applied Systems Health Management Conference*, Virginia Beach, Virginia, May 2011.)

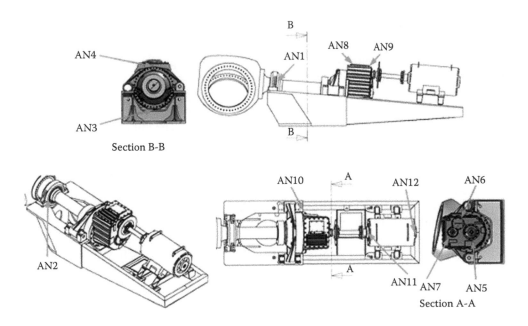

FIGURE 18.5 Vibration sensors' positions on drivetrain components in Condition Monitoring System. (From Sheng S. and Veers P. *NREL Mechanical Failures Prevention Group: Applied Systems Health Management Conference*, Virginia Beach, Virginia, May 2011.)

TABLE 18.4

Notation and Description of the Sensors in Figure 18.5

Sensor Label	Description
AN1	Main bearing radial
AN2	Main bearing axial gear radial
AN3	Ring gear radial 6 o'clock
AN4	Ring gear radial 12 o'clock
AN5	LSS radial
AN6	ISS radial
AN7	HSS radial
AN8	HSS upwind bearing radial
AN9	HSS downwind bearing radial
AN10	Carrier downwind radial
AN11	Generator drive end radial
AN12	Generator non-drive end axial

Source: Sheng S. and Veers P., *Presented at the Mechanical Failures Prevention Group: Applied Systems Health Management Conference 2011 Virginia Beach*, Virginia, May 10–12, NREL 2011.

does not suffer from the inherent limitations of vibration analysis. A piezoelectric crystal in the sensor converts the stress wave amplitude into an electrical signal, which is then amplified and filtered to remove unwanted low-frequency sound and vibration energy.

Stress Wave Energy (SWE) technique is based on sound rather than motion measurement. It is an excellent indicator of the overall health of WT drive systems because it provides direct measurement and comparison of the amount of friction and impact occurring within the machine. SWE measurements, however, provide a quantitative measure of friction and shock events during the entire life cycle of the machine.

Alternative CMS exist to monitor *oil condition*—count wear particles, determine ferrous contents in oil, and oil viscosity. Drivetrain monitoring system uses vibration, proximity detection, bearing temperature, and oil particle counting. Particle counting sensor accompanied by a software package for recording and storing the collected data [23] counts oil debris, both ferrous and nonferrous particles, and divides each type of particle into five bins. The minimum detectable ferrous particle size is 50 μm and the minimum nonferrous particle size is 150 μm. The sensor also can be integrated with the SKF CMS *WindCon* [24] or *SwanTech SWANWind* CM package [22].

Another option for monitoring oil condition is the *Kittiwake package*, which includes one sensor suite [25] composed of three types of sensors. The sensor suite is capable of monitoring the lubricant condition in terms of total ferrous debris in parts per million, relative humidity as a percentage, and oil quality (changes with the level of such contaminants as soot, oxidation products, glycol, and water) on a customized scale.

Another unit from *Kittiwake* [26] was the particle content sensor, which was used to count oil debris, both ferrous and nonferrous particles, and divide each particle type into five bins. The minimum detectable ferrous particle size is 40 μm and the minimum nonferrous particle size is 135 μm. The data collected by the sensor suite and particle content sensor are wirelessly transmitted through a cellular modem to a server located in the United Kingdom and can be viewed through a web browser.

System monitoring *rotor status*—blade imbalance, ice detection, and blade damage—is developed by *Moog/Insensys*. It includes blade mechanical and aero imbalance identification with FBGs on blades, as well as tower load measurement with FBGs.

Another CMS based on the use of fiber optic for monitoring *blade condition* is developed by Micron Optics and Sandia National Laboratories [27]. Systems that use SCADA data analysis for CM of WTs are developed by University of Strathclyde in the United Kingdom [28].

CMS solution for WTs, such as SKF WindCon [24], includes one of the vibration-based CM techniques. It typically has eight accelerometers of different sensitivities mounted onto the main components of the turbine drivetrain. The system includes sensors, data export, analysis, and lubrication. The system monitors health through vibration sensors and can be linked in the turbines SCADA data.

The data are analyzed by the software which can detect issues in operation such as bearing condition, shaft deflection, and foundation weaknesses. The system enables CM on an unlimited number of turbines and turbine data points. Sensors and software combine to continuously monitor and track several operating conditions, such as: unbalanced propeller blades, misalignment, shaft deflections, mechanical looseness, foundation weakness, bearing condition, gear damage, generator rotor/stator problems, resonance problems, tower vibrations, blade vibrations, electrical problems, and inadequate lubrication conditions.

The system continuously monitoring *corrosion* inside monopile [29], which consists of the corrosion rates, corrosion potential and oxygen concentration sensors installed inside the monopole, was tested for two years at the North Sea. Corrosion management inside the monopoles has been based on the assumption that no oxygen is present within the confined space of the monopile. The monitoring-setup at Belwind Wind Farm has been able to detect that the level of oxygen inside stays high. Based on installed electrodes that are lowered inside the monopile structure, measurements of corrosion rates at multiple depths are conducted.

The survey of existing commercially available CMS around the world in the last decade are provided in Ref. [2]. According to this survey, by 2014 there were

- 27 systems primarily based on drivetrain vibration analysis
- 1 system using Motor Current Signature Analysis, Operational Modal Analysis, and AE techniques
- 4 systems solely for oil debris monitoring
- 1 system using vibration analysis for WT blade monitoring
- 3 systems based on fiber optic strain measurement in WT blades, mast, and foundation

For example, Japanese company NTN developed a CMS for WTs, which can detect the failure of main bearings, gearboxes, and many mechanical components of WTs in their early stages [30]. One feature of CMS is that users can monitor from a remote location. This system includes a data acquisition module, data management software, and monitoring and analysis software for a client PC.

Different sensors for vibration (acceleration sensors), rotation (proximity sensors), current (AC, DC current sensors), temperature (thermocouples), and so on can be used in this system and changing of the number of channels can also be accomplished easily.

NREL conducted an extensive research, analysis, and testing of multiple CMS and sensor packages [31], which allowed to come up with the following conclusions:

1. The spectrum analysis of the *vibration signal* (or stress waves) can distinguish between healthy and damaged gearboxes, and, to a certain extent, pinpoint the location of damaged gearbox components. The diagnosis can determine which stage of the monitored gearbox has damage, but it may not be able to specify which bearing or gear, since several bearings or gears may have the same characteristic fault frequencies.
2. The *stress wave* amplitude histogram appears to be effective for detecting gearbox abnormal health conditions.
3. *Oil cleanliness* level can be used to control and monitor WT gearbox run-in. The typical run-in interval observed in the GRC tests was in the 1 + h range for each load level.
4. *Oil debris* particle counting is effective for monitoring gearbox component damage, but is not effective for pinpointing damage location. Note that readings are affected by sensor

mounting locations, and similar particle counting trends can be obtained between the inline filter (main) and the off-line filter (kidney or side stream) loops.

5. A damaged gearbox releases particles at increased rates.
6. Oil CM results, specifically moisture, total ferrous debris, and oil quality, indicate that oil total ferrous debris appears indicative of gearbox component damage. More data are required to evaluate the measurements of oil moisture and quality.
7. Periodic oil sample analysis may help pinpoint a failed component and support root cause analysis.
8. Electrical signature-based techniques so far have not distinguished between the healthy and damaged gearboxes.

Table 18.5 presents some of the CMS developed and available for installation on WTs around the world. The table shows CMS name, manufacturing company/country, and the section of the

TABLE 18.5

Wind Turbine Condition Monitoring Systems (CMS)

N	CMS	Manufacturer	Country	Drive Train	Rotor	Tower
1	OneProd Wind System	ACOEM/01db-Metravib	France	+		
2	WT-CMS Adapt.wind	Bently Nevada (GE)	USA	+		
3	PlantProtect	Beran Instruments	UK		+	+
4	WTAS (WT Analysis System) Type 3651	Brüel & Kjær Vibro A/S	Denmark	+	+	+
5	E-GOMS	Eickhoff	Germany	+		
6	Epro MMS	Emerson Process Management	USA	+		
7	FAG WiPro FAG-ProCheck-Wind	FAG Industrial Services GmbH	Germany	+	+	+
8	WinTControl	Flender Service GmbH	Germany			
9	SMP-8C	Gamesa	Spain	+		
10	E-Sentry System	Global Maintenance Technology	USA	+		
11	MetalSCAN Series 3000	Gas TOPS Ltd.				
12	Turbine Condition Monitoring (TMC®)	Gram & Juhl A/S	Denmark	+	+	+
13	AE System	Holroyd Instruments/	UK	+		
14	Metallic Contamination Sensor MCS 1000	HydacFiltertechnik GmbN	Germany	+		
15	BLADEcontrol® Ice Detector BID	IGUS ITS GmbH	Germany		+	
16	Metal Particle Sensor MPS 01.2	Internormen Technology GmbH	Germany	+		
17	RMS IDS Ice Detection System	Insensys Ltd.	UK		+	
18	Online Sensor Suite	Parker Kittiwake	UK	+		
19	VibroWeb XP	Prufteknik Condition Monitoring GmbH	Germany	+		
20	Winergy CDS	Rovsing Dynamics	Denmark	+		
21	Turbine Condition Monitoring (TCM)	Siemens Wind Power AS	Germany	+	+	+
22	WindCon	SKF Engineering & Research Centre B. V	Germany	+	+	
23	SWANwind	SwanTech	USA	+		
24	DriveMon Wind	Vatron	Austria	+		
25	Vestas Condition Monitoring System (VCMS)	Vestas Wind Systems A/S	Denmark			
26	Winergy CGS	Winergy AG (Siemens)	Germany			
27	WT-HUMS	WindSL	Israel	+		+
28	Omega-Guard®	μ-SEN GmbH	Germany	+		

TABLE 18.6

Physical Property Monitored and/or Techniques Used in CMS in Table 18.5

Physical Property/Technique	CMS (# According to Table 18.5)
Acoustic emission	4, 13, 22, 23, 27, 28
Oil cleanliness	1, 3, 4, 5, 7, 13, 18, 19, 20, 22, 27
Vibration	21
Strain	17, 21
Displacement	19
Temperature	1, 4, 6, 7,10, 19, 24, 27
Humidity	10
Pressure	10
Accelerometer	1, 2, 3, 4, 5, 6, 7, 9,10, 12, 15, 19, 20, 27
Tachometer	1, 2, 3, 4, 5, 7, 9, 12, 13, 17, 20, 21, 27, 28
Video	12

WT each system is monitoring [32,33]. Table 18.6 is an addition to Table 18.5 and it shows which property is monitored and/or which technique is used by each of the CMS presented in Table 18.5.

18.1.5 Wireless Condition Monitoring Systems for Wind Turbines

By 2015, practically none of CMS using *wireless sensors* to monitor various WT physical conditions was commercially available yet. However, many systems and application have been in the stages of study, development, and testing around the world.

For example, wireless radios using Frequency Hopping Spread Spectrum (FHSS) send critical vibration-sensing data to the O&M office, and such wireless data transmission has proven its reliability in industries such as military, oil and gas, and water/wastewater. Now this technology is available as a solution for vibration monitoring in WTs [34] (Figure 18.6). This system may be used for installation in remote locations and difficult environments, it can transmit real-time data up to 60 miles line-of-sight reliably.

With some wireless data radio providers, the investment in radio-frequency (RF) network is not only useful for vibration monitoring. The devices can be scaled to monitor wind speed coming from an anemometer, or the RPM of the actual engine itself within the turbine. With these radios, they can be expanded upon without having to add more devices [35].

The system for structural monitoring of WTs using wireless sensor networks (WSNs) is presented in Ref. [35]. In this study, WSNs have been installed in three operational turbines in order to demonstrate their efficacy in unique operational environment. The first installation was used to verify that *vibrational (acceleration)* data can be collected and transmitted within a turbine tower and that it is comparable to data collected using a traditional tethered system. In the second installation, the wireless network included strain gauges at the base of the structure and the data have been collected regarding the performance of the wireless communication channels within the tower. The final installation was on a turbine with embedded braking capabilities within the nacelle to generate an "impulse-like" load at the top of the tower.

Building and testing new CMS for WTs is a very active and prospective field; therefore, by the time this book will reach its audience, there might be many more CMS available commercially.

More complicated system developed in University of Strathclyde in Scotland, United Kingdom, for multiple parameters wireless monitoring described in Ref. [36] combines *electrical sensors, oil condition sensors, strain gauges, temperature, and vibration sensors* installed on every WT of offshore wind farm. The signals from wireless transmitters will send collected information to

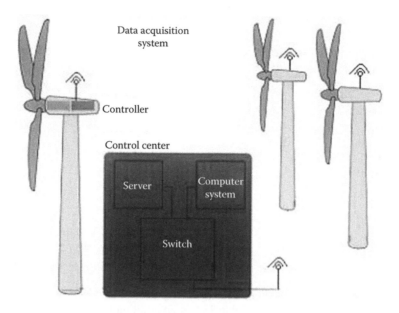

FIGURE 18.6 Wind turbines outfitted with wireless transceivers. (From Dianich B. Wireless technologies for wind turbine vibration monitoring, *Wind Turbine Condition Monitoring Workshop—NREL*, 2011, Online. Available: http://www.nrel.gov/wind/pdfs/day2_sessionviii_04_nrel_2011_bret_d_v2.pdf.)

onshore station, which in turn will transmit the data to headquarter via Internet for data analysis and decision-making on current physical condition of the most significant components of the WTs.

The performance evaluation of these systems was conducted in the laboratory settings and the real implementations and deployments in the field are projected.

Similar type of data transfer via wireless communication was suggested in Ref. [37] for CM of WT foundations, specifically in the area of foundation–tower interface used for large onshore turbines, which include the use of an embedded steel can (another type is use of a ring of bolts). It was determined that the embedded can style has been showing the most serious signs of failure.

The authors are proposing an inexpensive monitoring solution that actively monitors the structural integrity and measures displacements of the turbine, and reports its status to a remote technical center (RTC) or head office. Inspection of the displacement data and trending can enable technical personnel to improve the understanding of failures and allow the development of appropriate techniques to resolve them.

Numerous types of displacement sensors are available. However, many are not suitable for the climate within the turbine, are too costly or would pose difficulties during installation. The most suitable for the application on the foundations are IR, linear variable differential transformers (LVDTs), and Hall-effect sensors.

The communications between the sensors and the data aggregation device could be either wired or wireless. Wireless solution reduces the installation cost and eases deployment. Using this technology, devices can operate for more than 3 years with two AAA batteries reporting every 10 s, making it ideal for *SHM* applications.

Temperature monitoring of WT may be done with thermocouples, but with the development of new technologies, wireless SAW *(Surface Acoustic Waves)* temperature sensors have been applied for enhanced CM of WTs [38]. SAW sensors can be placed on rotating parts, like on the rotating outer bearing of a WTG. As the bearing condition is determined by its temperature (inner and outer ring), time, and stress, the turbine is equipped with remote real-time temperature control. On the outer bearing—rotating part—only a wireless sensor is applicable. Commercially available SAW sensors are not only wireless but also totally passive and maintenance-free sensors.

Based on SAW technology, these sensors operate with RF communication. They are not affected by harsh environment, electromagnetic fields, and harsh climate conditions. As they require no maintenance and feature infinite autonomy, they are well suited for offshore applications, where maintenance interventions are always very difficult and costly. More affordable and less complex than vibration analysis systems, they allow direct measurement on the most critical points. They, thus, provide earlier and more accurate indication of possible defect than classic sensors located on the engine casing.

WT blade health monitoring with *piezo ceramic-based wireless sensor* network was suggested and tested in China [39]. The system was developed for automated real-time health monitoring of WT blades. A distributed piezo ceramic transducer system was formed by embedding water-proof coated piezo ceramic patches in predetermined locations in the composite blades. Desired guided waves were generated by a selected piezo ceramic patch as an actuator to propagate through the entire composite blade. Wave responses were detected by other distributed embedded piezo ceramic patches. The wave propagation response in the WT blade will be attenuated due to acoustic impedance mismatch at the damaged area. Thus, the damage status inside the blade can be evaluated through the analysis of the sensor signals.

Various ideas and technologies using *energy harvesters* (EH) as sensors for SHM of WTs have been reviewed in Ref. [40]. The specific energy sources for harvesting in WTs include ambient mechanical vibrations, wind and aeroelastic vibrations, rotational kinetic energy, thermal energy, and solar energy. If the structure has a rich enough loading, then it may be possible to extract the needed power directly from the structure itself. Harvesting energy using piezoelectric materials by converting applied stress to electricity is most common. Other methods to harvest energy are electromagnetic, magnetostrictive, or thermoelectric generators.

In Ref. [41], an accumulated energy sensor (*piezoelectric EH*) mounted on the surface of the WT blade converts low-frequency vibrational strain energy from the blade to electrical charge that is subsequently stored to power an RF transmitter. The RF transmitter wirelessly communicates a single pulse to a centralized monitoring system in the turbine nacelle when sufficient electrical charge has been stored. The idea of this sensing approach is that the timing of data output from the RF transmitter, which is tied to the charging time, is indicative of the structural health. In a damaged blade, changes in the stiffness (associated with damage) will lead to a change in blade strain resulting in a change in the timing of the RF pulses.

Wireless SHM system was developed and tested on onshore large WT (GE 1.5 MW, 68 m in height) to collect the structural responses, and the system performance of the proposed wireless sensing system NTU-WSU was also evaluated in Taiwan [42]. The system monitored vibration motion of support structure and the soil structure interactions between the tower foundation and soil with multiple three-dimensional velocity sensors.

Wireless SHM system for WT blades based on the integration of WSNs and AE technology was suggested and developed in the United Kingdom [43]. As described earlier in this Chapter, AE technique is able to detect the defects in the blades under wind load while the WT is in service, since most types of blade failure cause detectable AE waves, including crack initiation and growth and crack opening. In this wireless SHM system, the AE data captured from the rotating blades is sent through wireless medium to the MIB510 serial gateway, which is attached to the WT tower (Figure 18.7). This data can then be transferred to the remote control unit via wired or wireless communication, such as using Wi-Fi or any other wireless techniques.

Different approach to determining structural problems with the blades is given in Ref. [44]. With increased demand of energy, the sizes of wind blades are getting bigger and bigger due to which WTs are installed very high above the ground level or offshore. This study in Korea proposes a single *laser displacement sensor* (LDS) system, providing evaluation of all of the rotating blades. The LDS system is able to measure deflection of the blade from the tower (Figure 18.8).

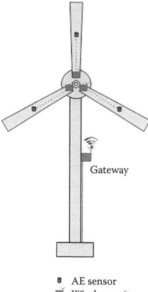

Gateway

🔋 AE sensor
📶 Wireless unit

FIGURE 18.7 Wireless structure monitoring system based on monitoring acoustic emission (AE) from wind turbine blades. (From Bouzid O. M. et al. *Journal of Sensors*, vol. 2015, Article ID 139695, 11pp., 2015, Online. Available: www.hindawi.com/journals/js/aa/139695/.)

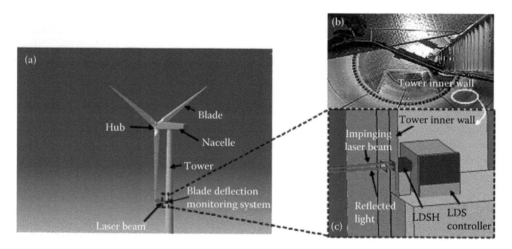

FIGURE 18.8 System for measuring deflection of the blade from the tower: (a) Installation of the blade deflection monitoring system within the tower; (b) Inside the actual wind turbine tower; (c) Inside wind turbine tower with laser displacement sensor (LDS) installation. (From Kim H.-C., Giri P., and Lee J.-R. A real-time deflection monitoring system for wind turbine blades using a built-in laser displacement sensor, *Proc. 6th European Workshop on Structural Health Monitoring (EWSHM 2012)*, vol. We.2.B.2, Dresden, Germany, 9pp., July 2012, Online. Available: www.ndt.net/article/ewshm2012/papers/we2b2.pdf. With permission.)

If the blade bolt loosening occurs, it causes deflection in the affected blade. If nacelle tilts, it will result in change of blade's position causing collision as the blades come close to the tower due to the nacelle tilt. If mass loss damage in the blade occurs due to the lightning it will result in rotational imbalance which may lead to collision accidents. The proposed system can identify such problems. The displacement is continuously monitored by impinging the laser beam of the noncontact LDS at the rotating blades in an operating condition of a WT system. Damage of the blade such as nacelle tilt, bolt loosening, or blade mass loss causes measurement irregularities or changes, indicating the detection of any possible damage. It allows any repair action to be taken readily which can prevent serious accidents.

The system is cost-effective and costs only about a thousand dollars. This wireless solution is achieved using Zigbee technology which operates in the industrial, scientific, and medical (ISM) radio bands, typically 2.4 GHz, 915 and 868 MHz. The output from the LDS is fed to the microcontroller which acts as an analog to digital converter which in turn is connected to the Zigbee transceiver module, which transmits the data. At the other end, the Zigbee reads the data and displays on the PC from where user can monitor the condition of wind blades.

18.2 TIDAL AND WAVE TURBINES

Tidal power technology was still in development stage by 2015, and no clear tidal turbine design has emerged as industry standard for extracting energy from tidal flow. The state of the art in turbine design includes vertical axis solutions, some with major structural and operational variations [45]. However, a common focus is the horizontal axis design, holding many similarities with a standard wind turbine (WT).

Maintenance on tidal turbines requires a lift operation to access the turbine above sea level. This can be a costly and lengthy procedure, resulting in prolonged periods of downtime. An effective CMS would therefore be of great benefit to this industry, allowing the health state of system components to be known, and allowing maintenance to be scheduled efficiently.

CM has already been well established for the wind industry. However, despite similarities between tidal and wind power turbine design, the operating environment is vastly different. Water is over 800 times denser than air and, despite slower flow rates (around 3 m/s compared to around 15 m/s for offshore wind), tidal flow has a much higher kinetic energy compared to wind flow [46]. This causes tidal turbines to operate with higher torque and thrust loading, inducing increased stress on the machine, particularly on the low speed stages of the drivetrain. Additionally, the marine environment provides other complications, such as corrosion and interaction with plant and animal life (see Chapter 10). Another complication is limited historical data of failures from tidal turbines required to implement CM techniques used as standard in the wind industry.

18.2.1 INITIAL STAGES OF TIDE TURBINE CONDITION MONITORING

Because of the lack of tide turbines failures historical data, anomaly detection technique was suggested and developed in Ref. [47] for identifying developing faults within tidal turbines with limited historical data. Using the data-mining methodology, key relationships between sensor data parameters from an operational tidal turbine were identified, describing the normal response of the turbine over variable operating conditions.

These trends were then defined using several modeling techniques, allowing for deviations from expected data patterns to be detected from live turbine data, alerting the operator to the possible onset of a fault. The use of sensor data from turbine components (such as the gearbox, generator, bearings, blades, etc.) can allow the onset of faults to be detected before they cause failure. This enables an efficient maintenance strategy to be employed, as maintenance can be scheduled to reflect to the known health of system components. The study in Ref. [47] was focused on data from the following sources: tri-axial generator vibration velocity, gearbox vibration velocity, bearing vibration velocity, bearing displacement, bearing temperature, generator rotor speed, and output power.

Though anomaly detection is useful as an initial stage of CM, it is only suitable for indicating if a deviation from the defined normal behavior has occurred. Specific failure modes cannot be identified through this method. Further stages of CM include diagnosis and prognostics.

### 18.2.2	CONDITION MONITORING VIA SENSOR NETWORK ON TIDAL TURBINE

#### 18.2.2.1	DeltaStream Condition Monitoring System

One of horizontal axis tidal turbines developed by DeltaStream works like a WT, but it uses water flow instead of wind flow. The water flow turns the blades of the turbine and electricity is obtained by means of the generator located in the nacelle [48]. It is primarily designed to be located on the seabed in areas with high tidal stream flows but could also be installed in suitable rivers and estuaries. When mounted in tidal areas it generates power during both the flow and ebb of the tide (see Figure 10.6). AC power is brought onshore from the DeltaStream unit through a submarine cable to its onshore power conversion and SCADA system.

Multiple sensors (total 78 devices) are installed on this turbine to monitor various physical conditions and parameters of the operating turbine. In this particular configuration, each group of three turbines is connected to a base station where there is a cable junction box. From this base station, the information is sent to the onshore base.

Power needed by the sensors is provided by a shore base station using cables and through an auxiliary transformer enclosure. This transformer is mounted in one of the vertical towers of the turbines. It provides energy for all auxiliary instruments and for the yaw system in the turbine [49]. Table 18.7 shows the location, the type of the sensors, and function of the monitoring devices. The structure

TABLE 18.7

Sensors Location, Type, and Function in DeltaStream Tidal Turbine CMS

Location	Type of Sensor	Monitoring Function	# Of Sensors
Blades (Rotor)	Strain gauge	Detection of defects or cracks	18[a]
	Accelerometer	Detection of defects due to destructive vibration	3[b]
Pintle beam	Strain gauge	Detection of defects	8
Gearbox cover	Acceleration sensor	Detection of any failure in the component of the gearbox	6
Inside the lubrication oil	Aqua sensor	Monitoring of the oil quality, warning when it is degraded	1
System (gearbox)	Metallic contamination sensor	Detection of the damage in the components of the gearbox	1
Gearbox shafts	Torque sensor	Detection of overloading of the shaft in the gearbox	1
	Speed sensor	To monitor the rotating speed of the shaft	1
Generator	Temperature sensor	To ensure the generator is working in the correct range of temperature	5
	Current sensor	To monitor the output power	1
Hydraulic system	Pressure sensor	To ensure the correct operated pressure, to detect oil leakage	10
	Linear transducer	Detection of failures in actuators	2
Inside the chassis	Leakage sensor	Detection of water	2
Structure frame	Strain gauges	To monitor frame stresses	8
Around the structure frame	Passive acoustics	To detect impact of marine animals	3
In the frame	CCTV	An underwater camera to monitor general condition	3
Water around the turbine structure	Acoustic Doppler Current Profile (ADCP)	To record the current velocity to monitor the performance of the turbine	3

[a]	Connected every 6 to a router to transmit to base station.
[b]	Connected to a router to transmit to base station.

is shared by the three turbines, thus, in total there are 200 sensors installed on DeltaStream tidal turbine. Most of them are located in the nacelle.

18.2.2.2 TidalSense Condition Monitoring System

Another example of CMS for tidal turbines is *TidalSense* CMS, which consists of a number of networks of *Macro Fiber Composite* (MFC) sensors and other sensors for data acquisition and a control system with software processing to evaluate the acquired data. TidalSense system can be broadly categorized as the system monitoring structural health for the purposes of incipient defect detection on tidal turbine blades.

The system includes active *Long Range Ultrasound Guide Waves* (LRUG) sensors and passive AE sensors or a combination of both. LRU technology, also known as *guided wave ultrasonic inspection*, uses arrays of ultrasound transducers to detect inconsistencies in long (high aspect ratio) components.

The transducer operates as a sensor and an actuator, "listening" for the initiation and propagation of defects, and using a guided response to identify and classify these defects. Flexible interdigitated MFC transducers are used as LRUG sensors in TidalSense system [50,51]. The system also includes other sensors (see Table 18.8) and allows data analysis in real time to detect damage. The system also controls the brakes of the tidal turbine to allow (or not) the blades movements.

The TidalSense CMS is designed so that it can be incorporated with the tidal turbine as it is being built and retro-fitted to ones already in the field, thus allows for maximizing cost benefit over the whole range of tidal turbines. The system wirelessly transmits data at 2 Mbps over distance of 40 m maintaining 100% signal integrity.

18.2.2.3 SeaGen Condition Monitoring System

The SeaGen-S tidal turbine manufactured by MCT (Marine Current Turbines, UK) incorporates twin horizontal axis rotors [52]. The rotors utilize an active blade pitching system, which limits structural forces during high flow conditions. This allows the use of blades that are highly efficient over the full range of tidal velocities, from initial cut-in through to rated flow. The tidal turbine is equipped with a web-based SCADA system.

This system offers remote control, a variety of status views, and useful reports from a standard Internet web browser. The status views present electrical, mechanical, meteorological, and tidal

TABLE 18.8
Sensors Location, Type, and Function in TidalSense CMS

Location	Type of Sensor	Monitoring Function	# Of Sensors
Turbine	Gyroscope sensor	Monitor position, velocity, and acceleration	1
	Long Range Ultrasonic Guide (LRUG) sensors	Detect damage by inducing a test wave in the structure, data analysis in real time	N/A
	Passive AE sensors	Detect damage in turbine	N/A
Buoyancy system	Gyroscope sensor	Monitor position, velocity, and acceleration	1
Housing box for electrical devices	Temperature, humidity, and dew point sensors	Monitor the box status	2
Mooring chain system	Load cell sensors	Measure the forces generated by the buoyancy system and the turbine on the mooring chain system	2
Turbine	RPM sensors	Measure the rotational velocity of the tidal turbine	N/A
Outside turbine	IP underwater camera	See the tidal turbine from surface	1

data, as well as operation, fault, and grid status. In addition to the Web SCADA system, the turbine is equipped with a web-based Turbine Condition Monitoring (TCM) system.

The TCM system carries out precise, continuous, real-time, condition diagnostics on main turbine components. The TCM system has various alarm levels, from informative through alerting level to turbine shutdown. No details of the system design and monitored conditions are available.

18.2.2.4   SKF Condition Monitoring of Bearings for Wind, Tidal, and Wave Turbines

SKF condition monitoring (SKF Insight) uses bearing-embedded sensors to monitor the critical parameters that are likely to lead to an early failure, for example, lubricant contamination or excessive loads or temperatures. By proactively eliminating such anomalies, the failure can be avoided. The same technology can be used in a more positive way: By verifying the integrity of an installation and giving a better understanding of the operating environment, a machine may be uprated to extend its life or power rating beyond the initial design [53]. SKF Insight puts together the technology that can monitor the actual conditions experienced by bearings in a particular application. Embedded sensors measure loads, lubrication conditions, speed, vibration, and temperature using power harvested from the application environment.

SKF algorithms and diagnostics interpret these data in terms of the severity of the conditions, or how far the operating conditions are departing from their original design condition. It can also identify excessive loads, duty excursions, lubricant contamination, and lubrication problems so that modifications can be made to the operating conditions to avoid damage before it occurs.

This package of sensors and algorithms has been named SKF Insight because that is what it provides—an insight into the operating conditions and how those are likely to affect the reliability of the installation. Various sensors could be installed into an SKF bearing [53]. For example, the Low-Frequency Seismic Sensor can be installed at bearing supports, where vibration often indicates major machinery issues. It is often installed on the bearing housing and measures vibrations within the range of 0.5 Hz to 1.0 kHz.

Intelligent wireless communication technology packaged inside the bearing enables it to communicate within environments where traditional Wi-Fi cannot operate. Bearings enabled with SKF Insight create smart networks, communicating through one another and via a wireless gateway to send information relevant to their condition for analysis. The gateway can be local to the machine or local to the plant. CMS designed and manufactured by SKF are applicable to tidal and wave turbines [54].

18.3   SOLAR MODULES

Reduced efficiency of photovoltaic (PV) panels depends on the environment and operating conditions. Damage from freeze/thaw cycles, lightning strikes, moisture intrusion, and micro vibrations within the building structures can seriously affect the integrity of the installation. In Chapter 9, the role of different internal and external causes of solar panel/cells leading to various kinds of deterioration has been described.

External causes include environment/climatic factors, such as corrosive gases, morning dews, rains, snow, lightning, windborne debris, and so on. Internal causes might include the presence of toxic chemicals, like cadmium and arsenic used in the PV production process, but these environmental impacts are minor and can be easily controlled through recycling and proper disposal.

Another internal cause of PV failures are "hot spots" (see Sections 9.2.7 and 9.2.8) that can eventually lead to severe overheating of the cell and its permanent damage. The difficulty in finding and tracking down such problems, and potential issues, that arise in solar systems installations is great.

These and other causes of the failure should be monitored during the operations and timely received information via monitoring system about developing problem in PV panel would be a reason for timely scheduled maintenance and repair.

18.3.1 USE OF INFRARED THERMOGRAPHY FOR INSPECTION AND TESTING OF SOLAR SYSTEMS

Hot spot heating can occur on a PV panel because of cell failure, interconnection failure, partial shading, and by having mismatched cells. When these problems occur, a cell in a string of cells becomes negatively biased and instead of producing electrical energy it produces heat energy.

These problems showed up as individual cells that were producing abnormal amounts of heat. Besides the obvious problem of that cell not producing any usable electrical energy, the cell also has the potential to degrade those around it. The majority of panels are constructed using plastic and other components such as holders, corners, and in some cases covers that can be susceptible to heat damage.

Over a long enough period of time, the damaged cell has the capacity to burn through its backing as well as cause the cells around it to become damaged and to overheat affecting the entire panel. When individual cells become damaged, due to defects in manufacturing or external incidents, the power output lowers and efficiency drops. Unfortunately, it is extremely difficult to see any of these problems with the naked eye and the only way to test the panels for efficiency is to take a voltage reading.

IR thermography is a valuable technology that can be used on a wide scale to pinpoint problems in large PV arrays. It is important for all of the cells to work together to efficiently produce the maximum output of electricity. IR thermography has the ability to see the heat differential between solar cells and can be used to determine whether any of those cells are damaged or defective. In some cases, the smaller solar sites can be walked along and scanned looking for these problems.

IR Thermography can be applied to determine the operational status of PV solar systems on a large scale using flight. Technique called *Solar Thermography* [55] is the use of an IR camera to inspect PV solar systems for problems that can cause damage to the cells, loss of efficiency, and fire hazards. Flying over the solar panel arrays allows gathering data about a large site in a time efficient manner. The ability to get a literal "bird's eye" thermographic view of the array allows for pinpoint accuracy down to the cell and significantly decreases the time spent finding and identifying problems. A series of problems which could be pinpointed with the aerial IR thermography are presented in Ref. [55].

Solar 800 kW array located on top of the Portland Habilitation Center in Portland, OR, is like many new solar sites that are being built either on top of a roof or high up in the air. This makes it impossible to look at it with IR camera from ground level and very difficult to scan while standing directly on top of it. Therefore, it is necessary to turn to flight to solve the problem.

Figure 18.9 shows a unique phenomenon that is peculiar to the panel manufacturing type in which instead of the standard practice of soldering the cell's connections together these connections

FIGURE 18.9 Strings of failed overheated solar cells seen with infrared (IR) thermography. (From Denio H. III and Denio H. II. *Proc. InfraMation*, 2011, Online. Available: http://www.oregoninfrared.com/sites/default/files/Solar%20Infrared%20Paper.pdf. With permission.)

are pressed and laminated. These "strings" seen in Figure 18.9 are actually a circuit of cells in series with each other. One of the pressed on connections shorted out and overheated causing the rest of the circuit to begin to overheat with it. Further investigation carried out at the ground level found bubbles in the lamination. Those panels in question were replaced and the problem repaired.

There are a number of various defects in solar panels which could be detected with IR inspection, they are summarized in Ref. [56]. Some of the serious defects in PV panels detected with IR inspection are shown in Figure 18.10.

Like many new technologies, *Unmanned Aerial Vehicles (UAVs)* (drones) started with military applications and gradually found their way into civilian life. Introduction of drones to the various services increased the capability of IR inspection of solar PV installation [57,58]. Carefully planned inspection and regular automated reporting solutions offered by the service companies help to maintain solar panel efficiency and reliable operation. In the *AIRX3 Photovoltaic* [57], the inspection sensor array comprises a specially designed High Definition (HD) inspection camera

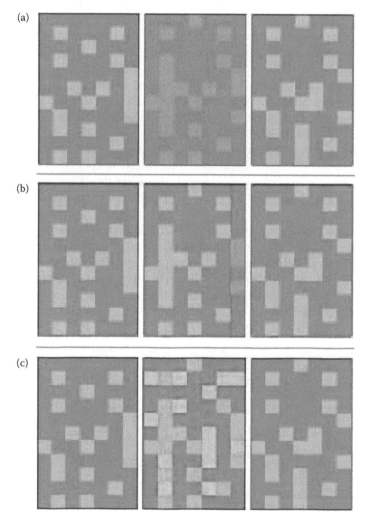

FIGURE 18.10 Infrared (IR) images of common photovoltaic (PV) panel defects: Panel overheating as compared with the other cells (a); overheating pattern caused by short circuit in a string of cells (b); "patchwork pattern" where randomly distributed individual cells are significantly hotter (c). (Modified from *Practical guide: Solar Panel Thermography*. Testo, Inc., Online. Available: http://www.murcal.com/pdf%20folder/15. testo_thermography_guide.pdf.)

mounted adjacent to a calibrated IR FLIR TAU 640 thermal camera mounted on an AscTec Falcon 8—*Remotely Piloted Aircraft System* (RPAS).

The system can be safely flown within 10 m of the panel surface, permitting detailed visual inspection or at greater distances to permit the harvested thermographic images to be assembled with HD image capture.

The data includes a Global Positioning System (GPS)-tagged identifier allowing precise positional recording of local hot spots or complete array defects. During the inspection process, images are transmitted to a main ground station for direct visualization and recorded for post processing.

The AscTec Falcon 8 is piloted by an AIRX3 qualified pilot and includes sensor arrays that can be remotely controlled and triggered by certified IR PV solar panel inspectors. Following the onsite procedures, specially prepared reports complete the turnkey IR thermal analysis subcontracted offer of service.

With support from investors such as Google, a new company called *Skycatch* [58] has developed a UAV that is used for remote monitoring and inspecting construction sites, mining operations, and farms. Now Skycatch is getting into the solar array inspection business. It plans to begin testing its technology on PV farms built by industry giants SolarCity and First Solar.

The Skycatch quadcopter weighs about 2.3 kg (5 pounds) and is armed with a variety of sensors, depending on the application. For solar farm inspection, it uses a Tau IR camera, which adds only 70 g to the overall weight. Autonomous navigation employs GPS and sonar, and the quadcopter can also be operated by remote control when necessary. It can fly with the speed 80 km/h (50 mph), reach altitudes of 120 m (400 ft), and maintain its position under windy conditions; its drift is about 30 cm (1 foot) in 11 m/s (25 mph) winds.

18.3.2 DEVELOPMENT OF REMOTE MONITORING SYSTEM FOR PV PANELS

Monitoring and recording of electrical and nonelectrical variables of the PV modules, the inverters, and the medium-voltage components could be an answer for monitoring and control of a solar power plant. A solution to this problem is based on installing WSN nodes with appropriate sensors monitoring the solar power panels. Such system proposed in Algeria [59] includes installation of a WSN node on each PV panel. Each node will control and monitor a PV panel and transmit the data to a gateway node, the information will be then forwarded to the central unit via Internet where the operator can monitor at distance the set of PV arrays via graphic user interface as human machine communication program.

Based on priority for faults detection, it is suggested to use light, temperature, humidity, and current sensors with I2C communication link. It also suggested using a dust sensor by optical sensing system. An IR emitting diode (IRED) and a phototransistor are diagonally arranged into this device. It detects the reflected light of dust in air. Several other publications present various ideas of monitoring solar power plant health based on wireless communication [60,61]. However none of these systems has yet been built and tested in the field.

REFERENCES

1. May Allan Forsith and McMillan David. A review of commercially available condition monitoring systems, *Engineering and Physical Sciences Research Council (EPSRC)*, 7, Online. Available: https://www.strath.ac.uk/media/departments/eee/iee/windenergydtc/Allan_May.pdf.
2. Crabtree C. J., Zappalá D., and Tavner P. J. *Survey of Commercially Available Condition Monitoring Systems for Wind Turbines*, Technical Report, Durham University School of Engineering and Computing Sciences and the SUPERGEN Wind Energy Technologies Consortium, May 2014, Online. Available: http://dro.dur.ac.uk/12497/1/12497.pdf?DDD10+ttsd23+dul4eg.
3. Antoniadou I., Dervilis N., Papatheou E. et al. Aspects of structural health and condition monitoring of offshore wind turbines, *Philosophical Transactions A*, January 2015, Online. Available: http://rsta.royalsocietypublishing.org/content/roypta/373/2035/20140075.full.pdf.

4. Tchakoua P., Wamkeue R., Ouhrouche M. et al. Wind turbine condition monitoring: State-of-the-art review, new trends, and future challenges, *Energies*, 7, 2595–2630, 2014, Online. Available: https://ideas.repec.org/a/gam/jeners/v7y2014i4p2595-2630d35349.html.

5. Wisznia R. Condition monitoring of offshore wind turbines, Master of Science Thesis EGI 2013:017, Stockholm, Sweden, 42pp., 2013, Online: Available: http://www.diva-portal.org/smash/get/diva2:606267/fulltext01.pdf.

6. Wind Turbine Condition Monitoring, *National Instruments*, April 2013, Online. Available: www.ni.com/white-paper/9231/en/pdf.

7. Rumsey M. A. and Paquette J. A. *Structural Health Monitoring of Wind Turbine Blades*, 2008, Online. Available: http://windpower.sandia.gov/other/SPIE-2008-6933-14.pdf.

8. Zaher, A., McArthur, S. D. J., and Infield, D. G., Online wind turbine fault detection through automated SCADA data analysis, *Wind Energy*, 12(6), 574–593, 2009.

9. Kim K., Parthasarathy G., Uluyol O., and Foslien W. Use of SCADA data for failure detection in wind turbines, *2011 Energy Sustainability Conference and Fuel Cell Conference*, Washington, D.C., August 7–10, 2011, Online. Available: http://www.nrel.gov/docs/fy12osti/51653.pdf.

10. Wiggelinkhuizen E. J., Verbruggen T. W., Braam H. et al. CONMOW: Condition monitoring for offshore wind farms, *European Wind Energy Association Conference (EWEC)*, Milan, Italy, 8pp., May 2007, Online. Available: http://www.ecn.nl/docs/library/report/2007/e07044.pdf.

11. Márquez F. P. G., Tobias A. M., Pérez J. M. P., and Papaelias M. Condition monitoring of wind turbines: Techniques and methods, *Renewable Energy*, 46, p. 169–178, 2012, Online. Available: http://lms.ctl.cyut.edu.tw/sysdata/user/3/10111612/blog/doc/4e4c1ed17a7f46e6/attach/285275.pdf.

12. *National Instruments Products for Wind Turbine Condition Monitoring*, November 2014, Online. Available: www.ni.com/white-paper/7676/en/pdf.

13. Jungert A. Damage detection in wind turbine blades using two different acoustic techniques. *The e-Journal of Nondestructive Testing*, 25, 2008, Online. Available: http://www.gastopsusa.com/knowledge_center_documents/1/DEWEK_2006.pdf.

14. Dupuis R. Application of oil debris monitoring for wind turbine gearbox prognostics and health management, *Annual Conference of the Prognostics and Health Management (PHM) Society*, Portland, Oregon, 15pp., October 2010, Online. Available: http://ftp.phmsociety.org/sites/phmsociety.org/files/phm_submission/2010/phmc_10_044.pdf.

15. Muller Jane Errichello Robert. Oil cleanliness in wind turbine gearboxes, *Machinery Lubrication*, 7, 2002, Online. Available: http://www.machinerylubrication.com/Read/369/wind-turbine-gearboxes-oil.

16. Feng Y., Qiu Y., Crabtree C. J. et al. Use of SCADA and CMS signals for failure detection and diagnosis of a wind turbine gearbox, *Scientific Proc. European Wind Energy Association Conference (EWEC)*, Brussels, Belgium, 2011, Online. Available: https://community.dur.ac.uk/supergen.wind/docs/publications/Feng,%20Qiu,%20Crabtree,%20Tavner,%20Long_EWEA2011presentation.pdf.

17. Daschner E. *Wind Turbine Lubrication Challenges and Increased Reliability*, DA Forum–Hannover Messe, 2011, Online. Available: http://www.faduc.com/wind-turbine-lubrication-challenges-hannover-apr2011_26715/#solomid.

18. Rumsey M. A. Condition monitoring and wind turbine blades, *Wind Turbine Reliability Workshop "Digging Down for Reliability,"* June 2009, Online. Available: http://windpower.sandia.gov/2009Reliability/PDFs/Day2-08-MarkRumsey.pdf.

19. Madsen T. Lind. *Corrosion Monitoring*, DTU Department of Mechanical Engineering, May 2012, Online. Available: http://www.thomaslindmadsen.dk/User_files/c81029225be8841e0eea2771111f3d00.pdf.

20. Van der Mijle Meijer Harald. Corrosion in offshore wind energy "a major issue," "Essential Innovations," Den Helder, Netherlands, February 2009, Online. Available: http://www.we-at-sea.org/wp-content/uploads/2009/02/3-Harald-vd-Mijle-Meijer.pdf.

21. Gong X. Online nonintrusive condition monitoring and fault detection for wind turbines, 2012. Electrical Engineering Theses and Dissertations. University of Nebraska-Lincoln, Paper 46, Lincoln, Nebraska, 2012, Online. Available: http://digitalcommons.unl.edu/cgi/viewcontent.cgi?article=1051&context=elecengtheses.

22. *Monitoring Wind Turbines Drivetrains*, Online. Available: http://famos.scientech.us/PDFs/StressWave_Web/MonitoringWindTurbinesDrivetrains.pdf.

23. *Macom Condition Monitoring*, Macom TechAlert 10 Brochure, Online. Available: http://www.macom.co.uk/ta10.pdf.

24. *Make the Most Out of Your Maintenance Resources*, SKF WindCon Brichure, November 2013, Online. Available: http://www.skf.com/binary/30-143837/SKF-Wind-Con-Bro-Update_6Nov13.pdf.

25. *Kittiwake Online Sensor Suite*, Online. Available: http://www.kittiwake.com/sites/default/files/MA-K19079-KW%20Iss%201%20Single%20Pages.pdf.

26. *Kittiwake Particle Content Sensor*, Online. Available: http://www.kittiwake.com/particle_content_sensor.htm.
27. Rumsey M., Nolet S., Neal B., and White J. Fiber optic sensors and the sensor blade. *Collaboration between Micron Optics and Sandia National Laboratories*, Online. Available: windpower.sandia.gov/2010BladeWorkshop/PDFs/1-E-2-Turner.pdf.
28. Quail F. and Feuchtwang J. Wind turbine condition monitoring activities at strathclyde university, *Engineering and Physical Sciences Research Council (EPSRC) Seminar "Wind Energy Systems"* (Doctoral Training Centre), Institute of Energy and Environment University of Strathclyde, Glasgow, Scotland, 2013, Online. Available: http://www.nrel.gov/wind/pdfs/day1_sessioniii_02_quail.pdf.
29. De Sitter G., Weijtjens W., Ingelgem V. et al. Foundation monitoring systems: Analysis of 2-years of monitoring at the north sea, *European Wind Energy Association (EWEA) Conference*, Barcelona, Spain, March 2014, Online. Available: http://proceedings.ewea.org/annual2014/conference/posters/PO_266_EWEApresentation2014.pdf.
30. Takeuchi A., Haseba T., Ikeda H. Application of condition monitoring system for wind turbines, *NTN Technical Review*, 80, 4, 2012, Online. Available: http://www.ntn.co.jp/english/products/review/pdf/NTN_TR80_en_015_018p.pdf.
31. Sheng S., Link H., LaCava W. et al. *Wind Turbine Drivetrain Condition Monitoring During GRC Phase 1 and Phase 2 Testing*, NREL Technical Report NREL/TP-5000-52748, October 2011, Online. Available: http://www.nrel.gov/docs/fy12osti/52748.pdf.
32. Elliott M. Reducing costs of offshore wind-Opportunities for condition monitoring, BVG Associates, 2013, Online. Available: http://www.merinnovateproject.eu/wp-content/uploads/martin-elliott-bvg-assoc-mer-innovate-170913.pdf.
33. *Condition Monitoring Systems (CMS)/Monitoring Bodies for CMS. DNV GL List of GL Renewables Certification (GL RC)*, 07.05.2015, Online. Available: http://www.gl-group.com/pdf/Condition_Monitoring_System_GL_RC_r2(1).pdf.
34. Dianich B. Wireless technologies for wind turbine vibration monitoring, *Wind Turbine Condition Monitoring Workshop – NREL*, 2011, Online. Available: http://www.nrel.gov/wind/pdfs/day2_sessionviii_04_nrel_2011_bret_d_v2.pdf.
35. Swartz R. A., Lynch J. P., Zerbst S., Sweetman B., and Rolfes R. Structural monitoring of wind turbines using wireless sensor networks, *Smart Structures and Systems*, 6(3), 183–196, 2010, Online. Available: http://www.rms-group.org/rms_papers/tamug_papers/other/lynch_cambridge_08.pdf.
36. Tachtatzis C., Harle D. A., Atkinson R. C. et al. Design and implementation of wireless sensor systems, *British Council Workshop*, Ho Chi Minh City, Vietnam, February 2014.
37. Currie M., Saafi M., Tachtatzis C., and Quail F. Structural health monitoring for wind turbine foundations, *Proc. Institution of Civil Engineers (ICE) – Energy*, 166(4), 162–169, September 2013, Online. Available: https://pure.strath.ac.uk/portal/files/30206386/Currie_et_al_2013.pdf.
38. *Wind Turbine Generator Bearing Wireless Temperature Monitoring*, SENSeOR, France, Online. Available: http://www.senseor.com/images/stories/download/Press_room/Press_releases/2012_04_SENSeOR_Wind_turbine_monitoring_with_wireless_sensors.pdf.
39. Song G., Li H., Gajic B. et al. Wind turbine blade health monitoring with piezoceramic-based wireless sensor network, *International Journal of Smart and Nano Materials*, 4(3), 150–166, 2013, Online. Available: http://www.tandfonline.com/doi/full/10.1080/19475411.2013.836577#.VblJHvlVhBc.
40. Davidson J. and Mo C. Recent advances in energy harvesting technologies for structural health monitoring applications, *Smart Materials Research*, Hindawi Publishing Corporation, vol. 2014, Article ID 410316, 14pp., 2014, Online. Available: http://www.hindawi.com/journals/smr/2014/410316/.
41. Lim D.-W., Mantell S. C., and Seiler P. J. Wireless structural health monitoring of wind turbine blades using an energy harvester as a sensor, *Proc. 32nd ASME Wind Energy Symposium*, National Harbor, Maryland, January 2014, Online. Available: http://www.aem.umn.edu/~SeilerControl/Papers/2014/LimEtAl_14AFM_WirelessSHMOfWindTurbineBlades.pdf.
42. Lu K.-C., Peng H.-C., and Kuo Y.-S. Structural health monitoring of the support structure of wind turbine using wireless sensing system, *Proc. 7th European Workshop on Structural Health Monitoring*, La Cité, Nantes, France, July 8–11, 2014, Online. Available: https://hal.inria.fr/hal-01020460/document.
43. Bouzid O. M., Tian G. Y., Cumanan K., and Moore D. Structural health monitoring of wind turbine blades: Acoustic source localization using wireless sensor networks, *Journal of Sensors*, Hindawi Publishing Corporation, vol. 2015, Article ID 139695, 11pp., 2015, Online. Available: www.hindawi.com/journals/js/aa/139695/.

44. Kim H.-C., Giri P., and Lee J.-R. A real-time deflection monitoring system for wind turbine blades using a built-in laser displacement sensor, *Proc. 6th European Workshop on Structural Health Monitoring (EWSHM 2012)*, vol. We.2.B.2, Dresden, Germany, 9pp., July 2012, Online. Available: www.ndt.net/article/ewshm2012/papers/we2b2.pdf.

45. Aly H. H. H. and El-Hawary, M. E. State of the art for tidal currents electric energy resources, *Proc. 24th Canadian Conference on Electrical and Computer Engineering (CCECE)*, pp. 1119–1124, Niagra Falls, ON, Canada, May 8–11, 2011.

46. Winter A. I. Differences in fundamental design drivers for wind and tidal turbines, *Proc. OCEANS, 2011, IEEE – SPAIN Conference*, Santander, Spain, June 6–9, 2011.

47. Galloway G. S., Catterson V. M., Love C., and Robb A. Anomaly detection techniques for the condition monitoring of tidal turbines, *Proc. Annual Conference of the Prognostics and Health Management (PHM) Society*, 12pp., 2014, Online. Available: https://pure.strath.ac.uk/portal/files/38273184/Galloway_etal_PHM2014_anomaly_detection_techniques.pdf.

48. *DeltaStream Tidal Energy Solution*, Tidal Energy Limited, Cardiff, Online. Available: http://www.tidalenergyltd.com/cms/wp-content/uploads/downloads/2012/10/DeltaStream_White_Paper_Aug12.pdf.

49. Sandra Royo Pérez. *Thermoelectric Energy Harvesting for Wireless Self Powered Condition Monitoring Nodes*, School of Engineering Department of Power and Propulsion, Cranfield University, May 2013, Online. Available: https://dspace.lib.cranfield.ac.uk/bitstream/1826/8049/1/Sandra_Royo_Perez_Thesis_2012.pdf.

50. TidalSense DEMO Report Summary *Demonstration of a Condition Monitoring System for Tidal Stream Generators*, September 2014, Online. Available: http://cordis.europa.eu/result/rcn/148019_en.html.

51. *TidalSense Specifications*, Online. Available: http://www.innotecuk.com/tidalsense/#tabs2.

52. SeaGen-S 2 MW, MCT Product Brochure, 8pp., July 2013, Online. Available: http://www.marineturbines.com/sites/default/files/FINAL_MCT_Product_Brochure_8pp_Seagen_UPDATE_E_HIRes.pdf.

53. Howieson D. The future of condition monitoring, *Evolution*, SKF, April 2014, Online. Available: http://evolution.skf.com/us/the-future-of-condition-monitoring/.

54. *Ocean Energy Solutions from SKF*, SKF PUB 74/S2 13634 EN• April 2013, Online. Available: http://www.skf.com/binary/21-242561/OceanCapabilities_13634-EN.pdf.

55. Denio H. III and Denio H. II. Aerial solar thermography and condition monitoring of photovoltaic systems, *Proc. InfraMation*, 2011, Online. Available: http://www.oregoninfrared.com/sites/default/files/Solar%20Infrared%20Paper.pdf.

56. *Practical Guide: Solar Panel Thermography*. Testo, Inc., Online. Available: http://www.murcal.com/pdf%20folder/15.testo_thermography_guide.pdf.

57. Photovoltaic Solar Panel Inspection, AirX 3 Visual Solutions, Online. Available: http://www.airx3.com/#!solar-panel-inspection/c7uc.

58. Lombardo Tom. *UAVs to Inspect Solar Farms*, Engineering.Com, May 2014, Online. Available: http://www.engineering.com/ElectronicsDesign/ElectronicsDesignArticles/ArticleID/7544/UAVs-to-Inspect-Solar-Farms.aspx.

59. Al-Dahoud A., Fezari M., and Belhouchet Fatma Zohra. Remote monitoring system using WSN for solar power panels, *Proc. First International Conference on Systems Informatics, Modelling and Simulation*, 15(5), 120–125, 2014, Online. Available: ijssst.info/Vol-15/No-5/data/5198a120.pdf.

60. Olita F. *Advanced Control and Condition Monitoring PV Systems*, Aaborg University Institute of Energy Technology, Denmark, June 2012, Online. Available: http://projekter.aau.dk/projekter/files/65386935/thesis.pdf.

61. Haq Irsyad Nashirul, *Line Condition Monitoring System of Solar Power using Web-based Wireless Sensor*, Institut Teknologi Bandung, Indonesia, January 2010, Online. Available: https://www.academia.edu/228911/on-line_condition_monitoring_system_of_solar_power_plant_using_web_based_wireless_sensor.

62. Sheng S. and Veers P. Wind Turbine Drivetrain Condition Monitoring—An Overview, *Presented at the Mechanical Failures Prevention Group: Applied Systems Health Management Conference 2011 Virginia Beach, Virginia*, May 10–12, NREL 2011.

19 Electrical Equipment Maintenance and Life Extension Techniques

As soon as new electrical equipment is installed in service, the aging process begins, with the rate depending on the application, environment, maintenance quality, and many other conditions that shorten the life of the equipment. Most of the electrical power equipment in use today have already deteriorated to different degrees; some remain in service for several decades. As soon as the performance of the electrical equipment begins to decline, knowledge about various life extension techniques for electrical equipment becomes very important. Aging electrical equipment are often retrofilled, retrofitted, or rebuilt, which means that either the parts or the whole unit is replaced with new ones. This is a valuable solution near the end of the equipment life or if the conditions are so bad that failure is practically unavoidable or if some parts are obsolete and there is no way to fix them. On the other hand, there is a continuous, affordable, and justified way of keeping the equipment in good and properly working conditions by providing a timely, thorough, and educated maintenance—the best way to extend the life of the equipment is starting as soon as is recommended by the manufacturer after installation. Different approaches to maintain electrical equipment as the means of extending its life in service are presented in this chapter.

19.1 MAINTENANCE STRATEGIES

The simplest maintenance strategy is based on the old philosophy: if it is not broken, do not fix it. It is called *corrective maintenance* (CM), which is based on restoring operation by fixing or replacing a component in case of failure, and it does not include any additional maintenance inspections. CM is usually the optimal solution at facilities where a failure of components has only a minor impact and it would be more expensive to monitor components' condition in service. In distribution systems, this kind of maintenance may be applied in the minor branches of an LV network.

Traditionally, electric utility maintenance strategies have been based on a fixed interval or number of operations; this is called *a time-based maintenance* (TBM), which even nowadays is quite a common maintenance policy used worldwide. Another term for this fixed interval maintenance is *periodic maintenance* (PM). Very often, TBM may be inefficient in controlling the lifetime of components. It is labor intensive and ineffective in identifying problems that develop between scheduled inspections. For those equipment in continuous operation but still available for planned periods, TBM is suitable, but not necessarily always. For example, a breaker used for capacitor or reactor switching is operated frequently and soon reaches thousands of cycles, in which case TBM is recommended. But other breakers may only operate once or twice per year. Moreover, TBM may not be cost-effective, which was the cause of the development of different approaches to keep electrical power system in a healthy condition.

Maintenance based on equipment condition and not on the number of operations or years in service provides better judgment on what is needed and when. As more and more techniques that monitor equipment's physical conditions became available, a new strategy was developed—*condition-based maintenance* (CBM)—a strategy that takes advantage of the information obtained through condition monitoring. In using this strategy, maintenance is based on the needs and priorities. CBM tasks are based on an evaluation of the equipment condition by performing periodic or continuous (online) equipment condition monitoring. Predictive maintenance (PdM) is another name for CBM.

The ultimate goal of CBM/PdM is to perform maintenance at a scheduled point in time when the maintenance activity will be most cost-effective and before the equipment loses performance or fails. This is in contrast to TBM, where a piece of equipment gets maintained whether it needs it or not. The "predictive" component of PdM stems from a goal of predicting the future trend of the equipment's condition. This approach uses the principles of statistical process control to determine at what time in the future it will be very appropriate to carry out maintenance activities. Most PdM inspections are performed while the equipment is in service, thereby minimizing disruption of the normal system operations. Adoption of PdM can result in substantial cost saving and a higher system reliability.

When the condition of the component is combined with the importance of the component from a functional point of view, another type of maintenance strategy is introduced—*reliability-centered maintenance* (RCM). RCM is not a single method, but allows comparison of different maintenance methods, of which the most cost-effective one can be chosen without compromising on reliability. RCM can also be seen as a process where the role of other maintenance strategies (CM, TBM, CBM) is optimized. For example, using PdM can enable better outage scheduling, flexibility of operation, a more efficient management of spare parts, and so on [1]. With RCM, the consequences of a failure are additionally evaluated. RCM is the right amount of maintenance for the right equipment at the right time. Unfortunately, sometimes, RCM is reduced to a fourth principle: no maintenance. This is the least desirable option and leads to costly disturbances in the long run.

The role of RCM at the system level is to balance the spending in maintenance activities with their effect on system performance. Usually, the aim is to maximize the result with regard to system reliability or outage cost reduction. In distribution networks, the RCM process includes identifying the critical system levels from the reliability point of view and selecting suitable strategies for different levels. For assessing the criticality of different components and system levels, standard reliability analysis methods are used. Some guidelines, with typical fault statistic data, are given in IEEE Standard 4931997, "IEEE Recommended Practice for the Design of Reliable Industrial and Commercial Power Systems (Gold Book)" [2]. New maintenance strategies may be developed based on finding an optimal balance between the maintenance cost and the ability to maintain sufficient reliability in the system [3].

19.2   MAINTENANCE AS A LIFE EXTENSION TECHNIQUE

19.2.1   Time-Based Maintenance

TBM (or PM) remains one of the most often used techniques to keep electrical equipment in proper working conditions. Original equipment manufacturer (OEM) user manuals usually have a special maintenance section where the intervals for maintenance actions are specified. For example, all leading switchgear manufacturers provide detailed maintenance guidelines for their switchgear. If switchgear is maintained according to these guidelines, it can be expected to last for over 40 years.

Unfortunately, the importance of maintenance is often underestimated until it is too late, that is, until a fatal or serious failure occurs. Some of the reasons for none or poor maintenance are a lack of product knowledge and insufficient funds. TBM is usually performed on critical system components, such as primary transformers and primary substation switchgear. For safety reasons, security codes usually require some inspections to be made on a regular basis, thus leading to the mandatory use of TBM, such as the inspection of earthings in primary and secondary substations.

Recommended maintenance intervals for the most important components are given in Ref. [1]. For example, the most frequent maintenance interval for transformers and CBs should be 1 year, minor overhaul is recommended to be done every 5 years, while major overhaul should be performed once every 7 years for transformers, and every 8–10 years for CBs. PM always includes a number of tests, which are specified in Ref. [4].

19.2.2 Maintenance of Power CBs

19.2.2.1 Molded Case Circuit Breakers (MCCBs)

MCCBs are designed to require little or no routine maintenance throughout their normal lifetime; however, the maintenance of MCCBs deserves special consideration because of their importance for routine switching and for the protection of other equipment. Electric transmission system breakups and equipment destruction can occur if an MCCB fails to operate because of a lack of periodic/preventive maintenance. The need for maintenance of CBs is often not obvious as CBs may remain idle, either open or closed, for long periods of time. Breakers that remain idle for 6 months or more should be made to open and close several times in succession to verify proper operation and remove any accumulation of dust or foreign material on moving parts and contacts.

MCCB should be exercised at least once per year. This manual exercise helps to keep the contacts clean, due to their wiping action, and ensures that the operating mechanism moves freely. This exercise, however, does not operate the mechanical linkages in the tripping mechanism. The only way to properly exercise the entire breaker operating and tripping mechanisms is to remove the breaker from service and test the overcurrent and short-circuit tripping capabilities. A stiff or sticky mechanism can cause an unintentional time delay in its operation under fault conditions.

19.2.2.2 Low-Voltage Circuit Breakers

LV CBs operating at 600 VAC and below should be inspected and maintained every 1–3 years, depending on their service and operating conditions. More frequent maintenance and inspection are necessary if LV CBs are exposed to high humidity and high ambient temperature, and a dusty, dirty, or corrosive atmosphere. Older equipment and the breakers that provide frequent switching or fault operations also require more frequent maintenance. A breaker should be inspected and maintained if necessary whenever it has interrupted current at or near its rated capacity [5].

The manufacturer's instructions for each CB that usually describe general procedures should be carefully followed in the maintenance of LV air CBs. The breakers should be carefully inspected. Insulating parts, including bushings, should be wiped clean of dust and smoke, the alignment and condition of the movable and stationary contacts should be checked and adjusted, and all other required procedures should be performed according to the OEM's manual.

19.2.2.3 Medium-Voltage Circuit Breakers

There are many varieties of MV CBs based on different technologies, such as using live tank technology with vacuum, air, oil, or SF_6 as the interrupting media, and dead tank technology with oil, SF_6, and so on. This is also true for HV CBs. MV CBs which operate in the range of 600–15,000 V should be inspected and maintained either annually or after every 2000 operations, whichever comes first. This maintenance schedule is recommended by the applicable standards to achieve the required performance from the breakers. For MV air CBs, all OEM's manuals provide the requirements for a proper maintenance. It includes cleaning of the insulating parts, including the bushings, checking the alignment and condition of movable and stationary contacts and their adjustment according to the manufacturer's data, and cleaning and lubricating the operating mechanism exactly as described in the instruction book. Worn parts should be replaced.

The maintenance of MV oil circuit breakers (OCBs) includes thoroughly cleaning the tank and other parts that have been in contact with the oil. An important part of the oil-filled MV breakers is the testing of the dielectric strength of the oil, which is defined as the maximum voltage that can be applied across the fluid without electrical breakdown (see ASTM D300-00). If it is <22 kV, then the oil should be filtered or replaced. The oil should also be filtered or replaced whenever a visual inspection shows an excessive amount of carbon, even if the dielectric strength is satisfactory.

During these procedures, it is important to be sure that the gaskets are undamaged and all nuts and valves are tightened properly to prevent oil leakage [5,6].

19.2.2.4   High-Voltage Circuit Breakers

Most manufacturers recommend external and internal inspections at intervals ranging from 6 to 12 months. However, with proper external checks, frequent internal inspections may be avoided. The inspection schedule should be based on the interrupting duty imposed on the breaker. It is recommended to make a complete internal inspection after the first severe fault interruption. If internal conditions are satisfactory, progressively more fault interruptions may be allowed before an internal inspection is made. Average experience indicates that up to five fault interruptions are allowed between inspections on HV CBs of voltage class 230 kV and above, and up to 10 fault interruptions are allowed on CBs rated under 230 kV. Normally, not more than 2 years should elapse between external inspections or 4 years between internal inspections [5].

An internal inspection should include all items applied to an external inspection, in addition to that the breaker tanks or contact heads should be opened and the contacts and interrupting parts should be inspected. There are several problems which should be looked into during internal HV breaker inspections [5], such as the tendency for carbon or sludge to form and accumulate in interrupters or on bushings. Among other issues is the tendency for interrupter parts or barriers to burn or erode. The bushing gaskets may leak moisture into breaker insulating material, among other issues [5].

19.2.2.5   SF$_6$ Gas CBs

Some of the maintenance issues are determined by the SF$_6$ gas effect on humans. Although, in pure state, SF$_6$ gas is nontoxic, it can exclude oxygen and cause suffocation. If the normal oxygen content of air is reduced from 21% to <13%, suffocation can occur without warning. Therefore, CB tanks should be purged out after opening them. Toxic decomposition products are formed when SF$_6$ gas is subjected to an electric arc. The decomposition products are metal fluorides and form a white or tan powder. Solid arc products, if they come into contact with the skin, may cause irritation or possible painful fluoride burn. Toxic gases are also formed, and they have the characteristic odor of rotten eggs. Therefore, maintenance personnel should not breathe the vapors remaining in a CB where arcing or corona discharges have occurred in the gas, and should thoroughly wash off all the solid SF$_6$ decomposition products. The faulted SF$_6$ gas from the CB should be evacuated and flushed with fresh air before one starts working on the CB [5].

19.2.3   Periodic Lubrication of the Power CB

Almost all power CBs require periodic renewal of lubrication in their operating mechanism. There are several factors determining the frequency at which lubrication should be renewed [7]: the continuous current rating of the CB, the number of close–open operations since the most recent renewal, the time elapsed since the most recent renewal, and the CB's operational environment. For LV power CBs, the minimum number of close–open operations that a breaker must be able to accomplish before requiring service is specified in IEEE Standard C37.16 [8]. The lubrication in a CB's mechanism needs to be renewed after the number of operations specified in that standard.

The OEMs are allowed to suggest a greater number of operations than the number given in the manufacturing standard [8], depending on the continuous current rating of the breaker. For example, although an 800 A-rated CB is required by the manufacturing standard to endure 500 operations before service is needed, a manufacturer's instruction book might indicate that an 800 A-rated breaker would require renewal of lubrication after 1750 operations.

If a CB operates only a few times each year, a 500- or 1750-operation count might never happen within the useful life of the breaker. A requirement for renew lubrication still stands because all lubrication materials deteriorate when exposed to air. This deterioration is accelerated by environmental conditions such as elevated air temperature or the presence of airborne contaminants (see Chapter 6).

Most users launch the programs to lubricate critical CBs based on an established time interval. Malfunctions have been attributed to a failure of lubrication for CBs having as little as 5 years of normal service. The lubrication points that typically require critical attention are usually specified in the OEM user manual and these requirements should be followed. Lubrication of the bearing surfaces of the HV CB operating mechanism should be performed according to the manufacturer's manual; however, excessive lubrication should be avoided as oily surfaces collect dust and grit and get stiff in cold weather, resulting in excessive friction.

There is a large variety of materials that are used to lubricate a CB's mechanism. More than one type of lubrication material might be used in the same mechanism at different specific points. Additionally, the material that is recommended for renewal of lubrication is sometimes not the same material that was installed at the factory. It is important to use the material that is specified in a CB's OEM manual to avoid incompatibility problems (see Chapter 6).

Current-carrying components also require periodic lubrication, and OEMs specify lubrication products and terms of lubrication in instruction manuals. Current-carrying components of a CB might include main contacts, primary-circuit finger clusters, and bus studs. Some current-carrying parts should never be lubricated, which is also specified in the manuals. Information on lubrication recommendations, including lubrication products and points and terms of lubrication of major manufacturers of electrical MV and LV equipment is summarized in Ref. [9].

Since many types of CBs do not operate often and lubricants tend to harden and deteriorate with time, one very important rule is established to keep the lubricant in proper working conditions—to *exercise* the mechanism as often as possible. It will prevent the CB's failure to open, which is mostly caused by dried lubricant.

19.2.4 Refurbishment or Reconditioning

For the breaker that is not functioning properly, the reconditioning procedure includes disassembling of the CB to its component level (racking mechanism, operating mechanism, frame, all electrical components, and current-carrying parts). All component parts should be inspected and tested in the beginning of this procedure. Cleaning should be thorough and part specific followed by lubrication of all parts as required by the OEM's user manual. Some specifics of cleaning and lubrication of current-carrying parts are described in Chapter 6 of this book. After assembly, the reconditioned CB should be inspected and tested again. An example is given in Ref. [10] of how an LV CB looks like before and after this procedure (Figure 19.1), a remarkable difference.

MV switchgear manufacturers have developed a number of alternative solutions to help customers find the best safety and life cycle management option for their equipment. Having more choices, each with different advantages and costs, enables customers to select the specific solution that best fits the overall business strategy of the plant.

One of these choices is refurbishment: the existing switchgear is fully overhauled and restored to "as new" condition. If one or more main elements of the switchgear are found worn, they are replaced with modern equivalents (retrofit). Elements with the highest maintenance cost and failure risk can be targeted specifically. A combination of these two solutions for an MV switchgear offers customers a life extension for their equipment at a low cost.

This option is attractive for customers with no imminent need to modernize, but wishing to maintain acceptable performance over a short- to medium-term life extension. Refurbishment of switchgear can take place in a number of ways: (a) refurbishment of the switching devices or (b) refurbishment of the complete panel. Switching devices such as CBs, contactors, and switches are typical candidates for refurbishment. The devices are removed from the switchgear, sometimes on a rotational basis in order to maintain continuity of supply, and returned to the manufacturer.

(a)

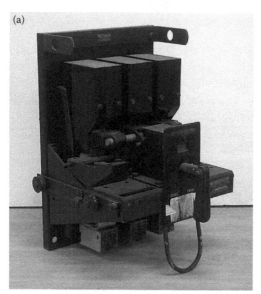

(b)

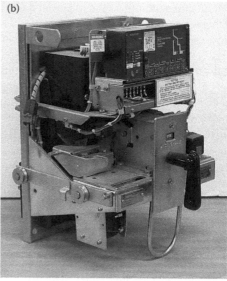

FIGURE 19.1   Westinghouse DB-25 CB before (a) and after reconditioning (b). (From Biggs J. *Circuit Breaker Life Extension*, Presented at Western Mining Electrical Association (WMEA) Meeting, Reno, NE, November 2004, Online. Available: http://www.wmea.net/Technical%20Papers/Circuit%20Breaker%20 Life%20Extension.pdf. With permission.)

For refurbishment, each breaker is completely disassembled to the smallest components. In general, bearings, cotter pins, and other selected components are totally replaced. All other parts are thoroughly checked for wear and damage and are replaced as needed. Each component is cleaned and replated if required. Product enhancements that have developed since the original manufacture are implemented to bring the breaker up to the latest production standards. When such a refurbished breaker leaves the factory, it has a new lease of life and a full new warranty [11].

19.2.5   Condition-Based Maintenance

The need for performing CBM using proper condition monitoring tools was revealed as early as in the 1960s by a study conducted for the development of a preventive maintenance program for Boeing 747. The study's purpose was to determine the failure characteristics of aircraft components [12]. The study was, at the request of the DoD (USA), documented and published in 1978 [13]. It was found that a relatively small part of all components (11%) had clear aging characteristics which would enable a scheduled overhaul, that is, predetermined preventive maintenance. The rest of the components (89%) did not show such aging characteristics, which led to a conclusion that they were random failures which are, thus, not applicable to scheduled overhauls.

Similar conditional probability curves within the manufacturing industry have been determined showing that only 30% of all components have clear aging characteristics, and that this percentage decreases as complexity and technology increase [14].

Evidently, when planning appropriate maintenance schedules, the aging feature of a component is not the best approach, and in some applications not even possible, thereby introducing CBM and condition monitoring as a reasonable solution to the issue [15]. CBM is defined as preventive maintenance based on performance and/or parameter monitoring and the subsequent actions. Therefore, CBM is a maintenance approach that utilizes condition monitoring in order to monitor the conditions over time and then uses its results in order to help to decide whether maintenance actions are required.

According to Ref. [16], CBM tasks are performed to determine whether a problem exists in the monitored item, how serious the problem is, and how long the item can be run before failure. CBM tasks also serve to detect and identify specific components in the items that are degrading, and diagnose the problem.

19.2.5.1   HV Switchgear

The development of novel comprehensive diagnostics for HV equipment became an integrated part of condition maintenance concepts. A combination of maintenance and measurement activities is covered by the application of CBM technology. One approach to HV switchgear CBM is used for reducing the maintenance costs without worsening the technical condition of the switchgear. It is achieved by focusing the maintenance and measurement activities on its critical components alone.

The benefits of the CBM approach increase considerably for significantly aged switchgear with a relatively high failure frequency, requiring high current maintenance costs and a long outage time for maintenance [12]. There are more indirect benefits of the CBM approach such as the use of the data on measured physical conditions as evidence in case of liability and extending the equipment lifetime by applying trend analyses of condition data.

Development of the CBM strategy is specific for each type of HV electrical equipment, because in each case, different aging processes are involved in the aging of switchgear and therefore in the shortening of lifetime. In Ref. [16], such a CBM strategy was developed for HV switchgear for which the average age of CBs was about 35 years. An aging model of CB components is required in order to determine which components are of great importance to be maintained. As a part of the program, an attempt was made to determine which factors affect the condition of insulation material in HV CBs [17]. An insulation material called "Coqolite" was applied to a large number of this 35-year-old switchgear in the Netherlands and is defined as a critical component because of the replacement cost and the consequences for environmental safety in case of failure. It was found that the moisture level in oil influences the aging process of Coqolite, which requires a more frequent monitoring of moisture level in oil. The location of the substation also aggravates the aging of Coqolite because of the diversity of groundwater in different regions in the Netherlands. Paying extra attention to the moisture level of the oil of CBs in areas with high groundwater level should help extend the technical lifetime of critical components such as Coqolite. Reconditioning the oil above a moisture level of 11 ppm in these areas can be looked at as a precautionary measure for slowing down the aging process of the insulation.

A general review of aging parameters and ways to control them can enable to reduce maintenance activities, which lead to in a decrease of total outage time and consequently to a cost reduction per switchgear bay. Further cost reduction may be achieved through analysis of the stored condition data and optimization of the maintenance and monitoring program. On the other hand, extra costs are induced by the implementation of CBM, the use of software applications, increased engineering activities, and additional measurement equipment. However, the costs per switchgear may significantly decrease if the total number of switchgear involved in the CBM program increases [12].

19.2.5.2   MV Switchgear

An application of CBM to the MV switchgear with the use of online temperature monitoring is described in Refs. [18,19]. A wireless temperature monitoring system was installed on two power plants to monitor the temperature of electrical contacts of aged equipment, which suffered multiple violent thermal failures in the past. The temperature data were continuously collected in the database and analyzed together with load data to determine any abnormalities in temperature behavior.

There could be two types of possible uses of the continuous temperature monitoring of electrical contacts. In one scenario for some unspecified reason, the temperature of the contact in good condition may suddenly increase above a predetermined level and the system issues an alarm requiring an immediate maintenance action to prevent thermal failure. The probability of such an event occurring is relatively low. On the other hand, a continuous observation of the temperature and analysis of the trends in temperature over a long period of time may allow to develop a reliable diagnostic tool for the contacts' physical condition, which will justify a scheduling of the maintenance of the specific unit.

Almost 4 years after temperature monitoring began, the temperature on the power contact exceeded 100°C. Since the monitoring system's high-temperature limit was preset at 95°C, this event triggered an alarm informing plant maintenance personnel that immediate measures should be taken. The breaker was removed from service, and the contact of the breaker were cleaned and maintained. The breaker was returned to service after proper testing.

At another power plant, a wireless temperature monitoring system (WTMS) was installed in conjunction with a power monitoring system. This utility experienced multiple overheating problems on main breakers in the past, which resulted in violent thermal failures. Temperature monitoring sensors have been installed on two main breakers, on the surface of the insulation as close as possible to the bolted connections under insulation. After some strong changes in thermal behavior have been recorded, load conditions for the year of monitoring were analyzed. It was found that the power monitor recorded a severe overload of the breakers (for up to 20%) during at least 5 months during the previous summer and fall.

It was concluded that continuous overload of the breakers could be the major cause of irreversible damage done to the contacts' physical conditions. After being informed about observations and warned that further continuous overload may cause serious damage of the breakers, plant personnel did not load either breaker above the maximum rated current [19]. In this case, online continuous monitoring of the temperature of current-carrying parts allowed to avoid thermal failure of the breakers, saving significant assets, and provided justified maintenance activity, which is the meaning and goal of CBM.

19.3   CBM METHODOLOGY AND LIFE MANAGEMENT

19.3.1   DISTRIBUTION POWER TRANSFORMERS

With the restructuring of the electricity sector into a profit-oriented business entity, CBM strategies are focused on cost-effective solutions through effective asset management. For power transformers, an effective CBM program is crucial as more transformers have reached their service life. Diagnostic techniques are able to reveal a degradation state for effectively incorporating a transformer life management program. To provide CBM of distribution transformers, it was suggested to determine the so-called transformer health index (THI) [20]. The CBM program requires a proper selection of the diagnostic techniques, which are divided into three practical levels or tiers, subject to the availability of instruments and financial limitations.

19.3.1.1   Level 1 Diagnostic Techniques

According to the CBM program, Level 1 diagnostic techniques are applied on energized units to assess the presence of faults, quality of the insulating oil, degradation level of insulating paper as well as physical, thermal, and operating performance of the transformers. These tests include *dissolved gas analysis (DGA)* in the transformer main tank, with the results analyzed for key gases and total dissolved combustible gas (TDCG) concentrations as well as key gases and TDCG rate of generation in accordance with IEEE Standard C57.104 [21] and IEC 60599 [22] limits and interpretation.

According to *oil quality analysis (OQA)*, oil samples are tested for breakdown voltage, water content, acidity, and power factor. These tests are the basic routine tests for mineral insulating oil in accordance with IEC 60422:2005 [23], and are sufficient to indicate the condition of the insulating oil. Since the life of the cellulosic material is directly related to the life of the transformer, analysis of the furanic compound in oil or *furfural analysis (FFA)* should also be performed. By measuring the quantity and types of furans present in a transformer oil sample, the insulation's overall degree of polymerization (DP) and remaining life estimation can also be done with a high degree of confidence.

Additional diagnostic techniques include inspection of transformer *physical conditions and operating performance*, which includes mechanical damage, corrosion, and oil leaks. This also includes a review of the transformer maintenance and performance historical records (on-load tap changer (OLTC) maintenance records and records of transformer internal and external tripping). *Infrared thermography* is also a part of the diagnostic as a technique indicating the thermal problems caused by overheating of conductors or deteriorated contacts. Infrared scanning is applied on the external components such as tanks, bushings, radiators, and cooling systems.

19.3.1.2 Level 2 Diagnostic Techniques

If the techniques included in Level 1 could not definitively classify some transformers as normal, then Level 2 diagnostic techniques should be used:

Transformer turns ratio measurement measured at every tap position of each phase against calculated nameplate values.

Winding resistance measurement (DC test) performed at every tap position to detect broken conductor strands, loose connections, and bad contacts in the tap changer

Dielectric dissipation factor/tan delta measurement characterizing the amount of dielectric power loss which will produce heat in the insulation during transformer operation that can cause deterioration of the insulation.

Excitation current measurement which detects short-circuited turns, core and winding problems, and poor tap changer contacts [24].

Insulation resistance (IR) and polarization index (PI) determining the presence or absence of harmful insulation contamination, degradation, and failure

19.3.1.3 Level 3 Diagnostic Techniques

Finally, if both sets of diagnostic tests still could not definitively classify some transformers as normal, Level 3 diagnostic techniques are applied:

Frequency response analysis (FRA) is used to detect faults such as short-circuited turns, movement, mechanical deformation, or displacement to windings or core and loose turns.

Partial discharge (PD) measurement is used to detect the presence of PD activity, which is commonly related to moisture in the insulation, cavities in solid insulation, metallic particles, and gas bubbles generated due to some fault condition. A significant increase in PD level can provide an early indication of a failure.

19.3.1.4 Transformer Health Index

To capture and quantify the results of multiple diagnostic tests, the concept of THI was formulated in Ref. [25], with the value of THI ranging from 1 to 100. It was used further to provide a qualitative indication of the condition of individual transformer accounting for the transformer age. Then each THI is ranked based on the Transformer Condition-Based Ranking to determine the next recommended maintenance actions. Out of 707 transformers, 49% have been found in "good" conditions ($85 \leq THI \leq 100$), 17% in "fair" condition ($55 \leq THI < 85$), 29% in "poor" condition ($10 \leq THI < 55$), and 5% in "very poor condition" ($THI < 10$).

It was found that a number of in-service transformers aged 5 years and below were diagnosed as in "poor" conditions, with the oil having a considerably high concentration of gases that relate to overheating. Most common problems found for transformers aged above 30 years were paper degradation and deterioration in physical conditions. Based on the problems found from the assessment, the corrective actions were recommended and carried out.

Based on the values of THI, the CBM actions have been planned accordingly to the transformers found in conditions ranging from "fair" to "very poor." Thirty transformers were selected for replacement or refurbishment, and 138 transformers have been scheduled for regular monitoring of gas concentration and content of furans. Additional monitoring of thermal and loading conditions has been specified for 55 transformers, and 133 transformers should have the oil reconditioned.

This is a very valuable example of developing the CBM concept, which provided a specific approach to the assessment of electrical equipment conditions and allowed taking maintenance actions where and when they were needed the most.

19.3.2 Power Cable Systems

Power cables are the critical component of an electrical distribution system, and they must operate as long as possible according to reliability and safety standards. A vital part of the maintenance program is proper diagnostic testing to identify defects that result in a system failure. It helps to predict when these defects may induce failure in the cable system. The test should be economically justified within the CBM concept and should not cause additional degradation to the system under test (see Chapter 8).

19.3.2.1 Cable Deterioration Diagnostics

Several diagnostic methods are available today that provide early identification of weak components of the cable system while the system remains energized. These methods can locate the degraded components of the system and determine the degree of degradation. This is essential for maintaining system reliability. As a part of CBM, cost saving is achieved by prioritizing the replacement only of the weak sections of the cable system and rehabilitating the aged, extruded cables. The degradation phenomena of power cables (see Chapter 8) are associated with PD and moisture ingress, which lead to thermal runaway failures. PDs are usually present at advanced stages of insulation degradation.

The majority of cable failures in an extruded cable system are related to water treeing, which leads to failure when they convert to electrical trees [25]. Once a water tree is converted to an electrical tree, the time to failure normally is very short, because the initiated electrical tree propagates rapidly through the already weakened dielectric material. Electrical trees can also be formed as a result of PDs in large voids in cables, terminations, and joints.

Diagnostic technology called CableWISE [26] can detect, locate, and assess cable degradation. In this technique, radio frequency (RF) sensors are placed over the exterior of an energized shielded cable. RF signals emitted by the cable system while in service are detected, amplified, and then recorded for being analyzed later. Data analysis allows to locate the deterioration, which emits RF signals, and to determine the significance of deterioration. This type of deterioration diagnostics is beneficial compared with other diagnostic techniques (DC HIPOT, very low frequency [VLF]) that apply overvoltage stresses to the cable during the test, with the possibility of inducing failure during the test or causing premature failure at a later time (Section 17.2).

The assessed condition of each cable component also depends on the type of cable insulation, influence of installation conditions, and previous condition assessments. The diagnostic technique provides the utilities with the guidance on managing assets properly and making educated and justified decisions on maintenance actions.

19.3.2.2 CBM and Life Extension of Power Cables

After the deterioration sites are located by diagnostic techniques in a power cable system, CBM actions may include various methods to improve cable physical conditions with a focus on the weak sections of the cable. Technologies that can be used to extend the life of extruded cables have been in use since the mid-1980s [25]. Initially, nitrogen was flowed through the strands to dry the cable core and keep the water out, but it was expensive and high maintenance was required to continuously feed the nitrogen to the cable. Research findings [25] showed that a by-product of crosslinking process, acetophenone, hinders water tree formation in cross-linked polyethylene (XLPE) insulation. Acetophenone was then used for some time in the field instead of nitrogen, but this technique for extending the life was also discontinued in the field due to its several drawbacks.

In the late 1980s, a new fluid based on silicone technology was developed and has been in use in the field effectively since then. This life enhancement fluid offers superior speed of treatment, increased cable life under harsh conditions of elevated temperature, and many other advantages [27].

Additional research conducted over several years has led to the development of a new fluid called CableCURE especially suited for larger, hotter running cables such as feeder cables. CableCURE is made up of 100% water-reactive cable treatment fluids. Via molecular interaction, CableCURE eliminates water from the void as well as water diffusing through the cable. When the fluid comes in contact with water, it increases in molecular weight by polymerization, filling the cavities containing the water tree with a more viscous water-resistant fluid, thus preventing further tree growth. The main benefit of CableCURE cable rejuvenation is that it significantly extends the economically useful lifespan of an underground power distribution cable, deferring the cost of cable replacement by decades. A significant improvement in cable system reliability is achieved, thus leading to fewer, if any, outages due to dielectric breakdown.

19.4 MAINTENANCE OF ELECTRICAL EQUIPMENT EXPOSED TO CORROSION AND WATER

19.4.1 Water-Damaged Electrical Equipment

Guidelines on how to handle electrical equipment that has been exposed to water through flooding, firefighting activities, and hurricanes are outlined in the National Electrical Manufacturers Association (NEMA) document [28] and in the recommendations of manufacturers and safety committees [29–31].

Electrical equipment exposed to water can be extremely dangerous if reenergized without proper reconditioning or replacement. Reductions in integrity of electrical insulation due to moisture, debris lodged in the equipment components, and other factors can damage electrical equipment by affecting the ability of the equipment to perform its intended function. Damage to electrical equipment can also result from flood waters contaminated with chemicals, sewage, oil, and other debris that will affect the integrity and performance of the equipment. Ocean water and salt can be particularly damaging due to the corrosive and conductive nature of the saltwater residue.

Switches and LV protective components such as MCCB and fuses, within assemblies such as enclosures, panel boards, and switchboards are critical to the safe operation of distribution circuits. By exposure to water and to the minerals and particles which may be present in the water, the ability of the components to protect these circuits is adversely affected. Such exposure can affect the overall operation of the mechanism through corrosion, the presence of foreign particles, and removal of lubricants. The condition of the contacts can be affected and the dielectric insulation capabilities of internal materials can be reduced. Exposure to water can cause corrosion and insulation damage to support structures, busways, wiring, electromechanical or electronic relays, and meters.

It is just a part of all possible damages, which could affect the electrical distribution equipment exposed to water, including corrosion, loss of lubrication, and deterioration of insulation quality. According to Ref. [28], most of water-damaged equipment, such as electronically controlled and solid state contactors and starters, components containing semiconductors and transistors, overload relays, MCCB and switches, and fuses, should be completely replaced. Some of manual and magnetic motor controllers and motor control centers (MCCs) may possibly be reconditioned.

In the case of LV *and MV CB and switches*, the operation of the mechanism can be impaired by corrosion, by the presence of particles such as silt (fine sand, clay, or other material carried by running water and deposited as a sediment) and by the removal of lubricants. The dielectric properties of insulation materials and insulators will degrade. For air CB, the condition of the contacts can be affected. Electronic trip units in LV power CB as well as the functionality of electronic protective relays and meters will be impaired.

It is possible that water-damaged LV and MV breakers could be maintained and returned to service by, for example, replacing contacts in air CB. However, the refurbishing of water-damaged power equipment, including cleaning, drying, and lubrication should be performed in close consultation with the manufacturer. However, the electronic trip units of LV power CB, and electronic protective relays and meters in any power equipment should be discarded and replaced, or at least returned to the manufacturer for inspection and possible refurbishment.

Exposure of *transformers* to water can cause corrosion and insulation damage to the transformer core and winding. The ability of the transformer to perform its intended function in a safe manner can also be impaired by debris and chemicals which may be deposited inside the transformer during a flood. Water and contaminates also can damage transformer fluids. All dry-type transformers regardless of kVA ratings and all dry-type control circuit transformers should be completely replaced. Some liquid-filled and cast-resin transformers may possibly be reconditioned by trained personnel in consultation with manufacturer.

When any *wire or cable product* is exposed to water, all metallic components (such as the conductor, metallic shield, or armor) are subject to corrosion that can damage the component itself and/or cause termination failures. If water remains in MV cable, it could accelerate insulation deterioration, causing premature failure. Wire and cable listed for only dry locations may become a shock hazard, when energized, after being exposed to water. Any wire or cable listed for dry locations only, such as type NM-B cable, as well as any cable that contains fillers, such as polypropylene, paper, and so on should be replaced if it has been exposed to water.

The guidelines [28] also are given for handling of other water-damaged electrical equipment, such as wiring devices, ground fault circuit interrupters (GFCIs), surge protectors, motors, and so on.

19.4.2 Electrical Equipment in Nuclear Industry

The U.S. Department of Energy (DOE) and Electric Power Research Institute (EPRI), in cooperation with nuclear power plant utilities and the Nuclear Energy Institute (NEI, formerly NUMARC, the Nuclear Management and Resources Council) have prepared guidelines for aging evaluations of nuclear power plant equipment for life extension considerations, including electrical equipment [32].

Aging Management Guidelines (AMGs) provide guidance on performing detailed evaluations of aging mechanisms and aging management strategies applicable to critical equipment groups. AMGs have been published for different groups of power equipment, including electrical switchgear [33]; transformers; power distribution equipment [34]; MCC [35]; and other equipment such as battery chargers, stationary batteries, inverters, and uninterruptible power supplies (UPS), and HV, MV, and LV cables.

In these documents, various stressors and aging mechanisms are discussed as well as failure modes. Typical stressors in nuclear plants environment are high or low temperature, radiation, humidity, and contaminants, and these factors are electrical, mechanical, chemical, thermal, or electrochemical in nature. Stressors acting over time produce aging mechanisms that ultimately can cause component degradation. Typical aging mechanisms for electrical equipment include wear, lubricant deterioration, corrosion, and so on. Operational demands, environmental conditions, failure data, and industry operations and maintenance history should always be considered to determine the significance of the electrical equipment aging mechanisms.

Effective management of aging mechanisms is discussed in these guidelines, including common maintenance, inspection, testing and surveillance techniques, or programs. An overview of the technical issues and recommended action to resolve these issues identified in AMGs for various types of electrical equipment are shown in Table 19.1. These technical issues may be of concern and the guidelines identify potential resolution strategies. The relatively small number of identified issues indicates a generally effective set of industry maintenance programs for the electrical components evaluated.

The guidelines for electrical switchgear define commonly used aging management actions: periodic visual inspection, measurement of various component properties (such as electrical resistance or contact tolerance), adjustment, operability and functionality checks, lubrication, and component

TABLE 19.1

AMG and Technical Issues for Electrical Equipment

Type of Equipment	Technical Issues	Resolution	AMG
Electrical switchgear	Overcurrent trip devices, lubrication practices, switchgear in "harsh" environments, and/or subject to a high number of operating cycles may warrant additional attention	Review and evaluate adequacy of existing maintenance and surveillance programs and alter if necessary	SAND93-7027
Power and distribution transformers	Insulation, insulating fluid, bushing contamination, transformers in "harsh" environments, and/or subject to a high number of operating cycles may warrant additional attention	Review and evaluate adequacy of existing maintenance and surveillance programs and alter if necessary	SAND93-7068
Motor control centers	MCCB, MCCs in "harsh" environments, and/or subject to a high number of operating cycles may warrant additional attention	Review and evaluate adequacy of existing maintenance and surveillance programs and alter if necessary; perform root cause analyses	SAND93-7069
Battery chargers, inverters, and uninterruptible power supplies	Inverter and UPS inspection frequency may need to be increased	Increase inspection frequency of inverters and UPS to match that of chargers	SAND93-7046
Stationary batteries	Seismic vulnerability of aged 1E lead–acid batteries may be a concern	Disassembly inspection of artificially aged and actual sample cells, seismic testing	SAND93-7071

Source: Adapted from *Aging Management Guideline for Commercial Nuclear Power Plants—Motor Control Centers,* Sandia National Laboratories Report SAND93-7069, February 1994.

replacement. Less commonly used aging management programs include infrared thermography, acoustic vibration analysis, and component refurbishment.

Evaluation results demonstrated that existing nuclear plant maintenance activities and programs are largely effective at managing the aging of switchgear components. In a few instances, additional actions may be taken to improve the effectiveness of the maintenance programs. For example, since overcurrent trip devices and other electrical devices caused a high percentage of the failures, an added consideration may be warranted for these units. Inadequate or degraded lubricant appears to have a strong correlation to instances of switchgear failure. Overhaul refurbishment may be required to effectively address lubricant degradation.

Switchgear components may require additional attention if they (1) are exposed to accident temperature, steam, or radiation conditions; (2) are exposed to temperatures in excess of 40°C (104°F) for extended periods; (3) are subject to a high number of operating cycles in relation to the manufacturer's design cyclic rating; or (4) are subject to environments containing high levels of dust or contaminants.

For power transformers commonly used, aging management actions include visual inspection, periodic observation of temperatures and levels (where applicable), and measurement of component properties (such as dielectric strength, oil viscosity, etc.). Degradation of transformer solid insulation and insulating fluid, as well as contamination of bushing external surfaces, appear to be closely related to the more damaging types of failures experienced by power and distribution transformers, and hence may warrant added consideration.

For MCCB, external operating handle mechanisms and starter contactor auxiliary contact mechanisms caused comparatively large fractions of the total number of MCC component failures, and therefore may require additional consideration.

REFERENCES

1. The present status of maintenance strategies and the impact of maintenance on reliability, Report of the IEEE/PES task force on impact of maintenance strategy on reliability, reliability, risk and probability applications subcommittee, *IEEE Transactions on Power Systems*, 16(4), 638–646, November 2001.
2. IEEE Gold Book™ Standard 493™-2007, *IEEE Recommended Practice for the Design of Reliable Industrial and Commercial Power Systems*, IEEE Publications, New York, 383pp., 2007.
3. Lehtonen M. On the optimal strategies of condition monitoring and maintenance allocation in distribution systems, *9th International Conference on Probabilistic Methods Applied to Power Systems (PMAPS)*, Stockholm, Sweden, June 11–15, pp. 1–5, 2006.
4. American National Standard for Maintenance Testing Specifications for Electrical Power Equipment and Systems, ANSI/NETA MTS-2011, 272 pp., 2011, Online. Available: http://www.iemworldwide.com/pdf/ansi-neta-mts-2011.pdf.
5. *Maintenance of Power Circuit Breakers, Facilities Instructions, Standards and Techniques*, United States Department of the Interior Bureau of Reclamation, V. 3–16.
6. Brikci F. and Nasrallah E. Maintenance of MV & HV power circuit breakers, *Electric Energy Magazine*, Online. Available: http://www.electricenergyonline.com/?page=show_article&mag=35&article=276.
7. Sprague M. J. Service-life evaluations of low-voltage power circuit breakers and molded-case circuit breakers, *IEEE Transactions on Industry Applications*, 37(1), pp. 145–152, January/February 2001.
8. IEEE C37.16 *Standard for Prefer Redratings, Related Requirements, and Application Recommendations for Low-Voltage AC (635 V and below) and DC (3200 V and below) Power Circuit Breakers*, IEEE Publications, New York, pp. C1–26, 2009.
9. *Square D® Services Lubrication Manual, Lubrication Instructions for Major Manufacturers' Low Voltage and Medium Voltage Circuit Breakers, Switches, and Switchgear*, 6th Edition, pp. 1–84, SqD Data Bulletin 0100DB0802, 2008.
10. Biggs J. *Circuit Breaker Life Extension*, Presented at Western Mining Electrical Association (WMEA) Meeting, Reno, NE, November 2004, Online. Available: http://www.wmea.net/Technical%20Papers/Circuit%20Breaker%20Life%20Extension.pdf.
11. New life for old switchgear, in *Power Services, Special Report*, ABB Review, ABB Ltd Publishing, Zurich, Switzerland, pp. 30–34, September 2004.
12. Overman R. and Collard R. The complimemtary roles of reliability-centered maintenance and condition monitoring, *Proceedings of the 18th International Maintenance Conference (IMC-2003)*, Clearwater Beach, FL, pp. 1–6, December 2003.
13. Nowlan F. S. and Heap H. F. *Reliability-Centered Maintenance*. Report Number AD-A066579, United States Department of Defense, 520pp., December 1978.
14. Page R. Maintenance management and delay reduction. *Maint. Asset Manage.* 17(1), 5–13, 2002.
15. Bengtsson M. Supporting implementation of condition based maintenance: Highlighting the interplay between technical constituents and human and organizational factors, *Int. J. Technol. Hum. Interact.* 4(1), 48–74, January/March 2008.
16. Mobley R. Predictive maintenance techniques, In *An Introduction to Predictive Maintenance*. Chapter 6, Butterworth-Heinemann/Elsevier Science, Amsterdam, Holland, pp. 99–113, 2002.
17. Groot E. R. S., Gulski E., van Dam A., and Wester F. J. Successful implemented condition based maintenance concept for switchgear, *CIRED 17th International Conference on Electricity Distribution*, Barcelona, Spain, Paper 14, pp. 1–7, May 12–15, 2003, Online. Available: www.cired.be/CIRED03/reports/R%201-14.pdf.
18. Livshitz A., Chudnovsky B. H., Bukengolts B., and Chudnovsky B. A. On-Line temperature monitoring of power distribution equipment, *Proc. of IEEE PCIC Conference*, Denver, CO, pp. 223–231, September 2005.
19. Chudnovsky B. H. Electrical contacts condition diagnostics based on wireless temperature monitoring of energized equipment, *Proceedings of the Fifty-Second IEEE Holm Conference on Electrical Contacts*, Montreal, Canada, pp. 73–80, September 2006.
20. Yang G., Young Z., Talib M. A., and Ahmad R. H. TNB experience in condition assessment and life management of distribution power transformers, *CIRED 20th International Conference on Electricity Distribution*, Prague, 8–11, June 2009.
21. IEEE Standard C57.104-2008. *IEEE Guide for the Interpretation of Gases Generated in Oil-Immersed Transformers*, IEEE Publications, New York, pp. C1–C27, 2009.

22. IEC Standard 60599. *Mineral Oil-Impregnated Electrical Equipment in Service—Guide to the Interpretation of Dissolved and Free Gases Analysis*, Ed. 2.1 B, Consolidated Edition, Publication of International Electrotechnical Commission (IEC), Geneva, Switzerland, 69pp., May 2007.

23. IEC Standard 60422. *Mineral Insulating Oils in Electrical Equipment—Supervision and Maintenance Guidance*, Ed. 3.0, Publication of International Electrotechnical Commission (IEC), Geneva, Switzerland, 85pp., October 2005.

24. Horning M., Kelly J., Myers S., and Stebbins R. *Transformer Maintenance Guide*, 3rd Ed., Transformer Maintenance Institute, published by S.D Myers, Inc., Tallmadge, OH, 437pp., 2004.

25. *Estimation of Life Expectancy of Polyethylene-Insulated Cables, Final Report*, EPRI Report EL 3154, January 1984.

26. Srinivas N. N. Condition based maintenance strategy tools for power cable system, *CIRED 20th International Conference on Electricity Distribution*, Prague, pp. 8–11, June 2009.

27. Stagi W. and Chatterton W. Cable rejuvenation-past present and future, *Proceedings of Jicable per C7.2.14*, Versailles, France, pp. 858–861, 2007.

28. *Guidelines for Handling Water-damaged Electrical Equipment*, Publication of National Electrical Manufacturers Association (NEMA), Rosslyn, VA, 2005.

29. *Guidelines for Handling Water-damaged Electrical Equipment*, Publication of National Electrical Manufacturers Association (NEMA), Rosslyn, VA, pp. 1–6, 2006.

30. *Water Damaged Electrical Distribution and Control Equipment, Data Bulletin* 0110DB0401R07/11, Schneider Electric, 2011, Online. Available: http://www2.schneider-electric.com/resources/sites/SCHNEIDER_ELECTRIC/content/live/FAQS/177000/FA177865/en_US/Water%20Damage%20 0110DB0401.pdf.

31. *Recovery of Water Damaged Electrical Equipment, Atomic Power Review*, Online. Available: http://www.atomicpowerreview.blogspot.com/2011/03/recovery-of-water-damaged—electrical.html.

32. Nakos J. T. and Rosinski S. T. *Research on U.S. Nuclear Power Plant Major Equipment Aging*, SAND94-2400C, U.S. DOE, Online. Available: http://www.osti.gov/bridge/servlets/purl/10190825-J7vdeV/webviewable/.

33. *Aging Management Guideline for Commercial Nuclear Power Plants—Electrical Switchgear*, Sandia National Laboratories Report SAND93-7027, July 1993.

34. *Aging Management Guideline for Commercial Nuclear Power Plants—Power and Distribution Transformers*, Sandia National Laboratories Report SAND93-7068, May 1994.

35. *Aging Management Guideline for Commercial Nuclear Power Plants—Motor Control Centers*, Sandia National Laboratories Report SAND93-7069, February 1994.

20 Life Extension Techniques, Inspection and Maintenance of Renewable Energy Equipment

20.1 EXTENDING LIFE OF WIND, TIDAL, AND WAVE TURBINES

Wind turbines (WTs) are typically designed for a 20-year life (see Chapter 16). However, the design life is a theoretical parameter and is also called the planned service life. The load-bearing capacity of the structure and suitability for use must be demonstrated for this period. At the end of the 20-year period, there is generally an interest in continuing to operate the WT beyond this period. From the other hand, extended life of wine turbines also lead to higher O&M costs and a greater risk of costly structural failures and increased safety risks.

Technical aspects of WT extended life are defined by the condition that further operation can only be tolerated if safe operation can be guaranteed during the extended service life period. Suitable procedures and guidelines are currently being developed to enable the continued operation of WTs which have reached or will soon reach their 20-year service life. In the longer term, these procedures will naturally also be available for WTs which only reach the end of their 20-year service life in a few years or further into the future [1]. The alternative is to decommission the turbine at the end of its design life and nonexistence of additional financial returns.

Life extension may be considered as a natural consequence of the service conditions. There are three wind load zones or classes according to the wind class under Standard DIN IEC 61400 or the wind load zone under the Deutsches Institut für BauTechnik (DiBt) Standard guidelines. According to DIN IEC 61400, in high wind (zone 3), wind speed is 27.5 m/s; medium wind (zone 2), it is 25 m/s; and low wind (zone 1), it is 22.5 m/s and less. If a WT at a location is actually exposed to lower average wind speeds or lower turbulence or has lower availability, then the components are subjected to correspondingly lower operating loads. These components will not be fatigued and further operation of the WT will be possible from a technical standpoint. It was estimated that if actual average wind speed at the location was within wind load zone 2, then potential extension after 20 years of operation can be from 3 to 6 years. If actual average wind speed at the location was within wind load zone 1, then potential extension after 20 years of operation can be more than 10 years [1].

There are many other factors that would shorten or extend actual service life of WT, including wind turbulence, WT technical availability, and so on.

Lifespan of tidal and wave turbines is also limited by the survivability of its components, such as turbine blades and gearboxes. Many of these components do not have a predicted lifespan longer than 20 years. Therefore, the search for the means of extending operational life of tidal and wave turbines and their components is also a subject of many efforts.

20.1.1 EXTENSION OF WT LIFE: MODELING APPROACHES

There could be various approaches to life extension of WTs, one of which is called a risk-based approach to life extension, which according to Ref. [2] can limit exposure to catastrophic failures and provide attractive financial returns. Under sponsorship by EPRI, DNV KEMA has developed cost models for three life extension methods and compared the internal rate of return (IRR) against a base case of a 20-year life.

The three approaches targeted were inspections, modified operations, and advanced controls. The modeling approaches considered the costs and revenue over the life of the turbines, including the cost of preventing or predicting structural fatigue failure and the cost of normal component wear out. The turbines were evaluated for extended lifespans of 22–35 years, depending on the approach.

The most attractive approach to life extension in this study was suggestion to use inspections to make decision about when and what to repair, replace, or decommission. In the targeted inspections approach, structural components are regularly inspected for cracks to predict structural failure. Inspections cost about $6300 annually per turbine, plus the cost of downtime. The model assumes decommissioning for turbines where cracks were present in the tower, foundation, mainframe, or hub, but replacement of cracked blades.

Another way to extend the life of WT is the modified operations approach, which can be used to reduce loads by derating or curtailing during high loading events. According to Ref. [2], the benefits were modeled as reduction in design driving fatigue loads from 5% to 9% at the expense of 1%–3% annual energy production loss, suggesting that operations are modified over the entire life of the turbines.

The third way of life extension modelled in Ref. [2] is an advanced controls approach, which can also reduce turbine loads over those experienced under nominal or older control schemes. There are two control options: lidar-based control and state estimation, a purely algorithm-based control approach. Lidar-based control requires a capital investment for the lidar device (estimated at $120,000 per turbine) plus annual O&M costs.

For all three approaches, the results of the study show an increase in *IRR* over the base case of decommissioning at year 20. Although the approaches were modeled separately, greater benefits may be achieved through combining approaches.

The major findings in Ref. [2] are

- The scenarios with the most years of life extension yield the largest financial benefit. Even a few extra years of life extension can increase IRR over the base case.
- Life extension through targeted inspections shows the largest financial gains. Additionally, this measure can be undertaken later in life, while measures to reduce loads must be initiated early to have a noticeable impact.

20.1.2   EXTENDING LIFE OF MAJOR WT COMPONENTS

The service life of some of the WT's main components such as gearboxes, frames, and blades were the first to be affected. Fatigue failures can damage some parts of the structure leading to, in some cases, a sudden collapse of the machine.

Depending on the case of each turbine, customers could choose between decommissioning the machine(s), dismantling them, or replacing them with larger and/or newer equipment. However, due to financial constraints, and technical and legal impediments, these two options are economically unfeasible for most.

Currently, WTs with more than 70,000 MW in the world have been running for more than 5 years. The real challenge exists in upgrading these turbines to make the wind farms profitable beyond their original useful lifetime, without the need for any subsidies. In the last 15 years, the whole industry has drastically increased its technical know-how and operational experience, now making a longer operation lifetime possible.

20.1.2.1   The WT Life Extension Program

The WT life extension program is offered by Gamesa Services, which is a global production and service company [3]. Gamesa Services has production centers in Spain and China, as the global production and supply hubs, while maintaining its local production capacity in India, the United States, and Brazil. The WT life extension program offered by Gamesa consists of a series of structural reforms and a monitoring system designed to prolong the useful lives of wind turbine generators made by Gamesa and also by other manufacturers beyond of the original design specifications.

20.1.2.1.1   Reliability-Centered Maintenance

One process being used successfully by Gamesa to prolong the useful life of WT is the RCM, a process that has been used for more than 50 years in other industrial sectors such as electrical nuclear, aeronautics, rail, and aerospace (see Section 19.1 "Maintenance Strategies"). It involves studying the failure modes of each component and the possible consequences of their failure on a more complex system. An RCM optimizes maintenance tasks, defining predictive, preventive, and corrective actions, and finally determining when a component should be upgraded with the latest state of the art design.

20.1.2.1.2   Reconditioning

Another way to extend the life of WTs is the reconditioning of its major components, which involves extending the useful life of blades, gearboxes, and generators, mainly by enhancing their constituent parts or by replacing these elements with the latest technological advances. In some cases, as a preventive measure, or during repairs, the damaged parts are replaced with the same original element. Reconditioning a turbine gearbox can involve redesigning the gear geometry, reinforcing the crown gear, replacing the existing bearings with a different design, and changing auxiliary systems connected to the gearbox to improve working conditions. This significant redesign can double the gearbox operating life for less than 20% of the cost of a new gearbox. It can also reduce maintenance costs.

20.1.2.1.3   Preventive and Corrective Actions

The useful life extension program involves investing in preventive and corrective activities when it is necessary to keep WTs working for 30 years, so an immediate replacement of the existing components is not required. When large components do fail, reconditioned components or the new model may be recommended instead of repairing the original ones.

20.1.2.2   WT Gearboxes: Life Extension Techniques

WT operators have a pressing need to extend the life of their gearboxes and bearings due to premature failures. The importance of reliability operations forces operators to make difficult predictions and decisions regarding future failure risk for warranty renewal and maintenance and operations planning.

To predict and extend the life of their gearbox, SunEdison and First Wind, WT operators in the United States, use prognostic system DigitalClone Live, developed by Sentient Science [4,5]. Sentient Science developed a new generation of autonomous, predictive, condition maintenance, monitoring systems to monitor the life of WTs. WT operators using Sentient Science prognostic data models, sensors, and processors can monitor and predict the remaining life of WT gearboxes (WTG).

With DigitalClone Live, WT operators now know gearbox failure rates and locations across its fleet, including critical bearing and gear components listed for inventory and supply chain management. Sentient Science is monitoring 218 WT gearboxes for First Wind with prognostic-based remaining useful life calculations per gearbox.

Using prognostic technique, WT operators can quantify their fleet's future failure risk under the real operating conditions for each turbine "As Is" and analyze how "What If" scenarios such as recommended changes in duty cycle, components, oil treatments, and service intervals could extend their fleet's life. In addition, a return-on-investment calculator developed by Sentient and SunEdison enables asset managers to control how their decisions today would lead to the longest gearbox life and most attractive return-on-investment.

The multiphysics prognostic DigitalClone model integrates with SCADA and sensor operational data to calculate the time until cracks will initiate in the component materials being used and when those cracks will cause mechanical failure. Then, the model updates the failure risk based on operational changes and alerts users of how to control their asset life and performance.

20.1.3 EXTENDING LIFE OF MONOPILE FOUNDATION STRUCTURES OF WIND, TIDAL, AND WAVE TURBINES

Foundations for WT generators (WTG) and/or offshore substations (OSS) are designed to support the structure for at least 20–25 years, the desired operational life. But in practice, the real state of health of the structures is not always certain. New concepts, assumptions regarding soil conditions and loads as well as unforeseen events, all have an influence on how the structure or some of its key components degrade. This in turn influences the real state of health of the foundations.

The only way of having a continuous view on the foundation's state of health together with its evolution in time is by equipping it with a smart, multisensor monitoring system that not only continuously collects essential data but also translates these data into indicators quantifying the various aspects of the state of health of the structure. The structural integrity monitoring would be an important tool to guide O&M operations and also could provide means to significant turbine life extension.

Zensor, Belgium company, developed autonomous and robust monitoring solutions for the foundation structures of offshore WTs as well as transformer stations [6]. Continuous corrosion monitoring of offshore WTs provides information on corrosion rates and potentials, and other related to corrosion parameters of foundation structures. This information may help to schedule condition-based maintenance (CBM) in timely manner, thus improving structural health and providing significant WT life extension.

20.1.3.1 Offshore Monopile WTs: Monitoring Corrosion

First time a permanent *corrosion* monitoring device was installed by Zensor at an offshore WT in front of the Zeebrugge coast in March 2012. It was Zensor PermaZEN corrosion sensor to follow up corrosion activity in a monopile foundation, which continuously monitors oxygen concentration and corrosion rates.

The latest Zensor CMSs include various subsets of relevant sensors to target relevant phenomena influencing the structural health of the *offshore* foundations, such as

- Corrosion activity
- Grout integrity
- Displacements and deformations
- Fatigue and dynamic behavior (natural frequencies, damping, etc.) of structures and components
- Impact
- Inclination
- Stresses and strains
- Scour
- Cathodic protection (impressed current: ICCP, and sacrificial anodes: sacrificial anode cathodic protection [SACP])
- Dissolved species, oxygen levels, temperatures, and pressures

20.1.3.2 Onshore Monopile WTs: Structural Integrity

For onshore WT foundations and towers, Zensor offers specific monitoring packages focusing on all aspects of structural integrity. Sensors are placed at various locations inside the foundation block and tower to monitor:

- Microcracks in concrete
- Water uptake in concrete

- Disbonding between steel faces and concrete
- Voids in concrete
- Disbonding between flanges and grout
- Integrity issues between different pours
- Fatigue life of wall segments and bolts
- Inclination
- Changes in damping and natural frequencies
- Mechanical shocks
- Relative movement between tower and foundation
- Thermal events

For each farm, a full, specified solution is provided, including the engineering, design, and construction. Collected data are presented in a live, web-based graphical user interface (GUI) and the results are further treated to gain insight in the evolution in the structure. The real age of individual structures is benchmarked to design assumptions. As such the risk profile can be further refined, maintenance can be optimized.

20.1.3.3   Extending Life of Major Tidal and Wave Turbine Components

Zensor solutions for monitoring are also available for tidal and wave power turbines and foundations [7]. Zensor offers sensor-based monitoring solutions for the prolonged follow-up of the integrity of energy production devices that rely on tidal and wave energy. Both foundation structures as energy harvesting devices are within scope. Aspects considered are as follows:

- Material loss through corrosion and abrasion
- Cracks and disbonding related to grouts and/or concrete
- Fatigue life
- Inclination
- Operational aspects
- Cathodic protection
- Overloads and other undesired events

Data are recorded and fed into a set of intelligent algorithms that generate reports and GUIs indicating the various parameters relevant for the follow-up of the state of health of the structures. As such, the O&M team has a hands-on tool for planning maintenance works, and management has a continuous view on the key performance indicators (KPIs) indicating the real state of health and risk profile of the structures and installations.

20.2   INSPECTION AND TESTING OF WIND, WAVE, AND TIDAL TURBINES

20.2.1   INSPECTION AND TESTING OF WTs

20.2.1.1   WT Components Inspection in the Field

When in production, various components of WTs should pass an extensive testing to comply with the national and international standards. For example, blades are going through multiple nondestructive testing (NDT), such as automated track scanner (AMS-14, AMS-57), ultrasound crawling scanner (AMS-46), and ultrasound manual scanning (MWS-6). Towers should pass fast automated ultrasound test (AMS-41) and flexible ultrasound scanner (AGS-2) [8].

When in operation, there are various possibilities for damage prevention of WT components. The most direct way of damage prevention is regular inspection and NDT testing of key components of the WTs [9]. There are some differences in defining which components of integrated system are the most critical ones. For a WT, the components identified as most critical are the gearbox, generator, and rotor blades.

The following mechanical components, which are estimated to be the most safety critical are [9,10]:

- Gearbox
- Generator
- Rotor blades
- Primary (low-speed) shaft
- Secondary (high-speed) shaft and the rotor hub
- Bolted joint between the blades and the hub
- Bolted joint between the steel tower and the foundation
- Steel tower

Some other miscellaneous mechanical components, such as the yaw motor, the blade adjustment, and the machine foundation are associated with the lower risks. The majority of components are suitable for dedicated programs for recurrent inspection and NDT, by which the probability of serious and costly failures can be significantly reduced [9].

During the life time of wind farms and WTs, periodic inspections are recommended every 2 or 4 years [10]. However, periodic inspections for offshore wind farms are statutory and have to be performed every 4 years. For onshore wind farms, the safety-related equipment has to be inspected periodically by an accredited inspection body. Inspections of other parts of onshore WTs are performed according to the individual contract and/or to the authorities' standards.

All European WT inspections are carried out according to IEC 61400 standards. For offshore wind farms in Germany, for example, periodic inspection services are performed also in accordance with German Federal Maritime and Hydrographic Agency (BSH) standards and all other common wind standards (GL, DNV, etc.) [11].

German technical expert forum German Wind Energy Association (BWE) developed and approved principles for the "Recurring Periodic Inspection of Wind Turbine Generator Systems" [12]. In this document, the scope of WT inspection is defined. It states, for example, that "During the inspection, the supporting structure (tower and accessible areas of the foundations), machine and rotor and essential components shall be examined and inspected especially for abnormalities and unacceptable deviations from the required conditions (e.g., damages, cracks, unexpected wear, corrosion, backlash, noises, condition of lubrication, leaks, tilting, misalignment, resonance behavior, imbalance, etc.)."

20.2.1.2 NDT Techniques for Turbine Component Inspection

NDT offers different techniques which are used or could be used for the inspection of WTs such as vibration analysis, thermography, x-ray imaging, AE, and UT [9,13]. More detail information about testing and monitoring techniques is given in Chapter 18 "Testing and monitoring techniques applied to renewable energy equipment."

Vibration analysis is the most known technology applied for CM, especially for rotating equipment. The types of sensors depend on the frequency range (from 0.01 to 100 kHz) relevant for the monitoring. The physical conditions of materials inspected/tested with acoustic technique are mainly focused on crack detection and growth. Commonly used acoustic methods are normally off-line and not suitable for online CM of WTs. Exception might be the usage of optical fuses in the blades and acoustic monitoring of structures.

Thermography is often applied for monitoring and failure identification of electronic and electric components. Typical thermographic applications involve the introduction of a controlled thermal load on the object of interest. Variations in the thermodynamic properties of the object then produce surface temperature patterns which can be detected with an IR imaging system. Hot spots, due to degeneration of components or bad internal contact, can be identified in a simple and fast manner. This method is particularly sensitive to flaws near the inspected surface.

AE technique is suitable for detecting the possible failure of blades in the real time. Acoustic monitoring has a strong relationship with vibration monitoring, though there is a principal difference. While vibration sensors are rigidly mounted on the component involved and register the local motion, the acoustic sensors perform registration of the component response. They are attached to the component by a flexible glue with a low attenuation. See Chapter 18 for more information.

These sensors are successfully applied for monitoring of bearing and gearboxes. In the case of AE application for the testing of a fiberglass structure, the source of emission is cracking of matrix and fibers. The presence of AE in a fiberglass structure subjected to an external load indicates a local failure in some part of the structure.

20.2.1.3 Inspection of Gearboxes and Bearings

It is generally accepted that gearboxes are the main problem for WTs because of their complexity and multiple moving parts which are possible weaknesses (e.g., bearings). Gearboxes generate a lot of failures and therefore they require more maintenance as well as careful inspections [10]. Several times per year, maintenance teams have to take care of the gearbox.

Premature failure of gearboxes and bearings due to contact fatigue is common and very costly. Continuous CM can help to detect damage prior to failure, thus helping to avoid costly, unplanned stops. Modern research on contact fatigue has demonstrated the importance of the surface conditions for contact fatigue failure of gears and bearings. These findings clearly indicate that securing higher and improved surface quality is a way of prolonging the lives of these components.

Gearbox inspection goal is to perform a general assessment of the gearbox's condition. Inspection of the gearbox via a flexible video-endoscope on all visible bearings, cogs, pinions, and sprockets may reveal multiple defects. An example of damaged WT gearbox found with remote visual inspection [14] is shown in Figure 20.1. It was most likely caused by metal debris floating around in the gearbox/oil.

Other examples are given in Chapter 13 "Wind Turbine Gearboxes and Bearings: Contamination and Failure Modes." In this examples (Figures 13.3 through 13.6), roller bearing-damaged surfaces and gear tooth micropitting have been found during inspection.

According to Ref. [10], the inspection program may include the following procedures:

- Oil sampling and analysis, including spectrometry, ferrometry, and counting the particles
- Conditions of gearbox, including cleanliness, painting, oil level, litters added, gear stay, bushings, and torque arms conditions
- Oil conditions such as foam presence, color, smell, presence of sour or rotten odor, and oil viscosity
- Tightness of seals on main shaft rotary connector, output shaft, sealing surfaces of covers, split line flanges, and gearbox oil lines
- Gearbox's cooling system and oil lines cooling systems

A gearbox has a known acoustic signature and the WT where it is integrated has been certified for a given and known emission noise level. When rollers, bearings, or teeth of gear wheels are deteriorated, this also affects the acoustic signature of the deteriorated gearbox. Therefore, during inspection, it should be confirmed that no abnormal noise is heard from the gearbox/bearings.

20.2.1.4 Inspection of Generator

Two main categories of failures are known regarding the generators: winding failure, which may result from defective insulation systems or poor winding design, and mechanical failure because of early bearing fatigue, which may result from poor lubrication.

FIGURE 20.1 Fracture on low-speed gear tooth which was found during a gearbox inspection using a Borescope. (CONCO makes its mark in the renewable energy sector, *Energize RE: Renewable Energy Supplement*, 2, 4–6, June 2014. With permission.)

The quality of the insulation is affected over the years because of functional stress. Root causes of defective insulation could be of electrical, mechanical, and chemical origin. The latter is mainly due to the proximity of chemical products like oil, corrosive vapors as well as dust, which affects the performance of insulation materials. There are many other factors that affect insulation: fluctuation of temperature and pollution of the environment (moisture, particles, and deposits in hot and wet areas). These causes lead to deterioration of the insulation properties.

By a lack of regular inspections and therefore appropriate corrective measures, the sum of these root causes may lead to incidents and thus loss of production. It is strongly recommended to perform periodic insulation tests on the generator (windings stator and rotor) in order to ensure a safe operation and to avoid incidents with losses of production. These tests allow detecting early defects or premature aging of the insulation properties of the generator's windings [10].

A generator inspection should be performed according to the scope of work determined by a wind energy operating or servicing company. For example, at Deutsches Windenergie-Institut (DEWI), wind energy service provider in Germany [10] during inspection and testing procedure performs the following investigations:

- Verification of HSS coupling
- Verification of the HSS alignment and calculation of the compensations to realign the coupling train

- Determination if abnormal noise is heard from the bearings. A video-endoscopy and/or grease analysis of the front and rear bearings can be done if requested
- Inspection of generator's cooling system
- Testing of insulation test of the generator (stator and rotor)

20.2.1.5   Inspection of Rotor Blades

Rotor blades are manufactured with GRP, also known as glass fiber-reinforced plastic (GFRP) and epoxy resin with an exterior gel coat. The blade interior structure consists mainly of a spar reinforced with rigid polyurethane foam encased in GRP as well as additional interior strength via "sandwich sheets" in built-up layers. Rotor blades are a highly stressed part of a WT because of the constant wind contact. They need regular inspection to evaluate their structural safety by experienced experts.

There are several types of damages according to the localization of the deviations. Inside the blade, these damages are cracks formed at the bonding resin; missing adhesive; discontinuities on the sandwich; delaminations within GFRP; and problems in the bonding, waves, air inclusions, and so on.

Outside the blades on the surface, the damages include erosion on the blade surface, defects in laminate (spalling, flaking, and cavities), deficient bond at the bonding surfaces, and cracks. Damages produced by *lightning strikes* could be not repairable and would require a replacement of the blade.

It is strongly recommended to take care of the blades by performing yearly blade inspections as well as making sure that repairs and cleaning are part of the preventive maintenance program implemented by the manufacturer or other maintenance service company mandated by the owner or its operator.

During blades inspection, both interior and exterior sides of each blade and its root region should be checked for air inclusions, cracks, lightning damages, erosion wears, delamination, defective bonds, and other quality problems. The condition of the joint seals, lightning protection systems, and additional aerodynamic parts should be checked as well as existing rain deflectors or possible repaired areas. The lightning protection system can be verified with a measurement of the ohmic resistance.

During the blade inspection, an important issue of moisture should be addressed. It is necessary to prevent the *accumulation of moisture* in the GRP. Such accumulation of moisture in the GRP leads to mechanical forces in the material which finally lead to a serious damage. When moisture penetrates different areas of the blade interior, it may cause imbalances within the blade structure. Moisture can penetrate the blade structure via previous structural cracks or as a result of surface damage caused from previous lightning strikes. Blade repairs that result in surface porosity may lead to future water access.

The worst case is a combination of lightning impact on a blade and residual internal moisture. The high temperatures generated by lightning can lead to the accumulated moisture instantaneously changing into steam by a thermal expansion effect inside the blade. After such a phenomenon, the following damages are observed: delamination, burst bonding, trailing edge cracking, detached blade pieces, debonding, longitudinal cracks, spar separation, fires due to the presence of hydraulic fluids/lubricants, and partial or complete blade destruction [10].

The aim of regular blade inspections is to reduce the risk by determining the issues and taking necessary actions immediately. The combination of valuable blade inspection programs and appropriate preventive maintenance works may significantly extend useful life of WT blades.

According to Ref. [12], "The rotor blades shall be inspected at close proximity, both externally and internally (if accessible) with regard to relevant damages to the surface and for structural defects in the body of the blade (e.g., cracks on bonded bridges, delamination, etc.)." A very roughly estimated testing interval for the internal damages is 4 years [9].

Currently, regular *visual inspection*, assisted by technical instruments, from the ground is probably the best and most accessible alternative for controlling surface damages. A seemingly promising technique for discovering internal damages during operation is *thermographic* testing. To detect delamination and microfractures in composite materials with thermal imaging cameras, a method called *pulse thermography* is often employed. The composite material is excited with the use of a lamp. The thermal imaging camera is then used to monitor the thermal distribution through the material. Differences in the speed at which parts of the rotor blade heat up or cool down indicate damage as found in Refs. [15–17].

The NDT testing using pulse-echo ultrasound to detect internal damages in WT blades has been developed in Ref. [18] and optimized for *in situ* WT blade inspection. The system is designed to be light weight so it can be easily carried by an inspector climbing onto the WT blade for *in situ* inspection. A software system has been developed to control the automated scanning and show the damage areas in a 2D/3D map with different colors so that the inspector can easily identify the damage areas.

Numerous NDT techniques for rotor blades are still under development. A review of recent research and development in the field of damage detection for WT blades is presented in Ref. [19]. This paper reviews frequently employed sensors, including fiber optic and piezoelectric sensors, and four promising damage detection methods, namely, transmittance function, wave propagation, impedance, and vibration-based methods.

20.2.1.6   Use of Vibration Diagnostics for Periodic Inspection and Testing

Vibration diagnosis of drive train (gearbox and generator) is an important part of periodic inspection and testing of WTs. CM on the drive train is usually associated with both periodic inspection and video-endoscopy of the gearbox. This inspection allows to verify the technical condition of the complete machine, which provides good information for the vibration investigations such as cracks in rotor blades and foundations, problems on components around and inside the gearbox, or any excessive wear of components of the WT.

A vibration diagnosis provides an accurate knowledge of the kinematic chain condition, which allows to avoid the preventive replacement of expensive turbine components sometimes required by insurance companies [10]. Therefore, vibration testing allows to determine and deploy the most suitable maintenance actions.

Vibration diagnostic may be performed via continuous CM, based on fixed data acquisition and analysis system, software, and wireless communication with suitable sensors which collect data and issues an alarm as soon as the component characteristics start to change. It could be also performed periodically, based on a "checkup" strategy. A mobile data acquisition and analysis system with its measuring instruments is implemented along the kinematic chain from the main bearing to the generator rear bearing for a temporary troubleshooting activity. With this method, possible installation and material faults can be detected already during the first diagnosis by comparing the characteristic frequency lines of the measured spectra.

The vibration diagnosis provides early fault detection and allows early detection of potential damages of the rotor bearings, generator, or gearbox as well as incipient failures and damages on the gear wheels. Various defects of rotor blade, drive train, lubrication, as well as electrical problems, resonance problems, and abnormal tower vibrations can be detected with a periodic vibration diagnosis.

Among detectable blade problems, there are unbalanced propeller blades, blade vibrations, small misalignments of a blade angle, and first blade wears like delamination. Vibration diagnostics can detect problems with the drive train such as misalignment, shaft deflections, mechanical looseness, bearing condition, gear damage, and generator rotor/stator problems.

20.2.1.7   Inspection of Bolted Joints, Welds, Tower, and Foundation

Bolted joints are probably the most safety critical element in a WT. It is recommended that the bolts are tap tested at least annually. This simple and cheap testing technique has proven to be surprisingly effective in crane applications which also contain several safety critical joints. UT or removal of the bolts for surface testing (aviation procedure) can be an alternative whenever extra concern about safety prevails.

Estimative fracture mechanics analyses show that recurrent testing of selected *welds in steel tower* should be effective to maintain safe operation. Magnetic particle testing with a 4-year interval seems reasonable as a first estimate, with complementary visual inspection with a more regular interval.

Onsite, all components of the turbines (blades, tower, and subsea components) are getting through detailed inspection on regular base. NDT inspections include visual inspection provided by UAV helicopters/drones and use of portable and manual ultrasound scanner (ATS-2 and MWS-6). Several important properties of tower are inspected and one of them is very important for monopile offshore WT—*corrosion inspection* of foundation [8]. Pitting and general corrosion may lead to decrease of the monopile foundation wall thickness, which is inspected and measured using ultrasound technology, manually or with automated devices. These inspections allow determine coating and base metal damages to prevent catastrophic failures by applying various corrosion protection means and techniques (see Chapter 12 "Wind Energy Equipment Corrosion"). Both the primary and secondary axle of WTs also should regularly be visually inspected for corrosion and pits that could start a fatigue process.

During periodic inspections of WT, according to Ref. [12], the foundation "shall be examined with regard to the required earth surcharge (load), condition of the surfaces in the visible area, concrete spalling and concrete cover, condition of the concrete backfill and the sealing joint, inadequate water drainage, etc. The tower structure shall be examined for damages with regard to stability (e.g., corrosion, cracks, spalling in the supporting steel/concrete structures, deformation, gaps, faulty welded joints)."

20.2.1.8   Other WT Periodical Inspection Tasks

For complicated integrated system as WT, there are many other components and properties to inspect on regular base. Here are some recommendations stated in Ref. [12]: "The external lightning protection equipment on the rotor, motor, and tower inclusive of the connection to the foundation earth electrode shall be examined to ensure they are free of defects. The system controls and the electrical equipment will undergo a visual inspection with regard to all connections, fasteners, condition of insulation, proper installation, discoloration, and accumulation of dirt. A visual and slackness inspection shall be performed as a minimum for scheduled, pre-tensioned bolted connections."

20.2.2   Inspection and Testing of Wave and Tidal Turbines

Wave and tidal turbines operate under harsh marine conditions; they are exposed to multiple environmental factors that lead to deterioration of the materials turbine components are made of [20]. For example, some of the tidal power plants contain special hydrofoils and hydro-buoys made of composites: multilayered GFRP used for the skin and multilayered carbon fiber-reinforced plastic (CFRP) used for the main spar. The constructions made from such composites should be periodically tested against the faults.

One of the NDT techniques which enable to detect the defects both during manufacturing and in service inspection is based on application of ultrasonic-guided waves [13,21,22]. The expected types of possible defects inside hydrofoil are similar to the defects found in aging composite WT blades [13,21]:

- Delamination between the skin and the adhesive layer
- Delamination between the main spar and the adhesive layer
- Adhesive joint failure between the skins along the leading and the trailing edges
- Internal multiple delamination or splitting between the layers of the skin and the main spar

Disbonding between the inner shell and bonding flange of the tidal turbine support is not uncommon. Other defects include delamination between the laminate plies that comprise the support structure and structural damage due to impact.

The study in Ref. [22] was performed to select the modes of ultrasonic-guided waves to be used for inspection of the most critical regions of composite components of the hydrofoils used in tidal power plants and to determine the parameters of their excitation, propagation along the sample, and to identify the regions of coverage. It was estimated that in order to use the fundamental modes, the frequency of operation below 100 kHz should be used for inspection of the skin and even lower frequency should be used for inspection of the main spar.

20.3 MAINTENANCE TECHNIQUES OF RENEWABLE ENERGY EQUIPMENT

Maintenance goal is to support the operability of technical systems. There are different approaches toward maintenance strategies (see Section 19.1). Suitability of the different maintenance approaches could differ for application in the maintenance process for power generation units in wind, wave, and tidal offshore arrays for energy production.

A definition of the different maintenance approaches and terms mentioned is given in the European standard EN 13306:2010 "Maintenance-Maintenance Terminology" accepted in many European countries and Britain [23]. A comprehensive discussion of those approaches and their applicability in offshore wind projects can be found in Ref. [24]. Some of the maintenance strategies used in renewable energy field will be discussed in the next sections.

20.3.1 Planned Preventive Maintenance (PPM)

The objective of preventive maintenance is to replace components and refurbish systems that have defined useful lives, usually much shorter than the projected life of the WT. Tasks associated with scheduled maintenance fall into this category.

These tasks include periodic inspections of the equipment, oil and filter changes, calibration and adjustment of sensors and actuators, and replacement of consumables such as brake pads and seals. Housekeeping and blade cleaning generally fall into this category. The specific tasks and their frequency are usually explicitly defined in the maintenance manuals supplied by the turbine manufacturer. Costs associated with planned maintenance can be estimated with reasonable accuracy, but can vary with local labor costs and the location and accessibility of the site. Scheduled maintenance costs are also dependent on the type and cost of consumables used [25].

To provide PPM [26], it is required to develop a program to ensure that regular checks are made on critical components, and the likelihood of possible future component failures is identified (in conjunction with CM).

There are multiple issues that need to be considered when developing a PPM program. They include information provided by the manufacturers; frequency of maintenance activities; and definition when the maintenance is to be performed, for example, time of year, day/night, and during peak or off-peak generation. There should be documentation defining maintenance activities to be performed; tools and equipment to be used, including calibration requirements; availability of competent personnel; disciplines required; specialist knowledge; and employed personnel or contractors.

Many other documents should be prepared and available, as well as various appliances and equipment necessary for safely performing complicated maintenance procedures [26]. The number of failures will be reduced by planned maintenance.

20.3.2 Predictive Maintenance and SCADA

Predictive maintenance is a rising trend in the wind industry. Predictive maintenance, which is well known to more mature industries, uses high-tech CM technologies and may be the most cost-effective of all O&M strategies as it reduces maintenance costs and breakdown frequency, increases machine life and productivity, and reduces spare parts inventories and the use of overtime [27].

CMSs monitor the status of all components subject to wear. In a wind system, the gearbox, bearings, and generator are steadily monitored and data are archived on the basis of the acoustic frequencies measured. Automated analysis typically draws up a comparison between the ideal and actual situation. In the event of a discrepancy, it signals a preventive service operation to be planned before any damage gets serious (see Chapter 18).

With CM, turbine inspections and services may be planned in advance, rather than as a reaction to damage done. The data recorded by the CM system for each WT may be evaluated in a database containing stored plant history. The historical data on individual turbine types and product families are evaluated in addition to the values of the individual turbines. With CM procedures evolving, CM will become part of the utility's overall monitoring program, including SCADA monitoring, performance monitoring, fault categorization, failure tracking, and spare parts management.

SCADA systems aid in improving turbine performance, particularly in cold climates. The technology may be used to detect trailing- and leading-edge ice formation. It can initiate different control systems, such as pitch, stall, and active stall control. Also, rather than relying on visual inspections, SCADA systems can detect when ice has melted, so the turbine may resume normal operation.

20.3.3 Condition-Based Maintenance (CBM)

CBM is also based on information delivered by CM systems, which use remote sensing and analytics to predict gearbox failures and other maintenance requirements. These systems reduce the need to physically visit WTs, which are often in remote locations and require climbing or rising to heights of 100 feet or more.

They measure and monitor the physical operation of gearboxes and other moving parts that are subject to changing loads and highly variable operating conditions that create high mechanical stress on WTs. Measurements may include vibration, strain, acoustics, temperature, as well as voltage, current, and electrical power [27]. See also Chapter 18 for more information.

For example, if CM reports that a component is experiencing vibrations outside its normal specifications, project managers can take predictive maintenance steps to inspect the component and initiate repairs or replacements before a breakdown occurs.

> *In-service vibration analysis* is an effective technique for assessing the condition of the gears and bearings, identifying problems early, and recommending an early solution. The analysis can detect gear damage such as scoring and abrasive wear, bearing defects, such as cage fractures and ring damage, alignment errors, looseness, imbalance, and resonance areas.
> *Laser cladding* (a process that applies powdered metal to a surface with a laser) is one repair technique that can restore damaged parts instead of replacing them. *Oil flushing* is also important because clean lubricants support reliable WT performance. The use of CM can also lead to improved O&M logistics and planning, for instance, advanced warning of impending issues of multiple turbines in one area can allow for advanced scheduling of downtime and equipment needed to perform repairs.

Utility experiences with yaw bearings and teeth have shown that repairing the teeth can be especially costly. They need good lubrication to hold up against static and dynamic loads and stress during WT operation. Lubrication supports smooth rotation for the orientation of the nacelle under all weather conditions. Some WT manufacturers now use self-lubricating gliding elements instead of a central lubrication system. Despite state-of-the art design, the bearings and teeth are subject to wear and need to be regularly inspected, adjusted, and replaced as needed.

Utilities are monitoring pitch control performance in wind projects. Depending on wind speed, a turbine's pitch system turns the blades into or out of the wind direction. The system typically adjusts the blades a few degrees every time the wind changes. This keeps the rotor blades at the optimum angle to maximize output for all wind speeds. Accumulated performance data will show how important this is, and when a maladjustment really is a problem to fix.

20.3.4   REACTIVE UNSCHEDULED MAINTENANCE (RUM)

Many circumstances, such as severe weather conditions, can arise, which require workers to carry out tasks beyond their normal work experience and/or which are more than usually hazardous by their nature. These tasks are called *reactive maintenance, unscheduled maintenance, failure-related maintenance,* and so on. Records of all breakdowns should be kept to influence future-planned maintenance policy revisions, training, and designs. Arrangements should be considered to manage unknown and changeable situations as a result of breakdowns, major component failure, and resultant damage.

A certain amount of unscheduled maintenance must be anticipated with any project. Commercial WTs contain a variety of complex systems that must all function correctly for the turbine to perform—rarely are redundant components or systems incorporated. Failure or malfunction of a minor component will frequently shut down the turbine and require the attention of maintenance personnel [25].

Unscheduled maintenance is usually very costly, and the costs can be separated into direct and indirect costs. The direct costs are associated with the labor and equipment required to repair or replace, with the component costs themselves, and with any consumables used in the process. The indirect costs result from lost revenue due to turbine downtime.

Labor costs are driven by the difficulty of accessing and working on the components. With the exception of some switchgear and power conversion equipment, most of the turbine equipment is accessed by climbing the tower. For safety reasons, a two-person crew is generally required for any up-tower activity. In remote locations, access to the turbine itself may be difficult and limited by weather.

Working conditions can be in extreme temperature conditions and may be curtailed by high winds. Some turbines are equipped with hoists and rigging equipment, but in general, all tools and equipment, in addition to spares, must be lifted into the nacelle. Space is limited inside the nacelle and working positions may be awkward. Work outside of the nacelle, including transitions into the hub on some turbines, requires working with a safety harness and lanyards.

Labor cost estimates for major component replacement are developed from experience. Although some major components may be reworked *in situ*, this is not generally the case, and replacement will require a crane to dismantle the drive train, and several personnel in addition to the crane operator. The equipment and procedures for disassembling the rotor or drive train are established during assembly. The actual cost, however, may vary due to accessibility to the turbine site, equipment availability, and wait time during high-wind conditions [25].

20.3.5   PLANNED MAINTENANCE AND LUBRICATION

Turbine lubrication is one important part of the process of maintaining equipment at wind power farms. Planned maintenance can also include services such as filter changing and the torqueing of bolts, while unplanned maintenance can also include electrical component failures and part replacements [28].

Because oils and greases for WT bearings are expected to continuously maintain optimal performance under extreme conditions, they need to meet higher standards than comparable lubricants for other industries. While bearing surface destruction in WTs is not much different from that observed in other equipment, the frequency and severity of the destruction is considerably greater. Blade root bearings, main shaft bearings, and yaw bearings are lubricated with grease, while the gearbox is lubricated with oil. Even in a well-designed and lubricated gearbox, oil film breakdown can occur during events that overstress components [29].

Keeping a WT's gearbox properly lubricated is important in extending the life of a WT. The gearbox is not the only part of the turbine that requires lubrication. The generator also requires lubrication, and there are lubrication points on the blades. Wind tower blades have bearings that will essentially feather the blade so operators can optimize the blade angle to match wind speed. The main shaft bearing also requires grease for lubrication, as well as the drivetrain and yaw and pitch drives. The turbines also use a hydraulic system that is used to provide a braking mechanism for a unit, but can also be used for hydraulic pitch control on the blades.

Turbine oil to be used in WT gearboxes are usually a fully synthetic oil designed to withstand the conditions WTs may be subject to, whether that is extreme temperatures or the potential of corrosion from saltwater for offshore WTs. The type of oil that is used in a turbine's gearbox in other parts of a WT is generally designated by the OEM for the units. More about lubrication of WTs is given in Chapter 14 "Wind Turbine Lubrication."

Lubrication during the maintenance should be provided by qualified personnel. A lubricant monitoring program can help prevent gear and bearing failures by showing when maintenance is required. Lubricant monitoring should include spectrographic and ferrographic analysis of contamination, particle counts and analysis of acidity, viscosity, and water content. Used filter elements should be examined for wear debris and contaminants.

The maintenance procedure may increase contamination of the oil. All maintenance that involves opening the gearbox or lubrication system should be performed with good housekeeping procedures. Oil should be added from a filter cart connected to the gearbox with quick connect couplings to minimize contaminants in the new oil and to minimize contaminant ingression during the transfer [30].

Some oil manufacturers not only offer different oils for different parts of the turbine, developing specializing lubricants for different parts of a WT, but also attempt to make multipurpose products when there would be no decline in quality. ExxonMobil, for example, produces Mobilgear SHC XMP 320 for turbine gearboxes as well as several other products for WTs, including Mobil SHC Grease 460 WT for use in main, pitch, and yaw bearings and Mobil SHC 524 lubricant for use in hydraulic systems (see Chapter 14 "Wind Turbine Lubrication").

Typical relubrication cycles for WT suggested in Ref. [31] are the following:

- Every 6 months—most greases including gear teeth and slew bearings
- Every 6 months—hydraulic and yaw and pitch gears check and top off oil
- Every 3–5 years—change oil in main gearbox, if filled with synthetic oil

Over a life of a WT, it will use 10 times the amount of lubricant for maintenance than the original OEM fill.

As turbine technology continues to develop and companies produce larger WTs, companies producing the lubrication also need to produce lubricants to keep up with the industry. The larger turbines being produced by companies require oil that can handle the extra stress created by the extra size. In renewable energy field, expectations of new generation of oils are high that it may lead to extension of oil drain intervals from 8 to 12 months to up to 3 years [32].

20.3.6 MAINTENANCE OF TIDAL AND WAVE TURBINES

Marine current turbines are not a completely new technology as they are similar in many aspects to WT, which are already well developed and commercialized. Therefore, maintenance techniques could be similar to the ones applicable to offshore WTs. However, there are several significant differences that increase the cost of maintenance of tidal and wave turbines. One of the main expenses in tidal power is maintenance—the saltwater corrodes the metal parts, and generators are hard to reach because they are often deep underwater.

Solution to bring the maintenance cost down may be by using CMSs [33], though there is no standard CM technique available that can provide details of the tidal blade integrity. TidalSense Demo is specifically making technology for an emerging method of generating tidal power known as tidal stream generation, where turbines are placed underwater in places where the tidal current moves quickly.

The project, which began in February 2012, uses long-range ultrasonic sensors to pick up any signs of wear and tear on the underwater turbines and then alert engineers so they can decide whether to take action. The CMS being developed by the project constantly checks the integrity of the exposed rotor blades, meaning that operators can save money by fixing problems quickly, and by adapting their designs if part of the rotor is particularly prone to breakage.

Maintenance costs associated with a tidal turbine farm are defined by a number of factors such as the size of the turbine farm, location, the layout of the turbine farm, and the efficiency at which electricity can be generated, with the latter often a function of local conditions [34]. A closely spaced layout will make maintenance easier by exposing less cable and reducing travel time by boat between turbines, but will reduce the efficiency of the turbines due to wake effects between the adjacent turbines. Conversely, maintenance costs will increase as the spacing of tidal current turbines increases due to the increase in exposed cables and distance between turbines, but the efficiency of the turbines is increased due to decreased wake interactions.

20.4 INSPECTION AND MAINTENANCE OF SOLAR PANELS AND PLANTS

20.4.1 Maintenance Procedures for Solar Panels

Solar panels maintenance includes different types of maintenance procedures, which are similar to the procedures applied to other renewable and traditional power technologies [35].

20.4.1.1 Preventive Maintenance
Scheduling and frequency of preventive maintenance is set by the operations function and is influenced by a number of factors, such as equipment type, environmental conditions (marine, snow, pollen, humidity, dust, wildlife, etc.) of the site, and warranty terms. Scheduled maintenance is often carried out at intervals to conform to the manufacturer recommendations as required by the equipment warranties.

20.4.1.2 Corrective Maintenance
It is required to repair damage or replace failed components. It is possible to perform some corrective maintenance such as inverter resets or communications resets remotely; also, less urgent corrective maintenance tasks can be combined with scheduled, preventive maintenance tasks.

20.4.1.3 Condition-Based Maintenance
CBM is the practice of using real-time information from data loggers to schedule preventive measures such as cleaning or to prevent corrective maintenance problems by anticipating failures or catching them early. Because the measures triggered by condition are the same as preventive and corrective measures, they are not listed separately. Rather, CBM affects when these measures occur, with the promise of lowering the frequency of preventive measures and reducing the impacts and costs of corrective measures.

20.4.2 Inspection and Maintenance of Solar Power Plants

Inspection and maintenance of solar power plants is scheduled on annual base. The inspection requirements placed on operators of PV power plants are very stringent and includes checking all the operating hardware and coordinates with the certified subcontractors and the utility company involved [36,37].

The inspection package can include visual inspection, performance test, removal of organic and inorganic contaminants, computerized digital error diagnosis, and temperature analysis, as well as a review of maintenance protocols. Defective modules are checked for typical errors, including cell fractions, insulation faults, hot spots, delaminating, and discoloration.

During inspection, the system is tested using insulation measurement, inspected for the presence of mechanical imperfections, such as loose connections and corrosion that can lead to ground faults and outages. Inspection services can include optical scanning, cleaning of the inverter, thermography for detection of hot spots to the network analysis, and upper shaft assessment.

High grass on the site and bushes and trees in the immediate vicinity can have a substantially negative impact on output. PV site maintenance therefore involves cutting the grass and weed clearance. Maintenance also includes cleaning PV modules.

The inverter is the heart of the solar power plant. The inverter station must have proper heating and cooling systems. Regularly exchanging the filter mats is also recommended, along with software updates and the cleaning of the inverter's components.

20.4.2.1   Dealing with Snow on PV Panels

Snow may seriously affect the productivity of solar panels. A study in Ref. [38] showed that power losses from snow can be as high as 15% in places like Truckee, CA, or as low as 0.3%–2.7% for a highly exposed 28° tilt roof mount system in Germany. Yearly yield losses are averaged around 3%. Snow is a variable issue, highly dependent on PV location. Unfortunately, there are not many ways to deal with it.

On residential-sized arrays, it could be worth one's time to simply shovel snow off of a few panels, as long as the array is not scratched. On a large commercial-sized array, this is very impractical. The reason for this is that there is little one can do to address snowfall. The snow problem may be resolved by itself, which happens when the temperature rises enough for snow starting to melt.

When it is hot enough out, the bottom part of the snow that is in contact with the array will melt first. Then, if the panel is at an angle, this small volume of snow (now water) will cause the entire block of snow to slide off the array. This will happen fastest when the sun is out. Now, when the sun is not out, and if it is too cold for the snow to melt at all, the snow will not naturally slide off the array. In this instance, because it is so cold, it is highly likely the sun is not out anyways, and the panels will not be able to generate power. In this case, it is not worth the effort to get the snow off the panels because it is cloudy and power output will be low anyway. Spraying water at snow-covered panels could get it off, but in the process the water can freeze and damage panels [39].

20.4.2.2   PV Panels Cleaning

PV panels installed either at industrial size solar plant or on a rooftop of the house in urban area get soiled or polluted if they are close to a major pollution source. Dust, leaves, and bird excrement can fall on PV panels, which lead to a reduced performance. Though some PV panels can be cleaned naturally by rain, snow, and wind, this is not true for panels close to farms, manufacturing plants, railroads, or other pollution sources. And it is especially not true for PV installations in desert areas, where experts say sand is a major problem [40].

After cleaning the panels, the researchers determined that panel productivity was reduced by 3.1%–13.8% per panel due to the dirt, with the average reduction in performance being 7.6%. The study was originally published in 1998, but reaffirmed later in other investigations done in 2007 and 2008. Energy loss actually may range as high as 25% in some areas according to the NREL. Individual dealers have reported losses as high as 30% for some customers who failed to ever clean their panels.

The most unusual cleaning method of cleaning solar panels is a self-cleaning technology originally developed at NASA for use in lunar and Mars missions, and it is not available commercially yet. The cleaning works by embedding very thin or transparent electrodes in the transparent cover

(a)

(b)

FIGURE 20.2 GEKKO PV cleaning robots (developed by Neiderberger Engineering and built and by Serbot): Rooftop PV cleaning robot (a) and Solar Farm cleaning robot (b). (Courtesy of SERBOT AG, with permission.)

over the solar panel. When a set of cyclical voltages is applied to the wires, it creates an electrostatic force that drives the particles to the edges of the pane. Within 2 min, the process removes about 90% of the dust deposited on the solar panel [40].

Another new technology is the use of the robots to clean PV panels, for example, the ones manufactured by Swiss company Serbot AG. The system for rooftop cleaning is shown in Figure 20.2a. This system works both in horizontal and vertical ways and has a remote control operated by specially designed software. It has a performance level of 350–400 m²/h and uses 0.5–1.2 L of water per minute and used on rooftop solar panels.

Another robot—GEKKO Solar Farm—was developed specifically for the cleaning of large free-field solar parks. It is shown in Figure 20.2b.

The challenge in PV panels cleaning is defined by the dimensions. The rows of panels in industrial installations are miles long, which makes an efficient use of cleaning tools a must. Furthermore, even a small layer of dirt leads to large performance losses. With a cleaning capacity of up to 2000 m²/h, GEKKO Solar Farm fulfills this requirement. In addition, thorough cleaning is guaranteed thanks to the proven use of demineralized water combined with multiple rotating brushes. The width of the cleaning brush can be adapted to the individual installation by fitting the cleaning robot with three, four, or five brushes. The GEKKO Solar Farm can be used up to an incline of 30° and can traverse gaps of up to 60 cm.

There is a different solution of cleaning PV panes in arid and desert regions, and also anywhere in the world where pressures on water resources threaten or may threaten water supplies. Israeli company Ecoppia has developed a high-tech means of cleaning and maintaining solar PV panels

FIGURE 20.3   Ecoppia E4 is water free and autonomous PV panel cleaning robot. (From http://www.ecop
pia.com/ecoppia-e4., *Ecoppia* 2016, all rights reserved. With permission.)

on a utility scale. This system comes in the form of a cloud-based solar robotics platform that is not
only highly efficient and effective but also energy independent and water free (Figure 20.3).

Fully automated E4 solar panel cleaning platform has been proven in the field to remove 99%
of the dust and other obscuring materials accumulated on solar panels each day. Ecoppia's self-
sustaining E4 solar robotics platform is equipped with its own solar PV panels, making it energy
independent. Rather than relying on water, the system relies on three key elements to clean solar
panels much more efficiently and effectively than conventional means: gravity (angle of inclination
of PV panels), air flow, and a microfiber fabric that has been proven to remove particles down to
several microns in size [41].

The technique for nonindustrial size PV systems cleaning is the use of high-pressure sprays
which are sprays with commercial anionic detergents. Effective cleaning solutions employ chemicals which reduce surface tension, lower the cost, are capable of being handled and mixed by
automated equipment, and are nontoxic, safe, and biodegradable. Detergent-based solutions are
normally able to restore the surface to 98% of its original reflectance. There are of sources less high-
tech ways of cleaning panels and variety of cleaning solutions, some of them may protect the panels
from accumulating additional dirt. However, some dirt that deposits on the panels contains fat and
cannot be easily removed, and bird dropping cannot be washed with rain [39].

Yet another option is the use Automatic Solar Panel Cleaning Systems (Heliotex™) to automatically wash and rinse solar panels [42]. Once the system is installed and programmed, it requires no
further attention except the occasional refilling of the soap concentrate and replacement of the water
filters. The system can be programmed to wash and rinse as often as is necessary for specific area.

The system attaches nozzles to solar panels. Water from an existing source, such as a hose bib
on a private residence, creates the water necessary for the system. The system can be attached to
an existing hose bib or plumbed in with a copper or plastic line. This system can be adapted to any
panel array configuration. Whether 20 panels or 20,000 panels, the system is scalable to the specific
requirements.

For desert and arid areas with high solar irradiance and limited or no access to cleaning water,
it is necessary to consider dry-type cleaning methods. Several types of dry (low water) cleaning
methods have been developed, such as active dry cleaning with brush (rotary or fixed) and passive
cleaning using various coating [43].

A comprehensive overview of PV soiling problems, primarily those associated with "dust" (sand)
and combined dust–moisture conditions that are inherent to many of the most solar-rich geographic

locations worldwide is provided in Ref. [44]. The review and evaluation of the key contributions to the understanding, performance effects, and mitigation of soiling problems spans a technical history of almost seven decades.

In the study in Ref. [45], a method to evaluate dust cleaning solution and to test the impact of a cleaning device on the performance of PV modules was developed. Investigation on the cleaning impact was performed by using the modules of six well-known producers. In this study, evaluation of the PV panels was done before and after a number of cleaning cycles in accelerated aging simulation. There was a difference in the impact for different cleaning modules on PV cells reflectance which indicates abrasion of coating. However, no mechanical impact was found for all six cleaning modules.

REFERENCES

1. Jürgen, H. Extended operation of wind turbines beyond their planned service life, *Wind Energy Report*, Germany, 2012, Fraunhofer Institute for Wind Energy and Wind Energy Technology (IWES), Germany, pp. 79–83, June 2013.
2. *Onshore Wind Turbines Life Extension*, EPRI Report #1024004, October 01, 2012.
3. *Life Extension Program*, Gamesa Services, Spain, Online. Available: http://www.gamesacorp.com/recursos/doc/productos-servicios/operacion-y-mantenimiento/life-extension-eng.pdf.
4. Life Extension with DigitalClone® Prognostics, case study, *Sentient Science*, April 2015, Online. Available: http://www.industrialinternetconsortium.org/case-studies/Sentient_SunEdison_case_study_2.pdf.
5. First Wind Inc. "DigitalClone Live™" drivetrain life extension technology to monitor the life of its clipper and GE wind turbines, *Sentient Science*, December 2013, Online. Available: http://sentientscience.com/documents/pressrelease-121513.pdf.
6. *Full Foundation Monitoring*, Zensor BVBA, Brussels, Belgium, Online. Available: http://www.zensor.be/node/40.
7. *Tidal and Wave*, Zensor, Online. Available: http://www.zensor.be/node/64.
8. *Wind Power Services*, FORCE Technology, Norway AS, 27pp., Online. Available: http://windcluster.no/files/2014/12/Leiv-Late-WCN-Forjukstreff-2014-Force-technology.pdf.
9. Isaksson E. and Dahlberg M. Damage prevention for wind turbines. *Phase 2 – Recommended Measures*, Elforsk Report 11:18, July 2011, Stockholm, Sweden, Online. Available www.elforsk.se/Rapporter/?download=report&rid=11_18_rapport_screen.pdf.
10. Hilario C. Wind turbine inspection, a strategic service? *DEWI Magazine*, 39, 56–65, August 2011, Online. Available: http://www.dewi.de/dewi_res/fileadmin/pdf/publications/Magazin_39/08.pdf.
11. *Periodic Inspections for Wind Farms*, TÜV Rheinland, Online. Available: https://www.tuv.com/media/corporate/industrial_service/Periodic_inspections-wind_farms-TUV_Rheinland.pdf.
12. Principles for the "Recurring Periodic Inspection of Wind Turbine Generator Systems," *BWE 2012 Technical Experts Forum*, Bundesverband Wind Energy, German Wind Energy Association, 9pp., October 2012, Online. Available: https://www.wind-energie.de/sites/default/files/attachments/page/sachverstaendigenbeirat/gb-grundsaetze-fuer-die-wiederkehrende-pruefung-2012-10-17.pdf.
13. Raišutis R., Jasiūnienė E., Šliteris R., Vladišauskas A. The review of non-destructive testing techniques suitable for inspection of the wind turbine blades, *Ultragarsas/Ultrasound*, 63(2), 26–30, 2008, Online. Available: www.ndt.net/article/ultragarsas/63-2008-No.2_04-Raisutis.pdf.
14. CONCO makes its mark in the renewable energy sector, *Energize RE: Renewable Energy Supplement*, 2, 4–6, June 2014, Online. Available http://www.ee.co.za/wp-content/uploads/2014/06/energize-revol2-pg-4-6.pdf.
15. Inspecting wind turbines with FLIR thermal imaging cameras, *FLIR Commercial Systems B.V.*, Online. Available: http://support.flir.com/appstories/AppStories/Electrical&Mechanical/Wind%20turbine%20inspections.pdf.
16. Krstulović-Opara L., Branko K., Endri G., and Željko D. The application of pulse heating infrared thermography to the wind turbine blade analysis, *Proceedings of the 16th International Conference on Composite Structures (ICCS 16)*, Porto, Portugal, 2pp., 2011, Online. Available: http://paginas.fe.up.pt/~iccs16/CD/121-160/156KRSTULOVIC.pdf.
17. Meinlschmidt P. and Aderhold J. Thermographic inspection of rotor blades, *Proceedings of the 9th European NDT Conference (ECNDT)*, Tu-1.5.3., Berlin, Germany, 9pp., 2006, Online. Available: www.ndt.net/article/ecndt2006/doc/Tu.1.5.3.pdf.

18. Guoliang Y., Neal B., Boot A. et al. Development of an ultrasonic NDT system for automated in-situ inspection of wind turbine blades, *Proceedings of the 7th European Workshop on Structural Health Monitoring*, La Cité, Nantes, France, July 2014, Online. Available: www.ndt.net/article/ewshm2014/papers/0156.pdf.

19. Li Dongsheng, Ho Siu-Chun M., Song Gangbing et al. A review of damage detection methods for wind turbine blades, *Smart Materials and Structures*, 24(3), 24, February 2015, Online. Available: http://iopscience.iop.org/article/10.1088/0964-1726/24/3/033001/pdf.

20. Boisseau A., Davies P., Choqueuse D. et al. Seawater ageing of composites for ocean energy conversion systems, *Applied Composite Materials*, 19(3), 459–473, June 2012, Online. Available: http://iccm-central.org/Proceedings/ICCM17proceedings/Themes/Behaviour/AGING,%20MOIST%20&%20VISCOE%20PROP/F1.2%20Boisseau.pdf.

21. Makaya K., Burnham K., and Tuncbilek K. Structural assessment of turbine blades using guided waves, *Proceedings of the Non-Destructive Testing Conference (NDT 2010)*, Cardiff, United Kingdom, pp. 225–236, September 2010, Online. Available: www.tidalsense.com/publications/pdfs/ESIA11_TWI_Submission1.pdf.

22. Raišutis R., Mažeika L., Samaitis V., Jankauskas A., Mayorga P., Garcia A., Correa M., and Neal B. Application of ultrasonic guided waves for investigation of composite constructional components of tidal power plants, *Proceedings of the 12th International Conference of the Slovenian Society for Non-Destructive Testing "Application of Contemporary Non-Destructive Testing in Engineering,"* Portorož, Slovenia, pp. 277–284, September 2013, Online. Available: www.ndt.net/article/ndt-slovenia2013/papers/277.pdf

23. Walford C. A. *Wind Turbine Reliability: Understanding and Minimizing Wind Turbine Operation and Maintenance Costs*, Sandia Report SAND2006-1100, 27pp., March 2006, Online. Available: http://prod.sandia.gov/techlib/access-control.cgi/2006/061100.pdf.

24. Standard EN 13306 Edition: 2010-10-01 Maintenance—Maintenance terminology, Online. Available: https://shop.austrian-standards.at/Preview.action;jsessionid=984CB13E259BE76F2F65BDCA28CCCDCC?preview=&dokkey=373702&selectedLocale=en.

25. Deliverable 6.1: Best practice guidelines for offshore array monitoring and control with consideration of offshore wind and oil & gas experiences, DTOcean, Document No. DTO_WP6_ECD_D6.1 Rev: 6.0, 37pp., Published July 24, 2014, Online. Available: http://www.dtocean.eu/content/download/22766/156991/file/DTO_WP6_ECD_D6.1.pdf.

26. *Guidelines for Onshore and Offshore Wind Farms*. Health & Safety in the Wind Energy Industry Sector, Renewable UK, September, Issue 1, 2010, Online. Available: http://c.ymcdn.com/sites/www.renewableuk.com/resource/collection/AE19ECA8-5B2B-4AB5-96C7-ECF3F0462F75/Offshore_Marine_HealthSafety_Guidelines.pdf.

27. Establishing an In-House Wind Maintenance Program, Wind and Water Power Program, US Department of Energy, December 2011, Online. Available: http://www1.eere.energy.gov/wind/pdfs/wind_om_report.pdf.

28. Martino J. Wind turbine lubrication and maintenance: Protecting investments in renewable energy, *Renewable Energy World Magazine*, May 2013, Online. Available: http://www.renewableenergyworld.com/articles/2013/05/wind-turbine-lubrication-and-maintenance-protecting-investments-in-renewable-energy.html.

29. Extending wind turbine gearbox life with lubricants, *TMC News*, May 2013, Online. Available: http://www.tmcnet.com/usubmit/2013/05/25/7161084.htm.

30. Muller J. and Errichello R. Oil cleanliness in wind turbine gearboxes, *Machinery Lubrication*, Online. Available: http://www.machinerylubrication.com/Read/369/wind-turbine-gearboxes-oil.

31. Brazen D. Lubrication of wind turbines, *Kansas Renewable Energy*, 2009, Online. Available: www.kcc.ks.gov/energy/kwrec_09/presentations/A2_Brazen.pdf.

32. Barr D. Modern wind turbines: A lubrication challenge, *Machinery Lubrication*, 9, 2002, Online. Available: http://www.machinerylubrication.com/Read/395/wind-turbine-lubrication.

33. Deighton B. Sensors to bring down the cost of tidal energy, *Horizon Magazine*, October 2013, Online. Available: http://horizon-magazine.eu/article/sensors-bring-down-cost-tidal-energy_en.html.

34. Li Y. and Florig H. K. Modeling the operation and maintenance costs of a large scale tidal current turbine farm, *Proceedings of the Oceans Conference*, Boston, MA, September 2006 Online. Available: http://www.lumina.com/uploads/main_images/Modeling%20the%20Operation%20and%20Maintenance%20Costs%20of%20a%20Large%20Scale%20Tidal%20Current%20Turbine%20Farm.pdf.

35. Keating T. J., Walker A., Ardani K. *Best Practices in PV System Operations and Maintenance*, Solar Access to Public Capital (SAPC) Working Group, March 2015, NREL, 2015, Online. Available: http://www.nrel.gov/docs/fy15osti/63235.pdf.

36. Operation and Maintenance Services for Solar Power Plants, Phoenix Solar, UK, Online. Available: http://www.phoenixsolar-group.com/business/us/en/about-us/services/Operations-and-maintenance. html.
37. AEG Power Solutions improves solar plant reliability with PV Care Check, *Sun & Wind Energy Magazine*, January 29, 2015, Online. Available: http://www.sunwindenergy.com/photovoltaics-press-releases/ aeg-power-solutions-improves-solar-plant-reliability-pv-care-check.
38. Andrews, R. W., Pollard A., and Pearce J. M. The Effects of Snowfall on Solar Photovoltaic Performance, *Solar Energy*, 92, 84–97, 2013, Online. Available: http://energy.sandia.gov/wp-content/ gallery/uploads/38-Andrews-Characterizing-Snow-Losses.pdf0__.pdf.
39. Shapiro D., Robbins C., and Ross P. *Solar PV Operation & Maintenance Issues*, Desert Research Institute, Online. Available: http://www.dri.edu/images/stories/editors/receditor/Solar_PV_Article.pdf.
40. Widman M. Let the light shine through, *PV Magazine*, (01), 2012, Online. Available: http://www.pv-magazine.com/archive/articles/beitrag/let-the-light-shine-through-_100005421/501/.
41. Burger A. *Cloud-Based Solar Robotics Platform Cleans PV Panels Without Water*, November 2014, Online. Available: http://www.triplepundit.com/2014/11/cloud-based-solar-robotics-platform-cleans-pv-panels-with out-water/.
42. Solar Panel Cleaning Services, Online. Available: http://www.solarpanelcleaningsystems.com/solar-panel-cleaning-services.html#pl.
43. Bkayrat R., Aldawsari A., and Al Zain H. *Best Practices for Mitigating Soiling Risk on PV Power Plants*, First Solar, Saudi Arabia-Jeddah, December 2014, Online. Available: saudi-sg.com/2014/files/ session/B-66.pdf.
44. Sarver T., Al-Qaraghuli A., and Kazmerski L. L. A comprehensive review of the impact of dust on the use of solar energy: History, investigations, results, literature, and mitigation approaches, *Renewable & Sustainable Energy Reviews*, 22, 698–733, June 2013.
45. Weber T., Ferretti N., Schneider F. et al. Impact and consequences of soiling and cleaning of PV modules, *PV Module Reliability Workshop (PVMRW)*, February 2015, Denver, USA, Online. Available: http://www.nrel.gov/pv/performance_reliability/pdfs/2015_pvmrw_105_weber.pdf.

Index

A

Milton Keynes UK
Ingram Content Group UK Ltd.
UKHW050458071024
449327UK00015B/425